Examining the Causal Relationship Between Genes, Epigenetics, and Human Health

Oscar J. Wambuguh
California State University - East Bay, USA

A volume in the Advances in Bioinformatics and Biomedical Engineering (ABBE) Book Series

Published in the United States of America by
IGI Global
Medical Information Science Reference (an imprint of IGI Global)
701 E. Chocolate Avenue
Hershey PA, USA 17033
Tel: 717-533-8845
Fax: 717-533-8661
E-mail: cust@igi-global.com
Web site: http://www.igi-global.com

Copyright © 2019 by IGI Global. All rights reserved. No part of this publication may be reproduced, stored or distributed in any form or by any means, electronic or mechanical, including photocopying, without written permission from the publisher. Product or company names used in this set are for identification purposes only. Inclusion of the names of the products or companies does not indicate a claim of ownership by IGI Global of the trademark or registered trademark.
    Library of Congress Cataloging-in-Publication Data

Names: Wambuguh, Oscar J., 1959- author.
Title: Examining the causal relationship between genes, epigenetics, and
  human health / by Oscar J. Wambuguh.
Description: Hershey PA : Medical Information Science Reference, [2019] |
  Includes bibliographical references and index.
Identifiers: LCCN 2018043704| ISBN 9781522580669 (hardcover) | ISBN
  9781522580676 (ebook)
Subjects: | MESH: Epigenesis, Genetic--genetics | Genetic Diseases,
  Inborn--genetics
Classification: LCC RB155 23 | NLM QU 475 | DDC 616/.042--dc23 LC record available at https://lccn.loc.
gov/2018043704

This book is published in the IGI Global book series Advances in Bioinformatics and Biomedical Engineering (ABBE) (ISSN: 2327-7033; eISSN: 2327-7041)

British Cataloguing in Publication Data
A Cataloguing in Publication record for this book is available from the British Library.

All work contributed to this book is new, previously-unpublished material. The views expressed in this book are those of the authors, but not necessarily of the publisher.

For electronic access to this publication, please contact: eresources@igi-global.com.

Advances in Bioinformatics and Biomedical Engineering (ABBE) Book Series

Ahmad Taher Azar
Benha University, Egypt

ISSN:2327-7033
EISSN:2327-7041

Mission

The fields of biology and medicine are constantly changing as research evolves and novel engineering applications and methods of data analysis are developed. Continued research in the areas of bioinformatics and biomedical engineering is essential to continuing to advance the available knowledge and tools available to medical and healthcare professionals.

The **Advances in Bioinformatics and Biomedical Engineering (ABBE) Book Series** publishes research on all areas of bioinformatics and bioengineering including the development and testing of new computational methods, the management and analysis of biological data, and the implementation of novel engineering applications in all areas of medicine and biology. Through showcasing the latest in bioinformatics and biomedical engineering research, ABBE aims to be an essential resource for healthcare and medical professionals.

Coverage

- Prosthetic Limbs
- Computational Biology
- Molecular Engineering
- Data Analysis
- Neural Engineering
- Orthopedic Bioengineering
- DNA Structure
- Robotics and Medicine
- Molecular Simulations
- Nucleic Acids

IGI Global is currently accepting manuscripts for publication within this series. To submit a proposal for a volume in this series, please contact our Acquisition Editors at acquisitions@igi-global.com or visit: https://www.igi-global.com/publish/.

The Advances in Bioinformatics and Biomedical Engineering (ABBE) Book Series (ISSN 2327-7033) is published by IGI Global, 701 E. Chocolate Avenue, Hershey, PA 17033-1240, USA, www.igi-global.com. This series is composed of titles available for purchase individually; each title is edited to be contextually exclusive from any other title within the series. For pricing and ordering information please visit https://www.igi-global.com/book-series/advances-bioinformatics-biomedical-engineering/73671. Postmaster: Send all address changes to above address. ©© 2019 IGI Global. All rights, including translation in other languages reserved by the publisher. No part of this series may be reproduced or used in any form or by any means – graphics, electronic, or mechanical, including photocopying, recording, taping, or information and retrieval systems – without written permission from the publisher, except for non commercial, educational use, including classroom teaching purposes. The views expressed in this series are those of the authors, but not necessarily of IGI Global.

Titles in this Series

For a list of additional titles in this series, please visit: www.igi-global.com/book-series

Expert System Techniques in Biomedical Science Practice
Prasant Kumar Pattnaik (KIIT University, India) Aleena Swetapadma (KIIT University, India) and Jay Sarraf (KIIT University, India)
Medical Information Science Reference • ©2018 • 280pp • H/C (ISBN: 9781522551492) • US $205.00 (our price)

Nature-Inspired Intelligent Techniques for Solving Biomedical Engineering Problems
Utku Kose (Suleyman Demirel University, Turkey) Gur Emre Guraksin (Afyon Kocatepe University, Turkey) and Omer Deperlioglu (Afyon Kocatepe University, Turkey)
Medical Information Science Reference • ©2018 • 381pp • H/C (ISBN: 9781522547693) • US $255.00 (our price)

Applying Big Data Analytics in Bioinformatics and Medicine
Miltiadis D. Lytras (Deree - The American College of Greece, Greece) and Paraskevi Papadopoulou (Deree - The American College of Greece, Greece)
Medical Information Science Reference • ©2018 • 465pp • H/C (ISBN: 9781522526070) • US $245.00 (our price)

Comparative Approaches to Biotechnology Development and Use in Developed and Emerging Nations
Tomas Gabriel Bas (University of Talca, Chile) and Jingyuan Zhao (University of Toronto, Canada)
Medical Information Science Reference • ©2017 • 592pp • H/C (ISBN: 9781522510406) • US $205.00 (our price)

Computational Tools and Techniques for Biomedical Signal Processing
Butta Singh (Guru Nanak Dev University, India)
Medical Information Science Reference • ©2017 • 415pp • H/C (ISBN: 9781522506607) • US $225.00 (our price)

Handbook of Research on Computational Intelligence Applications in Bioinformatics
Sujata Dash (North Orissa University, India) and Bidyadhar Subudhi (National Institute of Technology, India)
Medical Information Science Reference • ©2016 • 514pp • H/C (ISBN: 9781522504276) • US $230.00 (our price)

Applying Business Intelligence to Clinical and Healthcare Organizations
José Machado (University of Minho, Portugal) and António Abelha (University of Minho, Portugal)
Medical Information Science Reference • ©2016 • 347pp • H/C (ISBN: 9781466698826) • US $165.00 (our price)

Biomedical Image Analysis and Mining Techniques for Improved Health Outcomes
Wahiba Ben Abdessalem Karâa (Taif University, Saudi Arabia & RIADI-GDL Laboratory, ENSI, Tunisia) and Nilanjan Dey (Department of Information Technology, Techno India College of Technology, Kolkata, India)
Medical Information Science Reference • ©2016 • 414pp • H/C (ISBN: 9781466688117) • US $225.00 (our price)

701 East Chocolate Avenue, Hershey, PA 17033, USA
Tel: 717-533-8845 x100 • Fax: 717-533-8661
E-Mail: cust@igi-global.com • www.igi-global.com

To my family: my wife Josephine, whose love, patience, understanding and goodwill was a constant source of inspiration and advancement. To our daughter Anna (for her love, endearment and understanding); and to our son Jackson (for his humility and understanding) during the many hours I spent doing research, writing, and reviewing this textbook.

Table of Contents

Preface viii

Acknowledgment xi

Section 1
Understanding Life and Science

Chapter 1
Introduction to Life 1

Chapter 2
Life and Its Chemical Foundations 19

Chapter 3
Cellular Basis of Life 56

Section 2
Nature, Structure, and Functions of Genes

Chapter 4
Cell Division: The Cell Cycle, Mitosis, and Meiosis 93

Chapter 5
Genes: Modes of Inheritance 115

Chapter 6
Genes: Inheritance Patterns in Humans 128

Chapter 7
Genes: Structure, Replication, and Organization 145

Chapter 8
Genes: How They Work 162

Chapter 9
Genes: Expression and Regulation 186

Chapter 10
Genes: Recombinant DNA Technology.. 205

Section 3
Genes, Disorders, and Human Health

Chapter 11
Epigenetics.. 239

Chapter 12
Genomics and Genetic Testing 269

Chapter 13
Disorders of the Human Circulatory System 288

Chapter 14
Cancers.. 325

Chapter 15
Digestive, Ear/Nose/Throat, and Eye Disorders 361

Chapter 16
Endocrine and Immune System Disorders.. 399

Chapter 17
Muscle, Connective Tissue, and Neonatal Disorders 425

Chapter 18
Human Nervous System Disorders 468

Chapter 19
Genetics and Public Health.. 529

Appendix... 547

Glossary 554

About the Author 596

Index.. 597

Preface

As humans, we are fascinated to learn that life—with all its complexity from the molecular to organ systems levels, unravels from one molecule of life called deoxyribonucleic acid or DNA. This molecule encodes all the important features of life as we understand it. It is amazing that humans did not know how this molecule even looks like until about 65 years ago in 1953. Right from the time a microscopic sperm and egg in humans meet in the dark corridors of the Fallopian tubes in females to form a single fertilized egg (zygote), DNA direct all cellular processes of embryogenesis to make a functional human being.

From the time of DNA discovery, the next big event occurred 50 years later in 2003 when the sequencing of human DNA to elucidate the genes responsible for life was completed. Since then, scientists have continued unraveling our genetic mystery with new questions followed by increased research and collaboration. We know that to understand how genes work, we need to not only understand how DNA gene sequences dictate, but also the importance of other factors that control how genes are expressed. This is the exciting field of epigenetics which studies the chemical activation or deactivation of genes without altering the original DNA sequence. The word *epigenetic* has been defined by many authors to literally mean "in addition to changes in genetic sequence." Epigenomics is developing field that studies the genome-wide distribution of such epigenetic changes. As defined by the National Human Genome Research Institute (n.d.), "The epigenome consists of chemical compounds that modify, or mark, the genome in a way that tells it what to do, where to do it, and when to do it." Now we can understand why a cell in the liver expresses a different set of genes compared to a cell in the eye—although they both inherited the same original DNA from the sperm and egg. Chemical modifications completely shut-off all genes not required in the normal light sensory functions of the eye cell.

As discussed in Chapter 12, there are many epigenetic processes from the simplest one (methylation and chromatin modifications) to more complex ones like ubiquitination, sumolyation, miRNA, and siRNAs. These are natural processes essential to life—but they can also go awry affecting how cells function and causing health or behavioural problems. Although the epigenome can be passed on to offspring, studies indicate that during gametogenesis and after fertilization, the cell reprograms and erases all epigenetic modifications. Scientists are still learning more about epigenetic processes especially from studying twins living in different environments who show differentially expressed genes due to different epigenetic modifiers.

A variety of diseases presented in this book are caused by changes in the DNA gene sequences or changes in chromosomes number or integrity. Examples include haemophilia, cystic fibrosis, sickle-cell anaemia, Down Syndrome and Fragile-X syndrome among many others. We also know that a number of illnesses, behaviours and other health indicators are very closely linked to epigenetic mechanisms. These include cancers, cognitive malfunctioning, cardiovascular, autoimmune, respiratory and neurobehavioral

Preface

illnesses. There are many known or suspected drivers of epigenetic processes like heavy metals like lead and mercury, pesticides, tobacco smoke, petroleum-based emissions, PCBs, hormones, microbes, and medications. For instance, the drug *azacytidine*, approved by the FDA for the treatment of the blood disease myelodysplastic syndrome which can lead to leukaemia, turns on genes that had been shut off by DNA methylation. Unfortunately, while turning on hundreds of genes, it also turns off hundreds of other genes causing known side-effects like nausea, anaemia, vomiting and fever.

To enhance its utility, this book is divided into three major sections:

- **Section 1: Understanding Life and Science.** This section starts by defining life and its characteristics, its diversity, it introduces the theory of evolution as proposed by Charles Darwin in 1859, and discusses how scientists investigate research problems in the "Process of Science". The section next presents a discussion of life's chemical foundations, and the last section of the chapter explores the cellular basis of living organisms. It is hoped that this background will give readers a working foundation of concepts they need to comprehend what genes are, and how they work covered in Section 2.
- **Section 2: Nature, Structure, and Function of Genes.** In this section, it is hoped that readers will understand and appreciate what genes are, how they are transmitted from one generation to the next, how they work, how they are controlled, and how they can be engineered or manipulated to achieve specific human goals. Such goals could be for example, eliminating disease conditions, improving crop productivity, increasing pest resistance, cleaning up toxic environments, or "biopharming".
- **Section 3: Genes, Diseases, and Human Health.** This is like the "applied" section of the book where we will explore specific genetic diseases, their symptoms, genetic roots, and current therapies. It is hoped that this section will give readers information connecting genes and genetic susceptibility to disease. The latest scientific information gathered from currently published studies have been used in writing this section to ensure congruency with current research in the field. The section starts with a discussion of epigenetics and epigenetic mechanisms, through genomics and gene testing to coverage of specific genetic disorders, syndromes or diseases. Apart from the chapter on cancers which affect multiple systems (Chapter 14), and the one on neonatal diseases (Chapter 17), all other disorders are organized by human organ systems: circulatory system, digestive, ENT, eyes, skin, connective tissue, endocrine, skeletal, muscular, immune, and nervous systems.

The book ends with a chapter on genetics and public health. Since most genetic diseases in the realm of public health are an interplay of different genetic, lifestyle and environmental factors, public health specialists' roles are steadily evolving. With greater emphasis to the importance of molecular and cellular mechanisms in health and disease, and possible impact of this knowledge on individual and population health; public health practitioners are challenged to combine traditional public health goals of health education and promotion, outreach, and decreasing health disparities and equity; with safe and effective health interventions and diagnostics that meet individual health needs. A discussion on the critical role of public health and public policy in guiding public health practice and decision making, role of Big Data, personalized medicine and biomarkers, plus the bioethical, legal and social considerations concludes this final chapter.

Today, we know that there are thousands of known associations between genetic variants and complex human phenotypes. For example, although genome-wide association studies and other studies have demonstrated the association of more than 15,000 single nucleotide polymorphisms with a complex disease or trait—these association mechanisms remain unknown requiring continued research. Luckily, the rate of novel discoveries to understand the genetic architecture of complex genetic traits and diseases and to provide new insights into normal physiology and disease pathophysiology is rapidly increasing.

Being the first edition, this book will be an evolving publication as we learn more from the expanding fields of genomics, epigenomics, proteomics, pharmacogenomics, genome-wide association studies, genome wide interaction studies, next generation sequencing (whole genome sequencing and whole exome sequencing), gene expression profiling, and the many public health and policy initiatives that will surely accompany this renewed understanding of our genomes.

I greatly welcome any comments or suggestions from readers to help improve future editions of this text.

With warm regards,

Oscar J. Wambuguh
California State University – East Bay, USA
Summer 2019

REFERENCES

National Human Genome Research Institute. (n.d.). Retrieved from https://www.genome.gov/glossary/index.cfm?id=529

Acknowledgment

I would like to acknowledge and express gratitude to my "Dream Team" of four post-baccalaureate biomedical science students noted below. Their assistance in compiling the glossary of terms, end-of-chapter quizzes and "Thought Questions", and ensuring the references were properly formatted was surely invaluable.

Cindy Lam

Francine Sanvictores

Liane Hoang

Marian Banh

I would like to them to know that this textbook would not have been completed if it was not for their dedicated commitment to the cause. I greatly appreciate their diligence and patience in the many long and tireless hours they invested to get this work completed. I would like to wish each one of them success and excellence as they pursue their careers in the health professions. To my "Dream Team", a very big "Thank you!"

I would like to profoundly thank Ms. Colleen Moore, Editorial Assistant for IGI Global, for her assistance throughout the development process of this textbook. Her informative and constructive feedback to my many inquiries and ensuring that I received timely comments and suggestions from the textbook reviewers, is greatly appreciated. I also thank the textbook reviewers for their valuable contributions in the improvement of quality and presentation of the chapter contents.

Section 1
Understanding Life and Science

PROLOGUE

Living organisms share many characteristics that are similar be they bacteria, worms, plants or humans. This first chapter of the section lays out the common characteristics that unify life, and the variety of forms it has taken over many years of evolution (diversity). The mechanism through which evolution works to produce this diversity of lifeforms is presented as proposed by Charles Darwin in 1859. How scientists investigate research problems is discussed in the "process of science".

The second chapter explores the very basic elements that comprise living organisms and presents life's chemical foundations with major and minor elements of life, chemical bonds that are important in life, the chemical reactions of life, and the macromolecules that make life possible.

The third chapter in the section focuses on the basic unit of life: the cell. The components of cells—organelles are explored, how cells communicate with each other, and the various synthesis and breakdown metabolic reactions that occur in living systems.

This cellular basis of life will provide a good working foundation of concepts needed to better understand DNA and genes covered in Part II.

Chapter 1
Introduction to Life

ABSTRACT

This chapter introduces what life is and what characterises it, discusses its diversity, then introduces the theory of evolution. It ends with a section discussing how scientists investigate research problems called the "Process of Science." Living organisms share very defined characteristics—the sum of which make the "wholeness" we call life. Carl Linnaeus proposed a binomial system of classification where each organism's scientific name has two distinct parts: the genus and the species. Charles Darwin formulated the theory of natural selection which explains how evolution works. Humans share a lot in common with other living organisms, but there are features that make us distinctly human not shared with any other living organisms. The practice of science uses a carefully formulated series of steps in investigating problems. Science cannot explain everything especially philosophical questions that involve issues of right and wrong.

CHAPTER OUTLINE

1.1 Defining Life
1.2 Life's Defining Characteristics
1.3 Life's Diversity on Earth
1.4 Charles Darwin and Theory of Natural Selection
1.5 The Human Condition
1.6 How Science Works
1.7 Could Science explain Everything?
Chapter Summary

DOI: 10.4018/978-1-5225-8066-9.ch001

LEARNING OUTCOMES

- Understand how life is defined
- Classify life in its various diverse forms
- Explain the theory of natural selection and diversity of life on Earth
- Summarize the features that make us human
- Demonstrate how the process of science works
- Understand that science cannot explain everything

1.1 DEFINING LIFE

Each one of us will have at some point in life asked the question, "What is Life". We seek to understand what "being alive" means. What features make us different from a table? What do we do that a table cannot? Is life all the same, for example, life in a bacterium, life in ant and the life in a human being? Certainly, we see a thread connecting all three organisms: bacterium, ant, human. We need to understand this thread.

Living organisms share very defined threads or characteristics—the sum of which make the "wholeness" we call **life**. These threads differ in form but all perform similar functions. For example, humans utilize oxygen from the air to help break down food. Bacteria living in anaerobic conditions (without oxygen) are also still able to break down food. Humans produce young ones in **sexual reproduction**, and so do **bacteria** using **asexual reproduction**.

1.2 LIFE'S DEFINING CHARACTERISTICS

This section discusses the characteristics that collectively define what biologists consider living.

Organization

All living organisms are organized in units as simple as **cells** (like bacteria) to complex multi-cellular organisms with trillions of cells (like humans). Biologists do not consider anything below a cell living as in the case of **viruses**. Cells function as units with very well-defined order and an enclosing boundary (**cell membrane**), **genetic material**, and an **aqueous solution** made up of **sugars, ions, salts, amino acids**, and water. Some of the earliest known forms of life are single cells—which have remained that way as life evolved over millions of years to more complex life forms we see today.

In multi-cellular organisms, cells organize together to form **tissues** (as in skin tissue), which in turn form **organs** (as in heart); organs work together in **organ-systems**; and systems make a whole functioning individual organism (as in a wolf). (Figure 1)

Figure 1. Levels of biological organization (simplest levels to the biosphere)
Source: Image used under license from Shutterstock.com

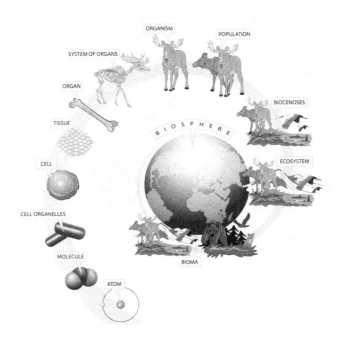

Growth and Development

Living organisms have genetic material primarily in form of **deoxyribonucleic acid (DNA)** which directs the process of development (Figure 2). In this molecule, instructions for growth and development are contained. For example, in humans, after fertilization of the egg by a sperm, the tiny fertilized egg (or zygote) begins a process of cell division that transforms and develops into a functional multi-cellular organism (**embryo**). Likewise, when we eat protein foods such as meat, instructions in our DNA direct the production of an **enzyme** known as pepsin to begin the digestion of proteins in our stomach.

Utilization of Energy

Living organisms must have a source of **energy** to grow and develop. This energy comes from an organism's ability to digest food, extract the energy in food, and use this energy to fuel its activities or store it for future use (Figure 3). As you read this sentence, your brain cells are fuelled by **glucose** reaching them through the blood. The glucose came from the breakfast or lunch you ate today, was digested, and its nutrients absorbed into the blood in the small intestines. When an organism's activities do not utilize all the energy absorbed from food, the body stores the excess in form of short-term **glycogen** (in muscles and liver) or long-term **sub-cutaneous fat** (under the skin as **adipose tissue**).

Figure 2. DNA directs the process of development. DNA carried by gametes carries the recipe or instructions of life and directs a fertilized egg from the 2-cell stage to a developing embryo.
Source: Image used under license from Shutterstock.com

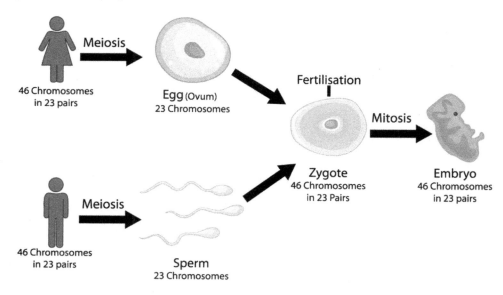

Figure 3. Interdependence between organisms
Source: Image used under license from Shutterstock.com

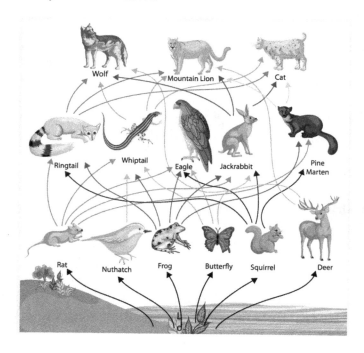

Introduction to Life

Reproduction

All living organisms produce young ones to propagate their kind—an **inborn** process that guarantees the survival of the **species**. Simple organisms reproduce with no exchange of genetic material by splitting into two (asexual). For example, bacteria simply divide into two daughter cells in a process called **binary fission**. In higher organisms, sexual reproduction occurs where specialized cells called **gametes** are produced whose sole goal is to unite into a new organism. In humans, gametes are produced by the **ovaries** (called **eggs**) and **testes** (called **sperms**).

Response to the Environment

Living organisms must be able to respond to their surroundings in order to survive. Factors such as light, temperature, wind, pH, food, and water are **stimuli** important in the survival of an organism. For example, bacteria respond to heat or cold by moving away from the source. Humans respond to heat by sweating and **vasodilation** or to cold by shivering and **vasoconstriction**. To do this, organisms have **receptors** organized in **cooperative lattices** (as in bacteria) or **sensory cells** (as in human taste buds).

Homeostasis

The ability to maintain stable internal operating environments is important in higher organisms. Such stability helps buffer fluctuations that would affect the organism's ability to operate in a variety of habitats from deserts to polar regions. For simple organisms like bacteria, if chemical conditions change, they are likely to die. Organisms like humans buffer chemical fluctuations using **negative feedback** loops that work to maintain stable operating conditions. For example, after a meal, our blood sugar level rises above the normal range of 70-99 milligrams per decilitre (mg/dl) of blood. Two hours after the meal the blood sugar level can rise to above 140 mg/dl. This high blood sugar reaches the brain which sends signals to the pancreas to release **insulin** into the blood. Insulin circulates through the blood to reach specific receptors at surfaces of cells. After recognition and binding between cell surface receptors and insulin, cells start up-taking glucose from the blood slowly bringing it to the stable level. On the other hand, if you skipped lunch bringing your blood sugar to below the normal range, the brain sends a different signal (**glucagon**) which targets the liver. The liver converts stored glycogen into glucose raising the level of sugar in the blood until it reaches the normal range. This is an example of **homeostasis** (homeo = constant; stasis = conditions) (Figure 4)

Interdependency

Organisms are not self-sufficient and depend on the resources in their environment to survive. For example, bacteria need food and some break down dead organisms for nutrition in a process called **decomposition**. Caribou eat vegetation to survive. Plants need water and nutrients from their environment to sustain themselves. Wolves hunt caribou for food. Humans use plants and animals as food sources. This inter-dependency creates distinct feeding levels of organisms like **decomposers** (bacteria and fungi); **detritus feeders** (insect larvae, beetles and termites); **producers** (primarily green plants); **consumers** (**herbivores**, **carnivores** and **omnivores**). (Figure 5)

Figure 4. Feedback mechanisms which regulate biological processes. The minus (-) signs are negative feedback loops from the thyroid to the hypothalamus-pituitary axis cancelling TRH and TSH production.
Source: Image used under license from Shutterstock.com

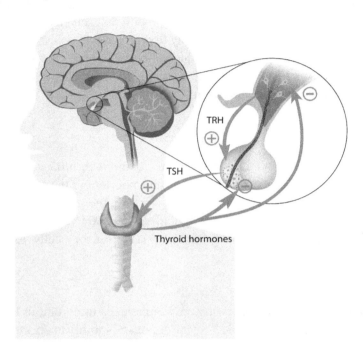

Figure 5. The organism's physical environment. This includes a source of chemical energy (sunlight), nutrients (vegetation or other consumers), and recycling of nutrients after death (nutrient cycling).
Source: Image used under license from Shutterstock.com

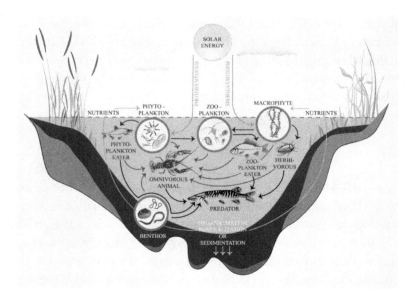

1.3 LIFE'S DIVERSITY ON EARTH

Considering the variety of living organisms on Earth, a way was needed to differentiate them depending on observable features of similarities and differences. Living organisms are classified as either prokaryotes (*pro* = coming before; no organized nucleus) or eukaryotes (with a distinct nucleus) depending on how complex their cells are organized. For example, simple single-celled organisms like bacteria have many things shared in common like single cell, cell structure, modes of life, etc. Likewise, they have habitat differences where each survives best. Some bacterial species like *Salmonella*, *Escherichia*, and *Staphylococcus* are found almost everywhere. Others survive better in areas characterized by harsh living conditions: very acidic (acidophiles), extremely hot (thermophiles), areas loaded with methane gas (methanogens), or very salty water environments (halophiles).

In the 18th Century, a Swedish scientist, Carl Linnaeus proposed a binomial system of classification where each organism's scientific name has two distinct parts: the *Genus* and the *species*. By taxonomic convention. the scientific name of an organism is always *italicized*. Although the Linnaeus system has been modified over the years by scientists, it is still widely used to this day. This taxonomic classification identifies eight groups starting from the most inclusive (Domain) to most exclusive (species). Thus, an organism like the American brown ("grizzly") bear is named *Ursus arctos* (Figure 6). It shares the genus (*Ursus*) with other bears like black bear (*Ursus americanus*), and polar bear (*Ursus maritimus*). It shares the same family (Ursidae) with other bears like sloth bears and giant pandas; and order (Carnivora) with other carnivores like mountain lions, coyotes and wolves. As a mammal which feeds its young milk produced in mammary glands, it shares the same Class (Mammalia) with humans; shares the same Phylum (Chordata – those with a vertebral column or spine) with humans; and as an animal, the same Kingdom (Animalia) with humans; and lastly, it shares the same Domain (Eukarya) with humans.

Currently, there are six recognized **Kingdoms** grouped into three **Domains**: Archaebacteria, Bacteria and Eukarya. The most common (regular) prokaryotes are classified together in one Kingdom called **Eubacteria**. The prokaryotic group usually found in very harsh environmental conditions are classified together in another Kingdom called **Archaebacteria**. (Figure 7)

Figure 6. Classifying living organisms. American brown bear (Ursus arctos).
Source: Image used under license from Shutterstock.com

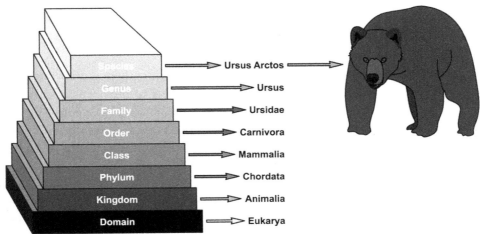

Figure 7. The three domains and six kingdoms of life
Source: Image used under license from Shutterstock.com

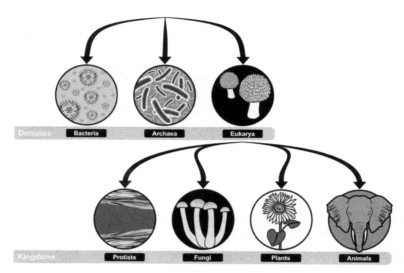

The eukaryotes are classified in four distinct Kingdoms: **Protista**, **Fungi**, **Plantae** and **Animalia**. Organisms considered protists occur in a variety of forms that may not be closely related with each other. Recent genetic advances have revealed group differences unknown to science before making some protists more closely related to plants, animals and fungi. Some use cilia or flagella for motion. Scientists however have retained these organisms in this Kingdom comprised primarily of single-celled organisms like paramecium, protozoa (like amoeba and malaria parasite *Plasmodium*), and *Euglena*.

The fourth Kingdom is the Fungi made up of organisms which are primarily decomposers of organic matter. This includes organisms such as common bread mould (*Rhizopus* spp.), mushroom, and yeast. The fifth Kingdom is the Plantae comprised of green plants all of which produce their own food from inorganic raw materials through the process of **photosynthesis (producers)**. The Animalia form the last Kingdom made of **invertebrates** (soft-bodied or boneless animals like crustaceans, annelids, and insects) and **vertebrates** (animals with bones including fish, amphibians, birds, reptiles, birds, and mammals).

1.4 CHARLES DARWIN VOYAGE AND THE THEORY OF NATURAL SELECTION

Thinking about the above diversity of life, one cannot help wonder: how did all this diversity come about? There had to be something responsible for the huge variety of life forms. That's what preoccupied foremost, an English naturalist named Charles Darwin in the late 19th Century (Figure 8). As the ship's naturalist, Charles participated on a trip around the world that took him from England to South America, Galapagos Islands, Australia, New Zealand, southern tip of Africa, and back to England between 1831-1836. On his trip Charles encountered a variety of living organisms all very well-adapted to their environments. He collected many specimens to take home with him. His observations and experiences on finches found both on the mainland of South America and on the islands of Galapagos were particularly noteworthy (Figure 9). Charles noted that the various mainland species of finches differed

Figure 8. Charles Darwin (1809-1882)
Source: Image used under license from Shutterstock.com

Figure 9. The various species of Galapagos finches Darwin studied
Source: Image used under license from Shutterstock.com

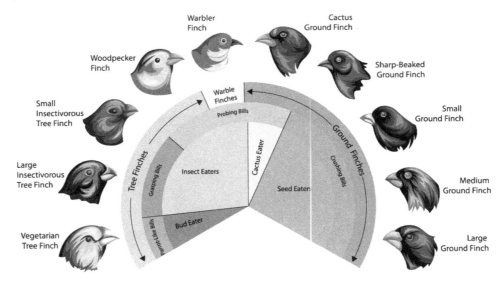

a lot in shapes and sizes of beaks from those on Galapagos islands. This observation greatly challenged his thinking on how organisms became adapted to the environments where they lived. He held many discussions with his peers particularly his colleague Alfred Wallace who was working on the same ideas from his trips to Indonesia. Using collaborative evidence from fossil records, a theory on how species physically changed over time was beginning to emerge.

Before exploring how Charles formulated a theory that could explain how evolution works, let us examine **artificial selection**. We are familiar with the huge variety of dogs today from the small Chihuahua to the Great Dane. They are all dogs with dog traits and features. Propagation of these breeds is motivated by human desire, experimentation or marketing. If there are more German Shepherds in demand on the market, a dog breeder will selectively propagate those dog breeds more often than others. The same with horse breeders breeding champion horse breeds. Crop breeders do the same. In artificial selection, breeders follow cues motivated by the prevailing circumstances as they propagate breeds that are on demand. Over time in the population, the breeds selected for increase in numbers.

The **Theory of Natural Selection** proposed by Charles Darwin follows the same reasoning. Many organisms like frogs, turtles, and fish lay thousands of eggs every season. We know that not all those individuals survive to maturity. Since in this case we do not have an artificial "human propagator", how are winners and losers chosen by the "forces of nature"?

The theory of natural selection has four main tenets (*variation, adaptive, selection, increase in frequency*) as discussed below (Figure 10).

1. **Variation:** The first tenet of the theory focused on the observation that individuals in a population are not similar in every way. Individuals have variation in features that can be passed along to offspring (that is, heritable). For example, we humans differ in many such traits like tongue rolling, widow's peak, hitchhiker's thumb, height, how fast we can run, attached or loose ear lobes, etc. The same is the case in frog, fish or turtle offspring.
2. **Adaptive:** Some of those features may be more adaptive in particular environments. Imagine a situation where all humans lived in a wild environment with hunters like bears, lions and wolves. These animals depend on speed to catch their prey. If we assume that to defend from these preda-

Figure 10. An illustration of natural selection by the peppered moth in Industrial Europe as a result of landscape changes brought about by air pollution
Source: Image used under license from Shutterstock.com

tors humans could only run away (much like deer); humans who could not run as fast would always fall victim, would die and not have a chance to reproduce. This means that in that environment, *running fast* is an adaptive feature of survival.
3. **Selection:** Over time, it is conceivable that humans who do not run fast in environments full of predators would gradually disappear. The predators would be considered the *natural agents of change* selecting the individuals who would survive to reproduce in the next generation.
4. **Increase in Frequency:** After several generations, slower humans will decrease in numbers after every generation. There just won't be enough of them who survive to reproduce and pass their features to the next generation. Eventually those features will be eliminated from the population. We can now say that *natural selection has occurred* in this population. Let's think further of the case of hunters like cheetahs and their favourite prey: Thomson's gazelles. If slower individual gazelles are gradually eliminated from the population, how are cheetahs able to capture any prey since (presumably) all remaining individuals are sprinters? Interestingly, in an environment like that of cheetahs and gazelles, speed is of essence both ways. Gazelles must run faster–in turn, cheetahs must run even faster to catch them. This is what biologists call an *evolutionary arm's race* that never ends (more technically referred to as **coevolution**).

Another good example to illustrate natural selection is **antibiotic** resistance, a common challenge in healthcare today. Although antibiotics kill most bacteria, the challenging medium might *by chance* selectively favour an individual bacterium which has developed an inheritable change (**mutation**) that allows it to overcome (resist) the chemical environment. Over time, individuals with this change increase in frequency in the population. At this time, we say that the bacteria have developed resistance towards an antibiotic like penicillin.

Charles therefore explained that natural selection, through adaptation to the existing environment ensures that only the fittest individuals can pass their features or traits to the next generation. Natural selection thus, is a mechanism that explains the evolution of species into the wide diversity or organisms we see today. One can now understand how the finches Charles studied in South America and Galapagos Islands were similar, yet with enough differences tuned by the environments where they lived, making them separate species.

Creationism vs. Evolution

Neo-Darwinists argue that natural selection and random mutations can mimic the powers of intelligent design without the need of an external designer or Creator. However, not everyone believes that the theory of natural selection (as proposed by Darwin) explains the origins of life and species diversity, particularly in humans. Several peer-reviewed literatures have been put together by scholars of intelligent design and can be read online (www.discovery.org). In the United States for instance, public opinion polls show that nearly three-quarters of Americans believe that God was involved in man's creation (Swift, 2017). Many have argued about "intelligent design" suggesting that evolution was guided by a divine hand; and therefore, creationism and evolution both complement, and should be taught together in school curricula. Although the US Supreme Court has ruled against intelligent design, it continues to be taught in many schools across the country. Interestingly, a Gallup poll in 2017 found that higher education may be associated with whether people believe in creationism or evolution: the poll found that about 21% of those with a postgraduate education believed in it while it was at 48% for those with only a high school diploma. Amongst postgraduates, only 31% believed that evolution occurred without God's guidance, while only 12% of high school graduates believed in it (Swift, 2017).

1.5 THE HUMAN CONDITION

As humans, we share a lot in common with other living organisms (section 1.2); other chordates (spine); other mammals (bearing live young, ability to produce milk for young); and with our primate relatives (elaborate brain, upright posture, grasping hands, stereoscopic vision). However, there are features that make us distinctly human not shared with any other living organisms.

These include:

1. **Advanced Brain:** Humans do not have the largest brains like sperm whales do; but brains with complex analytical and verbal skills, capacity for innovative thinking, and complex social and cultural behaviour.
2. **Muscle and Bone Arrangement:** Muscles in our hands and the nerves that operating them provide for a wide variety of manual dexterity unmatched in other species. We also have opposable thumbs. For example, immobilize your thumb and see how difficult it would be to write, hold a tool, or do other tasks.
3. **Walking Upright as the Chief Way for Locomotion:** Our closest primate cousins can walk upright but still use their forelimbs in locomotion especially in the thick vegetation where most live.
4. **Development of Speech:** This ability allows distinct formulation of words when speaking.
5. **Naked Skin:** Humans have less hair covering the body than most other species which, with the associated advanced brain, allowed the development of clothing.
6. **Ability to Make Fire:** This was a major advancement in human evolution allowing humans to stay in colder climates, cook food and defend against predators.
7. **Survival After Reproduction Ceases:** Only in humans do females survive much longer after reproduction ceases at menopause. Most other animals reproduce until they die. Culture and social bonding in human extended families allow older females to provide care and support which perhaps ensures success of the species as a whole.

1.6 HOW SCIENCE WORKS

What separates science from religion is the use of a carefully formulated series of steps in investigating problems. Unlike in religion which encourages blind faith, questioning nature is a major feature of science. Things must be tested, evidence evaluated, experiments designed and conducted, data collected in a specific way, and analyses conducted in accordance with established scientific procedures.

The process of science follows six steps in making inquiries about nature or things (Figure 11).

Introduction to Life

Figure 11. The process of science (scientific method). Hypothetical-deductive reasoning which allows scientists to investigate problems, design experiments, collect data, analyse it, and then make conclusions.
Source: Image used under license from Shutterstock.com

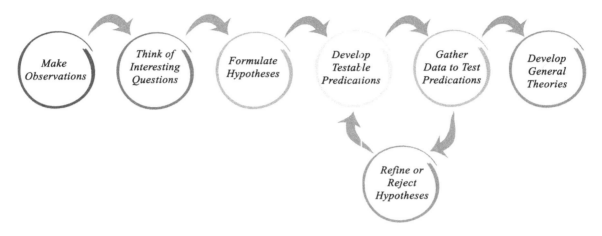

Make Observations

This identifies a specific phenomenon(a) or problem(s) in nature. For example:

1. Stella's vehicle is not starting.
2. Wolves always hunt in packs.
3. Young people love to party.

Develop Hypotheses

Once a problem or phenomenon is identified, the next step is to develop answers to the problem or provide "educated guesses". There can be several hypotheses. In the above examples, the hypotheses would be:

4. Stella's vehicle is out of gasoline. Stella's vehicle battery is dead. Stella's vehicle's starter is broken. Stella's ignition is out.
5. Wolves hunt in packs to maximize prey capture. Wolves hunt in packs to reduce injury from prey. Wolves hunt in packs because they are social animals.
6. Young people like to party because they like to socialize. Young people like to party to avoid other responsibilities. Young people like to party because once they become adults they will rarely party.

Make Predictions

Think of predictions about what you expect to happen if the hypotheses are correct.

7. Once we put gasoline (or battery, or starter, or ignition coil) in Stella's vehicle it will start.
8. Wolves hunting in packs almost always capture prey (or rarely sustain injuries). Social animals never hunt unless they are in a pack.
9. Young people in parties do nothing else except socialize. Young people do not accept any responsibilities. Adults rarely party.

Devise Experiments

Experimentation tests the accuracy of the predictions made. In the above Stella's case, fresh gasoline, or a good battery, or starter or ignition is the experiment. In the wolf pack, the experiment field observations should provide the basis for this prediction: do packs almost always succeed in capturing prey? A biologist can record instances of pack hunts that are successful versus pack hunts that are not successful. A study can also be conducted to investigate whether social animals like wolves (or lions or hunting dogs) ever hunt alone away from the pack.

In designing scientific experiments, it is often standard practice to have two groups of subjects: a **control group** and the **experimental group**. Some of the controls in the examples given above could be an empty gasoline tank versus a tank with gasoline; or non-working versus good working equipment in Stella's vehicle. In wolves, pack hunts that are unsuccessful or rarely observing injured wolves in pack hunts or successful lone wolf hunts. In young people, evidence of other activities occurring in parties apart from socializing; young people accepting other responsibilities instead of partying; or evidence of adults partying as much as young people.

In more typical cases of control versus experimental groups, we can think of a new experimental drug that has been found to reduce a tumour in three months. An experiment can be set with 50 subjects (women with same size tumour in their breast tissue). Half of the women (25) are the control and are given a control pill or a **placebo**. A placebo has all the other ingredients of the pill (like sugar, starch, saline solution or buffers) except the active drug treatment. In the "gold standard" testing of new experimental drugs, both patients and researchers do not know which half of the population receives the placebo or experimental treatment. This is often referred to as a "**double-blind experiment**" which ensures the researchers have no bias in the objective analysis of the results from the two groups. "Double" because neither the 50 subjects nor the researchers know which group received what.

Analyses of the Results

This includes assessing what the data collected means. Analyses may use statistical tests to evaluate **significance levels** which are based on **probability**. Such tests might include **correlation analysis**, **Student t-tests**, **analysis of variance, regression**, etc. Typically, most experiments in biology use the 95% significant level which means that the results observed could occur by chance only 5% of the time. In most medical drug trials, the higher significance level of 99% is sought (only 1% due to chance) for obvious reasons: it could result is severe complications or death of a subject. Statistical procedures yield measures like **mean** (average), **median, mode, standard deviation**, and **standard error**.

Introduction to Life

Publication of Findings

Typically, researchers share their findings in annual scientific meetings, conferences, symposiums, or in publications like **peer-reviewed** scientific journals, scientific reports, academic books or online scientific forums/sites. Results published or presented must provide enough information for the experiment to be reproduced elsewhere.

1.7 COULD SCIENCE EXPLAIN EVERYTHING?

Often with today's technology and use of the Internet, artificial intelligence, machine learning, smart phones, smart homes, smart/self-drive cars, smart TVs, advanced computer technology, nanotechnology, satellites, global positioning systems, and drone technology, humans might be tempted to believe that science can do or explain everything. We have learned a lot today than we did a decade ago and this advancement of technology, innovative thinking, creativity and capacity building is likely to continue into the future.

Despite this continued advancement, there are still many questions science cannot provide answers. Does God exist? If so, who is God? Does life have a meaning or purpose? Are human lives more valuable than a bee's or spider's? Should humans choose not to have children? Should humans pray? Should we **clone** humans? Should humans create **designer babies** with the traits we desire? Should we create **genetically modified organisms** like genetically modified foods or "**pharm-animals**"? Should we recycle? Should we dump wastes in the seas and oceans? These questions delve in the realm of philosophy that deals with **ethics**—issues of right or wrong.

Governments, corporations, non-governmental organizations, academic institutions, medical institutions, religious organizations, membership associations, cultural groups, parents, and individuals, are left to offer guidelines and/or make decisions about what is ethical to do. Often, governments will enact legislation prohibiting certain things. For example, according to United States Department of State (2018), US law does not allow the cloning of human beings for reproductive, therapeutic or experimental purposes. Another example is how the Vatican has for a long time used its global power and influence restricting certain reproductive rights to people (use of condoms in HIV/AIDs prevention or in women access to contraception and abortion) (The Vatican, 2018).

CHAPTER SUMMARY

Living organisms share very defined characteristics—the sum of which make the "wholeness" we call life. The characteristics that collectively define what biologists consider living include complex organization, growth and development, utilization of energy, reproduction, response to the environment, homeostasis and organism interdependency.

Considering the variety of living organisms on Earth, humans needed a way to differentiate them depending on observable features of similarities and differences. Living organisms are classified as either prokaryotes or eukaryotes depending on how complex their cells are organized. Carl Linnaeus proposed a binomial system of classification where each organism's scientific name has two distinct parts: the genus and the species. There are three Domains (Archaebacteria, Bacteria and Eukarya) and six Kingdoms (Eubacteria, Archaebacteria, Protista, Fungi, Plantae and Animalia).

Charles Darwin formulated the theory of natural selection which explains how evolution works. It has four main tenets—variation, adaptation, selection, and increase in frequency. Organisms are constantly adapting to their environment as conditions change.

Humans share a lot in common with other living organisms but, there are features that make us distinctly human not shared with any other living organisms. These include: advanced brain, muscle and bone arrangement, walking upright, development of speech, naked skin, ability to make a fire, and female survival after reproduction ceases.

The practice of science uses a carefully formulated series of steps in investigating problems. These include making observations, developing hypotheses, making predictions, designing experiments, analysing the results, and publishing findings.

Science cannot explain everything especially philosophical questions that that involve issues of right and wrong. In such cases humans depend on guidance from both governments non-governmental, and religious organizations.

End of Chapter Quiz

1. Which of the following statement defining living organisms is *false*?
    a. All organisms are organized in units of cells
    b. Living organisms have genetic material primarily in the form of DNA
    c. Living organisms must have a source of energy
    d. All organisms must reproduce with exchange of genetic material
    e. Living organisms must be able to respond to the environment
2. Homeostasis is
    a. solely dependent on the external environment
    b. a regulatory process that can only be seen in eukaryotes
    c. the ability to maintain stable internal environments
    d. dependent only on negative feedback
    e. dependent only on positive feedback
3. Which scientist proposed a binomial system of classification?
    a. Charles Darwin
    b. Robert Koch
    c. Thomas Edison
    d. Isaac Newton
    e. Carl Linnaeus
4. What are the three Domains of life?
    a. Protista, Fungi, Bacteria
    b. Bacteria, Eukarya, Archaebacteria
    c. Eukarya, Protista, Bacteria
    d. Plantae, Bacteria, Fungi
    e. Animalia, Bacteria, Protista

Introduction to Life

5. Eukaryotes are classified in four distinct Kingdoms namely
   a. Protista, Fungi, Plantae, Animalia
   b. Fungi, Prokaryotes, Protista, Animalia
   c. Animalia, Eukarya, Plantae, Fungi
   d. Plantae, Protista, Animalia, Fungi
   e. None of the Above
6. Charles Darwin proposed which theory?
   a. Meiosis Theory
   b. Combustion Theory
   c. Biogenesis Theory
   d. Theory of Natural Selection
   e. Spontaneous Generation Theory
7. What is the correct order of classification of living organisms?
   a. Phylum, Kingdom, Class, Family, Order, Genus, Species
   b. Kingdom, Phylum, Class, Order, Family, Genus, Species
   c. Kingdom, Phylum, Family, Class, Order, Species, Genus
   d. Kingdom, Family, Phylum, Class, Order, Family, Genus, Species
   e. None of the above
8. Which of the following is not part of the four tenets of the Theory of natural Selection?
   a. Connection
   b. Variation
   c. Adaptive
   d. Selection
   e. Increase in Frequency
9. Which of these features distinctly make us human?
   a. Advanced brain
   b. Muscle and Bone arrangement
   c. Development of speech
   d. Naked Skin
   e. All of the above
10. Which of the following is the correct way to write the scientific name of the African lion?
    a. *Panthera leo*
    b. Panthera leo
    c. *panthera leo*
    d. Panthera *leo*
    e. *Panthera Leo*

Thought Questions

1. Give 3 examples of how homeostasis plays a role in the human body.
2. Explain the four principles of the theory of natural selection.
3. Scientists conduct series of steps in order to investigate a problem. Design an experiment and include all of the required scientific procedures.

4. What are the features that make us distinctly human and are not shared with any other living organisms? Explain each feature in detail.
5. Can science explain everything? Why or why not?

REFERENCES

Discovery.org. (2018). *Peer-Reviewed Articles Supporting Intelligent Design.* Retrieved from https://www.discovery.org/id/peer-review/

Swift, A. (2017). In US, Belief in Creationist View of Humans at New Low. *News.gallup.com*. Retrieved from https://news.gallup.com/poll/210956/belief-creationist-view-humans-new-low.aspx

The Vatican. (2018). *Encyclical Letter Humanae Vitae of the Supreme Pontiff Paul VI.* Retrieved from http://w2.vatican.va/content/paul-vi/en/encyclicals/documents/hf_p-vi_enc_25071968_humanae-vitae.html

US State Department. (2018). *Views of the United States on cloning (September 23, 2002).* Retrieved from https://www.state.gov/s/l/38722.htm

KEY TERMS AND DEFINITIONS

Complex Organization: Levels of organization classified from the simplest living organisms to the most complex organisms.

Development: The process of growing and become more mature or advanced.

Domains: A taxonomy that classifies three different cellular life forms; Archaea, Bacteria and Eukarya.

Eukaryotes: Multicellular or unicellular organisms with membrane bound organelles.

Growth: The developmental process by which an organism matures or increases in size.

Homeostasis: The ability of the body to maintain stable conditions internally as the external environment changes.

Hypothesis: An assumption or a possible explanation that can later be scientifically tested.

Interdependency: Organisms that rely on resources in their environment to survive.

Kingdoms: The second highest taxonomy under Domain, that classifies and separates the following into smaller groups called phyla: Eubacteria, Archaebacteria, Protista, Fungi, Plantae, and Animalia.

Life: A combination of characteristics of an organism that consist of homeostasis, organization, metabolism, growth, response to stimuli, adaptation, and reproduction.

Natural Selection: The process that favors the survival and reproductive success of organisms that are best adjusted to the environment.

Prokaryotes: Single-celled organisms that lack membrane-bound organelles.

Reproduction: The ability of organisms to produce offspring.

Response to Stimuli: The ability of a living organism to detect a stimulus and respond accordingly.

Scientific Process: The hypothetical-deductive stepwise procedure used in science to make inquiries, formulate hypothesis, design experiments, collect and analyze data, and make conclusions.

Utilization of Energy: The ability of organisms to consume and transform energy into work.

Chapter 2
Life and Its Chemical Foundations

ABSTRACT

This chapter focuses on the chemical foundations of life. Matter is made up of elements classified into major and minor elements. Elements are made up of atoms which in turn are made up of sub-atomic particles called protons, electrons and neutrons. Chemical bonds are unions of electron structures when atoms lose, gain or share one or more electrons with other atoms. Water is important to life and has unique properties making it ideal to life on Earth. The pH of a substance is a measure of the balance between H^+ and OH^- ions ranging from 0 to 14 on a log scale. Most metabolic reactions that maintain life occur in living organisms involve five types of chemical reactions. Macromolecules are large complex molecules made up of repeating units (monomers) of sometimes the same molecule or of different molecules joined together by chemical bonds to form very long chains (polymers).

CHAPTER OUTLINE

2.1 Matter and Chemical Elements
2.2 Chemical Bonds and Life
2.3 Water, pH Balance and Life
2.4 The Central Role of Carbon
2.5 Chemical Reactions Governing Life
2.6 Life's Macromolecules
    2.6.1 Carbohydrates
    2.6.2 Lipids
    2.6.3 Proteins
    2.6.4 Nucleic Acids
Chapter Summary

DOI: 10.4018/978-1-5225-8066-9.ch002

Life and Its Chemical Foundations

LEARNING OUTCOMES

- Understand the nature of matter and chemical basis of life
- Explain chemical bonds and bonds important to life
- Understand the role of water in life
- Demonstrate the bonding behaviour of carbon and its importance in life
- Illustrate the nature and the different types of macromolecules in life

2.1 MATTER AND CHEMICAL ELEMENTS

Physicists define matter as anything that occupies space (volume) and has mass. Mass is different from weight in that it does not change depending on where you are. For example, in outer space, people experience 'weightlessness'; but they still have the same mass as when on Earth. Matter includes air, soil, water, bacteria, plants, animals, clouds, snow, ice, etc. but not energy (as in the electromagnetic spectrum). Matter is made up of elements, which are comprised of **atoms**. Atoms in turn are made up of smaller sub-atomic particles.

In living organisms, we identify *major elements and minor elements*. The major group includes four central elements carbon, nitrogen, oxygen and hydrogen, which make up to 96% of the *dry weight* of organisms. The minor group includes elements required in minor proportions by living organisms and comprising about 4% of the dry weight of organisms. These are elements like iron, calcium, copper, zinc, iodine, and phosphorous. Remember, *"wet weight"* is the weight you get on the scale, which includes watery fluids in your body. If all water were to be removed from your body entirely, you would get the dry weight. In research, scientists typically use the dry weight of organisms, which does not fluctuate as much as the wet weight.

All named elements are organized together in a *periodic table*. Most periodic tables of the elements common in many undergraduate college lecture halls list 104 elements identified by chemists. The list is comprised of 94 *natural elements* occurring on Earth, with the remainder being *synthetic elements* made in chemical laboratories around the world. Elements are made up of atoms. Atoms combine to form molecules, which in turn combine to form chemical compounds. Inside atoms are the sub-atomic particles called *protons, electrons* and *neutrons* (Figure 1). Both protons and neutrons have mass and are found in the nucleus of the atom, while electrons are weightless and are conceived as occupying a three-dimensional space around the nucleus in distinct shells or levels. Protons are positively charged particles counterbalanced by the negatively charged electrons. Thus, the numbers of protons always equal the numbers of electrons making an atom electrically neutral. Neutrons have no charge.

Structurally, electrons are best visualized as occupying defined "rings" around the nucleus like planets around the sun as seen here for hydrogen and helium (Figure 2).

The lowest level is near the nucleus and can hold a maximum of two electrons while the outer shells second, third and fourth all arranged in concentric circles, can hold a maximum of eight electrons each (Figure 3). This electron arrangement or configuration is often referred to as the *2:8:8:8 "octet rule"*. The rule focuses on the tendency of atoms to bond in ways that maximize the numbers of their electrons in the outermost shell to eight electrons as seen in inert gases.

Life and Its Chemical Foundations

Figure 1. Model of an atom showing sub-atomic particles
Source: Image used under license from Shutterstock.com

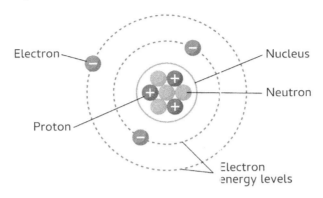

Figure 2. Simplified models of a hydrogen (H) and helium (He) atom. The hydrogen atom has one proton in the nucleus and one electron rotating around the nucleus. The helium nucleus consists of 2 neutrons and 2 protons; while two electrons rotate around the nucleus. Source: Image used under license from Shutterstock.com

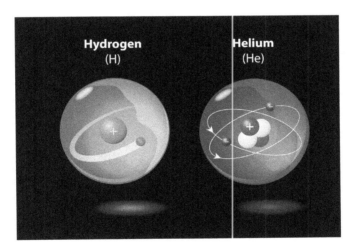

For our purposes, this simplified version will suffice but it is important to note that there is more complexity in shell structure. For instance, each shell is made up of orbitals organized in different planes. The first shell is made of up one *s orbital*, while the second shell is made up of one *2s*, and three *2p* orbitals (Figure 4). The shell to the very outside of the atom is referred to as the *outermost shell* while the one next to the nucleus is the *innermost shell*. Common elements like hydrogen, calcium, chlorine, and nitrogen use the electrons in their outermost shell to make *chemical bonds* or "unions" with other elements.

The 104 elements made up of gases, metals and non-metals are organized by chemists in a very specific way which factors their chemical behaviour. This arrangement is often referred to as the *Periodic Table* of the elements (Figure 5). On the table you can see that it is made up of names and letters with numbers. Each element is given a specific symbol and number.

Figure 3. Electron distribution in four elements. Note the different energy levels or shells. Source: Image used under license from Shutterstock.com

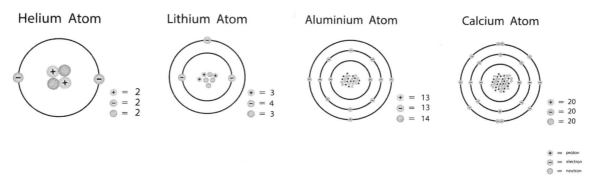

Figure 4. Electron shells and orbitals in the element Nitrogen. Electron shells are made of orbitals in different planes and each orbital can hold a maximum of 2 electrons. For instance, the first shell is made up of the 1s orbital with 2 electrons. The second shell has 4 orbitals (2s and three 2p orbitals).
Source: Image used under license from Shutterstock.com

For example, the element hydrogen at far left has the symbol H and number 1; helium at far right has symbol He and number 2. Lithium (Li) is number 3, beryllium (Be) 4, boron (B) 5, carbon (C) 6, nitrogen (N) 7, oxygen (O) 8, fluorine (F) 9 and neon (Ne) 10. What do these numbers mean? They represent an element's atomic number, the number of protons or electrons in its nucleus (remember, the #s of protons = #s of electrons).

Elements with one electron in their outermost shell are put together at far left (group 1) as seen for hydrogen (H), lithium (Li), sodium (Na), and potassium (K). Those with two electrons are next in group 2: beryllium (Be), magnesium (Mg), and calcium (Ca). Most elements in the two groups are metals except for hydrogen which occupies this position because its chemical behaviour mirrors those of metals. Groups 3-12 consist of elements referred to as the *"transition metals"* due to their unique chemical

Life and Its Chemical Foundations

Figure 5. The periodic table of the elements
Source: Image used under license from Shutterstock.com

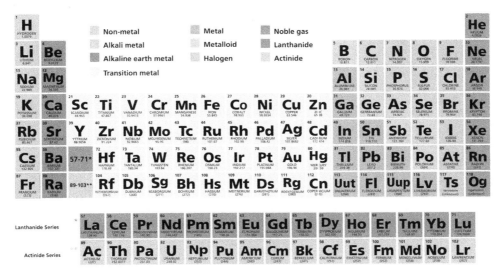

behaviour (Figure 2.4). Some are ions, others catalysts, while others can make bonds with other elements using electrons in shells other than just the outermost shell. The transition metals do not also follow the typical octet rule as illustrated by the first transition element Scantium (Sc) with 21 electrons arranged as 2:8:9:2. Gold (Au) has 79 electrons arranged as 2:8:18:32:18:1. Groups 13-17 generally are to the right on the transition metals and usually elements that are mostly non-metals and gases. Note that this is not exclusive as Al in position 13 is a metal. Group 18 elements occupy a special group and are often referred to as *inert gases*. They tend not to react with other elements. We will see why as we discuss chemical bonding in section 2.2 below.

Atoms often occur in multiple variants with one form stabilized in nature. These variants are called *isotopes*. Such alternate and unstable forms have the same numbers of electrons and protons but different numbers of neutrons. Some isotopes are radioactive and gradually break or decay by releasing radiation in form of *alpha, beta or gamma rays* to become the most stable form of the element. For example, carbon has three variants: carbon-12; carbon-13 (both stable forms) and carbon-14 (radioactive) (Figure 6).

This isotope property of emitting radiation has been useful to scientists in a number of ways most notably in medicine:

1. They have been used in radioactive dating for specimens found buried deep in the Earth's crust like *fossils*.
2. They have been used in research as *tracers*. For example, scientists can use carbon-14 incorporated in carbon dioxide to find out what happens to the gas once it is absorbed in green plants.
3. Medical scans – radioactive elements are often injected into organisms targeting a particular organ like thyroid gland. Using a device sensitive to the radiation emitted and connected to a computer, a 3-D scan of the thyroid can be viewed on the computer screen showing any abnormalities in growth.

Figure 6. Isotopes of carbon
Source: Image used under license from Shutterstock.com

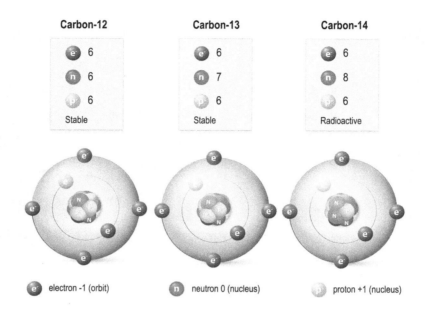

4. MRI (magnetic resonance imaging) – Radioactive substances like glucose with carbon-14, are often injected into the blood stream. Using computers, powerful magnetic fields and radioisotopes, images of body systems and organs can be generated on a computer screen. For example, in a vehicle accident that involved head injury, an MRI scan will allow detection of neuronal areas metabolically inactive (dead zones), which are then mapped to specific brain functions.
5. Radiotherapy – radiation emitted by radioisotopes has also been used in medicine to destroy wayward cells like those found in cancer.
6. PET (positron emission tomography) – in PET, the focus is not anatomy or structure but function. Radioisotopes are injected into the blood stream, inhaled or swallowed. These isotopes then accumulate in structures targeted and emit radiation, which can be detected by a computer and imaged. PET is used to study cellular function like tissue (or cell) oxygen uptake, blood flow, or therapeutic progress of a treatment plan.

2.2 CHEMICAL BONDS AND LIFE

Chemical bonds are unions of electron structures in atoms often using the outermost shells. They are formed when atoms lose, gain or share one or more electrons with other atoms. There are generally three types of bonds important in the understanding of living organisms: **ionic**, **covalent** and **hydrogen**. Atoms tend to react with others when their outermost shells do not have the maximum number of electrons using the 2:8:8:8 rule. Let's examine the first element in the periodic table, hydrogen. Hydrogen has only one shell that can hold a maximum of two electrons. Since it has an atomic number of 1, it has a single

Life and Its Chemical Foundations

electron revolving around a single proton in the nucleus (see Figure 2 above). Hydrogen then will form chemical bonds that will stabilize it. This could happen in two ways for hydrogen: it could lose that lone electron or share an electron with another atom.

$H - e^- \longrightarrow H^+$
(atom)   (ion or proton)

$H + H \longrightarrow H_2$
(atoms)   (molecule)

In the first reaction, the H atom loses an electron thereby becoming an ion or proton. The positive charge means it has an extra proton (positive charge) which has no counterbalancing electron in its structure. In this form, the H ion is stable but due to its charge, it is in a charged reactive state. In the second reaction, each H atom contributes an electron to make two which are shared on the first shell. The electrons revolve singly around each atom successively balancing the proton in each nucleus. Consider another atom like chlorine (Cl). Cl has the atomic number of 17 which, following the octet rule, has three shells with a 2:8:7 electron configuration (see Figure 9 below).

Ionic Bonds

Chemical bonds formed by loss or gain of electrons are called *ionic bonds*. A typical example is found in common salt, sodium chloride (Figure 7).

Sodium (Na) has 11 electrons (2:8:1). To be stable, there are two choices: it can get another element to donate 7 electrons or opt to lose its lone electron. Energetically, it is more favorable for the atom to lose the lone electron and become an ion.

Figure 7. Atomic bonding
Source: Image used under license from Shutterstock.com

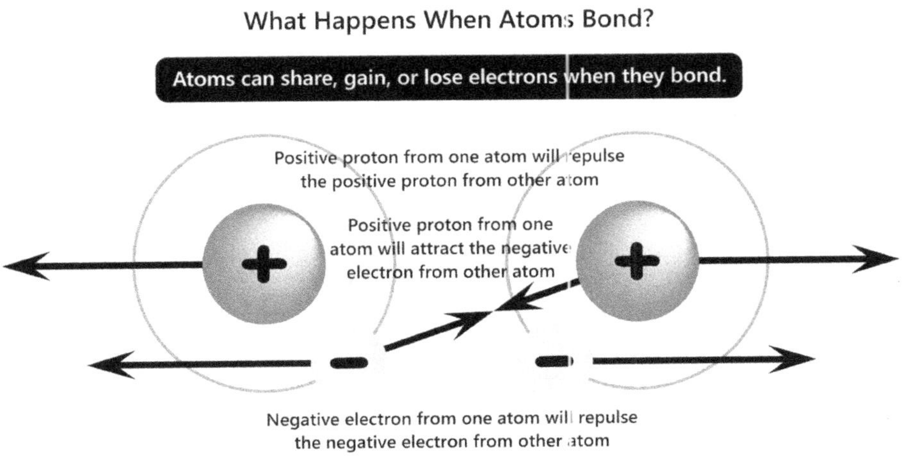

What Happens When Atoms Bond?

Atoms can share, gain, or lose electrons when they bond.

Positive proton from one atom will repulse the positive proton from other atom

Positive proton from one atom will attract the negative electron from other atom

Negative electron from one atom will repulse the negative electron from other atom

Na-e⁻ ⟶ Na⁺
(atom 2:8:1) (ion 2:8)

To be stable, chlorine on the other hand with a 2:8:7 configuration, would rather gain an electron than lose 7 electrons. It therefore becomes negatively charged – why? Because by "accepting" an electron, it now has an *extra* electron in its configuration not counterbalanced by a proton in the nucleus.

Cl+e⁻ ⟶ Cl⁻
(atom 2:8:7) (ion 2:8:8)

Sodium chloride is made up of the two ions Na⁺ and Cl⁻ held together by ionic bonds. The oppositely attracting charges make these bonds very strong considering that water can boil at 100°C and still not break the bonds!

Now practice forming ionic bonds with these elements and show how their configurations will change from atoms to ions as in the above examples: a) Magnesium (atomic # 12); Calcium (atomic # 20); and Aluminum (atomic # 13). Explain the configuration you got for Mg^{2+}, Ca^{2+} and Al^{3+}. In general, metals (and hydrogen) tend to form ionic bonds by losing electrons. Likewise, non-metals and gases in group 17 (known as *halogens*) tend to form ionic bonds by gaining electrons.

Figure 8. Electron transfer and ionic bonding. The attraction between oppositely charged atoms (or ions), is an ionic bond. An ionic bond can form between any two oppositely charged ions like Na⁺ and Cl⁻.
Source: Image used under license from Shutterstock.com

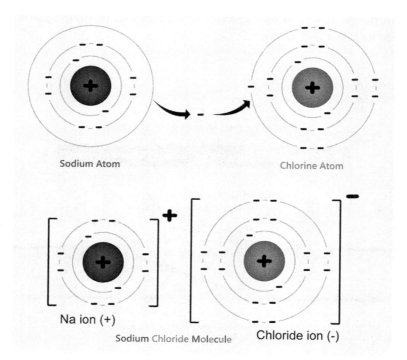

Life and Its Chemical Foundations

Covalent Bonds

When atoms unite their outer electron shells to share electrons as illustrated above for hydrogen, the bond holding the atoms together is called a *covalent bond*. Let's look at another example of atoms coming together to form single, double and triple covalent bonds (Figure 9). Did you notice that to maintain a stable oxygen molecule, O atoms must contribute two electrons each?

O+O $\rightarrow$ O_2
(atoms 2:6/2:6) (molecule 2:8)

Now consider nitrogen (2:5). How many electrons will be shared by the two nitrogen atoms? As you have seen in the case for H, O and N, covalent bonds can be single bonds as in H, double bonds as in O, or triple bonds as in N.

Now let's consider the case for water, H_2O. Hydrogen has a lone electron in its shell while oxygen has six electrons in its outermost shell. As discussed above in formation of covalent bonds, the H atoms will donate one electron each to the shared pool and O atom will donate two electrons—one shared with each H (Figure 10).

$2H_2 + O_2$ $\rightarrow$ $2H_2O$
(gas molecules) (water molecule)

Figure 9. Covalent bonding in Cl, O, and N molecules. Note the sharing of electrons creating a single (one electron pair), double (two electron pairs) and triple (three electron pairs) covalent bonds.
Source: Image used under license from Shutterstock.com

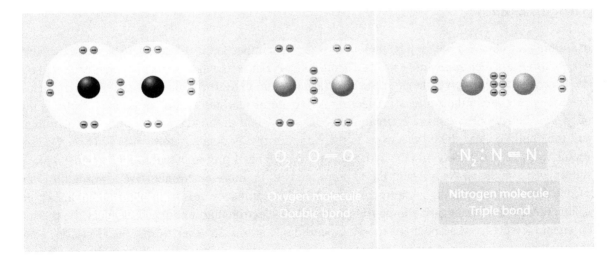

27

Figure 10. Covalent bonding in the water molecule
Source: Image used under license from Shutterstock.com

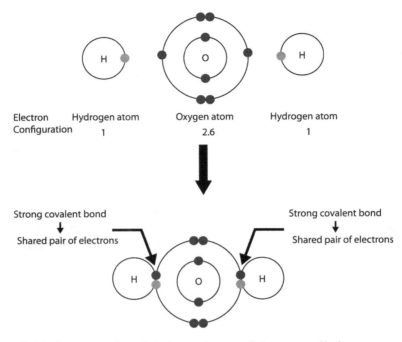

Hydrogen Bonds

In examining the covalent bonding behaviour of water, it is important to differentiate between two terms: *electropositive* and *electronegative* atoms. Electronegative atoms tend to attract covalently shared electrons. Electrons are negative, hence the word "negative" in the term. This *pull* or attraction is caused by the many protons in their nucleus. Electropositive atoms tend to "let go" or donate electrons more readily. This behaviour creates *polarity* in the covalent bond as compared to *non-polar* covalent bonds where the "pull" on the shared electrons is the same for the associated atoms like in H_2 or O_2 gases. Remember, in the water molecule the hydrogen atoms each has 1 proton in their nuclei against 8 protons in the oxygen's nucleus. As a result, the O atom tends to be more electronegative towards the shared electrons than H, thus "pulling" or attracting the shared electrons towards itself most of the time (Figure 9a). This creates a weak negative force (due to the extra electrons) around the O atom, while the same phenomenon creates a weak positive charge around the H atoms. In other words, the H protons has no counterbalancing electrons most of the time creating this weak positive charge around each. In liquid water, the weak negative charge around O tends to attract a neighbouring H atom; while the weak positive charge around H atoms attract a neighbouring O atom (Figure 9b). This weak attraction is called a *hydrogen bond* and occurs in many other molecules as we will see in the chapters ahead.

Life and Its Chemical Foundations

Figure 11. a) Hydrogen bonds between water molecules. The charged regions around the oxygen and hydrogen atoms in a water molecule are due to its polar covalent bonds. b) Oppositely charged regions of neighbouring water molecules are attracted to each other, forming hydrogen bonds. Each molecule can hydrogen-bond to multiple partners, and these associations are constantly changing in liquid water.
Source: Image used under license from Shutterstock.com

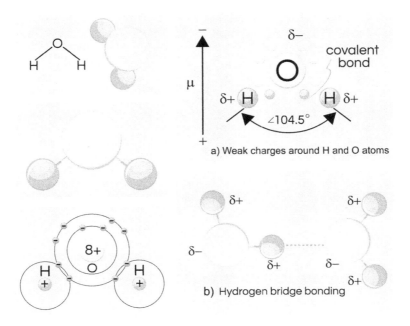

2.3 WATER, PH BALANCE AND LIFE

Water is often associated with life since 60-75% of living organisms are made up of water. For humans, depending on the surrounding conditions, we can live from a few hours to seven days without water. Under extreme heat humans can lose up to 1.5 litres/hour at which rate dehydration sets in and blood volume lowers and sweating ceases. Since sweating cools off the body in "naked "humans (as opposed to panting in more hairy animals like cats and dogs); the body overheats—which, together with the lower body fluid levels, can result in death within a few hours.

A number of water properties make it an ideal substance maintaining life.

1. **High Heat Capacity:** When water is heated, most of that energy is absorbed by the extensive hydrogen bonding (see Figure 9b above). If sufficient energy is absorbed, the hydrogen bonds break and water molecules separate as when steam forms with individual water molecules dissociating. The high boiling point of water at $100°C$ is evidence of this ability of water to take in a lot of heat energy without "breaking up". Similarly, when heating stops, and water begins to cool, enough energy must be lost to the environment for the hydrogen bonds to reform between the molecules. That's the reason why water is so useful in cooling organisms—as water evaporates, it carries with it a lot of heat energy. We also use water to cool engines in factories or in our vehicles too (we add ethylene glycol or ant-freeze to prevent water crystallizing to form ice).

Figure 12. How salt (Na⁺ Cl⁻) dissolves in water. Note the spheres of hydration forming around each of the atoms effectively separating them (dissolving in solution).
Source: Image used under license from Shutterstock.com

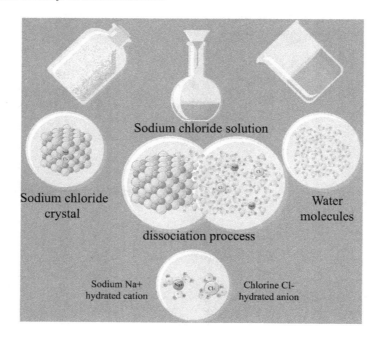

2. **Polarity:** The presence of positive and negative charges in water makes it a weak conductor of electricity and causes it to react with many other substances that have charges. For example, salt dissolves in water as the charged water ions (H+ and OH-) form hydration spheres around Na⁺ and Cl⁻ ions (Figure 12). In organisms with blood like humans—this watery medium which carries materials around the body has dissolved in it tens of mineral salts important in body functions like Ca^{2+}, Fe^{2+}/Fe^{3+}, Zn^{2+}, etc. Non-polar substances do not dissolve in water due to lack of charges within the molecule, e.g. oil and water cannot mix.

3. **Universal Solvent:** Water dissolves hundreds of substances chemicals and nutrients and due to its polarity making it ideal for washing, cleaning, cooking, making chemical solutions, etc. Our own kidneys are able to get rid of many substances because as blood passes through the body it carries with it substances we may no longer need like excess ions, drug metabolites, toxic substances and poisons.

4. **Adhesion and Cohesion:** Water tends to adhere to surfaces that are polar. For example, when we pour water on our skin, it does not all fall off—some remains adhered to the skin. This is because the skin always contains polar substances particularly salts excreted through sweating, which react with water. Cohesion on the other hand is the ability of water molecules to "stick together" or co-here. This is due to the attraction of each molecule to the neighbouring one. The forces of adhesion and cohesion allow tall trees to transport water against gravity from the roots to the leaves at the very top. For instance, costal redwoods in California can soar up to a height of 380-400 feet. These same forces also allow some organisms to literally "walk" on water. Very light water creatures like water striders can speed on the surface of water without disturbing or breaking the molecules. The

water molecules at top surface are attracted to those below by hydrogen bonds creating *surface tension* on the surface. The same reason why water in a glass looks curved at the top (meniscus) to the human eye. This curve is caused by water molecules clinging to the glass surface while at the same time, being attracted to adjacent water molecules (surface tension) causing the curve seen in a meniscus.

pH is a measure of the balance between H^+ and OH^- ions and ranges from 0 to 14 measured on a *logarithmic scale* (Figure 13). This means a pH of 3 is 1000 times as acidic as a pH of 6. The term pH means "power of hydrogen" in aqueous solutions and since element symbols are capitalized, the "H" is in upper case. A pH of zero means all ions present are H^+ (acidic) while a pH of 14 means only OH^- ions are present (alkaline or basic). When $H^+ = OH^-$ balance, the pH = 7.0. Pure water (distilled or dH_2O) is made up of only the two types of ions which is why it is useful in laboratory work where introduction of other substances in water can produce errors in procedures. The pH ion variation in fluids affect the solubility and behaviour of substances in it. For example, our stomachs produce the enzyme pepsin which starts the digestion of proteins. Pepsin will only be activated by an acidic environment from its inactive form pepsinogen. The pH of the human stomach therefore varies from 1-5. On the other hand, in the small intestine the pH is less acidic ranging from 6-7.5.

In homeostatic organisms like humans, pH must be maintained within a specific range in various tissues for proper cellular function. Substances called *buffers* are used to prevent wide fluctuations in pH. A common buffer used in the body is the bicarbonate ion ($HCO3^-$) buffer system. The system balances three substances depending on environment: carbon dioxide, CO_2; bicarbonate ions, HCO_3^-; and carbonic acid, H_2CO_3. In acidic situations where there are more H^+ (or vice versa), the bicarbonate ions combine with the H^+ ions in *reversible reactions* (denoted by the symbol $\rightleftharpoons$) to create a weak carbonic acid as shown by the equations below.

More H^+ ions favour this reaction: $HCO_3^- + H^+ \rightleftharpoons H_2CO_3$ (uptake of H^+)

More OH^- ions favour this reaction: $CO_2 + H_2O \rightleftharpoons H_2CO_3 \rightleftharpoons HCO_3^- + H^+$ (release of H^+)

2.4 THE CENTRAL ROLE OF CARBON

Living organisms are made up of carbon as one of the four main elements (C, N, H and O). The element carbon (atomic # 6) has been placed around the middle of the periodic table of the elements (ignoring the transition metals). It has a very versatile bonding behavior with the electronic configuration of 2:4. Carbon atoms can form single, double or triple covalent bonds with a variety of atoms including itself making organic compounds. For example, it can form four covalent bonds with hydrogen forming methane, CH_4 or ethane, C_2H_6. It can also form long chains (branched and linear or unbranched) made up of many repeating units; and three-dimensional structures most of which can serve as backbones for anchorage by many other atoms or groups of atoms. Some example include butane, octane, 2-methylpropane, cyclohexane, benzene, ethyl benzene, chlorobenzene, fatty acids, glucose (Figure 14).

When only carbon and hydrogen are in the molecule, they form molecules called hydrocarbons which tend to be very stable molecules and not very reactive e.g. butane, propane, and ring forms like cyclobenzene. When special groups of atoms called *functional groups* bond the carbon chain, they can

Figure 13. The pH scale and pH values of some aqueous solutions
Source: Image used under license from Shutterstock.com

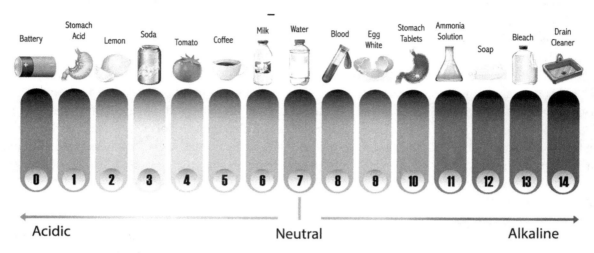

form macromolecules like carbohydrates, lipids and proteins. Functional groups can comprise a variety of elements including carbon, hydrogen, oxygen, nitrogen, phosphorous, and sulphur. They often determine the properties of the adjoining carbon skeleton and therefore, the chemical behavior of the whole molecule. Examples of functional groups include hydroxyl, aldehyde, ketone, carboxyl, amino, phosphate and sulfhydryl. Such groups can be bonded to a carbon skeleton to form molecules like ethanol, fatty acids, acetic acid, glucose, estrogen, and cholesterol (Figure 15).

Figure 14. How the element carbon forms molecular chains of varying length
Source: Image used under license from Shutterstock.com

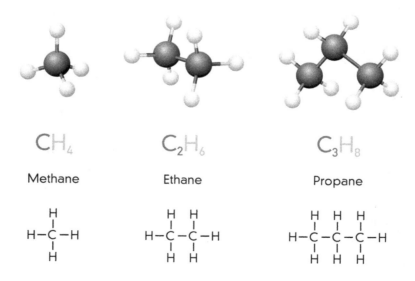

Life and Its Chemical Foundations

Figure 15. Some common functional groups

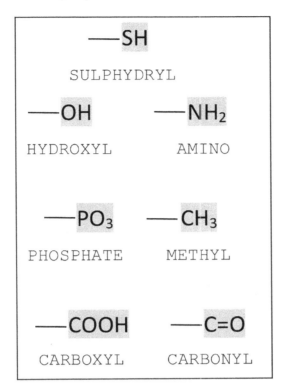

2.5 CHEMICAL REACTIONS GOVERNING LIFE

Thousands of metabolic reactions occur in living organisms to maintain life. The majority of them involve five types of chemical reactions as outlined below.

1. **Molecular Rearrangement:** In this case, while numbers and types of the atoms do not change, a reaction occurs when atoms in the reactants rearrange forming *isomers*. A good example is the reaction that changes glucose (formula $C_6H_{12}O_6$) to fructose (formula $C_6H_{12}O_6$). The atoms in the 3-D ring isomeric structures are rearranged (see Figure 16 below).

$$glucose\ (C_6H_{12}O_6) \rightarrow fructose\ (C_6H_{12}O_6)$$

2. **Electron Transfer:** Reactions involving electron transfer occur in atoms that form ions when they lose or gain electrons. The loss or gain of electrons by atoms is explained in section 2.2 (ionic bonds) above.

e.g. $Cl + e^-Cl \rightarrow V + Cl^-$

Figure 16. Monosaccharides (simple sugars)
Source: Image used under license from Shutterstock.com

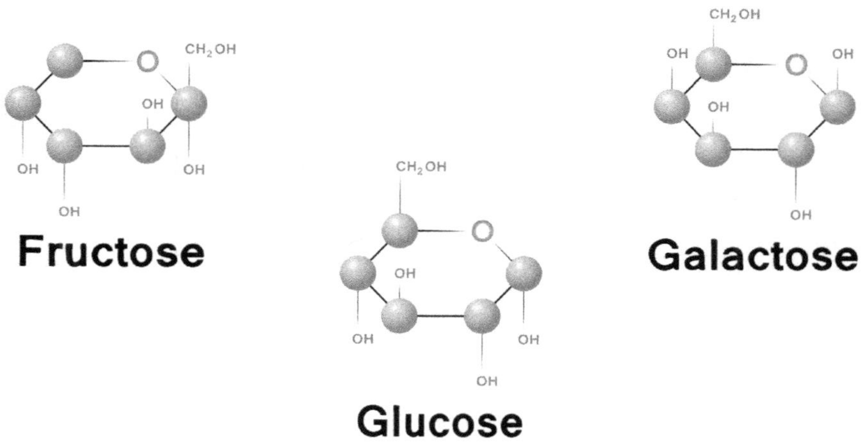

Fructose

Galactose

Glucose

3. **Functional Group Transfer:** In these types of reactions, the functional group component of the reacting molecules literally *trade* places in the reaction as shown below for the functional groups phosphate (PO_4^{3-}) and carbonate (CO_3^{2-}). This reaction is between phosphoric acid and limestone (chalk).

e.g. $2H_3PO_4 + 3CaCO_3 \rightarrow Ca_3(PO_4)_2 + 3H_2CO_3$

4. **Condensation (Dehydration):** These are reactions that occur with loss of water typically forming more complex molecules. Metabolic reactions that form macromolecules our body needs like proteins or lipids produce water as a by-product (Figure 17).
5. **Hydrolysis:** This is the opposite of dehydration with the addition of water cleaving molecules to form simpler molecules. Animals digest their foods using enzymes which operate in a watery medium which is used to hydrolyse macromolecules (Figure 18).

2.6 LIFE'S MACROMOLECULES

The name *macromolecule* represents large complex molecules made up of repeating units of (sometimes) the same molecule or of different molecules joined together by chemical bonds. Some are straight chain complex molecules while others are highly branched. Macromolecules form the basis of life on Earth made up of carbon atoms bonded to other atoms like hydrogen, oxygen, nitrogen, sulphur, phosphorous, and others. The foods made by green plants are complex macromolecules made from simple ingredients like water and carbon dioxide. These foods are then eaten by animals and broken into simpler molecules that are easily absorbed in the blood and used for energy, maintenance, or for building new macromolecules the body needs. In turn, some of these animals are eaten by others as a source of food.

Life and Its Chemical Foundations

Figure 17. Dehydration synthesis (condensation). Loss of water.

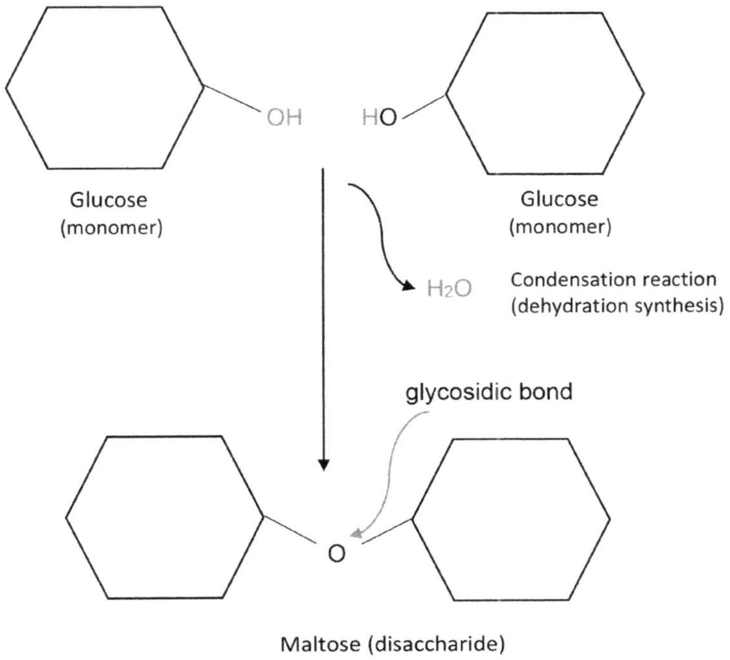

Maltose (disaccharide)

Figure 18. Hydrolysis reaction. Addition of water.

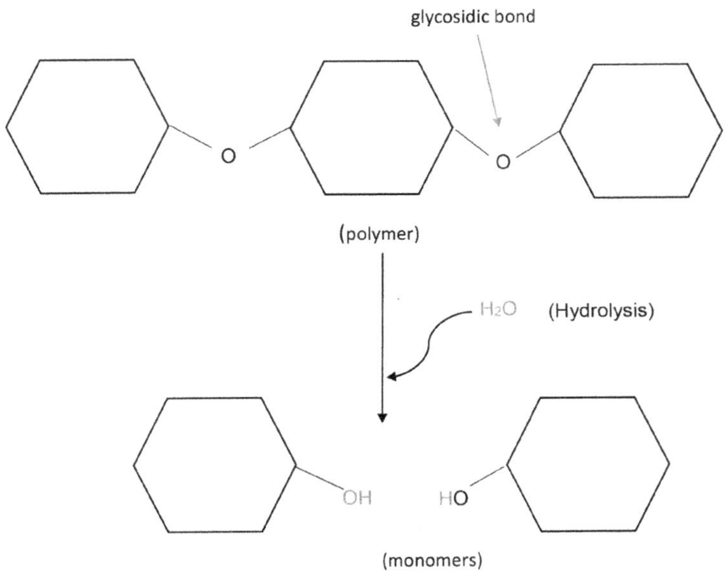

2.6.1 Carbohydrates

These macromolecules are perhaps the most abundant on the planet. Think of all those extensive cornfields or the huge wheat farms in the heartland states of the United States like Iowa, Ohio, Nebraska and Kansas. Likewise, the rice-fields of the Asia-Pacific Region in countries like China, India, Thailand and the Philippines. Carbohydrates are formed in green plants through the process of photosynthesis using sunlight as a source of energy.

$$CO_2 + H_2O \xrightarrow[\text{green leaves}]{\text{sunlight energy}} C_6H_{12}O_6 + O_2$$

Animals primarily utilize carbohydrates as a source of energy in a reverse process called *cellular respiration* where cells use oxygen gas to break the macromolecules into water and carbon dioxide.

$$C_6H_{12}O_6 + O_2 \xrightarrow[\text{cells}]{\text{enzymes}} CO_2 + H_2O$$

Carbohydrates occur in three main groups: monosaccharides (mono = one; saccharide = sugar), disaccharides and polysaccharides.

Monosaccharides

The monosaccharides are the simplest carbohydrates or sugars. Examples include the cell's most favourite source of energy—glucose. Others include fructose, mannose, galactose and ribose (see Figure 14a). When complex molecules like starches are digested in the body, they yield glucose molecules. Glucose and fructose are abundant in natural foods like fruits, berries, nectar, and honey. However, most people get a lot of additional simple sugars added in foods as high-fructose corn syrup in boxed cereals, fruit and sports drinks, desserts and candy.

Disaccharides

These are sugars made up of two monosaccharide units (*di* = two). When two units of glucose (or monomers) join through dehydration synthesis, they form the disaccharide maltose (also known as malt sugar used in beer making). The two units are joined by a covalent bond called a *glycosidic bond* (or glycosidic linkage, Figure 19). Glucose and fructose form sucrose, the common table sugar. Glucose and galactose form lactose, the sugar often found in milk. Though not always uniformly agreed, disaccharides are often included in the category of sugars called *oligosaccharides* which are made up of from 2-9 simple sugars.

Life and Its Chemical Foundations

Figure 19. Disaccharides two simple sugars joined by a glycosidic bond
Source: Image used under license from Shutterstock.com

Sucrose

Lactose

Maltose

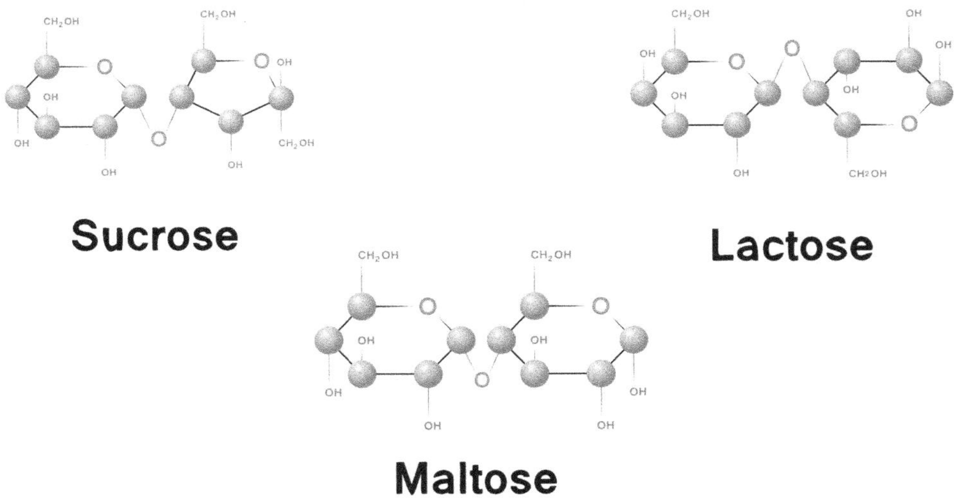

Polysaccharides

These are complex carbohydrates made up of at least ten simple sugars (monomers) to tens of thousands of repeating units (*poly* = many) all joined by glycosidic bonds (polymers) (Figure 20). Perhaps the most well-known is starch, a straight-chain polymer consumed by humans in corn, wheat, potatoes and rice. Another common one in the body is glycogen, a highly branched storage sugar in animals found abundantly in the muscles and the liver (Figure 21).

Figure 20. Starch is complex polysaccharides used for storage in plants
Source: Image used under license from Shutterstock.com

Amylopectin

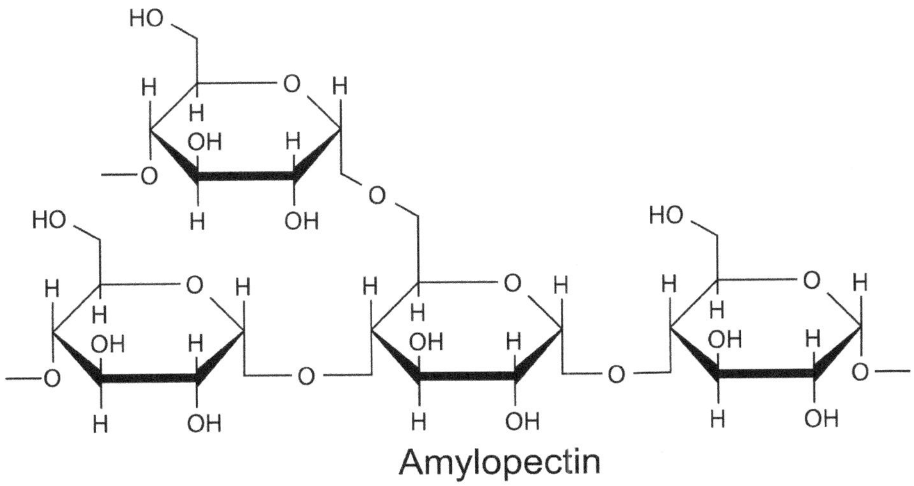

Figure 21. Glycogen is a complex polysaccharide used for storage in animals
Source: Image used under license from Shutterstock.com

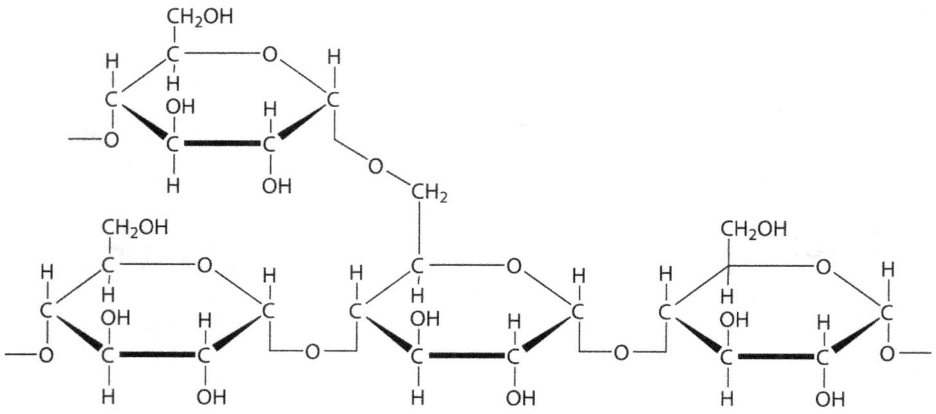

Other polysaccharides include:

1. **Fibres Like Cellulose (Figure 22) and Pectins:** Structural sugars found in plants abundant in vegetables, fruits, grains, and legumes. These are made of thousands of straight-chain glucose repeating units joined by glycosidic linkages.
2. **Gums:** Often sticky plant secretions useful in the food industry as emulsifiers, stabilizers and thickeners;
3. **Lignin:** The hardened, woody parts of plants often useful in the production of lumber and furniture in industry;
4. **Chitins:** Structural sugar reinforced with nitrogen found in *exoskeletons* of insects like beetles and grasshoppers and in crustaceans like crabs and lobsters (Figure 23).

2.6.2 Lipids

Lipids can occur in solid (*fats*) or liquid (*oils*) forms. They form a class of macromolecules used in the body as a source of energy, for storage and as a structural component in cells. Lipids are a rich source of energy due to their complex structure and presence of many high energy yielding bonds. They are generally non-polar and therefore insoluble in water. There are three generally recognized classes of lipids: triglycerides, phospholipids and sterols (Figure 24).

Triglycerides

These are the most abundant lipids in the body and richest source of energy. Animals store it under the skin as *adipose tissue*. Organisms that live in very cold climates like polar bears, penguins and seals often have a thick layer of adipose tissue which insulates the body by conserving body heat. Since triglycerides are insoluble in water, once absorbed in the intestines, they are carried by special units in the blood known as *chylomicrons* - the least dense of the family of blood vehicles called *lipoproteins*. The

Life and Its Chemical Foundations

Figure 22. Structure of cellulose, a complex polysaccharide. Cellulose is the main component of cell walls in plants which creates fibre.
Source: Image used under license from Shutterstock.com

Figure 23. Chitin – structural sugar reinforced with nitrogen found in insects and crustaceans
Source: Image used under license from Shutterstock.com

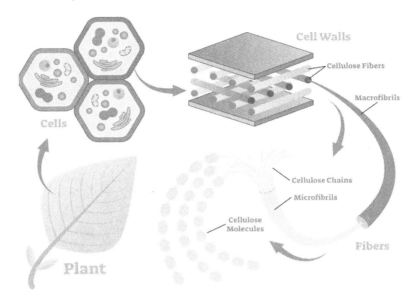

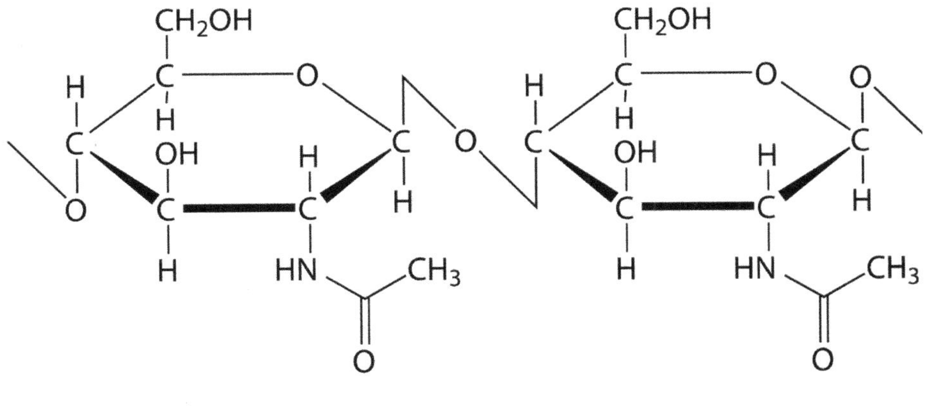

tri part of the name means that a triglyceride is made on *three* units of *fatty acids*, joined to a molecule of *glycerol* by a covalent bond known as *ester bond* (Figure 25).

Fatty acids are long hydrocarbon chains with either single or double bonds between carbon atoms and a carboxyl functional group at one end. Fatty acids with all single carbon bonds are referred to as *saturated fatty acids*. Saturation means all possible bonding positions with the carbon atom have been exhausted or taken by hydrogen atoms. Saturated fatty acids molecules pack very closely creating a very stable structure, that solidifies in nature (Figure 26a). Solidified fatty acids are called **fats** and are abundant in animal products like meats, milk, eggs, lard, butter and cheeses. Some plants like coconut and palm have saturated fatty acids found in coconut and palm oil.

Figure 24. Three types of lipids

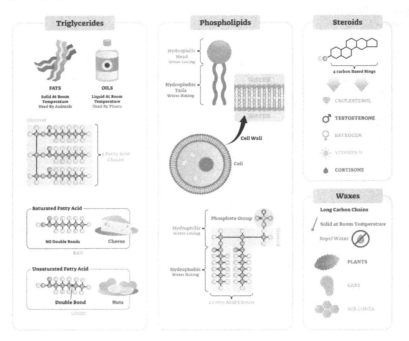

Figure 25. The synthesis and structure of a triglyceride. Note the three fatty acids and the glycerol molecule.
Source: Image used under license from Shutterstock.com

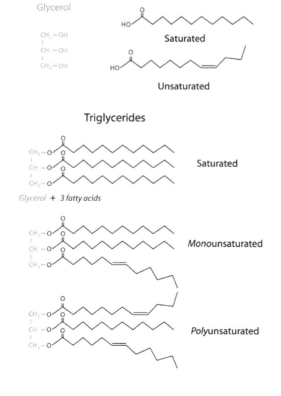

Life and Its Chemical Foundations

Figure 26. a) Unsaturated and b) saturated fatty acids. Unsaturated fatty acids can have single double bonds (monounsaturated) or multiple double bonds (polyunsaturated).
Source: Image used under license from Shutterstock.com

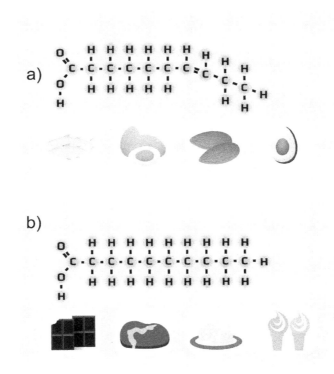

Fatty acids with double bonds are referred to as unsaturated fatty acids because the double bonds can be broken by adding hydrogen atoms (hydrogenation). The double bonds prevent close packing of the molecules due to "kinks" or "bends" in their structure making them "fluidy" and therefore liquids called *oils*. Fatty acids with single double bonds are often referred to as *monounsaturated fatty acids* (Figure 26b). Monounsaturated fatty acids include canola, safflower, and olive oils and can also be found in almonds, avocado and peanuts. Fatty acids with more than two double bonds are called *polyunsaturated fatty acids* (Figure 26b) which include corn, soybean, sunflower, cottonseed and flaxseed oils. The omega-3, omega-6, and omega-9 family of fatty acids are all unsaturated.

To enhance texture, consistency and increase shelf-life industry often saturate oils through the chemical process of adding hydrogen atoms (partial or full *hydrogenation*) to the unsaturated fatty acids. This alters the positioning of the hydrogen atoms from the *cis* to the *trans* position (Figure 27). The double bonds are also removed as hydrogen atoms "occupy" the bonding positions on carbon in full hydrogenation. These modified oils are often called *trans*fats. Hydrogenated products include margarine, potato/corn/tortilla chips, and baked goods like cakes, pies, and crackers.

Most of the fats we eat are a mixture of saturated, mono- and poly-unsaturated fatty acids. When we eat more saturated fats or hydrogenated fats the liver makes more very low-density *lipoproteins* (*VLDLs*) which circulate in the blood distributing fats to cells for use and for storage. They now become low-density lipoproteins (*LDLs*), which as they circulate and continue losing fats, they pick *cholesterol*

Figure 27. The process of commercial hydrogenation. Hydrogenation alters the positioning of the C and H atoms in healthy unsaturated fats (cis position) creating transfats (trans position).

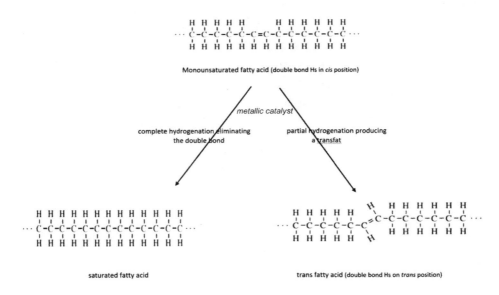

in the blood for distribution to cells for synthesis reactions and for deposit (Figures 28). This deposited cholesterol over time can slowly accumulate forming plaques in the blood vessels—a condition known as *atherosclerosis* (a form of *arteriosclerosis*). Plaques are made up of cholesterol, calcium, triglycerides and other blood components (Figures 29). When we eat more unsaturated fats, the liver makes more

Figure 28. Lipoproteins. With different densities, VLDL, LDL and HDL have different proportions of triglycerides, cholesterol and protein.
Source: Image used under license from Shutterstock.com

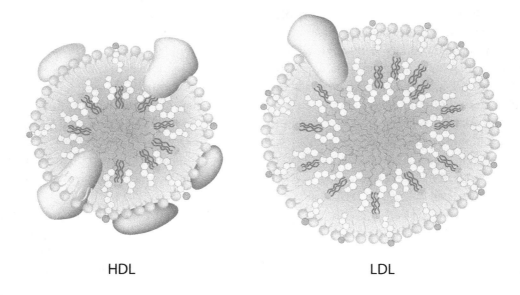

Figure 29. Formation of plaques in small arteries (a process known as arteriosclerosis) Source: Image used under license from Shutterstock.com

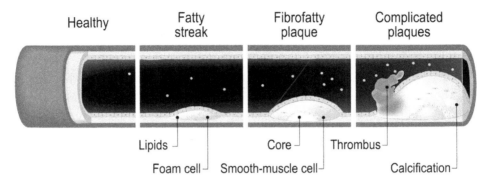

high-density lipoproteins (*HDL*s, Figure 28) which pick cholesterol in the blood and transport it from cells to the liver for destruction and disposal. That is the reason why the ratio of bad cholesterol (LDL) to good cholesterol (HDL) is so important in routine health screening. High cholesterol levels are attributed to risks for coronary heart disease.

Phospholipids

These macromolecules form the structure of cells and are integral components of cell membranes. They are formed by two fatty acids, glycerol, a phosphate group, and a molecule of choline joined by an ester bond. They are often characterized as having two non-polar "tails" and a polar "head", a property known as *amphipathic*. The polar head is made up of the phosphate group (negative charge) and choline (positive charge) (see Figure 24 above); while the non-polar tails are long-chain fatty acids. The polar head is *hydrophilic* (water loving) and the non-polar tails *hydrophobic* (non-water loving) (Figure 30). When phospholipids are put in water mixed with oil, phospholipids aggregate to form *bilayers* after settling or roundish assemblages called *micelles* if agitated. The polar heads orient to interact with water while the non-polar tails orient away from water (Figure 31).

Sterols

These are lipids with carbon skeleton composed on four interconnected rings made from cholesterol (Figure 32). Our liver typically makes enough cholesterol (800-1500 milligrams per day) for body needs: making vitamin D, steroid hormones like oestradiol (oestrogen) and testosterone (Figure 33), and bile salts important in emulsification of fats in the duodenum. However, we get a substantial part of it from our diet as cholesterol is naturally found in animal products like eggs, dairy products, poultry, fish and shell-fish.

Figure 30. Phospholipids have a hydrophilic head and hydrophobic tail (amphipathic)
Source: Image used under license from Shutterstock.com

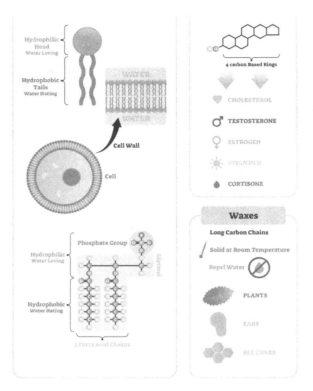

Figure 31. Phospholipids form bilayers and micelles in an aqueous environment
Source: Image used under license from Shutterstock.com

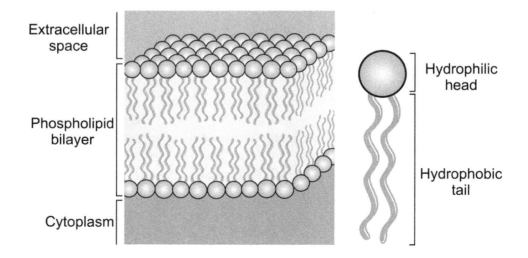

Life and Its Chemical Foundations

Figure 32. Structure of cholesterol – the precursor of steroid hormones like oestradiol and testosterone
Source: Image used under license from Shutterstock.com

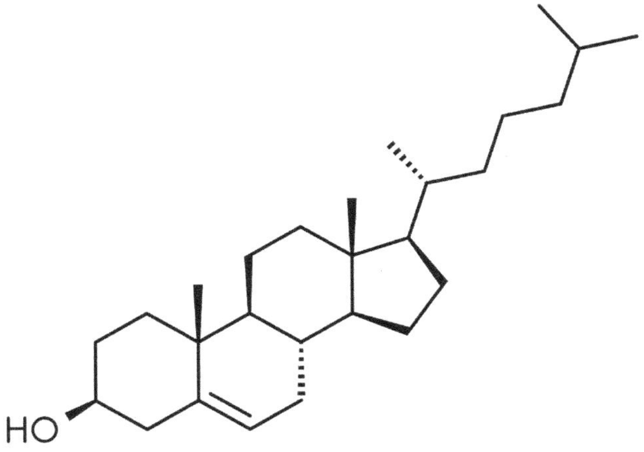

Figure 33. Structure of oestradiol and testosterone
Source: Image used under license from Shutterstock.com

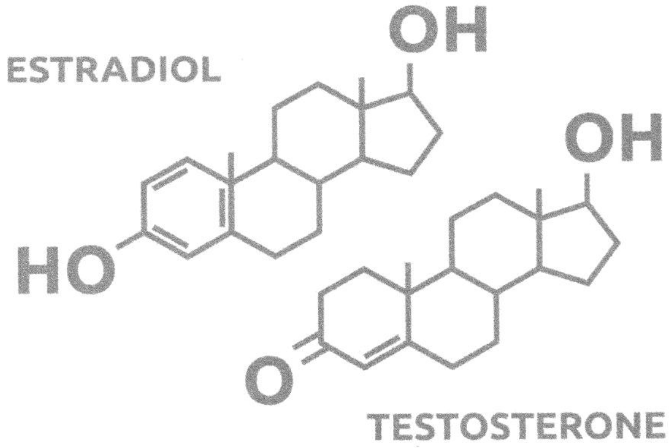

2.6.3 Proteins

This class of macromolecules has many roles in the body as food, hormones, enzymes, antibodies, transport molecules, and as structural materials. They are built from small monomers called *amino acids* which join through covalent bonds known as *peptide bonds* making proteins or polypeptides. An amino acid is made up of the amino functional group, an acid group, a hydrogen atom and an *"R" group* all covalently bonded to a central carbon atom (Figure 34). The "R" group can be polar, non-polar, acidic or basic in nature. There are 20 naturally occurring amino acids (Figure 35), nine of which are considered *"essential amino acids"* and must be present in our diets; the other 11 can be made by *transamination* reactions in the liver. Once assembled, proteins often fold extensively to form three-dimensional structures or configurations that give proteins their specific functions.

Figure 34. The structure of an amino acid
Source: Image used under license from Shutterstock.com

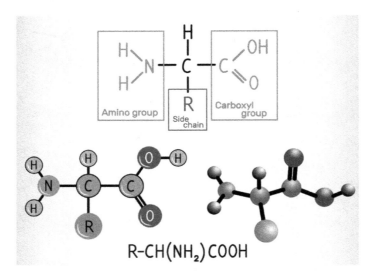

Figure 35. The 20 amino acids are grouped here according to the properties of their side chains (R-groups)
Source: Image used under license from Shutterstock.com

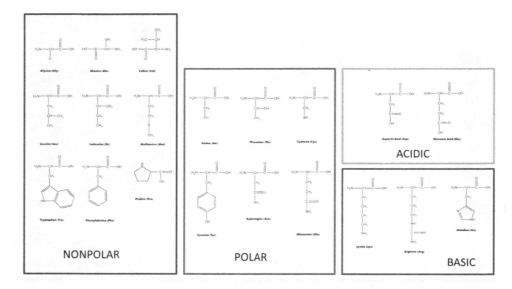

Primary Structure

The simplest protein structure is a long assemblage of amino acids in peptide bonds (Figure 36). It should be noted that this is the most basic protein structure and a functional protein will only develop at higher levels on interaction as seen in tertiary and quaternary structures discussed below.

Life and Its Chemical Foundations

Figure 36. The simplest level (primary) is a string of amino acids joined by peptide bonds Source: Image used under license from Shutterstock.com

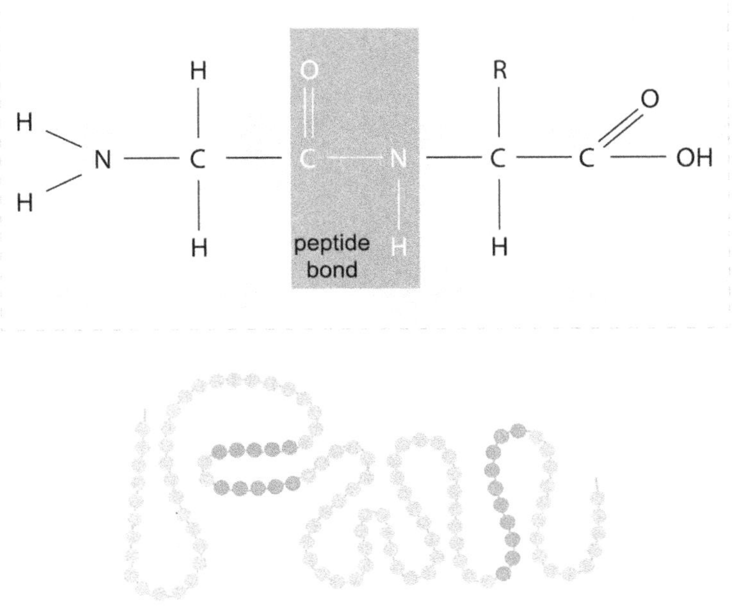

Secondary Structure

More often a protein will fold upon itself through interactions of neighbouring amino acids in this second level of protein structure. Hydrogen bonds stabilize the protein in two forms: alpha helices and beta chains. Alpha helices resemble the protein folding as a coil or a spring; while in beta chains the protein folds similar to way pieces of ham fold onto one another (see Figure 37). Examples of a protein with alpha helices is keratin found in nails, hair, skin and horns. Spider silk has many beta sheets.

Tertiary Structure

With more extensive folding, twists and turns and more bonding interactions, a protein can assume a 3-D level of conformation. Protein folding is assisted by special proteins called *chaperonins*. There are four forces that stabilize tertiary structure: *hydrogen bonds, hydrophobic interactions, disulphide bridges*, and *ionic bonds*. The non-polar portions of the protein may cluster together shielding them from other parts of the protein or solution stabilizing them. Ionic bonds form between charged portions of the protein stabilizing those parts. Disulphide bridges (S—S) form between two neighbouring sulphur atoms in the protein and stabilize the shape at those places. As in secondary structure, hydrogen bonds between polar parts of the molecule can develop stabilizing the protein into certain shape (Figure 37). Examples of proteins with the tertiary structure are enzymes which assume a 3-D configuration critical to their function (see Chapter 3, section 3.5 below).

Figure 37. Higher levels of protein structure. Secondary, tertiary and quaternary levels have a variety of bonds (H bonds, disulphide bridges, ionic bonds, hydrophobic interactions) that maintain these conformations as seen here.
Source: Image used under license from Shutterstock.com

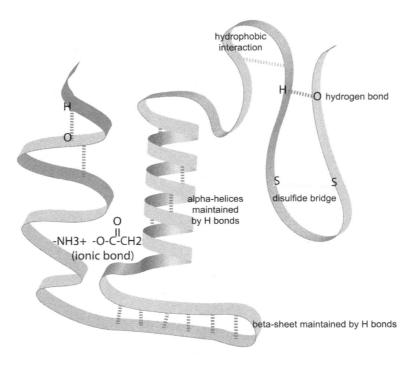

Quaternary Structure

In the fourth level, more than one polypeptide chain is involved. The arrangement of the various protein subunits and their interactions with each other form a functional 3-D complex. The same four forces as tertiary structure (hydrogen bonds, hydrophobic interactions, disulphide bridges, and ionic bonds) stabilize the various subunits in the quaternary structure. Examples of proteins with the quaternary structure are hemoglobin (a blood oxygen carrier protein, Figure 38) plus collagen and elastin found in skin and connective tissue.

- *Modified Proteins:* These are modified proteins with additional components like a lipid or carbohydrate and serve different functions in the body. For instance, lipoproteins carry cholesterol in the blood while *glycoproteins* and *glycolipids* are part of cell membrane structure (see Figure 6, Chapter 3).
- *Protein Denaturation:* The loss of 3-D shapes in proteins due to changes in pH, heat, UV light and chemicals can result in the disruption of bonds that hold the structure together. This is seen in egg hardening when eggs are boiled or cooked. The egg protein albumin loses its 3-D shape and hardens as seen here (Figure 39). The curdling of milk forming semi-solid globules in milk is the denaturing of the milk protein casein. Think of the reason why we usually do not iron (or put

Life and Its Chemical Foundations

Figure 38. Haemoglobin, a quaternary structure protein made up of four poly-peptide chains (1-4)
Source: Image used under license from Shutterstock.com

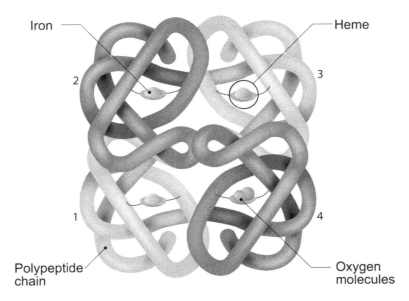

Figure 39. Loss of 3-D configuration (called denaturation). Boiling an egg damages the protein albumin which is now denatured.
Source: Image used under license from Shutterstock.com

in the dryer) woolen clothes, or even hang them to dry (drying them flat on the ground is the recommended procedure). Also, think of the reason why the skin sags, collapses and form wrinkles with increasing age. The skin's integrity is maintained by the protein fibers collagen and elastin (Figure 40).

Figure 40. Human skin and effects of aging

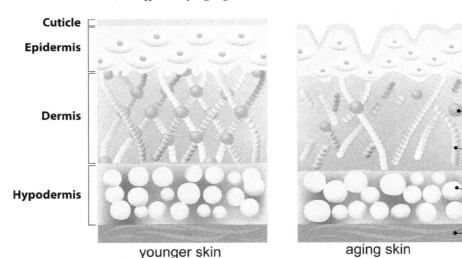

2.6.4 Nucleic Acids

These important macromolecules are the basis of heredity by carrying genetic information of organisms across generations. They include deoxyribonucleic acid (DNA) and ribonucleic acid (RNA). Both macromolecules work in harmony as cells use the genetic information carried on DNA, transferring it to RNA for the various cell metabolic functions. Nucleotides are made of single monomers called nucleotides joined together by *phosphodiester bonds*; and are made up of a sugar, a phosphate, and a nitrogenous base (Figure 41).

DNA and RNA (Figure 42) differ in several important ways:

1. DNA contains the sugar deoxyribose which contains one less oxygen than in RNA which has the sugar ribose;
2. DNA is double-stranded while RNA is single-stranded;
3. Both DNA and RNA contain three bases in common (adenine, guanine and cytosine), but a different fourth base—DNA has the base thymine, RNA has uracil;
4. Although the two macromolecules often work together in cells on a regular basis, DNA is the carrier of genetic information in most organisms; RNA carries genetic information only in some viruses.

CHAPTER SUMMARY

Physicists define matter as anything that occupies space (volume) and has mass. Matter is made up of elements classified into major and minor elements. The periodic table of the elements is lists 104 elements identified by chemists. Elements are made up of atoms which in turn are made up of sub-atomic particles called protons, electrons and neutrons. Protons are positively charged, and electrons negatively charged. Neutrons have no charge. Electrons occupy defined "rings" around the nucleus. Electrons are

Life and Its Chemical Foundations

Figure 41. Formation of a phosphodiester bond in nucleic acids. Nucleotides are made up of sugar, phosphate, and a nitrogen base. The phosphodiester bond is formed between the phosphate of one nucleotide and the sugar of the next one.
Source: Image used under license from Shutterstock.com

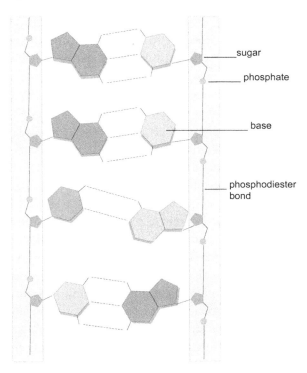

Figure 42. Differences between DNA and RNA
Source: Image used under license from Shutterstock.com

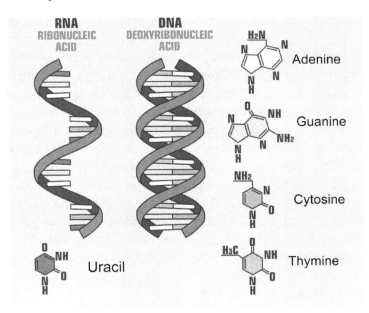

organized in different shells or energy levels following the 2:8:8:8 "octet rule". Atoms often occur in multiple variants called isotopes. Some of the isotopes are radioactive and gradually break or decay by releasing radiation. This property makes these isotopes useful in research as tracers or in medical imaging and radiotherapy.

Chemical bonds are unions of electron structures when atoms lose, gain or share one or more electrons with other atoms. Three types of bonds important in living organisms are ionic, covalent and hydrogen. Ionic bonds result when atoms lose or gain electrons. When atoms unite their outer electron shells to share electrons, they form covalent bonds. Hydrogen bonds do not involve loss, gain or sharing of electrons. They occur in situations where atoms which are already in covalent bonds exert different forces on the shared electrons. Electronegative atoms tend to pull the shared electrons towards their nucleus while electropositive atoms tend to donate the shared electrons more readily. This creates a weak negative force around the electronegative element and a weak positive charge around the electropositive atom.

Water is important to life and has unique properties making it ideal to life on Earth. These include polarity, high heat capacity, universal solvent, adhesive and cohesive properties. The pH of a substance is a measure of the balance between H^+ and OH^- ions and ranges from 0 to 14 measured on a logarithmic scale. Pure water (distilled) has a balanced number of H^+ and OH^- ions making its pH=7. Where there are no OH^- ions in a solution, is it is a pure acid (pH=0); and a purely alkaline when there are no H^+ (the pH=14). Living organisms use buffers like bicarbonate ion (HCO_3^-) to mitigate wide fluctuations in pH. Carbon has a very versatile bonding behaviour and can form single, double or triple covalent bonds with a variety of atoms including itself making organic compounds. Hydrocarbons are stable and less reactive compounds made up of only carbon and hydrogen atoms in long chains. Functional groups are special groups of atoms bonding a carbon chain and can be made up of a variety of atoms.

The majority of metabolic reactions that maintain life occur in living organisms involve five types of chemical reactions including molecular rearrangement, electron transfer, functional group transfer, condensation (or dehydration) and hydrolysis. Macromolecules are large complex molecules made up of repeating units (monomers) of sometimes the same molecule or of different molecules joined together by chemical bonds to form very long chains (polymers). The most abundant of them are the carbohydrates which occur in three forms: monosaccharides, disaccharides, or polysaccharides all containing glycosidic bonds.

Lipids can occur as fats which are saturated and solid in nature or as oils which are unsaturated and fluid in nature. The industrial process of hydrogenation can convert unsaturated fats to saturated. The types and amounts of lipids we eat have implications in human health. Three different types of lipids each with ester bonds are triglycerides, phospholipids and sterols. Triglycerides are the most abundant lipids in the body and richest source of energy and are carried in the body as lipoproteins. Phospholipids form the structure of cells and are integral components of cell membranes and are amphipathic (both hydrophilic and hydrophobic ends in their structure). Sterols are lipids derived from cholesterol and include vitamin D, steroid hormones like oestrogen and testosterone, and bile salts useful in digestion.

Proteins are a class of macromolecules made up of amino acids in peptide bonds and have multiple roles in the body as hormones, enzymes, antibodies, transport molecules, or part of the body structure. There are 20 naturally occurring amino acids. Once assembled, proteins often fold extensively to form three-dimensional structures or configurations that give proteins their specific functions. Four levels of protein structure are described: primary, secondary, tertiary, and quaternary. The body also has modified proteins serving different functions in the body.

Life and Its Chemical Foundations

Proteins can be denatured with the loss of their 3-D conformations because of changes in pH, heat, UV light and chemicals. Nucleic acids are macromolecules important in heredity carrying genetic information across generations. There are two types DNA and RNA which differ in several respects.

End of Chapter Quiz

1. The four main elements characterising life include all of the following EXCEPT?
   a. Carbon
   b. Nitrogen
   c. Oxygen
   d. Hydrogen
   e. Phosphorus
2. Chemical bonds formed by loss or gain of electrons are called
   a. ionic bonds
   b. covalent bonds
   c. hydrogen bonds
   d. polar bonds
   e. metallic bonds
3. An element has 35 protons. How many electrons will it have in its outermost shell?
   a. 2
   b. 7
   c. 8
   d. D.18
   e. 6
4. A pH of 4 is how many times more acidic than a pH of 6?
   a. 1
   b. 10
   c. 100
   d. 1,000
   e. 10,000
5. Water is an ideal substance in life because of which properties?
   a. High heat capacity
   b. Polarity
   c. Universal solvent
   d. Adhesion and cohesion
   e. All of the above
6. Rank the following bonds in order of relative bond strength: hydrogen/van der Waals, covalent, ionic, dipole interactions, London dispersion forces.
   a. ionic > covalent > hydrogen/van der Waals > dipole interactions > London dispersion forces
   b. covalent > ionic > hydrogen/van der Waals > dipole interactions > London dispersion forces
   c. covalent > ionic > dipole interactions > hydrogen/van der Waals > London dispersion forces
   d. ionic > covalent > London dispersion forces > hydrogen/van der Waals > dipole interactions
   e. hydrogen/van der Waals > ionic > covalent > London dispersion forces > dipole interactions

7. Which statement comparing DNA and RNA is *false*?
   a. DNA contains the sugar deoxyribose which contains one less oxygen than in RNA which has the sugar ribose
   b. DNA is single stranded; RNA is double stranded
   c. Both DNA and RNA contain three bases in common (adenine, guanine and cytosine), but a different fourth base—DNA has the base thymine, RNA has uracil
   d. Although the two macromolecules often work together in cells on a regular basis, DNA is the carrier of genetic information in most organisms; RNA carries genetic information only in some viruses
   e. None of the above
8. Which lipoprotein is increased in production when we ingest more unsaturated fats?
   a. Very low-density lipoproteins (VLDLs)
   b. Low density lipoproteins (LDLs)
   c. High density lipoproteins (HDLs)
   d. B & C
   e. A & B
9. Which of the following are properties of the cell membrane?
   a. Amphipathic
   b. Hydrophilic head and hydrophobic tail
   c. Bilayers
   d. Micelles
   e. All of the above
10. The pH of a solution is 7. Which of the following best demonstrates the pH when H^+ are added to the same solution?
    a. 2
    b. 7
    c. 7.5
    d. 10
    e. E.14

Thought Questions

1. What can break down cell membranes and how?
2. What bonds exist in each of the protein structures?
3. Although foods made with *trans*fats are some of the tastiest around, why is it not advisable to make them a major part of our human diet?
4. When human hair is wrapped around curlers with hair chemicals in a salon, and mild heat applied, hair is curled in new positions. Explain what happens in curled hair. (Hint: Hair is made up of the protein keratin).
5. Explain the difference between ionic, covalent, and hydrogen bonds.
6. If a solution is 5000 times as basic as water, what is the pH of the solution?

Life and Its Chemical Foundations

KEY TERMS AND DEFINITIONS

2:8:8:8 Octet Rule: A rule in electron configuration that focuses on the tendency of atoms to bond in a way that mimics the number of valence electrons in inert gases.

Atoms: Smallest unit of matter.

Chaperonins: A protein that aids the assembly and folding of other protein molecules in living cells.

Chemical Bonds: A union between atoms in a molecule.

Covalent Bond: A type of chemical bond involving the sharing of electrons between atoms.

Ester Bond: A bond formed from a dehydration reaction between a carboxylic acid and an alcohol.

Fatty Acids: Long hydrocarbon chains with either single or double bonds between carbon atoms and a carboxyl functional group at one end.

Glycosidic Bond: A covalent bond between the anomeric carbon of a sugar and an alcohol or amine of another molecule formed by a dehydration reaction.

Hydrocarbon: A compound consisting of hydrogen and carbon atoms.

Hydrogen Bond: A weak chemical bond involving the interaction of a hydrogen atom.

Ionic Bond: A type of chemical bond formed between oppositely charged particles and electrostatic attraction.

Isotopes: Atoms that share the same number of protons but differ in the number of neutrons.

Lipoproteins: A lipid and protein complex that aids in the transport of lipids throughout the lymph and blood.

Macromolecules: Large complex molecules made up of repeating units of the same or different molecules.

Particulate Matter: The mixture of solid and liquid particles suspended in the air.

Peptide Bonds: A type of covalent bond formed through a condensation reaction of two amino acids.

Periodic Table: A table of chemical elements arranged by respective characteristics/properties, including atomic number, mass, electronegativity, etc.

pH: A logarithmic scale that quantifies the acidity or basicity of an aqueous solution.

Phosphodiester Bond: The bond that forms between two phosphate groups of nucleotides.

Chapter 3
Cellular Basis of Life

ABSTRACT

To qualify as living, units of life called cells must be identifiable, distinct, and demonstrate most or all the qualities of life. Cells tremendously vary in size from about 0.5-500 micrometers. The smallest known single cells are those of bacteria while most higher organisms have multiple cells differentiated and functioning together as a single system. Communication in cells involves cell signaling, reception, transduction, and response. Signals received at the surface of the cell from other cells, or from blood or tissue fluid must be transferred to various parts of the cell and a cell response initiated. Cells actively take in raw materials which they use to function and perform maintenance activities. Collectively these activities are called cellular metabolism catalysed by enzymes. To avoid chaos in the body, cells maintain control of what reactions are needed all the time, needed only certain times or needed very rarely. This chapter explores the cellular basis of life.

CHAPTER OUTLINE

3.1 Cell Theory and Types of Cells
3.2 Structure of Cells
3.2.1 Animal and Plant Cells
3.2.2 Unique to Plant Cells
3.3 Cellular Communication and Transportation
3.4 Humans: Complex Cellular Systems
3.5 Metabolic Functions in Cells
Chapter Summary

DOI: 10.4018/978-1-5225-8066-9.ch003

Cellular Basis of Life

LEARNING OUTCOMES

- Understand the cell theory and the basic types of cells
- Demonstrate the various cellular structures
- Differentiate between plant and animal cells
- Explain how communication both outside and within cells
- Outline the various organ systems in complex organisms like humans
- Explain what metabolism means and the role of enzymes and cofactors.

3.1 CELL THEORY AND TYPES OF CELLS

Although life is based on chemistry and the various complex chemical reactions that characterize it, none these are components are holistic enough to be considered living by biologists. To qualify as living units of life must be identifiable, distinct and demonstrate most or all of the qualities of life discussed in Chapter 1 (section 1.2). The **Cell Theory** outlines three broad elements that characterize all units of life.

1. The smallest units of life occur at the **cell** level of organization
2. All living organisms are made up of one or more cells
3. New cells arise only from existing cells

Cells tremendously vary in size from about 0.5-500 micrometers. The smallest known single cells of bacteria vary in length from 0.5-2 micrometers. For comparison, the smallest frog eggs visible to the human eye as clusters clinging to vegetation in freshwater are about 1 millimeter in diameter. The largest cells are ostrich eggs which are about 5-6 inches in diameter (Figure 1). As noted in section 1.3, bacteria (**prokaryotes**) have existed on Earth as single cells for millions of years. Most higher organisms (**eukaryotes**) have multiple cells differentiated and functioning together as a single system.

At their most basic level using a **compound light microscope**, cells are seen as having a fluid-filled interior called **cytosol** which is bounded by an outside **plasma membrane** that delineates the cell boundary. The cytosol of the cell is about 65-70% water and contains dissolved substances and other products of the chemical reactions occurring both within and outside the cell. Towards the middle part of the cell is the control center or **nucleus** which carries the genetic information important in cellular operations (Figure 2).

If you take a cotton swab or toothpick and lightly scrape the inside of your cheeks, then put the contents onto a slide, add a stain like methylene blue which is absorbed by the cells and increases contrast, cheek cell basic structure is clearly visible (Figure 3). However, using an **electron microscope**, cells have a very complex structure with many "small organs" called **organelles** distributed into the cytosol (see below). The cytosol including the organelles is called the **cytoplasm**. Organelles are made up of membranes and perform specific functions in the cell—much like what organs like the heart do in the organism.

Figure 1. The size range of cells. Most cells are between 1 and 100 μm in diameter and are therefore visible only under the light microscope.
Source: Image used under license from Shutterstock.com

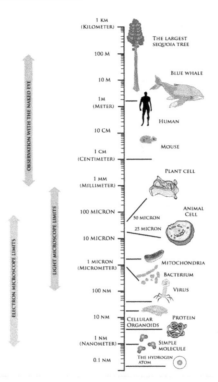

3.2 STRUCTURE OF CELLS

3.2.1 Animal and Plant Cells

The section details the plasma membrane structure, explores the nature of organelles and the nucleus, plus their important functions in the cell (Figures 4 and 5).

Plasma Membrane

Membranes serve as cell gatekeepers regulating all materials entering and leaving the cell and as mediators of cell-to-cell interactions. The plasma membrane is comprised of a phospholipid bilayer interspersed with **membrane proteins** (Figures 6). The hydrophilic (polar) heads of the phospholipids align towards the inside and outside of the cell packing the hydrophobic (non-polar) tails in the middle. This arrangement occurs because the cell's cytosol contains mainly water (polar) and the cell's outside is bathed or surrounded by tissue fluid or blood which is predominantly water. There are two types of membrane proteins: those attached on the inside of membrane (**peripheral proteins**) and those which partly or wholly transverse the membrane (**integral proteins**). Some of these proteins may be modified glycoproteins or glycolipids.

Cellular Basis of Life

Figure 2. Generalized structure of a cell under a light microscope - the plasma membrane, nucleus and cytosol are visible
Source: Image used under license from Shutterstock.com

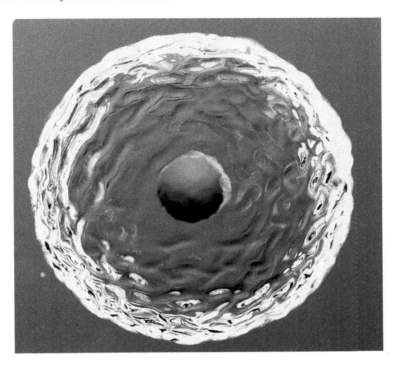

Figure 3. Human cheek cells. Note the stained nucleus.
Source: Image used under license from Shutterstock.com

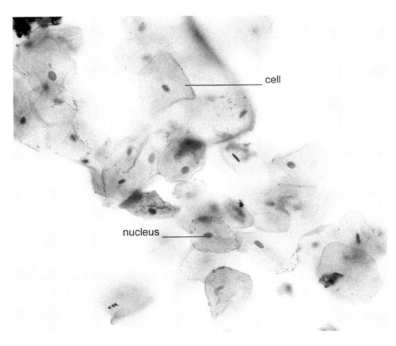

Figure 4. The basic structure of an animal cell as seen through an electron microscope. Note the numerous organelles in the cytosol.
Source: Image used under license from Shutterstock.com

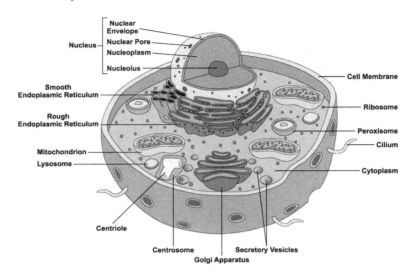

Figure 5. The basic structure of a plant cell as seen through an electron microscope. Note the green chloroplasts.
Source: Image used under license from Shutterstock.com

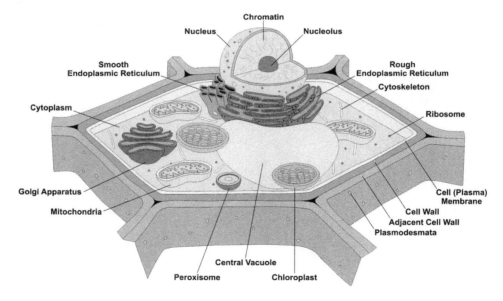

Figure 6. Structure of the plasma membrane. Note the many membrane proteins.
Source: Image used under license from Shutterstock.com

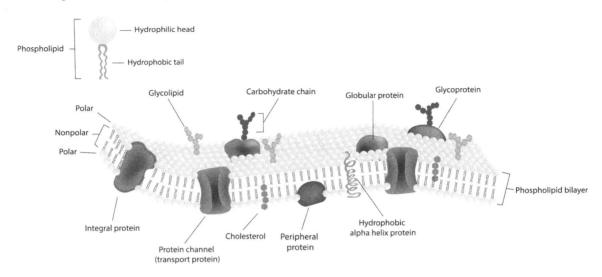

Figure 7. Cholesterol maintains fluidity of the plasma membrane. See text for explanation.
Source: Image used under license from Shutterstock.com

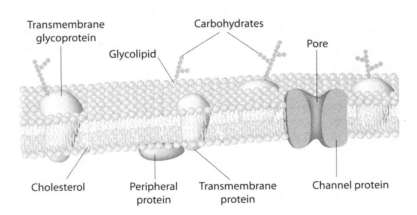

Membrane proteins serve many functions in the cells including:

1. **Transport:** Molecules constantly shuttle in and out the cell; proteins serves as a passageways or "highways".
2. **Receptors:** Have binding sites which recognize chemical signals in the fluid outside the cell, e.g. hormones.
3. **Enzymes:** Involved in catalyzing reactions important to the cell.

4. **Recognition Badge:** The proteins give the cell a sense of identity; this is reason the body reacts to foreign cells.
5. **Attachment:** Some proteins attach one cell to its neighbor or to its environment thereby anchoring it in place.

The plasma membrane also contains other molecules serving different functions. For example: 1) Glycoproteins are involved in cell anchoring, production of mucosal lubricants and other protective secretions. 2) Glycolipids are important in signal reception and other cell-to-cell interactions. 3) Cholesterol is found in the plasma membrane serving to regulate the fluidity (not too fluid or too viscous) of the membrane. Since the phospholipids unsaturated fatty acid chains with kinks move about (are "fluidy"), cholesterol helps maintain membrane firmness and sturdiness. At low temperatures cholesterol prevents the clumping and possible crystallization of phospholipid saturated fatty acid tails by packing closely, thus helping maintain the integrity of the membrane. Cholesterol also helps in the anchorage of membrane proteins allowing them to function smoothly (Figure 7).

Nucleus

The nucleus is the control centre of the cell in that it contains the instructions for cellular function carried on DNA. The nucleus is made up of the nuclear envelope - a membranous boundary made up of phospholipids and with numerous nuclear pores (openings); **chromatin** – made of DNA and associated proteins; the **nucleolus** – a dark body inside the nucleus; and the **nucleoplasm** – a fluid-filled space that bathes the nucleolus (Figure 8).

Just over two meters of DNA is tightly packed and coiled around blocks of protein known as **histones** in **nucleosome** units called chromatin (Figure 6; Schones & Zhao, 2011). This tight packing allows DNA to neatly fit within the nucleus. The nucleolus is the site for the manufacture of organelles called **ribosomes** (see below). Molecules made inside the nucleus leave through the nuclear pores to the cytoplasm of the cell (see Figure 9).

Figure 8. Nucleus structure and chromatin
Source: Image used under license from Shutterstock.com

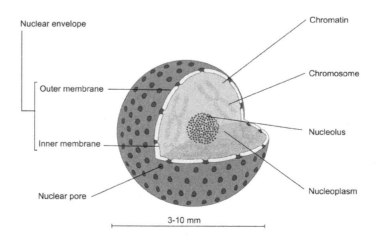

Cellular Basis of Life

Figure 9. Cellular organelle components
Source: Image used under license from Shutterstock.com

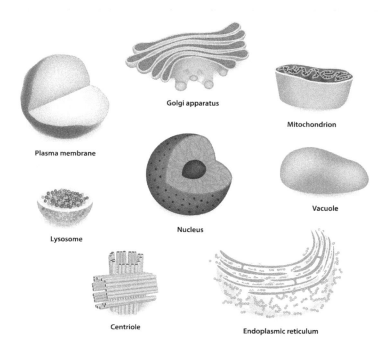

Ribosomes

The cell uses these organelles in the manufacture of proteins (protein synthesis). Some ribosomes are freely floating in the cytoplasm where they make proteins meant for internal cell use (cytosolic proteins, see Figure 4). Ribosomes attached to membrane structures called **endoplasmic reticula** manufacture proteins destined for outside the cell (export proteins) (Figure 7). Before ribosomes can make proteins, they need to become functional. When a larger ribosome subunit joins a smaller subunit, the complex makes a *protein synthesis-ready* functional ribosome (covered in Chapter 8).

Endomembrane System

- **Endoplasmic reticulum (ER):** This is a set of folded membranous structures which can have attached ribosomes (now called **rough ER**); or are smooth with no ribosomes attached (**smooth ER**). The two types of ER work as a set with molecules from the rough section passing through the smooth section on their way for further processing in other structures in the endomembrane system including Golgi body, vesicles, and the plasma membrane (see Figure 9). The rough ER is the site for the synthesis of proteins destined for export out of the cell to other parts of the body. Examples of such proteins in humans include hormones like insulin which is made in the rough ER of the pancreas, excreted out of the pancreas cells into the surrounding tissue fluid and blood system. The blood distributes the insulin to all body cells. Cells have protein receptors for insulin in their plasma membranes and once they recognize it, channel proteins in the plasma membrane

Figure 10. Lysosomes contain hydrolytic enzymes for intracellular digestion
Source: Image used under license from Shutterstock.com

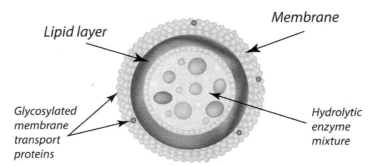

allow glucose to enter cells. The smooth ER is involved in the manufacture of lipids, the metabolism of some carbohydrates and in the liver drug/poison detoxification. Some of the lipids manufactured include steroid hormones oestradiol and testosterone in the smooth ER of the ovaries or testes respectively. Poisons like antibiotics are removed from the blood in the smooth ER of liver and sent to the kidneys for elimination—hence the typical smell in the urine of someone taking antibacterial drugs.

- **Golgi Apparatus:** The **Golgi complex** is another set of flattened membranes whose function is to modify products reaching it from the ER carried by small, round membrane structures called **vesicles**. In the Golgi complex are a variety of enzymes which help in the manufacture of new products like glycoproteins, lipoproteins or glycolipids. Such products are then packaged and shipped out through vesicles which fuse with the cell membrane and "dump" their products into the surrounding tissue fluid in a process known as exocytosis (see Figure 13 below).

Figure 11. The cytoskeleton. Made up of protein microfilaments (actin), intermediate filaments (keratin) and microtubules (tubulin).
Source: Image used under license from Shutterstock.com

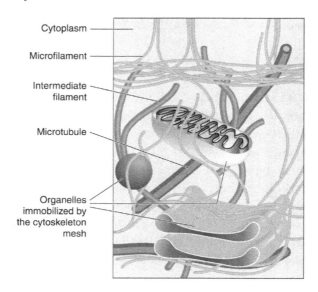

Cellular Basis of Life

Lysosomes

Just like any establishment with a cleaning crew responsible for emptying our trash cans and cleaning rooms to ensure steady operations; these roundish organelles are thought to bud-off from the Golgi complex to become independent organelles. They have a low pH due to the many **hydrolytic enzymes** they contain involved in intracellular digestion of many cell molecules (Figure 10). For example, immune cells like macrophages have lots of lysosomes to digest bacteria; cancerous cells; old, dying or dead cells; and other molecules identified by the cell as foreign. To maintain cellular operations at peak efficiency, cellular components like aging organelles are broken down and materials recycled in the lysosomes. In cases where normal healthy cells need to be destroyed (as in infected or injured cells) cells initiate self-destruction in a process known as apoptosis. A good example of this apoptotic process is the monthly menstruation of female mammals where the uterine endometrial tissue breaks down after ovulation. This breakdown only occurs if the egg is not fertilized—in which case, the endometrial layer, well primed for implantation of the fertilized egg, is no longer needed.

Peroxisomes

These are small round membranous structures (similar to lysosomes) in the cell involved in other cellular functions (see Figures 4 and 5). They contain enzymes involved in the oxidative degradation (using O_2) of macromolecules in the cell with the by-product hydrogen peroxide (H_2O_2, hence their name). Liver peroxisomes help in detoxification of alcohol and other toxic substances. Since H_2O_2 is toxic to the cell, it is converted by other enzymes in the peroxisome into harmless water and oxygen as shown.

$$2H_2O_2 \rightarrow 2H_2O + O_2$$

Mitochondria

These are membranous structures which are the cell's "power" factories where glucose is oxidatively broken into carbon dioxide and water in the process of cellular respiration (see Figure 4 and 5). This process releases energy locked in the covalent bonds of molecules like glucose which is then harnessed into a form usable to the cell in form of **triphosphate molecules** of adenosine or guanosine making ATP or GTP. Like human economic systems using dollars, the cells favourite currency is stored ATP or GTP molecules.

Centrioles

These structures only found in animal cells are made up of protein **microtubules** arranged in cylindrical patterns (see Figure 4 and 5). They microtubules attach to **chromosomes** and move them during the process of **cell division** when a cell makes daughter cells as discussed in the next chapter.

Cytoskeleton

These structures form the "skeleton" of the cell helping it maintain its integrity. The cytoskeleton is made of a meshwork of fibres called **microtubules** (made up of the protein tubulin) and **microfilaments** (made up of proteins actin and keratin, Figure 11) interspersed throughout the cell. Also found in the cytoskeleton are ATP powered motor proteins that aid in cell and organelle movement. A typical case is when organisms with cilia or flagella use these motor proteins to move microtubules thereby moving the whole organism around as in paramecium or flagellates. The cytoskeleton provides anchorage to organelles and offer cellular support and also maintaining cellular shape.

3.2.2 Unique to Plant Cells

There are three structures found only in plant cells including **cell wall**, **central vacuole** and **plastids** (see Figure 5 above).

- **Cell wall:** The cell wall made up of the complex polysaccharide cellulose and found just outside the plasma membrane. It gives plant cells firmness, rigidity and shape and is perforated by pores called **plasmodesmata** that allow exchange of materials between plant cells (Figure 5).
- **Central Vacuole:** The central vacuole is a fluid-filled structure occupying up to 75-80% of the cell's interior. They help in the cell's waste disposal process, provides cell support, and is used for the storage of important molecules like nutrients, and in cell growth. Often plant parts that are distasteful to animals have molecules stored in their central vacuole which repel animals (Figure 5).
- **Plastids:** Plastids are organelles used by plant cells to store sugars (called **leucoplasts**) or contain colored pigments with special functions in plants (**chromoplasts**). Chromoplasts like **chloroplasts** help in absorption of sunlight which powers their role as "sugar factories" in the plant. **Carotenoids** are other plant pigments which also help in sunlight absorption and make plants more visible to pollinators (Figure 5).

3.3 CELLULAR COMMUNICATION AND SIGNALLING

Cells stay put securely anchored in place and are also in constant communication with each other. The ability of cells to respond to their immediate environment requires that communication within them is facilitated in a timely and very specific way. These properties require that cell membranes of **epithelial cells** be equipped with junctions that make this possible as discussed below.

Gap Junctions

These structures allow cells to communicate with each other are called **gap junctions**. These junctions connect the cytoplasm of a cell with its neighbors and occur within membrane proteins channels which allow passage of small molecules, sugars and other ions. The opening and closing of gap junctions is regulated by membrane proteins depending on the situation. For example, when cells need to work in congruence as in contraction of heart cardiac muscle cells, gap junctions allow for electrical communication needed to synchronize the contractions.

Cellular Basis of Life

Tight Junctions

The contents of each cell need to be contained without leakage or loss of any kind. **Tight junctions** made up of protein strands make this fastening possible by uniting cell membranes of adjacent cells. However, these junctions serve as selectively permeable barriers with cells allowing specific molecules to enter or leave. A fine example of tight junction function is how gut contents stay in the lumen of gut while at the same time the products of digestion are absorbed into the blood.

Adhering Junctions

For cells to stay put, their internal structural cytoskeletal elements are tethered to **adhering junctions** made up both intracellular anchor and transmembrane anchor proteins. This is analogous to how rivets hold two pieces of iron sheets together. Such junctions are abundant in cells often subjected to sever mechanical stress as in cells that constantly need to contract and expand to function, e.g. muscle and heart cells.

Movement In-and-Out of Cells

Cells are constantly exchanging materials with their neighboring environment be it other cells, blood or tissue fluid. The process can occur in several different comprised of simple diffusion, facilitated diffusion, endocytosis, exocytosis, osmosis, passive and active transport.

- **Endocytosis:** "Endo" = "to the inside"; "cytosis" = "cell process". Cells are constantly taking in substances like molecules from their surrounding environment. These substances enter the cell by a receptor-mediated uptake or can push against the plasma membrane forming pockets inside the

Figure 12. Movement in and out of cells: endocytosis
Source: Image used under license from Shutterstock.com

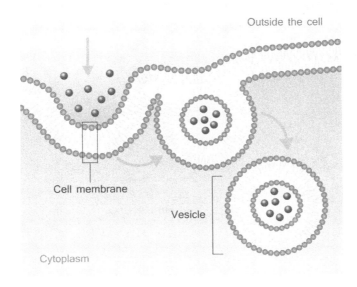

cell which become sealed off as vesicles enclosing the substances and enhancing the transportation of materials the cell needs (Figure 12). Organisms like bacteria use **phagocytosis** (a form of **endocytosis**) to ingest large particles like food items. **Pinocytosis** is another form of endocytosis where cells uptake molecules or fluids into the cell.

- **Exocytosis:** "Exo" = "to the outside". This is the reverse of the endocytosis where materials leave the cell in vesicles which fuse with the plasma membrane and empty their contents into the environment surrounding the cell (Figure 13).
- **Diffusion:** This is the process where substances move down a concentration gradient from a region where they are abundant (high concentration) to a region where they are scarce (low concentration). The best example would be how molecules from an air freshener diffuser gradually diffuse through the air to reach all parts of a room (Figure 14). In water, one can visualize this process when a drop of blue dye put on top of a glass full of water. The ink drop gradually diffuses throughout the glass creating a blue liquid. (Figure 15).
- **Osmosis:** This process occurs in plants and is the movement of water from a region where it is in high concentration to a region where it is scarce through a selectively permeable membrane. A good example is how plant cells uptake water from the soil. Water in the soil is more abundant than in plant roots, so plant root cells absorb water from the surrounding soil particles where it is in high concentration. The molecules pass through the selective filter of the root cells (plasma membrane) and enter the cells, then transported to the plants leaves by special water conducting vessels in the plant (Figure 16).
- **Passive Transport:** Water soluble substances can readily diffuse through the interior of channel or carrier proteins in a process called **facilitated diffusion**. The carrier or channel protein is the facilitator. (Figure 17).

Figure 13. Movement in and out of cells: exocytosis
Source: Image used under license from Shutterstock.com

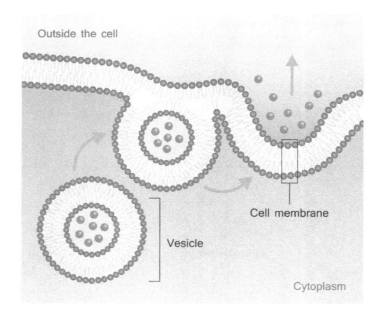

Figure 14. Diffusion of air freshener molecules in air
Source: Image used under license from Shutterstock.com

Figure 15. Diffusion of ink molecules in an aqueous medium
Source: Image used under license from Shutterstock.com

Figure 16. Osmosis. This is the movement of water molecules from a region where they are in high concentration (low solute - sugar in this case) to a region with low concentration. Source: Image used under license from Shutterstock.com

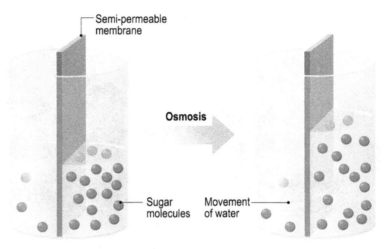

- **Active Transport:** Often in the cell, carrier proteins in the plasma membrane need to be energy boosted or primed for reaction through phosphorylation by a molecule of ATP. In this process the phosphorylated carrier changes its conformation to allow entry or exit of substances in/out of the cell. (Figure 18).
- **Coupled Transport:** When two substances are simultaneously transported across the cell one way (symport) or opposite ways (antiport). As one molecule moves along its concentration gradient the energy difference can be used to move a second molecule at the same time. A good example is where H^+ are concentrated on one side of the membrane and are facilitated across the

Figure 17. Facilitated diffusion. Integral (carrier) proteins facilitate passage of molecules down their concentration gradient either into or out of the cell.
Source: Image used under license from Shutterstock.com

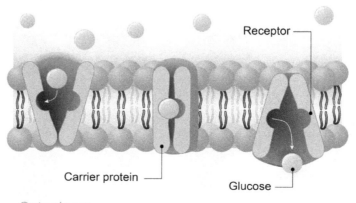

Cellular Basis of Life

Figure 18. Active transport. A membrane carrier protein needs an energy boost from ATP to change its conformation and transport molecules into and out of the cell. The donation of a phosphate by ATP is called phosphorylation.
Source: Image used under license from Shutterstock.com

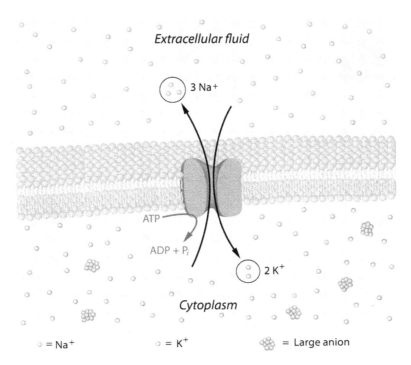

membrane to the other side. Another molecule like sucrose can link up with H^+ ions as they shuttle across through the channel protein.

- **Tonicity:** This term refers to the osmotic pressure gradient of water in solutions that are separated by a semi-permeable membrane. In defining tonicity, we will use the term "**solvent**" to mean water and "**solute**" as ions like Na^+ Cl^- or sugar molecules. The pressure gradient determines which direction water molecules will flow depending on the solute concentrations on both sides of the membrane (Figure 19). In isotonic solutions, the concentration of solutes is the same on both sides. For example, saline solution is often used in medical situations to counter a patient's dehydration. Saline solution is isotonic to the blood in patients. In cells subjected to a hypotonic solution in a beaker for example, there is less solute in the beaker (hence more water) than in the cell (hence less water). Water enters the cell resulting in swelling to the point of bursting and cell death. Cells put in a beaker with a hypertonic solution (more solute, less water) will lose water shrinking them to death (Figure 19). The best example to demonstrate both hypotonic and hypertonic processes is to use pealed potato cells. Prepare two beakers: in one put pure water (distilled water); in the other put water plus two teaspoons of salt and stir until well dissolved. Now get a fresh potato, peal it and cut two rectangular pieces (chips). Set a timer and dip each rectangle in the beakers. After 30 minutes, remove the potato chips. Describe what you observe?

Figure 19. Tonicity in cells. Water movement in isotonic, hypotonic and hypertonic environments. The size of the arrow indicates the volume of molecules moving either way. See text for further explanation.
Source: Image used under license from Shutterstock.com

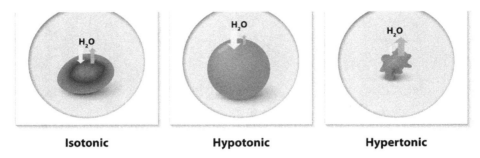

Cell Signal Reception, Transduction and Response

These processes are primarily facilitated by a variety of molecules including proteins and energy rich nucleotides (ATP and GTP); and also by ions like Ca^{2+}. Signals received at the surface of the cell from other cells, or from blood or tissue fluid must be transferred to various parts of the cell and an output (cell response) initiated. This molecular interaction process, often referred to as the **bio-signaling** or **signal transduction** involves three stages: signal reception, transfer (or transduction), and resulting cellular response (Figure 20).

Figure 20. Cell signalling. Three major steps are involved: signal reception, transduction, and cellular response.
Source: Image used under license from Shutterstock.com

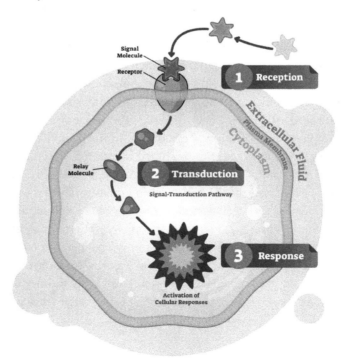

Cellular Basis of Life

Signal Reception

Signals reaching cells can be received at the cell surface (plasma membrane) or in the cytoplasm depending on the nature of the signal (often referred to as a **ligand**). If the signals are lipids like steroid hormones (e.g. estradiol or testosterone), they easily diffuse through the plasma membrane to the cytoplasm. How is this possible? Remember, the plasma membrane is primarily made of phospholipids. Like begets like, enabling the steroid hormone signals to interact with the plasma membrane seamlessly. Once in the cytoplasm however, steroid hormones must be recognized by other molecules to initiate transduction (Figure 21).

Non-steroid signals like protein hormones (e.g. insulin), catecholamines (e.g. hormone epinephrine) or eicosanoids (e.g. prostaglandins involved in inflammation) must be recognized at the plasma membrane surface by receptors. Specific protein receptors embedded in the cell membrane bind the signals, enabling conformational changes inside channel or transport proteins which transport the signal to the cytoplasm. There are three types of protein receptors for non-steroidal signals: G-protein coupled receptors (GPCR), receptor tyrosine-kinases & ion channel proteins. Non-steroid signals activate other molecular or ionic mediators in the cytoplasm.

- **G-proteins:** After signal binding, the GPCR activates GTP binding proteins called **G-proteins** located near the inside surface of the plasma membrane. A GTP molecule is hydrolyzed in the

Figure 21. Interaction of a steroid hormone with an intracellular protein receptor in the cytoplasm and activation of genes in the nucleus
Source: Image used under license from Shutterstock.com

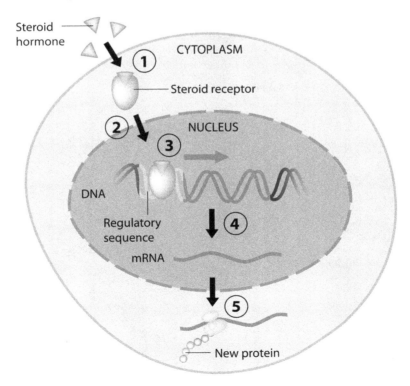

process and the energy released allows the GTP binding protein to activate yet another inactive membrane protein (Figure 22).

The activated protein initiates a cell response to the initial signal in one of two main ways:

1. **Receptor Tyrosine Kinases (RTKs):** According to Lemmon and Schlessinger (2010) humans have about 58 known RTKs all having similar architecture. RTKs exists in the cytoplasm as inactive monomers until they bind a signalling molecule that unites them into dimers. They have an extracellular ligand-binding domain, a transmembrane protein, and an internal domain the mediates several downstream signalling molecules through phosphorylation (Kriete & Eils, 2014). RTKs ligands can include cytokines, growth factors, and hormones. Once dimerized, several molecules of ATP are hydrolyzed to ADP and P. The phosphate attaches to the dimer (phosphorylated) creating an activated dimer complex. This in turn activates specific relay proteins which initiate the cell response to the signal (Figure 23).

2. **Ion Channel Proteins:** Channel proteins are integral proteins that span the plasma membrane repeatedly and selectively allow ions like potassium, sodium, chloride, etc. to move in and out of the cell. Ion movement is a function of gradients that can either be chemicals (ion concentration) or electrical (voltage). For instance, in bacteria channel proteins are made of four subunits assembled together forming a channel with pores at either end. The walls of the channel are structured in a way that stabilizes the transported ions being filtered creating selectivity. The pore loops are also made of different amino acids enabling each channel type to conduct specific ions. (Purves et al., 2001). For example, channels that transport potassium ions (K^+) will not allow calcium ions (Ca^{2+}) ions

Figure 22. A G-protein coupled receptor (GPCR). Guanosine triphosphate, GTP (an energy carrier akin to ATP) provides the energy required to activate the G-protein.
Source: Image used under license from Shutterstock.com

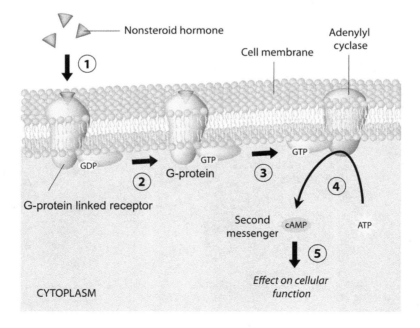

Cellular Basis of Life

Figure 23. Activation of Receptor Tyrosine Kinases (RTKs) in signal transduction. Note that energy required for signalling is donated by ATP.

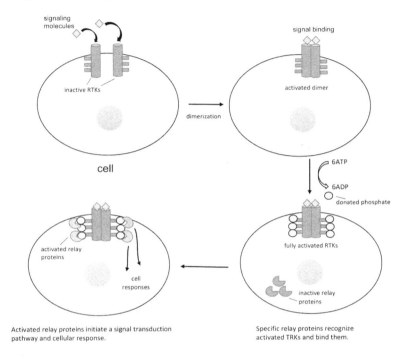

to enter the channel. In channels that are electrically regulated, a sensor made of a transmembrane domain with positive charged amino acids detects differences in membrane electrical potential. This change alters the orientation of the transmembrane protein by rotating it in a way that effectively opens the pore. Alternatively, channels that close because of the electrical changes have amino acid sequences that plug the pore, curtailing ion transport. (Purves et al., 2001). (Figure 24).

Signal Transduction Systems and Cellular Response

Signal transduction pathways vary depending on the type of molecule involved and nature of response needed by the cell. Cellular responses are varied and can include genomic processes, metabolic processes, cell cycle regulation and morphogenesis.

There are several systems of signal transduction systems: GPCRs, G-proteins, effector proteins, and receptor proteins for 2nd messengers (protein kinases A, G, C and calmodulin kinase) as depicted below:

$$GPCR \to G-Protein \to Effector\ Protein 2^{nd} \to Messenger \to Cellular\ response$$

- **Steroid Hormone Signaling:** In the cytoplasm, steroid hormones bind a protein receptor forming a hormone-receptor complex which propagates to the nucleus through the nuclear envelope to begin the cell response to the signal (see Figure 21). For steroid hormones, the activated hormone-receptor complex binds to specific DNA "acceptor sites" (on the **promoter**) assisted by several

Figure 24. Ion channel proteins. In response to a signal, proteins gates open to allow passage of molecules into and out of the cell.
Source: Image used under license from Shutterstock.com

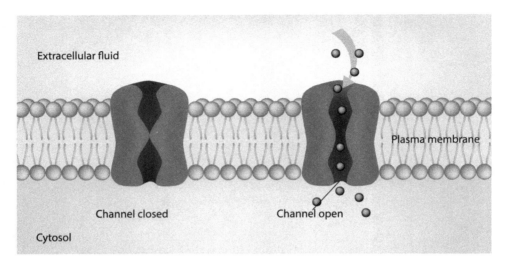

molecules like steroid response elements, DNA-binding "acceptor" proteins, and transcription factors (Landers & Spelsberg, 1992). Steroid hormones regulate **transcription** of genes by initiating or inhibiting **gene expression**. For example, imagine a young man as he reaches puberty. His testes start producing the male hormone testosterone which circulates in the blood to all cells of the body. Specific target cells in the larynx, shoulders, chin and upper lip among other places in the body have receptors that recognize the hormone. Following the above process of reception and transduction, the signal reaches its target, the DNA. In boys, genes that control voice, widening of the hips, sperm production, beard and moustache are transcribed transforming the boy into a young man with secondary sexual characteristics like a deeper voice, wider shoulders and more hair growth. The same case happens in females at puberty following the female hormone estradiol: widening of the hip region, ovulation, breast development and fat padding in many parts of the body.

- **Non-Steroid Hormone Signaling:** Transduction involves several protein molecules in a phosphorylation cascade and other multiple mediators (broadly called **second messengers**). Note that a signaling molecule like a hormone is considered the **first messenger** (see Figure. 22). Such second messengers are ions (like Ca^{2+}), lipid metabolism products (like diacylglycerol (DAG), inositol triphosphate (IP_3) and phosphatidyl-inositol-biphosphate (PIP_2)), nucleotides (like cyclic AMP or cyclic GMP), and a variety of protein mediators that relay, adapt, transduce, anchor, integrate, modulate or amplifier signals (Kulinsky & Kolesnichenko, 2005a).
  - **Protein Kinase Phosphorylation Cascade:** Protein kinases (PKs) are enzymes which phosphorylate proteins with serine and threonine amino acid residues. They also target other substrates like enzymes, ion channels, translational factors, and structural proteins. PKs are regulated by 2nd messengers such as cAMP, cGMP, DAG, Ca_{2+}-calmodulin, PIP_3 and other kinases called tyrosine kinases (these phosphorylate amino acid tyrosine). PKs regulate many intracellular processes like the induction of metabolic enzymes enhancing or decreas-

ing synthesis and amplifying catabolic activities of the cell (Kulinsky & Kolesnichenko, 2005b). Examples of such cellular processes include epinephrine activation of glucogenesis in "Fight or flight" emergency situations, or in gene activation by a growth factor.
  - **Phosphatidylinositol System:** A complex system involving G-protein coupled receptors, G-proteins, four second messengers (PIP_2, IP_3, DAG, and Ca^{2+}), enzymes (for the synthesis and degradation of these molecules), and two protein kinases (calmodulin and C kinase) (Kulinsky & Kolesnichenko, 2005a).
- **Calcium Ions:** A flexible and universal second messenger Ca^{2+} is used in activating or inhibiting many body processes. The protein calmodulin (CM) is often (but not always) the main receptor for Ca^{2+} ions and forms a Ca^{2+}-CM complex which binds protein kinases producing a variety of cellular effects including cell division, muscle contraction, gene transcription, fertilization, cell proliferation, and more. Research has also found out that Ca^{2+} increase in cells damages them bringing about apoptosis or cell death (Kulinsky & Kolesnichenko, 2005a).
- **Inositol Triphosphate (IP_3):** A GTP bound G-protein activates the enzyme phospholipase C whose substrate is PIP_2. PIP_2 is cleaved into two 2nd messengers DAG and IP_3. IP_3 opens Ca^{2+} channels in the endoplasmic reticulum and plasma membrane allowing Ca^{2+} influx into the cytosol.
- **Phosphatidyl-Inositol-Biphosphate (PIP_2):** This 2nd messenger produces a number of cellular responses including regulating channel and ion transporters like Ca^{2+}, K^+, Na^+, and ryanodine channels; regulating cytoskeletal structure; movement of vesicles in the cytosol; and exocytosis. PIP_3 also acts as a 2nd messenger and is thought to be involved in coupling cell signalling to actin polymerization. Research indicates that the levels of PIP_3 in unstimulated neutrophils (white blood cells) are very low (50 nM in resting cells) but they increase dramatically (2 μM) within 10s of stimulation with chemoattractant which closely coincides with the kinetics of actin polymerization (Howard & Oresajo, 1985; Stephens et al., 1991).
- **Diacylglycerol (DAG):** DAG as a second messenger recruits protein kinase C (PKC) which phosphorylates and alters activity of some ion channels and cytoplasmic enzymes. In the nucleus, PKC phosphorylates histones and activates specific transcription factors thereby stimulating gene expression promoting cell proliferation/differentiation, embryogenesis, and angiogenesis among other effects (Kulinsky & Kolesnichenko, 2005a).
  - **cAMP and cGMP systems:** A GPCR activates a GTP bound G-protein which activates the relay protein adenylyl cyclase. Adenylyl cyclase after hydrolysis ATP to form cAMP which is now the 2nd messenger. The main receptor for cAMP is protein kinase A (PKA) which dissociates after binding cAMP into free catalytic subunits. PKA subunits phosphorylates target proteins including enzymes, structural and matrix synthesis components. Some subunits enter the nucleus regulating genomic processes. cGMP resulting from GTP hydrolysis acts similarly to cAMP targeting gated ion channels and protein kinases. (Figure 25).

3.4 HUMANS: COMPLEX CELLULAR SYSTEMS

Four basic types of cells make up the human body: epithelial cells, connective tissue, muscle tissue and nervous tissue (Figures 26, 27, 28). **Epithelial tissue** cells cover internal and external surfaces of organisms and come in a variety of forms. Simple epithelia are single layers of cells of various shapes (squamous, cuboidal, or columnar) occurring in structures like the mouth. Stratified epithelial cells are

Figure 25. cAMP as a second messenger activating other proteins
Source: Image used under license from Shutterstock.com

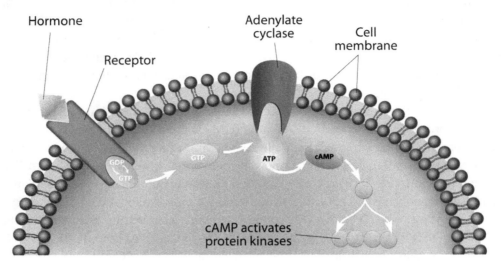

Figure 26. Connective animal tissue
Source: Image used under license from Shutterstock.com

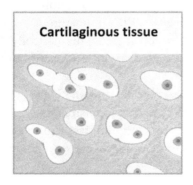
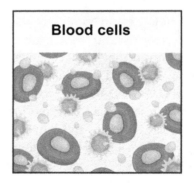
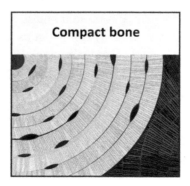
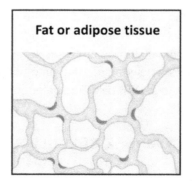

Cellular Basis of Life

Figure 27. Nervous animal tissue (neuron)
Source: Image used under license from Shutterstock.com

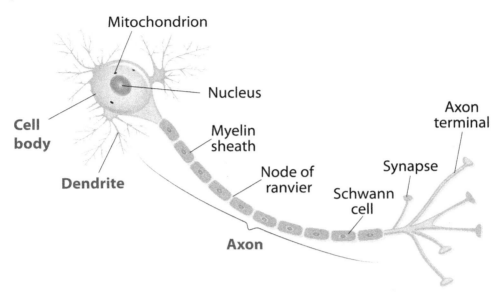

Figure 28. Muscle animal tissue
Source: Image used under license from Shutterstock.com

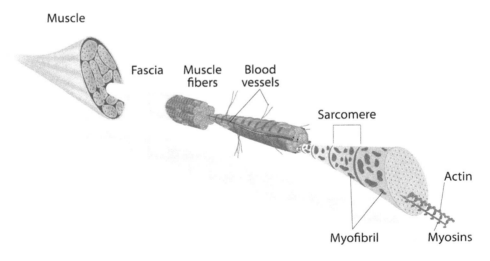

in multiple layers for cells that need to withstand some sort of force (mechanical or chemical) like skin cells covering the human body or human bladder/urethral cells.

Connective tissue literally "connects" many body systems and is widely distributed in the body including cartilage, bone, adipose tissue and blood. **Muscle tissue** can occur in form of skeletal (e.g. biceps), smooth (gut lining) or cardiac (heart) muscle tissues. **Nervous tissue** is made up of neurons and is found in the nervous system (brain, spinal cord and peripheral nerves).

During embryonic development, three germ layers form all the various body systems in vertebrate organisms which includes humans. The outermost layer (**ectoderm**), differentiates into the surface ectoderm which forms the epidermis, hair and nails. The neuroectoderm forms the nervous system which is further differentiated into the central nervous system (brain and spinal cord) and the peripheral nervous system (nerves in arms, hands, legs and feet). The middle layer (**mesoderm**) produces the dermal, skeletal, muscular, reproductive, circulatory and urinary systems. The innermost layer (**endoderm**) gives rise to two main tubes which share an anterior chamber called the pharynx. Extensions emanating from the pharynx form the thyroid, parathyroid, and thymus glands. The first is the digestive tube or gut with branching internal organs like the lungs, pancreas, and gall bladder. The second is the respiratory tube which forms the trachea, bronchi and the lungs.

3.5 METABOLIC FUNCTIONS IN CELLS

To do their job cells are actively taking in raw materials (energy and nutrients) which they use to function (build, store, break and eliminate substances) and perform maintenance activities like repairing injured tissue, programmed cell death or cell replacement. These activities are collectively called **cellular metabolism**. Plant absorb water and mineral salts from the soil and make energy with the help of sunlight in the leaves where carbon dioxide diffuses from the air in the process of **photosynthesis** (photo = light energy; synthesis = building up). The energy made is stored in high energy covalent bonds of molecules like sugar in many plants; lipids in certain plants like castor oil, palm oil and coconut; and proteins in leguminous plants like beans and lentils. Eating plant products like fruits and vegetables provides not only energy but major and minor minerals required for proper cellular metabolic functions.

Metabolism typically involves building up reactions (**anabolism**) and breaking down reactions (**catabolism**). These reactions are often complex, and many involve a series of steps where substrates are converted to final products through several intermediate substrates by **enzymes**. Enzymes are organic, usually proteins made in the cell and have a catalytic role of speeding up chemical reactions in living cells. Catalysts can be organic (enzymes) inorganic (metal ions). Reactions in living cells can involve linear, branched or cyclic pathways. In addition to enzymes, several other molecules are involved in metabolic pathways including **cofactors** (**vitamins** and metallic ions like Fe^{2+}) and energy carriers like ATP.

Enzymes

In living cells, enzymes are very specific by recognizing specific substrates (Figures 29, 30). For example, we eat a variety of complex macromolecules like carbohydrates, proteins or lipids. These large polymers cannot be absorbed into our blood stream in this form. They need to be broken down in smaller molecules called monomers like glucose, amino acids or fatty acids. These can easily diffuse from the lumen of the small intestine into the blood and transported to cells. Since enzymes are proteins, they have a 3-D configuration with a unique and specific groove or crevice called an **active site** (Figures 29, 30). This is the site where interaction between the enzyme and substrate occurs. Biologists propose an **induced-fit model** of enzyme-substrate interactions. According to this model, substrates do not perfectly fit into the active site. After the enzyme recognizes and binds the substrate, conformational changes in the active site enhance the enzyme-substrate complex for a better fit. Once this close proximity is established, interac-

Figure 29. An enzyme has an active site which recognizes and binds to a specific substrate Source: Image used under license from Shutterstock.com

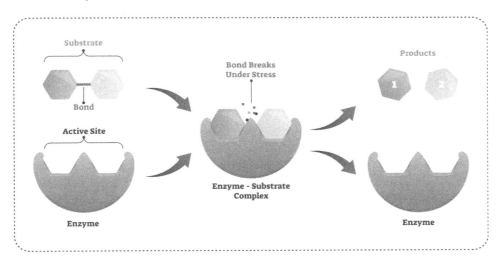

Figure 30. When enzyme binds the substrate it forms an enzyme-substrate complex and converts the substrates to products
Source: Image used under license from Shutterstock.com

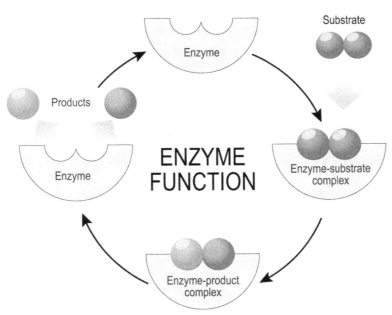

tions between the active site and substrate strains and rearranges the substrate(s) in ways that form new products (Figures 29, 30). Substances can also be combined in new ways to make different products.

Enzymes are not used up in a reaction much like a cab or taxi ferrying passengers and dropping them off before returning to original location. Like taxis we need to call or flag down, enzymes cannot initiate reactions, they only speed them up. Some examples of enzymes and their substrates are:

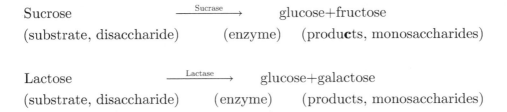

Borrowed from physics, biologists envision that enzymes lower the **activation energy** of a reaction. What does this mean? Imagine you are carrying a dead car battery to a shop to get a new one. If the car shop is located at the top of a flight of stairs with thirty steps, you will burn a lot of calories to get the battery to the shop and you will perhaps be sweating when you reach the shop. The sweat comes from oxidation reactions occurring in your body (working muscles, pumping heart, breathing lungs and better blood circulation). This expended energy is the activation energy you needed to get that battery to the shop. Reactions occur by increased collisions between the reactant molecules called kinetic energy. That is why it is easier to dissolve sugar in warm tea compared to iced tea. The hot tea molecules are moving faster in solution and colliding more often than in iced tea. Thus, in iced tea you will need to stir longer to dissolve the sugar. If the activation energy of a reaction is lower, it will occur faster just like if the flight of stairs had only ten steps, you will spend less energy carrying the battery up to the shop.

Enzymes make it possible for biochemical reactions to occur in living organisms by lowering the activation energy the molecules need to react. Note that these reactions would still occur—but very, very slowly, making life impossible. For instance, think of what happens if you encounter an emergency: while driving behind another motorist, you suddenly notice the vehicle swerve and hit a guard rail, the vehicle rolls over. What happens in you? Your **epinephrine** hormone levels rise causing a cascade of events in your body: your brain races, heart rate increases, lung capacity increases, breathing becomes faster and deeper, and your muscles get a big surge of glucose to energize your muscles and brain. Without enzymes, it is doubtful how fast you would respond to such an emergency.

Reactions occurring in the body are very sensitive to the body conditions of substrate concentration, enzyme concentration, pH, and temperature.

Let's consider each in turn.

- **Substrate Concentration:** Imagine you have not been able to take any food since breakfast and it is now 3 pm. Then as you walk, you remember you have some crackers in your bag. You check and find only 2 small crackers in a zipper bag and eat them. After eating, your gut enzymes will digest the crackers quickly since we have more enzymes in your gut that we have crackers (substrate). This reaction will be completed quickly because the substrate molecules are far fewer than the enzyme molecules available (Figure 31). Another example is the taxi business. The vehicles are the enzymes and people are the substrates. The taxis spend most of their time idle when there are only a few people needing them. The rate at which taxis with people are on the road ("reaction rate") will slow down because there are only so many people needing a ride.
- **Enzyme Concentration:** Now imagine the opposite of the above situation. You attended someone's birthday party and have eaten so much of the varieties of food and goodies available at the party. You are now stuffed. Your gut stores most of this food in the stomach. There is a valve (pyloric sphincter) regulating how much can enter the upper small intestine (duodenum) as diges-

Cellular Basis of Life

tion continues in the duodenum. You stay full longer. This is a situation where there are so many substrate molecules that the enzyme molecules available can only handle so many at any one time. As more substrates are converted to products, the enzyme molecules become available to bind more substrate molecules keeping the reaction rate up until all substrate is converted to products and your stomach empties. You will feel hungry after several hours which can last up to 12 hours depending on how much was ingested. In the taxi example, now imagine a ballgame just ended. There are just more people than the available taxis can handle. The rate at which taxis with people are on the road ("reaction rate") will slow down until the taxis drop off people and return to pick more (Figure 32).

- **Temperature: Homeotherms** (warm blooded animals) like humans maintain stable temperature conditions n their body giving cells optimal operating conditions. Cold-blooded organisms (**poikilotherms**) like reptiles are usually sluggish and slow in cold weather until they can absorb heat from their surroundings and their bodies reach optimal operating temperature. That's the reason a rattlesnake will rarely strike at you in early mornings when it is cold. Enzymatic reactions that maintain life operate at temperatures of about $98.6°F$ in humans (Figure 33). Metabolism will still occur slightly below or above this temperature in humans, but these reactions are not operating optimally. Why do cells operate within this narrow temperature range? Because the enzymes and other organic molecules important in life are destroyed by colder and warmer temperatures. This is much like how a protein's 3-D conformation is destroyed (denaturation) as discussed in Chapter 2 (section 2.6.3).

pH: Like temperature, the pH of body tissues needs to be maintained within a certain range which (unlike temperature) varies depending on the part of the body. Since pH means presence of H^+/OH^- ions, certain enzymes operate optimally at pHs that are low (acidic), neutral or high (alkaline). For example, the human gut is a good example of these variations in pH. In the mouth, the enzyme salivary amylase

Figure 31. The effect of substrate concentration on enzyme activity

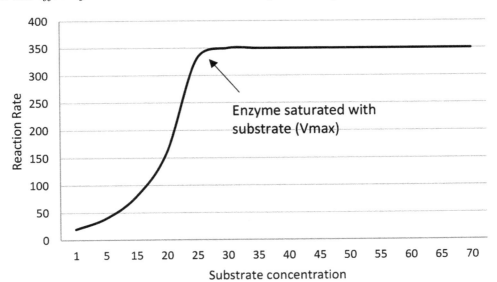

Figure 32. The effect of enzyme concentration on enzyme activity

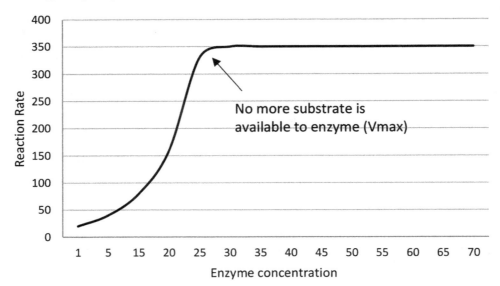

Figure 33. The effect of temperature on enzyme activity

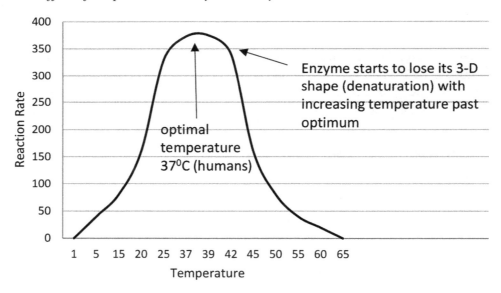

operates optimally under near neutral conditions; while in the stomach, the enzyme pepsin is activated by hydrochloric acid from pits in the stomach wall. In the upper intestine (duodenum), most enzymes from the pancreas, intestines operate optimally under alkaline conditions (Figure 34). Variations in pH can disrupt enzyme structures because of protonation/hydroxylation (or addition of H^+/OH^- ions) which disrupt bonding in the charged amino acids that are part of the enzyme structure.

Control of Enzyme Activity

The thousands of metabolic reactions in the body do not occur all at the same time. To avoid chaos in the body, cells maintain control of what reactions are needed all the time (like heart contraction), needed only certain times (as when food is eaten) or needed vary rarely (in emergency situations). For example, your body does not make hormones like insulin until they are needed after food is ingested; cells generally produce antibodies when they are invaded by foreign organisms; and unless there is a drastic change in ambient temperature, **thermoreceptors** in our skin do not keep sending signals to the brain when stable conditions predominate. That's why you feel very cold and chilly when you step out of a warm room in the winter to your car parked outside.

Cells control metabolic reaction by inhibition or activation. The normal action of an enzyme was depicted as shown above (Figure 30). However, when the enzyme needs to be regulated, certain substances are used by cells to stop enzyme action by having a competing substrate for an enzyme's active site called **competitive inhibition** (Figure 35a). For example, the poison cyanide acts by competitively blocking the active site of the enzyme cytochrome-C oxidase important in the electron transport chain that generates energy (ATP) in cellular respiration. In **non-competitive inhibition**, enzymes have two sites: active and another site away from the active site (called **allosteric site,** Figure 35b). Substances binding the allosteric site alter the shape of the enzyme and its active site such that the enzyme can no longer recognize its substrate. Many drugs against microbes act allosterically as seen in bacteria, where the antibiotic penicillin bind the allosteric site of the enzyme responsible for forming bacterial cells walls, causing death of the microbe. **Allosteric activation** is the binding of an activator to the allosteric site changing the shape of the active site in a way that enhances the enzymes ability to recognize and bind the substrate. In **feedback inhibition** (a type of non-competitive inhibition), the end product in a

Figure 34. The effect of pH on enzyme activity. Note that different enzymes have different optimal pHs depending on where they found in the body.

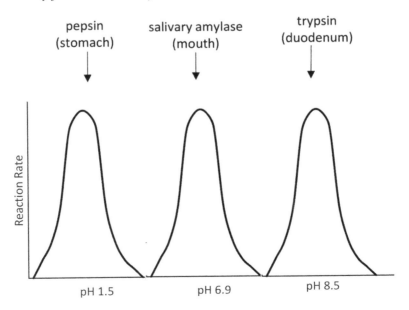

Cellular Basis of Life

Figure 35. a) Normal enzyme activity b) competitive inhibition c) non-competitive inhibition

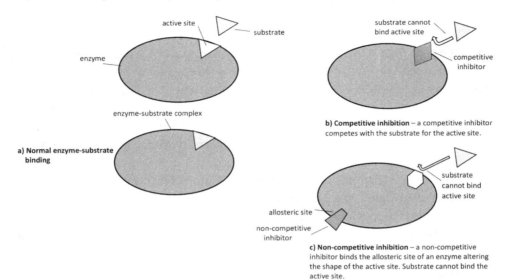

series of metabolic steps binds the allosteric site of the enzyme altering the shape of the enzyme's active site. A good example is the production of the amino acid isoleucine in the liver. Isoleucine binds the first enzyme in the chain stopping its own production (Figure 35c). Since allosteric binding is reversible, when the levels of isoleucine drop, the enzyme regains its active configuration and binds substrate.

- **Cofactors:** Cofactors include organic molecules called **coenzymes** (primarily vitamins) and inorganic metal ions (like Fe^{2+}, Fe^{3+}, Zn^{2+}). The cofactors assist enzymes in catalysis by acting as transfer agents during reactions. Although not substrates, cofactors bind the active site of an enzyme when a substrate binds and act as intermediate carriers of electrons, atoms or functional groups. The best example is seen in cellular respiration action of NAD^+ (B-vitamin niacin) and FAD^+ (B-vitamin riboflavin). In this metabolic pathway, NAD^+ accepts hydrogen atoms and electrons and delivers them to the electron transport chain where energy (ATP) is harnessed. ATP is a high energy carrier molecule in cells with its energy primarily held in the three energy-rich phosphates in the molecule (Figure 36).

CHAPTER SUMMARY

To qualify as living, units of life must be identifiable, distinct and demonstrate most or all the qualities of life. The Cell Theory outlines three broad elements that characterize all units of life including the cell as the basic unit of life and cells arising from existing cells. Cells tremendously vary in size from about 0.5-500 micrometers. The smallest known single cells of bacteria have existed on Earth as single cells for millions of years; while most higher organisms have multiple cells differentiated and functioning together as a single system.

Cellular Basis of Life

Figure 36. ATP, the high energy carrier molecule used by cells. When ATP donates a phosphate, it becomes ADP. After losing a second phosphate, it becomes AMP.
Source: Image used under license from Shutterstock.com

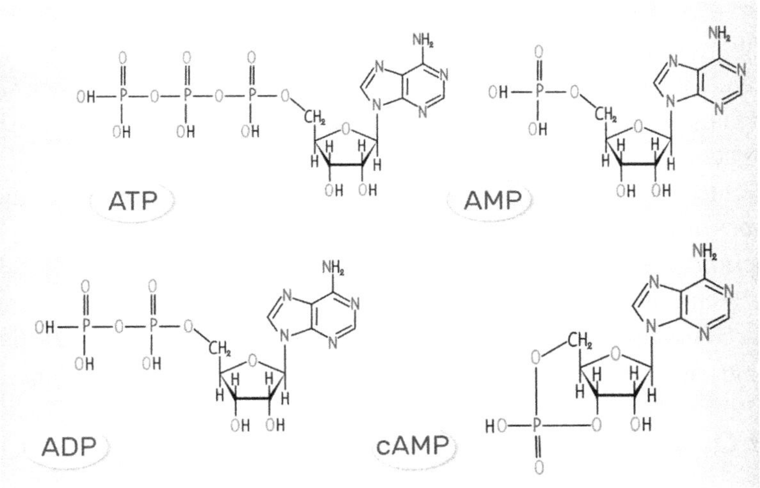

At very low magnification cells have a fluid-filled interior called cytosol which is bounded by an outside plasma membrane that delineates the cell boundary. Around the middle of the cell is the control center or nucleus which carries the genetic information important in cellular operations. At higher magnification, cells have a very complex structure with many "small organs" called organelles distributed into the cytosol made up of membranes and perform specific functions in the cell. Such organelles include ribosomes, endoplasmic reticulum, Golgi body, peroxisomes, lysosomes, mitochondria, and centrioles. The cytoskeleton maintains the cell shape and structure. Unlike animal cells, plant cells have a cell wall, central vacuole and plastids. Cells stay put securely anchored in place and are also in constant communication with each other due to presence of gap, tight, and adhering junctions. Materials exchange in and out of cells occur through several processes of endocytosis, exocytosis, diffusion, facilitated diffusion, passive transport, active transport, and co-transport.

The processes of cell signaling, reception, transduction and response are primarily facilitated by a variety of molecules like proteins and energy rich nucleotides (ATP and GTP), and by ions like Ca^{2+}. Signals received at the surface of the cell from other cells, or from blood or tissue fluid must be transferred to various parts of the cell and an output (cell response) initiated. Steroid hormones readily diffuse through the plasma membrane into the cytoplasm where they bind a protein receptor forming a hormone-receptor complex which propagates to the nucleus through the nuclear envelope activating specific genes. Non-steroid hormone signaling involves several protein molecules in a phosphorylation cascade and other multiple mediators called second messengers including ions, lipid metabolism products, nucleotides, and a variety of protein mediators that relay, adapt, transduce, anchor, integrate, modulate or amplifier signals.

Four basic types of cells make up the human body: epithelial cells, connective tissue, muscle tissue and nervous tissue. During embryonic development, three germ layers form all the various body systems in vertebrate organisms which includes humans.

Cells actively take in raw materials which they use to function and perform maintenance activities. Collectively these activities are called cellular metabolism. Metabolism typically involves building up and breaking down reactions. Metabolic reactions are often complex, and many involve a series of steps where substrates are converted to final products through several intermediate substrates by enzymes. Enzymes are usually proteins, very specific and speed up chemical reactions in living cells. With their 3-D configuration and an active site an enzyme recognizes and binds a substrate to form the enzyme-substrate complex in an induced-fit process. Enzyme catalysis is possible due to the lowering the activation energy of the reaction.

The rates of chemical reactions occurring in the body are very sensitive to the concentrations of substrate and enzymes available, the pH, and the temperature. To avoid chaos in the body, cells maintain control of what reactions are needed all the time, only certain times or needed very rarely. Cells control metabolic reaction by inhibition or activation. Cofactors which organic and inorganic molecules assist enzymes in catalysis by acting as transfer agents during reactions. They bind the active site of an enzyme when a substrate binds and act as intermediate carriers of electrons, atoms or functional groups.

End of Chapter Quiz

1. The Cell Theory states that
    a. the smallest units of life occur at the cell level of organization
    b. all organisms are made of one or more cells
    c. new cells arise only from existing cells
    d. A & B
    e. All the above
2. The three structures found only in plant cells are
    a. mitochondria, ribosomes, Golgi apparatus
    b. cell wall, chloroplast, plasma membrane
    c. cell wall, central vacuole, plastids
    d. central vacuole, plasmodesmata, mitochondria
    e. peroxisome, lysosome, centrosome
3. Which type of transport requires a carrier or a channel protein without the use of energy?
    a. Passive transport
    b. Active transport
    c. Coupled transport
    d. A & B
    e. A & C
4. The three stages to the signal transduction pathway are
    a. first phase, second phase, final stage
    b. signal reception, transfer, cellular response
    c. introduction, transduction, activation
    d. acute, chronic, terminal
    e. none of the above
5. Which of the following statements is false?
    a. Cholesterol is found in the plasma membrane

Cellular Basis of Life

    b.   Cholesterol helps maintain membrane firmness and sturdiness
    c.   At high temperatures cholesterol prevents the clumping and possible crystallization of phospholipid saturated fatty acid tails by packing closely, thus helping maintain the integrity of the membrane
    d.   Cholesterol regulates the fluidity of the membrane
    e.   Cholesterol helps in the anchorage of membrane proteins allowing them to function smoothly

6. Membrane proteins serve many functions such as
   a. binding sites that recognize chemical signals
   b. transporting molecules in and out of the cell
   c. giving the cell a sense of identity
   d. A & B
   e. all of the above

7. During the embryonic development, there are 3 germ layers. Which of the following is in the correct order starting from the outermost layer?
   a. Ectoderm, endoderm, mesoderm
   b. Mesoderm, endoderm, ectoderm
   c. Endoderm, ectoderm, mesoderm
   d. Ectoderm, mesoderm, endoderm
   e. Endoderm, mesoderm, ectoderm

8. Which of the following term has an incorrect description?
   a. Nucleolus: a dark body inside the nucleus
   b. Nucleoplasm: a fluid filled space that bathes the nucleolus
   c. Chromatin: made of RNA and associated proteins
   d. Nuclear envelope: a membranous boundary made up of phospholipids
   e. Nucleus: control centre of the cell that contains instructions for cellular function

9. Centrioles
   a. are found in both plant and animal cells
   b. are only found in animal cells
   c. are only found in plant cells
   d. sometimes found in plant cells
   e. none of the above

10. These structures allow cells to communicate with one another.
    a. Gap junctions
    b. Tight junctions
    c. Adhering junctions
    d. A & B
    e. All of the above

11. Which of the following statements is incorrect?
    a. Pinocytosis is another form of exocytosis where cells uptake molecules into the cell
    b. Phagocytosis is a form of exocytosis that ingests large particles
    c. A & B
    d. Exocytosis is where materials leave the cell in vesicles that fuse with the plasma membrane and empty their contents into the environment

    e. Endocytosis is taking in substances from the surrounding environment
12. Which of the following is not a 2nd messenger?
    a. Ca^{2+}
    b. Diacylglycerol (DAG)
    c. Cyclic AMP or cyclic GMP
    d. Hormones
    e. Inositol triphosphate (IP_3)

Thought Questions

1. Explain all the functions of membrane proteins.
2. Why do immune cells such as macrophages contain numerous lysosomes?
3. How are cells able to communicate with one another? For example, cells need to work in harmony in order to contract the cardiac muscle cells.
4. Why is important to use 0.9% NaCl solution IV infusion for treatment of fluid loss?
5. Explain what is wrong with the signal transduction pathway of a patient with cystic fibrosis.

REFERENCES

Kriete, A. & Eils, R. *Computational Systems Biology: From Molecular Mechanisms to Disease* (2nd ed.). Sunderland, MA: Sinauer Associates.

Kulinsky, V. I., & Kolesnichenko, L. S. (2005a). Molecular Mechanisms of Hormonal Activity. I. Receptors. Neuromediators. Systems with Second Messengers. *Biochemistry (Moscow)*, 70(1), 24–39. doi:10.100710541-005-0049-8 PMID:15701047

Kulinsky, V. I., & Kolesnichenko, L. S. (2005b). Molecular Mechanisms of Hormonal Activity. II. Kinase Systems. Systems with Intracellular Receptors. Transactivation of STS. *Biochemistry (Moscow)*, 70(4), 391–405. doi:10.100710541-005-0130-3 PMID:15892606

Landers, J. P., & Spelsberg, T. C. (1992). New concepts in steroid hormone action: Transcription factors, proto-oncogenes, and the cascade model for steroid regulation of gene expression. *Critical Reviews in Eukaryotic Gene Expression*, 2(1), 19–63. PMID:1543897

Lemmon, M. A., & Schlessinger, J. (2010). Cell signaling by receptor-tyrosine kinases. *Cell*, 141(7), 1117–1134. doi:10.1016/j.cell.2010.06.011 PMID:20602996

Purves, D., Augustine, G. J., Fitzpatrick, D., Katz, L. C., LaMantia, A.-S., McNamara, J. O., & Williams, S. M. (Eds.). (2001). *Neuroscience* (2nd ed.). Sunderland, MA: Sinauer Associates.

Cellular Basis of Life

KEY TERMS AND DEFINITIONS

Active Site: The specific site of an enzyme, in which binding of a certain substrate catalyzes a particular reaction.

Cell: The basic unit of structure and function in all living organisms.

Cell Theory: All living things are made up of one or more cells.

Connective Tissue: One of four basic tissue types in the body; supportive tissue such as bone, cartilage, and fat tissue that serve protection and support functions.

Cytosol: The aqueous component of a cell that make up the cytoplasm, proteins, and other cellular structures.

Enzymes: Super selective catalysts used to lower the activation energy of a chemical reaction.

Epithelial Tissue: One of four basic tissue types in the body; covering or lining tissue that serves to protect, support, secrete, and absorb (i.e., skin).

Muscle Tissue: One of four basic tissue types in the body; the three type include skeletal, cardiac, and smooth.

Nervous Tissue: One of four basic tissue types in the body; composed of specialized cells that function to receive stimuli and conduct impulses throughout the body (i.e., nerve cells, neurons).

Organelles: Specialized structures found in the cytoplasm of a living cell that carry out specific functions.

Section 2
Nature, Structure, and Functions of Genes

PROLOGUE

There are seven chapters in this section of the book which take readers through the process of cell division to the latest recombinant DNA technology. The foundation of life begins with the process of inheritance where cells faithfully divide and transmit their genes to their offspring. The processes of mitosis and meiosis are covered first in chapter 4.

The following chapter (5) explores patterns of inheritance illustrating the work of Gregor Mendel and how it formed the basis of understanding genetic inheritance. The next chapter (6) explores specific inheritance patterns in humans and introduces some common human traits, nature of chromosomes, karyotyping and screening of birth defects, genetic linkage, and specific human genetic disorders like haemophilia, cystic fibrosis, Down syndrome, fragile-X syndrome, and more.

Chapter 7 examines DNA structure and how it replicates during cell division, ending with a section on correcting DNA mistakes, the end of DNA molecule, and mitochondrial DNA. How genes work is explored in chapter 8 and the process of genetic mutations concludes the chapter. The regulation of gene expression and control is examined in chapter 9 with operon examples from lower organisms. The chapter ends with a section of gene control in higher organisms.

The last chapter in this section covers biotechnology by taking readers through restriction enzymes, cloning vectors, DNA libraries, genome sequencing, amplification, fingerprinting, gene therapy, applications of biotechnology, and concludes with some ethical and safety considerations of this technology.

Chapter 4
Cell Division:
The Cell Cycle, Mitosis, and Meiosis

ABSTRACT

Cells divide for three main reasons: growth and development, replace worn-out or injured cells, and reproduction of offspring. Cell division is part of the cell cycle divided into five distinct phases. The diploid state of the cell is the normal chromosomal number in species. During sexual reproduction, the cell's chromosome number is reduced to a haploid state to ensure constancy in chromosome number and thus continuation of the species. The process of cell division is controlled by regulatory proteins. Mitosis occurs in all body cells and is divided into four phases. Meiosis, which occurs in only the germ cells involved in reproduction, divides the chromosomes in two rounds termed meiosis I and meiosis II (reduction division). The human lifecycle starts with gametogenesis, the process that forms gametes which then combine to form a zygote. The zygote quickly becomes an embryo and develops rapidly into a foetus. This chapter explores cell division.

CHAPTER OUTLINE

4.1 Why Cells Divide
4.2 The Cell Cycle
4.3 Regulation of Cell Division
4.4 Mitosis
4.5 Meiosis and Sexual Reproduction
4.6 The Human Life Cycle
4.7 Mitosis and Meiosis Compared
Chapter Summary

DOI: 10.4018/978-1-5225-8066-9.ch004

LEARNING OUTCOMES

- Explain the reasons why cells divide and what is accomplished.
- Understand that actual cell division is part of a bigger picture – the cell cycle.
- Describe the mechanics of the cell cycle.
- Comprehend that cell division is highly regulated by cellular molecular signals.
- Understand that some cells remain in a dividing state forever while others remain arrested forever.
- Explain the two different types of cell division (mitosis and meiosis) and various phases of each.
- Understand the role of mitosis in living organisms.
- Recognize that meiosis involves reduction division, genetic reshuffling, and sexual reproduction.
- Understand the human life cycle and how mitosis and meiosis fit in.
- Be able to compare and contrast the two types of cell division.

4.1 WHY CELLS DIVIDE

Living organisms are composed of basic units of organization called **cells**. Some organisms remain as single cells (for example, bacteria and some **protozoans** like the malaria causing *Plasmodium falciparum*); while others differentiate into complex multicellular organisms (for example, worms, spiders, dogs, plants, humans) that are made up of billions of cells that function together as an organism. Why do cells need to divide? There are three main reasons why cells divide: (1) for growth and development; (2) to replace old, worn-out or injured cells (both of these occur in a type of cell division known as **mitosis**), and (3) for sexual reproduction (in a type of cell division known as **meiosis**).

1. Growth and Development

Many sexually reproducing animals start life as fertilized eggs and must grow, develop, and mature into adulthood. This growth and development means that cells must be able to divide and increase in number with subsequent specialization into functional cells, tissues, systems, and eventually individual organisms. Growth and development of the fertilized egg continues in stages until the young mature enough to be born. To illustrate, the human fertilized egg or **zygote** divides repeatedly within the first 24 hours to form a 'ball of cells' usually referred to as a **morula** which transforms into a **blastula** (or **blastocyst**) by end of first week (Figure 1).

In humans, this 'ball' continues to divide and develop in the next few weeks to a well differentiated and recognizable human **foetus** by four to five months (Figure 2).

Continued growth and development in the next several months fully develops the foetus until it is ready for birth as a **neonate** (Figure 3). After birth, the neonate, now an **infant**, continues to grow and develop until maturity. In adulthood, cell growth continues (albeit relatively more slowly) until death occurs. This continued growth and development is a defining aspect of life as new blood cells are formed, bones are dissolved and rebuilt, hair grows, the organism adds or loses weight, and many other metabolic functions that characterize living organisms occur.

Cell Division

Figure 1. Animal zygote development. Note the various stages fertilized egg, 2-cell stage, 4-cell stage, and blastula (blastocyst).
Source: Image used under license from Shutterstock.com

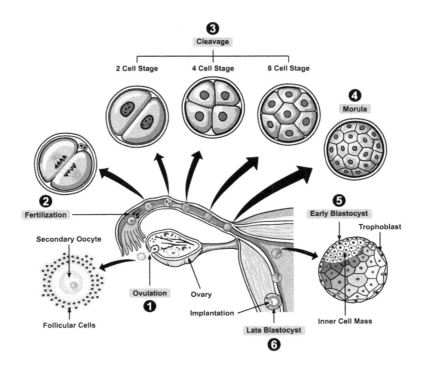

2. Replacement of Old, Worn-Out or Injured Cells

Just like vehicles need their old and worn-out parts replaced to continue functioning properly; in the normal process of living, cells are constantly getting worn-out and must be replaced with new ones. Such replacement comes from cells constantly dividing as need arises to keep the organism functioning at peak capacity. For example, as we chew food, the **epithelial cells** lining the outermost layer of the mouth are constantly sloughed off and swallowed with food. We usually do not sense this experience as it happens without triggering any nervous sensation. Think of situations where investigators might follow a suspect closely until he or she consumes a drink and throws away the container. Such containers are often a harvest for hereditary material (worn-out mouth epithelial cells) for investigators. Due to their importance in the absorption of nutrients which nourish and maintain the body, gut cells usually will divide constantly to keep the system functioning at peak capacity.

However, when we consume hot foods or fluids, we feel the hot foods sensation which often burns deep in the epithelial layer (injuring healthy epithelial cells) sensitizing the nerves and triggering either the swallow or spit response we are so familiar with. Such burns usually cause pain for a few days until they heal (replaced with new epithelial cells). You might also be familiar with body bruises that result from a bike fall on the hands, elbows, knees, and legs. These are slow to heal as underlying epidermal cells gradually replace the injured surface cells until the now blackened injured residues fall off. Another example: if you were to examine a carpenter's or machinist's palms, you would notice that they are

Figure 2. Human foetus development. Note the various stages of development from fertilized egg to nine months.
Source: *Image used under license from Shutterstock.com*

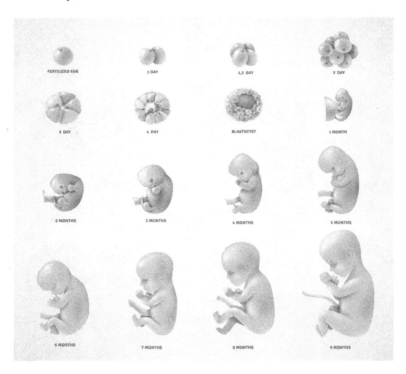

Figure 3. A neonate or new-born
Source: *Image used under license from Shutterstock.com*

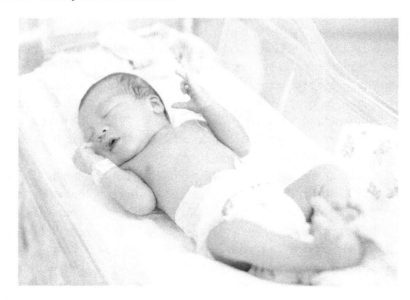

Cell Division

often rough and scaly. Why? This occurs because the constant pressure and rough handling caused by the nature of a carpenter's or machinist's job squeezes and destroys many palm surface epidermal cells triggering underlying cells to divide. The repeated pressure also causes 'epithelial layering' as more cells are produced to handle the external pressure, sometimes producing thick layers of epithelium, which perhaps is why a carpenter's or machinist's handshake is so unforgettable.

3. Reproduction

As recognized by the **cell theory** advanced by scientists in the mid-17th century, cells can only arise from other cells. This early recognition indicates that to perpetuate their kind, cells must have the capacity to reproduce themselves. This is a relatively easily achievable process in less complex, single-celled organisms like bacteria or **protists** like amoeba. Bacteria can divide from a single microscopic cell to a mass of bacterial cells visible to the naked eye in 24 hours. This makes single cells organisms like bacteria very handy as guinea pigs in scientific studies due to this short generation time. That also explains why bacteria can quickly develop resistance to antibiotics in a relatively short time. Some asexually reproducing organisms – that is, those organisms that do not form **gametes** that unite to make young – like hydra, can simply form a young one by growing a bud that eventually detaches from the parent organism to live independently. Many higher organisms like aphids also use **asexual reproduction** to make young that can utilize resources that are abundant but temporary (for example plant nectar, vernal pools).

In higher organisms like humans, the process of cellular reproduction is more complicated as cells are organized in a complex pattern that compartmentalizes genetic material from the rest of the cell and has many membrane-bound **organelles** which are like small "organs" within cells (see Figure 4, Chapter 3). To produce young, higher organisms form gametes that must then unite to form an individual organism. In humans, this occurs in form of gametes produced as sperms in males and eggs or ova in females. Such gametes go through a complex process of cell division that reduces their genetic material (which

Figure 4. How DNA is organized and packaged by histones in the nucleus of the cell to make chromosomes
Source: Image used under license from Shutterstock.com

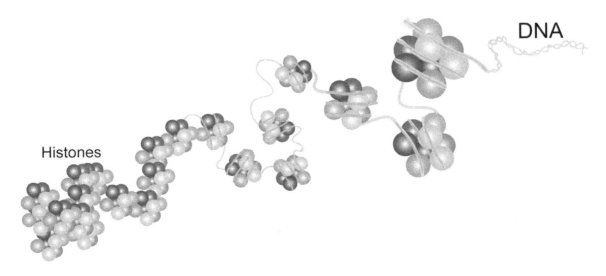

is organized in structures called **chromosome** located in the nucleus, Figure 4). Once mature, gametes are stored in male animals as sperms until mating occurs when they are introduced into the female (as in reptiles, birds, and mammals) or released directly on eggs (as in fishes or amphibians). In female animals, eggs mature and are stored either in unfertilized state (as in amphibians and fishes); or fertilized state (as in reptiles, birds and some mammals like the duck-billed platypus) for short periods of time until they are released or laid. In more advanced mammals, mating often occurs in seasonally defined periods of time termed breeding seasons when females are in **oestrous** and therefore, more receptive to males. During mating sperms are introduced into the female body to fertilize internally released eggs forming a zygote, the first stage in the formation of a new organism.

4.2 THE CELL CYCLE

When a cell divides, it passes along its **hereditary material** (deoxyribonucleic acid [DNA]), which must first be copied to make two identical copies. DNA contains all the information or instructions that a cell needs to do its metabolic duties. A cell also must make identical copies of all its **cellular machinery** made up of organelles (like ribosomes, mitochondria, endoplasmic reticulum, etc., see Figure 4, Chapter 3) and increases the volume of its **cytoplasm,** which houses the organelles, enzymes, ions, and other molecules necessary for cellular metabolism. Cell division therefore includes two main processes: division of the DNA (mitosis or meiosis) and cytoplasm division (**cytokinesis**). DNA is localized in the nucleus of the cell associated with proteins called **histones** (which help in its packaging and organization) forming **chromatin**. Very tightly packaged, chromatin is organized in units called **chromosomes** (Figure 4). For instance, human chromatin is organized into **23 chromosome pairs** in most body cells (called a **diploid** state); one set from each parent. The two chromosomes in a pair are termed **homologous** ("homo" means "same"), because they carry the same set of genes although in different varieties. For example, although the two chromosomes might carry the gene that determines eye colour, one chromosome might carry a gene that determines blue colour while the other one carries a gene that determines brown colour. In human reproductive structures or **germ cells** (testes in males and ovaries in females) the diploid state is reduced to a **haploid** state (or half the number of chromosomes) for reproduction (see section 4.5)

Cell division is part of a bigger process called the **cell cycle** which can take as short as a few hours t a day) to as long as several months depending on the type of cell involved. The cycle has five main phases or stages (1) **G_1 phase** (or Gap 1); (2) **S phase** (or Synthesis); (3) **G_2 phase** (or Gap 2); (4) **Mitotic or Meiotic phase** (M); and (5) **Cytokinesis** (Figure 5). The first four stages comprise the cell division phase called **interphase**, which takes over 90% of the cell cycle.

- **G_1 Phase**: During this first stage, most of what is occurring in the cell is growth (or increase in size) as the cell prepares for cell division and the cytoplasm doubles in size. It is typically the longest part of the cycle and can last 10 hours to over three months.
- **S Phase:** Perhaps the most important stage of the cycle, the synthesis phase is when DNA is duplicated and can last up to 8-10 hours. This stage is referred to as the "synthesis" phase as new genetic material is built from raw materials in the cell.
- **G_2 Phase:** This is usually a short stage lasting about 4 hours when the cell finalizes its growth before committing to the next stage. It is during this phase that chromosomes become super-coiled

Cell Division

Figure 5. The cell cycle. The cell cycle is organized into distinct phases: Interphase (made up of G1, S, G2), M (mitotic/meiotic phase), and cytokinesis.
Source: Image used under license from Shutterstock.com

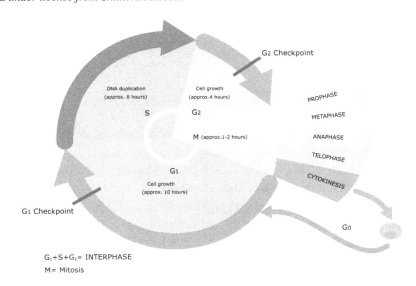

(condense) and shorten, organelles like mitochondria are copied and the cell proteins begin forming a spindle (these are protein microtubules that attach and move chromosomes during the M phase).
- **Mitotic or Meiotic Phase:** Lasting about 1-2 hours, mitosis which occurs in most body cells is divided into four separate stages often referred to as PMAT (**prophase, metaphase, anaphase, and telophase**) each characterized by separate chromosomal events (Figure 5). Meiosis, which only occurs in germ cells for reproductive purposes, also is divided into four separate stages and divides the nuclear material two times (hence is broken into PMAT I and PMAT II).
- **Cytokinesis:** "Cyto" stands for "cell" while "kinesis" means "movement". This is the shortest stage of the cycle lasting just a couple of minutes and divides the cytoplasm into two making two separate (daughter) cells, thus completing the cycle.

After cytokinesis, a cell can either enter G_1 phase, starting the cycle all over again, or enter a state known as G_0 state (or "non-dividing" state). Some body cells remain in a slow "routine" dividing state; for example, think of our skin cells before a bruise occurs. Although the cells are constantly dividing to replace worn out skin cells, this is often done as a matter of routine. However, when injury that damages some of the epidermal cells occurs, molecular signals (called **growth factors**) in the cell *prioritize* and *promote rapid cell division* in underlying epidermal cells to initiate healing. Other body cells remain in the G_0 state stage forever. Think of muscle cells - once fully developed in adult animals, they never divide again. So, what happens in muscle building? Muscle cells are special kinds of cells that can increase in size by increasing the size of tiny components called myofibrils as is often the case in muscle builders. Muscles also can decrease in size as myofibrils shrink as happens when muscle builders stop lifting weights or lose stamina and agility with advancing age. The same can be said of brain neurons, which remain in G_0 state too after the individual reaches adulthood. Any damage that occurs to any part of the central nervous system (CNS) often is never repaired as in the case of nervous tissue damage that occurs to people who have had vehicle accidents, overused drugs, or are suffering from dementia. However, recent research in this area is providing insights that indicate the possibility for neuron self-repair after damage. With time, such research will allow better understanding of cell structure, function, and repair mechanisms.

4.3 REGULATION OF CELL DIVISION

As we think about cell cycle regulation, let's first imagine of an analogous situation. You are at a 4-way traffic intersection on a busy highway. Green means "go"; amber means "slow down or delay crossing the intersection"; and red means "stop". Traffic is regulated by light signals, which have sensors that can detect traffic volume headed in any direction. These sensors send the signals to a "control centre," which then activate different light signals for varying durations of time depending on traffic density. That is why when you are approaching an intersection that is clear (mostly late at night) and you are turning, the green turn signal activates immediately your vehicle enters the zone served by the turn sensor.

The structure, integrity, and functioning of living organisms cannot be maintained without some degree of control of when cells can divide, delay division, or even stop dividing altogether. To achieve this, the body uses a variety of **growth factors** to regulate the process. Some examples of the hundreds of growth factors the body uses include mitosis promoting factor (described below), fibroblast growth factor, nerve growth factor, epidermal growth factor, and many others. Remember the skin bruises we discussed previously? The body needs to "know" when to trigger underlying skin epidermal cells to initiate cell division to begin the healing process. So, as in the light signal example, there are three principal **control checkpoints** controlling each of the four stages in the cell cycle: G_1, G_2 and M checkpoints (see Figure 5). Functioning similarly to the traffic sensors above, two better known regulatory proteins are involved in this control: **cyclin-dependent protein kinase (CDK)** and **cyclin**. The sensitivity of these proteins is dependent on the condition of the cell at any one time and both will interact at the specific check-point to delay the cycle, continue with it or terminate it (resting G_0 state).

How Do the Two Protein Regulators Function Together to Control the Cycle?

Much like the yeast added to flour in the making of bread to produce carbon dioxide that forces the dough to rise; cyclins are regulatory proteins that bind CDKs to activate them, hence the name "CDK". Cyclin's name is derived from the cyclic manufacture and destruction that characterizes these regulators within the cell. Once associated the **"CDK-cyclin complex"** then activates or inactivates specific cellular enzymes and proteins by adding a phosphate group (PO_4^{3-}) to amino acids that are part of their structure. For example, CDKs can add phosphate groups to amino acids comprising nuclear membrane microfilaments and to other microtubule associated proteins triggering specific events in the cell cycle.

What Happens at the G_2 Checkpoint?

Like a traffic signal, a 'checkpoint' is a control point where molecular signals in the cell regulate what happens at key stages of the cell cycle (see Figure 5). Such control might arrest the process of cell division altogether, delay it until the cell's conditions necessitate it (for example, if a cell receives signals

Cell Division

from growth factors as happens in injury), or allow it to proceed depending on the cell's needs at that time. The G_2 check point occurs at the end of the G_2 phase and regulates cycle entry into the M phase when mitosis occurs. As the cell slowly accumulates cyclin during this phase its concentration reaches a certain threshold, allowing it to bind CDK thus forming a **mitotic promoting factor (MPF)**. It is called "MPF" because the association between the kinase and cyclin creates an active complex (CDK-cyclin complex) that serves as a signalling molecule promoting mitosis, the stage following G_2. Depending on what the cell is primed to do, the amount of MPF generated will determine the next sequence of events. If the cell cycle is destined to terminate at the G_2 checkpoint (that is not enter the M phase), the amount of MPF available will not **phosphorylate** enough proteins to trigger (among other events) the beginning of nuclear membrane disintegration that precedes the binding of spindle filaments onto now condensing chromosomes during mitosis. The cycle then terminates. However, if enough MPF molecules are activated (for example by formation of more "CDK-cyclin complexes"), the MPF reaches a threshold that now allows the cell to enter the M phase, and the nuclear envelope disappears.

Experiments have shown that the rhythmic fluctuation of MPF activity and cyclin concentration in the cell is closely associated with the onset and termination of mitosis. What events would trigger such cell cycle termination? Such events might include for example, the cell's reception of regulatory message signals that "tell" it that the need for additional cells is over (as in situations when an injury has completely healed, or the cell is preparing to enter a G_0 [non-dividing] state).

- **G_1 and M Checkpoints:** The G_1 checkpoint regulates the cell's entry into the DNA synthesis phase where copies of the cell's DNA are made. The decision to proceed, delay, or halt this synthesis is determined at the G_1 checkpoint by the same control mechanisms discussed for G_2 checkpoint previously. The same occurs at the M checkpoint where the cell is entering the G_1 phase to begin the cell growth process. Regulatory proteins interact similarly to determine the events that occur at this point.

4.4 MITOSIS

As noted in section 4.2, mitosis is part of the five phases of the cell cycle when chromosomes are separated from each other and packaged to daughter cells. The M phase is often divided into 4 phases due to the distinguishable events that characterize each stage (prophase, metaphase, anaphase and telophase; Figure 6).

However, it must be emphasized that this is one continuous process that proceeds rapidly in actively dividing cells as seen on these onion hair root cells captured at various stages of mitosis (Figure 7).

A Closer Look at a Dividing and Non-Dividing Human Chromosome

Let us imagine a human cell in a non-dividing state. Although chromosomes are invisible at this stage through a light microscope, the 23 homologues (one from the female and the other from the male) are arranged in the nucleus of the cell in no particular order. Each of the 23 pairs would appear as a single chromosome from one parent similar in size to the other one from the other parent making 46 chromosomes. When **karyotype**s of human chromosomes are made, each pair is arranged according to size to a matching set of 23 homologous pairs (Figure 8).

Figure 6. Diagrammatic representation of the various mitotic stages - interphase, prophase, metaphase, anaphase, and telophase (PMAT)
Source: Image used under license from Shutterstock.com

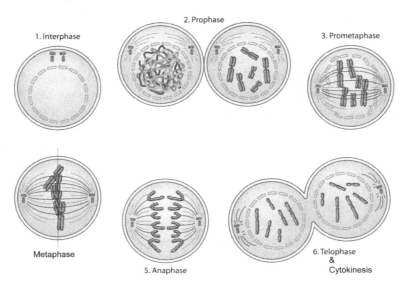

Figure 7. Mitosis in onion cells. Note the various distinct stages captured in this image. Source: Image used under license from Shutterstock.com

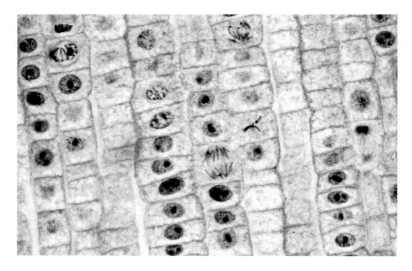

Now imagine a human cell in a dividing state (that would have to be after the S phase of the cell cycle when DNA has duplicated). Each of the 23 pairs of chromosomes would appear as a set of separate "twosomes" held together as "sisters". These daughter chromosomes assume a different name, **chromatids** or "baby chromosomes" as long as they remain attached to each other. The sisters are more closely attached at a region that looks like a "waist" and called a **centromere** which is not equidistant (more to towards

Cell Division

Figure 8. A human karyotype. Note that on this karyotype, the 23 chromosomes in human cells are arranged according to their length from the longest to the shortest.
Source: Image used under license from Shutterstock.com

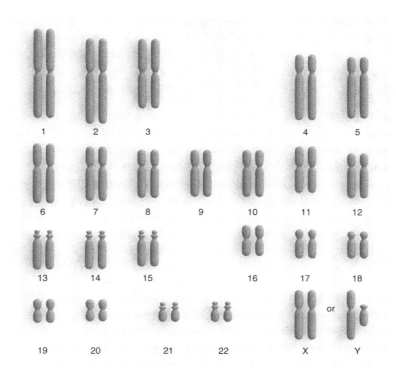

Figure 9. Homologous chromosomes. Each divides into two sister chromatids held together by a centromere.
Source: Image used under license from Shutterstock.com

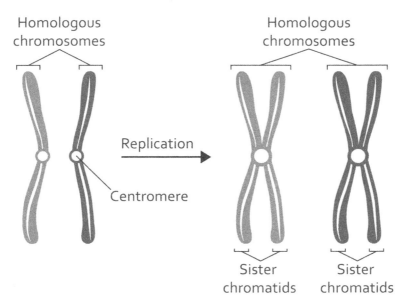

the top half than the bottom half). The centromere has an attachment point with a protein bundle known as a **kinetochore** where the spindle fibres involved in cell division attach (Figure 9).

- **Prophase:** As can be seen in the diagram, during this phase the now very condensed sister chromatids are visibly joined along their length with the centromere region very distinct (see Figure 6). The mitotic spindle (microtubules formed by the now replicated **centrioles**) begins to form moving the centrioles to opposite sides of the cell. The nuclear membrane separating the genetic material from the rest of the cytoplasm begins to collapse and the nucleolus disappears.
- **Metaphase:** The longest phase of mitosis, the centrioles by now have completed their migration to each pole of the cell. The spindle microtubules begin attaching to each set of chromatids at their kinetochores such that by end of this phase, all sets are attached to the spindles from each pole. Other microtubules (non-kinetochore) attach to each other from opposite sides of the cell. Each set of chromatids are aligned *individually* at the equator (cell mid-point, often referred to as the **metaphase plate**) of the cell. In humans, that would mean 46 sets of chromatids (or replicates) all lined up across the middle of the cell.
- **Anaphase:** As the spindle on each side of the cell shortens, the chromatids are separated along their length, each becoming a mature chromosome. Each chromosome, beginning with the centromere (where the spindle attaches), moves to opposite sides of the cell. The cell elongates as **non-kinetochore** microtubules lengthen preparing the cell cleavage. Each side of the cell now has a complete replica of the cell's set of chromosomes.
- **Telophase:** Kinetochores disappear and a nuclear envelope reforms around each set of chromosomes from collapsed fragments forming two daughter nuclei at each end of cell. The nucleolus reappears, the chromosomes now become less condensed and a cleavage furrow in animal cells begins to divide the cytoplasm marking the end of mitosis. In plant cells, a cell plate achieves the separation of the two daughter cells.
- **Cytokinesis:** The cleavage furrow now completes and separates the two identical daughter cells bringing the cell cycle to a close. Depending on the condition of the cell at this point, each daughter cell can begin a new cycle or enter the G_0 state until activated by growth factors.

4.5 MEIOSIS AND SEXUAL REPRODUCTION

As noted in Chapter 1 (section 1.1c), meiosis is the process of cell division that occurs in specific cells called **germ cells** for the purposes of producing young in many organisms (Figure 10). **Sexual reproduction** assures that the offspring get a mixture or reshuffled genes from both parents creating hybrids. That is one reason sisters or brothers are never quite alike in many respects even if they are produced by the same parents. That meiosis is a process that has evolved in many organisms means there must be important advantages to the species for **evolution** to have favoured the investment of resources required to complete process. The process of forming gametes takes resources away from other necessary needs in the life of an organism like securing food, preventing predation, and fighting off diseases. Forming gametes is one issue, ensuring organisms meet and mate and that fertilization occurs is another one full of many life-threatening risks for many organisms. Think of the risks many males like elk take to ensure their gametes are passed on to the next generation. A male elk invests plenty of resources to maintain

Figure 10. Meiosis in general
Source: Image used under license from Shutterstock.com

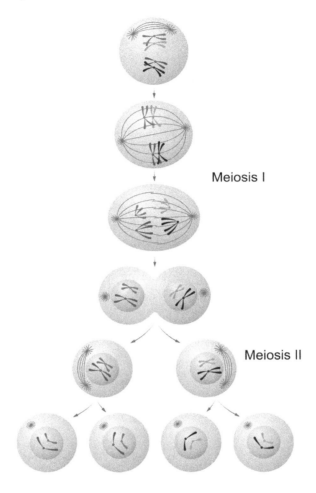

perfect health with highly developed horns and increased **testosterone** levels as the **rutting** season approaches. During this time **territories** have to be created, defended from other males with sometimes very brutal fighting, and females secured for mating. The more a male elk mates, the more genes it passes to the next generation, which becomes a measure of that individual's **fitness**. The elk population benefits from perpetuating genes belonging to individuals that are healthy, perhaps larger and in top condition; this enhances species survival because offspring will be healthier, fit and more likely to reproduce in the next generation.

The most important concept one needs to remember about meiosis is that it is a two-round process that divides the cell and its chromosomes exactly two times; the two rounds being referred to as **meiosis I** and **meiosis II** (see next section). For example, during meiosis I in human cells with a chromosome number of 46, the cell proceeds through the cell cycle and enters the M I phase divided into prophase I, metaphase I, anaphase I, and telophase I. At the end of this first round, there are two cells with half the original number of chromosomes (from 46 in original cell to 23 in each daughter cell). Without a break in the cycle, each daughter cell continues through prophase II, metaphase II, anaphase II, and telophase II. At the end of the second round, each of the two daughter cells divides into two creating four daughter cells each with 23 chromosomes.

Why 23 chromosomes, which is what each daughter cell had at end of meiosis I? Remember we have two homologues in the cell (one from each parent) and that each homologue preparing to divide has already been duplicated during the S phase and is joined with its sister (sister chromatids). In a human cell, this means the cell now has 46 x 2 sister chromatid sets. Now, with this in mind, meiosis I separates the homologues from each other creating two cells each with *one of the 23 homologues* but not the other and each *still joined with its sister* (if you want, you can visualize this as each daughter cell technically having two identical sets of 23 homologues). Recall that when homologues are not in the same cell we refer to them as being in a haploid state (23 in this case). Meiosis II then separates these sister chromatids from each other thus ending with four daughter cells with just one sister chromatid (now considered a full-fledged chromosome – in humans, that is 23 singles).

Why does meiosis divide the cell two times? This process, also referred to as **reduction division,** ensures continuity of the species. If this did not happen (for example) in humans, and gametes ended up with 46 chromosomes each; then after fertilization, the resulting zygote would have 92 chromosomes (46×2^1). In the next round of reproduction, the offspring would have double that number (46×2^2). In five generations, that number would swell to an offspring having 46×2^5 chromosomes. This is what meiosis reduction division precisely prevents.

Meiosis I

At the end of **interphase,** which precedes meiosis (just like in mitosis), the cell has duplicated its chromosomes and the organelles and doubled its amount of cytoplasm. In a human cell, this means the cell has 46 x 2 chromosomes with each of the 46 chromosomes joined with its sister.

- **Prophase I:** Taking 90% of meiosis, the cell accomplishes several tasks during this phase (Figure 11). As in mitosis, chromosomes begin to condense, a spindle begins to form as centrioles move to opposite poles of the cell, and the nuclear envelope starts dismantling. The rest of prophase I events described in what follows do not occur in mitosis. Each of the pairs (still joined to its sister) with its homologue (also joined to its sister) are now called **bivalents** and they make a "*four-some*" set called a **tetrad**. This pairing is called **synapsis** and is achieved by chromosomes aligning themselves together and held by proteins along their lengths (Figure 12).

Let's now review the meaning of these events. Remember it was noted earlier that sexual reproduction produces hybrids with a combination of features from both parents? To achieve this combination parental chromosome (homologues) must somehow exchange parts (genes) before gametes are formed (one source of the differences between sisters or brothers!). Synapsis then ensures that this exchange is achieved. The exchange of genes between two non-sister chromatids, referred to as **crossing over,** now occurs forming X-shaped contact points called **chiasmata** (recall crossing over must be between non-sisters, as the sister chromatids are copies of each other with identical genes as shown in Figure 12). Once crossing over is complete, the homologues slightly move apart but remain joined at the chiasmata, thus they are still visible as tetrads. The exchange of genes is generally not extensive and in humans occurs

Figure 11. The stages of meiosis I (PMAT I)
Source: Image used under license from Shutterstock.com

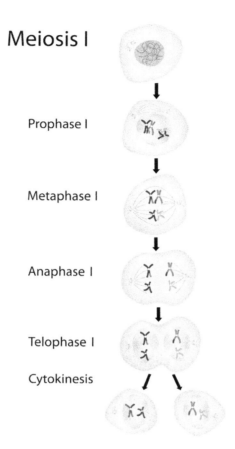

at about 2 to 3 cross-over events per pair. As prophase ends, spindle fibres attach onto the kinetochores at the centromeres of each homologous pair in the tetrad.

- **Metaphase I:** As in mitosis, chromosomes align at the equator of the cell during this phase with one major difference. In mitosis, chromosomes *align themselves individually* along the middle. In metaphase 1, chromosomes, still in tetrads joined *align themselves in pairs randomly* with one chromosome in each homologous pair facing each pole (Figure 11). Recall this alignment in mitosis is followed by separation between sister chromatids to each pole. In metaphase I, the same is achieved, the difference is that the *homologues that separate from each other* (rather than sisters as in mitosis) creating a haploid state at each pole. The random orientation of the homologues at the equator is another source of the variation seen in gametes as the maternal and paternal chromosomes assort without influencing each other; that is, independently. The first variation was in the reshuffling of genes during crossing over which again adds to the differences seen in sisters or brothers. To illustrate this independent alignment of maternal and paternal chromosomes at the equator, let's consider the human cell. There are 23 tetrads aligned at the equator in metaphase I.

Figure 12. Chromosome crossing-over during prophase I. Crossing-over is usually more extensive than is illustrated here.
Source: Image used under license from Shutterstock.com

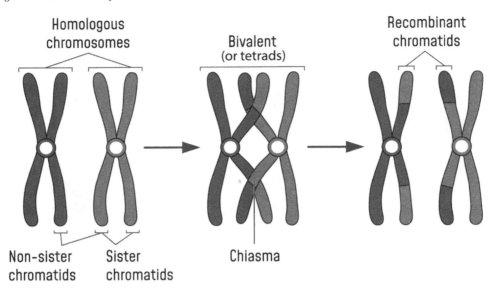

On each side of the equator is either a maternal or a paternal homologue attached to the spindle in each tetrad. The arrangement of these 23 homologues along the equator is random, that is during one round we might have chromosome 1 maternal, chromosome 2 paternal, chromosome 3 maternal, chromosome 4 paternal, and so on until chromosome 23. In another round we might have chromosome 1 paternal, chromosome 2 maternal, chromosome 3 maternal, chromosome 4 maternal, and so on until chromosome 23 in all sorts of combinations.

- **Anaphase I:** Homologous chromosomes separate (but each still joined with its *modified and thus genetically dissimilar sister* – remember, due to the exchange of genes during prophase) as their chiasmata cohesion proteins break down as each is pulled towards the sides of the cell by the shortening spindles. At this point, one can see how the haploid state arises at the end of meiosis I – the homologues have separated and once they do, the cell is no longer in a diploid state!
- **Telophase I:** Now separated into two clusters on each side of the cell, a nuclear envelope re-forms around each nucleus with cytokinesis occurring concurrently to create two daughter cells as seen in mitosis. Each daughter cell is haploid and in humans it would contain 23 individual homologues (each paired to its genetically dissimilar chromatid).

Meiosis II

After an interval that varies in duration in different cells, a short interphase follows in each of the two daughter cells but no chromosomal duplication as the cells prepare to enter the second meiotic phase. The sequence of events during this second phase of meiosis is very similar to mitosis – so, if you understood mitosis, this will be like a repeat only that we refer to the various stages as prophase II, metaphase II, anaphase II and telophase II (Figure 13).

Figure 13. The stages of meiosis II (PMAT II)
Source: Image used under license from Shutterstock.com

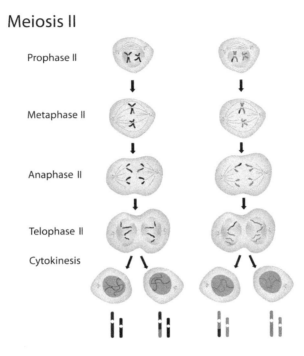

- **Prophase II:** With the just formed nuclear envelope disintegrating, the centrioles begin separating to each side of the cell forming a spindle. By late prophase II each chromosome is attached to the spindle and begins moving towards the equator.
- **Metaphase II:** Chromosomes, each attached to the spindle and joined with its sister chromatid, now align themselves at the equator individually along the whole length. In humans, this can be visualized as having 23 individual chromosomes lining up next to each other along the entire length of the cell mid-section.
- **Anaphase II:** The proteins holding the sisters at the centromere collapse as the spindle pulls the sisters towards the poles forming a cluster. At this point of separation each sister chromatid becomes a fully-fledged chromosome (after separation, the term chromatid is now not applicable).
- **Telophase II and Cytokinesis:** The nuclear envelope re-forms around each set of chromosomes, which by now begin decondensing, is followed by cytokinesis, which separates the two cells, completing meiosis II. Four daughter cells now result which in humans develop into gametes, sperms in males and ova in females. Each is haploid with 23 single chromosomes (see Figure 13).

4.6 THE HUMAN LIFE CYCLE

The process of meiosis is referred to as **gametogenesis** in humans (separated into **oogenesis** in females and **spermatogenesis** in males). After mating occurs in humans, gametes unite to form a **zygote** which develops rapidly through cell division (mitosis) to become an **embryo** (Figure 14); then as it differen-

Figure 14. Human life process forming an embryo

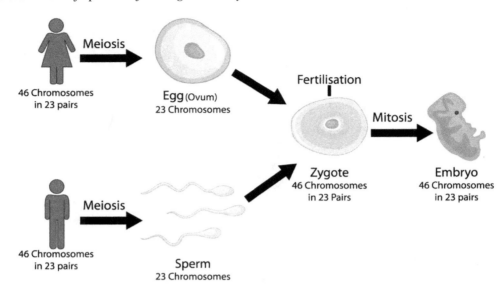

tiates, a **foetus** and when newly born, a **neonate**, then **infant** after few days. The infant's growth and development continues through mitosis (as discussed in section 4.1) into a sexually mature individual at puberty, when the cycle resumes.

4.7 MITOSIS AND MEIOSIS COMPARED

Although the processes of mitosis and meiosis both result in the division of the original cell to daughter cells, there are several marked differences in the events characterizing each process. These differences are apparent at several levels: where they occur in the body, genetic composition of daughter cells, number of daughter cells formed, number of divisions involved, alignment of chromosomes at metaphase and behavior of chromosomes during prophase (Table 1 and Figure 12).

Table 1. Comparison between mitosis and meiosis

Event	Mitosis	Meiosis
Location in the Body	Occurs in all body cells	Occurs only in germ cells during gamete formation
Genetic Composition of daughter Cells	Produces clones or genetically similar cells	Produces genetically different cells
Number of Daughter Cells	Produces two diploid daughter cells	Produces four haploid daughter cells
Chromosome Alignment at Metaphase	Chromosomes align individually at the equator	Chromosomes align at the equator as homologues or tetrads
Chromosome Behaviour during Prophase	Homologous chromosomes do not undergo synapsis	Homologous chromosomes undergo synapsis and exchange genes
Number of Divisions	Involves only one division	Involves two divisions: I and II

Cell Division

Figure 15. Mitosis and meiosis compared
Source: Image used under license from Shutterstock.com

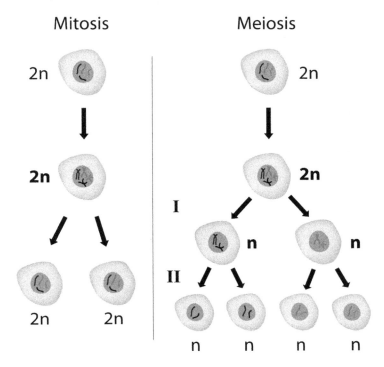

CHAPTER SUMMARY

Cells divide for three main reasons: growth and development, replace worn-out or injured cells, and reproduction of offspring. Cell division is part of a cycle known as the cell cycle that is divided into five distinct phases G_1, S, G_2, M, and cytokinesis. Chromosomes carrying hereditary information of the organism are localized in the cell's nucleus and are super-coiled chains of DNA held together by proteins called histones. The diploid state of the cell is the normal chromosomal number in species.

During sexual reproduction, the cell's chromosome number is reduced to a haploid state to ensure constancy in chromosome number and thus continuation of the species. The process of cell division is carefully controlled by growth factors that serve as regulatory proteins at primarily three control points in the cell cycle: G_1, G_2, and the M checkpoints. Two well-known types of regulatory proteins are Cyclin-dependent Protein Kinases (CDKs) and Cyclins.

The first kind of cell division, mitosis, occurs in all body cells as part of the cell cycle and is divided into four phases: prophase, metaphase, anaphase, and telophase. Mitosis produces two daughter cells of similar genetic make-up (clones). Meiosis, which occurs in only the germ cells involved in reproduction, divides the chromosomes in two rounds termed meiosis I and meiosis II (reduction division). During meiosis I, homologous chromosomes exchange genes before they are separated from each other to create two haploid daughter cells. During meiosis II, which is similar to mitosis, sister chromatids in each daughter cell are separated into two daughter cells. Thus, meiosis produces four daughter cells of different genetic make-up. The human lifecycle starts with gametogenesis, the process that forms gametes which then combine to form a zygote. The zygote quickly becomes an embryo and develops rapidly into a foetus which matures and is born as a neonate (infant after a few days) which develops through mitosis into a sexually mature individual at puberty, when the cycle repeats.

End of Chapter Quiz

1. Why do cells divide?
    a. For growth and development
    b. To replace old, worn-out or injured cells
    c. For sexual reproduction
    d. For energy generation
    e. All except D
2. How many pairs of chromosomes do humans have?
    a. 46
    b. 23
    c. 22
    d. 24
    e. 64
3. Which of following cell cycle is incorrectly stated?
    a. During the G1 phase, the cell increases in size and it is the shortest part of the cycle
    b. S phase is when DNA is duplicated. It is also referred to as the "synthesis" stage
    c. During the G2 phase, the chromosomes condense and shorten, organelles are copied and the cell proteins begin forming a spindle.
    d. Mitotic or Meiotic phase constitutes four separate stages known as prophase, metaphase, anaphase and telophase.
    e. Cytokinesis divides the cytoplasm into two making two separate daughter cells
4. The exchange of genes between two non-sister chromatids is referred to as
    a. synapsis
    b. homologue sisters
    c. crossing over
    d. gene swapping
    e. chiasmata
5. Which statement is incorrect about mitosis and meiosis?
    a. Mitosis involves only one division and whereas meiosis involves two divisions
    b. Mitosis produces two diploid daughter cells whereas meiosis produces four haploid daughter cells
    c. Mitosis occurs in all body cells, whereas meiosis occurs only in germ cells during gamete formation
    d. Mitosis produces genetically different cells whereas meiosis produces clones or genetically similar cells
    e. During metaphase in mitosis, chromosomes align individually at the equator. In meiosis, chromosomes align at the equator as homologues or tetrads

Cell Division

6. Which of the following displays the correct order of the phases during mitosis?
   a. Anaphase, cytokinesis, metaphase, prophase, telophase
   b. Prophase, anaphase, metaphase, telophase, cytokinesis
   c. Cytokinesis, metaphase, prophase, anaphase, telophase
   d. Prophase, metaphase, anaphase, telophase, cytokinesis
   e. Anaphase, prophase, metaphase, telophase, cytokinesis
7. What is the main function of the *S phase* of Cell division?
   a. Cell growth
   b. Chromosomes become supercoiled and shorten
   c. Production of new genetic material and DNA duplication
   d. Division of cytoplasm into two daughter cells
   e. Duplication of important organelles
8. What are the two protein regulators that function together to control the cell cycle?
   a. Cyclin dependent protein kinase (CDK) and cyclin
   b. Troponin and tropomyosin
   c. Cyclin and ubiquitin
   d. Cyclin dependent protein kinase (CDK) and ubiquitin
   e. None of the above
9. Which of following is a property of Mitosis, but not Meiosis?
   a. Occurs only in germ cells during gamete formation
   b. Produces two diploid daughter cells
   c. Produces genetically different cells
   d. Homologous chromosomes undergo synapsis and exchange genes
   e. Involves two divisions: I and II
10. Synapsis is achieved by _____?
    a. duplicating chromosomes
    b. formation of a spindle as centrioles move to opposite poles of the cell
    c. alignment of chromosomes at the centre of the cell
    d. chromosomes aligning themselves together and are held by proteins along their lengths
    e. the formation of an X-shaped contact points after the exchange of genes between two non-sister chromatids

Thought Questions

1. Why would investigators be interested in a cup that their suspect has drank from?
2. Why are bacteria often used in scientific studies?
3. Trisomy 21 is a condition in which an individual has an extra copy of chromosome 21. What happened during mitosis and meiosis in an individual with trisomy 21?
4. Why is reduction division important?
5. Compare and contrast mitosis and meiosis.

KEY TERMS AND DEFINITIONS

Cell Theory: All living things are made up of one or more cells.

Centromere: A structure that links two chromatids together to form a chromosome.

Chromatids: Half of an identical copy of a replicated chromosome.

Chromosomes: Units that contain chromatin specific for each species (e.g., humans have 46 chromosomes).

Crossing Over: The exchanging of genes between two non-sister chromatids that occurs during prophase I.

Deoxyribonucleic Acid (DNA): A type of nucleic acid that contains an organism's hereditary information; composed of adenine, guanine, thymine, and cytosine bases.

Evolution: The idea of change in the genetic composition over time.

Gametogenesis: The process of the gamete production.

Germ Cells: Reproductive cells (ovaries in females and testes in males).

Growth Factors: Molecular signals that prioritize and promote rapid cell division.

Meiosis: A type of cell division that results in germ cells for sexual reproduction.

Mitosis: A type of cell division that results in two identical copies of the original cell.

Mitotic Promoting Factor (MPF): A checkpoint that regulates the transition of cells from G2 to M phase.

Reduction Division: The first cell division in which the number of chromosomes is reduced by half.

Synapsis: The process of homologous chromosome pairing during the initial phases of meiosis.

Chapter 5
Genes:
Modes of Inheritance

ABSTRACT

Charles Darwin proposed a mechanism that explained how organisms evolved, but nothing was known on the form those heritable changes took, what they were, and how they were passed from one generation to another. Gregor Mendel started experiments using the common pea studying easily observable characteristics. Genes are units of DNA carrying information about specific traits. Any given trait has two alleles that may be the same (homozygous) or different (heterozygous). Mendel's first theory studied one trait (monohybrid) in peas. Mendel's second theory studied two traits (dihybrid). Two examples of inheritance do not conform to Mendel's theories. In incomplete dominance, the expression of both alleles results in an intermediate phenotype. In codominance, both alleles are expressed resulting in a new phenotype comprised of both alleles. Mothers who are Rh^- but bearing a Rh^+ baby will need a Rh immunoglobulin shot to counteract the formation of antibodies against a future foetus. This chapter explores genes.

CHAPTER OUTLINE

5.1 Heritable Traits
5.2 Genetics Terminology
5.3 The Work of Gregor Mendel
    5.3.1 1st Theory
    5.3.2 2nd Theory
5.4 Beyond Mendel's Work
    5.4.1 Incomplete Dominance
    5.4.2 Codominance
Chapter Summary

DOI: 10.4018/978-1-5225-8066-9.ch005

LEARNING OUTCOMES

- Explain heritable traits and give examples in humans.
- With examples, be able to explain the terminology used in genetics.
- Summarize the experimental work of Gregor Mendel.
- With examples, ability to explain genetic inheritance not consistent with Mendel's work.

5.1 HERITABLE TRAITS

In his famous book *On the Origin of Species* published in 1859, Charles Darwin proposed a mechanism which explained how organisms evolved. By gradually accumulating and preserving heritable changes which conferred survival advantages to the species, species genetically changed over time as they become better adapted to their environments. Darwin did not know what form those heritable changes took, what they were, and how they were passed from one generation to another. Little was known then about genes, DNA, heredity and proteins.

In the mid-1800s (1854-1856) Gregor Mendel, trained in mathematics and physics, started experimenting with the common pea studying easily observable characteristics like colours of pods, seeds and petals, stem length, flower location, and shapes of pea seeds. The garden pea is easily cultivated and pollinated. Mendel's experimental results were presented in 1865 to an audience in Natural Science Society of the then Brunn (today's Brno, Czech Republic). These experiments proved that heritable traits are carried in units that are not changed across generations. Today, Mendel's work (often referred to by modern biologists as classical genetics) is credited as laying the mathematical groundwork of the science of genetics.

5.2 GENETICS TERMINOLOGY

Before we explore the experiments Mendel did on garden peas, we will define some genetics terminology.

- **Genes:** Units of information on DNA about specific traits, e.g., flower colour.
- **Locus (Plural Loci):** The actual location or address of a gene on a chromosome.
- **Allele:** Alternative forms of the same gene responsible for a given trait.
- **Dominant:** The allele whose effects completely mask the expression of the other.
- **Recessive:** The allele whose effects show up only when dominant allele is absent.
  - *Note: Upper-case are letters used for dominant traits; while the recessive traits assume the lower-case letters used for dominant traits.*
- **Homozygous:** Alleles same, e.g. *AA or aa*
- **Heterozygous:** Alleles different, e.g. *Aa*
  - Homozygous Dominant would be *AA*
  - Homozygous Recessive *aa*
  - Heterozygous *Aa*
- **Genotype:** The actual alleles present in an individual, e.g. *Aa or aa*
- **Phenotype:** The physically observable traits in the individual like height or flower colour.

- **Generations:**
  - P (Parental)
  - F_1 (First Filial)
  - F_2 (Second Filial)

5.3 THE WORK OF GREGOR MENDEL

5.3.1 Theory of Segregation

Mendel's work involved the common garden pea or green pea. More by luck than design, he selected true breeding purple and white flowered plants to act as the parents in the first breeding generation. He followed the inheritance of this one trait (colour of flowers) often called **monohybrid inheritance**. All flowering plants have both the female and male parts in the same flower. The female part is called the **pistil** made up of the stigma and ovary; while the male part is called the **stamens** made up of anthers which produce pollen grains. To prepare for his first breeding cycle, he used scissors to remove all stamens in the purple flower to avoid contamination. Next, he used a small brush to rub the stamens of the white flowered plant onto the receptive part of the pistil (the stigma). Fertilization in plants occurs when the pollen grain germinates on the stigma and extends through it to reach the ovary where fertilization of the ovules occurs (Figure 1).

Figure 1. Fertilization in plants. The male pollen grain lands on the stigma, germinates and extends through the style to the ovary where it fertilizes the ovule.
Source: Image used under license from Shutterstock.com

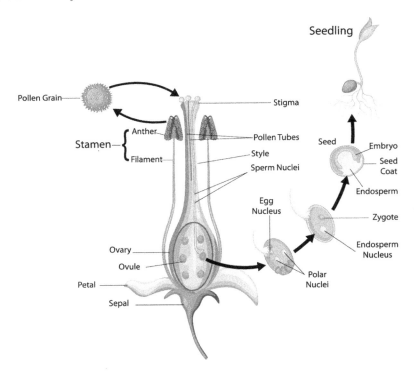

The first filial generation (F_1s) had all purple coloured flowers (Table 1). The purple trait was therefore completely dominant to the recessive white trait. In the second parental generation, he used the F_1s as the parents as shown in Table 2 and Figure 1. The F_1s produced two types of gametes: one carrying the purple colour trait (P), and the other the white colour trait (p). Using a **Punnet square** to compute the F_2s, three-quarters of the offspring displayed the purple colour trait while one-quarter displayed the white colour trait. This is a **phenotypic ratio** of 3:1; and a **genotypic ratio** of 1:2:1.

In the production of gametes during meiosis, the two alleles for each character (purple and white) **segregate** from each other and end up in different gametes. Thus, in the genotype Pp, 50% of the gametes get the allele *P*; the other 50% get the allele *p* (Table 2). This is the basis for Mendel's theory of gamete segregation.

It is difficult to tell apart plants that have the genotypes PP (homozygous dominant) from those with the genotype Pp (heterozygous dominant) since they both display the purple colour trait due to the presence of the dominate allele P. A **testcross** is usually conducted to tell PP and Pp apart where each parent is crossed with the homozygous recessive parent, pp (Table 3). If all F_1s resulting from a test cross are all purple, then the parents are PP; if half of the F_1s are purple and half white, then the parents are Pp.

Table 1. Mendel's theory of segregation using purple and white-flowered peas

(Parents)	P_1	PP	x	pp
(Gametes)	G_1	P	x	p
(Offspring)	F_1		Pp (purple flowered)	
	Which trait is dominant?			

Figure 2. Crossing Mendel's purple and white flowered peas (1st theory). A cross between tall and short plants is also shown.
Source: Image used under license from Shutterstock.com

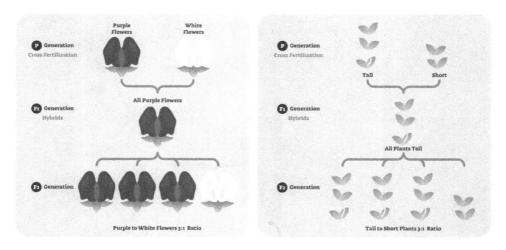

Genes

Table 2. Mendel's theory of segregation crossing the F1 plants

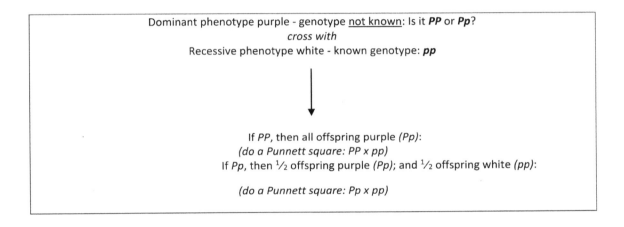

Table 3. The testcross

```
Dominant phenotype purple - genotype not known: Is it PP or Pp?
                    cross with
Recessive phenotype white - known genotype: pp
                        ↓

If PP, then all offspring purple (Pp):
(do a Punnett square: PP x pp)
If Pp, then ½ offspring purple (Pp); and ½ offspring white (pp):

(do a Punnett square: Pp x pp)
```

5.3.2 Theory of Independent Assortment

In his second round of experiments, Mendel studied the inheritance of two traits simultaneously. He used yellow round pea seeds and green wrinkled seeds as the parents in the first generation. The yellow and round traits completely dominate the wrinkled and green traits. This is referred to as **dihybrid inheritance** since two traits are used. Thus, the genotypes of the parents were YYRR and yyrr as shown (Table 4).

Q: Are these two traits (Y & R) transmitted from parent to offspring as a package, or are they inherited independently of each other?

The F_1s resulting were all yellow and round (YyRr) since the dominant traits are found in this genotype. Crossing the F_1s in the next generation, Mendel found the ratios shown in Table 5. The parent YyRr produces four types of gametes: the regular YR and yr and two **recombinants** Yr and yR. Why the recombinants? Remember during the formation of gametes in meiosis, chromosomes reshuffle their

Table 4. Mendel's theory of independent assortment crossing true breeding dihybrid plants

P₁	YYRR	x	yyrr
G₁	YR	x	yr
F₁		YyRr *(write phenotype)*	

Table 5. Mendel's theory of independent assortment crossing the F1 plants

P₂	YyRr	x	YyRr
G₂	YR, Yr, yR, yr *(recombinant gametes Yr and yR after cross-over)*		
F₂	*Make a Punnett square*		
Result:	9:3:3:1 Phenotypic Ratio as follows:		
	9/16 yellow round		
	3/16 green round *(recombinant genotype)*		
	3/16 yellow wrinkled *(recombinant genotype)*		
	1/16 green wrinkled		

genes and crossover which results in sets of genes exchanged between the chromosomes. This is how the gene carrying the yellow trait separates from the gene carrying the round trait forming a new gamete carrying yellow and wrinkled traits. The same for the gene carrying the green trait separating from the wrinkled to form the new gamete with green and round traits.

However, despite the crossing over process, the number of gametes with recombinant traits are always relatively fewer than those with the regular traits. As a result, the F_2s have a 9:3:3:1 ratio made up of 9/16 plants with the dominant traits yellow and round; 3/16 with yellow wrinkled (recombinant); 3/16 with green round (recombinant); and 1/16 with the recessive traits green wrinkled.

During meiosis, genes on chromosomes are not transmitted to gametes as a package (for example YR and yr). Instead, they are sorted during cell division for distribution to gametes independently of how they appear on the chromosome. This gene independence is what results in recombinants during crossing over producing gametes like Yr and yR. If the independence of genes on chromosomes was lacking (that is genes assorted together or dependently) the F1s (YyRr) would only produce gametes of two types: YR and yr. This is the basis of Mendel's Theory of Independent Assortment.

Although Mendel, through experimentation, laid the foundation for the principles of inheritance, he died in 1884. His work could not be fully comprehended as there was still a lot unknown about cell division, genes, and chromosomes. It was not until the middle-to-late of the following century that the structure of DNA (the molecule which carries genes) was discovered and Mendel's work was considered by geneticists as foundational.

Table 6. Incomplete dominance in snapdragon flowers

P_1	Red flower (RR)	x	White flower (WW)
G_1	R	x	W
F_1		RW *(pink flowers)*	

5.4 BEYOND MENDEL'S WORK

Mendel work involved peas that differed in specific traits (true breeding plants) like flower colour, seed shape, seed colour, etc. Mendel could not tell by looking at the plants whether a purple flowered plant was homozygous or heterozygous dominant (PP or Pp). In cases discovered later, clearly visible flower colour traits somehow seemed to "blend" into new flower colours.

5.4.1 Incomplete Dominance

There are several examples of traits that are inherited in an incomplete dominant way. To name a few examples: sickle-cell red blood cells trait, Tay-Sachs Disease, skin colour in humans, straight hair and curly hair in humans and flower colour in snapdragon flowers. In snapdragon flowers, petals can either be red or white. When red-flowered plants and crossed with white-flowered plants the F1s are all pink! (Table 6). In this case, the gene for red colour contributes red pigment and the gene for white colour contributes the white pigment. *None* of the genes contribute enough pigment to overcome the effects of the other, so red and white colours create a new colour pink. Note that the letters we use for each trait do not follow the typical upper-case letter for dominant trait and lower-case letter for the recessive. For both traits, we use the upper-case letters—because none dominates the other (that is, incomplete dominance).

Crossing the F_1s results in a genetic ratio of 1 red (RR), 2 pink (RW) and 1 white (WW) as shown below (Table 7).

Table 7. Crossing the F1s in snapdragon flowers

P_2	RW		x		RW
G_2		R, W	x	R, W	
F_2			(do Punnett square)		
			RR, RW, RW, WW		
			(Genotypic ratio 1:2:1)		

5.4.2 Codominance

In codominant alleles, *both* alleles are expressed (contrast that to the incomplete dominant situation where none of the alleles contribute enough pigment to offset the other pigment). Each allele contributes enough effect individually to be expressed in the organism. The ABO blood group system in humans is a perfect example of **codominance**.

Humans can have either of four blood types or phenotypes: A, B, AB and O (Figure 3). The phenotype is expressed as result of a gene than codes for an **antigen** attached to a glycolipid at the surface of red blood cells and also in other body fluids like saliva and tears. Both A and B alleles are dominant to O which is recessive. People with blood type A have the "A" allele, the genotype of which is often expressed as AA ($I^A I^A$) or AO ($I^A i$). Type A blood has **antibodies** against the B antigen. People with blood type B have the "B" allele and the genotype BB ($I^B I^B$) or BO ($I^B i$). Type B blood has antibodies against the A antigen. People with blood type AB have both alleles "A" and "B" and the genotype $I^A I^B$. Type AB blood has no antibodies against either antigen. Since the A phenotype and the B phenotype both appear, these alleles are said to be codominant. People with blood type O have both recessive alleles and no antigens (genotype ii). Type O blood has antibodies against both A and B antigens (Figure 3).

Another antigen of importance in blood-group typing is the Rhesus protein. A person who has the protein is Rh$^+$ and one lacking the factor is Rh$^-$. Thus, a person can be O$^+$ or O$^-$; A$^+$ or A$^-$; B$^+$ or B$^-$; and AB$^+$ or AB$^-$. The Rh factor is a protein that can be found on the surface of red blood cells. People whose

Figure 3. Codominant traits as seen in the human ABO blood group system
Source: Image used under license from Shutterstock.com

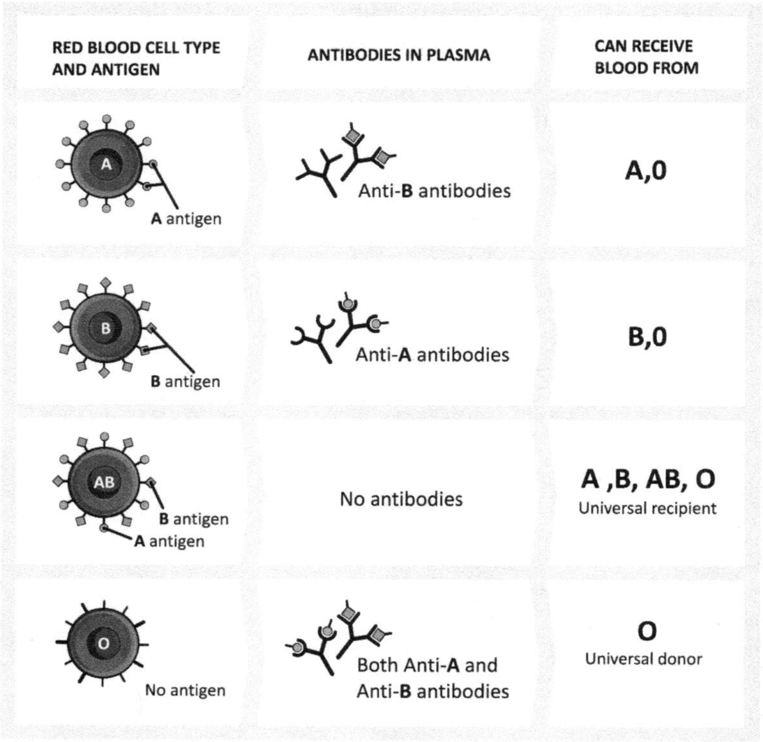

RED BLOOD CELL TYPE AND ANTIGEN	ANTIBODIES IN PLASMA	CAN RECEIVE BLOOD FROM
A antigen	Anti-B antibodies	A, O
B antigen	Anti-A antibodies	B, O
B antigen, A antigen (AB)	No antibodies	A, B, AB, O Universal recipient
No antigen (O)	Both Anti-A and Anti-B antibodies	O Universal donor

cells have this protein are Rh positive (Rh⁺), those without the protein are Rh negative (Rh⁻). The Rh factor is inherited and the foetus can inherit it from the father or the mother. During pregnancy, problems can occur if a mother is Rh negative and the foetus is Rh positive—a situation called **Rh incompatibility**. If the blood of a Rh-positive foetus gets into the bloodstream of the Rh-negative mother, her body will fight it by making Rh **antibodies** (Figure 4). These antibodies can cross the placenta and destroy the foetus's blood—a reaction that can lead to serious health problems, even death in a foetus or new-born. During pregnancy, a mother and her foetus usually do not share blood, but sometimes a small amount of blood from the foetus can mix with the mother's blood.

This can occur with any of the following:

- During labour and birth
- Birth genetic screening through amniocentesis or chorionic villus sampling (CVS)
- Bleeding during pregnancy
- Attempts before labour to manually turn a foetus from a breech presentation (a baby is considered breech when he/she is positioned to come out feet first - instead of head)
- Trauma to the abdomen during pregnancy

Rh incompatibility can lead to a type of anaemia in the foetus in which red blood cells are destroyed faster than the body can replace them. Without enough red blood cells, the foetus will not get enough oxygen and in some cases can die from anaemia.

Health problems usually do not occur during a Rh-negative woman's first pregnancy with a Rh-positive foetus because her body does not have a chance to develop a lot of antibodies. However, if preventive treatment is not given during the first pregnancy and the woman later gets pregnant with a

Figure 4. Rhesus factor in pregnancy. See text for explanation.
Source: Image used under license from Shutterstock.com

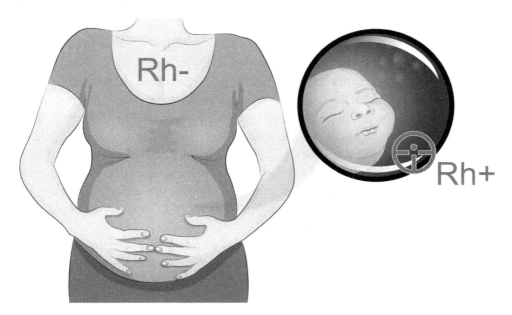

Rh-positive foetus, she can make more antibodies putting a future foetus at risk. A Rh-negative woman can also make antibodies after a miscarriage, ectopic pregnancy, or induced abortion. If such a woman gets pregnant after one of these events (and has not received treatment), a future foetus may be at risk of problems if it is Rh positive.

The goal of preventive treatment is to stop a Rh-negative woman from making Rh antibodies in the first place. This is done by finding out if the mother is Rh negative early in pregnancy (or before pregnancy) and, if necessary, giving them medication to prevent formation of antibodies. When a Rh-negative woman has not already made antibodies, a medication called **Rh immunoglobulin** (RhIg) can be given. RhIg stops the body from making antibodies, which can prevent severe foetal anaemia in a future pregnancy. RhIg is given to Rh-negative women:

- At around week 28 of pregnancy;
- Within 72 hours after the birth of a Rh-positive baby;
- After an ectopic pregnancy or a first-trimester miscarriage or abortion;
- After invasive procedures, such as amniocentesis, CVS, foetal blood sampling, or foetal surgery;
- If there is bleeding during pregnancy;
- If there is trauma to the abdomen during pregnancy;
- If there have been attempts to manually turn a foetus from a breech presentation.

People with type O blood are often referred to as **universal donors** since they can donate blood to all the other blood types (observing the Rhesus factor). Those with AB blood type are often referred to as **universal recipients** and can receive blood from any of the other blood types (again observing the Rhesus factor). In the US population, blood types O positive and A positive dominate (38% and 34% respectively) while blood type AB negative is the rarest (1%).

CHAPTER SUMMARY

Charles Darwin proposed a mechanism which explained how organisms evolved but nothing was known on the form those heritable changes took, what they were, and how they were passed from one generation to another. Gregor Mendel started experiments using the common pea studying easily observable characteristics. Genes are units of DNA carrying information about specific traits and their location on DNA is called locus. Any given trait has two alleles that may be the same (homozygous) or different (heterozygous). When one allele dominates the other, it is called the dominant allele and the other the recessive allele. In genetic notation, geneticists typically use upper case letters of the dominant trait to symbolize the both the dominant allele and recessive allele. A phenotype is the easily observable expression of the allele (like flower colour) while the genotype signifies the actual alleles responsible for the phenotype. Since the time of Mendel, the parental (P), first filial (F1), and second filial (F2) generations notation has been adopted.

Mendel's first theory of segregation studied one trait (monohybrid inheritance) in peas and found that during meiosis, the two alleles for each character segregate from each other and are packaged in different gametes. Mendel's second theory of independent assortment studied two traits (dihybrid inheritance). He found out that during meiosis, genes on chromosomes are not transmitted to gametes as

Genes

a package. Instead, they are sorted during cell division for distribution to gametes independently of how they appear on the chromosome.

Mendel work involved peas that happened to be true breeding and differed in specific traits. In cases discovered later, certain traits seemed to "blend" forming new phenotypes not present in the parental types. In incomplete dominance, the expression of both alleles results in an intermediate phenotype absent in the parental generation. In codominance, both alleles are expressed resulting in a new phenotype comprised of both alleles. Mothers who are Rh negative but bearing a Rh-positive baby will need a Rh immunoglobulin shot to counteract the formation of antibodies against a future foetus. The shot is also needed for mothers who have had an ectopic pregnancy, a miscarriage, an abortion, have bled during pregnancy, have experienced trauma during pregnancy, or have had invasive procedures performed.

End of Chapter Quiz

1. Incomplete dominance are traits that are inherited in an incomplete dominant way. An example of incomplete dominance includes
    a. sickle cell red blood cells trait
    b. Tay-Sachs disease
    c. skin colour in humans
    d. straight or curly hair in humans
    e. all of the above
2. People with type O blood are often referred to as _____, whereas people with type AB blood are often referred to as _____.
    a. universal donors; universal recipients
    b. universal recipients; universal donors
    c. universal carriers; universal recipients
    d. blood warriors; universal donors
    e. none of the above
3. An example of homozygous alleles is
    a. aa
    b. AA
    c. Aa
    d. AaAa
    e. A & B
4. A testcross is conducted to tell PP (homozygous dominant) and Pp (heterozygous dominant) apart where each parent is crossed with the homozygous recessive parent, pp. If half of the F1s are purple and half white, then the parents are
    a. homozygous dominant
    b. heterozygous dominant
    c. homozygous recessive
    d. heterozygous recessive
    e. not enough information is provided

5. The parental generation consisted of YYRR (yellow, round pea seeds) crossed with yyrr (green, wrinkled pea seeds) genotypes. The resulting F1 generation were all yellow and round YyRr. When the F1 generations were crossed with each other, this resulted in the F2 generation having a ratio of
   a. 1:2:1
   b. 9:6:1
   c. 9:3:3:1
   d. 15:1
   e. 3:1
6. When red flower plants are crossed with white flower plants the F1s are all pink. This is an example of
   a. Punnett square
   b. complete dominance
   c. codominance
   d. incomplete dominance
   e. testcross
7. The ABO blood group systems in humans is a perfect example of
   a. Punnett square
   b. complete dominance
   c. codominance
   d. incomplete dominance
   e. testcross
8. A cat has blue eyes and white fur coat. The blue and white traits are examples of
   a. phenotypes
   b. genotypes
   c. dominant
   d. recessive
   e. heterozygous
9. A locus
   a. is a unit of information on DNA about specific traits
   b. is the actual location or address of a gene on a chromosome
   c. is an unfixed position on a chromosome
   d. is observable traits in the individual
   e. none of the above
10. A rabbit has long ears. Long ears are dominant to short ears. A cross between a heterozygous long eared rabbit with a short-eared rabbit would produce offspring that are
    a. 100% long eared
    b. 50% long eared, 50% short eared
    c. 25% long eared, 50% medium eared, 25% short eared
    d. 100% short eared
    e. Not enough information to conclude

Genes

Thought Questions

1. In Poodles, the genes for cream fur colour and black fur colour show codominance. Heterozygous poodles have a Dalmatian pattern of both black and cream fur coats. Imagine a Dalmatian patterned (black and cream) Poodle puppy that has a black furred father. What could be the colour of the puppy's mother? Explain your reasoning.
2. An organism has four pairs of alleles EeFfGgHh in its diploid cells. How many genotypically different kinds of haploid cells can it produce?
3. In some plants, long leaves depend on a dominant gene, and a purple flower is a heterozygous gene for blue and red flowers (Pp). What percentage of the offspring would be purple and long, if you cross a long (heterozygous) and purple plants with a short and purple plant?
4. A mother's blood type is A and the father's blood type is B. What are all the possible blood types of their children?
5. The colour of a certain bird is determined by two alleles. When two green birds are crossed, the phenotypic ratio are as follows, 25% blue birds, 50% green birds, 25% yellow birds. What are the parental generation?

KEY TERMS AND DEFINITIONS

Allele: Alternative forms of the same gene responsible for a given trait.
Codominance: A genetic effect in which both alleles of are phenotypically expressed in an organism.
Dihybrid Inheritance: Inheritance of two different characteristic traits.
Genes: Units of information about specific traits on DNA.
Genotype: The actual alleles present in an individual.
Genotypic Ratio: The ratio used to describe differences in allele frequency from a cross or breeding event.
Heterozygous: Terminology used to describe different alleles.
Homozygous: Terminology used to describe the same alleles.
Locus: The location of a gene on a chromosome.
Monohybrid Inheritance: Inheritance of one characteristic trait.
Phenotype: The physically observable traits in an individual.
Phenotypic Ratio: The ratio used to describe differences in observable traits from a cross or breeding event.
Punnett Square: Visual diagram used to predict the offspring, genotype, and phenotype ratio of a particular cross or breeding event.
Recombinants: Novel genotypes that result from the reshuffling of genes and the crossover of chromosomes during meiosis.
Testcross: A genetic cross between a known homozygous recessive parent and a suspected heterozygous parent to determine specific genotypes (Pp or PP).

Chapter 6
Genes:
Inheritance Patterns in Humans

ABSTRACT

There are several commonly observed human traits. Chromosomes are composed of DNA tightly packed and coiled around blocks of protein known as histones. Humans have a total of 46 chromosomes made up of 1 pair of sex chromosomes and 22 pairs of autosomes. There are several minimally invasive procedures for detecting birth defects in humans during early pregnancy (prenatal diagnosis). Gene linkage occurs when genes that are close together are inherited as a set in autosomes or sex chromosomes. The interrelationships of parents and their children across many generations is called a family tree or pedigree and can be derived by tracing traits across several generations. Sex-linked disorders are genetic defects carried by the X chromosome. Autosomal disorders can be recessive or dominant. Disorders can result from the alterations in chromosome number in autosomes or sex chromosomes. When the integrity of chromosomes is affected, other disorders occur. This chapter explores this aspect of genes.

CHAPTER OUTLINE

6.1 Common Human Traits
6.2 Nature of Chromosomes
6.3 Human Karyotyping and Detection of Birth Defects
6.4 Sex Determination
6.5 Gene Linkage and Linkage Maps
6.6 Sexually-Linked Genes
6.7 When Genes and Chromosomes Change
    6.7.1 Sex-linked disorders
    6.7.2 Autosomal Disorders (Recessive)
    6.7.3 Autosomal Disorders (Dominant)
    6.7.4 Chromosomal Abnormalities

DOI: 10.4018/978-1-5225-8066-9.ch006

Genes

        6.7.4.1 Alterations in Chromosome Number (Autosomes)
        6.7.4.2 Alterations in Chromosome Number (Sex Chromosomes)
        6.7.4.3 Alterations in Chromosome Structure
Chapter Summary

LEARNING OUTCOMES

- Compare observable genetic traits in humans
- Explain the structure, nature and composition of chromosomes
- Summarize the human blood karyotyping process
- Illustrate the process of sex determination in humans
- With examples demonstrate gene linkage, sex-linked genes and linkage maps
- Outline the process of gene mutation with examples

6.1 COMMON HUMAN TRAITS

Let's examine some examples of common **human traits**. Check these traits and how they apply to you and your family.

1. **Hand Clasping:** Clasp your hands together without effort (that is naturally without thinking about the process). Check which thumb is lying on top of the other. Is it the right or left thumb? Interlacing fingers is consistent throughout life in humans (right fingers over left or vice versa). Several genes are involved in the expression of this phenotype.
2. **Eye Color:** Examine the eye of your colleague in class (the colored portion or iris). Is it blue, grey, amber, green, brown, hazel or other color? Eye color is determined by multiple genes.
3. **Dimples:** When you make a facial expression as in smiling, do your cheeks or those of your colleagues have a small part where skin is indented? Dimples are controlled by a single gene locus but other genes may have influence.
4. **Earlobes:** Check or feel your ears at lower lobe. Is it free or connected to the side of face? Free earlobes are determined by multiple genes with the phenotype being a continuum from free earlobes to attached earlobes and everything in-between.
5. **Tongue Rolling:** Try to roll your tongue at both sides upwards. Can you? This trait is determined by a single gene locus with the roller being dominant to non-roller.
6. **Mid-Digit Hair:** Check your middle finger segments for hair. Is there any? Research shows mid-digit hair is determined by recessive alleles although environmental factors may influence the trait.
7. **Freckles:** Most common on the face, freckles represent overproduction of melanin pigment in those cells. Freckles have been attributed to allele variants of one specific gene, the MC1R.
8. **Widows Peak:** Now look at your colleague's forehead. Does the hairline have an "M" shape or the hairline runs straight across the face? Researchers think several genes may be involved in determining this trait.

9. **Hitch-Hiker's Thumb:** Bend your thumb backwards at the top segment. How far does it go? The hitch-hiker's thumb is a phenotype where the top segment of the thumb bends significantly lower creating a large angle between the two segments. Studies indicate this trait varies in a continuum from straight to hitch-hiker's thumb and is thought to be controlled by multiple genes.

6.2 NATURE OF CHROMOSOMES

As noted in Chapter 3 (section 3.2.1), chromosomes are composed of DNA and extend about two meters in length in humans. They are tightly packed and coiled around blocks of proteins known as histones making chromatin. A chromosome has a constriction point called the **centromere**, which divides the chromosome into two sections—the "p" and "q" arms (Figure 1). The actual location of the centromere varies and is not equidistant to the length of the chromosome. Thus, the **"p" arm** is usually shorter than the longer **"q" arm.** This gives each chromosome its characteristic shape. Humans have a total of 46 chromosomes made up of two types of chromosomes, the **sex chromosomes** (1 pair named X and Y) and **autosomes** (22 pairs). Half of them (23) are contributed by the egg (mother) and the other half (23) by the sperm (father) restoring the normal chromosome number in humans (see Figure 14, Chapter 4).

6.3 HUMAN KARYOTYPING AND DETECTION OF BIRTH DEFECTS

The number of chromosomes, their physical appearance and their integrity can be assessed by a process called **karyotype analysis**. Cells are isolated and treated with chemicals which stimulate mitosis; they are then stained using a dye to enhance visibility. The stained chromosomes are then observed and pho-

Figure 1. The centromere which divides a chromosome into "q" and "p" arms
Source: Image used under license from Shutterstock.com

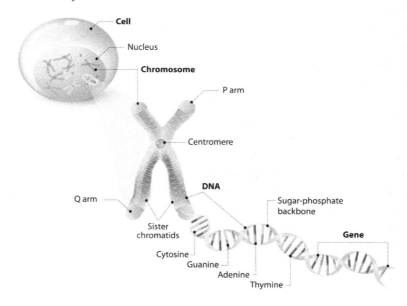

tographed. Since they are dividing cells, chromosomes appear as two sister chromatids joined around the mid-section. Similarly-sized and shaped chromosomes are arranged in pairs (one from each parent). For example, in humans, chromosome pair number 1 is the longest, while chromosome pair number 22 is the shortest (Figure 8, Chapter 4).

There are several "minimally invasive" procedures for detecting birth defects in humans during early pregnancy (prenatal diagnosis) including ultrasound, amniocentesis, foetal blood sampling, biopsy of foetal skin, chorionic villus sampling (CVS), and fetoscopy (Deka et al., 2012). Most prenatal genetic diagnosis is based on foetal tissue obtained by minimally invasive methods involving amniocentesis (>15 weeks), chorionic villus biopsy (>11 weeks), and fetoscopy (>18 weeks). These invasive methods involve minimal risks of miscarriage estimated to be 1–3% (Kumar, 2015).

- **Ultrasound:** Also referred to as sonography, ultrasound uses high frequency sound waves produced by a transducer. It then uses sound wave echoes to develop screen images of body tissues and organs like liver, heart, blood vessels, ovaries, kidneys, etc. for their integrity and consistency. It is also used in detecting birth defects, size of baby, baby positioning, delivery dates and presence of twins (Szabo, 2014).
- **Amniocentesis:** Early in pregnancy (about 15-20 weeks), a small sample of the amniotic fluid (the fluid which cushions the developing embryo in the womb), is drawn using a syringe. The fluid is then centrifuged, and foetal cells extracted and cultured for several weeks. Karyotyping is then performed on those cells (Figure 2).
- **Chorionic Villus Sampling (CVS):** At about 11-12 weeks of pregnancy, a small placental chorionic villi tissue is removed for analysis. A needle or a catheter can be guided by ultrasound through the birth canal to remove the sample. Chorionic villi are small finger-like projections in

Figure 2. Amniocentesis. Diagnostic testing detects chromosomal abnormalities in foetus.
Source: Image used under license from Shutterstock.com

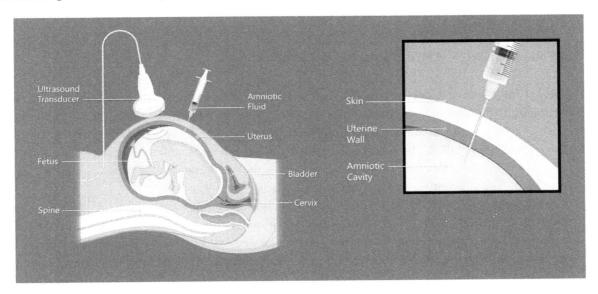

the placenta with the same genetic profile as foetal cells. One advantage of the CVS over amniocentesis is it is done earlier in pregnancy and the results are obtained sooner giving parents time to make decisions about the pregnancy.

- **Fetoscopy:** At about 18 weeks of pregnancy, an endoscope (as small as 1 millimeter in size) is guided by ultrasound through a small incision in the mother's abdomen to the amniotic sac. Fetoscopy provides access to the placental surface, umbilical cord, and foetus. Using ultrasound tomography, visual images can be generated to assess the condition of the foetus. (Deka et al., 2012).
- **Pre-Implantation Diagnosis (PID):** For couples who have difficulty in conceiving babies, both parents provide gametes in an in-vitro fertilization environment (lab or clinic). Fertilization is done under carefully controlled conditions. As the fertilized egg continues to divide, a karyotype is generated from cells derived from **morula** stage of differentiating embryo, and analysis done.
- **Foetal Blood Sampling (FBS):** Blood from the foetus is drawn using an ultrasound-guided needle directly from the foetus or umbilical cord. Blood can then be tested for anaemia or genetic defects (Moise, 2009).

6.4 SEX DETERMINATION

In humans, sex determination occurs at fertilization depending on whether an X or Y chromosome was inherited from the father (See Box 1). Research shows that the Y chromosome has a **male-specific region** (MSY) which harbours several active genes including the **sex determining region** (SRY) region; and the testis-determining factor (TDF). The SRY gene is thought to code for transcription factors that bind DNA and trigger differentiation of **Sertoli cells** which promote the formation of the primordial germ cells that eventually become the testis (Karkanaki et al., 2007). According to Karkanaki et al., (2007), sex determination is an interplay of multiple genes not only located in the Y chromosome but also in the X chromosome or autosomes 9, 11 and 17.

6.5 GENE LINKAGE AND LINKAGE MAPS

- **Gene Linkage:** Gene loci are closely associated with the probability of the genes being inherited together. The closer the genes are on the same chromosome, the more likely they will be inherited as a set—a phenomenon called **linkage** (Figure 3a). The farther apart they are, the higher the chances they will not be inherited together (not linked) and can be separated by homologous recombination which occurs through crossing-over during prophase I of meiosis. Genes located on different chromosomes are not linked (Figure 3b).
- **Linkage Maps:** To find the location (locus) of a gene, information about gene linkage can be used from experimental crosses. The probability of a recombination between two genes is proportional to the distance separating the genes. Looking at the percentage frequency of different genes being inherited together; linkage maps can be created of the relative distances between the genes. The distances between genes are expressed as map units, with one map unit being the effective distance to achieve a 1% recombination frequency.

Genes

Box 1.

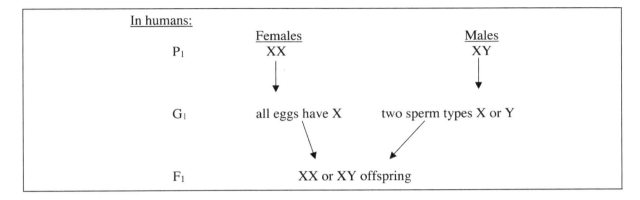

Figure 3. a) Linked b) unlinked genes

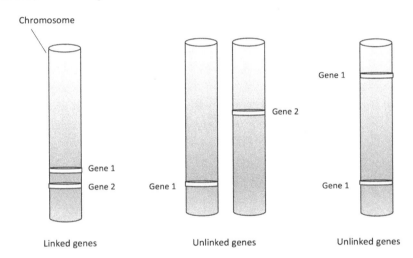

Let's look at an example. Assume the observed recombination frequencies between three gene pairs are as follows: *tr–p* 13%; *p–bn* 16.7%; and *tr–bn* 23%. Can you determine the order of the three genes *tr, p* and *bn*? The genes *tr* and *bn* frequency of 23% is less than the sum of *tr-p* and *p-bn* (13 + 16.7 = 29.7%). Put in another way, the crossing overs between *tr-p* are 56% as often as those between *tr-bn*; likewise crossing overs between *p-bn* are 72% as often as those between *tr-bn*. The crossing over between *p-bn* would cancel the first one (*tr-p*) reducing the observed frequency between *tr-bn*, while contributing to the frequencies between each of the pairs of genes, *tr-p* and *p-bn*. Thus, the combined frequency of *tr-p* and *p-bn* (29.7%) is closely related to the distance between the two genes. A map locating gene *p* somewhere between *tr* and *bn* would most likely be in line with this data (Figure 4).

Figure 4. Linkage maps. These are constructed from recombination frequencies between different genes. Distances between genes are expressed as map units (1 map unit = 1% recombination frequency).

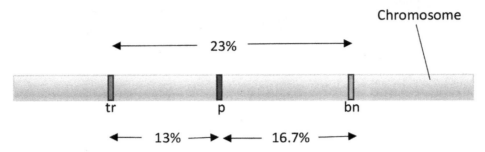

6.6 SEXUALLY-LINKED GENE INHERITANCE

Genes carried by the X chromosome are referred to as sex-linked since chromosomes X and Y are important in sex determination. The X chromosomes carries about 1400 genes while the Y carries about 200 genes most important of which is the cluster of genes in the MSY. Genes located on the X chromosomes can be linked (closer together) or unlinked (farther apart) (Figure 5).

Figure 5. Inheritance of sex-linked genes on the X-chromosome
Source: Image used under license from Shutterstock.com

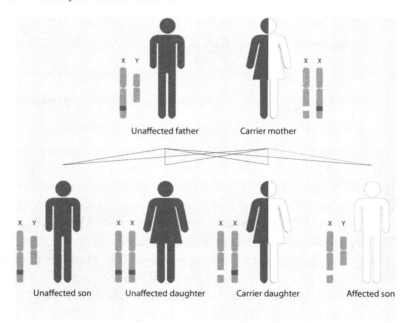

Genes

Box 2.

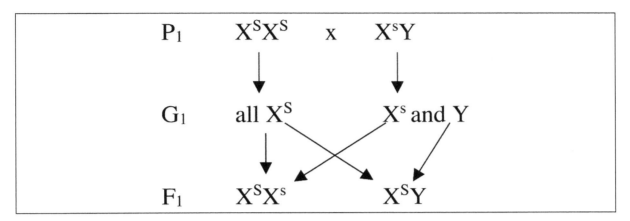

6.7 WHEN GENES AND CHROMOSOMES CHANGE

- **Tracing Genetic Disorders:** The interrelationships of parents and their children across many generations is called a **family tree** or **pedigree** and can be derived by tracing traits across several generations. Such pedigree analysis allows a better understanding of some common genetic disorders in humans. This way, a **personal wellness profile** can be generated for individuals. The profile measures your personal and family history of illnesses and medical conditions, your lifestyle practices, your general health status and guidance in achieving and maintaining good health.

6.7.1 Sex-Linked Disorders

Since human males have only one X chromosome, a disorder carried by the X will not be masked by the other chromosome since the other chromosome is a Y. In females however, where there are two XX chromosomes, one of them can mask any disorder carried by the other X. Hence, males express sex-linked disorders more often than females (see Figure 5).

Let's look at sex-linked inheritance examples. The first one (a) has been done to illustrate the gametes (G_1) and offspring (F_1) of a cross between normal mother and father with the disorder.

Find out the G_1s and F_1s for both (b) and (c) below.

1. First, we have a normal mother and father with disorder with these genotypes (see Box 2)

2. Second, we have a normal father but a carrier mother with these genotypes:

$X^SY \times X^SX^s$

3. Third, we have a carrier mother and a father with the disorder with these genotypes:

$X^SX^s \times X^sY$

135

Figure 6. Autosomal disorders
Source: Image used under license from Shutterstock.com

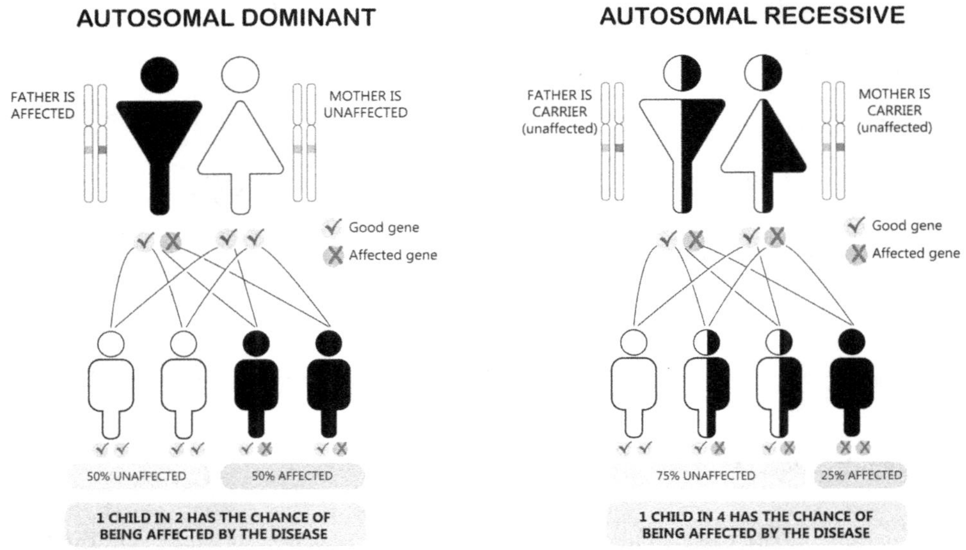

Three common examples of sex-linked traits are:

1. **Hemophilia:** Individuals with the genotype X^H and X^h lack a clotting factor in blood and clots form very slowly which could result in bleeding to death.
2. **Red-Green Colour-Blindness:** Individuals with the genotype X^C and X^c lack certain receptors on their retina which are sensitive to the red-green wavelengths of the electromagnetic spectrum.
3. **Duchenne Muscular Dystrophy (DMD):** Individuals with DMD have a faulty structural protein **dystrophin** that causes muscle cells to weaken and explode due to cellular pressure build up. This results in poor muscle development.

6.7.2 Autosomal Disorders (Recessive)

These disorders follow a typical Mendelian inheritance pattern discussed in Chapter 5. Only individuals inheriting two recessive alleles show the condition. Those with one copy of the recessive allele are carriers and are not affected by the condition (Figure 6).

1. **Cystic Fibrosis:** Individuals have mutations in the CFTR gene resulting in a faulty receptor protein which transports chloride ions (Cl^-). As a result, there is clogging in organs like lungs and pancreas leading to chronic infection, bronchitis, and poor nutrient absorption. This condition is common in people of European descent.
2. **Tay-Sachs Disease:** Individuals have a mutation in the HEXA gene and cannot produce a key brain enzyme (beta-hexosaminidase A) that metabolizes lipids in lysosomes. Lipids therefore accumulate in the brain damaging neurons (causing seizures, blindness, and motor performance). This condition is common in peoples of Ashkenazic Jew descent from central Europe.

Genes

3. **Sickle-Cell Anemia:** Individuals with this condition are carriers with the heterozygous genotype (Ss). Their hemoglobin is a mix of red blood cells with normal hemoglobin (S) and others with a crescent shape due to abnormal hemoglobin (s) (sickle-cell). Individuals with the homozygous dominant genotype (SS) are normal while those of homozygous recessive (ss) do not survive. Sickling of red blood cells causes premature breakage resulting in anaemia. Sickle-celled red blood cells bind oxygen poorly and in extreme physical stress or in areas with low oxygen levels (higher elevations for example) individuals face severe difficulty transporting oxygen to active body tissues and organs. The condition is common in peoples coming from tropical environments of Africa, Asia, South America, and several islands in the Pacific.
4. **Phenylketonuria (PKU):** These individuals have a mutation on the PAH gene responsible for the enzyme phenylalanine hydroxylase required for metabolism of the amino acid phenylalanine.
5. **Albinism:** Individuals lack pigmentation in the skin, hair and eyes and develop vision problems. A gene that codes for an enzyme involved in the metabolic pathway that makes the pigment melanin produces a non-functional form of the enzyme - thus, no melanin is deposited.

6.7.3 Autosomal Disorders (Dominant)

1. **Huntington's Disease:** This is a degenerative disease of the nervous system and can affect individuals from childhood to an advanced age. Symptoms typically appear in midlife (ages 30-50) as neurons in the motor control regions of the brain gradually deteriorate. The genetic cause of Huntington's Disease is a mutation of the HD gene which makes the protein huntingtin. The mutation is caused by bases C, A, and G on DNA repeating many more times than is normal interfering with gene function in HD patients (NINDS, 2018).
2. **Polydactyly:** Individuals develop extra fingers or toes on either little finger or on thumb. It is caused by a mutation on the GLI3 gene which controls patterning of limbs during the early stages of embryonic development. Studies indicate that the GLI3 protein controls other genes by activating (turning them on) or repressing (turning them off).
3. **Achondroplasia:** These individuals have abnormally short arms and legs relative to other body parts. For males, the average height is 4 feet, 4 inches and for females, the average height is 4 feet, 1 inch. The term means "without cartilage formation". During embryonic development, the skeleton is formed from cartilage (a flexible material) in a process called **ossification** where bone formation occurs. The genetic fault is in the FGFR3 gene which codes for a receptor protein involved in bone growth regulation; and as a result, bones of the arms and legs fail to develop normally (NINDS, 2018).

6.7.4 Chromosomal Abnormalities

6.7.4.1 Alterations in Chromosome Number (Autosomes)

As discussed in Chapter 4 (section 4.2, cell cycle), normal body chromosome number is referred to as diploid. During cell division, some cells end up with complete extra sets of chromosomes. Cells with one additional set of chromosomes (23 + 23 + 23) for a total of 69 chromosomes, are called **triploid**. In **tetraplo**id, cells have two additional sets of chromosomes (23 + 23 + 23 + 23) for a total of 92 chromosomes. Embryos with such extra sets of chromosomes often spontaneously terminate early in life. In

Figure 7. Down syndrome karyotype (Trisomy-21)
Source: Image used under license from Shutterstock.com

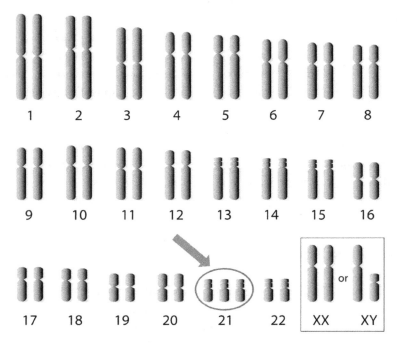

Down Syndrome, there is an extra autosome on chromosome #21 (**trisomy 21**) which occurs because of chromosomal **nondisjunction**. Nondisjunction occurs during cell division when the spindle breaks prematurely preventing chromosome #21 from separating from its homologue or sister chromatid (Figure 7). The extra genes disrupt the normal course of embryonic development producing the characteristic features and susceptibility to other health conditions. Individuals have characteristic facial features, short stature, heart defects, sexually under-developed and sterile, and with poor mental development.

6.7.4.2 Alterations in Chromosome Number (Sex Chromosomes)

1. **Klinefelter Syndrome:** This condition is caused by an extra X chromosome brought about by nondisjunction producing individuals with either XXX or XXY karyotype (Figure 8). In males, the extra set of genes on the X chromosome interfere with the normal development of the testes significantly reducing the levels of the male hormone testosterone. Individuals are sterile and feminine. In females, the extra X chromosome appears to have little effect on their development producing normal healthy females. Variants of Klinefelter syndrome have more than one extra set of the X chromosome (XXXX or XXXY) and have more severe signs and symptoms of the syndrome.
2. **Turner Syndrome:** Occurring in females only, individuals with this syndrome have only one copy of the X chromosome in their cells due to nondisjunction (**monosomy X**). Studies indicate the presence of one gene called **SHOX** important for bone development and growth whose loss leads to short stature and skeletal abnormalities. This results from an egg cell with no X chromosome (O type). If O fuses with a Y sperm the resulting zygote fails to develop since humans cannot survive without the genes on X chromosome.

Figure 8. Klinefelter syndrome
Source: Image used under license from Shutterstock.com

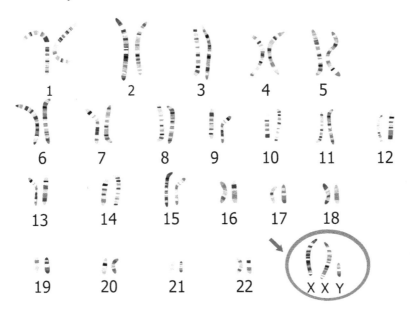

However, if O fuses with an X sperm, the XO individual develops into a sterile female, short stature, extra folds of skin on neck (webbed neck), and immature sex organs. Individuals are of normal intelligence but face developmental delays, and non-verbal learning disabilities (Figure 9). Today estrogen replacement therapy helps Turner syndrome girls develop secondary sexual characteristics.

3. **Jacob's Syndrome:** When the Y chromosome fails to separate from its partner (due to nondisjunction) during cell division, some gametes become YY. If they fuse with an X egg cell they produce an individual with the karyotype XYY (Figure 10). These individuals are fertile males of normal appearance with normal sexual development. They may however have mild mental impairment, are taller than average, and exhibit higher risk of social, behavioral and emotional problems than unaffected individuals.

6.7.4.3 Alterations in Chromosome Structure

1. **Cri-du-Chat Syndrome:** This condition occurs when the short (p) arm of chromosome #5 is structurally damaged by environmental factors like radiation or chemicals, and DNA repair enzymes are unable to fix it. Studies indicate the loss of multiple genes in the damaged section—with the loss of one gene (CTNND2), associated with severe intellectual disability. Individuals have delayed mentally development, small head (microcephaly), low birth weight and unusual facial features with a cry that sounds like the meowing of a distressed cat.
2. **Fragile-X Syndrome:** This condition occurs when a section of DNA with the bases C, G, and G is repeated up to 40 or more times. This expanded CGG section disrupts (turns off) the FMR1 gene

Figure 9. Turner syndrome (XO)
Source: Image used under license from Shutterstock.com

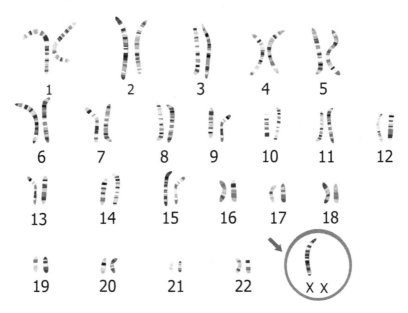

Figure 10. Jacob's syndrome (XYY)
Image used under license from Shutterstock.com

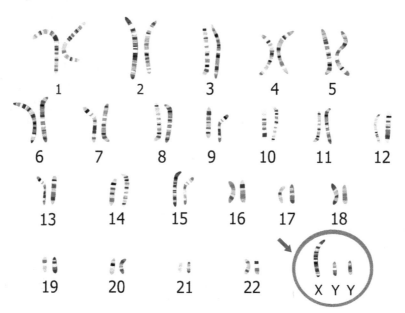

responsible for a protein that regulates the production of other proteins. This disrupts nervous system functions. Individuals with this syndrome have poor cognitive and delayed speech development with autism spectrum disorders that affect social interactions. They have characteristic features including flat feet, unusually flexible fingers, large ears, long and narrow face, and a prominent jaw and forehead. Males often display more severe symptoms than females.

Later chapters with a more comprehensive coverage of the genetic foundations of human diseases appear in Part III below (Genes, Diseases, and Human Health).

CHAPTER SUMMARY

Some common human traits hand clasping, eye color, free or attached earlobes, tongue rolling, mid-digit hair, freckles, widows peak, and hitch-hiker's thumb.

Chromosomes are composed of DNA tightly packed and coiled around blocks of protein known as histones. They have a constriction point called the centromere, dividing the chromosome into two sections—a shorter "p" arm and a longer "q" arm. Humans have a total of 46 chromosomes made up of two types of chromosomes, the sex chromosomes and autosomes. The number of chromosomes, their physical appearance and their integrity can be assessed by a process called karyotype analysis which can be used to detect birth abnormalities.

There are several minimally invasive procedures for detecting birth defects in humans during early pregnancy (prenatal diagnosis) including ultrasound, amniocentesis, foetal blood sampling, biopsy of foetal skin, chorionic villus sampling (CVS) and fetoscopy. In humans, sex determination occurs at fertilization depending on whether an X or Y chromosome was inherited from the father. The Y chromosome has a male-specific region (MSY) which harbours several active genes including the sex determining region (SRY) region; and the testis-determining factor (TDF).

Gene loci are closely associated with the probability of the genes being inherited together. The closer the genes are on the same chromosome, the more likely they will be inherited as a set—a phenomenon called gene linkage. Using the percentage frequency of different genes being inherited together helps in the creation of linkage maps showing the relative distances between the genes. These distances are expressed as map units, with one map unit being the effective distance to achieve a 1% recombination frequency. Genes carried by the X chromosome are referred to as sex-linked. Genes on sex chromosomes or autosomes can be linked (closer together) or unlinked (farther apart).

The interrelationships of parents and their children across many generations is called a family tree or pedigree and can be derived by tracing traits across several generations. This produces a personal wellness profile which measures an individual's personal and family history of illnesses and medical conditions, lifestyle practices, general health status and guidance in achieving and maintaining good health.

Sex-linked disorders are genetic defects carried by the X chromosome. Since human males have only one X chromosome, a disorder carried by the X will not be masked by the other chromosome, Y; hence, males express sex-linked disorders more often than females. Three common examples of sex-linked traits are hemophilia, red-green color blindness and Duchenne Muscular Dystrophy. Recessive autosomal disorders follow typical Mendelian inheritance pattern where only individual inheriting two recessive alleles show the condition. Those with one copy of the recessive allele are carriers and are not affected by the condition. Several examples of such genetic conditions include cystic fibrosis,

Tay-Sachs Disease, sickle-cell anemia, phenylketonuria, and albinism. Dominant autosomal disorders include Huntington's Disease, polydactyly, and achondroplasia. Disorders that result from the alterations in chromosome number in autosomes include Down Syndrome where there is an extra autosome on chromosome #21 often called trisomy 21 resulting from chromosomal nondisjunction. Disorders that result from the alterations in chromosome number in sex chromosomes include Klinefelter, Turner, and Jacob's syndromes. When the integrity of chromosomes is compromised, disorders like Cri-du-Chat Syndrome and Fragile-X syndrome result.

End of Chapter Quiz

1. Which of the following statements is *false*?
    a. Chromosomes are composed of DNA
    b. Chromosomes have constriction points called centromeres
    c. Centromeres divide the chromosome into two sections-- the "p" and "q" arms
    d. The location of the centromere is equidistant to the length of the chromosome
    e. Humans have two types of chromosomes, sex chromosomes and autosomes
2. Minimal invasive procedures for detecting birth defects include
    a. amniocentesis
    b. foetal blood sampling and biopsy of foetal skin
    c. chorionic villus sampling
    d. fetoscopy
    e. all of the above
3. Ultrasound is used to detect
    a. birth defects
    b. size of the baby
    c. baby positioning
    d. presence of twins
    e. all of the above
4. The closer the genes are on the same chromosome, the more likely they will be inherited as a set is a phenomenon called
    a. gene linkage
    b. gene connection
    c. gene correlation
    d. gene interconnection
    e. gene relationship
5. Assume the observed recombination frequencies between three gene pairs are as follows: *cn-b 2%, vg-cn 5%, b-vg 19%*. Determine the order of three genes.
    a. b-cn-vg
    b. b-vg-cn
    c. cn-b-vg
    d. vg-b-cn
    e. none of the above
6. The interrelationship of parents and their children across many generations is called
    a. transplacental relationship

Genes

      b.   family tree
      c.   investigation
      d.   pedigree
      e.   B & D

7. What is an example of sex-linked traits?
   a. Haemophilia
   b. Red-Green colour-blindness
   c. Duchenne muscular dystrophy
   d. All of the above
   e. A & B only
8. Which of the following is a dominant autosomal disorder?
   a. Tay-Sachs Disease
   b. Sickle-cell anaemia
   c. Huntington's Disease
   d. Albinism
   e. Phenylketonuria
9. Individuals with the Fragile-X Syndrome have characteristic features such as
   a. flat feet, flexible fingers, large ears, long and narrow face, prominent jaw and forehead
   b. poor cognitive and delayed speech development
   c. small ears, wide face, arched feet
   d. A & B
   e. B & C
10. Which of the following statements is *true* about Klinefelter Syndrome?
    a. In females, the extra X chromosomes has a significant effect on their development
    b. It is caused by an extra X chromosome producing individuals with either XXX or XXY karyotype
    c. Both males and female with Klinefelter are sterile.
    d. In males, the extra set of genes on the X chromosome increases the levels of testosterone
    e. Variants of Klinefelter syndrome have more than one extra set of the Y chromosome

Thought Questions

1. Genes A, B, C, and D are located on the same chromosome. The recombination frequencies (RF) are: A-B = 8%, A-D = 11%, D-C = 53%, B-C = 34%, and A-C = 42%. What is the order of the genes on the chromosome? Show your work.
2. Draw the pedigree of a mother who is heterozygous for colour blindness and a father who is recessive for colour-blindness, and they have 2 girls and 1 boy.
3. Explain what happens in an individual with Down's Syndrome.
4. If a female has curly hair and all her children also have curly hair, would you say that the gene for curly hair is sex-linked? Why or why not?
5. What is the difference between autosomal dominant and autosomal recessive disorder?

REFERENCES

Karkanaki, A., Praras, N., Katsikis, I., Kita, M., & Panidis, D. (2007). Is the Y chromosome all that is required for sex determination? *Hippokratia, 11*(3), 120–123. PMID:19582205

National Institute of Neurological Disorders and Stroke (NINDS/NIH). (2018). *Huntington's Disease: Hope Through Research.* Retrieved from https://www. ninds.nih.gov/

National Institutes of Health (NIH). (2018). *Genetics Home Reference: Your Guide to Understanding Genetic Conditions.* Retrieved from https://ghr.nlm. nih.gov/condition

Szabo, T. L. (2014). *Diagnostic Ultrasound Imaging: Inside Out.* San Diego, CA: Academic Press.

KEY TERMS AND DEFINITIONS

"p" Arm: The shorter section of the chromosome when it is divided by the centromere.

"q" Arm: The longer section of the chromosome when it is divided by the centromere.

Autosomes: Twenty-two pairs of chromosomes that exist in the human genome and are not sex chromosomes.

Chromosomes: Units that contain chromatin specific for each species (e.g., humans have 46 chromosomes).

Karyotype Analysis: A process used to observe the number and appearance of chromosomes in order to detect birth disorders.

Linkage: Genes that are closer together on the same chromosome are more likely inherited together.

Pedigree: A diagram used to determine the lineage and probability of inheriting genes in humans and direct ancestors.

Personal Wellness Profile: A questionnaire that is used to assess an individual's health lifestyle.

Sex Chromosomes: The 23rd pair of chromosomes, X and Y, that determines sex and sexual characteristics of an organism.

Sex Determining Region: Also known as testis-determining factor, it is a gene on the Y chromosome that gives rise to male phenotypes, such as the testes.

Chapter 7
Genes:
Structure, Replication, and Organization

ABSTRACT

Two types of nucleic acids, DNA and RNA, carry genetic information of organisms across generations. Many researchers are credited with the early work that laid the foundation of the discovery of the structure of DNA. During cell division, the cell replicates its DNA and organelles during the synthesis (S) phase of the cell cycle. Four main steps are involved in the processes of replication. DNA replication errors and cells have evolved a complex system of fixing most (but not all) of those replication errors proofreading and mismatch repair. With repeated cell division, the DNA molecule shortens with the loss of critical genes, leading to cell death. In gonads, a special enzyme called telomerase lengthens telomeres from its own RNA sequence which serves as a template to synthesize new telomeres. Although most DNA is packaged within the nucleus, mitochondria have a small amount of their own DNA called mitochondrial DNA. This chapter explores this aspect of genes.

CHAPTER OUTLINE

7.1 Discovery of Nucleic Acids
7.2 The Elucidation of DNA Structure
7.3 How DNA Makes Copies (Replication)
7.4 When DNA Makes Mistakes Copying
7.5 End of the DNA Molecule
7.6 Mitochondrial DNA
7.7 DNA Organization in the Nucleus
Chapter Summary

DOI: 10.4018/978-1-5225-8066-9.ch007

LEARNING OUTCOMES

- Outline the nature and types of nucleic acids.
- Explain how DNA was discovered.
- Summarize the process of DNA replication.
- With examples demonstrate how DNA corrects mistakes during replication.
- Explain why DNA shortens after each replication.
- Explain inheritance of mitochondrial DNA.
- Describe how DNA is organized in the nucleus of a cell

7.1 DISCOVERY OF NUCLEIC ACIDS

As introduced in section 2.6.4, **nucleic acids** are the basis of heredity and carry genetic information of organisms across generations. They are of two types: deoxyribonucleic acid (DNA) and ribonucleic acid (RNA). In the mid-1800s, a physician of Swiss origin named Friedrich Miescher was doing research on lymphoid cells (white blood cells). Using pus drained from surgical wounds as a result of infection, he extracted white blood cells from which he isolated a substance he called "**nuclein**". He found out that "nuclein" was made of hydrogen, nitrogen, oxygen, and phosphorous. This "nuclein" is the "nucleic acid" in names DNA and RNA.

"Nuclein" was closely related to the work of Gregor Mendel (section 5.3), who had demonstrated that specific traits in garden peas (like flower colour and shape of pea seeds) were passed along to offspring in packages. Today, we know these packages are the genes. How genes related to nucleic acids was not obvious and took time to unravel. In the 1930s, an English scientist Frederick Griffith was working on pneumonia-causing bacteria in mice. When he mixed parts of the virulent strain with that of a harmless strain, the harmless strain somehow became virulent. He proposed that the virulent strain passed a "transforming" substance to the harmless strain. Later in 1943, an American scientist, Oswald Avery, working on disease-causing bacteria also showed that bacteria could transfer the ability to cause disease to other harmless bacterial strains—which in turn could pass it to the next generation. These experiments demonstrated that the "transforming" substance was the nucleic acid DNA. Nucleic acids are made up of nucleotides which are comprised of a 5-carbon sugar, a phosphate group (PO_4^{3-}), and four bases (thymine, cytosine, adenine, and guanine).

7.2 THE ELUCIDATION OF DNA STRUCTURE

Many researchers are credited with the early work that laid the foundation of the discovery of the structure of DNA. In 1948, Linus Pauling, an American theoretical physical chemist, using **X-ray diffraction** techniques found out that proteins assume an alpha-helix shape spiraled like a spring coil. In the mid-1950s, Alfred Hershey and Martha Chase, doing experiments on viruses, discovered that when viruses infected bacteria, they injected DNA. Around the same time, Columbia University biochemist, Erwin Chargaff discovered two features of DNA: 1) that the amount of guanine bases was equal to those of cytosine bases; while those of adenine were equal to those of thymine; 2) the four bases A, T, C and G

Genes

while constant for all living organisms, varied in amounts in each species. Note that two pairs of bases have similar structures—A and G both have two carbon-nitrogen rings (**purines**); while C and T have a single carbon-nitrogen ring (**pyrimidines**) (Figure 1).

Around 1951, Rosalind Franklin, a British physical chemist was working on X-ray diffraction technology and applied it to the study of DNA. Examining X-ray diffraction images of DNA, she determined that the DNA molecule existed in a helical conformation. Her DNA X-ray diffraction pictures paved the way for further work by the American geneticist James Watson and British physicist Francis Crick, who in 1953, would be awarded the Nobel prize for DNA discovery (Figure 2). In recognition of the work of Franklin, the Rosalind Franklin University of Medicine and Science in North Chicago, Illinois in the United States was so named.

Watson and Click pondering over the previous work by Pauling, Chargaff and DNA image results from Franklin, and extensive analysis and modelling, eventually figured out the structure of DNA. They gathered that DNA was made of two strands, with the 3' to 5' directions reversed on either side (see Figure 1). That is, the two strands ran in opposite directions (anti-parallel) and twist together forming a double-helix. The two strands of are held together at their bases by hydrogen bonds: bases at the middle, and the sugar-phosphate forming the backbone (Figure 3). Like a spiral staircase, the sugar-phosphate in DNA equate to the staircase headrails, and the bases to the staircase steps. The **3' OH carbon end** of one nucleotide joins the next at the **5' carbon end** where the phosphodiester bond forms, and so on down the length of the DNA chain. The numbers stand for the five carbons in the sugar deoxyribose as shown (Figure 4).

Figure 1. Nucleotides are the building blocks of nucleic acids (DNA and RNA). Here you can see that DNA is made up of a 5-carbon sugar (deoxyribose; RNA has ribose), a phosphate group and four nitrogenous bases (A, C, T, G; RNA substitutes T with U). Note the reversed 3' to 5' direction on both sides.
Source: Image used under license from Shutterstock.com

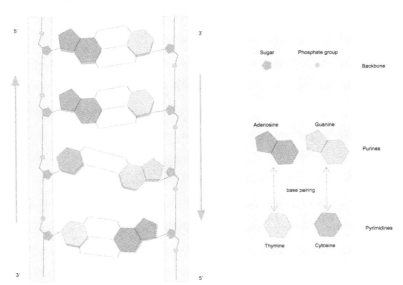

Figure 2. Commemorative plaque for DNA discovery by James Watson and Francis Crick in 1953
Source: Image used under license from Shutterstock.com

Figure 3. The structure of DNA with nitrogen bases at the middle and a sugar-phosphate backbone
Source: Image used under license from Shutterstock.com

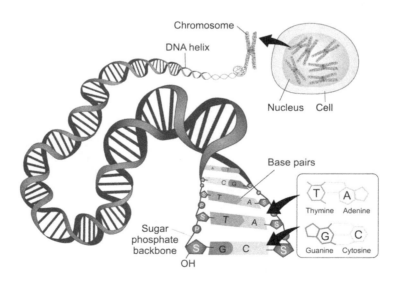

Figure 4. The five carbons in the sugar deoxyribose
Source: Image used under license from Shutterstock.com

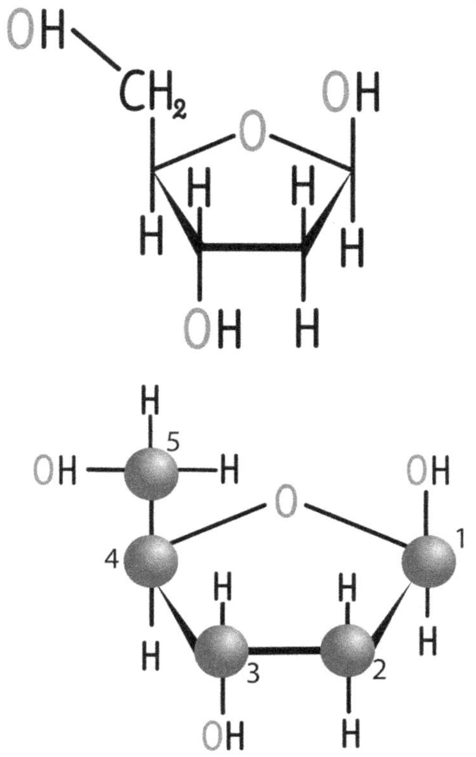

7.3 HOW DNA MAKES COPIES (REPLICATION)

As cells go through cell division to form daughter cells, DNA and organelles are copied or replicated during the synthesis (S) phase of the cell cycle (see Chapter 4, Figure 5). This section of the chapter will outline the process by which a copy of DNA is made from the existing template. Although the understanding of the process of replication was first obtained from research on the prokaryotic bacterium *Escherichia coli* (*E. coli*); the process is well known in higher organisms (eukaryotes).

There are four main steps in the processes of replication (Figure 5):

1. Unwinding of the DNA strands at a specific location known as the replication fork.
2. Preparation of a primer onto which nucleotides are added.
3. Assembly of the new DNA strands.
4. Packaging of new DNA into daughter cells.

- **Unwinding of DNA Strands:** At specific regions called **replication origins, initiator proteins** bind DNA at several places. This forms a **protein-DNA complex** with the DNA wrapped around the proteins. This complex now binds the enzyme **DNA helicase** which breaks the hydrogen

Figure 5. DNA Replication Process. Unwinding of the DNA, the leading and lagging strands, synthesis of an RNA primer, and Okazaki fragments.
Source: Image used under license from Shutterstock.com

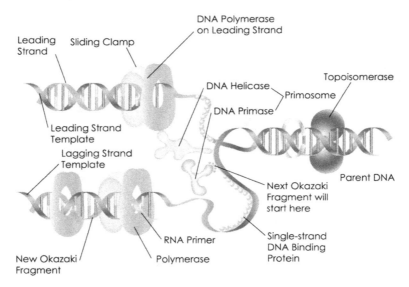

bonds holding the two strands, thereby unwinding it creating a replication fork. Special **binding proteins** attach to the DNA to prevent the two strands from re-joining. With this unwinding of the DNA, the strands ahead of the replication strain and twist further. This strain is relieved by another enzyme called topoisomerase which is just ahead of the fork (Figure 5). Remember, the two strands hydrogen bond instantly much like what two pieces of Velcro would do.

- **Creation of the Primer:** Another enzyme, **DNA primase** attaches to the helicase-DNA complex (creating a **primosome**) and creates a small piece of RNA called a **primer** in the direction away from the replication origin. This is the foundation piece onto which nucleotides will be added in the new strand (Figure 5). Completion of this step leads to the building of another protein-DNA complex creating a second replication fork.
- **Assembly of the New DNA Strands:** The protein complexes in the two forks now proceed in opposite directions from the replication origin along each template strand until the whole length of DNA is replicated. In *E. coli*, the two replication forks add about 500-1000 nucleotides per second taking less than one hour to duplicate the small circular bacterial genome. In more complex higher organisms where DNA is tightly packed as chromatin, the process is more complex. Research from the early 1960s using **autoradiography** showed that the replication forks in human cells move at only 50 nucleotides per second. At this rate, replication of the human genome would take many hours. To speed up the process, studies indicate that in eukaryotes, several pairs of replication forks are in operation simultaneously until all DNA is replicated.

The primer now completed, an enzyme called **DNA Polymerase III** wraps around the unwound strand and adds free nucleotides onto the primer from the surrounding nucleoplasm. These nucleotides are delivered by the blood to tissues from the assimilated nutrients in the lumen of the small intestine. Since the polymerase can only add nucleotides to the 3' OH side, one strand is synthesized continuously (**leading strand**) in the 5' to 3' OH direction (Figure 5). Replication in the fork proceeds in one direction as the replication bubble moves along the DNA length. Since the two strands are antiparallel, a different mechanism of replication occurs on the strand oriented in the opposite direction (**lagging strand**) since nucleotides can only be added in the 5' to 3' OH direction.

Within the time it takes the DNA Polymerase to replicate the leading strand in the replication fork; replication must be completed on the lagging side to keep the bubble moving along at a steady pace. On the lagging strand as a result, DNA is replicated in sections. Thus, several primers are synthesized with several polymerases adding free nucleotides onto them (Figure 5). This discontinuous synthesis produces short stretches of replicated DNA called **Okazaki fragments** in honour of two of the Japanese scientists (Okazaki et al., 1968). Since the replicated pieces have both RNA pieces (primers) and DNA, another enzyme **DNA Polymerase I** recognizes them and removes them, filling the gap with DNA complimentary to the template. However, this polymerase cannot join the DNA pieces to the existing replicated strand. Yet another enzyme, **DNA ligase** recognizes the missing phosphodiester bonds between nucleotides and joins the DNA fragments into one continuous strand. This way, DNA replication on the leading and lagging strands is completed during the period the fork is open. The replication fork moves along until the whole of the DNA is replicated.

The newly replicated strands base-pair with the parent strands forming the new DNA that each daughter cell inherits. The mode of replication is called **semi-conservative replication** (Figure 6).

7.4 WHEN DNA MAKES MISTAKES COPYING

DNA replication is not without errors. Errors occur since the flexibility of the DNA double helix can accommodate slightly misshaped pairings like joining A to G instead of T (Crick, 1966). Insertions or deletions can also cause errors in replication. DNA Polymerase makes mistakes once every 100,000 nucleotides every time a cell divides. This could mean 120,000 mistakes in the human genome with an estimated 6 billion base pairs. In the course of evolution, cells have evolved a complex system of fixing

Figure 6. Semi-conservative mode of DNA replication
Source: Image used under license from Shutterstock.com

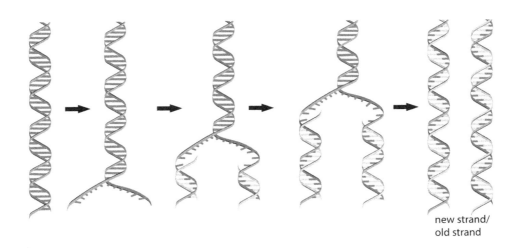

new strand/
old strand

most (but not all) of those replication errors. Mistakes fixed immediately following replication are done by the process of **proof-reading**, where the polymerase recognizes the error, fixes it, and continues with replication. Errors corrected after replication is over are called **mismatch repair**. These errors are easily recognized by the polymerase because they cause structural deformities due to incorrectly matched base pairs (e.g. A-G or C-T). As discussed in Chapter 8, errors that cannot be fixed during replication become established in the cell as **mutations** and can be passed along to offspring (Fray, 2008).

7.5 END OF THE DNA MOLECULE

Since DNA Polymerase III adds nucleotides only on the 3' OH end, there is no way of completing the 5' ends of daughter DNA strands on the lagging strand. For bacteria with a circular DNA, this is no problem. However, for eukaryotes with linear DNA, this presents a big problem. Once the replication fork reaches the end of a DNA molecule, there is no space to produce a primer needed for the last Okazaki fragment at the tip of a linear DNA. While the leading strand is synthesized completely, the lagging strand is not because the fork dissociates following completion of replication for the leading strand. As a result, lagging strand is shortened by one Okazaki fragment. Repeated replication therefore results in shorter and shorter strands and essential genes could be lost! To avoid this, special repetitive expendable and non-coding sequences called **telomeres** (Figure 7) are found on the ends of DNA comprised of TTAGGG nucleotides (up to 100-1000 pairs). In such cases, the fork remains intact long enough to complete the synthesis of the terminal Okazaki fragment. Shortening of telomeres protects organisms from cancer by limiting the number of cell divisions. Tumour cells have been shown to have very short

Figure 7. Telomeres are short sequences found at ends of chromosomes
Source: Image used under license from Shutterstock.com

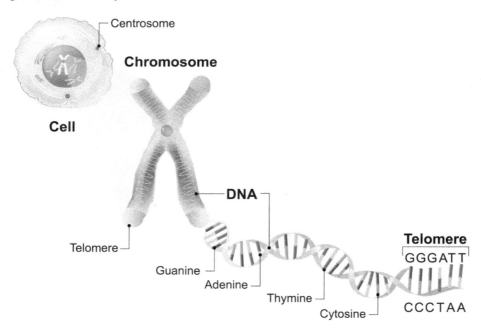

telomeres. In evolutionary sense, this termination of cell division is an adaptive "yardstick" preventing cells from uncontrolled cell proliferation or cancer.

In germ cells (gonads), a special enzyme called telomerase lengthens telomeres from its own RNA sequence that serves as a template to synthesize new TTAGGG telomere segments. However, telomerase is not present in most cells of multicellular organisms like humans, and older humans have shorter DNAs that probably limits tissue lifespan and age of organism (Figure 8). Experiments have shown that human fibroblast cells isolated in the lab with ample nutrients continue dividing for a limited period (sixty times) followed by programmed-cell death or apoptosis.

7.6 MITOCHONDRIAL DNA

Mitochondria are organelles that make energy for the cell (section 3.2.1, Chapter 3). Although most DNA is packaged in chromosomes within the nucleus (nuclear DNA), mitochondria also have a small amount of their own DNA called **mitochondrial DNA** (mtDNA). It is thought that early eukaryotic organisms lacking certain genes they needed for survival, engulfed bacterial cells in a process called **endosymbiosis**. Over time as organisms evolved, some of these genes were incorporated into the nuclear genome while others remained in the mitochondria.

Mitochondrial DNA is estimated to contain 37 genes essential for normal mitochondrial function. Thirteen of those provide instructions for making enzymes involved in oxidative phosphorylation. Dur-

Figure 8. DNA shortens after each successive cell division. See text for explanation.
Source: Image used under license from Shutterstock.com

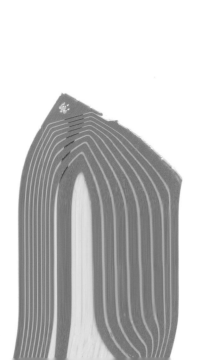
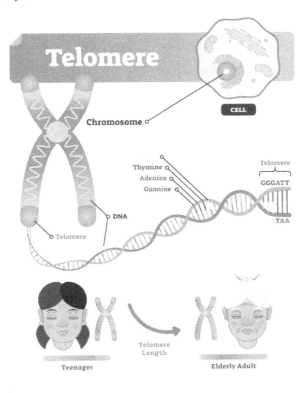

ing this process, oxygen is used to breakdown glucose yielding energy for the cell in form of ATP. The remaining genes provide instructions molecules associated with gene function including transfer RNAs and ribosomal RNAs (covered in chapter 8). If mtDNA has genes which cause disease, they can be transferred to offspring through the egg.

The inheritance pattern from parents to offspring (two scenarios) would occur as shown in Box 1.

7.7 DNA ORGANIZATION IN THE NUCLEUS

With human DNA up to six feet long and with trillion of cells making up an adult human, over many years of evolution, there had to be a mechanism of storing this DNA in the confines of each cell nucleus. As you undoubtedly have noticed folding away your holiday season lights from the Christmas tree, this task requires time, effort and energy. Often, we get so fed up untangling the twisted lights that some people throw them away. To avoid this fate, the DNA in the nucleus is associated with **histone** and **non-histone proteins**. Together they make chromatin.

There are several types of histone proteins labelled H1, H2A, H2B, H3, and H4 (Van Holde, 1988); while the non-histone proteins are numerous often greater than 1000. Both histone and non-histone proteins are associated with DNA in coiling and super-coiling (Figure 9). The basic repeating functional and structural unit of chromatin is known as a **nucleosome**, made up of eight histone proteins and about 146 base pairs of DNA (Wolffe, 1999). Remember from section 4.2 that chromosomes are made of

Box 1.

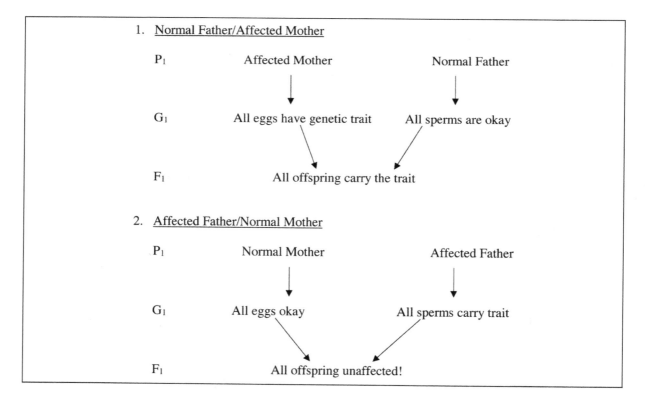

DNA packaging by several types of histone proteins in the nucleus
used under license from Shutterstock.com

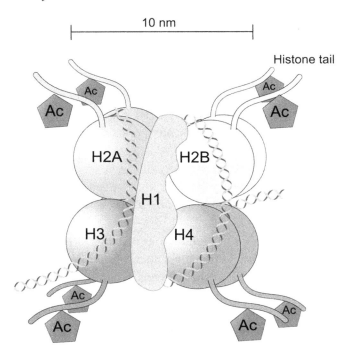

Since each chromosome has many repeating units of nucleosomes, they typically look like beads under an electron microscope (Olins & Olins, 2003).

indicate that during transcription and replication where access to the DNA strands is needed, mechanisms of loosening up this compact structure momentarily (Venkatesh, 2015; Smith & 2005). To make chromatin more accessible, cells enzymatically modify or remodel histones by methyl, acetyl or phosphate group. This process is reversible, thereby preserving the integrity omatin on completion of transcription or replication.

ER SUMMARY

of nucleic acids, DNA and RNA carry genetic information of organisms across generations. cids are made up of nucleotides which are comprised of a 5-carbon sugar, a phosphate group, bases. Many researchers are credited with the early work that laid the foundation of the discovery of the structure of DNA. They include Linus Pauling, Alfred Hershey, Martha Chase, Erwin and Rosalind Franklin. In 1953, James Watson and Francis Crick announced the discovery of cture. They found that DNA was made of two antiparallel strands, held together at their bases gen bonds, and twisting together to form a double-helix.

cell division, the cell replicates its DNA and organelles during the synthesis (S) phase of the Four main steps are involved in the processes of replication: unwinding of the DNA strands, on of a primer, assembly of the new DNA strands, and packaging of new DNA into daughter cells. The process of replication involves many types of enzymes and factors. The main enzyme is called DNA Polymerase III, which wraps around the unwound DNA strand and adds free nucleotides onto the primer. Since the polymerase can only add nucleotides to the 3' OH side, the strand synthesized continuously in the 5 to 3' OH direction. is called the leading strand. A different mechanism of replication occurs on the lagging strand which is oriented in the opposite direction. Several primers are synthesized onto which free nucleotides are added producing short stretches of replicated DNA called Okazaki fragments. The enzyme

DNA Polymerase I removes the primers while another enzyme, DNA ligase, joins the DNA fragments into one continuous strand. The newly replicated strands base-pair with the parent strands in a semi-conservative mode of replication.

DNA replication errors occur due to the flexibility of the DNA double helix or through insertions or deletions. Since mistakes occur once every 100,000 nucleotides every time a cell divides, cells have evolved a complex system of fixing most (but not all) of those replication errors proof-reading and mismatch repair.

Only able to add nucleotides to the 3' OH end, there is no way DNA Polymerase III can add nucleotides on the 5' end of the lagging strand. Following a succession of cell divisions, the DNA molecule becomes shorter and shorter with the loss of critical genes, leading to cell death. Cells that are actively dividing have special repetitive and expendable non-coding sequences called telomeres found on DNA molecule. In reproductive cells (in the gonads), a special enzyme called telomerase lengthens telomeres from its own RNA sequence which serves as a template to synthesize new telomeres. However, telomerase is not present in most cells of multicellular organisms like humans; thus, older organisms have shorter DNAs with increasing age, which limits tissue longevity. Although most DNA is packaged in chromosomes within the nucleus (nuclear DNA), mitochondria also have a small amount of their own DNA called mitochondrial DNA. With an estimated 37 genes, some provide instructions for making enzymes involved in oxidative phosphorylation, while the rest provide instructions molecules associated with gene function including transfer RNAs and ribosomal RNAs.

With human DNA up to six feet long and with trillion of cells making up an adult human, cells have developed a complex mechanism of storing DNA in the nucleus. DNA is associated with proteins called histones making chromatin. The basic repeating functional and structural unit of chromatin is known as a nucleosome, made up of eight histone proteins and about 146 base pairs of DNA. However, during transcription and replication, cells have mechanisms of loosening up this compact structure momentarily to make chromatin more accessible.

End of Chapter Quiz

1. Nucleic acids are made up of nucleotides which are comprised of
    a. 5-carbon sugar
    b. phosphate group
    c. four bases (thymine, cytosine, adenine, guanine)
    d. all of the above
    e. B & C

ich of the following bases are correctly paired?
   A. A and G both have two carbon-nitrogen rings (purines); while C and T have a single carbon-nitrogen ring (pyrimidines)
   B. A and G both have a single carbon-nitrogen ring (pyrimidines); while C and T have two carbon-nitrogen rings (purines)
   C. A and C both have two carbon-nitrogen rings (purines); while G and T have a single carbon-nitrogen ring (pyrimidines)
   D. A and T both have two carbon-nitrogen rings (purines); while G and C have a single carbon-nitrogen ring (pyrimidines)
   E. None of the above

ich of the following statements about DNA is *false*?
   A. DNA has a double helix structure
   B. The two strands run in the same direction
   C. The two strands are held together at their bases by hydrogen bonds: bases at the middle, and the sugar-phosphate forming the backbone
   D. 3' OH end of one nucleotide joins the next at the 5' carbon end
   E. None of the statements are false

at are the main steps in the processes of replication?
   A. Unwinding of the DNA strands at a specific location known as the replication fork
   B. Preparation of a prime onto which nucleotides are added
   C. Assembly of the new DNA strands
   D. Packaging of new DNA into daughter cells
   E. All of the above

at happens when DNA makes mistakes during replication?
   A. They are fixed immediately following replication by a process of proofreading
   B. They are easily recognized by the polymerase because they cause structural deformities
   C. Errors that are fixed during replication become mutations and cannot be passed along to offspring
   D. A & B
   E. A & C

omeres
   A. lengthen as you get older
   B. are special repetitive, expandable and non-coding sequences on DNA
   C. are made up of six-base sequence (GGGTTA) nucleotides
   D. A & B
   E. B & C

of these researchers are credited with the early work that laid the foundation of the structure
NA except
   A. Linus Pauling
   B. Erwin Chargaff
   C. Louis Pasteur
   D. Rosalind Franklin
   E. Martha Chase

8. The thought of early eukaryotic organisms engulfing bacterial cells to obtain certain genes for survival is known as
   a. Spontaneous Generation Theory
   b. Endosymbiosis
   c. Exocytosis
   d. Germ Theory
   e. Miasma Theory
9. DNA ligase is an enzyme that
   a. recognizes hydrogen bonds and cleaves the DNA strand creating a fork
   b. ligates the leading and lagging strands
   c. recognizes the missing phosphodiester bonds between nucleotides and joins the DNA fragments into one continuous strand
   d. replicates the leading strand in the replication fork
   e. wraps around the unwound strand and add nucleotides onto the primer
10. The basic repeating functional and structural unit of a chromatin is called
    a. nucleosome
    b. non-histone proteins
    c. primosome
    d. primer
    e. diffraction

Thought Questions

1. Summarize the and draw the process of DNA replication.
2. Why do telomeres shorten as we age?
3. Compare and contrast the inheritance pattern from parents to offspring regarding these two scenarios
   a. Normal Father/Affected Mother
   b. Affected Father/Normal Mother
4. Is DNA replication always continuous? Why or why not? Explain your reasoning.
5. Describe the structure and bonding of purines and pyrimidines.

REFERENCES

Annunziato, A. (2008). DNA Packaging: Nucleosomes and Chromatin. *Nature Education*, *1*(1), 26.

Bednar, J., Horowitz, R., Grigoryev, S., Carruthers, L., Hansen, J., Koster, A., & Woodcock, C. (1998). Nucleosomes, linker DNA, and linker histones form a unique structural motif that directs the higher-order folding and compaction of chromatin. *Proceedings of the National Academy of Sciences of the United States of America*, *95*(24), 14173–14178. doi:10.1073/pnas.95.24.14173 PMID:9826673

Crick, F. H. S. (1966). Codon-anticodon pairing: The wobble hypothesis. *Journal of Molecular Biology*, *19*(2), 548–555. doi:10.1016/S0022-2836(66)80022-0 PMID:5969078

I. S. (1966). Codon-anticodon pairing: The wobble hypothesis. *Journal of Molecular Biology*, –555. doi:10.1016/S0022-2836(66)80022-0 PMID:5969078

., Wang, Y., & Allis, C. D. (2003). Histone and chromatin cross-talk. *Current Opinion in Cell* 5(2), 172–183. doi:10.1016/S0955-0674(03)00013-9 PMID:12648673

. R., & Burgoyne, L. A. (1973). Chromatin sub-structure. The digestion of chromatin DNA spaced sites by a nuclear deoxyribonuclease. *Biochemical and Biophysical Research Communications*, *52*(2), 504–510. doi:10.1016/0006-291X(73)90740-7 PMID:4711166

R. D. (1974). Chromatin structure: A repeating unit of histones and DNA. *Science*, *184*(4139), doi:10.1126cience.184.4139.868 PMID:4825889

Mader, A. W., Richmond, R. K., Sargent, D. F., & Richmond, T. J. (1997). Crystal structure leosome core particle at 2.8 A resolution. *Nature*, *389*(6648), 251–260. doi:10.1038/38444 05837

1974). Subunit structure of chromatin. *Nature*, *251*(5472), 249–251. doi:10.1038/251249a0 22492

R., Okazaki, T., Sakabe, K., Sugimoto, K., & Sugino, A. (1968). Mechanism of DNA chain Possible discontinuity and unusual secondary structure of newly synthesized chains. *Proceedings National Academy of Sciences of the United States of America*, *59*(2), 598–605. doi:10.1073/ .598 PMID:4967086

L., & Olins, D. E. (1974). Spheroid chromatin units (v bodies). *Science*, *183*(4122), 330–332. 26cience.183.4122.330 PMID:4128918

E., & Olins, A. L. (2003). Chromatin history: Our view from the bridge. *Nature Reviews. r Cell Biology*, *4*(10), 809–814. doi:10.1038/nrm1225 PMID:14570061

Gross-Bellard, M., & Chambon, P. (1975). Electron microscopic and biochemical evidence matin structure is a repeating unit. *Cell*, *4*(4), 281–300. doi:10.1016/0092-8674(75)90149-X 22558

2008). DNA Replication and Causes of Mutation. *Nature Education*, *1*(1), 214.

L., & Peterson, C. L. (2005). ATP-dependent chromatin remodeling. *Current Topics in Developmental Biology*, *65*, 115–148. doi:10.1016/S0070-2153(04)65004-6 PMID:15642381

J. O., & Kornberg, R. D. (1975). An octamer of histones in chromatin and free in solution. ngs of the National Academy of Sciences of the United States of America, *72*(7), 2626–2630.)73/pnas.72.7.2626 PMID:241077

le, K. E. (1988). *Chromatin: Springer Series in Molecular Biology.* New York: Springer-Verlag k, Inc.

de, K. E., Sahasrabuddhe, C. G., & Shaw, B. R. (1974). A model for particulate structure in n. *Nucleic Acids Research*, *1*(11), 1579–1586. doi:10.1093/nar/1.11.1579 PMID:10793713

Watson, J. D., & Crick, F. H. S. (1953). Molecular structure of nucleic acids. *Nature*, *171*(4356), 737–738. doi:10.1038/171737a0 PMID:13054692

Wolffe, A. P. (1999). *Chromatin: Structure and Function* (3rd ed.). San Diego, CA: Academic Press.

Woodcock, C. L. (2005). A milestone in the odyssey of higher-order chromatin structure. *Nature Structural & Molecular Biology*, *12*(8), 639–640. doi:10.1038/nsmb0805-639 PMID:16077725

Woodcock, C. L., Safer, J. P., & Stanchfield, J. E. (1976). Structural repeating units in chromatin. I. Evidence for their general occurrence. *Experimental Cell Research*, *97*(1), 101–110. doi:10.1016/0014-4827(76)90659-5 PMID:812708

KEY TERMS AND DEFINITIONS

3' OH End: Refers to the 3' carbon of DNA and RNA sugar backbone that has a hydroxyl group attached to it.

5' Carbon End: Refers to the 5' carbon of DNA and RNA backbone that has a phosphate group attached to it.

Apoptosis: Programmed cell death.

Autoradiography: A technique that utilizes X-rays to visualize radioactively labeled molecules.

Binding Proteins: Proteins that act as a link to bring two molecules together.

Chromatin: A DNA and protein complex that forms a chromosome.

DNA Helicase: An enzyme that initiates the unwinding (or unfurling) of DNA during DNA replication.

DNA Ligase: Enzyme that joins the fragmented nucleotides together via a phosphodiester bond.

DNA Polymerase I: Enzyme that has proofreading ability, where it removes nucleotides from the 5' end and replace with the correct DNA.

DNA Polymerase III: A holoenzyme that catalyzes the synthesis of the complementary strand of DNA in the 5' to 3' end direction.

DNA Primase: Enzymes that create RNA primers complementary to DNA strands.

Endosymbiosis: A form of relationship in which one cell lives inside another to form and acts as a single organism.

Initiator Proteins: Proteins that bind to the replicator and signal to begin replication.

Lagging Strand: A new DNA strand that is synthesized discontinuously in the 5' to 3' direction.

Leading Strand: A new DNA strand that is synthesized continuously in the 5' to 3' direction.

Mismatch Repair: A repair system that corrects any errors in base pairing.

Mitochondrial DNA: Genetic material (DNA) in the mitochondria that carries the code to convert chemical energy into ATP.

Mutation: An alteration in the DNA sequence due to an insertion, deletion or rearrangement of a single base pair or fragments.

Nucleic Acids: Polymers that are made of nucleotides and form either deoxyribonucleic acid (DNA) or ribonucleic acid (RNA).

...ein: A term which was previously used to describe a nucleic acid.
...eosome: DNA wrapped around histones, forming a repeating unit of chromatin.
...zaki Fragments: Term used to label the synthesized DNA fragments on the lagging strand.
...er: Short RNA sequences created from DNA primase.
...osome: A protein complex involved with DNA primases to synthesize RNA primers.
...freading: Editing and correcting the errors that enzymes may have made while they were creat... ...hter strands.
...ein-DNA Complex: A structure of protein and DNA that have bound together to regulate cell ...on.
...ication Origins: The starting point for DNA replication to begin its process.
...-Conservative Replication: A type of DNA replication in which the two new strands of DNA ...d each contain an original strand and a new strand.
...merase: An enzyme that extends the telomeres in chromosomes to reverse shortened telomeres.
...meres: Protective chemical structures located at the end of a chromosome to prevent the loss ...

...y Diffraction: A method that determines the molecular structure of a substance using a beam ...s which diffract or spread out in specific patterns.

Chapter 8
Genes:
How They Work

ABSTRACT

Genes are regions on DNA that contain the instructions for making specific proteins. In humans, genes vary in size from hundreds of DNA bases to over 3 million base pairs. From DNA to proteins, two steps are involved. Transcription is accessing the gene and reading the instructions therein in the nucleus producing as a single strand of RNA called messenger RNA (mRNA). Translation is reading the instructions on mRNA to assemble the specified proteins on the surface of ribosomes. Genetic mutations are heritable, small-scale alterations in one or more base pairs that damage DNA. Although new mutations introduce new variation, these are constantly removed from populations. Mutations can arise naturally during DNA replication or can be caused by environmental factors like chemicals or radiation. They can be harmful, neutral, or beneficial to the organism and are generally of five types: point mutations, frameshift mutations, transposons, transitions, and transversions. This chapter explores this aspect of genes.

CHAPTER OUTLINE

8.1 Connecting Genes and Proteins
8.2 Making a Transcript from DNA
8.3 How Enzymes Read DNA: Genetic Code
8.4 How the mRNA Transcript is Translated
8.5 When Genes Change: Genetic Mutations
Chapter Summary

DOI: 10.4018/978-1-5225-8066-9.ch008

LEARNING OUTCOMES

- ...trate how genes were connected to proteins by experimentation.
- ...ain the process of transcription to make messenger RNA.
- ...marize the elements of the genetic code.
- ...ain the process of translation.
- ... examples, demonstrate how mutations occur, types and their effects.

CONNECTING GENES AND PROTEINS

...1-1800s, Charles Darwin proposed the theory of evolution which explained a mechanism that ...rganisms to evolve as their environments change (section 1.4). This is what we call adaptation ...eritable—allowing advantageous changes to accumulate and benefit the population. **Genetic** ...n occurs because favorable genetic changes (mutations) accrue in on organism allowing it ...o existing conditions. The experimental work of Gregor Mendel on garden peas (Chapter 5, ...3) in the latter half of 1800 showed that genetic traits (like flower colour) are passed on to ...in distinct packages we today call genes. In the early 1900s scientists continued working to ... how genes worked. How do genes allow organisms to adapt? How do genes pass traits they ...n parent to offspring?

...experiments by Beadle and Tatum (Cairns, 2003) provided experiential evidence connecting ... enzymes working on the bread mold *Neurospora crassa*. They demonstrated that mutant ...mbarded with X-rays were unable to grow in growth medium that was minimal in nutrients. ... when the mutant strains were put in a complete growth medium, they thrived. The normal strain ...ed or wild-type) can make all the nutrients it requires and missing from the minimal medium. ...othesized that the mutant strains had mutations in specific genes required in the synthesis of ...g nutrients. To find out exactly what genes had mutated in the mutants, follow-up experiments ...ochemical synthesis of the amino acid arginine were performed by Srb and Holowitz on *N*. ...orowitz, 1995). In the 3-step metabolic pathway required for arginine synthesis (see below) ...ant required the addition of a specific amino acid in the medium. It was hypothesized that the ...cked the gene coding for the enzyme required to make the added amino acid (see Box 1)

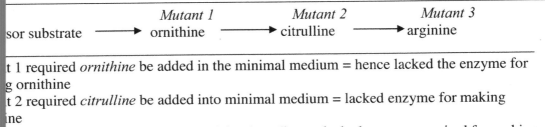

...sor substrate → *Mutant 1* ornithine → *Mutant 2* citrulline → *Mutant 3* arginine

...t 1 required *ornithine* be added in the minimal medium = hence lacked the enzyme for
...g ornithine
...t 2 required *citrulline* be added into minimal medium = lacked enzyme for making
...ine
...t 3 required *arginine* added into minimal medium = lacked enzyme required for making
...ne

The conclusion from these experiments was that the mutants had incurred faults in the genes required for making the enzymes needed at each step in the pathway giving way for the *one gene-one enzyme hypothesis*. As more information became available from research in later years—that not all proteins were enzymes, this hypothesis was modified to read *one gene–one polypeptide hypothesis*. Some proteins are used in the body for other purposes like building body structure (muscles), in hair, hormones, antibodies, etc.

Genes are regions on DNA that contain the instructions for making specific proteins. In humans, genes vary in size from 100s of DNA bases to over 3 million base pairs. The **Human Genome Project** was an international research effort tasked with determining the sequence of the human genome and identifying the genes that it contains. The effort began in 1990 and ended in 2003 funded largely by the US Government through the National Institutes of Health and the Department of Energy. The effort required collaboration between researchers in the US and other countries. It is estimated that humans have between 20,000 to 25,000 protein-coding genes which makes only about 1.5% of the genome for coding proteins, **rRNA**, and **tRNA**. The rest of the genome once known as "**junk DNA**" is today known to be crucial to the survival of the species. This non-coding DNA consists of **repetitive DNA** (transposons, ~44%; non-transposons, ~14%); **introns** (~5%); **regulatory sequences** (~20.4%); and other unique non-coding DNA sequences (~15%) (Figure 1).

The process by which genes direct the synthesis of proteins is like how we use a recipe book in cooking. The recipe must be accessed, instructions read, then implemented in specific steps. In class, I always use familiar examples regularly occurring in our bodies. Let's start with eating a sandwich for lunch. After we have chewed and swallowed, we usually forget about the food (unless the food causes illness – vomiting and/or diarrhea). The natural physiological process in the body takes over once food reaches the stomach. In the end, we know that the nutrients in the sandwich needs to be delivered to

Figure 1. Types of DNA sequences in the human genome. See text for explanation.

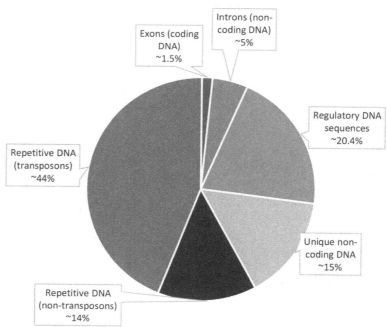

ich break them down for energy, use them for body maintenance and repair, and to build other olecules the body needs (e.g. antibodies or hormones). A good example is the making of insu- h the body needs to allow cells uptake glucose from the blood. Blood sugar can often rise to 20 mg per deciliter of blood after a meal. Cellular uptake of glucose lowers blood sugar to the atically maintained level of ~90 mg/deciliter of blood.

en, how do cells make insulin with instructions from their DNA in the pancreas? Well, we know lin is a protein hormone. The pancreas contains the same DNA as other cells of the body and ere in this DNA, are the **gene** instructions (recipe) for assembling the string of amino acids re- assemble this polypeptide (Figure 2). In the next few pages, we will see a very close interplay DNA and RNA in accomplishing this feat of accessing the instructions on DNA and using them functional proteins like insulin.

DNA to proteins, two steps are involved (Figure 3):

anscription: This involves accessing the gene recipe and reading the instructions therein in the cleus. The transcript is made as a single strand of RNA.

anslation: The transcript of instructions must leave the nucleus to the cytoplasm where it will translated into the assembly of the specified protein. Remember, protein synthesis occurs in the ribosomes or those attached on the rough ER (Chapter 3, section 3.2.1).

. *A gene is a unit of heredity carrying specific instructions for making a protein. DNA carries ds of genes which in humans vary in size from a few hundred base pairs to more than 2 million rs. Between 20,000 and 25,000 genes (~1.5%) were identified as protein-coding genes in humans uman Genome Project.*
age used under license from Shutterstock.com

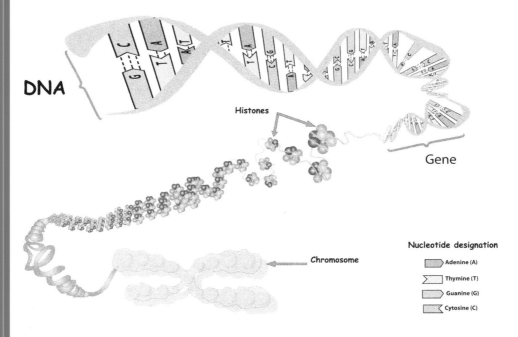

Figure 3. Overview of transcription and translation. Information carried by genes is transcribed then translated into proteins in the cell.
Source: Image used under license from Shutterstock.com

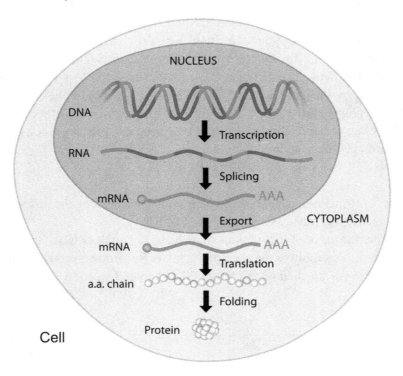

8.2 MAKING A TRANSCRIPT FROM DNA

As we noted in Chapter 1, one big feature of living cells is order. This orderliness prevents chaos from developing which would make cells not respond to signals they receive from different parts of the body. For example, in an extreme emergency, we often use *instinct* to respond to situations (rather than *active thinking*). This is as it should be considering that humans evolved not in cities—but in wild environments with other sometimes dangerous beasts. The time it takes for a human to decide—that is gathering the facts at hand to make a rational decision is too long for some situations like escaping from danger. Thus, we instinctively react and ran *away* as far as is possible from the present danger without considering we could be running *towards* other dangers! In the same vein, cellular processes are well coordinated in the body, and the need for insulin is communicated by the brain to the pancreas, which secretes insulin, and eventually blood sugar level is lowered.

The process of transcription begins when a gene called **TBP** codes for a **TATA binding protein** which binds the **promoter** region—a regulatory section near the beginning of a gene sequence. This regulatory region is marked with the repetitive bases thymine and adenine (TATA). Binding the TATA box is like announcing to other enzymes that gene reading must "start here" (Figure 3a). The chromatin structure is modified to allow an enzyme known as **RNA Polymerase II** to bind the promoter assisted by **transcription factors**. Transcription factors are an important family of proteins whose main roles are to activate or repress (curtail) gene expression (Figure 4). For example, the development of the fertilized

Initiation of transcription and the role of transcription factors in assisting RNA Polymerase II the promoter region of a gene
ge used under license from Shutterstock.com

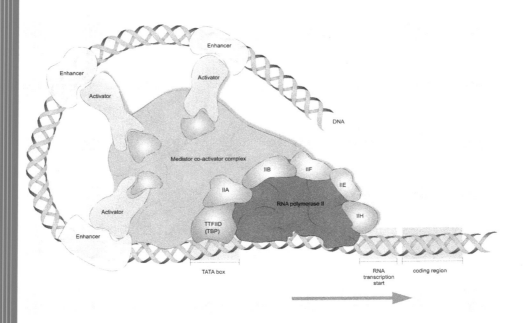

specialized embryo with many functionally different cells and tissues is directed by **homeotic** ich code for transcription factors.

Polymerase II binding forms a **pre-initiation complex** which is followed by an active confor- change in which a few bases at the start site are dissolved (see Figure 4). Dissolution requires of the ATP-dependent DNA helicase which initiates DNA unwinding. This enhances the po- of the template strand of the promoter towards the active site of the RNA Polymerase II (Hahn, ychik & Hampsey, 2002). Initiation of transcription then begins with the synthesis of the first diester bond of the RNA. After making about 30 RNA bases, the polymerase dissociates from oter and the rest of the transcription machinery to proceed with elongation of the RNA strand. A polymerase now begins the slow process of recruiting factors necessary for RNA chain syn- longation), RNA processing, RNA export and chromatin modification. Remember, for every T ountered in DNA by the RNA polymerase, the base U is added as the mRNA is synthesized. ngly, many of the transcription machinery at the promoter remain attached at the start site in a **complex**. This complex earmarks actively transcribed genes enabling the by-passing the slow of recruitment in additional rounds of transcription. (Hampsey, 1998)

elongation actively continues making a new strand of **precursor messenger RNA (pre-mRNA)** rminator signal is reached at the end of a gene sequence (Figure 5). The termination process ted by proteins which interact with RNA polymerase, the pre-mRNA transcript, and the DNA . Termination is complex, involving post translational modifications of the polymerase as well iation of several *trans*-acting factors (Aneshkumar, 2013).

Figure 5. The process of transcription which makes a messenger RNA (mRNA)
Source: Image used under license from Shutterstock.com

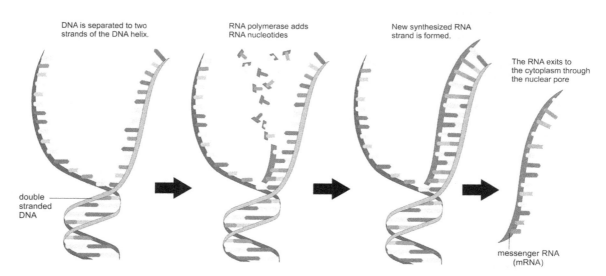

The high-molecular-weight nuclear pre-mRNA produced is called **hnRNA** (Figure 6). The next step after termination is processing the hnRNA into a **mature** mRNA to allow its export from the nucleus and facilitate translation. To ensure the pre-mRNA is stabilized and translation signalling factors added, two stages are involved: **pre-mRA splicing**, and 5' (upstream end), and 3' OH (downstream end), modification. Most mammalian pre-mRNAs contain non-coding sequences called **introns** and coding sequences called **exons**. In addition, there are usually other non-coding regions located at the 5' and 3' OH ends of the gene flanking the protein coding sequences. These sequences are known as *untranslated regions* (UTRs) because they are usually transcribed, but not translated. The 5' and 3' OH UTRs generally contain transcription start sites (TSS) and polyadenylation signals respectively, important in regulating a gene. (D'Souza et al., 2003)

Introns must be removed (spliced) by esterification reactions catalysed by a **spliceosome** and associated auxiliary proteins (see Figure 6). The spliceosome clips the intron and joins the two exons that border the intron. A spliceosome is composed of a complex of five subunits of **small nuclear ribonucleoprotein particles (snRNPs**, pronounced "snurps") including U1, U2, U4, U5 and U6 recruited after the completion of the pre-mRNA. snRNPs are a class of dynamic RNA-protein complexes that accumulate in the nucleus. (Maniatis & Reed, 1987; Kelly & Corbett, 2009; Suhana et al., 2015). During mRNA processing, some exons are spliced in different patterns in a process called **alternative splicing** (see Figure 2, Chapter 11). Alternative splicing plays important roles in the versatility of proteins derived from a single gene in different cell types or developmental stages (Inose et al., 2015).

Modification occurs at the two ends—5' and 3' OH of the pre-mRNA. On the 5' end modification occurs even before the RNA polymerase has completed transcribing the pre-mRNA. A **7-methylguanosine cap** is added on the 5' end catalysed by a nuclear enzyme, guanylyl transferase. On the 3' OH end, an enzyme known as **poly-A polymerase** adds a set of approximately 200 adenine (A) nucleotides making a **poly-A tail** in a process called **polyadenylation** (Figure 7). This pre-MRA processing (cap and tail)

... The precursor messenger RNA (also known as high-molecular-weight nuclear pre-mRNA) is ... introns and exons. snRNPs which make up the spliceosome help in pre-mRNA splicing which ... non-coding sequences (introns) leaving coding ones (exons).
Image used under license from Shutterstock.com

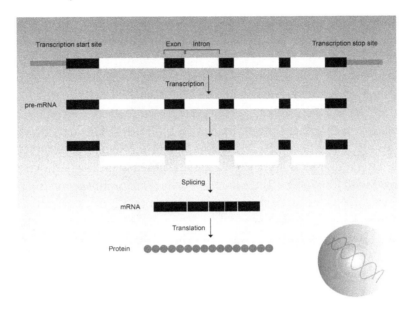

... RNA processing: Addition of the 5' GTP cap and poly-A-tail on the 3'OH end

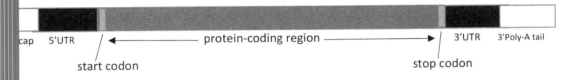

... protect the now **mature mRNA** from cellular degradation as it is exported to the cytoplasm, ... low initiation factors involved in translation to bind and begin the process of translation. ... mature mRNA is now ready for export to the cytoplasm. Studies in budding yeast indicate that ... protein complex, known as the ***t*ranscription and *ex*port (TREX) complex,** is assembled upon ... mRNA during transcription and is a crucial component in determining the efficiency of mature ... export (Reed & Cheng, 2005; Kelly & Corbett, 2009). Components of this TREX recruit **adap-**
teins which in turn assemble export receptors which bind the mRNA transcript. Two adaptor ... **TAP** and **CRM1** have been identified as nuclear export receptors for different classes of mRNA ... Inose et al., 2015). After the mRNA has recruited the appropriate export receptors, the result-
... **NA ribonucleoprotein (mRNP) complex** is translocated through nuclear pore complex to the ... sm. Here, the mRNP is remodelled to replace export factors with a set of proteins factors which ... translation.

8.3 HOW ENZYMES READ DNA: GENETIC CODE

All lifeforms existing on the planet (with minor exceptions) share the same standard **genetic code** (Zamudio & Jose, 2017). Crick (1968) in one of the initial articles describing the genetic code, used the term 'frozen accident' to describe the universality of the genetic code; that is, its apparent inability to accept new variations. Fifty years later, scientists have advanced significantly in understanding the molecular mechanisms that govern the genetic code. Nevertheless, questions remain regarding the origin and evolution of the code, particularly the observation that the code stopped incorporating new amino acids despite its redundancy in codon sequences. (de Pouplana et al., 2017)

The genetic code is seen as consisting of a syntax (symbols and rules); semantics (meaning of symbols or combinations thereof); and cipher (algorithm for decryption). ("The Genetic Code", n.d.). During translation, molecules recognize the bases on the mature RNA in sets of three (**triplets**) called **codons**. Since we know there are four bases A, C, T and G, using mathematics, the question becomes how many "words" can be formed with four bases. Since there are only 20 naturally occurring amino acids that can be assembled in a variety of ways to make poly-peptides (section 2.6.3, Chapter 2, and Table 1), each must be encoded for by a separate code "word". If the "words" were one letter long, only four "words" would be possible (A, T, C, G). That leaves 16 amino acids with no code! If the "words" were two letters long, then 4 x 4 = 16 "words" would be possible (for example, AG, UU, GC). That leaves four amino acids with no code! If the words are made of three letters, it can be deduced that there will be 4 x 4 x 4 (or 4^3) kinds of codons. We have more than enough now—so, what do the other 44 codons (64-20) do? Sixty-one codons specify amino acids, three (UAG, UAA, UGA) serve as "stop" signals, and one (AUG) codes for the amino acid methionine and serves as a "start" signal (Figure 8). There is more than one

Table 1. The 20 naturally occurring amino acids

Ala: Alanine
Asp: Aspartic acid
Glu: Glutamic acid
Cys: Cysteine
Phe: Phenylalanine
Gly: Glycine
His: Histidine
Ile: Isoleucine
Lys: Lysine
Leu: Leucine
Met: Methionine
Asn: Asparagine
Pro: Proline
Gln: Glutamine
Arg: Arginine
Ser: Serine
Thr: Threonine
Val: Valine
Trp: Tryptophan
Tyr: Tyrosine

The 64 codons comprising the Genetic Code. Note that some amino acids (like leucine) are [coded] by more than one codon (redundancy). See text for explanation.
[Ima]ge used under license from Shutterstock.com

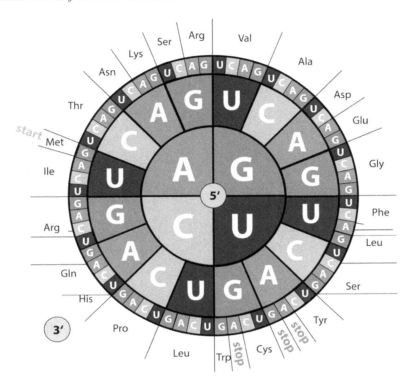

some amino acids—thus, redundancy! For example, lysine has two codons (AAA and AAG) [and leu]cine has six codons (UUA, UUG, CUU, CUC, CUA, and CUG). These signals are recognized [by molec]ules which direct them to perform different actions. This set of 64 codons comprises the ge[netic cod]e which was deciphered in 1961 by Marshall Nirenberg—for which he won the Nobel Prize in [Physiolog]y or Medicine in 1968.

HOW THE mRNA TRANSCRIPT IS TRANSLATED

[Two form]s of RNA polymerase (I and III) transcribe a number of tRNA and rRNA genes as well as other [small RN]A genes distributed through the genome (Aneeshkumar et al., 2013). The rRNAs are synthesized [from nuc]leolar DNA while tRNA is synthesized from regular DNA. Both RNAs are transported to the [cytoplasm] at the same time as the mature mRNA. In the cytoplasm, the rRNA is combined in equal parts [with prot]eins to form ribosomes. The 40S and 60S subunits join to form a functional ribosome which [attaches t]o the mRNA (Figure 9).

[Init]iation of Translation: Perhaps the most complex step of the translation process is initia[tion]. All the participants in the process need to be readied. These include the mature mRNA; the [40S] and **60S ribosomal subunits** in eukaryotes (note that in bacteria these are the 30S and 50S

Figure 9. The formation of the functional ribosome by joining the 40s and 60s sub-units. Note the three sites: A (aminoacyl) site, P (peptidyl) site, and E (exit). A mRNA is already bound to the ribosome ready for translation.
Source: Image used under license from Shutterstock.com

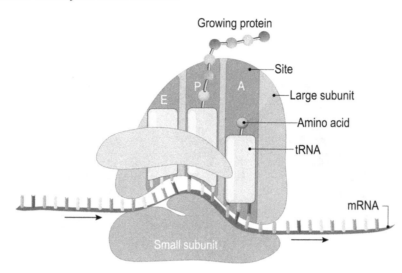

subunits); the first transfer RNA with the first amino acid already attached (Met-tRNA); and 11 translation initiation factors consisting of 24 independent gene products (Dever et al., 2016). The current estimate of the number of "basic" initiation factors of eukaryotes is >16 and growing (Browning & Bailey-Serres, 2015).

Before going through the process of translation, let's discuss the third type of RNA in more detail. We have the mature mRNA and the functional ribosome, now we need a molecule that will be able to recognize the codons on the mRNA and link them to the corresponding amino acid on the surface of the ribosome. This is the transfer RNA (tRNA) molecule. An enzyme called **aminoacyl-tRNA synthase** catalyses the addition of an amino acid to the tRNA. Since there are 20 amino acids, there are 20 aminoacyl-tRNA synthases. By recognizing the codons on the mRNA transcript, the synthases acetylate tRNA molecules (often termed "charging") with the appropriate amino acid at one end which corresponds to the anti-codon at the other end of the tRNA molecule. The tRNA molecule resembles a clover leaf with the **anticodon** at the bottom and the amino acid attachment site at top end (Figure 10).

- **Elongation:** The ribosome has three sites through which tRNAs move: **A (aminoacyl) site**, **P (peptidyl) site,** and **E (exit) site** (see Figure 9). A eukaryotic elongation factor binds the first aminoacyl-tRNA in a GTP-dependent manner and then directs the tRNA to the A site of the ribosome. Codon recognition by the tRNA triggers GTP hydrolysis by the elongation factor releasing it and enabling the aminoacyl-tRNA to be accommodated into the A site (Dever & Green, 2012). Translation has officially begun (Figure 11)! The first tRNA with the amino acid methionine then moves on to the P site, as the second tRNA with its amino acid enters the ribosome at its A site base-pairing with the codon next to AUG. The first tRNA then moves to the E site of the ribosome as the second tRNA enters the P site (Figure 11). The third tRNA next enters the A site. The

...0. The structure of transfer RNA (tRNA) molecule. Note the mRNA (anticodon) and the amino ...achment sites.
...mage used under license from Shutterstock.com

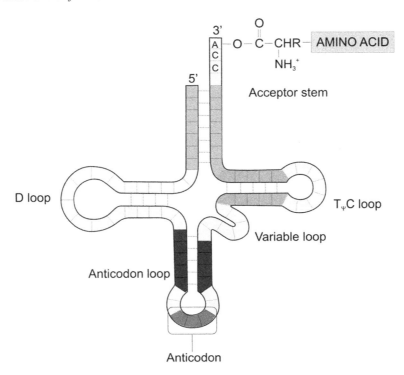

...**ptidyl transferase** centre of the ribosome catalyses (by dehydration synthesis) the addition of ... amino acid of the second tRNA to the first amino acid carried by the first tRNA (methionine) ...rming a peptide bond. The P site always holds the tRNA that is attached to the elongating poly-...ptide chain. The ribosome then proceeds to the next codon on the mRNA and the next tRNA ...se-pairs with the codon. The subsequent translocation of the mRNA by one codon shifts the ...ptidyl-tRNA to the P-site and the deacylated tRNA to the E-site, freeing the A-site for the next ...propriate charged-tRNA and continuation of the cycle (Browning & Bailey-Serres, 2015). This ...ocess continues until a stop codon is encountered.

...**rmination:** Once a stop codon has been reached (UAG, UAA, or UGA), obviously for stop ...dons, there are no tRNAs available to enter the ribosome. In eukaryotes two termination protein ...tors catalyse the process. The first factor is tRNA-shaped protein responsible for stop codon ...cognition and peptidyl-tRNA hydrolysis. It binds the ribosome triggering the polypeptide re-...ase. The second factor is a release factor. The ribosome dissociates from the mRNA and its two ...bunits separate. As noted by Dever and Green (2012), full dissociation of the ribosomal com-...ex will occur following termination in some cases; whereas in other cases, partial dissolution of ...complex will allow "reinitiation" events to follow. In cases where same protein transcribed on ...mRNA is required by the cell, several ribosomes attach to the mRNA and translate in sequence

Figure 11. Initiation of translation bringing together the mRNA, the tRNA and the now functional ribosome. The first tRNA moves to the E site of the ribosome as the second tRNA enters the P site and so on until all the mRNA is translated. See text for explanation.
Source: Image used under license from Shutterstock.com

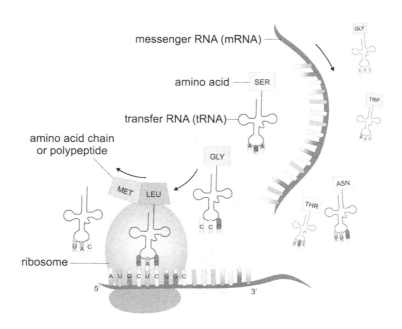

(called **polyribosomes**). They may appear as clusters, linear arrays, or rosettes. Reinitiation allows the continued translation of two or more transcripts without undergoing complete recycling between these events. A summary of how genes work (protein synthesis) is depicted in Figure 12.

Eventually, going to our protein hormone example, insulin is ready and is released into the blood and transported to all cells; then binds receptors at the cell surface allowing cells to open protein channels to uptake glucose from the blood. This re-establishes sugar homeostasis in the body. In class I usually liken the process of protein synthesis to our experiences after travelling to a foreign country. For instance, you visit Germany and meet with some German friends who don't know how to read or write in English. You also don't know how to read or write in German. You return home and two weeks later, you receive a post card in the mail written in German. What do you do? You need to either seek a friend who knows how to read in German or use an online dictionary or Google translator. The postcard can be likened to the mRNA transcript carrying a message. The dictionary is the genetic code. You now sit and translate the message on a couch. The couch or sofa is the ribosome and you are the tRNA. The translated message is your final polypeptide product, thus ending the process.

A summary of the process of protein synthesis
Image used under license from Shutterstock.com

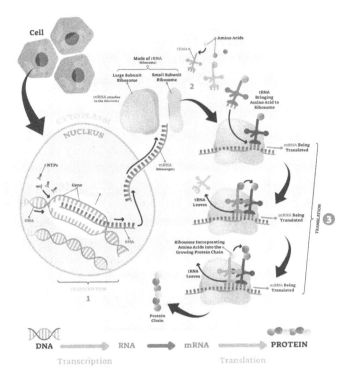

WHEN GENES CHANGE: GENETIC MUTATIONS

We discuss three types of gene mutations: single gene, chromosomal, and genome mutations. **Genetic mutations** are heritable, small-scale alterations in one or more base pairs that damage DNA. The change can affect a single nucleotide pair or larger segments of a chromosome. Genetic variation is the differences in the DNA of individuals in a population. Evolutionary biologists seek to understand how these differences are generated and maintained to create the enormous diversity we see today (Luo et al., 2016). Although new mutations introduce new variation, these are constantly removed from populations. Many studies indicate that some nucleotide sequences have more mutation rates than others. Such biased patterns of genetic changes are sources of non-randomness in the production of genetic variation. (Luo et at., 2016)

What causes mutations? Mutations can arise naturally (spontaneously) during DNA replication. They can also be caused by environmental factors like chemical mutagens (alkylating and deamination agents); particulate ionizing radiation (alpha and beta particles, etc.); non-particulate ionizing radiation (x-rays, gamma rays, etc.); and non-ionizing radiation (like ultraviolet [UV] rays).

Mutations can be harmful, neutral or beneficial to the organism. Harmful mutations affecting an organism's fitness and reproduction are removed by natural selection (section 1.4). To limit the number of mutations, cells routinely correct errors in DNA. Every so often, these repair mechanisms fail, allowing a small number of cells in a population to accumulate mutations more quickly than other cells. This "hypermutation" enables some cells to rapidly adapt to changing conditions to avoid the entire population from becoming extinct as seen in bacteria faced with a crippling antibiotic (Swings et al., 2017). Swings et al. study focused on populations of the bacterium *Escherichia coli* studying hypermutation after exposure to varying levels of alcohol which causes bacteria to experience extreme stress. Further experiments showed that individual populations of bacteria can alter the rate of mutation several times following changing alcohol stress levels. The experiments demonstrate that hypermutation occurs rapidly in these conditions and is essential for bacteria to adapt to the level of alcohol to avoid extinction (Swings et al., 2017).

Neutral mutations have no deleterious effect on the organism. For example, if there is a base change that converts the codon CUU to CUC, the effect is neutral or silent because both codons code for the *same* amino acid (leucine). This is a big advantage of the redundancy of the genetic code when a single amino acid can be coded for by multiple codons. Beneficial mutations occur when they increase the fitness of an organism allowing it to produce more offspring. For example, when bacteria develop resistance to an antibiotic like penicillin, the mutant individual(s) have a survival advantage in a penicillin environment. They are thus selected for and leave more offspring as the rules of natural selection dictate. Over time, these new mutants increase in frequency in the population. Genetic drift on the other hand, is the random increase in, or loss, of a genetic variant from a population over time (Huang et al., 2016). Mutation rates vary tremendously across species from viruses to humans and can evolve under conditions that allow the population to adapt to new environmental conditions by "hitchhiking" on beneficial mutations (Cobben et al., 2017).

Single gene mutations are generally of five types: **point mutations, frameshift mutations** (insertions or deletions—dubbed "*indels*"), **transposons, transitions,** and **transversions**.

- **Base Substitutions:** When one of the four bases is exchanged for another (like A to G), such change occurs at a single letter level, thus referred to as a *point* mutation. For example, in sickle-cell condition, one base sequence in the normal haemoglobin gene (at codon CTC) becomes CAC in the mutant. This base change changes the codon from GAG to GUG which in turn changes the amino acid glutamic acid in the normal haemoglobin to valine in the mutant (see below). In the above case, a protein is produced which is a non-functional—thus, we say it does not make the right sense, hence called a **missense mutation**. Had the substitution resulted in a stop codon (like UAA, UAG or UGA), protein synthesis would have terminated at that point resulting in a polypeptide with fewer than the correct number of amino acids in haemoglobin, hence called **nonsense mutation**.
Normal haemoglobin DNA: C A C G T G G A C T G A G G A C T C C T C
Normal haemoglobin protein: Val-His-Leu-Thr-Pro-<u>Glu-Glu</u>
Sickle-cell haemoglobin DNA: C A C G T G G A C T G A G G A C A C C T C
Sickle-cell haemoglobin protein: Val-His-Leu-Thr-Pro-<u>Val</u>-Glu
- **Frameshift Mutations (Insertions):** This occurs when a new base(s) is inserted into a DNA sequence. An insertion affects the way the sequence is read during translation. For example, consider the case below:

JEN LET HER DOG EAT THE BUN
… changes to JEN LET HERE DOG EAT THE BUN.

13. Disruption of corn kernel colours by transposition
Image used under license from Shutterstock.com

, if these three letter words made a protein, at the point of insertion, the words would become: T HER EDO GEA TTH EBU. This would change all amino acids after HER because the reading (frameshift) was altered.

 bases are inserted in one set (or sets) of three letters, it does not cause a change in the reading f the sequence but is likely to affect the function of the protein produced.

example:

LET HER DOG EAT THE BUN

hanges to JEN LET HER OLD DOG EAT THE BUN or

LET HER OLD TAN DOG EAT THE BUN or

LET HER OLD TAN DOG EAT THE HOT BUN

mples of diseases where insertions are the cause are Fragile-X syndrome and Huntington's Dis- both cases, there is an insertion of a trinucleotide repeat (one CAG triplet in Huntington's and 0 CGG triplets in Fragile-X syndrome).

meshift Mutations (Deletions): This happens when a letter or letters are deleted from the sequence. s will also affect the way the sequence is read as seen in the example below where the letter R eted.

LET HER DOG EAT THE BUN

hanges the reading frame to: JEN LET HED OGE ATT HEB affecting all amino acids after LET.

nsposons or **transposable elements** are mobile genetic DNA sequences found in the genomes y all eukaryotes. Moving spontaneously about the genome, they were discovered by Barbara tock in corn kernels in the 1940s—for which she won a Nobel prize in 1983 when she was 81 d. Transposons are repetitive DNA sequences that have the capability to move (transpose) from ation to another in genome. The movement of transposons can result in altered gene expression, mutations, induced chromosome rearrangements and, due to extensive copying, enlarged genome sizes. As a result, they are considered an important contributor in gene and genome evolution (Kazazian, 2004). As reported by Schnable et al. (2009), transposable elements represent the most abundant repeats in most plant genomes—for example, in maize's (*Zea may*) genome they constitute more than eighty-five percent. When transposons insert into a gene region, they usually inactivate that gene by interrupting the gene sequence, or its regulator sequence. In corn transposons disrupt the kernel colors. Due to activator-controlled transposition at different stages of seed development, the genes of corn kernels are capable of producing a variety of coloration patterns. (Figure 13)

Transitions are changes that alter the one-ring pyrimidines (C and T) from one (C) to the other (T) or vice versa (T) to (C). This can also happen to double-ring purines (A and G) when one is changed to the other. **Transversions** change one-ring pyrimidines (C and T) to double-ring purines (A and G) or the other way around (purine to pyrimidine). Transition rates are higher than transversion rates in animal genomes. Luo et al. (2016) report that transitions occurred more frequently than transversions; explaining that the process of transversion is probably more complicated than transition because it means changing one-ring to two-rings or a two-ring structure to one-ring.

Chromosomal mutations are structural changes where whole sections of a chromosome are changed or modified. There are four ways this can occur (see Box 2).

Genome mutations affect the chromosome number. The normal human chromosome number is a set of 23 pairs (one set from each parent making 46 chromosomes). Human genetic conditions affected by variations from this normal chromosome number because of nondisjunction can affect autosomes (Down Syndrome); and sex chromosomes (Klinefelter, Turner, Jacob, Cri-du-Chat, and Fragile X syndromes) as discussed in detail earlier in Chapter 6 (section 6.7.4).

Box 2.

```
Deletion – a section of a chromosome is lost.
    Original chromosome    ATT TGC GCC TAT TGT AAG
    New chromosome         ATT TGC G̶C̶C̶ T̶A̶T̶ TGT AAG

Duplication – an extra copy of a chromosome is added.
    Original chromosome    ATT TGC GCC TAT TGT AAG
    New chromosome         ATT TGC GCC TAT GCC TAT TGT AAG

Inversion – part of a chromosome order is reversed.
    Original chromosome    ATT TGC GCC TAT TGT AAG
                                   ←
    New chromosome         ATT CGT CCG TAT TGT AAG

Translocation – a section of one chromosome is added onto another chromosome.
    Original chromosome    ATT TGC GCC TAT TGT AAG
                               CCA GAA inserted here
                                         ↓
    New chromosome         ATT TGC CCA GAA GCC TAT TGT AAG
```

ER SUMMARY

...s evolve as their environments change, a process called adaptation which is passed on to off-
...ritable)—allowing advantageous changes to accumulate and benefit the population. Genetic
... occurs because favorable genetic changes (mutations) accrue in on organism allowing it to
...xisting conditions. Early experiments by Beadle and Tatum provided experiential evidence
...g genes and enzymes working on the bread mold *Neurospora crassa*. They concluded that
...ts had incurred faults in the genes required for making the enzymes needed at each step in
...ay giving way for the *one gene-one enzyme hypothesis* which was later modified to read *one
...polypeptide hypothesis* with more research.

...are regions on DNA that contain the instructions for making specific proteins. In humans, genes
...e from 100s of DNA bases to over 3 million base pairs. The Human Genome Project was an
...nal research effort tasked with determining the sequence of the human genome and identify-
...nes that it contains. It is estimated that humans have between 20,000 to 25,000 protein-coding
...ch makes only about 1.5% of the genome codes for proteins, rRNA, and tRNA. The rest of
...e once known as "junk DNA" is today known to be crucial to survival of the species. From
...roteins, two steps are involved. Transcription is accessing the gene and reading the instructions
...the nucleus producing as a single strand of RNA called messenger RNA (mRNA). Translation
...the instructions on mRNA to assemble the specified proteins on the surface of ribosomes. The
...transcription begins when a gene called TBP codes for a TATA binding protein which binds
...ter region announcing to other enzymes that gene reading must "start here". The chromatin
...s modified to allow an enzyme known as RNA Polymerase II to bind the promoter assisted
...iption factors.

...NA polymerase now begins the slow process of recruiting factors necessary for RNA chain
...(elongation), RNA processing, RNA export and chromatin modification. RNA elongation
...ontinues making a new strand of precursor messenger RNA (pre-mRNA) until a terminator
...eached at the end of a gene sequence. The next step after termination is processing the nascent
...A into a mature mRNA. Processing involves removing the non-coding sequences (introns)
...e coding ones (exons), a process catalysed by a spliceosome and associated auxiliary proteins.
...some is composed of a complex of five subunits of small nuclear ribonucleoprotein particles
... recruited with the completion of the pre-mRNA.

...her end of the mRNA (5' and 3' OH) 7-methylguanosine cap is added on the 5' end and a
...l on the 3' OH end. This pre-MRA processing (cap and tail) serves to protect the now mature
...om cellular degradation as it is exported to the cytoplasm, and to allow initiation factors in-
...translation to bind and begin the process of translation. The mature mRNA is now ready for
...the cytoplasm. Studies in budding yeast indicate that a multi-protein complex, known as the
...ion and *export* (TREX) complex, is assembled upon the pre-mRNA during transcription and is
...component in determining the efficiency of mature mRNA. Components of this TREX recruit
...roteins which in turn assemble export receptors which bind the mRNA transcript. After the
...s recruited the appropriate export receptors the resulting mRNA ribonucleoprotein (mRNP)
...s translocated through nuclear pore complex to the cytoplasm.

All lifeforms existing on the planet (with minor exceptions) share the same standard genetic code. During translation, molecules recognize the bases on the mature RNA in sets of three (triplets) called codons. There are 64 codons in the genetic code with sixty-one of them specifying amino acids, three (UAG, UAA, UGA) serve as "stop" signals, and one (AUG) codes for the amino acid methionine and serves as a "start" signal. More than one codon codes for some amino acids—thus, redundancy! For example, lysine has two codons (AAA, AAG) while leucine has six codons (UUA, UUG, CUU, CUC, CUA, CUG).

The RNA polymerase II transcribes other RNA molecules needed for the initiation of translation. These include tRNA and rRNA. The rRNAs are synthesized from nucleolar DNA while tRNA is synthesized from regular DNA. Both RNAs are transported to the cytoplasm at the same time as the mature mRNA. In the cytoplasm, the rRNA is combined in equal parts with proteins to form ribosomes. The 40S and 60S subunits join to form a functional ribosome which attaches to the mRNA. The ribosome has three sites through which tRNAs move the A, P and E sites. The first tRNA with the amino acid methionine moves on to the P site as the second tRNA with its amino acid enters the ribosome at its A site base-pairing with the codon next to AUG. The first tRNA then moves to the E site of the ribosome as the second tRNA enters the P site. The third tRNA next enters the A site. The peptidyl transferase centre of the ribosome catalyses by dehydration synthesis the addition of the amino acid of the second tRNA to the first amino acid carried by the first tRNA (methionine) forming a peptide bond. This process of tRNAs shuttling through the three ribosomal sites as the polypeptide is assembled continues until a stop codon is encountered.

Once a stop codon has been reached, there are no tRNAs available to enter the ribosome. In eukaryotes two termination protein factors catalyse the process. The first factor is tRNA-shaped protein responsible for stop codon recognition and peptidyl-tRNA hydrolysis. bind the ribosome triggering the polypeptide release. The second factor is a release factor. The ribosome dissociates from the mRNA and its two subunits separate. In cases where same protein transcribed on the mRNA is required by the cell, several ribosomes attach to the mRNA and translate in sequence creating what is known as a polyribosome appearing as clusters or rosettes.

Genetic mutations are heritable, small-scale alterations in one or more base pairs that damage DNA. This change can affect a single nucleotide pair or larger segments of a chromosome. Although new mutations introduce new variation, these are constantly removed from populations. Many studies indicate that some nucleotide sequences have more mutation rates than others. Mutations can arise naturally (spontaneously) during DNA replication or can be caused by environmental factors like chemicals or radiation. Mutations can be harmful, neutral or beneficial to the organism. Single gene mutations are generally of five types: point mutations, frameshift mutations (insertions or deletions—dubbed *"indels"*), transposons, transitions, and transversions. Chromosomal mutations are structural changes where whole sections of a chromosome are changed or modified. Genome mutations affect the chromosome number. The normal human chromosome number is a set of 23 pairs (one set from each parent making 46 chromosomes). Human genetic conditions affected by variations from this normal chromosome number because of nondisjunction can affect autosomes and sex chromosomes.

Chapter Quiz

Select the best fit description of genetic adaptation.
- Genetic adaptation occurs when favourable genetic changes such as mutations accrue in an organism allowing it to adapt to existing conditions
- Genetic adaptation allows organisms to de-evolve as their environment changes
- It is not heritable
- All of the above
- None of the above

Researchers of the Human Genome Project estimated that humans have approximately how many protein-coding genes?
- 5,000 to 10,000
- 10,000 to 15,000
- 20,000 to 25,000
- 30,000 to 35,000
- 40,000 to 50,000

A regulatory section near the beginning of a gene sequence is called
- initial region
- promoter region
- precursor region
- pre-initiation complex
- start complex

During translation, molecules recognize the bases on the mature RNA in sets of three called
- codons
- triplets
- tRNA
- 40S ribosomal subunits
- 60S ribosomal subunits

The start codon _____, codes for the amino acid methionine
- AUG
- UAG
- UAA
- UGA
- CUG

What two steps are required to make proteins from DNA?
- Elongation and Termination
- Transfiguration and Translation
- Transcription and Translation
- Initiation and Termination
- None of the above

7. What are causes of mutations?
   a. Chemical mutagens
   b. Particulate ionizing radiation
   c. Mutations can arise naturally (spontaneously)
   d. X-rays, gamma rays and ultraviolet or UV rays
   e. All of the above
8. Which of the following statements about mutation is *false?*
   a. Mutations are heritable, small-scale alterations in one or more base pairs that damage DNA
   b. Mutations are always harmful to the organism
   c. Mutations can be caused by environmental factors
   d. Point mutation is a type of a single gene mutation
   e. Harmful mutations affecting an organism's fitness and reproduction are removed by natural selection
9. If there is a base change that converts the codon CGA to CGC but codes for the same amino acid arginine, this is called
   a. nonsense mutation
   b. missense mutation
   c. frameshift mutation
   d. silent mutation
   e. transversion
10. Chromosomal mutations are structural changes where whole sections of a chromosome are changed or modified. Which of the following is not a chromosomal mutation?
    a. Deletion
    b. Duplication
    c. Inversion
    d. Translocation
    e. Addition

Thought Questions

1. Why do you think the Human Genome Project's goal of sequencing the entire human genome and identifying the genes that it contains was so important?
2. Using the given set of chromosomes, give examples of the following chromosomal mutations: deletion, duplication, inversion, and translocation.
   a. ATT TGC GCC TAT TGT AAG
3. What does RNA elongation continue to make and how does it know when to stop?
4. Give a brief description of step of the translation of the mRNA transcript.
5. Transposons are mobile genetic DNA sequences. How can the movement of these transposons affect our genome?

REFERENCES

(2001). Human genome published. *Nature Biotechnology*, *19*(3), 191. doi:10.1038/85582

...eri, A. G., Rijal, K., & Maraia, R. J. (2013). Transcription termination by the eukaryotic RNA ...se III. *Biochimica et Biophysica Acta*, *1829*(3-4), 318–330. doi:10.1016/j.bbagrm.2012.10.006 ...099421

..., K. S., & Bailey-Serres, J. (2015). Mechanism of Cytoplasmic mRNA Translation. *The Arabook / American Society of Plant Biologists*, *13*, e0176. doi:10.1199/tab.0176 PMID:26019692

...H. (1968). The origin of the genetic code. *Journal of Molecular Biology*, *38*(3), 367–379. ...16/0022-2836(68)90392-6 PMID:4887876

...na, L. R., Torres, A. G., & Rafels-Ybern, A. (2017). What Froze the Genetic Code? *Life (Chi-*..., *7*(14), 1–6. doi:10.3390/life7020014

...E., & Green, R. (2012). The Elongation, Termination and Recycling Phases of Translation ...otes. *Cold Spring Harbor Perspectives in Biology*, *4*(7), a013706. doi:10.1101/cshperspect. ...PMID:22751155

...E., Kinzy, T. G., & Pavitt, G. D. (2016). Mechanism and Regulation of Protein Synthesis in ...*myces cerevisiae*. *Genetics*, *203*(1), 65–107. doi:10.1534/genetics.115.186221 PMID:27183566

...(2004). Structure and mechanism of the RNA Polymerase II transcription machinery. *Nature ... & Molecular Biology*, *11*(5), 394–403. doi:10.1038/nsmb763 PMID:15114340

...M. (1998). Molecular Genetics of the RNA Polymerase II General Transcriptional Machinery. ...*logy and Molecular Biology Reviews*, *62*(2), 465–503. PMID:9618449

...M., & Cairns, J. (2003). The Centenary of the One-Gene One-Enzyme Hypothesis. *Genetics*, ...39–841. PMID:12663526

..., N. H. (1995). One-gene-one-enzyme: remembering biochemical genetics. *Protein Science: A ...on of the Protein Society*, *4*(5), 1017–1019. doi:.5560040524 doi:10.1002/pro

...Mukai, K., Ito, M., & Masuda, S. (2015). Gene Regulation through mRNA Expression. *Ad-...Biological Chemistry*, *5*(02), 45–57. doi:10.4236/abc.2015.52005

...T., & Reed, R. (1987). The role of small nuclear ribonucleoprotein particles in pre-mRNA ...*Nature*, *325*(6106), 673–678. doi:10.1038/325673a0 PMID:2950324

...childer, R. J., & Kimball, S. R. (2015). Role of Precursor mRNA Splicing in Nutrient-Induced ...s in Gene Expression and Metabolism. *The Journal of Nutrition*, *145*(5), 841–846. doi:10.3945/ ...3216 PMID:25761502

...& Cheng, H. (2005). TREX, SR proteins and export of mRNA. *Current Opinion in Cell Biol-*...), 269–273. doi:10.1016/j.ceb.2005.04.011 PMID:15901496

Schröder, H. C., Bachmann, M., Diehl-Seifert, B., & Müller, W. E. G. (1987). Transport of mRNA from Nucleus to Cytoplasm. *Progress in Nucleic Acid Research and Molecular Biology*, *34*, 89–142. doi:10.1016/S0079-6603(08)60494-8 PMID:3326042

The Genetic Code in Operation for Protein Construction. (n.d.). Retrieved from http:// hyperphysics. phy-astr.gsu.edu/hbase/Organic/gencode.html#c3

Woychik, N. A., & Hampsey, M. (2002). The RNA polymerase II machinery: Structure illuminates function. *Cell*, *108*(4), 453–463. doi:10.1016/S0092-8674(02)00646-3 PMID:11909517

Zamudio, G. S., & José, M. V. (2017). On the Uniqueness of the Standard Genetic Code. *Life (Chicago, Ill.)*, *7*(7), 1–8. PMID:28208827

KEY TERMS AND DEFINITIONS

60S Ribosomal Subunits: In eukaryotes, the larger subunit of the ribosome complex that is responsible for peptide formation.

7-Methylguanosine Cap: A cap that is added to the 5' end of the hRNA molecule that protects the mRNA from degradation.

Adaptor Proteins: Proteins that facilitate the binding of a mRNA transcript by recruiting appropriate signal components such as receptors. They form complexes with other proteins to regulate signal transduction pathways.

Aminoacyl-tRNA Synthase: Enzyme that catalyzes the addition of a corresponding amino acid to its tRNA.

Anticodon: A set of 3 nucleotides on a tRNA that correspond to a complementary codon in mRNA.

Codons: A set of 3 nucleotides or triplets on the mRNA that codes for a specific anticodon.

Exons: Coding regions of mRNA that are spliced together by spliceosomes during post-translational modification and will exit the nucleus.

Frameshift Mutations: Mutations that occur from the insertion or deletion of nucleotides and results in a shift in the reading frame.

Genetic Adaptation: Traits that result from natural selection and random variation, which give rise to favorable genetic changes in an organism.

Genetic Code: Standard set of rules and meanings of nucleotide triplets in which all life forms share.

Genetic Mutations: Heritable alterations in one or more base pairs that damage DNA and can affect a single nucleotide pair or larger segments of a chromosome.

Homeotic Genes: Highly conserved DNA sequences that code for specific transcription factors that regulate gene expression of specific anatomical structures.

Human Genome Project: An international project created in 1990 with the goal of sequencing and identifying the entire human genome.

Introns: Non-coding regions of mRNA that are removed by spliceosomes during post-translational modification.

DNA: Non-coding DNA sequences in the genome.

Mature mRNA: mRNA that has been spliced, processed, and is ready for exportation to the cytoplasm to begin the process of translation.

Missense Mutation: A type of point mutation that results in the substitution of one amino acid for another in the final polypeptide.

mRNA Ribonucleoprotein (mRNP) Complex: Protein complex containing the appropriate export proteins that is translocated through the nuclear pore to the cytoplasm.

Nonsense Mutation: A type of point mutation that results in the substitution of a stop codon for an amino acid in the final polypeptide.

Peptidyl Transferase: Enzyme that catalyzes the formation of peptide bonds in a growing polypeptide chain.

Point Mutations: Mutations that occur from one nucleotide change for another in DNA.

Poly-A Polymerase: Enzyme that catalyzes the addition of adenine nucleotides during post-transcriptional modification.

Poly-A Tail: A set of approximately 200 adenine nucleotides at the 3' OH end of a pre-mRNA.

Polyadenylation: The process by which poly-A polymerase adds on adenine nucleotides are adding to the 3'OH end of pre-mRNA.

Polyribosomes: A complex of several ribosomes that attach to mRNA and translate in sequence.

Precursor Messenger RNA (Pre-mRNA): A primary transcript that contains both exons and introns, and precedes the formation of mRNA.

Promoter: A specific region upstream of the DNA sequence that regulates transcription by binding RNA polymerase II.

Repetitive DNA: Short repeating sequences of DNA.

RNA Polymerase Ii: Enzyme complex that catalyzes the transcription of DNA by binding to the promoter.

Small Nuclear Ribonucleoprotein Particles (snRNPs): Uridine-rich RNA-proteins that interact with modified pre-mRNA and other proteins to form a larger complex known as the spliceosome.

Spliceosome: An enzyme that catalyzes esterification reactions in the removal of introns and the splicing of exons.

TATA Binding Protein: Protein that recognizes and binds to a specific region of repeating thymine and adenine bases.

Transcription Factors: Family of proteins that regulate (activate or repress) gene expression and transcription.

Transitions: A type of mutation in which pyrimidines are changed to another (i.e., C to T) or purines are changed to another (i.e., A to G). The transition rate is higher than the transversion rate in animal genomes.

Transposons: Repetitive DNA sequences that can insert or remove themselves from one location to another in the genome.

Transversions: A type of mutation in which pyrimidines are changed to purines, and vice versa.

Chapter 9
Genes:
Expression and Regulation

ABSTRACT

Gene expression patterns are dependent on their internal cell environment of their DNA, their immediate internal cell environment, and the integrity of their DNA. It also depends on the cell's external environment comprised of signals from other parts of the body including chemicals, nutrients, and/or mechanical stress. Gene regulation is achieved by a wide range of mechanisms that cells use to control whether genes are transcribed, when they are transcribed, and to regulate the quantity of certain proteins based on the cellular and/or environmental feedback. Proper regulation of gene expression is required by organisms to respond to continually changing environmental conditions. Some bacterial genes are transcribed as a unit under a regulatory system called an operon which contains functionally related genes. Three well studied operons include the lactose operon, histidine operon, and tryptophan operon. Gene regulation in higher organisms can occur at various stages from DNA level to protein assembly. This chapter explores this aspect of genes.

CHAPTER OUTLINE

9.1 Overview of Gene Expression
9.2 Gene Regulation in Lower Organisms (Prokaryotes)
    9.9.1 The Lactose Operon
    9.9.2 The Tryptophan Operon
9.3 Gene Regulation in Higher Organisms (Eukaryotes)
    9.3.1 Regulation at the DNA Level
    9.3.2 Regulation at Transcription Level
    9.3.3 Regulation at Post-Transcription Level
    9.3.4 Regulation by non-coding RNAs
Chapter Summary

DOI: 10.4018/978-1-5225-8066-9.ch009

...ING OUTCOMES

...lain the basis of gene expression and the need for regulation.
...trate using a bacterial model how operons work.
...marize how the tryptophan operon works.
...ine with examples how gene regulation occurs in higher organisms.

...ERVIEW OF GENE EXPRESSION

...s inherit their genomes from their parents. However, in the process or embryonic development, ...enes expressed (turned "on" and "off") in differentiating cells which are destined for different ...d organs? For example, with billions of cells making up our bodies, human life starts from ...ertilized egg cell with half the genes from one parent, and the other half from the other par-...wing rapid cell divisions of the fertilized egg, into a mass of cells, to a more defined morula, ...gastrula, and neurula stages, the fertilized egg develops and differentiates into an embryo. ...ls become the future nervous system, others become sensory cells, others liver cells, gut cells, ...cular cells, and so on. All these cells use the same original DNA but express different genes. ...ls (neurons) express genes involved in higher order operations like cognition; while sensory ...ess different genes like photosensitive retinal cells in the eye. This utilization of the same DNA ...ssing different genes involves gene regulation in the early stages of embryonic development.
...xpression patterns we see in cells is dependent on their internal cell environment (inherited ...eir immediate internal cell environment (neighbouring cells or tissues), and the integrity of their ...e to damage from physical and/or age-related factors). Cellular gene expression also depends ...xternal (outside cell) environment comprised of signals from other parts of the body includ-...icals, nutrients, and/or mechanical stress. For example, cells subjected to extreme mechanical ...e a carpenter's hands or people who walk barefooted for a long time) will develop additional ...e layers of cells in their epidermis.
...regulation is achieved by a wide range of mechanisms that cells use to control whether genes ...cribed; when they are transcribed; and to regulate the quantity of certain proteins based on the ...nd/or environmental feedback (Stefanski et al., 2016). Proper regulation of gene expression is ...by organisms to respond to continually changing environmental conditions (and host responses ...ous agents like bacteria) (Schiano & Lathem, 2012). We will first discuss gene regulation in ...ganisms (prokaryotes) then move over to higher organisms (eukaryotes).

...NE REGULATION IN LOWER ORGANISMS (PROKARYOTES)

...cterial genes are transcribed as a unit under a regulatory system called an **operon.** Three well ...perons include the **lactose operon, histidine operon,** and **tryptophan operon**. As explained ...below, the key regulatory element of transcription initiation in operons is called the **operator**, ...c DNA sequence which binds a regulatory protein called a **repressor**.

Why are operons so prevalent? Traditionally, genes in the same operon have similar expression patterns. This perhaps explains why operons tend to contain functionally related genes; and why genome rearrangements that would destroy operons are strongly selected against (Morgan et al., 2006).

9.2.1 The Lactose Operon

Our intestinal tracts have billions of bacteria and *Escherichia coli* (*E. coli*), due to its ease of growth and versatility has been extensively studied by researchers. It has a genome that encodes about 4300 proteins and was one of the first organisms to have its DNA sequenced in the lab (Esmaeili et al., 2015). Two well-understood model transcriptional regulatory systems for learning the basic principles of gene regulation are operons—perhaps the best studied being the tryptophan (trp) and lactose (lac) operons. Operons are clusters of co-regulated genes with related functions and often code for genes operating in the same functional pathway (Osbourn & Field, 2009). As a result, operons will include genes encoding enzymes of consecutive steps in metabolic pathways (Zaslaver et al., 2006) or genes encoding interacting proteins (Mushegian & Koonin, 1996; Huynen et al., 2000). The mechanism of gene regulation in *E. coli* using the lactose operon was illustrated decades ago by Jacob and Monod (1961) who won the Nobel prize for their work. Their research demonstrated how enzyme quantities can be controlled directly at the level of transcription (Stefanski et al. 2016). The bacterium thrives in a lactose-rich environment. Lactose being a disaccharide needs to be broken to the monosaccharides glucose and galactose which can be utilized by the bacterium. The genes required for the uptake and metabolism of lactose are only turned on when this lactose is present in high amounts in the growing environment. Glucose is the favourite carbon and energy source for *E. coli*, as well as for many other organisms. Although this bacterium can also feed on other sugars (like lactose), it only does so when glucose is absent (Narang & Pilyugin, 2008).

There are three structural genes in the lac operon of *E. coli*: lacZ, lacY and lacA coding for respectively β-galactosidase, permease, and thiogalactoside transacetylase. The genes in the *lac* operon code for the proteins necessary to transport and metabolize lactose. *E. coli* shows a lot of versatility and efficiency in the way it utilizes glucose and lactose (Narang & Pilyugin, 2008). The enzyme β-galactosidase breaks down lactose into either allolactose (approximately half of it) or into glucose and galactose (the other half). The second enzyme permease (a transmembrane protein) is necessary for the uptake of lactose. The third enzyme (thiogalactoside transacetylase) plays no role in the operon regulatory pathway. The other elements in the operon are the promoter and operator (Figure 1).

We will look at how the operon operates in three scenarios: 1) when glucose present, lactose absent; 2) when glucose is absent, but lactose is available; and, 3) when both glucose and lactose are present in varying amounts.

The question to ask is how this lac operon switching *mechanism* works in *E. coli* as the types of substrates available fluctuate. The lac operon has two regulatory pathways: it is subject to negative (repressor) and positive (CAP) regulation as discussed below.

1. **Glucose Present, Lactose Absent:** The **lac repressor** acts as a negative regulator (negative gene control) preventing the RNA polymerase from transcribing genes when lactose levels are low. The lac operon is OFF for there is no need for lactose enzymes. A repressor protein coded for by the regulatory gene LacI (outside operon, Figure 1) is activated to block the promoter by binding onto the operator in a lock-and-key mechanism. This binding causes the DNA loops out at that location

...king it inaccessible to the RNA-Polymerase. The operon is OFF and no transcription occurs (Figure 1).

Glucose Absent, Lactose Present: Lactose is converted to allolactose which acts as an **inducer** forming a complex with the repressor protein. The probability for the inducer binding the repressor depends on the inducer concentration inside the cell. The induction process is facilitated by the permease transporter, which moves the allolactose (inducer) into the cell. The allolactose-repressor complex causes a conformational change in the repressor and inactivates it. (Narang & Pilyugin, 2008). Now, the allolactose-repressor complex is unable to bind to the operator region of the lactose promoter, thus leaving the operator site accessible to the RNA polymerase. The polymerase can now bind the promoter to transcribe the 3 genes onto one mRNA molecule. The operon is ON and transcription occurs (Figure 1).

Both Glucose and Lactose Present (But Glucose Levels LOW): The second regulatory mechanism (positive regulation) in the lac operon is controlled by glucose. As the concentration of extracellular glucose decreases, the intracellular production of a protein activator known as **cyclic AMP (cAMP)** increases. cAMP binds to **cAMP receptor protein (CRP)** to form the cAMP-CRP complex called **catabolite activator protein (CAP)** (Figure 2, a). The CAP complex binds just upstream of the lactose promoter and by through positive regulation assists the RNA polymerase to bind promoter and start transcribing the genes. In this case, the lac genes are regulated by CAP which increases the affinity of RNA polymerase to transcribe the operon (operon ON). Thus, with low glucose and high lactose levels, CAP is bound to the promoter, and the repressor is inactivated in the presence of allolactose. (Schiano & Lanthem, 2012; Esmaeili et al., 2015; Stefanski et al., 2016)

Figure 1. The lac operon in E. coli. Note the three structural genes, the promoter and the operator. The operon is OFF when lactose is present; ON when lactose is absent.
Image used under license from Shutterstock.com

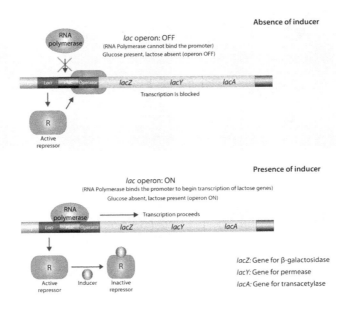

4. **Both Glucose and Lactose Present (But Glucose Levels HIGH):** With high glucose amount in the cell, the cAMP levels are low. As a result, CAP is not activated, and the RNA polymerase has no affinity to bind the promoter. No transcription occurs, operon is OFF (see Figure 2, b).

9.2.2 The Tryptophan Operon

Tryptophan (Trp) is an essential amino acid among the assemblage of the 9 required (of 20 total) amino acids in mammals. Tryptophan is generally synthesized by free-living prokaryotes, lower eukaryotes, and higher plants. The trp operon is functionally similar to the lac operon except in one feature: in the lac operon, lactose acts as an inducer promoting transcription; while in the trp operon, tryptophan acts as repressor, blocking transcription. Bacteria living in the gut will utilize the tryptophan they get in the host's diet to build the proteins they require.

The classical structure of the trp operon contains both the genes trpE, trpD, trpC, trpB and trpA coding for the five enzymes required for the synthesis of tryptophan, and the regulator region (Figure 3). Similar to the lac operon, a single mRNA is transcribed for all five enzymes. Next to the enzyme coding sequences is the regulatory region made up of the **promoter**, an **operator**, and a region referred

Figure 2. a) Lactose present but glucose scarce (cAMP levels HIGH activating CAP for transcription). b) Lactose present, glucose present (cAMP levels LOW, CAP inactive – no transcription)

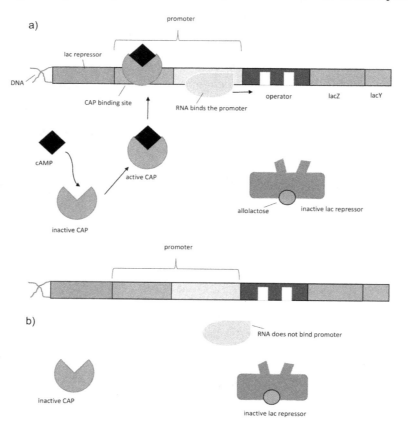

Tryptophan present; operon OFF. No transcription (upper diagram). b) tryptophan absent; transcription occurs (lower diagram).
ge used under license from Shutterstock.com

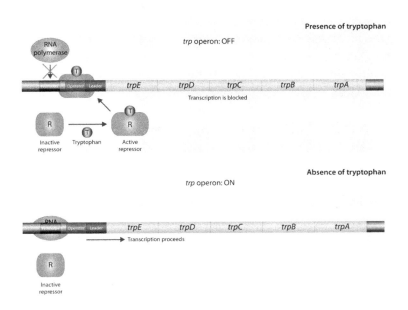

eader (with a section called the **attenuator** located in the leader) (Kagan et al., 2008). The codes for the leader region of the DNA.

am on the 5' end is the regulatory gene trpR which codes for the repressor protein whose expression varies in response to changes in the amount of free tryptophan available in the cell. In the lac operon where lactose induces transcription, tryptophan is active in repression rather duction.

operon order: trpR *promoter operator* trpL trpE trpD trpC trpB trpA

operon is regulated in two ways: through an "on-off" mechanism, and by a modulation system.

ptophan Present: The protein produced by the *trpR* gene—the trp repressor is unable to bind operator (as it is) unlike in the lac repressor. If the amounts of tryptophan present in cell are ificant, the repressor and the tryptophan molecule join forming an active repressor complex ch can now bind to the operator. This binding blocks the RNA polymerase from accessing the noter. Operon is OFF—no transcription occurs (see Figure 3a).

ptophan Absent: The repressor is inactive and without tryptophan to form an active repres- complex, the operator is unblocked giving RNA polymerase access to the promoter. Operon is —transcription occurs (see Figure 3b).

ounts of Tryptophan Intermediate: The trpL gene coding for the leader sequence has four ains 1-4, which control the operon in a process called **attenuation**. Domain 1 produces a short ide made up of 14 amino acids. Domain 3 can base pair with domain 2 or 4. When tryptophan ls are high, the translation of the short peptide is rapid as domain 2 associated with the ribosome making it unavailable. With no domain 2, domains 3 and 4 join, looping the DNA at that point and blocking RNA polymerase—no transcription occurs. If tryptophan levels are low, the translation of the short peptide is slow, domain 2 is not associated with the ribosome making it available. Domains 3 and 2 therefore join—no looping occurs and transcription proceeds. Domain 4 is critical to the process of attenuation because its absence (as seen in engineered bacteria with domain 4 missing) causes transcription to proceed even when levels of tryptophan are high in the cell. As a result, domain 4 is called the attenuator since it needs to be present to reduce or attenuate the transcription process when tryptophan levels are high.

9.3 GENE REGULATION IN HIGHER ORGANISMS (EUKARYOTES)

The evolutionary forces that selected for gene clusters (operons) are not fully understood. However, a few studies indicate that in the unpredictable environments most bacteria live, the amounts of transcribed molecules like mRNA may vary from 50 this hour to 100 or more in the next hour (Ray & Igoshin, 2012). According to the two authors, if the genes were split up into separate operons, the "noise" created in this process would be grossly inefficient and selection would be against noise amplification (Ray & Igoshin, 2012).

Operons are also found in eukaryotes, although their distribution is irregular perhaps indicating that operons may not be ancestral traits in eukaryotes (Lawrence 1999; Hastings 2005). In the early 1990s, structures with remarkable similarities to prokaryotic operons were found in the eukaryotic nematode *Caenorhabditis elegans* (Spieth et al., 1993; Zorio et al., 1994). Studies have indicated that about 15% of genes in *C. elegans* comprise at least 1,000 operons consisting of linked genes that are under the control of a single promoter and are transcribed as a single poly mRNA (Blumenthal et al., 2002). However, unlike in bacteria, this poly mRNA is then trans-spliced into mono mRNAs that are translated individually. (Osbourn & Field, 2009). Interestingly, several operon-like structures have been identified in Drosophila, but those lack the poly mRNA trans-splicing machinery. In most metabolic pathways in the plants studied, genes are generally not clustered as operons with notable exceptions. Functional gene clusters operating plant metabolic pathways have been reported in maize (cyclic hydroxamic acid pathway, Frey et a., 1995); in oat (*avenacin gene cluster*, Qi et al., 2004); and in rice (diterpenoid momilactone cluster, Wilderman et al., 2004). With the pressure for organisms to adapt to prevailing environmental conditions, selection forces will likely drive formation of new gene clusters—which can be viewed as units of strong gene interaction with tightening linkage between structural genes. Physical clustering may be advantageous functionally as it allows groups of genes to be regulated in a coordinated manner. (Stahl & Murray, 1966)

The overwhelming complexity of higher organisms means that gene regulation must occur at various stages from the initial level of reading DNA instructions, to the final level of protein assembly, and at many points in-between. The understanding of the regulation of gene expression has greatly been facilitated by recent advancements in single-cell and single-molecule imaging technologies. These observations have revealed highly dynamic aspects of transcriptional and post-transcriptional control in cells. Cell proliferation and differentiation depend on rigorously controlled gene activities.

Gene regulation in higher organisms can occur at various stages from DNA to protein. It can occur at each of the six steps of gene expression: 1) transcription; 2) mRNA processing control; 3) mRNA transport and localization; 4) translation; 5) protein activity; and (6) mRNA half-life (Inose et al., 2015).

These mechanisms will be covered more fully in chapter 11 (Epigenetics), but we will summarize the main epigenetic highlights in this section of the chapter.

4. Gene regulation by chromatin remodelling
Image used under license from Shutterstock.com

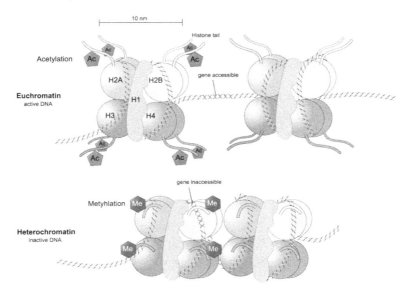

DNA Level Regulation

Modification

...tin remodelling is modification of the configuration of histone "spools," around which strands ... are wrapped—an architecture that allows DNA to be compressed into the nucleus of a cell. ... and non-histone proteins are modified in ways that affect the structure of chromatin and how ...le it becomes in transcription and DNA duplication. There are two types of chromatin: the less ...sed euchromatin where genes more accessible; and the more compressed heterochromatin where ... genes is blocked (Figure 4). Post-translational modifications of the histone tails regulate two ... processes, namely transcriptional activation and repression (Cohen et al., 2011). Chromatin ...ling thus, is a process through which changes in gene expression occurs without altering the ... DNA gene sequence.

...tion and Methylation

...tion means adding the acetyl group ($COCH_3$) to histones that form the foundation onto which DNA ... (Figure 5). Protein acetylation regulates DNA processes, including gene expression, replication, ...nd recombination (Kouzarides, 2007). In eukaryotic cells, it is well established that acetylation ...s histones and transcription factors, modulating gene expression and other DNA transactions ... et al., 2005). Methylation of DNA is the attachment of a methyl group (CH_3) to specific genes ...arly on the cytosine-phosphate-guanine (CPG) sequences. It acts as an on/off switch determining ...nes are activated and when they are repressed. Adding a CH_3 group represses gene expression ...5). Demethylation on the other hand is the removal of the CH_3 group from these genes which ...s them.

Figure 5. Acetylation and methylation of histone tails
Source: Image used under license from Shutterstock.com

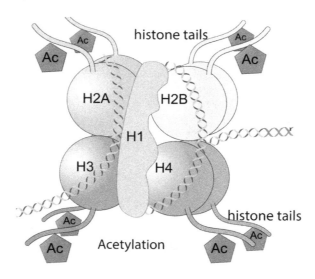

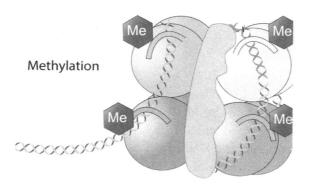

Other mechanisms that affect gene expression are discussed in detail in chapter 11. They include phosphorylation, beta-N-glycosylation, carbonylation, citrullination, ubiquitylation, sumoylation, biotinylation, ADP-ribosylation, crotonylation, proline isomerization, aspartic acid isomerization, N-formylation, propionylation, and butyrylation.

9.3.2. Regulation at Transcription Level

Transcription factors can either activate transcription or repress it. **Activators** are molecules which recognize and bind sequences called enhancer sites located upstream (distal) of the promoter region of the gene they regulate. **Repressors** bind specific sequences called **silencers** upstream of the promoter region of a gene and curtail gene transcription by preventing RNA polymerase from binding the promoter.

egulation at Post-Transcription Level

ulation can occur at various levels during mRNA processing depending on the needs of the
 one moment. These include alternative splicing, polyadenylation, and 5' capping. These pro-
re explained in Chapter 8 (section 8.2). Other mechanisms include mRNA export regulation,
lf-life regulation, tRNA fragments (tRFs) regulation, translation initiation, and degradation
ed protein (details in Chapter 11, section 11.2).

egulation by Non-Coding RNAs (ncRNAs)

wide surveys have revealed that a large portion of the eukaryotic genomes is transcribed into
g RNA (ncRNA) of two classes: structural and regulatory ncRNAs. Structural ncRNAs are
ousekeeping required for the normal function and viability of the cell. This includes tRNAs,
nRNAs, small nucleolar RNAs (snoRNAs), RNase P RNAs, and telomerase RNA (Prasanth &
2007). Regulatory ncRNAs are expressed at certain stages of development, during cell differ-
 or as a response to environmental stimuli. Depending on their length, ncRNAs can be further
to two short ones, a medium ncRNA, and a long ncRNA (Nie et al., 2012). Details of each of
hanisms are explained in Chapter 11 (section 11.2).

ER SUMMARY

ression patterns are dependent on their internal cell environment of their DNA, their imme-
rnal cell environment, and the integrity of their DNA. It also depends on the cell's external
ent comprised of signals from other parts of the body including chemicals, nutrients, and/or
al stress. Gene regulation is achieved by a wide range of mechanisms that cells use to control
enes are transcribed; when they are transcribed; and to regulate the quantity of certain proteins
 the cellular and/or environmental feedback. Proper regulation of gene expression is required
isms to respond to continually changing environmental conditions.
bacterial genes are transcribed as a unit under a regulatory system called an operon. Operons
ntain functionally related genes. Three well studied operons include the lactose operon, histidine
nd tryptophan operon. The genes in the *lac* operon code for the proteins necessary to transport
bolize lactose. The enzyme β-galactosidase breaks down lactose into either allolactose or into
and galactose. The second enzyme permease is a transmembrane protein required for the up-
ctose. The third enzyme (thiogalactoside transacetylase) plays no role in the operon regulatory
The other elements in the operon are the promoter and operator.
 glucose is present but lactose absent the lac operon is OFF for there is no need for lactose
 and no transcription occurs. If glucose absent and lactose present, an isomer of lactose (al-
) joins with a repressor to form the allolactose-repressor complex which prevents the repressor
ding the operator region of the promoter and the operon is on and transcription occurs. If both
and lactose present, but glucose levels low, positive gene regulation occurs. A protein activator
 cyclic AMP binds to cAMP receptor protein to form catabolite activator protein (CAP). This
assists the RNA polymerase to bind promoter to start transcribing the genes. On the other hand,

if both glucose and lactose present, but glucose levels high, the cAMP levels are so low that CAP is not activated resulting in no transcription occurs.

The trp operon is functionally similar to the lac operon except that in the lac operon, lactose acts as an inducer promoting transcription; while in the trp operon, tryptophan acts as repressor blocking transcription. The trp operon contains both the genes trpE, trpD, trpC, trpB and trpA coding for the five enzymes required for the synthesis of tryptophan, and the regulator region. There is also a regulatory region made up of the promoter, an operator, and a region referred to as the leader (the attenuator is in the leader region). The trp operon regulated in two ways: through an "on-off" mechanism, and by a modulation system called attenuation.

If tryptophan present, the trp repressor is unable to bind the operator blocking transcription. If the amounts of tryptophan present in cell are significant, the repressor and the tryptophan molecule join forming an active repressor complex which can now bind to the operator blocking the RNA polymerase—no transcription occurs. If tryptophan is absent, the repressor is inactive, thus the operator is unblocked giving RNA polymerase access to the promoter and transcription occurs. If the amounts of tryptophan are intermediate, the trpL gene codes for the leader sequence which has four domains which control the operon in a process called attenuation. Domain 4 is critical to the process of attenuation because its absence causes transcription to proceed even when levels of tryptophan are high in the cell. Thus, domain 4 is called the attenuator since it needs to be present to reduce or attenuate the transcription process when tryptophan levels are high.

Operons are also found in eukaryotes, although their distribution is irregular perhaps indicating that operons may not be ancestral traits in eukaryotes. Gene regulation in higher organisms can occur at various stages from DNA to protein: 1) transcription; 2) mRNA processing control; 3) mRNA transport and localization; 4) translation; 5) protein activity; and (6) mRNA half-life.

Post-translational modifications to histones are an integral component of epigenetic regulation. Chromatin structure is closely associated with the accessibility of DNA to transcriptional machinery, and hence gene activity. Histone and non-histone proteins are modified in ways that affect the structure of chromatin and how accessible it becomes in transcription and DNA duplication. Post-translational modifications of the histone tails regulate two opposite processes, namely transcriptional activation and repression.

Acetylation means adding the acetyl group ($COCH_3$) to histones that form the foundation onto which DNA is bound. Histone methylation involves transfer of a methyl group to histone tails. Phosphorylation is the addition of the phosphate (PO_4) group to the histone tails.

Studies show that binding of chromatin by nucleocytoplasmic proteins (beta-N-glycosylation) regulates transcription of genes especially those associated with metabolism and aging. Carbonylation is the formation of carbonyl groups by direct oxidation of basic amino acid residues—changes that affect gene expression. Citrullination involves the conversion of peptidyl arginine to citrulline and may lead to transcriptional repression. Ubiquitylation covalently attaches one or more ubiquitin molecules to a lysine residue. Ubiquitylation may participate in transcriptional activation, while on the other hand, it may be associated with transcriptional repression. Sumoylation involves small ubiquitin-related modifier polypeptides which join many cellular proteins, altering their interaction with other proteins regulating their function. Sumoylation has been implicated in controlling many important processes, including regulation of the cell cycle, transcription, nucleocytoplasmic transport, DNA replication and repair, chromosome dynamics, and apoptosis, as well as ribosome biogenesis.

...nylation is the addition of biotin (vitamin B7), a cofactor for four carboxylases involved in gene ...on and chromatin structure. ADP-ribosylation is a reversible covalent modification where the ...ose group from NAD^+ is transferred to a specific amino acid of an acceptor protein. The process ...many crucial cellular activities including cell differentiation, transcription, chromatin modifica-...A damage detection and repair, apoptosis, and carcinogenesis. Crotonylation is the addition of ...onyl group onto the amino group of a lysine residue. N-formylation may influence epigenetic ...isms governing chromatin states and could accumulate with age and contribute to the deregula-...hromatin which may affect gene function. Propionylation and butyrylation may be associated ...lular metabolic status and could regulate genes involved in energy metabolism.

...scription factors can either activate transcription or repress it. Activators are molecules which ...ze and bind enhancer sites in the promoter region of the genes they regulate. Often, they require ...tors to function adequately. Repressors bind silencer sites in the promoter region of a gene and ...RNA polymerase from binding the promoter. They require corepressors which facilitate their ...Gene regulation can occur at various levels during mRNA processing depending on the needs ...ell at any one moment. In humans, more than 90% of transcripts undergo alternative splicing, a ...that is precisely controlled in accordance to cellular/tissue/organ needs of the organism. Several ...isms have been identified that enhance 5' cap methylation, the final step in 5' cap generation, ...s that regulates gene expression. Data also suggests that de-capping the 5' end also regulates ...ction.

...native polyadenylation (APA) produces multiple isomeric forms of could affect gene expression ...ng on the developmental stage and tissue organization levels of the organism. The RNA export ...rminal centre-associated nuclear protein promotes mRNA export, and likely enables rapid changes ...expression by facilitating nuclear export of specific classes of mRNA. Some genes regulate their ...on level by controlling their mRNA half-life. mRNAs with a shorter half-life result in lower pro-...ression level—due to reduced translation. Most regulation is exerted during translation initiation, ...e AUG start codon is identified and decoded by the methionyl tRNA specialized for initiation. ...ed protein degradation via the ubiquitin–proteasome system controls a wide variety of cellular ...s, from transcriptional regulation and stress response to cell cycle regulation.

...ome-wide surveys have revealed that a large portion of the eukaryotic genomes is transcribed ...-coding RNA (ncRNA) of two classes: structural and regulatory. Structural ncRNAs are used ...ekeeping activities for the normal function and viability of the cell and include tRNAs, rRNAs, ..., snoRNAs, RNase P RNAs, and telomerase RNA. Regulatory ncRNAs include short RNAs ...NA, siRNAs, asRNAs, piRNAs); medium ncRNAs, long ncRNAs, paRNAs, crasiRNAs, and ...As. Non-coding RNAs regulate genes during transcription and/or and may be associated with ...tic gene control.

Chapter Quiz

...activate the genes need for the uptake and metabolism of lactose, what is required in the grow-...g environment?

- Low amounts of glucose
- High amounts of glucose
- Low amounts of lactose
- High amounts of lactose

e. The genes are always activated
2. Which of the following lac operons codes for the correct enzyme?
   a. lacZ codes for permease
   b. lac Y codes for β-galactosidase
   c. lacA codes for thiogalactoside transacetylase
   d. lacA codes for β-galactosidase
   e. lacZ codes for thiogalactoside transacetylase
3. There is an increase in production of cyclic AMP (cAMP) in which of the following lac operon scenarios?
   a. Glucose present lactose absent
   b. Glucose absent, lactose present
   c. Both glucose and lactose present (glucose in low levels)
   d. Both glucose and lactose present (glucose in high levels)
   e. All of the above
4. In the lac operon, lactose acts as an _____; while in the trp operon, tryptophan acts as a(n) _____.
   a. repressor; inducer
   b. inducer; repressor
   c. activator; inducer
   d. inducer; deactivator
   e. deactivator; repressor
5. The trp operon is regulated in which of the following ways?
   a. Through an enzyme coding sequence that is adjacent to the regulatory region
   b. Through an "on-off" mechanism and by a modulation system called attenuation
   c. By an operator and a leader
   d. By lactose acting as a repressor and tryptophan acts as a promoter
   e. None of the above
6. All of the following correctly describe properties between the lac operon and trp operon EXCEPT?
   a. Negative control can only be seen in the trp operon.
   b. A single mRNA is transcribed in each operon.
   c. Attenuation can only be seen in the trp operon.
   d. Allolactose-repressor complex turns on transcription whereas the tryptophan-repressor complex turns off transcription.
   e. Lactose acts as an inducer whereas tryptophan acts as a repressor.
7. _____ is called the attenuator because it needs to be present to attenuate the transcription process when tryptophan levels are high.
   a. Domain 1
   b. Domain 2
   c. Domain 3
   d. Domain 4
   e. Domain 5
8. Methylation _____ gene expression while demethylation _____ gene expression.
   a. activates; represses
   b. controls; represses

regulates; hinders
represses; activates
restrains; promotes

trp operon contains these genes
trpA
trpB
trpC
trpD
All of the above

regulation can occur at which of the following steps?
mRNA processing control
Translation
Repression of gene expression
A & B
C only

Questions

Compare and contrast the lactose operon with the tryptophan operon.
Describe the scenarios in which the lac operon is OFF.
What are the different stages that gene regulation can occur in higher organisms?
What is the order of the Trp operon?
What is the difference between acetylation and methylation?

REFERENCES

de S., Penalva, L. O., Marcotte, E. M., & Vogel, C. (2009). Global signatures of protein and mRNA levels. *Molecular BioSystems*, 5(12), 1512–1526. doi:10.1039/b908315d PMID:20023718

T., Frevel, M., Williams, B. R. G., Greer, W., & Khabar, K. S. A. (2001). ARED: Human element-containing mRNA database reveals an unexpectedly diverse functional repertoire of proteins. *Nucleic Acids Research*, 29(1), 246–254. doi:10.1093/nar/29.1.246 PMID:11125104

T., Evans, D., Link, C. D., Guffanti, A., Lawson, D., Thierry-Mieg, J., ... Kim, S. K. (2002). Analysis of *Caenorhabditis elegans* operons. *Nature*, 417(6891), 851–854. doi:10.1038/na- PMID:12075352

Poręba, E., Kamieniarz, K., & Schneider, R. (2011). Histone Modifiers in Cancer: Friends or *nes & Cancer*, 2(6), 631–647. doi:10.1177/1947601911417176 PMID:21941619

A., Pelletier, D. A., Hurst, G. B., & Escalante-Semerena, J. C. (2012). System-wide Stud- Lysine Acetylation in *Rhodopseudomonas palustris* Reveal Substrate Specificity of Protein nsferases. *The Journal of Biological Chemistry*, 287(19), 15590–15601. doi:10.1074/jbc. 2104 PMID:22416131

Esmaeili, A., Davison, T., Wu, A., Alcantara, J., & Jacob, C. (2015). PROKARYO: An illustrative and interactive computational model of the lactose operon in the bacterium *Escherichia coli*. *BMC Bioinformatics*, *16*(1), 311. doi:10.118612859-015-0720-z PMID:26415599

Frey, M., Kliem, R., Saedler, H., & Gierl, A. (1995). Expression of a cytochrome P450 gene family in maize. *Molecular & General Genetics*, *246*(1), 100–109. doi:10.1007/BF00290138 PMID:7823905

Glozak, M. A., Sengupta, N., Zhang, X., & Seto, E. (2005). Acetylation and deacetylation of non-histone proteins. *Gene*, *363*, 15–23. doi:10.1016/j.gene.2005.09.010 PMID:16289629

Ha, M., & Kim, V. N. (2014). Regulation of microRNA Biogenesis. *Nature Reviews. Molecular Cell Biology*, *15*(8), 509–524. doi:10.1038/nrm3838 PMID:25027649

Hastings, K. E. (2005). SL trans-splicing: Easy come or easy go? *Trends in Genetics*, *21*(4), 240–247. doi:10.1016/j.tig.2005.02.005 PMID:15797620

Hu, L. I., Chi, B. K., Kuhn, M. L., Filippova, E. V., Walker-Peddakotla, A. J., Bäsell, K., ... Wolfe, A. J. (2013). Acetylation of the Response Regulator RcsB Controls Transcription from a Small RNA Promoter. *Journal of Bacteriology*, *195*(18), 4174–4186. doi:10.1128/JB.00383-13 PMID:23852870

Huynen, M., Snel, B., Lathe, W. III, & Bork, P. (2000). Predicting protein function by genomic context: Quantitative evaluation and qualitative inferences. *Genome Research*, *10*(8), 1204–1210. doi:10.1101/gr.10.8.1204 PMID:10958638

Ingolia, N. T., Lareau, L. F., & Weissman, J. S. (2011). Ribosome Profiling of Mouse Embryonic Stem Cells Reveals the Complexity of Mammalian Proteomes. *Cell*, *147*(4), 789–802. doi:10.1016/j.cell.2011.10.002 PMID:22056041

Ivanov, P., Emara, M. M., Villen, J., Gygi, S. P., & Anderson, P. (2011). Angiogenin-induced tRNA fragments inhibit translation initiation. *Molecular Cell*, *43*(4), 613–623. doi:10.1016/j.molcel.2011.06.022 PMID:21855800

Jacob, F., & Monod, J. (1961). Genetic regulatory mechanisms in the synthesis of proteins. *Journal of Molecular Biology*, *3*(3), 318–356. doi:10.1016/S0022-2836(61)80072-7 PMID:13718526

Ji, Z., Lee, J. Y., Pan, Z., Jiang, B., & Tian, B. (2009). Progressive lengthening of 3′ untranslated regions of mRNAs by alternative polyadenylation during mouse embryonic development. *Proceedings of the National Academy of Sciences of the United States of America*, *106*(17), 7028–7033. doi:10.1073/pnas.0900028106 PMID:19372383

Kagan, J., Sharon, I., Beja, O., & Kuhn, J. C. (2008). The tryptophan pathway genes of the Sargasso Sea metagenome: New operon structures and the prevalence of non-operon organization. *Genome Biology*, *9*(1), R20. doi:10.1186/gb-2008-9-1-r20 PMID:18221558

Komander, D., Clague, M. J., & Urbe, S. (2009). Breaking the chains: Structure and function of the deubiquitinases. *Nature Reviews. Molecular Cell Biology*, *10*(8), 550–563. doi:10.1038/nrm2731 PMID:19626045

htt, A. R., Schor, I. E., Allo, M., Dujardin, G., Petrillo, E. & Munoz, M. J. (2013). Alternative A Pivotal Step between Eukaryotic Transcription and Translation. *Nature Reviews Molecular logy, 14*(3), 153-165. doi: 101038/nrm3525

des, T. (2007). Chromatin modifications and their function. *Cell, 128*(4), 693–705. doi:10.1016/j. 7.02.005 PMID:17320507

e, J. (1999). Selfish operons: The evolutionary impact of gene clustering in prokaryotes aryotes. *Current Opinion in Genetics & Development, 9*(6), 642–648. doi:10.1016/S0959-)00025-8 PMID:10607610

hmad, Y., Shlien, A., Soroka, D., Mills, A., Emanuele, M. J., ... Lamond, A. I. (2014). A pro-hronology of gene expression through the cell cycle in human myeloid leukemia cells. *eLife, 3,* doi:10.7554/eLife.01630 PMID:24596151

, T., Ohe, K., Katayama, T., Matsuzaki, S., Yanagita, T., Okuda, H., ... Mayeda, A. (2007). a: Sequence-Specific RNA-Binding Factor Causing Sporadic Alzheimer's Disease-Linked ipping of Presenilin-2 Pre-mRNA. *Genes to Cells, 12*(10), 1179–1191. doi:10.1111/j.1365-07.01123.x PMID:17903177

L., & Catic, A. (2016). Touch and go: Nuclear proteolysis in the regulation of metabolic genes er. *FEBS Letters, 590*(7), 908–923. doi:10.1002/1873-3468.12087 PMID:26832397

stein, R., & Roth, S. Y. (2001). Histone acetyltransferases: Function, structure, and catalysis. *Opinion in Genetics & Development, 11*(2), 155–161. doi:10.1016/S0959-437X(00)00173-8 1250138

ian, A. R., & Koonin, E. V. (1996). A minimal gene set for cellular life derived by comparison lete bacterial genomes. *Proceedings of the National Academy of Sciences of the United States ica, 93*(19), 10268–10273. doi:10.1073/pnas.93.19.10268 PMID:8816789

A., & Pilyugin, S. S. (2008). Bistability of the *lac* Operon During Growth of *Escherichia coli* on and Lactose + Glucose. *Bulletin of Mathematical Biology, 70*(4), 1032–1064. doi:10.100711538-9-7 PMID:18246403

, A. E., & Field, B. (2009). Operons. *Cellular and Molecular Life Sciences, 66*(23), 3755–3775. 00700018-009-0114-3 PMID:19662496

e, L. A., Schmeing, T. M., Maag, D., Applefield, D. J., Acker, M. G., Algire, M. A., ... Ramakrish-2007). The eukaryotic translation initiation factors eIF1 and eIF1A induce an open conformation)S ribosome. *Molecular Cell, 26*(1), 41–50. doi:10.1016/j.molcel.2007.03.018 PMID:17434125

T. V., Lorsch, J. R., & Hellen, C. U. T. (2007). The mechanism of translation initiation in eu-. In M. Mathews, N. Sonenberg, & J. W. B. Hershey (Eds.), *Translational Control in Biology licine* (pp. 87–128). Cold Spring Harbor, NY: Cold Spring Harbor Laboratory Press.

M. (2006). The anaphase promoting complex/cyclosome: A machine designed to destroy. *Nature Molecular Cell Biology, 7*(9), 644–656. doi:10.1038/nrm1988 PMID:16896351

Piccirillo, C., Khanna, R., & Kiledjian, M. (2003). Functional Characterization of the Mammalian mRNA Decapping Enzyme hDcp2. *RNA (New York, N.Y.)*, *9*(9), 1138–1147. doi:10.1261/rna.5690503 PMID:12923261

Price, M. N., Arkin, A. P., & Alm, E. J. (2006). The Life-Cycle of Operons. *PLOS Genetics*, *2*(7), e126. doi:10.1371/journal.pgen.0020126 PMID:16789824

Qi, X., Bakht, S., Leggett, M., Maxwell, C., Melton, R., & Osbourn, A. (2004). A gene cluster for secondary metabolism in oat: Implications for the evolution of metabolic diversity in plants. *Proceedings of the National Academy of Sciences of the United States of America*, *101*(21), 8233–8238. doi:10.1073/pnas.0401301101 PMID:15148404

Ray, J. C. J., & Igoshin, O. A. (2012). Interplay of Gene Expression Noise and Ultrasensitive Dynamics Affects Bacterial Operon Organization. *PLoS Computational Biology*, *8*(8), e1002672. doi:10.1371/journal.pcbi.1002672 PMID:22956903

Roth, S. Y., Denu, J. M., & Allis, C. D. (2001). Histone acetyltransferases. *Annual Review of Biochemistry*, *70*(1), 81–120. doi:10.1146/annurev.biochem.70.1.81 PMID:11395403

Sadakierska-Chudy, A., & Filip, M. (2015). A Comprehensive View of the Epigenetic Landscape. Part II: Histone Post-translational Modification, Nucleosome Level, and Chromatin Regulation by ncRNAs. *Neurotoxicity Research*, *27*(2), 172–197. doi:10.100712640-014-9508-6 PMID:25516120

Schiano, C. A., & Lathem, W. W. (2012). Post-Transcriptional Regulation of Gene Expression in *Yersinia* Species. *Frontiers in Cellular and Infection Microbiology*, *2*, 129. doi:10.3389/fcimb.2012.00129 PMID:23162797

Schwanhäusser, B., Busse, D., Li, N., Dittmar, G., Schuchhardt, J., Wolf, J., ... Selbach, M. (2011). Global quantification of mammalian gene expression control. *Nature*, *473*(7347), 337–342. doi:10.1038/nature10098 PMID:21593866

Sobala, A., & Hutvagner, G. (2013). Small RNAs derived from the 5′ end of tRNA can inhibit protein translation in human cells. *RNA Biology*, *10*(4), 553–563. doi:10.4161/rna.24285 PMID:23563448

Sonenberg, N., & Hinnebusch, A. G. (2009). Regulation of Translation Initiation in Eukaryotes: Mechanisms and Biological Targets. *Cell*, *136*(4), 731–745. doi:10.1016/j.cell.2009.01.042 PMID:19239892

Spieth, J., Brooke, G., Kuersten, S., Lea, K., & Blumenthal, T. (1993). Operons in C. elegans: Polycistronic mRNA precursors are processed by trans-splicing of SL2 to downstream coding regions. *Cell*, *73*(3), 521–532. doi:10.1016/0092-8674(93)90139-H PMID:8098272

Stahl, F. W., & Murray, N. E. (1966). The Evolution of Gene Clusters and Genetic Circularity in Microorganisms. *Genetics*, *53*(3), 569–576. PMID:5331527

Stefanski, K. M., Gardner, G. E., & Seipelt-Thiemann, R. L. (2016). Development of a *Lac* Operon Concept Inventory (LOCI). *CBE Life Sciences Education*, *15*(2), ar24. doi:.15-07-0162 doi:10.1187/cbe

Vera, M., Biswas, J., Senecal, A., Singer, R. H., & Park, H. Y. (2016). Single-Cell and Single-Molecule Analysis of Gene Expression Regulation. *Annual Review of Genetics*, *50*(1), 267–291. doi:10.1146/annurev-genet-120215-034854 PMID:27893965

, Abreu, R. de S., Ko, D., Le, S. Y., Shapiro, B. A., Burns, S. C., ... Penalva, L. O. (2010). signatures and mRNA concentration can explain two-thirds of protein abundance variation n cell line. *Molecular Systems Biology, 6*(1), 400. doi:10.1038/msb.2010.59 PMID:20739923

asinghe, V. O., McMurtrie, P. I., Mills, A. D., Takei, Y., Penrhyn-Lowe, S., Amagase, Y., ... A. (2010). mRNA Export from Mammalian Cell Nuclei Is Dependent on GANP. *Current* ?0(1), 25–31. doi:10.1016/j.cub.2009.10.078 PMID:20005110

n, P. R., Xu, M., Jin, Y., Coates, R. M., & Peters, R. J. (2004). Identification of *Syn*-Pimarae Synthase Reveals Functional Clustering of Terpene Synthases Involved in Rice Phytoelochemical Biosynthesis. *Plant Physiology, 135*(4), 2098–2105. doi:10.1104/pp.104.045971 299118

hen, C. Y., & Shyu, A. B. (1997). Modulation of the fate of cytoplasmic mRNA by AU-rich Key sequence features controlling mRNA deadenylation and decay. *Molecular and Cellular 7*(8), 4611–4621. doi:10.1128/MCB.17.8.4611 PMID:9234718

A., Mayo, A., Ronen, M., & Alon, U. (2006). Optimal gene partition into operons correlates with tional order. *Physical Biology, 3*(3), 183–189. doi:10.1088/1478-3975/3/3/003 PMID:17021382

A., Cheng, N. N., Blumenthal, T., & Spieth, J. (1994). Operons as a common form of chromoanization in C. elegans. *Nature, 372*(6503), 270–272. doi:10.1038/372270a0 PMID:7969472

RMS AND DEFINITIONS

ttors: Molecules that recognize and bind to enhancer sites located upstream of the promoter

uation: A modulation system in which transcription is halted before the operon genes are ed.

Receptor Protein (CRP): A regulatory protein that binds cAMP and causes a conformational CAP.

olite Activator Protein (CAP): The cAMP-CRP complex that allows CRP to bind tightly to ter to begin gene transcription.

etitive DNA Binding: A competition to bind DNA when there is overlap between enhancer er sequences.

AMP (Camp): An intracellular protein activator or second messenger in signal transduction cellular processes.

ncesome: A type of enhancer containing multiple sites for activator binding.

er: Protein that binds to the repressor, thus preventing the repressor from binding to the operator.

ttor/Insulator Protein: A group of elements found between the enhancer and the promoter.

Repressor: Protein that acts as a negative regulator and prevents transcription by RNA polyhen lactose levels are low.

se Operon: An inducible system that requires an inducer (lactose) to remove a repressor protein operator site in order to begin transcription of a gene.

Operator: The key regulatory element of transcription initiation in operons that is composed of a specific DNA sequence.

Operon: Structure consisting of a cluster of co-regulated genes transcribed as a single mRNA and plays a role in regulation when binding to repressors or inducers.

Positive Gene Regulation: Gene regulation that results in increased enzymatic activity in the presence of product; in contrast to negative gene regulation.

Promoter: A specific region upstream of the DNA sequence that regulates transcription by binding to RNA polymerase II.

Repressors: Molecules that bind to the operator to halt transcription of a particular gene.

Ribosomes: Organelles that synthesize proteins.

Silencers: Specific sequences upstream of the promoter region that prevent RNA polymerase from binding to the promoter, thus curtailing gene transcription.

Splicing Regulatory Elements: Elements of gene expression between transcription and translation that play a role in protein processing and degradation.

Transcription Factors: Family of proteins that regulate (activate or repress) gene expression and thus transcription.

Tryptophan Operon: A regulatory system that, in the presence of tryptophan, will turn off transcription by binding to the trp repressor, thus preventing RNA polymerase from accessing the promoter.

Chapter 10
Genes:
Recombinant DNA Technology

ABSTRACT

...ent of recombinant DNA technology has offered new opportunities for innovations to produce a ...ge of bioproducts in food and agriculture, health and disease, and environment. Biotechnology ...ized universally as one of the key enabling technologies for the 21st century forming the basis of ...ngineering where genes are isolated, modified, and inserted into organisms. The new CRISPR-...hnology has made it easier to make direct changes to a DNA strand called gene editing. In ap-...ences such as clinical medicine, biotechnology, forensics, molecular, and evolutionary biology, ...ng DNA has become an important tool. Gene therapy is a technique used to correct single gene ...s where a cloned normal gene is separated and inserted into a cloning vector. Biotechnology ...d for oversight and regulation in ways that makes its application and products safe for human ...operating within human ethical and social guidelines. This chapter explores recombinant DNA ...gy.

CHAPTER OUTLINE

...ence of Recombinant DNA
...ombinant DNA Tool-Kit
....1 Restriction Enzymes
....2 Cloning Vectors
....3 CRISPR Cas9 System
...A Libraries
....1 Genomic
....2 Complimentary DNA
....3 Randomized Mutant
...etic Sequencing

...18/978-1-5225-8066-9.ch010

10.4.1 DNA Microarrays
10.4.2 Next Generation sequencing
10.5 DNA Amplification and Fingerprinting
10.6 Genomics and Proteomics
10.7 Gene Therapy
10.8 Biotechnology and its Applications
10.9 Biotechnology Safety and Ethical Considerations
Chapter Summary

LEARNING OUTCOMES

- With examples, explain what recombinant DNA technology means
- Be able to illustrate what the DNA tool kits are comprise of
- Explain the meaning of DNA libraries and how they are made
- Outline the process of DNA amplification, fingerprinting, and gene therapy
- With examples demonstrate the process of gene sequencing
- Present information of how humans use biotechnology and its ethical and safety implications

10.1 SCIENCE OF RECOMBINANT DNA

In the last two decades or so, the advent of **recombinant DNA technology** has offered new opportunities for innovations to produce a wide range of bio-products. These products have wide applications in food and agriculture, health and disease (gene therapy, antibody production, drug metabolism research, vaccine development, medical nutrition), and environment (phytoremediation and energy applications) (Khan et al., 2016). DNA recombination is done through the modification of microorganisms, animals, and plants to yield biomedically useful products and medical therapies in a field of study called **transgenics.** Some of these products include biopharmaceuticals (like vaccines, growth hormones, antibodies, recombinant protein and enzymes), vitamins, organic acids, amino acids, glucosamines, solvents, anti-tumour agents, cholesterol-lowering agents, and a host of other compounds (Adrio & Gemain, 2010).

Recombinant DNA technology involves altering genetic material outside an organism to obtain enhanced and desired characteristics in living organisms or their products. DNA fragments with the desirable gene sequence from a variety of sources are inserted using several types of vectors like bacteria, **plasmids** or **retroviruses** (Figures 1, 2, 3).

The organism's genome is then manipulated by either the introduction of one or several new genes and regulatory elements (Figure 1); or by decreasing or blocking the expression of original genes through recombining genes and elements (Bazan-Peregrino et al., 2013). When different genes are incorporated in different plasmids as shown, it creates a plasmid library (Figure 2).

Cloning a gene of interest into a bacterial plasmid

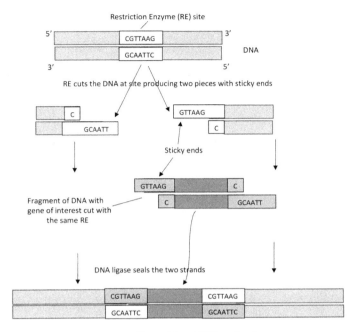

DNA plasmid library containing different genes of interest

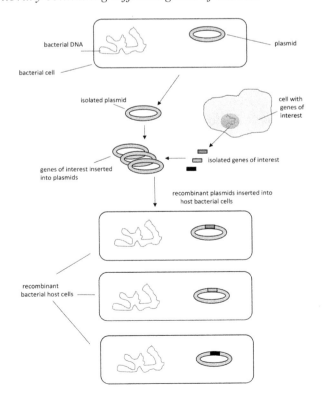

Figure 3. Viral library containing different genes of interest

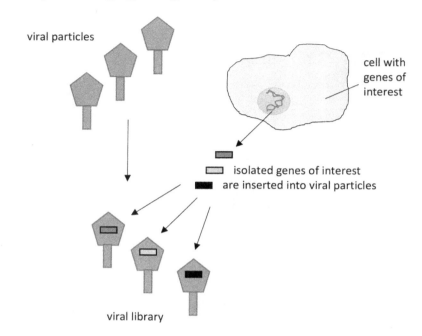

10.2 RECOMBINANT DNA TOOL-KIT

Recombination in microorganisms occurs through three main parasexual processes: **conjugation, transduction,** and **transformation**. Conjugation involves transfer of DNA via cell-to-cell contact; transduction occurs from host cell to recipient cell via mediation by a bacteriophage; while transformation involves uptake and expression of naked DNA by competent cells. Although competence can occur naturally, it can also be induced by changes in the physical and chemical environment; for example, cold calcium chloride treatment, protoplasting, electroporation and heat shock. (Adrio & Demain, 2010). Recombinant DNA forms the basis of **genetic engineering** where genes are isolated, modified and inserted into organisms. How are genes of interest isolated?

10.2.1 Restriction Enzymes

Restriction endonucleases (often abbreviated as REases) are the toolkits for all recombinant DNA operations worldwide. They have become the enabling tools of molecular biology, genetics and biotechnology, and made analysis at the most fundamental levels routine (Pingoud et al. 2014). They are what hammers are to carpenters or spanners to mechanics. Restriction endonucleases occur ubiquitously among prokaryotic organisms and their principal biological function is the protection of the host genome against foreign DNA, in particular bacteriophage DNA (Pingoud & Jeltsch, 2001).

There are more than 19,000 REases available (Roberts et al., 2010) of four main types: Type I, II, III and IV and almost all require a divalent metal cofactor (such as Zn^{2+} or Mg^{2+}) for activity. The metal ion is thought to stabilize the REase-DNA complex transition state by neutralizing the build-up of negative charge on the phosphorus. One REase (Type II), with over 3000 varieties available, is the best studied

4. Using restriction enzymes to make recombinant DNA

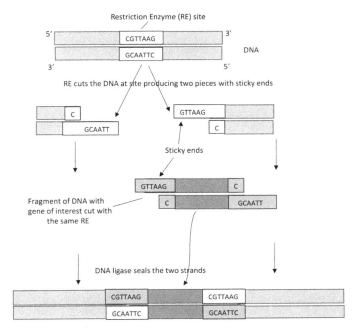

Recombinant DNA molecule

...useful as a tool for recombinant DNA technology (Wil et al., 2014). Some of the most common REases discovered since the 1970s are EcoRI, EcoRII, HindII, HindIII, SmaI, XmaI, NgoPII, PvuII, BamHI, FokI, Bg1I, Bg1II and MunI. Pingoud et al. (2014) contends that the discovery of II REases has produced dollars of economic activity, tons of jobs and careers and astounding ...s in our understanding and knowledge.

...ype II REase recognizes a sequence of about 4–8 base pairs in length, and in the presence of ...leaves the two strands of the DNA within or immediately adjacent to the recognition site to give ...osphate and a 3′ OH end. Once the REase binds DNA at the recognition site, conformational ...take place that constitute the recognition process and lead to the activation of the REase's cata-...ntres. The DNA phosphodiester bond joining nucleotides is cleaved producing ends with either ...OH overhangs referred to as "sticky ends" (Figure 4). This means they can snap back like Velcro ...they are complementary. For example, EcoRI recognizes and cleaves the sequence GAATTC ...), NgoPII the sequence GGCC (at G↓C), and XmaI the sequence CCCGGG (at C↓G).

Cloning Vectors

Viral Vectors

...re synthetically produced biological particles, in which the plasmid DNA carrying the therapeutic ...pression cassette is encapsulated or bound to a synthetic chemical compound and then released ...rget site upon delivery. Naturally occurring in bacteria, yeast, and some higher eukaryotic cells, ...ls are circular, double-stranded DNA molecules (Figure 5). They exist in a parasitic or symbiotic

Figure 5. A bacterial cell with a plasmid (independent circular piece of DNA)
Source: Image used under license from Shutterstock.com

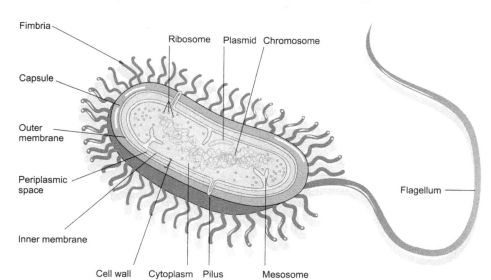

relationship with their host cell—they provide benefits to host cell and they are protected inside the host. Often, plasmids contain genes that encode enzymes that deactivate antibiotics, conferring survival benefits to the host cell. Whenever the host cell divides the plasmid DNA is also replicated therefore propagating benefits to future generations of the host cell.

In contrast to viral-derived vectors, non-viral systems are relatively easy to produce, and the risk for inflammatory complications is low (Chira et al., 2015). Non-viral systems are naked genetic material or complexed with a chemical compound. The plasmids used in recombinant DNA technology replicate in *E. coli* and engineered to optimize their use as vectors in DNA cloning. The non-viral vector once in the extracellular environment must maintain its integrity to achieve physical contact with the target cell. Research shows that the level of therapeutic gene expression depends of the type of promoter used to drive its expression and this is directly correlated with the efficiency of gene transfer *in vivo*. (Papadakis et al., 2004)

Viral Vectors

Viruses are appealing tools for therapeutic gene transfer because of their high transfection/transduction efficiency in wide range of human cells (Figure 6). Since they are pathogenic agents, attenuated versions are safely used in clinical applications. One way to attenuate viruses is to delete the viral genes that are responsible for replication, in which case the resulting vectors are replication-defective. Virus-derived vectors originate from different viral classes including **adenoviruses (Ad), adeno-associated viruses (AAV), retroviruses, lentiviruses, foamy viral vectors,** *Herpes simplex* **virus (HSV), poxviruses (PV) and hybrid vectors.** (Pushko et al., 2013; Zeltins, 2013). Approximately 70% of the vectors used in gene therapy clinical trials are represented by viral-based delivery systems (Ginn et al., 2013). However, due to past failures that has negatively impacted gene therapy, further optimization is needed to safely use viral vectors for future clinical proposes. Recent research seems to indicate that the ideal gene therapy

*Viral vectors. Useful vehicles for cloning genes.
Image used under license from Shutterstock.com*

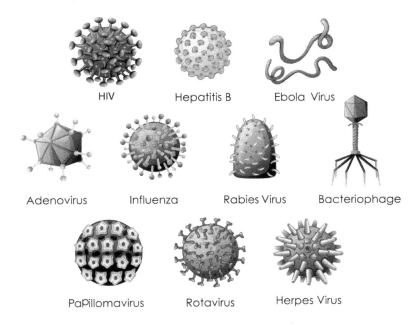

vector will be a compromise between non-viral and viral vectors. As a result, "convergent" hybrid vectors might represent promising tools for safe and efficient transfer of therapeutic genes, moving gene therapy a step closer to the clinic. (Chira et al., 2015)

CRISPR-Cas9 System

Gene editing systems exists in nature in bacterial cells where they capture bits of genetic material from viral invaders and use it to create short segments of DNA called **CRISPR arrays**. In the mid-1980s, Japanese researchers discovered unusual repeating sequences in the DNA of *E. coli* bacteria, later found in other bacterial species. These consisted of short sequences of genetic code and similar sequences in between. These palindromic repeats were given the CRISPR acronym derived from <u>C</u>lustered <u>R</u>egularly <u>I</u>nterspaced <u>S</u>hort <u>P</u>alindromic <u>R</u>epeats. Researchers also found CRISPR associated genes (Cas) which code for Cas enzymes. Just like the way vaccines help human immune cells "remember" infectious agents, the CRISPR arrays allow the bacteria to "remember" invaders if they invade again. The microorganisms make copies of short sections of viral DNA in their genome which are transcribed into RNA called CRISPR RNA (crRNA). When re-infected by similar viruses the bacteria produces segments of crRNA from the CRISPR arrays which target viral DNA. The crRNA forms a complex with a second piece of RNA called the trans-activating crRNA (tracrRNA), and CRISPR-associated protein 9 (Cas9), which are encoded in the bacterial genome. When this complex encounters viral (bacteriophage) DNA which is complementary to the crRNA sequence, it binds it. Cas9 is an endonuclease which cleaves double-stranded DNA, slicing through the viral DNA and preventing transcription, thereby disabling it. (Peddle and Van Haren, 2017)

Since it was discovered in 2012, researchers have been using the CRISPR-Cas9 technique to create a short **"guide" RNA** sequence whose sequence matches the damaged DNA they want to edit. The guide RNA then binds a Cas9 enzyme. They also create a "customized DNA" in a benign virus or plasmid and introduce it to the cell at the same time. Together, the "guide" RNA-Cas9 complex attaches to a specific sequence on the DNA and cuts (deletes) it at that target location (Figure 7). Cells are constantly monitoring the integrity of the genome and when they notice the cut, they attempt to repair it. They can either use their machinery for DNA repair comprised of polymerases, ligases, and nucleotides a process that could disrupt or delete the disease-causing gene. Cells can also correct the genetic fault if the "customized DNA" is available in the vicinity. Often, bacteria will use the "customized DNA" available to them thereby integrating the correct gene into the genome as researchers expect. (Lander, 2016; Komor et al., 2017)

CRISPR-Cas9 system has successfully been used to make direct changes to a DNA strand such as inserting or deleting specific sequences that code for proteins. Ryan et al. (2014) used the CRISPR-Cas9 system to create multiple DNA breaks simultaneously across the genome of yeast cells and joined 'barcoded' DNA or DNA for intact genes to these breaks. This avoids the need to use plasmids to introduce foreign DNA into cells. The CRISPR-Cas9 system has been welcomed by the scientific community because it is faster, cheaper, more accurate, and more efficient than other existing genome editing methods (GHR, 2018). Much of this enthusiasm centres on the clinical potential of CRISPR/Cas9 for treating human disease and editing the human genome (Lino et al., 2018). Genome editing CRISPR-Cas systems require a Cas9 endonuclease that is targeted to specific DNA sequences by a non-coding single guide RNA (sgRNA) (Jinek, et al., 2012). The Cas9-sgRNA ribonucleoprotein complex precisely generates double-strand breaks in eukaryotic genomes at sites specified by a 20-nucleotide guide sequence at the 5′

Figure 7. The Cas9-sgRNA ribonucleoprotein complex. See text for explanation.
Source: Image used under license from Shutterstock.com

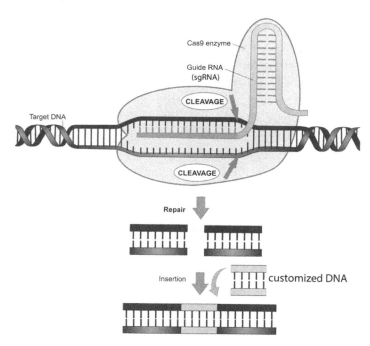

...he DNA strand (Sternberg et al., 2014) (Figure 7). Researchers are exploring the CRISPR Cas9 ...ting system to learn its applicability in single-gene defects (like haemophilia, cystic fibrosis and ...ell anaemia) and more complex diseases like schizophrenia, cancer, and heart disease.

... to the CRISPR-Cas9 system, there were two options for gene editing at the DNA level: zinc ...ucleases (ZFNs) or transcription activator-like effector nucleases (TALENs). Although research... had more time to tinker with genomes using these techniques, both techniques were marred by ...disadvantages. ZFNs were found to be low on-target efficiency and difficulty locating a potential ...te; while TALENs were very large and therefore difficult to deliver to the cells. For example, ...vith Hunter syndrome have a mutation that disables an enzyme called iduronate-2-sulfatase (IDS), ...ble for breaking down certain complex sugars in the body. When those sugars accumulate, they ...age organs including the lungs, heart and brain. Routinely, people with Hunter syndrome receive ...s of the IDS enzyme to replace the damaged enzyme. Unfortunately, the healthy enzyme is rapidly ... and patients need fresh dose frequently. Gene-editing enzymes called zinc finger nucleases have ...ed to insert a healthy version of defective genes into a region of the genome deemed to be safe. ...hers then used a virus to shuttle DNA that encodes the nuclease, and the healthy *IDS* gene into ...ls, where IDS is normally produced. (Carroll, 2011; Gupta & Musunuru, 2014; Ledford, 2018) ...panies like CRISPR Therapeutics are examples of corporations at the forefront of this latest gene ...echnology for the treatment of hemoglobinopathies like sickle-cell anaemia and beta-thalassemia ...ffect more than a quarter million people worldwide. The company receives patient blood stem ...m clinical trial sites in the field, edits them, and then ships them back to the trial sites. Patients ...dergo chemotherapy or irradiation to prepare their bone marrow for the transplantation of the ...lood stem cells. Hopefully, after this transplantation, they will produce healthier blood cells for ...ss, 2018).

...er alternative forms of CRISPR have since been discovered which can turn genes on and off ...CRISPR activation and interference (CRISPRa/i). This method can turn off a whole set of genes ... the understanding of how genes interact in complex diseases like heart disease, cancer, and ...er's. Despite the gene editing breakthrough, there are however several challenges to clinical use ...PR-Cas9. These include gene editing efficiency or precision editing, off-target effects (that is ...n the wrong places on the DNA), and heterogeneity of some diseases. Other concerns with the ... Cas9 gene editing include safety, ethical concerns especially in human editing of germ cell lines ...eggs). Such manipulations of the human genome could cause cancers or knock-off/deactivate ...Using CRISPR, organisms can be manipulated to create biological weapons. Today, CRISPR is ...ommended in clinical trials across the globe. (Bourzac, 2017; Peddle & MacLaren, 2017)

...NA LIBRARIES

...advancements in DNA synthesis and assembly techniques have enabled the production of highly ...ibraries with relatively even distribution of variants (Bradley, 2014). A DNA library (like a book ... is an assortment of DNA fragments that have carefully been cloned into vectors like bacteria ...es. These synthetic DNA libraries can store the entire genome of an organism (like a human ...o allow the sequences useful in the production of antibodies, enzymes, various other proteins, ...omes to be more thoroughly examined. To measure and evaluate DNA libraries, a procedure known as next generation sequencing (NGS) is the current state of the art for measuring large numbers of individual DNA sequences.

DNA libraries are of three types: Genomic DNA, complimentary DNA, and randomized mutant DNA. Let's review each of these.

10.3.1 Genomic

To prepare a genomic library, the entire DNA of an organism is cleaved with REases into tiny fragments producing millions of genomic DNA fragments. Once the sequences of interest are identified, they are inserted into different cloning vectors (plasmids, retroviruses, adenoviruses, etc.). The enzyme DNA ligase is used to join the two DNA types making recombinant DNA. If the vector carrying the DNA of interest is a plasmid it is introduced into bacterial cells for uptake. Bacteria are often treated in specific procedures (like heat or chemical treatments) to increase uptake rates. Now inside the bacterium, every time the bacterium replicates, new copies of the introduced DNA of interest is copied (a process called cloning, Figure 1). For viral vectors, those carrying the DNA of interest are identified and labelled for future use (Figure 3). A genomic library therefore is a set of clones that collectively contain all sequences (those of interest plus those not of current interest) in the original DNA.

10.3.2 Complimentary DNA

Complimentary DNA (cDNA) is made from mRNA transcripts that have undergone processing (splicing/5' capping/3' OH adenylation) to produce mature mRNA ready for translation. Copies of cDNA are extremely powerful tools for analysing the structure, organization, and expression of eukaryotic genes (Okayama & Berg, 1982). cDNA has been used to identify initiation, coding and termination sequences of mRNAs. In addition, their use as hybridization probes has allowed searching, isolation, identity, and characterization of the corresponding genes on chromosomal DNA (Okayama & Berg, 1982).

To prepare cDNA from mRNA, an enzyme called **reverse transcriptase** is utilized. This enzyme can reproduce a copy of DNA from mRNA. It is like going in "reverse" since mRNA comes from DNA. Since DNA is double stranded the enzyme DNA Polymerase is used to make a single DN copy into a double stranded DNA making cDNA (Figure 8). This can then be inserted into a cloning vector for use in research or biomedicine.

10.3.3 Randomized Mutant

Different from genomic and cDNA libraries, a randomized mutant library is created in the laboratory from simple molecules (*de novo* synthesis). The benefits of *de novo* synthesis are that different sequences (codons) can be incorporated into the DNA in specific locations. This produces variants of the original gene as a result of the sequence mixture in the DNA. Once this new DNA is inserted into a cloning vector, it produces a mutant library from which individual clones can be created and proteins expressed studied.

Today, biotechnology companies are using protein engineering techniques to screen for mutant libraries for novel proteins with "enhanced expression levels, solubility, stability, enzymatic activity, or interaction with desired binding partners" (GeneScript, 2018, p.1).

How complementary DNA (cDNA) is made from eukaryotic genes

```
DNA
exon  intron  exon  intron  exon  intron  exon  intron  exon
        ↓
                                                        mRNA
        ↓
DNA                                         processed mRNA
synthesis                                   with no introns
        Reverse transcriptase enzyme
        (makes a copy of DNA from mRNA)
        ↓
        DNA Polymerase synthesizes
        a second DNA strand
        ↓
        complete cDNA of gene without introns
```

GENETIC SEQUENCING

[...] sciences such as clinical medicine, biotechnology, forensics, molecular and evolutionary bi[ology ...] quencing DNA has become an important tool. The original sequencing technology was called [...] **equencing** in honour of the scientist who developed it in the mid-1970s—Frederick Sanger. [...] ry useful for sequencing short stretches of DNA but can be time consuming for sequencing [...] organism's genome. The preparation of microarrays was the prime procedure used in Sanger [...] g technology. Sequencing technology has been advancing since the human genome was first [...] lly sequenced on 2003. Today, **next-generation sequencing (NGS)** is emerging as a feasible [...] sequence an organism's genome in short period. NGS has allowed the sequencing of whole [...] as seen in **whole genome sequencing (WGS)** and **whole exome sequencing (WES)**.

DNA Microarrays

[...] **cid arrays (DNA microarrays)** are a group of technologies in which specific DNA sequences [...] deposited or synthesized in a 2-D (or 3-D) array on a glass microscope slide such that the DNA [...] tly or non-covalently attached to the surface (Figure 9). Such a DNA array is used to measure [...] ve concentrations of the nucleic acid species in solution with a mixture of fluorescent-labelled [...] cids by binding (through hybridization) of these "targets" to the "probes". DNA microarrays [...] to measure gene expression levels by simultaneously measuring the relative concentrations of [...] ferent DNA or RNA sequences. They are also used in combination with other procedures in

Figure 9. A DNA microarray. A DNA microchip looks like this chip packed with small single bases of about 20-25 bases.
Source: Image used under license from Shutterstock.com

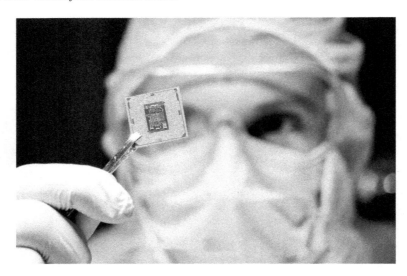

determining the binding sites of transcription factors and single-nucleotide-polymorphism (SNP) genotyping. (Horak & Snyder, 2002; Bumgarner, 2013)

Microarrays have several disadvantages. 1) They can only detect sequences that the array is designed to detect. 2) Arrays provide an indirect measure of relative concentration. 3) In complex genomes, it is often challenging to design arrays in which multiple related DNA/RNA sequences will not bind to the same probe on the array—that is, a sequence on an array that was designed to detect "gene Z", may also detect "genes W, X, and Y" if those genes have significant sequence homology to gene Z. (Bumgarner, 2013)

While microarrays were the leading technology in gene sequencing—in recent years, the emergence of next generation sequencing technologies combined with the rapid decrease in the cost of sequencing has now made sequencing for almost all assays cost competitive compared to microarrays (Bumgarner, 2013).

10.4.2 Next Generation Sequencing

Next-generation sequencing (NGS) was developed during the 2007-2017 decade and allows simultaneous sequencing of millions of DNA fragments without previous sequence knowledge (Kamps et al., 2017). Using NGS, an entire human genome can be sequenced within a single day. The implications and the impact of NGS in understanding the biological processes of diseases like cancer and in personalizing patient care are unprecedented. Once NGS has been performed, the client's genome can be stored for future evaluation and as new medical questions arise, the genomic data can be re-analysed. With improved software, increased knowledge and understanding about genome–phenotype linkages, and the development of computational analyses handling massive data, revaluating a patient's genome can help identify previously undetected correlations. (Kamps et al., 2017) Whole genome sequencing (WGS) can detect non-coding mutations, structural variants including copy number alterations, mitochondria mutations, and pathogen detection, as well as protein-coding mutations (Nakagawa & Fujita, 2018).

le exome sequencing refers to the sequencing of all the exons (protein coding sequences) in a genome. Whole exome sequencing is an ideal procedure for identifying mutations which af-
teins coded by genes, thereby resulting in genetic defects. Since the activity of genes are often
ed by sequences outside the exome, WES would not detect such mutations. This is where whole
sequencing becomes handy as it sequences the whole organism's genome (protein coding and
ing sequences). Technological advances and the lowering cost of DNA sequencing have made
ibility of WGS as a highly accessible clinical test for numerous indications feasible. There have
ny recent, successful applications of WGS in establishing the etiology of complex diseases and
therapeutic decision-making in tumour and non-tumour related diseases and in various aspects
ductive health. (Chrystoja & Diamandis, 2014). WGS has also been utilized in understanding
diversity and evolutionary analysis.

main disadvantage of NGS in the clinical setting is putting in place the required infrastructure,
computer capacity and storage, and the personnel expertise required to comprehensively analyse
rpret the subsequent data. In addition, the volume of data needs to be managed skilfully to extract
cally important information in a clear and robust interface (Behjati & Tarpey, 2013).

NA AMPLIFICATION AND FINGERPRINTING

mplification: *The chemical synthesis of genes from* oligonucleotides *de novo* used as the building
or enzymatic assembly, makes it possible to synthesize gene sequences and even whole genomes
et al., 2008). This technology is now more affordable than cloning genes and it enables the use of
genome databases to construct any intended target. Such targets can include synthetic enzymes,
genes, regulatory elements, and even entire pathways (Burbelo et al., 2010).

eloped in the 1980s by Kary Mullis, the nested **polymerase chain reaction (PCR)** is based on
ty of the DNA Polymerase ability to make copies of DNA given all the necessary ingredients of
nthesis: primers with a 3' OH end, template strand, free nucleotides under temperature regulated
ons. Conventional PCR is a cyclic process involving denaturation, annealing, and extension steps
bles the target sequences after each cycle (Shrestha et al., 2010). During this process millions of
DNA can be obtained from the original DNA (Figure 10). However, due to reaction inhibitions,
lation of pyrophosphate molecules, and end-product recombination (strand self-annealing), the
ocess soon plateaus making the end product slightly unreliable. As a result, new PCR techniques
ize "real-time" monitoring of the product made after each successive PCR cycle. **Amplicon**
ination has been identified as the most frequent cause of diagnostic false-positive results in PCR
cation (Bell & Ranford-Cartwright, 2002). **Real-time polymerase chain reaction (RT-PCR)** is
nd a more attractive technology because it does not need post-PCR analysis, resulting in shorter
und times and minimizing the risk of amplicon contamination (Fotedar et al., 2007). RC-PCR
s the increasing amplicon copies in real time after each cycle.

NA Fingerprinting: First described by Jeffreys et al. (1985), this novel method allowed for the
st time the ability to discriminate between humans, animals, plants and fungi on the individual
vel using DNA markers (Nybom et al., 2014). Jeffreys et al. (1985) discovered that certain re-
ons of DNA showed variations in the number of tandem repeats known as **variable number of**
ndem repeats (VNTRs or microsatellite DNA). VNTRs typically consist of repeated 10 to 60

Figure 10. The polymerase chain reaction (PCR). See text for explanation.
Source: Image used under license from Shutterstock.com

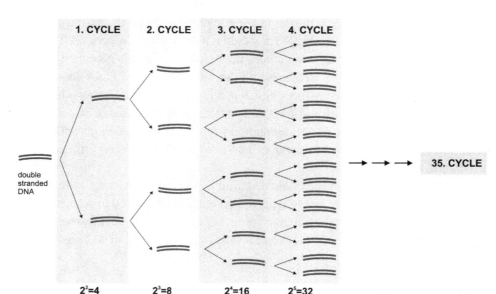

DNA base pair units. They are scattered throughout the human genome occurring on an average of every 10,000 nucleotides making them powerful genetic markers (Raina & Dogla, 2002). Once DNA is cleaved by REases, a set of DNA sequence fragments of varying length known as **restriction length fragment polymorphisms (RFLPs)**, are produced in the test tube. These are then put in gel-wells using a micropipette in a process called **gel electrophoresis** which is connected to electric current. DNA, comprised of many phosphates (PO_4) is negatively charged and will move towards the positive electrode when subjected to an electric current. The different RFLPs navigate through the small spaces within the gel. Smaller RFLPs migrate farthest while longer pieces do not migrate far. Over time, a series of RFLP marks are produced indicating the farthest a specific piece migrated. This is called a **DNA fingerprint**, indicating the number of repeated sequences differing from one individual to the other (Figure 11).

10.6 GENOMICS AND PROTEOMICS

Since the ACTG bases on DNA is just a list of the 3.2 billion bases in humans, it is necessary to the study how these bases determine whole sets of genes and how they function together in a recent field called **genomics**. Genomics is dedicated to the large-scale analysis of the properties of genomes (Conesa & Mortazavi, 2014). To identify genes, scientists use computer software to scan stored DNA sequences for transcriptional, translational, regulatory, splicing sites, start and stop signals. To determine the function of a gene, scientists can disable it and observe the consequences in the cell/organism. They can also introduce mutations and study the effects on cellular operations.

. *A DNA fingerprint produced from VTNR analysis. Here the defendant's blood is shown at ?ft, blood from clothes he/she was wearing (middle), and blood from the victim (extreme right).*

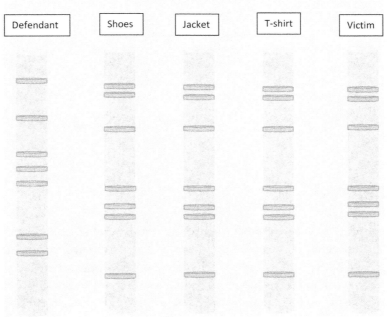

)s one of the several cutting-edge approaches for understanding global genes expression and :tional mechanisms is to study the proteins translated from those genes and that scientific known as **proteomics**. The word "proteome" is derived from PROTEins expressed by a ge- hich was coined by Wilkins in the mid-1990s. Analogous to genomics, the term "proteomics" the study and characterization of the complete set of proteins present at a given time in the dns et al., 1995). It is the proteins, not genes, that are responsible for the phenotypes of cells impossible to elucidate mechanisms of disease, aging, and effects of the environment solely ng only on the genome.

made, proteins undergo considerable modification post-translationally, they are translocated : cell, or can be synthesized or degraded. Thus, examination of the proteome of a cell is like snapshot" of the protein at any given time in this dynamic environment. The aim of proteomics y to identify all the proteins in a cell but also to create a complete three-dimensional (3-D) map l indicating where proteins are located. The growth of proteomics is a direct result of advances arge-scale nucleotide sequencing of expressed sequence tags and genomic DNA. (Graves & , 2002).

Figure 12. CRISPR technology

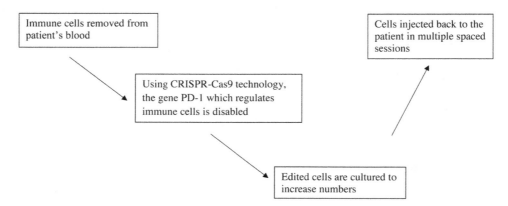

10.7 GENE THERAPY

This technique (also referred to as **eugenics**) is used to correct single gene disorders where a cloned normal gene (e.g. insulin gene) is separated and inserted in to a cloning vector like a retrovirus. The retrovirus is inserted into patient cells in vitro (cells are removed from patient and cultured in lab). The modified retrovirus is then allowed to insert its genome in the chromosome of patient's cells. These engineered cells are now injected into patient. When the cells divide, more copies of the gene are made, and the gene is expressed—eliminating the impacts of the disorder in the patient! With the advent of the new CRISPR technology (section 10.2.3 above), researchers are modifying human DNA in clinical trials to treat diseases such as sickle-cell anemia. In 2016, researchers in China used CRISPR technology to start treatment on a patient with non-small-cell lung cancer (see schematic).

Challenges remain however, due to gene interactions that are not well understood. For example, a specific gene edit may cause further problems downstream on the DNA where unrelated genes are turned off or might fail to work properly after the editing process.

There are primarily three reasons for using gene therapy. 1) Replacing a mutated gene that causes disease with a healthy copy of the gene. 2) Inactivating or "knocking out" a mutated gene that is functioning improperly. 3) Introducing a new gene into the body to help fight a disease. Several studies have shown that gene therapy can have very serious health risks, such as toxicity, inflammation, and triggering cancer. Despite the ~1800 gene therapy clinical trials reported world-wide, only ONE product has been approved to be used in clinical applications in the European Union—*Glybera*. It is used for treatment of lipoprotein lipase deficiency, by means of a virus which delivers the functional copy of the gene into the patient muscle cells (Figure 13).

10.8 BIOTECHNOLOGY AND ITS APPLICATIONS

Biotechnology is a utilization of knowledge, techniques and procedures rooted in a variety of disciplines including biochemistry, molecular biology, genetics, microbiology, and bioinformatics. The science of biotechnology involves working with living cells or cell products and finding ways in which some of these products can be used commercially by human in medicine, agriculture, environment, and industry.

3 The process of gene therapy in correcting a genetic disorder

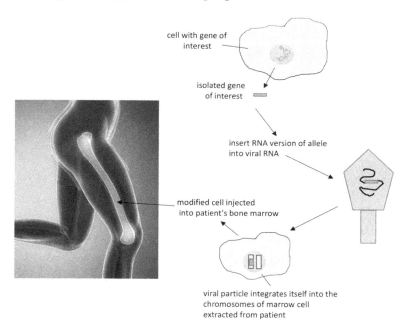

have used living organisms for generations from times past when we domesticated animals, d crops, and we continued improving our agricultural products (plants and animals) through eeding and hybridization techniques. However, not everyone is happy with the increasing avail- f genetically modified organisms (GMOs) in the market place (Figure 14). If we restrict the n to genetic modification of biological systems to generate goods and services, then it started ears ago when our ancestors first domesticated agriculture, intentionally selecting genetically or preferable crops and animals for domestication (McHughen, 2016).

y, modern technology has allowed continued biotechnological improvement through genetic nation, stem cell research, cell and tissue culture. Bacteria and yeasts have commercially been many biotechnological tasks, ranging from biofuel production, commodity chemical synthesis, oduction of industrial and biopharmaceutical proteins (Dragosits & Mattanovich, 2013). Ge- gineering strategies have been employed to tackle the environmental issues such as converting to biofuels and bioethanol, cleaning the oil spills, carbon, and other toxic wastes, and detecting nd other contaminants in drinking water, and are used in biomining and bioremediation (Khan 16).

dicine

ct of biotechnology to date has been most pronounced in the **pharmacogenomics** sector although ues to advance in the fields of gene therapy and genetic screening. Recombinant biological prod- e revolutionized modern medicine by providing both remarkably effective vaccines to prevent nd therapeutic drugs to treat a wide variety of unmet medical needs (Volkin et al., 2015). The s primarily centred on pharmaceutical development of three different types of biotechnology-

Figure 14. Opposition to transgenic plants or genetically modified organisms (GMOs)
Source: Image used under license from Shutterstock.com

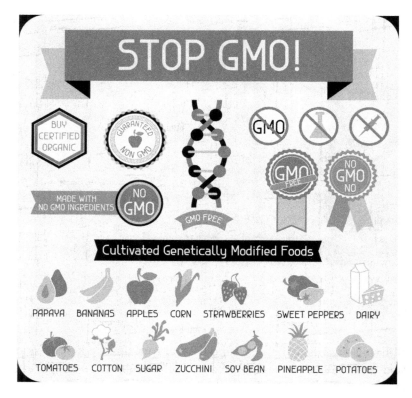

based product candidates: 1) **protein-based therapeutics**; 2) other biological molecules like peptides, polysaccharides, and DNA/RNA; 3) vaccine development from **macromolecular antigens**. Since the late 1990s, there has been a surge in development and commercialization of new vaccines (both new vaccine antigens; new formulations, and delivery technologies). These include polysaccharide-protein conjugates like pneumococcal and meningococcal vaccines; killed and live, attenuated viruses like hepatitis A, rotavirus and shingles vaccines; and recombinant protein technologies that includes virus-like particles (human papillomavirus or HPV vaccine) and a recombinant hemagglutinin flu vaccine. New formulations and delivery technologies have been introduced to improve and expand the utility of vaccines such as influenza (e.g., nasal and intradermal delivery, new adjuvants). Already approved vaccines have also been developed with new mixture formulations to simplify vaccination schedules and ensure compliance (e.g., measles, mumps, rubella. varicella, diphtheria, tetanus toxoid, acellular pertussis, hepatitis B and inactivated poliovirus vaccines). (Volkin et al., 2015)

The first genetically engineered recombinant protein approved by the US Food and Drug Administration (FDA) was synthetic human insulin. Today, sixteen categories of medications produced through biological processes involving recombinant DNA (referred to as biologics, biopharmaceuticals, or recombinant proteins) constitute a $108 billion market (Wuest et al., 2012). The term transgenics refers to the genetic engineering processes that remove genetic material from one species of plant or animal and add it to a different species. Glenn (2013) discusses several ways where the field of transgenics transformed the way we live today. Transgenics has been used to fortify foods with minerals and vitamins; producing

t glow in the dark using firefly genes; saving endangered species; xenotransplantation which
...splantation of living tissues or organs from one species to another; tissue engineering where
... organs can be grown in scaffolds using engineered stem cells; and "biopharms" where animals
... milk with therapeutic protein products. (Glenn, 2013)

...ture

...logy has enabled the generation of **genetically modified (GM) crops or transgenic plants,**
... crops into which one or several genes coding for desirable traits have been inserted through
... of genetic engineering. The genes may originate from the same or other species and or-
... unrelated to the recipient organism. Biotechnology confers several advantages in GM crops
... improved nutritional value, better food flavours, fresher produce, pesticide resistant crops,
...-resistant crops. In animals, the principal applications of biotechnology currently have been
... production of vaccines for human and animal diseases and animal disease diagnosis (Papin et
... In the livestock productivity and to improve the availability of nutrients from feeds, animal
... aids such as enzymes, probiotics, single-cell proteins and antibiotic feed additives are widely
... intensive production systems worldwide. Advanced biotechnology-based diagnostic tests also
... sease identification and disease-causing agents and to monitor the impact of disease control
... (FAO, 2018).

...ial Biotechnology

... **bioeconomy** that includes increasing reliance on biological processes and bio-based products
... element of the overall global sustainability transition (Johnson, 2017). Industry has applied the
... of modern molecular biology to improve the efficiency and reduce the environmental impacts
... industrial processes like textile, paper and pulp, and chemical manufacturing (Keener et al., n.d). The
... on of biotechnology-based tools to traditional industrial processes—also called **bioprocessing**
... to as **industrial biotechnology**. It includes the manufacture of bio-based products like fuels,
... and plastics from renewable feedstocks from diverse microbial systems from bacteria, yeasts,
... to marine diatoms and protozoa. A diverse group of companies are using such microorganisms,
... products (enzymes), and advances in genetic engineering as the foundation for these bioprocesses.
... et al., 2012)

...nmental Biotechnology

... into the potential of biological organisms like bacteria and many fungal and plant-based **biore-**
... **rs**, environmental biotechnology can more efficiently clean up many wastes and environmental
... ...ants than conventional methods greatly reducing our dependence on land-based solid and
... waste disposal. Bacteria have been engineered to clean up toxic waste and purify sewage. For
... a recent innovation using bacteria is the development of a microbial fuel cell called *BioVolt*,
... convert 2250 litres of sewage into enough clean water for at least 15 people in a day, while
... me time producing power to power itself (New Scientist, 2016). The bacteria liberate some
... as they respire, which are tapped to produce electricity.

Forensics

The collection of **forensic evidence** and the application of forensic sciences has become essential to criminal investigations and prosecutions (Peterson et al., 1987, Figure 10.14). According to Fisher (2004), forensic evidence fulfills several roles in criminal investigations. 1) Proving a crime has been committed or establish key elements of a crime; 2) Placing a suspect in contact with the victim or with the crime scene; 3) Establishing the identity of persons associated with a crime; 4) Exonerating an innocent person; 5) Corroborating a victim's testimony; and 6) Assisting in establishing the facts of what occurred.

In summary, biotechnology is recognized universally as one of the key enabling technologies for the 21st century, and confidence in this view stems from its position as a radical innovation, the impact it has had and will have on major global problems (disease, malnutrition, and environmental pollution); the promise it holds for achieving industrial sustainability (optimal use of renewable resources, amelioration of global warming, and introduction of clean or cleaner products and processes); and the increasing realization that it has become a mature technology capable of achieving economic competitiveness, generating new markets, and having wide industrial applicability (Keener et al. n.d,)

10.9 BIOTECHNOLOGY SAFETY AND ETHICAL CONSIDERATIONS

The primary goal of regulation is to protect our society, community and environment from harms. In our imperfect world, however, practical realities demand prioritization in the allocation of resources—human, temporal and financial—to regulate and manage only a portion of the spectrum of potential hazards. No nation has the resources to fully regulate everything for every risk, so a system of prioritization must be adopted everywhere. (McHughen, 2016)

The Animal and Plant Health Inspection Service (APHIS) branch of USDA, through its Biotechnology Regulatory Service office, administers regulatory authority under the Plant Protection Act of 2000. The Act has laid out strict lab procedures to prevent modified organisms from contaminating natural ecosystems. Many individuals seem to agree that GM organisms as food need to be clearly identified and labelled. As research and bioprocessing of products continues, both scientists and society need to deliberate about the ethical ramifications of the possibilities opened by scientific research. According to McLean (2000) and Glenn (2013), the science of ethics requires that humans justify their actions and account for their intentions. Reasons must be given why we do what we do, and our reasoning needs to address three main areas: 1) Incentives, or the ways that we encourage scientists to do particular kinds of research; 2) Intentions, or the goals of that research; and 3) Actions, or the potential applications of research results. (McLean, 2000)

Ethical Implications

Morality is characterized as the "norms about right and wrong in human conduct" (Beauchamp & Childress, 2013, p.3). We rightly judge all human conduct by the standards of morality (autonomy, nonmaleficence, beneficence, and justice) applicable to all persons in all places. For example, situations like: humans know not to lie, not to steal property, to keep promises, to respect the rights of others, not to kill or not to cause harm to innocent persons, and so on. (Christen et al., 2014). According to Hahm and Lee (2012), biomedical ethics (or bioethics and medical ethics) is an interdisciplinary study of the issues that result from advances in medical practices and research. It deals with issues that are [...] and controversial and new to society. As a result, society must provide solutions or judgments [...] effective and applicable—which has allowed the field to develop and progress through the con[...] efforts of experts from diverse backgrounds such as medicine, bioscience, ethics, theology, legal [...] and public policy. As noted by Hahm and Lee, biomedical ethics is a multi-dimensional way [...] communication where there are conflicting perspectives of the liberal and the conservative, ethical [...] and policy makers, and even between disciplines. The overarching goal is not to defeat the [...] with political tactics, but a collective way of reaching common understanding and solutions [...] problem. (Hahm & Lee, 2012)

According to McLean (2000) biotechnology ethics should focus on several primary questions:

- What benefits and what harms can be predicted for biotech innovations in both the research and application phases; and which courses of action will result in the best consequences overall?
- Who are the ethically relevant stakeholders, and what rights do they have? Which course of action protects those rights? Is human dignity respected?
- Which option treats everyone the same unless there is an ethically justified reason to treat them differently?
- Which course of action seeks the common good?
- Which option best develops virtues, and which virtues (such as trust and compassion), might be particularly relevant to biotech development and human health?

Other individuals might have concerns arising from gene modifications in procedures like gene [therapy], such as:

- Who has the right to examine someone else's genes?
- How can "good" and "bad" uses of gene therapy be distinguished?
- Who decides which traits are normal and which constitute a disability or disorder?
- Will the high costs of gene therapy make it available only to the wealthy?
- Would the widespread use of gene therapy make society less accepting of people who are different?
- Should people be allowed to use gene therapy to enhance basic human traits such as height, intelligence, or athletic ability?
- If transgenic or chimeric organisms (especially those with human genes) show some degrees on intelligence, do they have rights of protection?
- Will transgenic organisms be treated like regular (unmodified) organisms? Why are people opposed to GMOs?

CHAPTER SUMMARY

[Adve]nt of recombinant DNA technology has offered new opportunities for innovations to produce a [ran]ge of bioproducts in food and agriculture, health and disease, and environment. Some of these [...] include biopharmaceuticals, vitamins, organic acids, amino acids, and many other compounds. [Recomb]inant DNA technology uses DNA fragments with the desirable gene sequence inserted in vectors [like bact]eria, plasmids or retroviruses. Recombination in microorganisms occurs through three main parasexual processes: conjugation, transduction, and transformation. Recombinant DNA forms the basis of genetic engineering where genes are isolated, modified and inserted into organisms.

Restriction endonucleases or REases are the cleaving tools of molecular biology, genetics and biotechnology, and made analysis at the most fundamental levels routine. There are more than 19,000 REases available of four main types: Type I, II, III and IV which require a divalent metal cofactor like Zn^{2+} or Mg^{2+} for activity. There are two types of cloning vectors: non-viral vectors like plasmids, and viral vectors like adenoviruses and retroviruses.

Editing of genomes is natural for bacterial cells where they capture bits of genetic material from viral invaders and use it to create short segments of DNA called CRISPR arrays. These arrays allow the bacteria to "remember" previously encountered viruses. The process of chopping the viral particles I accomplished in bacteria by using an enzyme such as Cas9. This is the basis of the new gene editing tool called CRISPR Cas9. Researchers using the CRISPR Cas9 technique create a short "guide" RNA sequence that binds a Cas9 enzyme. CRISPR-Cas9 system has successfully been used to make direct changes to a DNA strand such as inserting or deleting specific sequences that code for proteins. Researchers are exploring the CRISPR Cas9 gene editing system to learn its applicability in single-gene defects and more complex diseases like schizophrenia, cancer, and heart disease.

A DNA library is an assortment of DNA fragments that have carefully been cloned into vectors like bacteria or viruses. These synthetic DNA libraries can store the entire genome of an organism (like a human being) to allow the sequences useful in the production of antibodies, enzymes, various other proteins, and genomes to be more thoroughly examined. DNA libraries are of three types: Genomic DNA, complimentary DNA, and randomized mutant DNA.

In applied sciences such as clinical medicine, biotechnology, forensics, molecular and evolutionary biology, sequencing DNA has become an important tool. The original sequencing technology called Sanger sequencing was very useful for sequencing short stretches of DNA but can be time consuming for sequencing a whole organism's genome. Sequencing technology has been advancing since the human genome was first successfully sequenced on 2003. Today, next-generation sequencing is emerging as a feasible method to sequence an organism's genome in short period. NGS has allowed the sequencing of whole genomes as seen in whole genome sequencing and whole exome sequencing.

Several techniques are available in sequencing: DNA microarrays where specific DNA sequences are synthesized in a 2-D or 3-D arrays on a glass microscope slide. DNA microarrays are used to measure gene expression levels by simultaneously measuring the relative concentrations of many different DNA or RNA sequences. However, they have several disadvantages by only detecting sequences that the array is designed to detect; providing only indirect measure of relative gene concentrations through expression; and lack of precision in binding specific sequences in closely related organisms. In recent years, the emergence of next generation sequencing technologies combined with the rapid decrease in the cost of sequencing has now made sequencing for almost all assays cost competitive compared to microarrays.

Though requiring elaborate infrastructure and personnel next-generation sequencing allows simultaneous sequencing of millions of DNA fragments without previous sequence knowledge such that an entire human genome can be sequenced within a single day. Whole exome sequencing refers to the sequencing of all the exons in a person's genome. Whole exome sequencing is an ideal procedure for identifying mutations which affect proteins coded by genes, thereby resulting in genetic defects.

Using short nucleotides from scratch has made it possible to synthesize gene sequences and even whole genomes. The nested polymerase chain reaction is based on the ability of the DNA Polymerase to

ies of DNA given all the necessary ingredients of DNA synthesis: primers with a 3' OH end, strand, free nucleotides under temperature regulated conditions. During this process millions d DNA can be obtained from the original DNA. Newer real-time polymerase chain reaction y is more attractive as it allows access to amplicon copies in real time after each cycle.

n regions of DNA show variations in the number of tandem repeats known as variable number repeats consisting of repeated 10 to 60 DNA base pair units scattered throughout the human Cleaving DNA with REases, produces a set of DNA sequence fragments of varying length restriction length fragment polymorphisms which through the process of gel electrophoresis a very specific pattern of migration through the gel called a DNA fingerprint.

udy how DNA bases determine whole sets of genes and how they function together is a new tudy called genomics. Genomics is dedicated to the large-scale analysis of the properties of Since it is the proteins, not genes, that are responsible for the phenotypes of cells, a new field ocusing on the proteins translated from genes is known as proteomics has emerged. The aim of s is not only to identify all the proteins in a cell but also to create a complete three-dimensional p of the cell indicating where proteins are located.

therapy is a technique used to correct single gene disorders where a cloned normal gene (e.g. ne) is separated and inserted in to a cloning vector like a retrovirus. After uptake, these engineered njected into the patient and when they divide, more copies of the gene are made, and the gene sed—eliminating the impacts of the disorder in the patient! There are primarily three reasons gene therapy including replacing a mutated gene, inactivating or "knocking out" a mutated introducing a new gene into the body to help fight a disease. Several studies have shown that apy can have very serious health risks, such as toxicity, inflammation, and triggering cancer.

chnology is a utilization of knowledge, techniques and procedures rooted in a variety of dis- including biochemistry, molecular biology, genetics, microbiology, and bioinformatics. The f biotechnology involves working with living cells or cell products and finding ways in which these products can be used commercially by human in medicine, agriculture, environment, and industry.

chnology is recognized universally as one of the key enabling technologies for the 21st century. , this scientific innovation has called for oversight and government regulation in ways that makes ation and products safe for human use and operating within human ethical and social guidelines. ary goal of regulation is to protect our society, community and environment from harms. As and bioprocessing of products continues, both scientists and society need to deliberate about al ramifications of the possibilities opened by scientific research.

Chapter Quiz

ich are the three main parasexual processes that occur in recombination in microorganisms?
  Transcription, translation, transformation
  Conjugation, transduction, transformation
  Conjugation, transduction, translation
  Conjugation, transcription, translation
  Transcription, transduction, transformation

2. Why are viruses appealing tools for therapeutic gene transfer?
   a. They are small and easy to manipulate
   b. They are not live
   c. They have high transfection/transduction efficiency in a wide range of human cells
   d. They are contagious and have high transmission rates
   e. None of the above
3. Type I, II, III and IV are the four main types of restriction endonucleases (REases) and require which of the following for activity?
   a. Short segments of DNA called CRISPR arrays
   b. Divalent metal cofactor like Zn^{2+} or Mg^{2+}
   c. Non-viral vector like plasmids
   d. Viral vector like adenoviruses and retroviruses
   e. C and D
4. What is the most frequent cause of diagnostic false-positive results in PCR amplification?
   a. End-product recombination
   b. Plateau of the PCR process
   c. Temperature regulated conditions
   d. Amplicon contamination
   e. Mechanical error
5. What is/are the primary reason(s) for using gene therapy?
   a. Replacing a mutated gene that causes disease with a healthy copy of the gene
   b. Inactivating a mutated gene that is not functioning
   c. Introducing a new gene into the body to help fight a disease
   d. All of the above
   e. A & B
6. Which is a disadvantage of microarrays?
   a. Array's provide an indirect measure of relative concentration
   b. It is difficult to ensure binding of DNA to the solid surface
   c. They can only detect sequences that the array is designed to detect
   d. A & D
   e. None of the above
7. What is produced once DNA is cleaved by REases?
   a. Cas9 endonuclease
   b. ACTG
   c. RFLPs
   d. CRISPR
   e. REase-DNA complex
8. To prepare cDNA from mRNA, which enzyme is needed to reproduce a copy of DNA from mRNA?
   a. Reverse transcriptase
   b. DNA polymerase
   c. Reverse ligase
   d. RNA transcriptase
   e. RNA polymerase

What are the three types of DNA libraries?
- Genomic DNA, complementary DNA, randomized mutant DNA
- Sequencing DNA, cloning DNA, translation DNA
- Genomic DNA, proteomics DNA, randomized DNA
- Chromosomal DNA, complementary DNA, reverse DNA
- None of the above

Bacteria that have been engineered to clean up toxic wastes are called
- superbugs
- bioremediators
- biobacteria
- bacteria helpers
- ecobacteria

Thought Questions

Name two advantages of synthetic DNA libraries that can store the entire genome of an organism?

What are the three primary reasons for using gene therapy? How can gene therapy be used for a patient with a genetic disorder such as Haemophilia (a disease caused by insufficient blood-clotting factors)?

You are the CEO of a big oil company and you find yourself having to deal with an accidental oil spill in the middle of the Pacific Ocean. How can bioremediators help you clean the oil spill?

List a few ways biotechnology is incorporated in your daily lives.

Why is it important to regulate biotechnology safety and consider the ethical issues?

REFERENCES

. (2016). *Bacteria made to turn sewage into clean water – and electricity.* Retrieved from https://newscientist.com/article/mg23130840-100-bacteria-made-to-turn-sewage-into-clean-water-and-ity/

. L., & Demain, A. L. (2010). Recombinant organisms for production of industrial products. *neered Bugs*, *1*(2), 116–131. doi:10.4161/bbug.1.2.10484 PMID:21326937

, U., Tsurupe, G., Segwagwe, A., & Obopile, M. (2014). Development and application of modern ural biotechnology in Botswana: The potentials, opportunities and challenges. *GM Crops and iotechnology in Agriculture and the Food Chain*, *5*(3), 183–194. doi:10.4161/21645698.2014. PMID:25437237

Peregrino, M., Sainson, R. C. A., Carlisle, R. C., Thoma, C., Waters, R. A., Arvanitis, C., ... r, L. W. (2013). Combining virotherapy and angiotherapy for the treatment of breast cancer. *Gene Therapy*, *20*(8), 461–468. doi:10.1038/cgt.2013.41 PMID:23846253

amp, T., & Childress, J. (2013). *Principles of Biomedical Ethics.* New York: Oxford University

Behjati, S., & Tarpey, P. S. (2013). What is next generation sequencing? *Archives of Disease in Childhood - Education and Practice, 98*(6), 236–238. doi:10.1136/archdischild-2013-304340 PMID:23986538

Bell, A. S., & Ranford-Cartwright, L. C. (2002). Real-time quantitative PCR in parasitology. *Trends in Parasitology, 18*(8), 337–342. doi:10.1016/S1471-4922(02)02331-0 PMID:12380021

Bourzac, K. (2017, September 27). Gene Therapy: Erasing sickle-cell disease. *Nature, 549*(7673), S28–S30. doi:10.1038/549S28a PMID:28953858

Bradley, L. H. (2014). High-quality combinatorial protein libraries using the binary patterning approach. *Methods in Molecular Biology (Clifton, N.J.), 1216*, 117–128. doi:10.1007/978-1-4939-1486-9_6 PMID:25213413

Bull, A. T., Marrs, B. L., & Kurane, R. (1998). *Biotechnology for clean industrial products and processes. Towards industrial sustainability.* Paris, France: Organisation for Economic Cooperation and Development.

Bull, A. T., Ward, A. C., & Goodfellow, M. (2000). Search and Discovery Strategies for Biotechnology: The Paradigm Shift. *Microbiology and Molecular Biology Reviews, 64*(3), 573–606. doi:10.1128/MMBR.64.3.573-606.2000 PMID:10974127

Bumgarner, R. (2013). Overview of DNA microarrays: Types, Applications and their future. *Current Protocols in Molecular Biology, 101*(1). doi:10.1002/0471142727.mb2201s101 PMID:23288464

Burbelo, P. D., Ching, K. H., Han, B. L., Klimavicz, C. M., & Iadarola, M. J. (2010). Synthetic biology for translational research. *American Journal of Translational Research, 2*(4), 381–389. PMID:20733948

Chira, S., Jackson, C. S., Oprea, I., Ozturk, F., Pepper, M. S., Diaconu, I., ... Berindan-Neagoe, I. (2015). Progresses towards safe and efficient gene therapy vectors. *Oncotarget, 6*(31), 30675–30703. doi:10.18632/oncotarget.5169 PMID:26362400

Christen, M., Ineichen, C., & Tanner, C. (2014). How "moral" are the principles of biomedical ethics? – a cross-domain evaluation of the common morality hypothesis. *BMC Medical Ethics, 15*(1), 47. doi:10.1186/1472-6939-15-47 PMID:24938295

Chrystoja, C. C. & Diamandis, E. P. (2013). Whole genome sequencing as a diagnostic test: challenges and opportunities. *Clinical Chemistry, 60*(5), 724–33. doi:. 2013.209213 doi:10.1373/clinchem

Conesa, A., & Mortazavi, A. (2014). The common ground of genomics and systems biology. *BMC Systems Biology, 8*(2Suppl 2), S1. doi:10.1186/1752-0509-8-S2-S1 PMID:25033072

Cyranoski, D. (2016). CRISPR gene-editing tested in a person for the first time. *Nature, 539*(7630), 479. doi:10.1038/nature.2016.20988 PMID:27882996

DeGrazia, D. (2003). Common morality, coherence, and the principles of biomedical ethics. *Kennedy Institute of Ethics Journal, 13*(3), 219–230. doi:10.1353/ken.2003.0020 PMID:14577458

Dragosits, M., & Mattanovich, D. (2013). Adaptive laboratory evolution – principles and applications for biotechnology. *Microbial Cell Factories, 12*(1), 64. doi:10.1186/1475-2859-12-64 PMID:23815749

B., Nelson, J. E., & Winters, P. (2012). Perspective on opportunities in industrial biotechnology ble chemicals. *Biotechnology Journal, 7*(2), 176–185. doi:10.1002/biot.201100069

A. J. (2004). *Techniques of crime scene investigation* (7th ed.). New York, NY: CRC Press.

Agricultural Organization (FAO). (2018). The State of Food and Agriculture 2003-2004. *What tural biotechnology?* Retrieved from http://www.fao.org /docrep/006/ Y5160E/y5160e07.htm

R., Stark, D., Beebe, N., Marriott, D., Ellis, J., & Harkness, J. (2007). Laboratory diagnos-ques for Entamoeba species. *Clinical Microbiology Reviews, 20*(3), 511–532. doi:10.1128/04-07 PMID:17630338

pt. (2018). *Mutant Libraries (Protein Engineering)*. Retrieved from https://www. genscript. etic_library.html

. G., Benders, G. A., Andrews-Pfannkoch, C., Denisova, E. A., Baden-Tillson, H., Zaveri, J., ... O. (2008). Complete chemical synthesis, assembly, and cloning of a Mycoplasma genitalium *Science, 319*(5867), 1215–1220. doi:10.1126cience.1151721 PMID:18218864

., Alexander, I. E., Edelstein, M. L., Abedi, M. R., & Wixon, J. (2013). Gene therapy clini-worldwide to 2012 - an update. *The Journal of Gene Medicine, 15*(2), 65–77. doi:10.1002/ PMID:23355455

M. (2013). *Ethical Issues in Genetic Engineering and Transgenics.* American Institute of Sciences. Retrieved from http://www.actionbioscience.org/ biotechnology/ glenn.html

R., & Haystead, T. A. J. (2002). Molecular Biologist's Guide to Proteomics. *Microbiology cular Biology Reviews, 66*(1), 39–63. doi:10.1128/MMBR.66.1.39-63.2002 PMID:11875127

H., & Lee, I. (2012). Biomedical Ethics Policy in Korea: Characteristics and Historical De-t. *Journal of Korean Medical Science, 27*(Suppl), S76–S81. doi:10.3346/jkms.2012.27.S.S76 661876

E., & Snyder, M. (2002). ChIP-chip: A genomic approach for identifying transcription factor bind-*Methods in Enzymology, 350*, 469–483. doi:10.1016/S0076-6879(02)50979-4 PMID:12073330

A. J., Wilson, V., & Thein, S. L. (1985). Hypervariable "minisatellite" regions in human DNA. *14*(6006), 67–73. doi:10.1038/314067a0 PMID:3856104

A. J., Wilson, V., & Thein, S. L. (1985, July). Individual-specific "fingerprints" of human DNA. *16*(6023), 76–79. doi:10.1038/316076a0 PMID:2989708

F. X. (2017). Biofuels, Bioenergy and the Bioeconomy in North and South. *Industrial Bio-y (New Rochelle, N.Y.), 13*(6), 289–291. doi:10.1089/ind.2017.29106.fxj PMID:29282380

R., Brandão, R. D., van den Bosch, B. J., Paulussen, A. D. C., Xanthoulea, S., Blok, M. J., & A. (2017). Next-Generation Sequencing in Oncology: Genetic Diagnosis, Risk Prediction and lassification. *International Journal of Molecular Sciences, 18*(2), 308. doi:10.3390/ijms18020308

K., Hoban, T., & Balasubramanian, R. (n.d.). *Biotechnology and its Applications.* Raleigh, NC: h Carolina Cooperative Extension Service, North Carolina State University.

Khan, S., Ullah, M. W., Siddique, R., Nabi, G., Manan, S., Yousaf, M., & Hou, H. (2016). Role of Recombinant DNA Technology to Improve Life. *International Journal of Genomics*, *14*(2405954). doi:10.1155/2016/2405954 PMID:28053975

Komor, A. C., Badran, A. H., & Liu, D. R. (n.d.). CRISPR-Based Technologies for the Manipulation of Eukaryotic Genomes. *Cell*, *169*(3), 559. doi:10.1016/j.cell.2017.04.005

Lander, E. S. (2016). The Heroes of CRISPR. *Cell*, *164*(1-2), 18-28. doi:.2015.12.041 doi:10.1016/j.cell

Ledford, H. (2018). *First test of in-body gene editing shows promise*. Retrieved from https://www.nature.com/articles/d41586-018-06195-6

Loenen, W. A. M., Dryden, D. T. F., Raleigh, E. A., Wilson, G. G., & Murray, N. E. (2013). Highlights of the DNA cutters: A short history of the restriction enzymes. *Nucleic Acids Research*, *42*(1), 3–19. doi:10.1093/nar/gkt990 PMID:24141096

McHughen, A. (2016). A critical assessment of regulatory triggers for products of biotechnology: Product vs. process. *GM Crops and Food: Biotechnology in Agriculture and the Food Chain*, *7*(3-4), 125–158. doi:10.1080/21645698.2016.1228516 PMID:27813691

McLean, R. M. (2000). *Thinking Ethically About Human Biotechnology*. Retrieved from https://www.scu.edu/ethics/focus-areas/bioethics/resources/thinking-ethically-about-hum an-biotechnology/

Nakagawa, H., & Fujita, M. (2018). Whole genome sequencing analysis for cancer genomics and precision medicine. *Cancer Science*, *109*(3), 513–522. doi:10.1111/cas.13505 PMID:29345757

Nybom, H., Weising, K., & Rotter, B. (2014). DNA fingerprinting in botany: Past, present, future. *Investigative Genetics*, *5*(1), 1. doi:10.1186/2041-2223-5-1 PMID:24386986

Okayama, H., & Berg, P. (1982). High-efficiency cloning of full-length cDNA. *Molecular and Cellular Biology*, *2*(2), 161–170. doi:10.1128/MCB.2.2.161 PMID:6287227

Papadakis, E. D., Nicklin, S. A., Baker, A. H., & White, S. J. (2004). Promoters and control elements: Designing expression cassettes for gene therapy. *Current Gene Therapy*, *4*(1), 89–113. doi:10.2174/1566523044578077 PMID:15032617

Papin, J. F., Verardi, P. H., Jones, L. A., Monge-Navarro, F., Brault, A. C., & Holbrook, M. R., ... Yilma, T. D. (2011). Recombinant Rift Valley fever vaccines induce protective levels of antibody in baboons and resistance to lethal challenge in mice. *Proceedings of the National Academy of Sciences of the United States of America*, *108*(36), 14926–14931. 10.1073/pnas.1112149108

Peterson, J. L., Ryan, J. P., Houlden, P. J., & Mihajlovic, S. (1987). The uses and effects of forensic science in the adjudication of felony cases. *Journal of Forensic Sciences*, *32*(6), 1730–1753. doi:10.1520/JFS11231J

Pingoud, A., & Jeltsch, A. (2001, September 15). Structure and function of type II restriction endonucleases. *Nucleic Acids Research*, *29*(18), 3705–3727. doi:10.1093/nar/29.18.3705 PMID:11557805

Pingoud, A., Wilson, G. G., & Wende, W. (2014). Type II restriction endonucleases—A historical perspective and more. *Nucleic Acids Research*, *42*(12), 7489–7527. doi:10.1093/nar/gku447 PMID:24878924

P., Pumpens, P., & Grens, E. (2013). Development of virus-like particle technology from ghly symmetric to large complex virus-like particle structures. *Intervirology*, *56*(3), 141–165. 159/000346773 PMID:23594863

., & Dogra, T. D. (2002). Application of DNA fingerprinting in medicolegal practice. *Journal dian Medical Association*, *100*(12), 688–694. PMID:12793630

, R. J., Vincze, T., Posfai, J., & Macelis, D. (2010). REBASE—a database for DNA restriction ification: Enzymes, genes and genomes. *Nucleic Acids Research*, *38*(suppl_1), D234–D236. 093/nar/gkp874 PMID:19846593

. W., Skerker, J. M., Maurer, M. J., Li, X., Tsai, J. C., Poddar, S., ... Cate, J. H. D. (2014). Selec- hromosomal DNA libraries using a multiplex CRISPR system. *eLife*, *3*, e03703. doi:10.7554/ 703 PMID:25139909

, H. K., Hwu, K., & Chang, M. (2010). Advances in detection of genetically engineered crops plex polymerase chain reaction methods. *Trends in Food Science & Technology*, *21*(9), 442–452. 016/j.tifs.2010.06.004

, M. (2018). Target, delete, repair. CRISPR is a revolutionary gene-editing tool, but it's not risk. *Stanford Medicine, Winter 2018*. Retrieved from https://stanmed. stanford.edu/2018winter/ R-for-gene-editing-is-revolutionary-but-it-comes-with-risks.html

g, F. M., & Raso, J. (1998). Biotech pharmaceuticals and biotherapy: An overview. *Journal of cy & Pharmaceutical Sciences*, *1*(2), 48–59. PMID:10945918

Fisher Scientific. (n.d.). *Genomic DNA and cDNA Libraries*. Retrieved from https://www.ther- r.com/us/en/home/life-science/cloning/cloning-applications/ library-construction.html

Ma, K., & Saaem, I. (2009). Advancing high-throughput gene synthesis technology. *Molecular ems*, *5*(7), 714–722. doi:10.1039/b822268c PMID:19562110

D. B., Hershenson, S., Ho, R. J. Y., Uchiyama, S., Winter, G., & Carpenter, J. F. (2015). Two s of Publishing Excellence in Pharmaceutical Biotechnology. *Journal of Pharmaceutical Sci- 04*(2), 290–300. doi:10.1002/jps.24285 PMID:25448882

, M. R., Sanchez, J. C., Gooley, A. A., Appel, R. D., Humphery-Smith, I., Hochstrasser, D. F., ams, K. L. (1995). Progress with proteome projects: Why all proteins expressed by a genome e identified and how to do it. *Biotechnology & Genetic Engineering Reviews*, *13*(1), 19–50. do 0/02648725.1996.10647923 PMID:8948108

D. M., Harcum, S. W., & Lee, K. H. (2012). Genomics in mammalian cell culture bioprocessing. *nology Advances*, *30*(3), 629–638. doi:.biotechadv.2011.10.010 doi:10.1016/j

A. (2013). Construction and characterization of virus-like particles: A review. *Molecular Bio- gy*, *53*(1), 92–107. doi:10.100712033-012-9598-4 PMID:23001867

KEY TERMS AND DEFINITIONS

Bioeconomy: An economy in which biological processes and biological based products are used to reduce environmental impacts and improve overall global sustainability.

Bioprocessing: The process and application of biotechnology-based tools as a means for commercial use.

Bioremediators: Biological organisms such as bacteria and fungi—microorganisms that are used to reduce the amount of environmental pollutants.

Biotechnology: The utilization and application of knowledge, techniques, and procedures from a variety of biology-related disciplines.

Complementary DNA (cDNA): DNA synthesized from mature mRNA transcripts that can be used to analyze the structure, organization, and expression of eukaryotic genes.

CRISPR Arrays: Short segments of DNA that allow bacteria to remember and recognize when an infectious agent invades again.

DNA Fingerprint: A unique pattern of repeated sequences, or RFLPs, obtained via gel electrophoresis, that can be used to identify an individual's DNA.

DNA Library: An assortment of DNA fragments that have carefully been cloned into vectors.

DNA Microarrays (Nucleic Acid Arrays): Specific DNA sequences that are deposited and bound to a solid surface such as a glass microscope slide.

Forensic Evidence: Evidence that is obtained by scientific methods and is often used in criminal and legal proceedings.

Gel Electrophoresis: A method used in the lab to separate DNA fragments based on their size.

Gene Therapy: The replacement of defective genes with functional genes or the introduction of new genes to prevent or treat diseases.

Genetic Engineering: The process of altering the characteristics of an organism by insertion of modified genes.

Genomics: A study of how bases on DNA determine whole sets of genes and how they function together.

Guide RNA: An RNA sequence used in CRISPR Cas9 technology to direct enzymes to specific DNA sequences.

Industrial Biotechnology: The implementation of biotechnology-based tools into traditional industrial processes as a means to improve and sustain global sustainability.

Next Generation Sequencing (NGS): A sequencing method used to measure large numbers of individual DNA sequences in a short period of time.

Pharmacogenomics: A sector within biomedicine and biotechnology that aims to revolutionize modern medicine by advancing pharmaceutical development in vaccines, therapies, and biological molecules.

Plasmids: Small circular pieces of double stranded DNA that can replicate independently of the host's chromosomal DNA, usually found in bacteria.

Polymerase Chain Reaction (PCR): A technique used to make copies of a target piece of DNA.

Protein-Based Therapeutics: A type of biotechnology-based product that aims to prevent and treat a variety of medical issues that have yet to be addressed.

Proteomics: A study and characterization of proteins expressed by a genome at one given time in a cell.

...Time Polymerase Chain Reaction (RT-PCR): A newer, faster technique of PCR amplification ...not require post-PCR analysis.

...nbinant DNA Technology: Biomedical technology that focuses on altering genetic material ...organism as a means to enhancing and improving characteristics inside living organisms or ...ucts.

...ction Endonucleases (REases): DNA-cleaving enzymes that biologically function to protect ...enome against foreign DNA.

...ction Length Fragment Polymorphisms (RFLPs): A set of DNA sequence fragments of ...ngth.

...viruses: A group of RNA viruses that replicate by producing a DNA copy of their genome ...s it into a host cell.

...se Transcriptase: An enzyme that produces a copy of DNA from mRNA.

...r Sequencing: The original sequencing technology used to sequence short stretches of DNA.

...genics: The process of isolating genetic material from one organism and introducing it into ...rganism's genome.

...ble Number of Tandem Repeats (VNTRs or Microsatellite DNA): Specific regions of DNA ...nstration variations in the number of tandem repeats.

...e Exome Sequencing: A sequencing method used to determine all the exons in a genome of ...sm.

...e Genome Sequencing: A sequencing method used to determine whole genomes of an organism.

Section 3
Genes, Disorders, and Human Health

PROLOGUE

In this section of the book, we will "connect the dots" from the knowledge and concepts gained in Section 1 and Section 2 to human genetic disorders. Some of the genetic disorders discussed are fairly common, while others are somewhat rare. The Genetic Disease Foundation estimates that there are over 6,000 genetic disorders that can be passed down across many generations of families, many of which are fatal or severely debilitating (GDF, 2018). Since we cannot be able to cover all these conditions, we will focus on about 85 disorders in the next several chapters covering most human physiological systems.

We start with a chapter on epigenetics (11) and genomics/genetic testing methodology (12); then move on to disorders of the human organ systems. These start with circulatory and respiratory systems (13), cancers (14), human digestive, ENT, and eye disorders (15), endocrine and immune systems (16), muscle, connective and neonatal disorders (17), and concludes with human nervous system disorders (18).

The last chapter examines the roles, challenges and opportunities of public health and public policy in the genetic information age.

Introduction to Genes and Disease

When genes become dysfunctional and make products (proteins) that are faulty, we talk of a gene mutation (see section 8.5 above for a detailed review of mutation types). In most cases, mutations cause genetic disorders that manifest as conditions, syndromes, or diseases because the specific metabolic pathway where the gene product is involved cannot be completed. Genes by themselves are not responsible for a disorder—these are caused by gene products not functioning properly.

Nevertheless, mutations are essential for generating variation in living organisms. How does this happen?

Mutations produce altered (new) genes that can give rise to products that allow organisms to become better adapted to an ever-changing environment. For example, when bacteria develop resistance to antibiotics, it is the development of new mutations after the administration of an antibiotic that allows some bacteria cells to survive and multiply. One way bacteria develop resistance is by using a small piece of

DNA existing separately from the bacterial genome (called a plasmid) which encodes resistance
to antibiotic. Plasmids are circular, double-stranded DNA molecules occurring naturally in bacteria,
and some higher eukaryotic cells. They exist in a parasitic or symbiotic relationship with their
host. Whenever the host cell divides the plasmid DNA is also replicated therefore propagating it to
generations of the host cells. Often, plasmids contain genes that encode enzymes that deactivate
antibiotics, conferring benefits to the host cell. Another way is through bacterial conjugation where
a bacterium with an advantageous mutated gene shares its DNA with another individual through a
bridge between the two cells.

Mutations can be inherited from a parent to a child (hereditary) or they can happen during a person's lifetime (acquired). Acquired mutations can be caused by environmental factors such as ultraviolet radiation from the sun. Acquired mutations that humans develop during their lifetime are in cells called somatic cells—the cells that make up most of the body. If a gene mutation is carried by the egg or sperm cells future generations can inherit the defective gene from their parents. Diseases can occur due to a mutation in a single gene or in a set of genes and as determined by the degree of mutation, they can be categorized into four main types:

Chromosomal diseases: These occur when the entire chromosome, or large segments of a chromosome, are missing, duplicated or otherwise altered. The best example is Down Syndrome where there is an entire copy of chromosome #21.

Single-gene disorders: These occur when an alteration occurs in a gene (monogenic) which now encodes for a non-functional protein. A good example is sickle-cell anaemia where a mutated form of the hemoglobin producing gene exists—resulting in some defective hemoglobin molecules which bind to oxygen very poorly. Their "sickle" shape tends to cause them to cluster as they travel in the blood system causing severe congestion which affecting breathing and metabolism.

Multifactorial disorders: The vast majority of medical conditions with a genetic link involve either the complex interaction of a number of genes or the complex interaction between genes and the environment. Good examples of multifactorial disorder are Type-2 diabetes, hypertension, heart disease, psychiatric illnesses (e.g. bipolar disorder), and some cancers

Mitochondrial disorders: These are often rare disorders caused by mutations in mitochondrial DNA located within the mitochondria. These are typically transmitted to offspring through the egg cell. Remember during fertilization the sperm only contributes the paternal DNA, while the egg contributes the maternal DNA and the cytoplasm of the fertilized egg (zygote).

Naming Genetic Conditions

Names given to genetic conditions described below are not in one standard way, they are often derived from one or a combination of sources.
Several criteria are used:

The basic genetic or biochemical defect that causes the condition (e.g., alpha-1 antitrypsin deficiency);
One or more major signs or symptoms of the disorder (e.g., hyper-manganesemia with dystonia, polycythemia, and cirrhosis);

- *The parts of the body affected by the condition (e.g., craniofacial-deafness-hand syndrome);*
- *The name of a physician or researcher, often the first person to describe the disorder (e.g., Marfan syndrome, named after Dr. Antoine Bernard-Jean Marfan);*
- *A geographic area (e.g., familial Mediterranean fever, which occurs mainly in populations bordering the Mediterranean Sea);*
- *The name of a patient or family with the condition (e.g. amyotrophic lateral sclerosis also called Lou Gehrig disease after the famous baseball player who had the condition).*

Chapter 11
Epigenetics

ABSTRACT

This chapter focuses on epigenetics: the study of stable, often heritable changes that influence gene expression but are not mediated by DNA sequence. These changes play crucial roles in chromatin state structure which influences processes such as gene expression, DNA repair, and recombination. Evidence indicates that epigenetic patterns are altered by environmental factors which are associated with disease risk including diet, smoking, alcohol intake, environmental toxicants, and stress. Studiers have linked environmental pollutants with epigenetic variations particularly changes in DNA methylation, histone modifications, and microRNAs. Growing data have linked epigenetic alterations with heavy metals, persistent organic toxicants, and water chlorination by-products. Studies focusing on the effects of air pollution in humans demonstrate an association between exposure to air pollution and DNA methylation. Several classes of pesticides can modify epigenetic marks, including endocrine disruptors, persistent organic pollutants, arsenic, several herbicides, and insecticides. This chapter explores epigenetics.

CHAPTER OUTLINE

- Epigenome and Epigenomics
- Epigenetic Mechanisms
  - 11.2.1 Regulation at the DNA Level
  - 11.2.3 Regulation at the Transcription Level
  - 11.2.4 Regulation at Post-Transcription Level
  - 11.2.5 Regulation by non-coding RNAs
- Epigenetics, Environment and Disease
  - 11.3.1 Physical Activity, Cancer, Diet and Chronic Diseases
  - 11.3.2 Heavy Metals
  - 11.3.3 Organic Toxicants
  - 11.3.4 Air Pollution
  - 11.3.5 Pesticides
- Summary

DOI: 10.4018/978-1-5225-8066-9.ch011

LEARNING OUTCOMES

- Define epigenetics and epigenomics
- Explain the various epigenetic mechanisms
- Summarize the effects of lifestyle habits and environmental pollutants in gene expression and disease.

11.1 THE EPIGENOME

Epigenetics is the study of stable, often heritable changes which influence gene expression but are not mediated by DNA sequence. These changes play crucial roles in chromatin state regulation which influences processes such as gene expression, DNA repair, and recombination (Berger et al., 2009). The most distinctive aspects of epigenetics are that these changes in expression occur without a change in the original sequence of the DNA which is the arrangements of the adenine, cytosine, guanine, and thymine (ACGT) bases that constitute specific genes (Robinson, 2016). The term "epigenomics" was coined to describe all of the chemical modifications that are added to the genome to regulate gene expression and activity. Epigenetic changes are inheritable, modifiable, or erasable in response to developmental cues or external and environmental stimuli. A better understanding of the epigenomic factors that contribute to many disease processes will lead to additional strategies for treatment in the future. For example, defects in epigenetic regulation have been linked to developmental defects, metabolic disorders and cancer in humans. The connection between the epigenome and more common complex diseases including psychosis, diabetes and asthma are also being discovered. (Murrell et al., 2005; Berger et al., 2009; Fingerman et al., 2013)

Epigenomics is the identification of all epigenetic modifications implicated in gene expression (Costa, 2010). It is increasingly becoming an important tool to better understand human biology in both normal and pathological states. Such changes occur on a genomic scale rather than a single gene. According to Toden and Goel (2014) it is essential to understand that genetics and epigenetics function interactively and cooperatively. The two processes can be equated to a computer system, in which genetics serves as the hardware, whereas epigenetics functions as computer software and provides readout of our genes and dictates whether our genes are expressed in a positive or negative manner. Epigenetic modifications are varied and diverse but can be grouped into four major classes: post-translational modification of histone proteins, chromatin conformation/accessibility, DNA modification, and non-coding regulatory RNA (Bernstein et al., 2007). The epigenome is the distribution of these epigenetic features throughout the genome and is (unlike the genome itself) dynamic and localization of epigenetic features is influenced by cell or tissue type, age, exposure to environmental stimuli or many other factors. Such factors complicate the definition of an organism's epigenome—but owing to the important roles epigenetic phenomena play in human health and development, the National Institutes of Health (NIH) in the United States launched the Roadmap Epigenomics Project in 2007 to combine whole genome epigenetic analysis with genetic sequencing (Fingerman et al., 2013). This has created a series of publicly available reference epigenome maps encompassing a wide array of cell lines, cell and tissue types from individuals at various developmental stages and health states. The National Centre for Biotechnology Information (NCBI) which is part

maintains the Epigenomics database as a repository for these data. Data have been collected from ...ge-scale studies such as the NIH Roadmap Epigenomics Project, ENCODE and modENCODE ...n smaller single laboratory studies. (Fingerman et al., 2011; Fingerman et al., 2013)

...PIGENETIC MECHANISMS

...genome undergoes extensive reprogramming events associated with embryogenesis and early ...ment during two key time-points in early gestation with the purpose of establishing cell- and ...ecific gene expression. The first period occurs at the primordial germ cells stage. In males, such ...ich eventually become sperm, methylation is initiated during gestation. In female primordial ...lls destined to become eggs, methylation occurs after birth in mature oocytes. The second pe- ...pigenetic reprogramming occurs at fertilization, which helps in the establishment of individual ...cific methylation patterns.

...e regulation in higher organisms can occur at various steps from chromatin (DNA) level to post-...onal level. Gene expression is comprised of six steps: 1) transcription; 2) mRNA processing; 3) ...transport and localization; 4) mRNA half-life; 5) translation; and, 6) protein activity. (Hajkova ...002; Jirtle & Skinner, 2007; Inose et al., 2015)

...will examine the main processes occurring at each of those levels in turn.

...gulation at the DNA level (histone modification processes *[methylation, acetylation, phosphoryla-...n, glycosylation, carbonylation, ubiquitylation, biotinylation, sumoylation, citrullination, ADP-...osylation, N-formylation, crotonylation, propionylation and butyrylation, plus also proline and ...partic acid isomerization]*, piRNAs, and epigenetic inheritance).

...gulation at transcription level (transcription factors, enhancers, activators and coactivators, ...ressors, and basal factors).

...gulation at post-transcription level (mRNA processing, mRNA degradation, translation initia-...n, protein processing and degradation).

...gulation by non-coding RNAs (microRNAs, siRNA, and RNAi).

...ulation at the DNA Level

...e Modification

...nslational modifications to histones are an integral component of epigenetic regulation. Histones ...n alkaline proteins composed of many amino acids with basic side chains (particularly lysine ...inine) (Grazioli et al., 2017). **Chromatin remodelling** is the modification of the configuration ...ne "spools," around which strands of DNA are wrapped, an architecture that allows DNA to be ...ssed into the nucleus of a cell. Chromatin structure is closely associated with the accessibility of ...transcriptional machinery, and hence gene activity (Sadakierska-Chudy & Filip, 2015). Histone ...nslational modifications regulate chromatin-templated processes through the modulation of chro-...ackaging and regulation of chromatin structure. Binding proteins specific to the post-translational ...ation are recruited and recognize modified (Kouzarides, 2007).

Chromatin is a very dynamic structure changing according to the regulatory mechanisms available that include histone-modifying, histone modification-recognizing, and histone modification-erasing proteins ("so-called writer, reader, and eraser proteins respectively") (Sadakierska-Chudy & Filip, 2015, p.172). Histone and non-histone proteins are modified in ways that affect the structure of chromatin and how accessible it becomes in transcription and DNA duplication. Post-translational modifications of the histone tails regulate two opposite processes, namely transcriptional activation and repression (Cohen et al., 2011). Chromatin remodelling thus, is another process through which changes in gene expression occurs without altering DNA gene sequences (see Figure 4, Chapter 9). Histone modifications consist of methylation, acetylation, crotonylation, ubiquitination, sumoylation, phosphorylation, hydroxylation and proline isomerization as discussed below. (Davie & Spencer, 1999; Houston et al., 2013; Kouzarides, 2007; Peterson & Laniel, 2004)

Methylation

Histone methylation involves transfer of a methyl group (CH_3) from a high-energy enzymatic donor to amino groups of lysine and arginine mainly on the H3 and H4 histone tails. Methylation can occur through exposure to specific ubiquitous chemicals such as bisphenol A or specific dietary components. In contrast to acetylation (below), methylation is more complex and can induce structural changes. The attachment of methyl groups is carried out by enzymes called arginine methyltransferases and the removal by a demethylase (Chang et al. 2007). Methylation can lead to activation or silencing of gene expression depending on its localization in the histone tails (Figure 1).

Acetylation/Deacetylation

Acetylation means adding the acetyl group ($COCH_3$) to histones that form the foundation onto which DNA is bound. Histone acetyltransferases catalyse the transfer of an acetyl group from a donor molecule like acetyl coenzyme A (acetyl-CoA), to a target lysine residue which forms part of a histone block. In eukaryotic systems, there are at least five classes of acetyltransferases. Proteomic studies show that hundreds of proteins are acetylated. (Crosby et al., 2010; Marmorstein & Roth, 2001; Roth et al., 2001). Protein acetylation regulates DNA processes, including gene expression, replication, repair, and recombination (Kouzarides, 2007). In eukaryotic cells, it is well established that acetylation regulates histones and transcription factors, modulating gene expression and other DNA transactions (Glozak et al., 2005). As the positively charged lysine residues can normally interact with the negatively charged phosphate groups on DNA, the presence of acetylation inhibits their DNA binding (Hu et al., 2013, Figure 1). Reversal of acetylation (deacetylation) occurs when acetyl groups are removed from lysine residues by two families of enzymes, Zn^+-dependent deacetylases or histone deacetylases and NAD^+-dependent sirtuins (Sadakierska-Chudy & Filip, 2015).

Histone methylation and acetylation mechanisms
Image used under license from Shutterstock.com

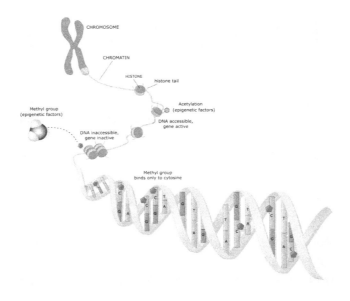

Phosphorylation

Phosphorylation is the addition of the phosphate (PO_4^{3-}) group to the histone tails of the amino acids serine, threonine, and tyrosine residues in four core histones (H2, H2A, H2B, and H3S), and H4S1 by a family of enzymes called kinases. Addition of the negatively charged PO_4^{3-} group can induce changes in the chromatin structure controlling several processes: mitosis, meiosis, DNA damage response, apoptosis, and gene expression (Cheung et al., 2000; Metzger et al., 2008).

Ubiquitylation (Ubiquitination)

Ub is a small protein of 76 amino acids. Ubiquitylation covalently attaches one or more ubiquitin proteins to a lysine residue (Zhang, 2003). Ubiquitylation is reversible, and the removal of Ub is achieved by enzymes called isopeptidases (Wilkinson, 2000). The role of ubiquitylation in transcription regulation (activation or repression) is still controversial because different studies provide contradictory findings. On the one hand, H2B ubiquitylation may participate in transcriptional activation (Zhang, 2003) and on the other hand, ubiquitylation of H2A at lysine is associated with transcriptional repression (et al., 2013).

Sumoylation

The **small ubiquitin-related modifier (SUMO)** is a polypeptide of less than 100 amino acids. SUMOylation is a lysine-targeted post-translational modification of a SUMO family of ubiquitin-like proteins which are covalently conjugated to target proteins altering their function (Henley et al., 2014). SUMO has been demonstrated to modify proteins throughout the cell. SUMO is joined to many cellular proteins, altering their interaction with other proteins regulating their function. More than 1,000 nucleoproteins undergo sumoylation (Hochstrasser 2009), and this pathway has been implicated in controlling many important processes, including regulation of the cell cycle, transcription, nucleocytoplasmic transport, DNA replication and repair, chromosome dynamics, and apoptosis, as well as ribosome biogenesis (Wang & Dasso 2009). Defective protein SUMOylation is implicated in a growing number of disorders of the nervous system.

Crotonylation

Tan et al. (2011) described lysine crotonylation as a post-translational modification of histones where the crotonyl group (C_4H_5O) is added onto the amino group of a lysine residue. Studies indicate that crotonylation is associated with active chromatin and is enriched at the promoters and enhancers of active genes in human somatic cells (Tan et al., 2011). The authors suggest that crotonylation affects chromatin structure and facilitates histone replacement which influences gene expression. Histone crotonylation was observed in kidney tissue, which might suggest it plays a role in epigenetic regulation of gene expression during kidney injury (Tan et al., 2011).

Proline Isomerization

The process where a compound is converted into another with the same molecular composition but with a different arrangement of atoms in three-dimensional space is called isomerization. Proline isomerization is one important way to achieve large conformational changes and reach various macrostates of multidomain proteins. Such changes resulting from proline switching from the trans state to cis state in a 180° flip are crucial in controlling protein activity in many biological processes including cell signalling, neurodegeneration, channel gating, and gene regulation. Different isomeric structures differ significantly in physical and chemical properties. It is still not well understood how isomerization of histone prolines contribute to transcriptional and epigenetic regulation. (Nelson et al., 2006; Monneau et al., 2013)

Aspartic Acid Isomerization

Aspartic acid isomerization is the post-translational modification found in histone H2B and was found to be an important factor in gene expression in mammalian brain and testis cells. This modification is emerging as an important factor in the development of autoimmune disease. (Doyle et al., 2013)

Hydroxylation

hydroxylation is a post-translational modification catalysed by enzymes called dioxygenases. Hydroxylation modification occurs on various amino acids, including proline, lysine, asparagine, and histidine (Zurlo et al., 2016). Hydroxylation can affect the enzymatic activity of certain and disrupting their interaction with direct activators or influencing the occurrence of other translational modifications that in turn affect their activity (Ploumakis & Coleman, 2015).

O-Glycosylation

enzymes, a transferase and a hydrolase are involved in this post-translational modification where than 500 nucleocytoplasmic proteins undergo glycosylation, binding on the hydroxyl group of threonine residues (Alfaro et al., 2012). Studies show that this binding of chromatin regulates option of genes especially those associated with metabolism and aging (Love et al., 2010).

Carbonylation

bound carbonyl groups are formed by direct oxidation of basic amino acid residues, including arginine and lysine. Protein carbonylation, an irreversible and non-enzymatic post-translational modification often used as a marker of oxidative stress. Carbonylation in histone proteins may mask the charges and thus affect the relaxation of chromatin and accumulation of transcription factors. changes affect gene expression (Sadakierska-Chudy & Filip, 2015). Due to oxidative modification, protein carbonylation has been associated with several age-related or metabolic diseases such as Alzheimer, Parkinson, Diabetes, Chronic lung disease (Weng et al., 2017).

Citrullination

Citrullination involves the conversion of peptidyl arginine to citrulline by the enzyme peptidyl-arginine deiminase (Bannister & Kouzarides, 2011). Citrullination may lead to transcriptional repression, but research is going on to elucidate the precise mechanism of action. Research indicates that protein citrullination, a calcium-driven post-translational modification, may play a causative role in the neurodegenerative disorders of Alzheimer's disease, multiple sclerosis (MS), and prion disease (Lazarus et al., 2015).

Biotinylation

Biotin is **vitamin B7** and acts as a cofactor for four carboxylases, which play essential roles in the metabolism of glucose, proteins, and fatty acids (Camporeale & Zempleni, 2006). Biotin is involved in gene regulation and chromatin structure when covalently attached to the amino group of lysine residues in histones (Zempleni et al., 2008). A growing body of evidence suggests that histone biotinylation plays important role in biological processes, including gene silencing, chromatin remodeling, cellular responses to DNA damage, genome stability, mitotic condensation of chromatin (Kothapalli & Zempleni, Filenko et al., 2011).

ADP-Ribosylation

ADP-ribosylation is a reversible covalent modification where the ADP-ribose group from **nicotinamide adenine dinucleotide** (NAD^+) is transferred to a specific amino acid of an acceptor protein like lysine, arginine, glutamate, aspartic acid, cysteine, asparagine, and phosphoserine (Hassa et al. 2006). The process is mediated by members of the ADP-ribosyltransferase family of enzymes. ADP-ribosyltransferases are primary sensors of DNA damage and catalyse the addition of ADP-ribose onto target proteins. Poly-ADP-ribosylation controls many crucial cellular activities including cell differentiation, transcription, chromatin modification, DNA damage detection and repair, apoptosis, and carcinogenesis (Masutani et al., 2005; Hassa et al., 2006; Rakhimova et al., 2017).

N-Formylation

N-formylation is a modification involving the reaction of deoxyribose oxidation products with histone proteins, which may influence epigenetic mechanisms governing chromatin states (Jiang et al. 2007). The authors report that lysine formylation could accumulate with age and could contribute to the deregulation of chromatin which may affect gene function.

Propionylation and Butyrylation

Two histone acetyltransferases use propionyl-CoA or butyryl-CoA as substrates to catalyse propionylation or butyrylation of lysine residues. Propionylation and butyrylation may be associated with cellular metabolic status and could regulate genes involved in energy metabolism. Although the biological functions of the two processes remain unknown, they may be involved in the epigenetic control of regulatory elements exposure to transcription factors. (Chen et al., 2007)

2. Regulation at Transcription Level

Transcription factors can either activate transcription or repress it (see Figure 4, Chapter 8). Activators are molecules which recognize and bind sequences called enhancer sites located upstream (distal) of the promoter region of the gene they regulate. Evidence suggests that activators require coactivators to do their job. Currently, over 400 coregulators (coactivators and corepressors) have been described. On binding the enhancers sites, DNA lops around triggering mediator proteins to interact with other proteins at the promoter region. Such proteins include basal factors required for the initiation of transcription by RNA polymerase. This enhances the rate of transcription of specific genes which may be located a distance away on the same or different chromosome. Activators are thus involved in **positive gene regulation.** Enhancer sites may contain multiple sites for activator binding creating an **enhancesome**. It should be noted that there is another sequence between the enhancer and the promoter called an insulator. If an **insulator protein** binds it, the activation of the promoter by the enhancer (already with activator) is blocked—no transcription occurs.

...sors bind specific sequences called **silencers** upstream of the promoter region of a gene and ...e transcription by preventing RNA polymerase from binding the promoter. Corepressors bind ... to facilitate their action. When enhancer and silencer sequences overlap, they compete to bind ...mpetitive DNA binding). If the enhancer binds first, transcription is initiated; if silencer binds ...scription is curtailed.

...lation at Post-Transcription Level

...lation can occur at various levels during mRNA processing depending on the needs of the cell ... moment. These include alternative splicing, polyadenylation, and 5' capping. These processes ...ained in Chapter 8 (section 8.2) and are reviewed below. The observation that protein and ...oncentrations correlate significantly has provided useful insights into the roles of translation ... in both prokaryotic and eukaryotic cells (de Sousa et al., 2009). Other studies suggest that the ...ip between mRNA and protein abundance is remarkably complex indicating that measurements ... levels alone cannot be relied upon to provide an accurate reflection of protein abundance in ...ations (Tony et al., 2014). As recent studies have shown, translation (and protein degradation) ...s important in regulating gene expression as are transcription (and transcript degradation) ...äusser et al., 2011).

...l molecular mechanisms can potentially affect translation and mRNA stability, including ... initiation, elongation, termination and even protein degradation. Research has shown several ... are involved post-transcription regulation: mRNA export regulation, mRNA half-life regula-...A fragments (tRFs) regulation, translation initiation, and degradation of translated protein. In ...an systems, translation and protein degradation are thought to contribute as much as ~40% ...iation in protein concentrations, while transcription and transcript degradation contribute a ...nount (Vogel et al., 2010). Ribosome profiling, a recently developed technique enables precise ...ents of translation of each mRNA in the cell. Small fragments of mRNA associated with a are ...nd quantitatively analysed by sequencing. This sequence information allows the calculation of ...ge number of ribosomes per mRNA, which is associated with translation efficiency for each ...ngolia et al., 2011).

...ive Splicing

..., more than 90% of transcripts undergo alternative splicing (Figure 2). It is precisely controlled, ... to the required tissue and organ function. Mutations in **splicing regulatory elements** (SREs) ...alternative splicing cause human genetic diseases and cancers (Kornblihtt et al., 2013). Aber-...ing has been implicated in causing up to 15% - 50% of genetic diseases in humans as seen in ...r's disease. In Alzheimer's disease, an oncogene product with a high affinity for binding pre-...esults in aberrant splicing that produces mRNA lacking exon 5. As a result, the protein product ...errant splicing (known as PS2V), accumulates in the brain of people with Alzheimer's disease, ...ause cell death (Manabe et al., 2007).

Figure 2. Alternate RNA splicing of a gene producing different mRNA molecules which are translated into different (but related) proteins

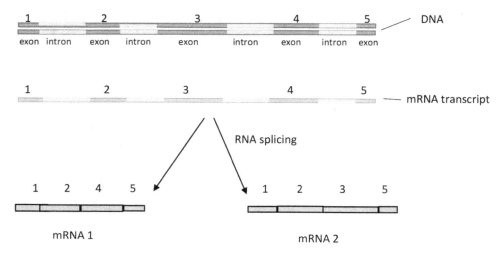

5' Cap Methylation

Several mechanisms have been identified that enhance methylation, the final step in 5' cap generation (see Figure 7 in Chapter 8. In yeast, studies indicate that there are enzymes that degrade mRNA harboring a methylation-defective cap. The observation that the 5' mRNA end can have different methylation levels implies the existence of a mechanism that regulates gene expression. It is yet unknown how a difference in methylation levels affects gene expression, and similarly what determines the methylation levels (Lutz & Moreira, 2011). Data suggests that de-capping the 5' end also functions as a gene regulatory step (Piccirillo et al., 2003).

Alternative Polyadenylation (APA)

This phenomenon produces multiple isomeric forms of mRNA because of the existence of multiple polyadenylation sites in one gene (see Figure 7, Chapter 8). APA often controls the fate of mRNA by diversifying the length of the mRNA 3' end (Lutz & Moreira, 2011). Recent **genome-wide analyses** have revealed that APA is observed widely throughout the genome, and that its pattern varies according to developmental stage and tissue organization (Ji et al., 2009).

mRNA Export Regulation

Recent research has revealed that gene regulation via mRNA export may be more important than previously recognized (Lutz & Moreira, 2011). RNA export factor germinal centre-associated nuclear protein (GANP) is known to promote TAP-dependent mRNA export (Wickramasinghe et al., 2010). It has been shown that GANP also promotes mRNAs that are highly expressed, short-lived, and highly enriched

al components of the gene expression machineries. Thus, GANP likely enables rapid changes
expression by facilitating nuclear export of specific classes of mRNA, by bridging target genes
nuclear pore. (Wickramasinghe et al., 2010)

Half-Life Regulation

enes regulate their expression level by controlling their mRNA half-life. mRNAs with a shorter
result in lower protein expression level—due to reduced translation (Xu et al., 1997; Bakheet et
). This system of control appears to be frequent in activities related to the cell cycle, develop-
d immune response. Such mRNAs usually have an AU-rich element (ARE) in their 3' UTR, a
e for microRNAs (miRNAs). mRNAs containing an ARE tend to be rapidly degraded because
rter poly-A tail.

Fragments (tRFs) Regulation

to their main function as amino acid carriers, recent studies indicate that tRNAs are used as
rs for several different classes of small RNAs. Studies by Sobala and Hutvagner (2013) have
hat tRNA molecules can be processed into small RNAs that are derived from both the 5' end
NA, called tRNA fragments (tRFs). Angiogenin is a stress-activated ribonuclease that cleaves
ithin anticodon loops to produce tRNA-derived stress-induced fragments like 5' tRNA. These
e been shown to inhibit the process of protein translation without the need for complementary
es in the mRNA. Some 5' tRNA halves specifically inhibit translation by a mechanism involving
ment of eIF4G, a protein involved in protein translation (Ivanov et al., 2011).

ation Initiation

cess of translation can be divided into initiation, elongation, termination, and ribosome recy-
ost regulation is exerted during translation initiation, where the AUG start codon is identified
ded by the methionyl-tRNA specialized for initiation (Met-tRNAi, see Figure 9, Chapter 8). In
base-pairing of the Shine-Dalgarno (SD) sequence (5'-AGGAGGU-3') located just upstream of
codon with the complementary anti-SD sequence recruits the 30S ribosomal subunit directly to
tion region of the mRNA. This means most translational control in bacteria involves modulating
ility of the SD sequence. In eukaryotes, by contrast, the start codon is generally identified by a
mechanism, where the small (40S) ribosomal subunit loaded with Met-tRNAi in a preinitia-
plex (PIC), binds to the mRNA near the 5' end and scans the 5' untranslated region (5'UTR)
UG codon. Consequently, RNA structures that impede the ability of ribosomes to interact with
R in single-stranded form, or subsequently to scan the 5'UTR, reduce the efficiency of initia-
coy AUG codons in the 5'UTR can also waylay scanning ribosomes to impede recognition of
ct start codon. (Passmore et al., 2007; Pestova et al., 2007; Sonenberg & Hinnebusch, 2009)

Degradation of Translated Protein

Regulated protein degradation plays a key role in sculpting the proteome during the cell cycle. Regulated protein catabolism via the ubiquitin–proteasome system (UPS) controls a wide variety of cellular functions, from transcriptional regulation and stress response to cell cycle regulation (Figure 3). The UPS acts as a disposal pathway and optimizes cellular functions by removing misshaped proteins. The system's 'surgical' role controls cell function by adjusting the abundance of regulatory proteins involved in numerous biological pathways, such as signal transduction, stress response, immunity, cell cycle regulation, and transcriptional regulation. (Maneix & Catic, 2016). Ubiquitination is achieved through the coordinated actions of ubiquitin-activating enzymes, ubiquitin-conjugating enzymes, and ubiquitin ligases. Furthermore, ubiquitination is a reversible protein modification which can be reversed by enzymes called deubiquitinases, which either trim or fully remove the ubiquitin chains that are attached to protein substrates, providing an additional layer of regulation to this dynamic process (Komander et al., 2009).

Figure 3. Protein degradation by a proteasome. Proteins to be degraded are often tagged with a small protein called ubiquitin.

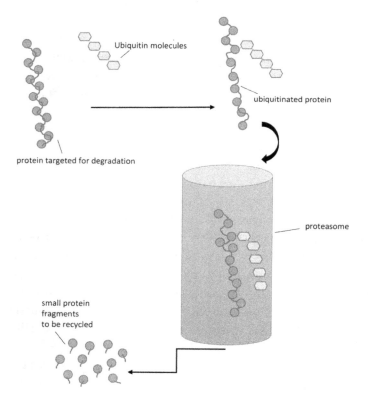

...lation by Non-Coding RNAs (ncRNAs)

...wide surveys have revealed that a large portion of the eukaryotic genomes is transcribed into ...g RNA (ncRNA) of two classes: structural and regulatory ncRNAs. Structural ncRNAs are ...ousekeeping required for the normal function and viability of the cell. This includes tRNAs, ...nRNAs, small nucleolar RNAs (snoRNAs), RNase P RNAs, and telomerase RNA (Prasanth ..., 2007).

...tory ncRNAs are expressed at certain stages of development, during cell differentiation, or ...nse to environmental stimuli. Depending on their length, ncRNA can be further divided into ...ps: two short ones (microRNAs [miRNA] and piwi-interacting RNA [piRNA]); a medium ...nd a long ncRNA (Nie et al., 2012). Non-coding RNAs can affect the expression of other ...e level of transcription or translation and play a role in chromatin regulation via interaction ...matin-modifying enzymes and transcription factors. Many studies have reported that several ...re associated with epigenetic regulation (Kaikkonen et al., 2011).

...IAs (miRNAs)

...nown class of ncRNAs, miRNAs regulate gene expression at the post-transcriptional level ...). The miRNA molecules repress translation of target mRNA resulting in decreasing levels ...pression. The group of miRNAs involved in epigenetic regulation are called "epi-miRNAs". ...JAs can influence epigenetic phenomena either by directly inhibiting enzymes involved in ...hylation, histone modifications, and chromatin remodelling, or by altering the availability of ... necessary for these enzymatic reactions. miRNAs can regulate the expression of genes that ... indirectly regulate epigenetic status, so that when the miRNA-epigenetic regulatory circuitry ...d, normal chromatin function may be impaired leading to various diseases. More than 1000 ...ave been identified, implying that the expression of more than 5% of mRNAs is affected by ... Microarray data suggests that almost one third of mRNAs are potential targets for miRNAs, ... that this perhaps represents a global function rather than a "fine-tuning" mechanism. (Iorio ...0; Ha & Kim, 2014)

...ntly described sub-class of miRNAs, called circulating miRNAs (cmiRNAs) mediate cell-cell ...-tissue cross-talk. These intracellular mediators are produced in various tissues both at baseline ...ponse to physiological and/or pathological stimuli and are secreted into the blood where they ...ivered to their respective recipient tissues to modulate gene expression (Grazioli et al., 2017).

...terfering RNAs (siRNAs)

...dicate that siRNA-mediated suppression of transcription is associated with histone and DNA ...n of mammalian cells, which target the promoter region (Morris et al., 2004; Castanotto et ...Suzuki et al., 2005).

Figure 4. Formation and function of microRNAs. The miRNA is formed from a non-coding region of DNA. It forms a complex with one or more proteins, then binds a complementary sequence of mRNA. miRNAs repress translation or degrade target mRNAs resulting in decreasing levels of gene expression.

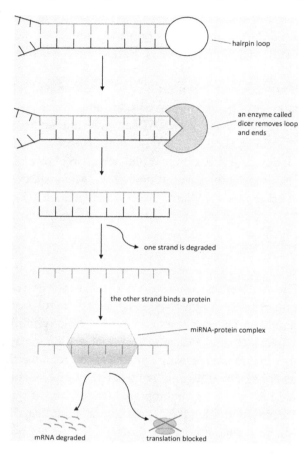

Antisense RNAs (asRNAs)

As described by Zhou et al. (2010), antisense RNAs (asRNAs) are single-stranded RNAs complementary to mRNA. They bind the mRNA and through DNA methylation, block translation.

Piwi-Interfering RNAs (piRNAs)

piRNAs are the largest class of small ncRNAs in vertebrates, and guide DNA methylation of transposable elements in foetal, male germ cells. This process recruits epigenetic factors to specific genomic sites which affects gene expression. (Zhou et al., 2010)

Non-Coding RNAs (lncRNAs)

...A transcription and processing are complicated processes in which the majority of lncRNAs are ...polyadenylated, and 5′-capped (as in protein-coding RNA). Unlike other ncRNAs, most lncRNAs ...lized in the nucleus, which suggests they may be involved in the regulation of chromatin. They ...ely guide chromatin-modifying complexes to specific genomic loci. Nuclear lncRNA molecules ...ely interact directly or indirectly with the components of chromatin-remodelling complexes (Nie ...012). In addition, lncRNAs may be involved in epigenetic gene silencing, i.e., **genomic imprint-** ...X chromosome inactivation (Ponting et al., 2009).

...ter-Associated RNAs (paRNAs)

...ss of ncRNAs include promoter-associated small RNAs involved in transcription are of four ...types: terminal-associated short RNAs, transcription start site-associated RNAs, transcription ...n RNAs, and promoter-upstream transcripts (Kaikkonen et al., 2011). Kapranov et al. (2007) ...s short paRNAs located near the promoter or transcription start side. Studies suggest that paRNAs ...te to transcriptional gene silencing and chromatin organization (Kanhere et al., 2010).

...mer Repeat-Associated Small Interacting RNAs (crasiRNAs)

...As are very important in centromere establishment as well as chromosome segregation. Studies ...that crasiRNAs facilitate the recruitment of adaptor proteins to the centromere-specific DNA ...ltimately methylates DNA. crasiRNAs could therefore be considered as regulatory elements in ...tic changes. (Lindsay et al., 2012)

...ere-Specific Small RNAs (tel-sRNAs)

...As have been found in telomere and sub-telomere regions in mammalian cells. Studies show ...sRNAs contain UUAGGG repeats that inhibit telomerase activity in vivo (Schoeftner & Blasco, ...As a result, tel-siRNAs could potentially act as sensors of chromatin status and mediators in the ...ic length control and telomeric heterochromatin formation (Cao et al., 2009).

PIGENETICS, ENVIRONMENT AND DISEASE

...an increasing body of evidence demonstrating that epigenetic patterns are altered by environmental ...known to be associated with disease risk including diet, smoking, alcohol intake, environmental ...ts, and stress (Relton & Smith, 2010). Research indicates that many factors can modulate the ...tic mechanisms associated with a variety of human diseases. However, since epigenetic changes ...ll, potentially cumulative, and they may develop over time, it may be difficult to establish the ...ffect relationships among environmental factors, epigenetic changes and diseases (Baccarelli & ...2009). There are more than 13 million deaths worldwide because of environmental pollutants, ...dies have linked environmental pollutants with epigenetic variations particularly changes in DNA methylation, histone modifications and microRNAs. Approximately 24% of diseases are caused by environmental exposures that can be averted through preventive measures. Growing evidence suggests that environmental pollutants may cause diseases via epigenetic mechanism-regulated gene expression changes. (Hou et al., 2012) Elucidating epigenetic processes and how they interact with one another to control gene expression has the potential to transform our understanding of how the phenotype of the cell is maintained and how it is affected in disease conditions (Baylin & Jones, 2011).

Epigenomic signatures can be propagated through cell division and an individual's epigenome often reflects his/her prenatal environmental exposure experience. Therefore, epigenomic profiling of individuals exposed to environmental pollutants might provide molecular archives of one's past or even prenatal environmental exposures. Numerous clinical and preclinical studies show that most of the epigenetic changes are reversible, which offers novel insights to develop new preventive and therapeutic strategies that might take advantage of molecules that modify the activities of epigenetic enzyme. DNA methylation patterns are reprogrammed genome wide, generating cells with a broad developmental potential which is critical for **genomic imprinting**. Reprogramming is likely to have a crucial role in establishing nuclear totipotency in normal development in the erasure of acquired epigenetic information, and in stem cell differentiation (Reik & Walter, 2001). While the genome is virtually identical in all diploid cells from the same individual, each tissue and potentially each individual cell or cell type, may exhibit a unique epigenomic profile.

11.3.1 Physical Activity, Diet, Cancer, and Chronic Diseases

Numerous studies report that modification of lifestyle factors, especially increasing physical activity levels, can influence the epigenetic patterns involved in human cancer, metabolic, cardiovascular and neurodegenerative diseases (Lee et al. 2012; Rea, 2017). For instance, studies in human cells demonstrate that moderate exercise up-regulates the methylation status of a protein that serves as an important mediator of an inflammatory signaling pathways (Grazioli et al., 2017). The protein's methylation pattern is associated with the level of pro- and anti-inflammatory cytokines during exercise, which regulates lymphocyte activation and differentiation (Bopp et al., 2010). These epigenetic mechanisms are thought to contribute to lowering of inflammation, thereby preventing the occurrence of diseases linked to a chronic inflammation. Physical activity also counteracts hypermethylation and hypermethylation associated with tumorous mutations in the genome eliminating any side effects for the patients. Grazioli et al. (2017) demonstrated that 12 weeks of low frequency, moderate intensity power training has the capacity to reduce the global DNA methylation in peripheral mononuclear cells of elderly subjects.

The epigenome appears to go through extensive remodelling in multiple malignancies and today, therapeutics targeting enzymes involved in these processes (DNA methyltransferases and histone deacetylases) are used in several cancer types (Baylin & Jones, 2011). Typically, loss of DNA methylation is observed on a global scale while gains in DNA methylation occur in a more specific manner, and these changes are accompanied by abnormal nucleosome positioning and chromatin modifications (Baylin & Jones, 2011). Epigenetics helps us understand that genetic or hereditary forms of most cancers are extremely rare and that most cancers can be realistically prevented or managed by making simple day-today changes in our diets. Studies indicate that 90% to 95% of these cancers are "sporadic" or "acquired" in nature, occur as a function of aging, and are intimately tied to lifestyle and dietary habits. The expression of cancer-associated genes in most of sporadic cancers is controlled by the epigenome and a heritable manner (Toden & Goel, 2014). However, these epigenetic changes are reversible and can easily be modified by dietary and environmental stimuli. Cancer is commonly associated with hypermethylation of tumour-suppressor genes (You & Jones, 2012). Studies suggest that exercise may reduce DNA methylation in certain tumor suppressor genes, thereby inhibiting tumor progression with positive effects on cancer survival (Zeng et al., 2012).

There is increasing evidence that epigenetic factors may partly explain the beneficial effects of exercise on the prevention and treatment of Type-2 diabetic patients and other metabolic disorders which also results in positive effects on metabolic outcomes in offspring (Barres & Zierath, 2011). Keller et al. (2011) found that aerobic exercise reduces the expression of various types of miRNAs in human skeletal muscle, 22% of which target genes that regulate transcription and 16% target genes involved in muscle metabolism, mostly in oxidative phosphorylation. Other studies indicate that physical exercise modulates histone H3 and H4 acetylation in different tissues promoting chromatin modification that can lead to selective activation or inhibition of specific genes related to cancer, muscle wasting or behavioral diseases. (McGee & Hargreaves, 2011; Zimmer et al., 2014; Philp et al., 2014; Lavratti et al., 2017)

Physical activity has positive impacts on a wide range of biological functions, counteracting those epigenetic mechanisms able to alter gene expression in cardiovascular diseases (Movassagh et al., 2010). As epigenetic modifications of specific genes implicated in neurological disorders (e.g. epilepsy, schizophrenia, Alzheimer and Parkinson diseases) central nervous system plasticity is subjected to epigenetic modifications induced by exercise. Lavratti et al. (2017) demonstrated a relation between exercise training and levels of global histone acetylation in people affected by neurodegenerative diseases. For example, showed that 90 days of combined exercise program (1 hour, 3 times/week, aerobic and strength training) induced significant histone H4 hypoacetylation in schizophrenia patients, reducing transcriptional activity and gene expression.

Diet and Cancer: Many dietary compounds, especially **phytochemicals**, are potent epigenetic modulators in cancer and other age-associated chronic diseases making these compounds potential targets of cancer prevention and treatment. Many dietary plant-derived compounds such as polyphenols (including flavanols and isoflavones), have anti-tumorigenic properties likely mediated through reversal of DNA methylation, leading to reactivation of these genes slowing down or preventing tumor growth. Another mechanism by which dietary compounds can influence epigenome is by regulating the expression of various noncoding RNAs, including micro-RNAs (miRNAs). Flavanols are a class of flavonoids that include forms of catechins commonly present in green tea, cocoa, and various fruits. Epigallocatechin-3-gallate, the major polyphenol found in green tea, is a potent antioxidant, has putative antitumorigenic properties, and has been shown to reverse hypermethylation of tumor suppressor genes including p16 and p15 in multiple cancers. Resveratrol is another polyphenol that belongs to the stilbene class and is found primarily in grapes and peanuts. The antitumorigenic effects of resveratrol have been investigated extensively and it has been shown to modulate DNA methylation in vitro. Curcumin is the major active principle found in the rhizomes of turmeric (a plant of the ginger family) and is used as a traditional medicine in India and Southeast Asia and shown to exert anti-inflammatory, antioxidative, antiproliferative, and proapoptotic activities in both in vivo and in vitro studies. Curcumin was identified as a potential hypomethylating agent in various cancers including leukemia, ovarian, and colon. (Toden & Goel, 2014)

11.3.2 Heavy Metals

Growing data have linked epigenetic alterations with heavy metal exposure like arsenic, nickel, cadmium, mercury, lead and chromium. (Takiguci et al., 2003; Filipic & Hei, 2004; Huang et al., 2008; Collotta et al., 2013)

Evidence suggests that exposure to arsenic cause alterations in histone modifications through DNA methylation both globally and in the promoter regions of certain genes. Nickel has been proposed to increase chromatin condensation and trigger de novo DNA methylation of critical tumour suppressor or senescence genes. Evidence on nickel-induced histone modifications includes increases of H3 dimethylation, loss of histone acetylation in H2A, H2B, H3 and H4, H3 phosphorylation, and increases of the ubiquitination in H2A and H2B. It is also associated with gene silencing due to histone deacetylation. Cadmium alters global DNA methylation, induces global DNA hypomethylation and prolonged exposure was shown to lead to DNA hypermethylation. It can also decrease DNA methylation in proto-oncogenes and promote oncogenes expression that can result in cell proliferation.

An established carcinogen, cadmium reduces genome methylation, inhibiting DNA methyltransferases by the metal latching onto the methyltransferase DNA binding domain. Lead is among the most prevalent toxic environmental metal whose long-term exposure has been shown to alter epigenetic marks. Studies suggest that changes in DNA methylation may represent a biomarker of past lead exposure. Pilsner et al. (2009) characterized genomic DNA methylation in the lower brain stem region in polar bears and reported an inverse association between cumulative lead measures and genomic DNA methylation level. Chromium has been associated with repressed gene expression, hypermethylation in lung cancer patients, and carcinogenic outcomes. Recent investigations have demonstrated that aluminium exposure can alter the expression of several miRNAs in human neural cells which corresponded to the decreased expression of complement factor H, a repressor of inflammation.

Exposure to mercury has been associated with brain tissue DNA hypomethylation in the polar bear and in after 48 or 96 hours of exposure to the chemical, hypermethylation of specific genes in Hg-treated mouse embryonic stem cells was observed (Arai et al., 2011).

11.3.3 Organic Toxicants

- **Benzene:** Exposure to benzene, an industrial chemical has been associated with increased risk of haematological malignancies, particularly with acute myeloid leukaemia and acute non-lymphocytic leukaemia. Studies on police officers and gas-station attendants (Hou et al., 2012) have shown that low-dose exposure to airborne benzene is associated with alterations in DNA methylation in blood DNA of healthy subjects which resembles those found in haematological malignancies. Treating human lymphoblastoid cells with benzene metabolites consistently show reductions in global DNA methylation. Research has also shown that benzene exposure induces hypermethylation of poly (ADP-ribose) polymerases-1, a gene involved in DNA repair. (Hou et al., 2012)
- **Bisphenol A (BPA):** An **endocrine disruptor** affecting reproductive systems, BPA is also a weak carcinogen associated with increased cancer risk in adult life through foetal exposures. BPA is used in industry as a plasticizer in epoxy resins for food and beverage containers, baby bottles and dental composites. In mice, BPA has been reported to cause colour shifts from viable yellow agouti to a yellow-coat phenotype which is associated with increased cancer rates, obesity and insulin resistance (Dolinoy et al., 2007). In breast epithelial cells treated with low-dose BPA,

...e expression profiling identified 170 genes with expression changes in response to BPA. For ...mple, the expression of lysosomal-associated membrane protein 3 was silenced due to DNA ...permethylation in its promoter (Weng et al., 2010).

...oxin: Dioxins and dioxin-like compounds come from three classes of compounds (chlorinated ...enzo-p-dioxins, chlorinated dibenzo-p-furans, and polychlorinated biphenyls) (Van den Berg ...l., 2006; Zikmund-Fisher et al., 2013). These are highly toxic substances which can cause can-..., reproductive and developmental problems, damage to the immune system, and can interfere ...h hormone functioning. More than 90% of human exposure is through food, mainly meat and ...ry products, fish and shellfish (EPA.gov, 2018). Also, products of combustion processes diox-...are toxin and a carcinogenic with bioaccumulation effects in natural food chains. One proposed ...hway to carcinogenesis is related to the powerful dioxin-induced activation of microsomal en-...mes, which might activate other pro-carcinogen compounds to active carcinogens. Alterations ...DNA methylation at multiple genomic regions have also been reported in splenocytes of mice ...ated with dioxin, due to dioxin's immune-toxicity.

...xahydro-1,3,5-Trinitro-1,3,5-Triazine (RDX): An explosive and a common ammunition con-...uent used in military and civil activities, RDX is usually found in soils and in water. RDX has ...n demonstrated to cause neurotoxicity, immune-toxicity and cancers plus altered expression of ...RNA in mouse brain and liver (Zhang & Pan, 2009).

...ethylstilbestrol (DES): This is a synthetic oestrogen which was used in the 1940s and 1960s to ...vent miscarriages in pregnant women (Laitman, 2002). A moderate increase in breast cancer ...k has been shown both in daughters of women who were treated with DES during pregnancy, ...well as in their daughters. Hsu et al. (2009) demonstrated that the expression of 9.1% of ~898 ...RNAs evaluated were altered in breast epithelial cells when exposed to DES.

...emicals in Drinking Water: Water chlorination often performed to purify drinking water ...killing microbes produces chlorination by-products such as triethyltin, trihalomethanes (like ...oroform), dichloroacetic acid (DCA), and trichloroacetic acid (TCA) (IARC, 1991). These ...micals have been shown to induce certain epigenetic changes. For example, rats chronically ...oxicated with triethyltin in drinking water showed development of cerebral oedema (Hou et al., ...2). Mice treated with DCA, TCA, and chloroform show global hypomethylation and increased ...ression of a proto-oncogene involved in liver and kidney tumours. In female mice liver, tri-...omethanes demonstrated carcinogenic activity by methylation in the promoter region of the ...C regulatory gene (codes for a transcription factor) reducing its activity consistent with their ...cinogenic activity (Coffin et al., 2002).

Air Pollution

...tudies of air pollution universally demonstrated an association between exposure to air pollution ...A methylation of circulating leukocytes (Baccarelli et al. 2010; Hou et al. 2012). Air pollution **...ulate matter** has been shown to cause decreased DNA methylation, global DNA hypermethyl-...sperm genomic DNA, a change that persisted after removal of environmental exposure. Inhaled ...haust particles induced hypermethylation of several sites of gene promoters correlated with ...in immunoglobulin-E levels. In steel workers, exposure to metal-rich particulate matter induced ...anges in the expression of two inflammation-related miRNAs. Evidence also shows extensive alterations of miRNA expression profiles in human bronchial epithelial cells treated with diesel exhaust particles. (Jardim et al., 2009)

Dynamic epigenetic patterns occurring during embryogenesis adapt embryos for further differentiation at the zygote stage and during primordial germ cells formation. Heijmans et al. (2009) have suggested that the epigenome may represent a molecular archive of the prenatal environment, via which the in-utero environment may produce serious ramifications on health and disease later in life. For example, prenatal exposure to cigarette smoke was associated with increased overall blood DNA methylation level in adulthood. There were decreased methylation level in adults and children prenatally exposed to smoking, and global DNA hypomethylation in new-borns with utero exposures of maternal smoking. Studies are on-going to provide additional information on the regions sensitive to the prenatal environment and may indicate developmental influences on human disease.

- **Nanoparticles:** Certain nano-sized compounds can induce an impaired expression of genes involved in DNA methylation reactions leading to global DNA methylation changes, as well as changes of gene specific methylation of tumour suppressor genes, inflammatory genes, and DNA repair genes, all potentially involved in cancer development (Stoccoro et al., 2013). Nanomaterials are in size comparable to biological structures and their small size enables them to be introduced in the biological systems via cellular uptake, they can pass cellular barriers making them potent carriers of drugs and other small molecules. **Nanotechnology** therefore is a promising field for broad variety of new biological and biochemical applications. In one study (Tuomela et al., 2013) gene expression of human cells exposed to ZnO and TiO_2 nanoparticles was analysed with microarrays to elucidate how these materials modulate transcription in different cell types. Results showed that 2703 genes were significantly differentially expressed in human macrophages upon exposure to 10 µg/ml ZnO nanoparticles, while in dendritic cells only 12 genes were affected. In T cell leukaemia-derived cell lines, 980 genes were differentially expressed. (Tuomela et al., 2013)

11.3.5 Pesticides

In vitro, animal, and human investigations have identified several classes of pesticides that modify epigenetic marks, including endocrine disruptors, **persistent organic pollutants,** arsenic, several **herbicides** and **insecticides** (Collotta et al., 2013). Growing evidence suggests that epigenetic events can be induced by pesticide exposures. Studies from animal models show that exposure to some pesticides such as vinclozolin and methoxyclor, induces heritable alterations of DNA methylation in male germline associated with testis dysfunction. They also affect ovarian function via altered methylation patterns. Dieldrin, a widely used organochlorine pesticide, has been shown to increase acetylation of core histones H3 and H4 in a time-dependent manner when acetylation was induced within 10 minutes of dieldrin exposure. This suggests that histone hyperacetylation is an early event in dieldrin-induced diseases.

Since the literature is not yet conclusive regarding the role of environmental contaminants in epigenetics (Terry et al. 2011); more controlled human experiments are necessary where the subjects' DNA methylation is measured before and after exposure to different air pollutants. Research linking epigenetics to environmental factors is rapidly increasing to a promising field that is expected to lead to clinical and public health interventions. The use of such controlled studies in environmental health should help to definitively link exposure to epigenetic marks, as well as to related phenotypes (Burris & Baccarelli, 2014). Future epidemiology studies will help elucidate whether the effects of environmental exposures genome are mitigated by positive changes in lifestyles or worsened by the interaction with other ... More studies on the epigenetic mechanisms involved in disease pathogenesis, particularly ... epigenetics in health and disease, their relationships with environmental exposures and the ... associated with the disease phenotype may help develop preventive and therapeutic strategies ..., 2012).

ER SUMMARY

...cs is the study of stable, often heritable changes which influence gene expression but are not ...by DNA sequence. These changes play crucial roles in chromatin state regulation which influ-...cesses such as gene expression, DNA repair, and recombination. The most distinctive aspects ...etics are that these changes in expression occur without a change in the original sequence of ... Epigenomics is the identification of all epigenetic modifications implicated in gene expres-...increasingly becoming an important tool to better understand human biology in both normal ...logical states.

...igenome is the distribution of these epigenetic features throughout the genome and is (unlike ...e itself) dynamic and localization of epigenetic features is influenced by cell or tissue type, ...sure to environmental stimuli or many other factors. Owing to the important roles epigenetic ...a play in human health and development, the National Institutes of Health (NIH) launched ...nap Epigenomics Project in 2007 to combine whole genome epigenetic analysis with genetic ...g. This has created a series of publicly available reference epigenome maps encompassing ...ray of cell lines, cell and tissue types from individuals at various developmental stages and ...tes.

...egulation in higher organisms can occur at various steps from chromatin (DNA) level to post-...al level. Gene expression is comprised of six steps: 1) transcription; 2) mRNA processing; 3) ...nsport and localization; 4) mRNA half-life; 5) translation; and 6) protein activity. Regulation ...A level includes histone modification processes [methylation, acetylation, phosphorylation, ...tion, carbonylation, ubiquitylation, biotinylation, sumoylation, citrullination, ADP-ribosylation, ...ation, crotonylation, propionylation and butyrylation, plus also proline and aspartic acid isomeri-...RNAs, epigenetic inheritance. Regulation at Transcription level includes transcription factors, ..., activators and coactivators, repressors, and basal factors. Regulation at post-transcription level ...mRNA processing, mRNA degradation, translation initiation, protein processing and degrada-...lation by non-coding RNAs includes microRNAs, siRNA, and RNAi.

...is an increasing body of evidence demonstrating that epigenetic patterns are altered by en-...tal factors known to be associated with disease risk including diet, smoking, alcohol intake, ...ental toxicants, and stress. Research indicates that many factors can modulate the epigenetic ...ms associated with a variety of human diseases. Studies have linked environmental pollutants with ...c variations particularly changes in DNA methylation, histone modifications and microRNAs. ...rous studies report that modification of lifestyle factors, especially increasing physical activity ...n influence the epigenetic patterns involved in human cancer, metabolic, cardiovascular and ...enerative diseases. The epigenome appears to go through extensive remodelling in multiple ...cies and today, therapeutics targeting enzymes involved in these processes are used in several ...pes. Epigenetics helps us understand that genetic or hereditary forms of most cancers are extremely rare and that most cancers can be realistically prevented or managed by making simple day-today changes in our diets. Epigenetic changes are reversible and can easily be modified by dietary and environmental stimuli. Cancer is commonly associated with hypermethylation of tumour-suppressor genes. Studies suggest that exercise may lower DNA methylation in certain tumor suppressor genes, thereby inhibiting tumor progression with potential effects on cancer survival.

Increasing evidence that epigenetic factors may partly explain the beneficial effects of exercise on the prevention and treatment of Type-2 diabetic patients and other metabolic disorders which also results in beneficial effects on metabolic outcomes in offspring. Physical activity has positive impacts on a wide range of biological functions, counteracting those molecular mechanisms able to alter gene expression in cardiovascular diseases and neurological disorders (e.g. epilepsy, schizophrenia, Alzheimer and Parkinson diseases). Many dietary compounds, especially phytochemicals, are potent epigenetic regulators in cancer and other age-associated chronic diseases making these compounds potential targets of cancer prevention and treatment.

Growing data have linked epigenetic alterations with heavy metal exposure like arsenic, nickel, cadmium, mercury, lead and chromium. Organic toxicants including benzene, bisphenol-A, dioxins, hexahydro-trinitro-triazine, diethylstilbestrol, water chlorination by-products have been shown to cause altered gene expression, DNA methylation and hypermethylation, neurotoxicity, immune-toxicity, and cancer.

Studies focusing on the effects of air pollution in humans demonstrate an association between exposure to air pollution and DNA methylation of circulating leukocytes. Air pollution of particulate matter has been shown to cause decreased DNA methylation, global DNA hypermethylation and extensive alterations of miRNA expression in human cells. Certain nano-sized compounds can induce an impaired expression of genes involved in DNA methylation reactions leading to global DNA methylation changes, as well as changes of gene specific methylation of tumour suppressor genes, inflammatory genes, and DNA repair genes, all potentially involved in cancer development

In vitro, animal, and human investigations have identified several classes of pesticides that modify epigenetic marks, including endocrine disruptors, persistent organic pollutants, arsenic, several herbicides and insecticides. However, the literature is not yet conclusive regarding the role of environmental contaminants in epigenetics, and more controlled human experiments are necessary. More studies on the epigenetic mechanisms involved in disease pathogenesis, particularly the role of epigenetics in health and disease, their relationships with environmental exposures and the pathways associated with the disease phenotype may help develop preventive and therapeutic strategies.

End of Chapter Quiz

1. Which of the following statement about epigenetics is false?
    a. It is the study of stable, heritable changes which influence gene expression but are not mediated by DNA sequence
    b. Epigenetics changes are inheritable in response to developmental cues or external and environmental stimuli
    c. The changes in expression occur with a change in the original sequence of DNA
    d. An understanding of epigenomic factors will lead to more methods to treat diseases
    e. All of the statements are true

Epigenetic modifications can be grouped into four major classes which are
- Post-translational modification of histone proteins, chromatin conformation/accessibility, DNA modification, and non-coding regulatory RNA
- Pre-translational modification of histone proteins, chromatin conformation/accessibility, DNA modification, and non-coding regulatory RNA
- Post-translational modification of histone proteins, chromatin conformation/accessibility, RNA modification, and non-coding regulatory RNA
- Pre-translational modification of histone proteins, chromatin conformation/accessibility, DNA modification, and non-coding regulatory DNA
- None of the above

Histone modifications consists of
- Ubiquitination
- Acetylation
- Methylation
- Phosphorylation
- All of the above

Protein hydroxylation is post-translational modification catalysed by enzymes called
- hydrolases
- dioxygenases
- catalases
- acetylases
- endonucleases

Due to oxidative modifications, protein carbonylation has been associated with age related or metabolic diseases such as
- Diabetes
- Diabetes and Alzheimer's
- Diabetes, Alzheimer's and Parkinson's
- Diabetes and Chronic lung disease
- Diabetes, Alzheimer's, Parkinson's and chronic lung disease

Biotin is _____ and acts as a _____ for four carboxylases, which play essential roles in the metabolism of glucose, proteins and fatty acids.
- vitamin B1; cofactor
- vitamin B1; corepressor
- vitamin B7; cofactor
- vitamin B7; corepressor
- vitamin B2; precursor

Transcription factors
- can only activate transcription
- can only repress transcription
- can either activate or repress transcription
- can activate and repress transcription simultaneously
- None of the above

8. _____ is a stress-activated ribonuclease that cleaves tRNA within anticodon loops to produce tRNA-derived stress-induced fragments
   a. Angiogenin
   b. Angiotensin
   c. Angiotensinogen
   d. Angiotensin II
   e. Angiogram
9. What are the two classes of non-coding RNAs (ncRNA)?
   a. Specific and non-specific ncRNAs
   b. Structural and regulatory ncRNAs
   c. Supportive and regulatory ncRNAs
   d. Structural and non-regulatory ncRNAS
   e. Sensitive and regulatory ncRNAs
10. Antisense RNAs are
    a. double-stranded RNAs complementary to mRNA
    b. single-stranded RNAs complementary to mRNA
    c. double-stranded RNAs complementary to tRNA
    d. single-stranded RNAs complementary to tRNA
    e. double-stranded RNAs complementary to rRNA

Thought Questions

1. If you have multiple histone modifications for one gene expression, does it matter which order the modifications occur? Explain why or why not.
2. Describe how physical activity and diet plays a role in protein methylation.
3. Describe how the promoter region is affected due to the exposure of heavy metals.

REFERENCES

Arai, Y., Ohgane, J., Yagi, S., Ito, R., Iwasaki, Y., Saito, K., ... Shiota, K. (2011). Epigenetic Assessment of environmental chemicals detected in maternal peripheral and cord blood samples. *The Journal of Reproduction and Development, 57*(4), 507–517. doi:10.1262/jrd.11-034A PMID:21606628

Baccarelli, A., & Bollati, V. (2009). Epigenetics and environmental chemicals. *Current Opinion in Pediatrics, 21*(2), 243–251. doi:10.1097/MOP.0b013e32832925cc PMID:19663042

Baccarelli, A., Tarantini, L., Wright, R. O., Bollati, V., Litonjua, A. A., Zanobetti, A., ... Schwartz, J. (2010). Repetitive element DNA methylation and circulating endothelial and inflammation markers in the VA normative aging study. *Epigenetics, 5*(3), 222–228. doi:10.4161/epi.5.3.11377 PMID:20305373

Barres, R., & Zierath, J. R. (2011). DNA methylation in metabolic disorders. *The American Journal of Clinical Nutrition, 93*(4), 897S–900. doi:10.3945/ajcn.110.001933 PMID:21289222

Baylin, S. B. & Jones, P. A. (2011). A decade of exploring the cancer epigenome—biological and translational implications. *Nature Reviews Cancer, 11*, 726–734. doi:10.1038/nrc3130

..., L., Kouzarides, T., Shiekhattar, R., & Shilatifard, A. (2009). An operational definition of ... *Genes & Development, 23*(7), 781–783. doi:10.1101/gad.1787609 PMID:19339683

..., Aumuller, E., Gnauck, A., Nestelberger, M., Just, A., & Haslberger, A. G. (2010). Epigenetic ... estrogen receptor expression and tumor suppressor genes is modulated by bioactive food com-... *nnals of Nutrition & Metabolism, 57*(3-4), 183–189. doi:10.1159/000321514 PMID:21088384

..., B. E., Meissner, A., & Lander, E. S. (2007). The mammalian epigenome. *Cell, 128*(4), 669–681. ...16/j.cell.2007.01.033 PMID:17320505

..., Baccarelli, A., Hou, L., Bonzini, M., Fustinoni, S., Cavallo, D., ... Yang, A. S. (2007). Changes ...ethylation patterns in subjects exposed to low-dose benzene. *Cancer Research, 67*(3), 876–880. ...58/0008-5472.CAN-06-2995 PMID:17283117

... Radsak, M., Schmitt, E., & Schild, H. (2010). New strategies for the manipulation of adaptive ...esponses. *Cancer Immunology, Immunotherapy, 59*(9), 1443–1448. doi:10.100700262-010-...MID:20361184

... H., & Baccarelli, A. A. (2013). Environmental Epigenetics: From Novelty to Scientific Dis-...ournal of Applied Toxicology, 34*(2), 113–116. doi:10.1002/jat.2904 PMID:23836446

... & Xu, X. (2010). Diet, epigenetic, and cancer prevention. *Advances in Genetics, 71*, 237–255. ...16/B978-0-12-380864-6.00008-0 PMID:20933131

... C., Ge, R., Yang, S., Kramer, P. M., Tao, L., & Pereira, M. A. (2000). Effect of trihalometh-...ell proliferation and DNA methylation in female B6C3F1 mouse liver. *Toxicological Sciences*, ...3–252. doi:10.1093/toxsci/58.2.243 PMID:11099637

... M., Bertazzi, P. A., & Bollati, V. (2013). Epigenetics and pesticides. *Toxicology, 307*, 35–41. ...16/j.tox.2013.01.017 PMID:23380243

... F. (2010). Epigenomics in cancer management. *Cancer Management and Research*, (2): 255–265. ...47/CMAR.S7280 PMID:21188117

... D. C., Huang, D., & Jirtle, R. L. (2007). Maternal nutrient supplementation counteracts bisphenol ...d DNA hypomethylation in early development. *Proceedings of the National Academy of Sciences ...ited States of America, 104*(32), 13056–13061. doi:10.1073/pnas.0703739104 PMID:17670942

... Z., Wang, Y., Ai, N., Hou, Z., Sun, Y., Lu, H., ... Yang, C. S. (2003). Tea polyphenol (-)-epi-...chin-3-gallate inhibits DNA methyltransferase and reactivates methylation-silenced genes in ...ll lines. *Cancer Research, 63*(22), 7563–7570. PMID:14633667

... N. A., Kolar, C., West, J. T., Smith, S. A., Hassan, Y. I., Borgstahl, G. E. O., ... Lyubchenko, ...11). The Role of Histone H4 Biotinylation in the Structure of Nucleosomes. *PLoS One, 6*(1), ...doi:10.1371/journal.pone.0016299 PMID:21298003

..., & Hei, T. K. (2004). Mutagenicity of cadmium in mammalian cells: Implication of oxidative ...age. *Mutation Research, 546*(1-2), 81–91. doi:10.1016/j.mrfmmm.2003.11.006 PMID:14757196

Fingerman, I. M., Zhang, X., Ratzat, W., Husain, N., Cohen, R. F., & Schuler, G. D. (2013). NCBI Epigenomics: What's new for 2013. *Nucleic Acids Research, 41*(D1), D221–D225. doi:10.1093/nar/gks1171

Fingerman, I. M., McDaniel, L., Zhang, X., Ratzat, W., Hassan, T., Jiang, Z., ... Schuler, G. D. (2011). NCBI Epigenomics: a new public resource for exploring epigenomic data sets. *Nucleic Acids Research, 39*(Database issue), D908–D912. doi:10.1093/nar/gkq1146

Franks, A. L., & Slansky, J. E. (2012). Multiple Associations Between a Broad Spectrum of Autoimmune Diseases, Chronic Inflammatory Diseases and Cancer. *Anticancer Research, 32*(4), 1119–1136. PMID:22493341

Gao, A., Zuo, X., Liu, Q., Lu, X., Guo, W., & Tian, L. (2010). Methylation of PARP-1 promoter involved in the regulation of benzene-induced decrease of PARP-1 mRNA expression. *Toxicology Letters, 195*(2-3), 114–118. doi:10.1016/j.toxlet.2010.03.005 PMID:20230882

Goel, A., Kunnumakkara, A. B., & Aggarwal, B. B. (2008). Curcumin as "Curecumin": From kitchen to clinic. *Biochemical Pharmacology, 75*(4), 787–809. doi:10.1016/j.bcp.2007.08.016 PMID:17900536

Grazioli, E., Dimauro, I., Mercatelli, N., Wang, G., Pitsiladis, Y., Luigi, L. D., & Caporossi, D. (2017). Physical activity in the prevention of human diseases: Role of epigenetic modifications. *BMC Genomics, 18*(S8Suppl 8), 802. doi:10.118612864-017-4193-5 PMID:29143608

Heijmans, B. T., Tobi, E. W., Lumey, L. H., & Slagboom, P. E. (2009). The epigenome: Archive of the prenatal environment. *Epigenetics, 4*(8), 526–531. doi:10.4161/epi.4.8.10265 PMID:19923908

Henley, J. M., Craig, T. J., & Wilkinson, K. A. (2014). Neuronal SUMOylation: Mechanisms, Physiology, and Roles in Neuronal Dysfunction. *Physiological Reviews, 94*(4), 1249–1285. doi:10.1152/physrev.00008.2014 PMID:25287864

Hou, L., Wang, S., Dou, C., Zhang, X., Yu, Y., Zheng, Y., ... Baccarelli, A. A. (2012). Air pollution exposure and telomere length in highly exposed subjects in Beijing, China: a repeated-measure study. *Environment International, 48*, 71–77. doi:.envint.2012.06.020 doi:10.1016/j

Hou, L., Zhang, X., Wang, D., & Baccarelli, A. (2012). Environmental chemical exposures and human epigenetics. *International Journal of Epidemiology, 41*(1), 79–105. doi:10.1093/ije/dyr154 PMID:22253299

Hsu, P.-Y., Deatherage, D. E., Rodriguez, B. A. T., Liyanarachchi, S., Weng, Y.-I., Zuo, T., ... Huang, T. H.-M. (2009). Xenoestrogen-Induced Epigenetic Repression of *microRNA-9-3* in Breast Epithelial Cells. *Cancer Research, 69*(14), 5936–5945. doi:.CAN-08-4914 doi:10.1158/0008-5472

Huang, D., Zhang, Y., Qi, Y., Chen, C., & Ji, W. (2008). Global DNA hypomethylation, rather than reactive oxygen species (ROS), a potential facilitator of cadmium-stimulated K562 cell proliferation. *Toxicology Letters, 79*(1), 43–47. doi:10.1016/j.toxlet.2008.03.018 PMID:18482805

IARC. (1991). Chlorinated drinking-water. *IARC Monographs on the Evaluation of Carcinogenic Risks to Humans, 52*, 45–141. PMID:1960849

Jardim, M. J., Fry, R. C., Jaspers, I., Dailey, L., & Diaz-Sanchez, D. (2009). Disruption of microRNA expression in human airway cells by diesel exhaust particles is linked to tumorigenesis-associated pathways. *Environmental Health Perspectives, 117*(11), 1745–1751. doi:10.1289/ehp.0900756 PMID:20049127

hang, L., Peng, V., Ren, X., McHale, C., & Smith, M. (2010). A comparison of the cytoge- erations and global DNA hypomethylation induced by the benzene metabolite, hydroquinone, se induced by melphalan and etoposide. *Leukemia*, *24*(5), 986–991. doi:10.1038/leu.2010.43 0339439

., Vollaard, N. B. J., Gustafsson, T., Gallagher, I. J., Sundberg, C. J., Rankinen, T., ... Timmons, J.). A transcriptional map of the impact of endurance exercise training on skeletal muscle phenotype. *of Applied Physiology*, *110*(1), 46–59. doi:10.1152/japplphysiol.00634.2010 PMID:20930125

A., Ho, S.-M., Hunt, P. A., Knudsen, K. E., Soto, A. M., & Prins, G. S. (2007). An Evaluation ence for the Carcinogenic Activity of Bisphenol A. *Reproductive Toxicology (Elmsford, N.Y.)*, 40–252. doi:10.1016/j.reprotox.2007.06.008 PMID:17706921

, C. J. (2002). DES exposure and the aging woman: Mothers and daughters. *Current Women's Reports*, *2*(5), 390–393. PMID:12215312

, C., Dorneles, G., Pochmann, D., Peres, A., Bard, A., de Lima Schipper, L., ... Elsner, V. R. Exercise-induced modulation of histone H4 acetylation status and cytokines levels in patients with hrenia. *Physiology & Behavior*, *168*, 84–90. doi:10.1016/j.physbeh.2016.10.021 PMID:27810494

, R. C., Buonora, J. E., Flora, M. N., Freedy, J. G., Holstein G. R., Martinelli, G. P., ... Mueller, 015). Protein Citrullination: A Proposed Mechanism for Pathology in Traumatic Brain Injury. *rs in Neurology*, *6*, 204. doi:.2015.00204 doi:10.3389/fneur

M., Shiroma, E. J., Lobelo, F., Puska, P., Blair, S. N., & Katzmarzyk, P. T. (2012). Impact of l Inactivity on the World's Major Non-Communicable Diseases. *Lancet*, *380*(9838), 219–229. 1016/S0140-6736(12)61031-9 PMID:22818936

. S., & Moreira, A. (2010). Alternative mRNA polyadenylation in eukaryotes: An effective or of gene expression. *Wiley Interdisciplinary Reviews. RNA*, *2*(1), 23–31. doi:10.1002/wrna.47 21278855

, S. L., & Hargreaves, M. (2011). Histone modifications and exercise adaptations. *Journal of l Physiology*, *110*(1), 258–263. doi:10.1152/japplphysiol.00979.2010 PMID:21030677

au, Y. R., Soufari, H., Nelson, C. J., & Mackereth, C. D. (2013). Structure and Activity of the l-Prolyl Isomerase Domain from the Histone Chaperone Fpr4 toward Histone H3 Proline Isom- n. *Journey of Biological Chemistry*, *288*(36), 25826–25837. doi:10.1074/jbc.M113.479964 23888048

sagh, M., Choy, M. K., Goddard, M., Bennett, M. R., Down, T. A., & Roger, S. (2010). Differen- IA Methylation Correlates with Differential Expression of Angiogenic Factors in Human Heart . *PLoS One*, *5*(1), e8564. doi:10.1371/journal.pone.0008564 PMID:20084101

A., Rowland, T., Perez-Schindler, J., & Schenk, S. (2014). Understanding the acetylome: Translat- geted proteomics into meaningful physiology. *American Journal of Physiology. Cell Physiology*, , C763–C773. doi:10.1152/ajpcell.00399.2013 PMID:25186010

Pilsner, J. R., Hu, H., Ettinger, A., Sánchez, B. N., Wright, R. O., Cantonwine, D., ... Hernández-Avila, M. (2009). Influence of Prenatal Lead Exposure on Genomic Methylation of Cord Blood DNA. *Environmental Health Perspectives*, *117*(9), 1466–1471. doi:10.1289/ehp.0800497 PMID:19750115

Pilsner, J. R., Hu, H., Ettinger, A., Sánchez, B. N., Wright, R. O., Cantonwine, D., ... Hernández-Avila, M. (2009). Influence of Prenatal Lead Exposure on Genomic Methylation of Cord Blood DNA. *Environmental Health Perspectives*, *117*(9), 1466–1471. doi:10.1289/ehp.0800497 PMID:19750115

Ploumakis, A. & Coleman, M. L. (2015). OH, the Places You'll Go! Hydroxylation, Gene Expression, and Cancer. *Molecular Cell, 58*(5), 729–741. doi:.2015.05.026 doi:10.1016/j.molcel

Rakhimova, A., Ura, S., Hsu, D., Wang, H., Pears, C., & Lakin, N. D. (2017). Site-specific ADP-ribosylation of histone H2B in response to DNA double strand breaks. *Scientific Reports*, *7*(43750). doi:10.1038rep43750 PMID:28252050

Rea, I. M. (2017). Towards ageing well: Use it or lose it: Exercise, epigenetics and cognition. *Biogerontology*, *18*(4), 679–691. doi:10.100710522-017-9719-3 PMID:28624982

Reik, W., Dean, W., & Walter, J. (2001). Epigenetic reprogramming in mammalian development. *Science*, *293*(5532), 1089–1093. doi:10.1126cience.1063443 PMID:11498579

Relton, C. L., & Smith, G. D. (2010). Epigenetic Epidemiology of Common Complex Disease: Prospects for Prediction, Prevention, and Treatment. *PLoS Medicine*, *7*(10), e1000356. doi:10.1371/journal.pmed.1000356 PMID:21048988

Robison, S. K. (2016). The political implications of epigenetics. *Politics and the Life Sciences*, *35*(2), 30–53. doi:10.1017/pls.2016.14

Ross, S. A. (2003). Diet and DNA methylation interactions in cancer prevention. *Annals of the New York Academy of Sciences*, *983*(1), 197–207. doi:10.1111/j.1749-6632.2003.tb05974.x PMID:12724224

Singh, B. N., Shankar, S., & Srivastava, R. K. (2011). Green tea catechin, epigallocatechin-3-gallate (EGCG): Mechanisms, perspectives and clinical applications. *Biochemical Pharmacology*, *82*(12), 1807–1821. doi:10.1016/j.bcp.2011.07.093 PMID:21827739

Stoccoro, A., Karlsson, H. L., Coppedè, F., & Migliore, L. (2013). Epigenetic effects of nano-sized materials. *Toxicology*, *33*(1), 3–14. doi:10.1016/j.tox.2012.12.002 PMID:23238276

Supic, G., Jagodic, M., & Magic, Z. (2013). Epigenetics: A new link between nutrition and cancer. *Nutrition and Cancer*, *65*(6), 781–792. doi:10.1080/01635581.2013.805794 PMID:23909721

Takiguchi, M., Achanzar, W. E., Qu, W., Li, G., & Waalkes, M. P. (2003). Effects of cadmium on DNA-(Cytosine-5) methyltransferase activity and DNA methylation status during cadmium-induced cellular transformation. *Experimental Cell Research*, *286*(2), 355–365. doi:10.1016/S0014-4827(03)00062-4 PMID:12749863

Terry, M. B., Delgado-Cruzata, L., Vin-Raviv, N., Wu, H. C., & Santella, R. M. (2011). DNA methylation in white blood cells: Association with risk factors in epidemiologic studies. *Epigenetics*, *6*(7), 828–837. doi:10.4161/epi.6.7.16500 PMID:21636973

, & Goel, A. (2014). The Importance of Diets and Epigenetics in Cancer Prevention: A Hope for the Future? *Alternative Therapies in Health and Medicine*, *20*(2), 6–11. PMID:25362211

S., Autio, R., Buerki-Thurnherr, T., Arslan, O., Kunzmann, A., Andersson-Willman, B., ... , R. (2013). Gene Expression Profiling of Immune-Competent Human Cells Exposed to En- Zinc Oxide or Titanium Dioxide Nanoparticles. *PLoS One*, *8*(7), e68415. doi:10.1371/journal. 8415 PMID:23894303

ates Environmental Protection Agency. (2018). *Learn about Dioxin*. Retrieved from https:// gov/dioxin/learn-about-dioxin

l, R. A. (2009). Is epigenetics an important link between early life events and adult disease? *Research*, *71*(1), 13–16. doi:10.1159/000178030 PMID:19153498

Huang, K., Kaunang, F. J., Huang, C., Kao, H., Chang, T., ... Lee, T. (2017). Investigation and ion of protein carbonylation sites based on position-specific amino acid composition and physi- al features. *BMC Bioinformatics*, *18*(S3), 66. doi:10.118612859-017-1472-8 PMID:28361707

I., Hsu, P.-Y., Liyanarachchi, S., Liu, J., Deatherage, D. E., Huang, Y.-W., ... Huang, T. H.-M. pigenetic influences of low-dose bisphenol A in primary human breast epithelial cells. *Toxi- d Applied Pharmacology*, *248*(2), 111–121. doi:10.1016/j.taap.2010.07.014 PMID:20678512

& Jones, P. A. (2012). Cancer Genetics and Epigenetics: Two Sides of the Same Coin? *Cancer*), 9–20. doi:10.1016/j.ccr.2012.06.008 PMID:22789535

Irwin, M. L., Lu, L., Risch, H., Mayne, S., Mu, L., ... Yu, H. (2012). Physical activity and breast rvival: An epigenetic link through reduced methylation of a tumor suppressor gene L3MBTL1. *ncer Research and Treatment*, *133*(1), 127–135. doi:10.100710549-011-1716-7 PMID:21837478

, & Pan, X. (2009). RDX induces aberrant expression of microRNAs in mouse brain and liver. *ental Health Perspectives*, *117*(2), 231–240. doi:10.1289/ehp.11841 PMID:19270793

Fisher, B. J., Turkelson, A., Franzblau, A., Diebol, J. K., Allerton, L. A., & Parker, E. A. (2013). t of misunderstanding the chemical properties of environmental contaminants on exposure case involving dioxins. *The Science of the Total Environment*, *447*, 293–300. doi:10.1016/j. 2013.01.030 PMID:23391895

P., Baumann, F. T., Bloch, W., Schenk, A., Koliamitra, C., Jensen, P., ... Elter, T. (2014). Im- ercise on pro inflammatory cytokine levels and epigenetic modulations of tumor competitive tes in non-Hodgkin-lymphoma patients-randomized controlled trial. *European Journal of ogy*, *93*(6), 527–532. doi:10.1111/ejh.12395 PMID:24913351

Guo, J., Takada, M., Wei, W., & Zhang, Q. (2016). New Insights into Protein Hydroxylation portant Role in Human Diseases. *Biochimica et Biophysica Acta (BBA) -. Revue Canadienne*, 208–220. doi:10.1016/j.bbcan.2016.09.004 PMID:27663420

KEY TERMS AND DEFINITIONS

Alternative Splicing: A phenomenon of splicing that results in different options or patterns for a given gene; a process mediated by introns, exons, and proteins that increases the complexity of gene expression.

Biotin (Vitamin B7): A vitamin B-complex that aids in carbohydrate and fat metabolism and in fatty acid and glucose synthesis.

Chromatin Remodelling: Regulated modification of chromatin to allow DNA to be compressed into the nucleus of a cell.

Competitive DNA Binding: A competition to bind DNA when there is overlap between enhancer and silencer sequences.

Endocrine Disruptor: Exogenous chemicals that alter or interfere with the body's normal endocrine function, thus further impairing homeostasis.

Enhancesome: A type of enhancer containing multiple sites for activator binding.

Epigenetics: The study of changes in an organism that influences gene expression but are not mediated by the original DNA sequence.

Epigenomics: The identification of the complete set of epigenetic modifications implicated in gene expression.

Genome-Wide Analyses: The process of measuring and identifying gene features of entire genomes.

Herbicides: A pesticide used to kill unwanted vegetation.

Imprinting: A phenomenon where of the two inherited copies of each gene (alleles), one of them is silenced by epigenetic processes such that only one of them (either from mother or father) is expressed.

Insecticides: A pesticide used to harm, kill or repel insects.

Insulator/Insulator Protein: A group of elements found between the enhancer and the promoter.

Nanotechnology: Field of research designed for the implementation of nanomaterials for new biological or biochemical applications.

Nicotinamide Adenine Dinucleotide: An important electron carrier involved in various redox reactions of living systems—for example, aerobic respiration.

Particulate Matter: The mixture of solid and liquid particles suspended in the air.

Persistent Organic Pollutants: Toxic organic chemicals that are resistant to environmental degradation and that persist and accumulate in adipocytes of living organisms.

Phytochemicals: Biologically active chemicals present in plants that play a role in human health and disease.

Positive Gene Regulation: Gene regulation that results in increased enzymatic activity in the presence of product; in contrast to negative gene regulation.

Silencers: Specific sequences upstream of the promoter region that prevent RNA polymerase from binding to the promoter, thus curtailing gene transcription.

Small Ubiquitin-Related Modifier (SUMO): A polypeptide of less than 100 amino acids that attach or detach to other proteins to modify their functions.

Splicing Regulatory Elements: Elements of gene expression between transcription and translation that play a role in protein processing and degradation.

Chapter 12
Genomics and Genetic Testing

ABSTRACT

This chapter focuses on the Human Genome Project (HGP), which determined that humans have between 20,000 to 25,000 protein-coding genes and only about 1.5% of the genome codes for proteins, rRNA, and tRNA. The remainder once referred as "junk DNA" is today known to be crucial to survival of the cell. Research indicates that genes are not contiguous, and some genes occur within the introns of other genes; some genes can overlap with each other either on the same or on different DNA strands with protein-coding and/or regulatory elements; plus, the vast majority of human genes undergo alternative splicing leading to different proteins being encoded by the same gene. Advances in genomics and gene testing technologies have created exceptional opportunities for the delivery of personalized medicine. Clinical genetic testing has been helpful in identifying gene variants associated with risks for a number of diseases and health conditions.

CHAPTER OUTLINE

- The Human Genome Project
- Genomic Research, Precision Medicine and Pharmacogenetics
- Gene Name and Cytogenetic Location
- Genetic Testing Technology and Methods
  - 12.4.1 Cytogenetic Testing
  - 12.4.2 Biochemical Testing
  - 12.4.3 Molecular Testing
- Genomics, Genetic Testing and Society
- Summary

LEARNING OUTCOMES

- Explain the work of the Human Genome Project and its research and health implications
- Outline the role of genomics in precision medicine and pharmacogenetics
- Illustrate with examples how genes are named
- Differentiate between the various genetic testing methods currently in use
- Summarize the implications of genomic data and gene testing in society

12.1 THE HUMAN GENOME PROJECT

The **Human Genome Project (HGP)** was an international research effort whose primary goal was to determine the sequence of the human genome and identify the genes it contains (introduced in chapter 8, section 8.1). Funded largely by the US Government through the National Institutes of Health and the Department of Energy and involving collaboration between researchers in the US and other countries, the effort started in 1990 and was completed in 2003. It is now known that humans have between 20,000 to 25,000 protein-coding genes which makes only about 1.5% of the genome coding for proteins, rRNA, and tRNA. The rest of the genome once known as "junk DNA" is today known to be crucial to survival of the species. This non-coding DNA consists of repetitive DNA (59%); introns and regulatory sequences (24%); and other non-coding DNA sequences (15.5%).

According to Naidoo et al. (2011), the HGP provided the raw DNA sequence that helped generate several secondary studies which have greatly improved our knowledge of the architecture and function of the genome. New insights have been gained in gene number and density, non-protein-coding RNA genes, pervasive transcription, high copy number repeat sequences and, evolutionary conservation (Naidoo et al., 2011). For example, genome-wide association studies have gradually increased in the last 15 years exploring complex human diseases and traits, and gene sequencing technologies have become more advanced. The advent of next-generation sequencing technologies has dramatically changed studies in structural and functional genomics. For instance, several microarray-based methods have been replaced by sequencing-based approaches such as ChIP-Seq, RNA-Seq, Methyl-Seq and CNV-Seq (Werner, 2010; Naidoo et al., 2011).

Other progress made from the reference genome generated by the HGP are several large-scale international projects which have contributed substantially to our understanding and knowledge of human genetics and genomics. Examples of such projects include the **International HapMap Project**; the **Encyclopaedia of DNA Elements (ENCODE) Project, the 1000 Genomes Project,** the **International Cancer Genome Consortium,** the **National Institute of Health (NIH) Roadmap Epigenomics Program,** and the **Human Microbiome Project.**

Research indicates that genes are not contiguous as earlier thought—some genes are known to occur within the introns of other genes; some genes can overlap with each other either on the same or on different DNA strands with shared coding and/or regulatory elements; plus, the vast majority of human genes undergo alternative splicing leading to different proteins being encoded by the same gene (Yu et al., 2005; Yang & Elnitski, 2008; Pan et al., 2008). Gene density has been shown to vary greatly between the human chromosomes; and so is uneven gene distribution within chromosomes. Some chromosomal regions have been reported to be gene-poor regions (often called 'gene deserts') with no protein-coding genes but may contain regulatory sequences (Ovcharenko et al., 2005).

...oding RNA genes are as widespread as they are diverse and may well exceed protein-coding ...erms of their number. Non-coding RNAs include tRNAs, rRNAs and snRNAs, but also regu-...As involved in sequence-specific transcriptional and post-transcriptional modulation of gene ...n. These include microRNAs, siRNAs, piwi-interacting RNAs, tiRNAs, transcription start ...iated RNAs [TSSa-RNAs], promoter upstream transcripts [PROMPTs], promoter-associated ...ASRs and PALRs] and longer non-coding RNAs (most of these were discussed in section ...ecently large intergenic non-coding RNAs (lincRNAs)—numbering at least 3,000 in the human ...ave been described. They represent a novel category of evolutionarily conserved RNAs with a ...ray of functions ranging from stem cell pluripotency to cellular proliferation. Some lincRNAs ...e involved in guiding chromatin-modifying complexes to specific genomic loci regulating gene ...n. (Guttman et al., 2009; Khalil et al., 2009; Loewer et al., 2010; Ørom & Shiekhattar, 2011) ...gh > 90 per cent of the human genome appears to be represented in nuclear primary transcripts, ...DE project has found that only 35-50 per cent of processed transcripts have so far been an-...genes, implying that many genes may not yet have been recognised as such (Gingeras, 2007). ...urn out that large numbers of unannotated transcripts collectively classified as transcripts of ...function (TUFs), may become functionally significant. Such transcripts include antisense ...s of protein-coding genes, isoforms of protein-coding genes, and transcripts that either overlap ...annotated gene transcripts or which are derived entirely from inter-genic regions. (Kapranov ...7; Dinger et al., 2009)

...GP revealed that repeat sequences account for at least 50 per cent of the human genome sequence ...as transposable elements, pseudogenes, simple sequence repeats, blocks of tandemly repeated ..., and segmental duplications or low copy number repeats (Naidoo et al. 2011). Pseudogene-...NA transcripts may harbour functional elements with a regulatory role, although research is ...n what proportion of the pseudogenes identified to date have retained or acquired a function ...on-coding RNAs. (Svensson et al., 2006; Zheng & Gerstein, 2009). Transposable elements ...bout 40 per cent of the human genome and constitute a major source of inter-individual struc-...ability. Some of these transposable elements have contributed gene-coding sequences to the ...nome; others have contributed functional non-coding sequence like regulatory elements and ...As. (Mills et al., 2007; Piriyapongsa et al., 2007; Xing et al., 2009)

...ding to Naidoo et al (2011), the elaborate transcriptional mechanisms in the genome, wide-...currence of overlapping genes and shared functional elements, males it extremely difficult to ...precisely and unambiguously where one gene ends and another one begins. Single nucleotide ...hisms (SNPs) are the most abundant type of genetic variation in the human genome, occurring ...ls of approximately one SNP to every kb of DNA sequence throughout the genome. This is ...ately equivalent to 3.5 million SNPs being carried by each individual genome. The DNA se-...f any two unrelated genomes are estimated to be about 99.9 per cent identical; the 0.1 per cent ...s mainly SNPs, and these are believed to be responsible for many of the phenotypic differences ...ng individuals in populations—for example, disease susceptibility, drug responses and physical ...as height. Most SNPs are predicted to be neutral and have no functional roles. However, their ...e in the human genome means they may serve as useful genetic markers in individuals. More ...0 SNPs have been reported to be associated with various human complex diseases and traits ...ing degrees of replication and success (Frazer et al., 2009; Naidoo et al., 2011).

12.2 GENOMIC RESEARCH, PRECISION MEDICINE AND PHARMACOGENETICS

Genomics is the study of the structure, function, evolution and mapping of genomes (Vlahovich et al., 2017). According to NIH's National Human Genome Research Institute (NHGRI), there are vast amounts of data about human DNA generated by the Human Genome Project and other genomic research, scientists and clinicians have more powerful tools to study the role that multiple genetic factors acting together and with the environment play in much more complex diseases. These diseases, such as cancer, diabetes, and cardiovascular disease constitute most of health problems in the United States. (NHGRI, 2018)

The human genome is highly variable, and each personal genome differs from the reference human assembly in ~ 3.5 million SNPs and 1000 large (>500 base pairs) copy number variations (CNVs). Apart from the typical Mendelian single gene variants of disease, SNVs as well as CNVs, in noncoding, conserved, or regulatory regions can confer disease. The small differences conferred by SNPs may help predict a person's risk of particular diseases and response to certain medications. The true challenge for personalized genomics is to identify disease-causing mutations among the approximately 3.0–3.5 million SNVs (on average) and ~1,000 CNVs in a given human diploid genome. (Gonzaga-Jauregui et al., 2012)

Translation of genomic information to clinical applications has shown great promise. **Pharmacogenomics** will allow the development of tailored drugs to treat a wide range of health problems, including cardiovascular disease, Alzheimer disease, cancer, HIV/AIDS, and asthma (GHR-NIH, 2018). For instance, in pharmacogenetics, the US Food and Drug Administration (FDA) has approved genotyping tests for the screening of genetic variants in candidate genes that influence the responses and adverse effects of several commonly used anticancer drugs. Genomics information has also been used to develop molecular-targeted cancer therapies. The discovery of the fusion protein (breakpoint cluster region ABL oncogene with a tyrosine kinase activity domain) ultimately led to the development of a molecular-targeted drug (*imatinib*) as a treatment for chronic myeloid leukaemia (CML) (Mauro et al., 2002). *Imatinib* is an inhibitor targeting the tyrosine kinase domain of the fusion.

Recent research has demonstrated that genomic technology, and particularly genome-wide (whole genome or exome) sequencing, can identify genetic causes of rare paediatric diseases much more effectively than conventional clinical and laboratory methods (Friedman et al., 2017). The Global Alliance for Genomics and Health is an international collaboration of more than 400 healthcare, research, disease advocacy, life science, and information technology institutions working together to promote human health through sharing of genomic and clinical data (GAGH, 2018).

Genome-based research is already enabling medical researchers to develop improved diagnostics, more effective therapeutic strategies, evidence-based approaches for demonstrating clinical efficacy, and better decision-making tools for patients and providers. Ultimately, it appears inevitable that treatments will be tailored to a patient's genomic makeup. Thus, the role of genetics in health care is starting to change profoundly and the first examples of the era of genomic medicine are with us today. (NHGRI, 2018) In the era of genome-wide association studies and whole-genome sequencing the great challenge lies in data interpretation and how genomic information can be used to discover new drugs or molecular biomarkers for clinical applications that will eventually translate into patient benefit. The ultimate goal of these studies is to improve the clinical management of patients and to bring about personalized medicine through the development of new therapeutic agents tailored to the individual, based upon their genetic information. Although progress made towards achieving these goals has been promising, many challenges in the translational phase remain. (Guttmacher et al., 2010; Altman et al., 2011)

GENE NAME AND CYTOGENETIC LOCATION

...go Gene Nomenclature Committee (HGNC) designates an official name and symbol for each ...uman gene. HGNC assigns a unique name and symbol to each human gene, which allows ef-...rganization of genes in large databanks, aiding the advancement of research. More than 13,000 ...stimated 20,000 to 25,000 genes in the human genome named. Geneticists use a standardized ...describing a gene's cytogenetic location. In most cases, the location describes the position of a ...ar band on a stained chromosome: 7q31.2. It can also be written as a range of bands, if less is ...about the exact location: 17q12-q21. The combination of numbers and letters provide a gene's ...s" (locus) on a chromosome.

...gene address made up of several parts:

...he chromosome on which the gene can be found – 1st number or letter represents the chromo-...me. Chromosomes 1-22 (the autosomes) are designated by their chromosome number; while the ...x chromosomes are designated by X or Y.

...he chromosome arm is the 2nd part of the gene's address, e.g., 7q is the long arm of chromosome ...and Xp is the short arm of the X chromosome.

...he 3rd part is the position of the gene on the p or q arm based on a distinctive pattern of light and ...rk bands that appear when chromosome is stained in a certain way – see dig. slide 10!

...he genes position is usually designated by two digits (representing a region & a band), some-...mes followed by a decimal point and one or more additional digits (representing sub-bands ...ithin a light or dark area).

...he number indicating the gene position increases with distance from the centromere; e.g. 14q21 ...epresents position 21 on the long arm of chromosome 14; 14q21 is closer to the centromere than ...4q25!

GENETIC TESTING TECHNOLOGY AND METHODS

...y evolving knowledge and technologies have led to wider use and availability of genetic testing, ... reduced costs of DNA sequencing and significantly reduced reporting times on genetic tests ...ovich et al., 2017). However, genomic research raises a number of important issues for research-...cluding the complexity of informed consent, sample and data storage, return of results, research ...ing children and privacy and confidentiality (Green & Guyer, 2011). **Research in genetic testing** ...d to learn more about the contributions of genes to health and to disease. Sometimes the results ...ot be directly helpful to participants, but they may benefit others by helping researchers expand ...nderstanding of the human body, health, and disease. Genetic screening is the use of *systematic* ...ms directed either at whole populations of *asymptomatic* individuals or at subpopulations in which ...known to be increased or in which the specific phase of life merits screening, such as in the case ...gnant women or new-borns. Although these tests may be chosen for recreational purposes (such ...estry testing), they may yield (sometimes unsought) medically relevant information. Commercial ...c testing or screening may be aimed at a broad range of disorders, ranging from traits and common ...ders to rare and serious disorders. (Becker et al., 2011)

Many potentially treatable conditions cannot be detected in infants using current new-born screening methods like amniocentesis and chorionic villus sampling. Most of these disorders result from genetic mutations (either inherited from one or both parents or arising *de novo* in the child) and could, in principle, be diagnosed shortly after birth by means of available genomic technologies. Such conditions include many early-onset seizure disorders, cardiac arrhythmias, cardiomyopathies, diseases of the blood or bone marrow, liver diseases and kidney disorders. Clinical laboratories currently employ molecular genetic technologies for a variety of purposes, including the identification of bacteria or viruses involved in a patient's infection and matching tissue antigens between a donor organ and a patient who requires organ transplantation. In addition, genetic testing is routinely performed by clinical labs in circumstances other than new-born. There are several different kinds of genetic tests that could be used in new-born screening. (Friedman et al., 2017) Depending on the condition being measured, three categories of genetic testing—cytogenetic, biochemical, and molecular—are available to detect abnormalities in chromosome structure, protein function, and DNA sequence, respectively.

12.4.1 Cytogenetic Testing

In this technique, chromosomes are examined to identify structural abnormalities. Through cytogenetic analysis, cells from various tissues in the body like bone marrow cells, blood cells or amniotic fluid foetal cells can be extracted and cultured. After several days of cell culture, chromosomes are fixed, spread on microscope slides, and stained for easier chromosomal identification. Each chromosome will have distinct bands making it easy to identify abnormalities. A technique called fluorescent in situ hybridization (FISH) is used to vividly paints chromosomes or portions of chromosomes with fluorescent molecules to identify chromosomal abnormalities like insertions, deletions, translocations, and/or amplifications). FISH is commonly used to identify specific chromosomal deletions associated with paediatric syndromes such as DiGeorge syndrome (a deletion of part of chromosome 22) and cancers such as chronic myelogenous leukaemia (a translocation involving chromosomes 9 and 22). (Smith & Hung, 2017; NCBI, 2018)

12.4.2 Biochemical Testing

Inborn errors of metabolism are often congenital and include a broad spectrum of defects of various gene products that affect intermediary metabolism in the body (Yu & Scott, 2006). Since genes make proteins, examining them instead of the gene provides insights into such "inborn errors of metabolism". Biochemical tests can be developed that directly measure enzyme activity, level of reaction metabolites (indirect measurement of enzyme activity), and the size or amount of protein (protein structure). Samples where such proteins are present are typically blood, urine, amniotic fluid, or cerebrospinal fluid. Due to the degradability and unstable nature of most gene products, this test requires prompt product collection, storage and shipping to the laboratory. Several technologies are available such as high-performance liquid chromatography, gas chromatography, mass spectrometry, and tandem mass spectrometry which enable both qualitative detection and quantitative determination of metabolites. (Rivera-Colón et al., 2012; NCBI, 2018)

and Genetic Testing

Molecular Testing

...ique works best when the gene sequence of interest is available, and the function of the protein ...n such that a biochemical test cannot be developed. Requiring very small sample amounts ...st can be performed on any tissue sample using several different molecular technologies like ...uencing, polymerase chain reaction-based assays, genomic hybridization, DNA chip and protein ...ys. Comparative genomic hybridization is a molecular cytogenetic method for analysing gains ...n DNA that are not detectable with routine chromosome analysis. Based on the proportion of ...tly-labelled patient DNA to normal-reference DNA this method can detect small deletions ...cations, but not structural chromosomal changes such as balanced reciprocal translocations ...ons or changes in chromosomal copy number. DNA microarray analysis (DNA chip analysis) ...s gene expression where molecules of mRNA bind or hybridize specifically to a known DNA ...gene or portion of from which it originated). When an array contains many DNA templates, the ... level of hundreds to thousands of genes from an individual patient sample can be measured ...omputer to detect the amount of mRNA bound to each site on the array. In protein microar-...is the amount of protein present in biological samples is quantified. Next, the hybridization ...d target proteins in a patient sample is measured against a reference sample (biomarker). The ...absence, increase, or decease of a particular protein can be an indicator of disease in a person. ...ple, analysis of the cerebrospinal fluid of a patient for amyloid beta or tau proteins may be used ...se Alzheimer's disease. (Ferrin et al, 2018; Hoff et al., 2018; NCBI, 2018)

...c tests are done on a small sample of tissue or fluids from the body like blood, cheek cell swabs, ...r, skin, tumors, or a sample of the amniotic fluid obtained by amniocentesis. The sample is sent ...atory that tests it for certain differences in the given genetic material. The laboratory provides ...report of the test results, usually to a health care provider who then talks with the patient about ... (GHR-NIH, 2018b). Genetic tests can be done to confirm a suspected diagnosis, to predict ...ility of future illness, to detect the presence of a carrier state in unaffected individuals (whose ...may be at risk), and to predict response to therapy. Genetic tests may be carried out in the pre-...a, either through pre-implantation genetic, chorionic villus sampling (CVS), or amniocentesis. ...ests may be carried out on adults for all of these indications. (GHR-NIH, 2018b)

...**either of the three** categories of genetic testing—cytogenetic, biochemical, and molecular, ...**several types of genetic tests that can be performed:**

...**gnostic Testing:** In situations where there are symptoms or manifestation of an illness, a diag-...ic test may be requested to precisely identify the cause of an illness or disease so that choices ...t treatment or health management can be made. For example, a person having seizures fre-...tly may be required to have a medical imaging test like a CT scan or MRI.

...**dictive (pre-Symptomatic) Genetic Test:** This test is ordered to find gene changes that increase ...rson's likelihood of developing diseases. The results of the test provide information about a ...on's risk of developing a specific disease like haemophilia; information useful in lifestyle and ...thcare choices.

...**rier Testing:** This test may be ordered for people who "carry" a change in a gene that is linked ...sease. Carriers may show no signs of the disease; however, they can pass on the gene change to ...r children, who may develop the disease or become carriers themselves. Some diseases require a gene change to be inherited from both parents for the disease to occur. A typical case is sickle-cell anaemia which is caused by a faulty haemoglobin. Individuals who are normal have the dominant allele genotype, SS; carriers have the heterozygous genotype, Ss; while those with the condition have the homozygous recessive genotype, ss.

4. **Prenatal Testing:** This is offered during pregnancy to help identify foetuses that have certain diseases. This includes several methods including ultrasound, amniocentesis, CVS and pre-implantation diagnosis, fetoscopy, and foetal blood sampling (see chapter 6, section 6.3 for a fuller discussion of these procedures).

5. **New-Born Screening:** This test is used to test babies one or two days after birth to find out if they have certain diseases known for developmental, genetic, and metabolic disorders in the new-born baby. Such conditions are treatable if detected early and can include conditions like phenylketonuria (inability to metabolize the amino acid phenylalanine); hemoglobinopathy (production of abnormal haemoglobin); cystic fibrosis (caused by a defective CTFR gene that makes a chloride channel protein); hearing loss; severe combined immunodeficiency (SCID) caused by deficient T-lymphocyte production; Duchenne muscular dystrophy (MDM) caused by a faulty muscle protein dystrophin; and eye and vision screens that could detect diseases like retinoblastoma (cancerous cells in the immature retina).

6. **Pharmacogenomic Testing:** Pharmacogenomics is the study of how genetic factors relate to interindividual variability of drug response (Kitzmiller et al., 2011). It provides gives information about how certain medicines are processed by an individual's body allowing the choice of medications that are optimal in a specific genetic profile. For instance, warfarin is used for the long-term treatment and prevention of thromboembolic events. Statins lower the concentration of low-density lipoprotein cholesterol (LDL-C), resulting in a relative-risk reduction of about 20% for each 1 mmol/L (39 mg/dL) decrement in LDL-C; they are one of the most commonly prescribed classes of drugs, but their side effects can limit their appeal. Tamoxifen is prescribed to prevent the recurrence of oestrogen-receptor-positive breast cancer, to treat metastatic breast cancer, to prevent cancer in high-risk populations, and to treat ductal carcinoma in situ. (Kitzmiller et al., 2011)

12.5 GENOMICS, GENETIC TESTING AND SOCIETY

Advances in genomics and gene sequencing technologies have created exceptional opportunities for the delivery of personalized medical care (Hamilton et al., 2016). Clinical genetic testing has been helpful in identifying gene variants associated with risks for a number of diseases and health conditions including heart disease, stroke, dementia, cancer, and Type-2 diabetes (Bellcross et al., 2012). These tests provide risk information that can guide decisions regarding disease prevention, diagnosis, and treatment for individuals seeking testing and their families. There has also been an upsurge in the availability of direct-to-consumer genetic tests that predict risks for multiple common diseases and traits and are advertised and sold directly to the general public, typically without the involvement of a healthcare provider (Frueh et al., 2011; Bellcross et al., 2012). Applications of genomic medicine is expected to substantially improve health outcomes including reduced incidences of illness (morbidity) and mortality.

As noted by Zhao and Grant (2011), genetic testing has been transformed by an explosion of genomic data, powerful new technologies and analytical approaches. It is now generally known that risk information generated from genome analysis for a range of diseases conditions will inform disease prevention

Bloss *et al.*, 2011). How useful genetic test information is in improving health outcomes remains le and may be a major obstacle to further translation or adoption (Hunter *et al.,* 2008; Khoury, Even if clinical utility of this information is demonstrated, barriers to full benefits still exists patients/consumers are unable to correctly interpret and understand the significance of genomic rmation, either in the specific context of health care or for one's overall sense of personal well-Haga et al., 2013).

udes of the public towards genetic testing for the risk of diseases (including cancer), are re-o be generally positive (Etchegary, 2014). For example, in a study carried out in the US, 97% of ants indicated that they were at least somewhat interested in the topic of genetic testing and the had positive attitudes about genetic research and showed approval of the use of genetic testing e detection (Haga et al., 2013). A Dutch survey found that 64% of participants believed genetic would help people to live longer. Nevertheless, there are also concerns amongst the public that est results could be used to discriminate against those with a genetic predisposition for illness genetic testing could result in people being labelled as having "good" or "bad" genes. (Haga et ; Henneman et al., 2013)

in factors may hinder some communities (especially minority populations) from access to genetic including low awareness of these tests; knowledge of genetic testing; services and resources e; language barriers; stigma associated with being at risk; fatalistic views of some of the diseases cer; anticipation of negative emotions; uncertainty about the information provided; and mistrust data would be used (Allford et al., 2013). The most agreeable reasons for genetic testing were to be helping people make decisions on screening; motivating self-examination; ability of results help to family and children; reducing concerns about cancer; reducing uncertainty; providing a personal control; and helping plan for the future (Hann et al., 2017). There was less agreement participants towards benefits of tests such as help in making important life decisions; providing nce; helping with cancer prevention; and making decisions about preventative surgery (Hann 017). When asked for reasons why they enrolled for genetic testing, participants cited family/ motives (62%), information (28%), and society (9%) (Kinney et al., 2006).

most frequently endorsed limitations or barriers to genetic testing/counselling included: anticipated d worry about offspring/relatives if test result is positive; anticipated personal emotional reaction fear, anger) if test result is positive; concern about family's reaction or impact on family; concerns nfidentiality; concern about jeopardizing/losing insurance; cost; and feeling unable to handle the tionally. Reported reasons for a lack of interest/intention to pursue testing included cost, time, at others would find out, belief the results could be wrong, worry about increased risk, concern scomfort, concern about discrimination, not wanting blood taken, logistical reasons, personal and having heard negative experiences of others who had undergone testing. (Hann et al., 2017)

adoption of **personalized medicine** will be driven, in part, by the public's understanding and n new clinical genetic applications (Syurina et al., 2011). Haga et al. (2013 conclude that it is lear what levels of knowledge of genetics are needed or desired to ensure informed decision and optimize the understanding of genomic risk. Certainly, studies on people's biased knowledge that more effort is needed to present the benefits, risks, and limitations of genetic testing to nformed decision making, both in the context of health care and in terms of one's overall sense nal well-being and social identity (Haga et al., 2013; Hamilton et al., 2016). Studies show that 5% of the American public are aware of direct-to-consumer (DTC) genetic testing, indicating eed and an opportunity for educating the public about the availability, utility, and limitations of these tests (Agurs-Collins et al., 2015). Ensuring that the public has a complete understanding of the characteristics of DTC genetic tests and can make appropriate, informed decisions about their use will require efforts from policy makers, clinicians, and researchers. Oversight from the US Food and Drug Administration (FDA) will be crucial to ensure that DTC genetic testing companies provide accurate, consistent and reliable results. Public health education and promotion agencies will need to develop better health communication and educational materials to establish knowledge among the general public about the characteristics, potential benefits, and limitations of DTC genetic testing and the risk information it provides. Health care providers will face increasing challenges of addressing concerns or questions about the low predictive power and uncertain clinical validity of DTC genetic testing, and with helping their patients to determine how this risk information may be relevant to their lives. DTC genetic testing is fraught with many challenges including inconsistency in disease risk estimates across companies, lack of established clinical utility and validity, and misleading web-based advertisements. (Lewis et al., 2011; Kolor et al., 2012; Agurs-Collins et al., 2015)

CHAPTER SUMMARY

The Human Genome Project (HGP), an international research effort whose primary goal was to determine the sequence of the human genome and identify the genes it contains; found out that humans have between 20,000 to 25,000 protein-coding genes and only about 1.5% of the genome codes for proteins, rRNA, and tRNA. The rest of the genome once known as "junk DNA" consisting of repetitive DNA (59%); introns and regulatory sequences (24%); and other non-coding DNA sequences (15.5%) is today known to be crucial to survival of the species. The advent of next-generation sequencing technologies has dramatically changed studies in structural and functional genomics. For instance, several microarray-based methods have been replaced by sequencing-based approaches such as ChIP-Seq, RNA-Seq, Methyl-Seq and CNV-Seq. Other progress made from the reference genome generated by the HGP are several large-scale international projects which have contributed substantially to our understanding and knowledge of human genetics and genomics.

Research indicates that genes are not contiguous and some genes occur within the introns of other genes; some genes can overlap with each other either on the same or on different DNA strands with shared coding and/or regulatory elements; plus, the vast majority of human genes undergo alternative splicing leading to different proteins being encoded by the same gene. Gene density and distribution varies between chromosomes; and some chromosomal regions have been reported to be gene-poor regions (often called 'gene deserts') with no protein-coding genes but may contain regulatory sequences. The HGP revealed that repeat sequences account for at least 50 per cent of the human genome sequence classified as transposable elements, pseudogenes, simple sequence repeats, blocks of tandemly repeated sequences, and segmental duplications or low copy number repeats. Transposable elements make up about 40 per cent of the human genome and constitute a major source of inter-individual structural variability. Some of these transposable elements have contributed gene-coding sequences to the human genome; others have contributed functional non-coding sequence like regulatory elements and microRNAs. The DNA sequences of any two unrelated genomes are estimated to be about 99.9 per cent identical; the 0.1 per cent comprises mainly SNPs, and these are believed to be responsible for many of the phenotypic differences noted among individuals in populations—for example, disease susceptibility, drug responses and physical traits such as height.

and Genetic Testing

...nics is the study of the structure, function, evolution and mapping of genomes. The human ... highly variable, and each personal genome differs from the reference human assembly in ~ 3.5 ...NPs and 1000 large (>500 base pairs) copy number variations (CNVs). Apart from the typical ... single gene variants of disease, SNVs as well as CNVs, in noncoding, conserved, or regulatory ... confer disease. The small differences conferred by SNPs may help predict a person's risk of ... diseases and response to certain medications. The true challenge for personalized genomics ...ify disease-causing mutations among the approximately 3.0–3.5 million SNVs (on average) ...0 CNVs in a given human diploid genome. Translation of genomic information to clinical ...ns has shown great promise. Pharmacogenomics will allow the development of tailored drugs ...wide range of health problems, including cardiovascular disease, Alzheimer disease, cancer, ...S, and asthma (GHR-NIH, 2018). Recent research has demonstrated that genomic technol-...particularly genome-wide (whole genome or exome) sequencing, can identify genetic causes ...ediatric diseases much more effectively than conventional clinical and laboratory methods. ...based research is already enabling medical researchers to develop improved diagnostics, more ...herapeutic strategies, evidence-based approaches for demonstrating clinical efficacy, and better ...making tools for patients and providers. Ultimately, it appears inevitable that treatments will ...d to a patient's genomic makeup.

...han 13,000 of the estimated 20,000 to 25,000 genes in the human genome have been named. ...s use a standardized way of describing a gene's cytogenetic location. In most cases, the location ...the position of a particular band on a stained chromosome: 7q31.2. It can also be written as a ...ands if less is known about the exact location: 17q12-q21. The combination of numbers and ...vide a gene's "address" (locus) on a chromosome. Rapidly evolving knowledge and technolo-...led to wider use and availability of genetic testing, greatly reduced costs of DNA sequencing ...icantly reduced reporting times on genetic tests. However, genomic research raises a number ...ant issues for researchers, including the complexity of informed consent, sample and data stor-...n of results, research involving children and privacy and confidentiality. Research in **genetic** ... used to learn more about the contributions of genes to health and to disease. Sometimes the ...y not be directly helpful to participants, but they may benefit others by helping researchers ...eir understanding of the human body, health, and disease. Although these tests may be chosen ...tional purposes (such as ancestry testing), they may yield (sometimes unsought) medically ...nformation. Commercial genetic testing or screening may be aimed at a broad range of disor-...ing from traits and common disorders to rare and serious disorders.

...potentially treatable conditions cannot be detected in infants using current new-born screening ...ike amniocentesis and chorionic villus sampling. Most of these disorders result from genetic ... (either inherited from one or both parents or arising *de novo* in the child) and could, in principle, ...sed shortly after birth by means of available genomic technologies. Such conditions include ...y-onset seizure disorders, cardiac arrhythmias, cardiomyopathies, diseases of the blood or bone ...iver diseases and kidney disorders. Clinical laboratories currently employ molecular genetic ...ies for a variety of purposes, including the identification of bacteria or viruses involved in a ...nfection and matching tissue antigens between a donor organ and a patient who requires organ ...ation. In addition, genetic testing is routinely performed by clinical labs in circumstances other ...born. There are several different kinds of genetic tests that could be used in new-born screening. ...g on the condition being measured, three categories of genetic testing—cytogenetic, biochemi-...molecular—are available to detect abnormalities in chromosome structure, protein function, and DNA sequence, respectively. Using either of the three categories of genetic testing, there are several types of genetic tests that can be performed: diagnostic testing, predictive and pre-symptomatic genetic tests, carrier testing, prenatal testing, new-born screening, and pharmacogenomic testing.

Advances in genomics and gene sequencing technologies have created exceptional opportunities for the delivery of personalized medical care. Clinical genetic testing has been helpful in identifying gene variants associated with risks for a number of diseases and health conditions including heart disease, stroke, dementia, cancer, and Type-2 diabetes. These tests provide risk information that can guide decisions regarding disease prevention, diagnosis, and treatment for individuals seeking testing and their families. Applications of genomic medicine is expected to substantially improve health outcomes including reduced incidences of illness (morbidity) and mortality. Attitudes of the public towards genetic testing for the risk of diseases (including cancer), are reported to be generally positive. Nevertheless, there are also concerns amongst the public that genetic test results could be used to discriminate against those with a genetic predisposition for illness and that genetic testing could result in people being labelled as having "good" or "bad" genes. Certain factors may hinder some communities (especially minority populations) from access to genetic services including low awareness of these tests; knowledge of genetic testing; services and resources available; language barriers; stigma associated with being at risk; fatalistic views of some of the diseases like cancer; anticipation of negative emotions; uncertainty about the information provided; and mistrust of how data would be used.

The adoption of personalized medicine will be driven, in part, by the public's understanding and interest in new clinical genetic applications. Studies on people's biased knowledge suggests that more effort is needed to present the benefits, risks, and limitations of genetic testing to ensure informed decision making, both in the context of health care and in terms of one's overall sense of personal well-being and social identity. Ensuring that the public has a complete understanding of the characteristics of DTC genetic tests and can make appropriate, informed decisions about their use will require efforts from policy makers, clinicians, and researchers. Public health education and promotion agencies will need to develop better health communication and educational materials to establish knowledge among the general public about the characteristics, potential benefits, and limitations of DTC genetic testing and the risk information it provides.

End of Chapter Quiz

1. The human genome only contains ____ percent of coding proteins.
    a. 1%
    b. 1.5%
    c. 10%
    d. 15%
    e. 20%
2. The genome sequences of any two unrelated genomes are estimated to be about ____ identical.
    a. 10%
    b. 25%
    c. 50%
    d. 75%
    e. 99%

cs and Genetic Testing

Which of these is true about "gene deserts"? They
- are overlapping genes
- may contain regulatory sequences
- are gene dense regions
- are non-coding genes
- Both B and D

"Junk" DNA
- includes repetitive DNA, introns, and regulatory sequences
- are part of the human genome known to be crucial to crucial to the survival of the species
- are part of the human genome known to have no effect on the survival of the species
- A and B
- A and C

The three categories of genetic testing are
- Cytogenetic, biochemical, and molecular
- Cytochrome, biochemical and molecular
- Developmental, biochemical and molecular
- Developmental, biogenetic, and molecular
- None of the above

What technique is used to paint chromosomes with fluorescent molecules to identify chromosomal abnormalities?
- Tandem mass spectrometry
- Fluoroscopy
- Fluorescent in situ hybridization
- Mass spectrometry
- B and C

Several methods of prenatal testing include
- amniocentesis
- chorionic villus sampling
- ultrasound
- foetal blood sampling
- all of the above

Pharmacogenetics is the study of how genetic factors relate to _____ variability of drug response.
- intraindividual
- interindividual
- inter-community
- intra-community
- none of the above

What are some limitations to genetic testing/counselling?
- Anticipated emotional reaction
- Cost
- Impact on family
- Concern about confidentiality
- All of the above

10. Who designates an official name and symbol for each known human genome?
    a. Gene Name Committee
    b. Binomial Nomenclature Committee
    c. Charles Linnaeus
    d. Hugo Gene Nomenclature Committee
    e. Health Nomenclature Committee

Thought Questions

1. Why is "junk DNA" so important and what role does it play in species survival?
2. Given the different types of genetic tests you have read about, which one do you think is the most practical and effective, and why?
3. How would the gene address be affected if an individual has more than two sex chromosomes?

REFERENCES

Agurs-Collins, T., Ferrer, R., Ottenbacher, A., Waters, E. A., O'Connell, M. E., & Hamilton, J. G. (2015). Public Awareness of Direct-to-Consumer Genetic Tests: Findings from the 2013 U.S. Health Information National Trends Survey. *Journal of Cancer Education: The Official Journal of the American Association for Cancer Education, 30*(4), 799–807. doi:10.100713187-014-0784-x PMID:25600375

Allford, A., Qureshi, N., Barwell, J., Lewis, C., & Kai, J. (2014). What hinders minority ethnic access to cancer genetics services and what may help? *European Journal of Human Genetics, 22*(7), 866–874. doi:10.1038/ejhg.2013.257 PMID:24253862

Altman, R. B., Kroemer, H. K., McCarty, C. A., Ratain, M. J., & Roden, D. (2011). Pharmacogenomics: Will the promise be fulfilled? *Nature Reviews. Genetics, 12*(1), 69–73. doi:10.1038/nrg2920 PMID:21116304

Becker, F., van El, C. G., Ibarreta, D., Zika, E., Hogarth, S., Borry, P., ... Cornel, M. C. (2011). Genetic testing and common disorders in a public health framework: How to assess relevance and possibilities. *European Journal of Human Genetics, 19*(S1), S6–S44. doi:10.1038/ejhg.2010.249 PMID:21412252

Bellcross, C. A., Page, P. Z., & Meaney-Delman, D. (2012). Direct-to-consumer personal genome testing and cancer risk prediction. *Cancer Journal (Sudbury, Mass.), 18*(4), 293–302. doi:10.1097/PPO.0b013e3182610e38 PMID:22846729

Bloss, C. S., Jeste, D. V., & Schork, N. J. (2011). Genomics for disease treatment and prevention. *The Psychiatric Clinics of North America, 34*(1), 147–166. doi:10.1016/j.psc.2010.11.005 PMID:21333845

Dinger, M. E., Amaral, P. P., Mercer, T. R., & Mattick, J. S. (2009). Pervasive transcription of the eukaryotic genome: Functional indices and conceptual implications. *Briefings in Functional Genomics & Proteomics, 8*(6), 407–423. doi:10.1093/bfgp/elp038 PMID:19770204

Etchegary, H. (2014). Public attitudes toward genetic risk testing and its role in healthcare. *Personalized Medicine, 11*(5), 509–522. doi:10.2217/pme.14.35 PMID:29758777

Kirakodu, S., Jensen, D., Al-Attar, A., Payyala, R., Novak, M. J., ... Gonzalez, O. A. (2018). ression analysis of neuropeptides in oral mucosa during periodontal disease in nonhuman *Journal of Periodontology*, *89*(7), 858–866. doi:10.1002/JPER.17-0521 PMID:29676776

A., Murray, S. S., Schork, N. J., & Topol, E. J. (2009). Human genetic variation and its contribu-nplex traits. *Nature Reviews. Genetics*, *10*(4), 241–251. doi:10.1038/nrg2554 PMID:19293820

, J. M., Cornel, M. C., Goldenberg, A. J., Lister, K. J., Sénécal, K., & Vears, D. F. (2017). newborn screening: Public health policy considerations and recommendations. *BMC Medical* , *10*(1), 9. doi:10.118612920-017-0247-4 PMID:28222731

W., Greely, H. T., Green, R. C., Hogarth, S., & Siegel, S. (2011). The future of direct-to-consumer enetic tests. *Nature Reviews. Genetics*, *12*(7), 511–515. doi:10.1038/nrg3026 PMID:21629275

Home Reference – National Institute of Health. (2018a). *What is pharmacogenomics?* Retrieved s://ghr.nlm.nih.gov/primer/ genomicresearch/pharmacogenomics

Home Reference – National Institute of Health. (2018b). *Genetic Testing: What It Means for lth and Your Family's Health*. Retrieved from www.genome.gov/ pages/health/patientspub-netictestingfactsheet.pdf

T. R. (2007). Origin of phenotypes: Genes and transcripts. *Genome Research*, *17*(6), 682–690. 01/gr.6525007 PMID:17567989

lliance for Genomics & Health. (2018). *Enabling genomic data sharing for the benefit of hu-th*. Retrieved from https://www.ga4gh.org

Jauregui, C., Lupski, J. R., & Gibbs, R. A. (2012). Human Genome Sequencing in Health ase. *Annual Review of Medicine*, *63*(1), 35–61. doi:10.1146/annurev-med-051010-162644 248320

D., & Guyer, M. S.National Human Genome Research Institute. (2011). Charting a course for medicine from base pairs to bedside. *Nature*, *470*(7333), 204–213. doi:10.1038/nature09764 307933

er, A. E., McGuire, A. L., Ponder, B., & Stefánsson, K. (2010). Personalized genomic informa-aring for the future of genetic medicine. *Nature Reviews. Genetics*, *11*(2), 161–165. doi:10.1038/ PMID:20065954

M., Amit, I., Garber, M., French, C., Lin, M. F., Feldser, D., ... Lander, E. S. (2009). Chro-nature reveals over a thousand highly conserved large non-coding RNAs in mammals. *Nature*,), 223–227. doi:10.1038/nature07672 PMID:19182780

B., Barry, W. T., Mills, R., Ginsburg, G. S., Svetkey, L., Sullivan, J., & Willard, H. F. (2013). owledge of and attitudes toward genetics and genetic testing. *Genetic Testing and Molecular rs*, *17*(4), 327–335. doi:10.1089/gtmb.2012.0350 PMID:23406207

Hamilton, J. G., Shuk, E., Arniella, G., González, C. J., Gold, G. S., Gany, F., ... Hay, J. L. (2016). Genetic Testing Awareness and Attitudes among Latinos: Exploring Shared Perceptions and Gender-Based Differences. *Public Health Genomics*, *19*(1), 34–46. doi:10.1159/000441552 PMID:26555145

Hamilton, J. G., Shuk, E., Arniella, G., González, C. J., Gold, G. S., Gany, F., ... Hay, J. L. (2016). Genetic Testing Awareness and Attitudes among Latinos: Exploring Shared Perceptions and Gender-Based Differences. *Public Health Genomics*, *19*(1), 34–46. doi:10.1159/000441552 PMID:26555145

Hann, K. E. J., Freeman, M., Fraser, L., Waller, J., Sanderson, S. C., Rahman, B., ... Lanceley, A. (2017). Awareness, knowledge, perceptions, and attitudes towards genetic testing for cancer risk among ethnic minority groups: A systematic review. *BMC Public Health*, *17*(1), 503. doi:10.118612889-017-4375-8 PMID:28545429

Henneman, L., Vermeulen, E., van El, C. G., Claassen, L., Timmermans, D. R. M., & Cornel, M. C. (2013). Public attitudes towards genetic testing revisited: Comparing opinions between 2002 and 2010. *European Journal of Human Genetics*, *21*(8), 793–799. doi:10.1038/ejhg.2012.271 PMID:23249955

Hoff, F. W., Hu, C. W., Qiu, Y., Ligeralde, A., Yoo, S. Y., Mahmud, H., ... Kornblau, S. M. (2018). Recognition of Recurrent Protein Expression Patterns in Pediatric Acute Myeloid Leukemia Identified New Therapeutic Targets. *Molecular Cancer Research*, *16*(8), 1275–1286. doi:10.1158/1541-7786.MCR-17-0731 PMID:29669821

Hunter, D. J., Khoury, M. J., & Drazen, J. M. (2008). Letting the genome out of the bottle--will we get our wish? *The New England Journal of Medicine*, *358*(2), 105–107. doi:10.1056/NEJMp0708162 PMID:18184955

Kapranov, P., Willingham, A. T., & Gingeras, T. R. (2007). Genome-wide transcription and the implications for genomic organization. *Nature Reviews. Genetics*, *8*(6), 413–423. doi:10.1038/nrg2083 PMID:17486121

Khalil, A. M., Guttman, M., Huarte, M., Garber, M., Raj, A., Rivea Morales, D., … Rinn, J. L. (2009). Many human large intergenic noncoding RNAs associate with chromatin-modifying complexes and affect gene expression. *Proceedings of the National Academy of Sciences of the United States of America*, *106*(28), 11667–11672. 10.1073/pnas.0904715106

Khoury, M. J. (2010). Dealing with the evidence dilemma in genomics and personalized medicine. *Clinical Pharmacology and Therapeutics*, *87*(6), 635–638. doi:10.1038/clpt.2010.4 PMID:20485318

Kinney, A. Y., Simonsen, S. E., Baty, B. J., Mandal, D., Neuhausen, S. L., Seggar, K., ... Smith, K. (2006). Acceptance of genetic testing for hereditary breast ovarian cancer among study enrollees from an African American kindred. *American Journal of Medical Genetics*, *140A*(8), 813–826. doi:10.1002/ajmg.a.31162 PMID:16523520

Kitzmiller, J. P., Groen, D. K., Phelps, M. A., & Sadee, W. (2011). Pharmacogenomic testing: Relevance in medical practice: Why drugs work in some patients but not in others. *Cleveland Clinic Journal of Medicine*, *78*(4), 243–257. doi:10.3949/ccjm.78a.10145 PMID:21460130

S., Cabili, M. N., Guttman, M., Loh, Y.-H., Thomas, K., Park, I. H., ... Rinn, J. L. (2010). Large [intergen]ic non-coding RNA-RoR modulates reprogramming of human induced pluripotent stem cells. [Nature] *Genetics*, *42*(12), 1113–1117. doi:10.1038/ng.710 PMID:21057500

M. J., O'Dwyer, M., Heinrich, M. C., & Druker, B. J. (2002). STI571: A paradigm of new agents [for canc]er therapeutics. *Journal of Clinical Oncology*, *20*(1), 325–334. doi:10.1200/JCO.2002.20.1.325 [PMID:1]1773186

E., Bennett, E. A., Iskow, R. C., & Devine, S. E. (2007). Which transposable elements are active [in the hu]man genome? *Trends in Genetics*, *23*(4), 183–191. doi:10.1016/j.tig.2007.02.006 PMID:17331616

N., Pawitan, Y., Soong, R., Cooper, D. N., & Ku, C.-S. (2011). Human genetics and genomics a [decade] after the release of the draft sequence of the human genome. *Human Genomics*, *5*(6), 577–622. doi:10.1186/1479-7364-5-6-577 PMID:22155605

[National] Human Genome Research Institute. (2018). *A Brief Guide to Genomics*. Retrieved from https://[www.ge]nome.gov /18016863/a-brief-guide-to-genomics/

[NCBI. (]2018). *Appendix I: Genetic Testing Methodologies*. Retrieved from https://www.ncbi. nlm.nih.[gov/boo]ks/NBK115548/

J. A., & Shiekhattar, R. (2011). Long non-coding RNAs and enhancers. *Current Opinion in [Genetics] & Development*, *21*(2), 194–198. doi:10.1016/j.gde.2011.01.020 PMID:21330130

[Ovcharen]ko, I., Loots, G. G., Nobrega, M. A., Hardison, R. C., Miller, W., & Stubbs, L. (2005). Evolution [and func]tional classification of vertebrate gene deserts. *Genome Research*, *15*(1), 137–145. doi:10.1101/[gr.3015]505 PMID:15590943

Shai, O., Lee, L. J., Frey, B. J., & Blencowe, B. J. (2008). Deep surveying of alternative splic[ing com]plexity in the human transcriptome by high-throughput sequencing. *Nature Genetics*, *40*(12), [1413–14]15. doi:10.1038/ng.259 PMID:18978789

[Piriyapo]ngsa, J., Mariño-Ramírez, L., & Jordan, I. K. (2007). Origin and evolution of human microR[NAs fro]m transposable elements. *Genetics*, *176*(2), 1323–1337. doi:10.1534/genetics.107.072553 [PMID:1]7435244

Colón, Y., Schutsky, E. K., Kita, A. Z., & Garman, S. C. (2012). The structure of human GALNS [reveals t]he molecular basis for mucopolysaccharidosis IV A. *Journal of Molecular Biology*, *423*(5). doi:10.1016/j.jmb.2012.08.020 PMID:22940367

., & Hung, D. (2017). The dilemma of diagnostic testing for Prader-Willi syndrome. *Transla[tional P]ediatrics*, *6*(1), 46–56. doi:10.21037/tp.2016.07.04 PMID:28164030

n, O., Arvestad, L., & Lagergren, J. (2006). Genome-wide survey for biologically functional [pseudog]enes. *PLoS Computational Biology*, *2*(5), e46. doi:.pcbi.0020046 doi:0.1371/journal

E. V., Brankovic, I., Probst-Hensch, N., & Brand, A. (2011). Genome-based health literacy: A new [challeng]e for public health genomics. *Public Health Genomics*, *14*(4-5), 201–210. doi:10.1159/000324238 [PMID:2]1734434

Vlahovich, N., Fricker, P. A., Brown, M. A., & Hughes, D. (2017). Ethics of genetic testing and research in sport: A position statement from the Australian Institute of Sport. *British Journal of Sports Medicine*, *51*(1), 5–11. doi:10.1136/bjsports-2016-096661 PMID:27899345

Werner, T. (2010). Next generation sequencing in functional genomics. *Briefings in Bioinformatics*, *11*(5), 499–511. doi:10.1093/bib/bbq018 PMID:20501549

Xing, J., Zhang, Y., Han, K., Salem, A. H., Sen, S. K., Huff, C. D., ... Jorde, L. B. (2009). Mobile elements create structural variation: Analysis of a complete human genome. *Genome Research*, *19*(9), 1516–1526. doi:10.1101/gr.091827.109 PMID:19439515

Yang, M. Q., & Elnitski, L. L. (2008). Diversity of core promoter elements comprising human bidirectional promoters. *BMC Genomics*, *9*(2), S3. doi:10.1186/1471-2164-9-S2-S3 PMID:18831794

Yu, C., & Scott, C. R. (2006). Human biochemical genetics: An insight into inborn errors of metabolism. *Journal of Zhejiang University. Science. B.*, *7*(2), 165–166. doi:10.1631/jzus.2006.B0165 PMID:16421978

Yu, P., Ma, D., & Xu, M. (2005). Nested genes in the human genome. *Genomics*, *86*(4), 414–422. doi:10.1016/j.ygeno.2005.06.008 PMID:16084061

Zhao, J., & Grant, S. F. (2011). Advances in whole genome sequencing technology. *Current Pharmaceutical Biotechnology*, *12*(2), 293–305. doi:10.2174/138920111794295729 PMID:21050163

Zheng, D., & Gerstein, M. B. (2007). The ambiguous boundary between genes and pseudogenes: The dead rise up, or do they? *Trends in Genetics*, *23*(5), 219–224. doi:10.1016/j.tig.2007.03.003 PMID:17382428

KEY TERMS AND DEFINITIONS

1000 Genomes Project: An international research effort launched between 2008 and 2015 that was responsible for generating the largest public catalogue of genetic variation and genotypic data.

Encyclopaedia of DNA Elements (ENCODE) Project: A public research association that started in 2003 with a main goal of identifying all functional elements in the genomes of humans and mice.

Hugo Gene Nomenclature Committee: A committee that sets the official standard for human genome nomenclature and associated genomic information. It aims to assign a unique yet meaningful name or symbol for each known human gene.

Human Genome Project: An international project created in 1990 with the goal of sequencing and identifying the entire human genome.

Human Microbiome Project: A research initiative established in 2008 under the United States National Institutes of Health that enabled the study of human microbiota and its role in the development of human health and disease.

International Cancer Genome Consortium: An international association established in 2007 with the aim of identifying, defining, and understanding the genomic changes of 25,000 untreated cancer cases.

International HapMap Project: An international project created in 2002 that contributed to the development of a haplotype map of the human genome.

and Genetic Testing

nal Institute of Health (NIH) Roadmap Epigenomics Program: An international consortium goal of creating accessible human epigenomic data that can be used to further integrate and current research in the scientific community.

nalized Medicine: The application of genomic advances and gene sequencing knowledge to- delivery of personalized medical care to improve health outcome and reduce patient mortality idity.

nacogenomics: A sector within biomedicine and biotechnology that aims to revolutionize modern by advancing pharmaceutical development in vaccines, therapies, and biological molecules.

Chapter 13
Disorders of the Human Circulatory System

ABSTRACT

This chapter focuses on genetic disorders affecting the human circulatory system. Genetic disorders can occur due to a defect in a single gene or in a set of genes. The body's circulatory system is made up of the heart and blood vessels (arteries, arterioles, veins, venules, and capillaries). The system carries both blood and lymphatic fluid in two circuits: pulmonary circulation (blood through the lungs for oxygenation) and systemic circuits (from the heart to all body parts). Fourteen disorders are presented in this chapter including sickle cell disease, Gaucher Disease, chronic myeloid leukaemia, Niemann-Pick Disease, haemophilia, atherosclerosis, ataxia telangiectasia, haemoglobinuria, thalassemia, William's syndrome, porphyria, long QT syndrome, and alpha-I-antitrypsin deficiency.

CHAPTER OUTLINE

13.1 Overview of the Circulatory System
13.2 Sickle-Cell Anemia
13.3 Burkitt Lymphoma
13.4 Gaucher Disease
13.5 Chronic Myeloid Leukemia
13.6 Niemann-Pick Disease
13.7 Hemoglobinuria
13.8 Hemophilia
13.9 Thalassemia
13.10 Atherosclerosis
13.11 William's Syndrome
13.12 Ataxia Telangiectasia
13.13 Porphyria

DOI: 10.4018/978-1-5225-8066-9.ch013

...rs of the Human Circulatory System

...ong QT Syndrome
...lpha-I-Antitrypsin Deficiency
...Summary

...NING OUTCOMES

...entify each genetic disorder affecting the circulatory system
...tline the symptoms of each disorder
...plain the genetic basis of each disorder
...mmarize the therapies available to treat each disorder

...VERVIEW OF THE CIRCULATORY SYSTEM

...ly's circulatory system can be compared to the major road networks that connect many states in ...ed States (often called "freeways"). Interstate-80 connecting New York state in the eastern US ...ornia in the west coast is a major conduit likened to the body's major artery: the aorta—which ...freshly oxygenated blood to all parts of the body. The circulatory system is made of the heart ...od vessels (arteries, arterioles, veins, venules, and capillaries).
...circulatory system carries both blood and lymphatic fluid. Blood is composed of plasma (water, ...s and wastes) and cells (erythrocytes, leukocytes and thrombocytes) (Figure 1).

...1. Components of blood. Note that blood plasma comprises 50% or more of blood. The most ...nt blood cells are red blood cells.
...mage used under license from Shutterstock.com

PLASMA
Constitutes between 52 and 62 percent of whole blood. It is a straw-colored fluid in which blood cells, proteins, and other substances are suspended and transported.

WHITE BLOOD CELLS
LEUKOCYTES - constitute less than 1 percent of whole blood. These attack and destroy potentially harmful foreign matter.

PLATELETS
THROMBOCYTES - constitute less than 1 percent of whole blood. These form clots, blocking blood from exiting wounds.

RED BLOOD CELLS
ERYTHROCYTES - constitute between 38 and 48 percent of whole blood. These cells keep tissue alive by bringing oxygen to it and taking carbon dioxide away.

Blood also carries a variety of substances like hormones, antibodies, gases, and enzymes and frequently also infectious agents like bacteria, viruses, etc. In the heart, blood is separated in two chambers: oxygenated blood in one chamber; and de-oxygenated blood in the other (Figure 2). Deoxygenated blood enters the right side of the heart from tissues through two major veins superior vena cava (from upper body) and inferior vena cava (lower body). It enters the right atrium, passes through the tricuspid valve to the left ventricle; then pumped to the lungs for oxygenation through the pulmonary valve in the pulmonary artery (the only artery carrying de-oxygenated blood). In the lungs, the high partial pressure of oxygen breathed in from the air favours dissociation of the carboxyhaemoglobin releasing carbon dioxide to form oxyhaemoglobin. Carbon dioxide is released to the air through the nostrils. Once oxygenated, blood returns to the heart through the pulmonary vein (the only vein carrying oxygenated blood) to the left atrium, then passes through the mitral valve to the left ventricle. The left ventricle then pumps the blood through the aortic valve to the aorta and then to all parts of the body. The passage of blood from the heart to the lungs is often referred to as pulmonary circulation; while blood passing through the aorta to all parts of the body is referred to as systemic circulation (Figure 2).

Blood reaching cells and tissues provides nourishment to cells and takes away wastes from the cells. Since not all blood drains from tissues due to leakage of plasma from the capillaries, some of it collects

Figure 2. Blood circulation in the human body
Source: Image used under license from Shutterstock.com

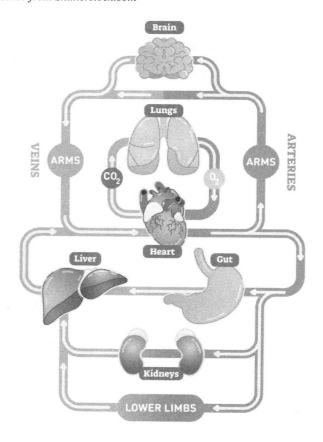

fluid and is drained in vessels of the lymphatic system. The lymphatic system is a subsystem of the circulatory system made up of vessels, tissues and organs (Figure 3). In addition to carrying lymphoid fluid, it also transports white blood cells which fight infectious agents.

SICKLE-CELL DISEASE (SCD)

Sickle-cell disease is an inherited chronic red blood cell disorder associated with significant morbidity and mortality and affecting 100,000 persons in the United States and millions worldwide (McGann et al., 2015). It currently has no established cure except in cases where patients have undergone successful bone marrow or stem-cell transplantation. Gene therapy for SCD is the ultimate goal for a cure, although it is not feasible at the present, significant strides have been made at the basic level to enable genetic correction of **hemoglobinopathies** (Steinberg, 2008).

Symptoms

Three types of complications can occur in SCD: pain syndromes and related issues; anaemia and its related consequences due to haemolysis; and, organ or tissues damage. Under low oxygen states, the sickle haemoglobin undergoes rapid intracellular polymerization, which damages the erythrocyte membrane and significantly alters both its shape and its flexibility while carrying oxygen to the body

Figure 3: The lymphatic system in humans
Image used under license from Shutterstock.com

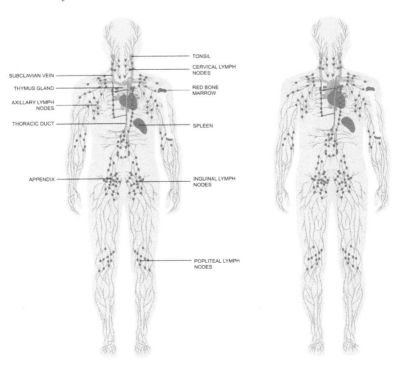

(Figure 4). These stiff and 'sickled' red blood cells have a decreased lifespan and result in both acute and chronic haemolysis. They also undergo a complex process known as vaso-occlusion, in which sickled erythrocytes and other circulating cells adhere to the vascular endothelium, aggregate together to disrupt blood flow especially in the small blood vessels, and cause lack of oxygen (hypoxia) to tissues and vital organs. This can cause brain damage, kidney/heart failure, anaemia, mental impairment, and other physiological complications.

Genetic Basis

Haemoglobin consists of four protein subunits (2 subunits of alpha-globin and 2 subunits of beta-globin). A gene referred to as HBB (haemoglobin, beta) provides instructions for making beta-globin, and a mutation in this gene produces an abnormal version of beta-globin known as haemoglobin S (HbS). Other mutations in the HBB gene lead to additional abnormal versions of beta-globin such as haemoglobin C (HbC) and haemoglobin E (HbE). (GHR, 2018) SCD includes a group of congenital haemolytic anaemias all characterized by the predominance of sickle haemoglobin (HbS) which distorts blood cells into a sickle shape. SCD is an autosomal recessive disease caused by a point mutation that changes the codon GAG → GTG in the haemoglobin gene on chromosome 11 resulting in the substitution of glutamic acid by valine at position 6 of the *beta*-globin polypeptide chain (Steinberg, 2009).

Inheritance of the sickle cell trait from two parent carriers would proceed as shown in Figure 5.

The frequency of carriers ($Hb^A Hb^S$) varies significantly around the world, with high rates associated with zones of high malaria incidence. The presence of both normal (A) and sickle shaped (S) hemoglobin molecules in carriers interferes with the multiplication of the malaria parasite *Plasmodium falciparum*—making carriers resistant to malaria; and therefore, with a selective advantage in malaria prone tropical regions. Natural selection then favors the existence of the heterozygous condition ($Hb^A Hb^S$) due to this advantage, and therefore the continuation of the sickle cell trait in human populations.

Figure 4. Normal and sickle-cell blood cells
Source: Image used under license from Shutterstock.com

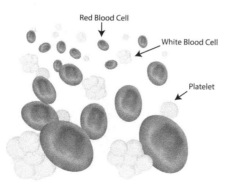
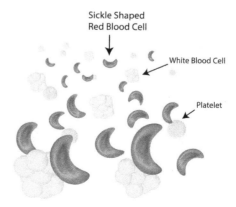

...5. Inhertitance of the sickle cell trait from two parent carriers

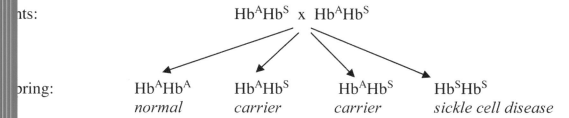

Parents: Hb^A Hb^S x Hb^A Hb^S

Offspring: Hb^A Hb^A   Hb^A Hb^S   Hb^A Hb^S   Hb^S Hb^S
             normal   *carrier*   *carrier*   *sickle cell disease*

Current Therapies

...S and Europe, new born hemoglobinopathy screening allows early preventive treatments such ...illin prophylaxis and provision of important immunizations, which together have significantly ...the early mortality. However, in many developing nations in the tropical regions of the world ...e sickle cell burden is greatest, early diagnostic capacities are lacking and childhood mortality is ... combination of fluids, painkillers, antibiotics, immunizations, and transfusions are used to treat ...ns and complications of SCD. Today, much of the pain experienced and treated at home is of ... moderate intensity and responses well to acetaminophen or **non-steroidal anti-inflammatory** ... NSAIDs) (Dampier & Brodecki, 2002). **Blood transfusion** is one of the three current major ...hes for the effective and promising therapeutic approaches for SCA in general; the other two ...hydroxycarbamide (also known as hydroxyurea), and **stem cell transplantation**. The transfu-...blood from normal donors to patients with SCA achieves the following two major goals: (1) ...ment of the oxygen-carrying capacity of blood and its delivery to tissues; and (2) dilution of ...ng sickled red cells in order to improve microvascular perfusion. Complications of blood transfu-...atients with SCA include the transmission of infectious disease, alloimmunization, haemolytic ...ion reactions (acute or delayed), allergic reactions, febrile reactions, volume overload, and iron ...l. Patients with SCD who are potential candidates for bone marrow or stem cell transplantation ...eceive irradiated cellular blood components.

...roxyurea, a once daily oral medication, has emerged as the primary disease-modifying therapy for ...d can be taken in infants from 9 months of age (McGann & Ware, 2015). Research on the drug for ...30 years demonstrates that hydroxyurea is a safe and effective therapy for SCA, although the drug ... underutilized for a variety of reasons. The primary benefits of hydroxyurea for SCA relate to its ...o increase foetal haemoglobin (HbF) levels, which inhibits intracellular HbS polymerization and ...s the sickling process within erythrocytes (Lebensburge et al., 2010). McGann and Ware (2010) ...e the importance of the global sickle cell community working together to improve both the length ...lity of life for hundreds of thousands of infants born with SCA each year in low-resource settings, ...y sub-Saharan Africa and India. On the positive side, a partnership was established between ...ators in North America and sub-Saharan Africa, to develop a prospective multi-centre research ...l designed to provide data on the safety, feasibility, and benefits of hydroxyurea for children with ...he Realizing Effectiveness Across Continents with Hydroxyurea (REACH) program will treat ...f 600 children age 1–10 years with SCA: 150 at each of four different clinical sites within sub-...Africa (Angola, Democratic Republic of Congo, Kenya, and Uganda). (McGann et al., 2016)

13.3 BURKITT LYMPHOMA (BL)

Considered the most aggressive form of lymphoma, this is a rare form of lymphocyte cancer accounting for about 2% of all lymphoma cancers. It is named after Denis Burkitt, the British surgeon who first identified the cancer in children in African the late 1950s. There are three types of Burkitt lymphoma: sporadic (non-African), immunodeficiency-associated and endemic (African). The endemic version is often diagnosed in children infected with malaria and/or the Epstein-Barr virus (EBV) but the pathogenic mechanism is unclear. The first evidence that all African patients with Burkitt lymphoma were infected with EBV was found in the mid-1960s with EBV found in tumour cells in virtually all Burkitt lymphoma cases (Henle & Henle, 1966). The growth-transforming function of EBV is now known to require co-expression of at least 6 latent infection viral genes that encode the nuclear antigens and the latent membrane protein-1 (Rowe et al., 2014).

Symptoms

The symptoms of BL vary greatly and depend on the disease's type and may spread or worsen quickly as the aggressive cancer advances. Patients with endemic Burkitt lymphoma may have swelling or disfigurement in the jaw or face; while patients with sporadic Burkitt lymphoma may have swelling or pain in the abdomen. The disease may also spread quickly to the central nervous system and brain, causing severe neurological symptoms, including paralysis. Other symptoms include swollen lymph nodes, night sweats, fever, fatigue, loss of appetite, and weight loss. (CTCA, 2018)

Genetic Basis

The most frequently mutated genes in Burkitt lymphoma are oncogenes *MYC* (40%) and *ID3* (34%). Other frequently mutated genes included the known suppressor genes *ARID1A*, *SMARCA4* and *TP53*, as well as the oncogene *PIK3R1* and *NOTCH1*. Chromosomal translocations involving the MYC gene located on chromosome 8 alter the pattern of MYC's expression –disrupting its usual function of controlling cell growth and proliferation. *(Love et al., 2012;* Schmitz et al., 2012)

Current Therapies

While Burkitt's lymphoma can be cured in developed countries in ~85% of cases occurring in younger patients using high-dose **chemotherapy** regimens, these regimens are unsafe in older patients, due to their immune suppression and logistical difficulties which may preclude effective delivery. New insights into BL pathogenesis may allow the clinical evaluation of drugs targeting the PI(3) kinase pathway, tonic BCR signalling, and cyclin D3/CDK6 in BL providing some hope. Ultimately, the rational combination of such targeted agents may provide more effective and less toxic treatment of BL worldwide. (Schmitz et al., 2012; Rowe et al., 2014)

13.4 GAUCHER DISEASE

Gaucher disease (GD) first described by Philippe Gaucher in 1882 as rare, autosomal, recessive genetic disease. The disease's incidence is around 1/40,000 to 1/60,000 births in the general population, reaching 1/800 births in the Ashkenazi Jewish population (Grabowski, 2008). Type-2 GD (<5% of cases in most countries) is characterized by early and severe neurological impairment starting in infants aged 3–6 months old. Death occurs before the third year of life, following massive aspiration or prolonged apnoea and the mean survival age is 11.7 months (range 2–25 months). Pulmonary symptoms and aspiration caused by Gaucher disease or the aggravation of respiratory conditions such as central apnoea is the cause of 50% of fatal cases (Mignot et al., 2006). Type-3 GD (also called juvenile) has severe forms with varying neurological signs including progressive epilepsy (16% of patients), cerebellar ataxia (20%–50% of patients), and **dementia** in some cases (Tylki-Szymańska et al., 2010).

Symptoms

Gaucher cells mainly infiltrate bone marrow, the spleen, liver, and other organs and are considered the main causes of the disease's symptoms. There are three GD types: 1, 2 and 3. Type-1 GD is usually distinguished by the absence of neurological impairment and is the most common form of the disease with a prevalence of 90%–95% in Europe and North America (Stirnemann et al., 2017). Its clinical presentation is variable, ranging from asymptomatic throughout life to early-onset forms presenting in childhood. The initial symptoms vary considerably, and patients can be diagnosed at any age. Symptoms include general fatigue, acute painful bone crises, delayed growth in children, enlarged spleen, focal lesions in the liver and spleen and development of gallstones.

Genetic Basis

Gaucher disease is a rare, autosomal recessive genetic disorder caused by mutations in the *GBA1* gene, located on chromosome 1. Faults in this gene lead to a markedly decreased activity of the lysosomal enzyme, glucocerebrosidase. Without the enzyme, the enzyme's substrate (glucosylceramide) accumulates in macrophages inducing their transformation into Gaucher cells. Patients with a heterozygous (or homozygous) mutation in the *GBA1* gene are now considered at risk for Parkinson's disease. Glucocerebrosidase deficiency results in the accumulation glucosylceramide in macrophages causing neurotoxicity in the brain. Glucosylceramide directly influences amyloid formation which then aggregate and form Lewy bodies in the nerve cells in Parkinson's disease.

Current Therapies

All GD patients require regular monitoring and once, treatment must generally be administered for life. There are currently two specific types of treatment for GD: **enzyme replacement therapy** (ERT), and **substrate reduction therapy** (SRT). The goal is to treat patients before the onset of complications. The principle of ERT is to supply the lacking enzyme in the cells, particularly the Gaucher cells. A recombinant glucocerebrosidase (Cerezyme®, Vpriv®, and Elelyso®). For SRT, the aim is to reduce excess cell glucosylceramide by decreasing its production. *Miglustat* (Zavesca®) is a glucosylceramide synthase inhibitor which reduces its biosynthesis of glucocerebrosidase in Gaucher cells. (Aerts et al., 2006; Lukina et al., 2010; Tylki-Szymańska et al., 2010; Stirnemann et al., 2017)

Other types of treatment can include **gene therapy**, that is, introducing the *GBA1* gene into hematopoietic cells and then injecting the corrected cells into patients (Dahl et al., 2015). Molecular chaperones can help the production of functional enzymes and can restore the intracellular activity of mutant GCase (Sanchez-Martinez et al., 2016). Symptomatic treatments include splenectomy, using level I, II and III analgesics, orthopaedic surgery for bone complications, and perhaps liver transplantation with severe liver disease progressing to fibrosis and liver failure (Tylki-Szymańska et al., 2010; Stirnemann et al., 2017).

13.5 CHRONIC MYELOID LEUKEMIA

Chronic myeloid leukemia (CML) is characterized by the unrestrained malignancy of **pluripotent bone marrow stem cells** that circulate in the blood causing enlargement of the spleen, liver, and other organs (Melo et al., 2003).

Symptoms

Up to 50% of patients are asymptomatic and diagnosed incidentally after routine laboratory evaluation. Clinical features are generally nonspecific: 1) Spleen enlargement is present in 46–76% of cases and may cause left upper quadrant pain or early satiety; 2) Fatigue, night sweats, symptoms of anaemia and bleeding due to platelet dysfunction may occur; 3) In less than 5% of patients are symptoms of hyperviscosity, generally seen when the white cell count exceeds 250,000 per microliter. The disease has chronic (most patients), accelerated, and blast phases. (Savage et al., 1997; Faderl et al., 1999; Melo et al., 2003; Thomson et al., 2015)

Genetic Basis

The hallmark of CML is the presence of a chromosomal abnormality called the *Philadelphia (Ph) chromosome* named after the city where it was first recorded. Translocation between the long arms of chromosomes 9 and 22 bring together two genes: the BCR (breakpoint cluster region) gene on chromosome 22 and the proto-oncogene ABL (Ableson leukemia virus) on chromosome 9. The translocation results in a *BCR-ABL* fusion gene and production of a BCR-ABL fusion protein with tyrosine kinase activity (Shtivelman et al., 1985; Lugo et al., 1990). Tyrosine kinase activity activates signal transduction pathways leading to uncontrolled cell growth sufficient for production of CML.

Current Therapies

Three tyrosine kinase inhibitors are now FDA-approved for initial treatment of chronic phase-CML: *imatinib, nilotinib* and *dasatinib*; all of which constitute adequate treatment options for patients with CML at the time of diagnosis. Induction chemotherapy is an option for lymphoid blast phase CML cases; while myeloid blast phase CML patients can be treated with tyrosine kinase inhibitors monotherapy.

13.6 NIEMANN-PICK DISEASE (NPD)

Albert Niemann, a German paediatrician first described NPD in 1914 in an Ashkenazi Jewish infant who presented with massive hepatosplenomegaly and a rapidly progressive neurodegenerative course that led to her death at 18 months of age. There are two types of NPD (types 1 and 2) (Schuchman & Desnick, 2017).

Symptoms

Type-A NPD patients exhibit hepatosplenomegaly and do not typically survive beyond the first year of life. A cherry-red spot is present in the retina (macula region) in ~50% of these infants. The disease is characterized by a rapidly progressive neurodegenerative course, with reduced pressure in the intraocular fluid in the eyeball (hypotonia), and failure to attain developmental growth milestones. For example, between 6 to 15 months development plateaus, followed by a rapidly progressive psychomotor deterioration, with most patients never able to sit independently. Most type-A infants do not survive beyond the third year of life. Type-B patients have no overt signs of central nervous system involvement, but hepatosplenomegaly may be profound and accompanied by signs of liver failure. Other common type-B disease manifestations include fatigue, bone and joint pain. (Hollak et al., 2012; Schuchman & Desnick, 2017)

Genetic Basis

NPD disorder is due to a continuum of phenotypes arising from mutations in the same *SMPD1* gene, more accurately defined as a single disorder with the enzyme acid sphingomyelinase (ASM) deficiency with three forms: acute neurological, chronic neurological, and chronic non-neurological. The SMPD1 gene is located within the chromosomal region 11p15.4, a hotspot region for imprinting within the human genome, and studies have shown that the *SMPD1* gene is preferentially expressed from the maternal chromosome (i.e., paternally imprinted). Types A and B NPD are inherited as recessive traits, and the degree of clinical involvement largely depends on the type of *SMPD1* mutations inherited. To date, more than 180 mutations have been found within the *SMPD1* gene causing types A and B NPD including point mutations (missense and nonsense), small deletions, and splicing abnormalities. ASM has its highest activity at reduced pH and catalyses the hydrolytic cleavage of sphingomyelin in lysosomes, producing phosphocholine and ceramide. Within lysosomes ASM interacts with other lipid hydrolases and performs an essential housekeeping function by maintaining proper sphingolipid homeostasis and participating in membrane turnover.

Current Therapies

Bone marrow transplantation (BMT) has been undertaken in several ASM-deficient NPD patients. Gene therapy has been extensively studied in mice using retroviral vectors, liver directed gene therapy using AAV vectors, and direct injection of gene therapy vectors with or without recombinant human ASM into the brain (Barbon et al., 2005; Dodge et al., 2005; Dodge et al., 2009). Recombinant human ASM **enzyme replacement therapy** has also been successful in several trials (McGovern et al., 2016). Based on these findings, the FDA completed trials to this therapy in 2015 with at least three additional clinical trials being recruited currently (USNLM, 2018). Since ERT is unlikely to impact the neurological features of the disease future research is focused on achieving widespread enzyme delivery to the brain, as well as developing alternatives to enzyme infusions, including gene therapy and small molecule approaches (Schuchman & Desnick, 2017).

13.7 PAROXYSMAL NOCTURNAL HAEMOGLOBINURIA (PNH)

PNH is a rare, clonal, hematopoietic stem cell disorder that manifests with a haemolytic anaemia from uncontrolled complement activation, bone marrow failure, and a propensity for thrombosis (DeZern & Robert, 2015). PNH is among the first diseases in which the role the complement system cascade plays in the pathogenesis is well-elucidated (Brodsky, 2008). The complement system is a part of the immune system consisting of plasma proteins that interact via three major pathways to form a membrane attack complex which causes membrane lesions, therefore lysing invading pathogenic cells.

Symptoms

Patients with PNH suffer from decreased red blood cells (anaemia); bone marrow failure owing to impaired blood formation (haematopoiesis); and the presence of blood haemoglobin in the urine (haemoglobinuria) and plasma (haemoglobinemia). Blood coagulation (thrombosis) is often a typical manifestation of PNH and is the leading cause of death in the disease (Hill et al., 2013). Thrombosis may occur at any site in PNH (venous or arterial) with intra-abdominal, cerebral, and hepatic vein thrombosis being the most common.

Genetic Basis

PNH is caused by a somatic mutation in *PIG-A* gene that leads to a marked deficiency or absence of two important GPI-anchored complement regulatory membrane proteins (CD55 and CD59) required for the biosynthesis of cellular anchors like glycosyl-phosphatidyl-inositol-anchored protein. CD55 regulates the formation and stability of enzymes called convertases, whereas CD59 blocks the formation of the membrane attack complex. PNH is associated with a high risk of major thrombotic (clot formation) events where erythrocytes become highly vulnerable to complement-mediated lysis owing to a reduction, or absence, of CD55 and CD59. PNH is an acquired genetic disorder resulting from mutations triggered by various environmental factors. An affected blood cell clone passes the altered PIG-A to all its descendants – red blood cells, white blood cells (including lymphocytes) and platelets.

Current Therapies

Eculizumab is a humanized monoclonal antibody that binds to C5 and inhibits its further cleavage. The drug decreases intravascular haemolysis, reduces thrombosis risk, and improves quality of life in PNH by inhibiting the formation of the membrane attack complex (Brodsky et al., 2008). *Eculizumab* is currently the only therapy approved by the FDA for PNH. Patients require close monitoring while on *eculizumab* treatment. Hematopoietic stem cell transplantation is the only curative therapy for PNH but is not recommended as initial therapy in the *eculizumab* era, given the risks of transplant-related morbidity and mortality (DeZern & Robert, 2015).

Disorders of the Human Circulatory System

13.8 HAEMOPHILIA

Haemophilia is a sex-linked inherited disorder resulting in a deficiency in certain clotting factors in the blood which affect the intrinsic clotting cascade required to convert prothrombin to thrombin, then fibrin (clot). Haemophilia A has a higher prevalence, occurring in about 1:5,000 male births, while haemophilia B, occurs in about 1:25,000. It is inherited as an X-linked recessive trait, although it can also be acquired in advanced age as a result of autoimmunity, cancer, or various metabolic disorders affecting both males and females (Sherman et al., 2017).

Symptoms

While the determination of which factor is missing is important for treatment, the clinical symptoms of haemophilia A and B are essentially comparable. The severity of X-linked haemophilia is dependent on the degree of residual clotting activity. Mild cases (5-40% activity) are typically asymptomatic outside of major trauma or surgery, whereas moderate cases (1-5% activity) are somewhat more vulnerable and may have prolonged bleeding even from minor injuries. However, severe haemophilia (<1% activity) brings additional complications. In addition to the difficulty responding to injury, these patients frequently develop spontaneous bleeds in capillary beds, particularly within joints. Over time, this causes significant chronic deterioration of the joints if not properly managed.

Genetic Basis

This X-linked bleeding disorder is caused by mutations in coagulation factor VIII (haemophilia A) or factor IX (haemophilia B). Haemophilia A is the most prevalent form representing 80% of haemophilia cases while haemophilia A represents 20% of cases. (Sabatino et al., 2012; Monahan, 2015; Rogers & Herzog, 2015)

Current Therapies

Of the two haemophilia types, gene therapy for type-B has been more successful, having advanced to multiple recent clinical trials. Primarily, this is due to the simplicity of factor IX compared to factor VIII. Currently, haemophilia A is treated by intravenous delivery of replacement clotting factor, either plasma-derived or recombinant which can be on demand or prophylactic (Gringeri, 2011). **Protein replacement therapy** carries the risk of deleterious immune responses against the therapeutic protein. As patients are not naturally producing clotting factor, the immune system can recognize the exogenous protein as a foreign antigen and form antibodies against the protein that prevent its function.

Gene therapy represents an appealing alternative to protein replacement therapy. Instead of repeated injections of protein, it would ideally involve a single injection that would induce long-term production of the defective clotting factor. A variety of mechanisms to introduce the transgene have been investigated, some of the most popular are recombinant viral vectors (**adeno-associated viruses (AAV)**, retroviruses and lentiviruses) and non-viral gene transfers. While a number of promising approaches for gene therapy for haemophilia have been elucidated, there are clearly numerous problems that still need to be addressed to develop approved gene therapies for both haemophilia A and B for use in humans. The first two Phase

I clinical trials in humans with severe haemophilia B, utilizing muscle- or liver- directed gene therapy with **AAV vectors**, were safe and successfully completed after having originally been approved based on the safety and efficacy demonstrated in several animal models (Sabatino et al., 2012). Multiple Phase I/II clinical trials are testing hepatic *in vivo* gene transfer with adeno-associated viral vectors in patients with severe haemophilia A, in some cases achieving normal factor VIII levels (Nienhuis et al., 2017).

13.9 THALASSEMIA

Thalassemias are a group of autosomal recessive disorders caused by reduction or lack of production of one or more of the globin chains that make up the haemoglobin (Hb) four subunits (Cao & Kan, 2013). Faulty synthesis of haemoglobin is caused by gene mutations, with thalassemia being among the commonest autosomal recessive disorders worldwide, prevalent in populations in the Mediterranean area, the Middle East, Central Asia, the Indian subcontinent, and the Far East (Modell & Darlison, 2008; Weatherall, 2010; Cao & Kan, 2013). In Southeast Asia, α-thalassaemia and β-thalassaemia are particularly prevalent (Fucharoen & Winichagoon, 2011).

Symptoms

Patients with thalassaemia disease commonly develop chronic haemolytic anaemia and ineffective red blood cell synthesis (**erythropoiesis**) due to imbalanced globin synthesis. The severity of anaemia is linked to the number of genetic aberrations, the specific combination of affected genes, other genetic and environmental modifiers and physiological stressors (Fucharoen & Winichagoon, 1987; 2011). Children with β-thalassaemia major appear healthy at birth due to presence of foetal haemoglobin. Defects in the globin genes result in impaired globin chain synthesis leading to the reduced haemoglobin content of red blood cells leading to chronic anaemia. Without treatment, the spleen, liver, and heart become greatly enlarged, and during the first decade of their life, they become pale, listless and fussy, and have a poor appetite. They grow slowly, often developing jaundice, with brittle, thin bones, osteoporosis and osteopenia, and face bones become distorted. Most of these thalassaemia major patients die in the paediatric age group with heart failure due to anaemia and infection, being the leading causes of death among untreated children.

Genetic Basis

Adult hemoglobin is composed of two alpha (α) and two beta (β) polypeptide chains. There are two copies of the haemoglobin α-gene located on chromosome 16 (HBA1 & HBA2), each encoding an α-chain. The haemoglobin β-gene (HBB) encodes the β-chain and is located on chromosome 11. In α-thalassemia, there is deficient synthesis of α-chains—the resulting excess of β-chains bind oxygen poorly, leading to a low concentration of oxygen in tissues (hypoxemia). Similarly, in β-thalassemia, there is a lack of β-chains and the excess α-chains can form insoluble aggregates inside red blood cells causing them (and their precursors) to die resulting in severe anaemia.

Disorders of the Human Circulatory System

Current Therapies

Patients with severe thalassemia phenotypes require regular blood transfusion to suppress erythropoiesis and maintain function, whereas patients with moderate and milder forms can be managed conservatively. Pharmacological agents such as hydroxyurea have been known to cause induction of foetal haemoglobin and reduce ineffective erythropoiesis, and thus may alleviate the symptoms in thalassaemia intermedia patients. Anecdotal reports also indicate that wheat grass juice can be helpful in raising the haemoglobin. For instance, a study by Marwah *et al.* (2004) reports that at least 8 patients consuming about 100 ml of wheat grass juice daily had their blood transfusion requirement falling by >25%. Other therapies include bone marrow haemopoietic stem cell transplants, use of available bone marrow/cord blood stem cell registries, and the drugs *deferiprone* and *deferasirox* are currently administered in India. A coordinated control programme was also developed in 1970s by a team of experts at the World Health Organization that emphasized political and financial support; improving curative services; prenatal diagnosis in couples who have given birth to an affected child and those identified to be at risk; prospective antenatal screening; community carrier screening; counselling and prenatal diagnosis; and a network of centres, and national/regional working groups (WHO 1994; Verma et al., 2011).

13.10 ATHEROSCLEROSIS

Atherosclerosis is a progressive inflammatory disorder underlying coronary artery disease (CAD or **arteriosclerosis**) and stroke, major causes of mortality and morbidity worldwide (Lusis, 2012). An estimated 7 million plus deaths are attributed to CAD annually (WHO, 2012; Mishra et al., 2016). In atherosclerosis, plaque gradually builds up inside blood vessels called arteries which carry oxygen-rich blood from the heart to all parts of your body. Coronary arteries are small arteries that supply the heart muscle with blood. If those arteries accumulate plaque, the heart muscle is starved of blood and oxygen and can cause heart attack, stroke, and could be fatal. Plaque, like that found in teeth, accumulates over time narrowing the diameters of the arteries—thus reducing blood flow. It is a mixture of calcium, cholesterol, cell debris, collagen, fibrin, and other substances found in blood. The key risk factors include genetic and/or environmental and include elevated levels of cholesterol, triglycerides in the blood, HBP, and exposure to cigarette smoke.

Symptoms

CHD symptoms may include sudden weakness, paralysis or numbness of the face, arms, or legs, especially on one side of the body, confusion, trouble speaking or understanding speech, trouble seeing in one or both eyes, problems breathing, dizziness, trouble walking, loss of balance or coordination, and unexplained falls, loss of consciousness, sudden and severe headache (NIH-NHLBI, 2018). The common clinical presentation of CHD is chronic stable chest pain or pressure (**angina**). The underlying mechanisms may include plaque formation in coronary arteries, spasm of normal or plaque containing arteries, left ventricular dysfunction due to prior acute myocardial cell death (**necrosis**) or weakened heart muscle (**ischaemic cardiomyopathy**). In addition to chest discomfort, shortness of breath, palpitations, fainting or fatigue may also be present and sometimes may be the only symptom. (Mishra et al., 2016). CHD is characterised by episodes of transient central chest pain often triggered by exercise, emotion or other forms of stress resulting from mismatch between blood demand and supply resulting in shortness of breath.

Genetic Basis

The pathology of CAD and stroke is complex involving multiple genetic and environmental factors which, until recently, have largely remained unelucidated. Understanding the genetic factors contributing to common forms would have important implications for prevention and treatment of the disease. The Human Genome Project, the **HapMap Project**, and subsequent genomic studies led to the identification of many of the common **single nucleotide polymorphisms** (SNPs) in the world's populations. The largest **genome-wide association studies (GWAS)** for CAD, the **CARDIoGRAM** study, included over 100,000 subjects of European descent, plus smaller CAD studies of Asian populations have also been reported (Samani et al., 2007; Peden et al., 2011; Schunkert et al., 2011). GWAS of large human populations have identified numerous novel genes/loci for atherosclerosis and related traits. Several the loci included well-known lipid genes such as LDLR and PCSK9, supporting the importance of low-density lipoproteins (LDL) in atherosclerosis. A few other genes such as CYPA1, CNNM2, and NT5C2, showed evidence of association with hypertension. One locus with a strong effect on both CAD and lipids contained 3 genes in association with one another (SELSR2, PSRC1 and SORT1), none of which were previously known to participate in lipid metabolism. In such CAD GWAS, a locus on chromosome 9p21 has proven to be most strongly associated with CAD. However, due to complexity of the disease important on-going and future tasks will seek to identify the causal genes and DNA variants at the GWAS loci; characterize the functions of the causal genes; and, extend the search to rare variants. (WTCCC, 2007; Schunkert et al., 2011; Lusis 2012; Lee et al., 2017)

Current Therapies

To open blocked or narrowed arteries, **coronary angioplasty**, is a procedure used to open blocked or narrowed coronary (heart) arteries. A small meshed tube called a stent in inserted into the artery to improve blood flow to the heart and relieve chest pain. In a type of surgery called **coronary artery bypass grafting,** arteries or veins from other areas in your body are used to bypass or go around narrowed coronary arteries. This will improve blood flow to heart, relieve chest pain, and even prevent a heart attack. To remove plaque build-up from the carotid arteries in the neck, a type of surgery called carotid endarterectomy is performed to restore blood flow to the brain, which could prevent a stroke. (NIH-NHLBI, 2018)

The mainstay treatment options for occlusive vascular disease are **stent insertion** or **bypass grafting**. However, significant complications such as stent thrombosis, re-stenosis or vein graft failure can arise (Levine et al., 1995). The molecular mechanisms that underlie these harmful processes need to be elucidated to help devise new therapies. In recent years, the role of oxidative stress, endoplasmic reticulum stress and mitochondrial dysfunction in promoting neointimal proliferation has been intensively studied (Tse et al., 2016). The roles of microRNAs in atherosclerosis have been the focus of recent research. These are non-coding RNAs involved in post-transcriptional regulation of genes by RNA silencing. Recent work has demonstrated microRNA regulation of flow-dependent vascular remodelling, and it appears that some are inducible by shear stress and play a protective role in atherosclerosis (Ha & Kim, 2014). Other microRNAs have anti-inflammatory properties by inhibiting the translation of several factors involved in inflammation. A recent study reports the use of polyethylene glycol-polyethyleneimine nanoparticles as vectors for microRNA delivery targeting E-selectin of inflamed endothelium of mice, which ameliorated endothelial inflammation and atherosclerosis. (Fang et al., 2010; Wang et al., 2016; Lee et al., 2017)

13.11 WILLIAM'S SYNDROME (WS)

WS is a rare congenital developmental disorder named after JCP Williams in 1961. WS is a neurogenetic disorder affecting human development and adult cognition. Affected individuals have mild to moderate intellectual disability, or learning difficulties, unique personality characteristics, distinctive facial features, and cardiovascular problems. About 1 in 7,500 to 20,000 people have WS (WSA, 2018).

Symptoms

Typical features of WS include dysmorphic craniofacial features (full lips, short nasal bridge and large forehead), high levels of calcium in the blood, hypertension, a heart (aorta) defect, and poor mental development (Collette et al., 2009). Cardiovascular complications are the major cause of death in these patients particularly when they have biventricular outflow obstruction. Children require close follow-up for blood pressure, thyroid functions, serum calcium levels and growth parameters (Kumar et al., 2014). WS individuals have a distinctive facial appearance and a unique personality that combines over-friendliness and high levels of empathy with anxiety (NIH-GARD, 2018).

Genetic Basis

WS is caused by a micro-deletion of chromosome band 7q11.23 involving about 24–28 genes and RNA transcripts. Several genes are included in this loss including CLIP2, ELN, GTF2I, GTF2IRD1, and LIMK1 which are involved in several body conditions. The ELN gene is associated with the connective tissue abnormalities and cardiovascular disease; visual-spatial tasks, unique behavioural characteristics, and other cognitive difficulties are associated with the other four. Lack of the GTF2IRD1 gene may be responsible for the distinctive facial features of WS. Interestingly, lack of NCF1 gene may be beneficial since research has showed that individuals whose NCF1 gene is intact have a higher risk of developing hypertension. (Mohan & Mohan, 2011; Cuenza & Adiong, 2015)

The physical and cognitive features associated with WS result in part from loss of one genomic copy of the deleted region. Other mechanisms contribute, including the effect of the deletion rearrangement on genes flanking the break point, and variations of DNA sequence, epigenetic mechanisms including genomic imprinting, parent-of-origin and tissue-specific effects, all of which may alter the expression of genes located on the non-deleted chromosome 7. Decreased gene expression in a given tissue may contribute disproportionately to the WS phenotype, and that subtle epigenetic effects on single genes or clusters of genes may contribute significantly to cognitive phenotypes. Genes in the WS region may regulate neighbouring gene expression and contribute to phenotype by multiple mechanisms. The expression of GTF2I in WS is related to the parental origin of the transmitted allele, lower when of paternal origin, which supports an epigenetic control mechanism and the hypothesis that GTF2I is paternally imprinted. Studies report that the gene GTF2I may play a key role in normal human brain development, and it appears to regulate the expression of other genes in the WS region. (Collette et al., 2009)

Current Therapies

Treatment for people with WS depend on the symptoms and severity in each individual. Strategies include feeding therapy for infants with feeding problems, early intervention programs and special education programs for children with varying degrees of developmental disabilities, behavioural counselling and/or medications for attention deficit disorder and/or anxiety, surgery for certain heart abnormalities, medications or diet modifications for hypercalcemia, orthodontic appliances or other treatments for malocclusion of teeth, gonadotropin-releasing hormone agonist for early puberty (NIH-GARD, 2018).

Children with WS have balance problems and weak muscle tone and **physical therapy** can greatly help individuals. **Occupational therapy** can be used to assist with fine motor skill development. **Speech therapy** is often for speech/language related issues since speech is often delayed in children with WS and articulation can be affected by muscle tone issues. Often children experience difficulties with processing information. (WSA, 2018; AAP, 2018)

13.12 ATAXIA TELANGIECTASIA (AT)

AT, also referred to as Louis-Bar Syndrome, was given its commonly used name by Elena Boder and Robert P. Sedgwick, who in 1957 described a familial syndrome of progressive cerebellar ataxia, oculocutaneous telangiectasia and frequent pulmonary infection (Rothblum-Oviatt et al., 2016).

Symptoms

AT is characterized by progressive cerebellar degeneration, telangiectasia, immunodeficiency, recurrent sinopulmonary infections, radiation sensitivity, premature aging, and a predisposition to cancer development, especially of lymphoid origin. Other abnormalities include poor growth, gonadal atrophy, delayed pubertal development and insulin resistant diabetes. It is important to note that AT is a complex disease and not all people have the same clinical presentation, constellation of symptoms and/or laboratory findings (e.g. telangiectasia is not present in all individuals with AT). The prevalence is estimated to be <1–9/100,000, although incidences as high as 1 in 40,000 and as low as approximately 1 in 300,000 have been reported (Swift et al., 1986).

Since not all children develop in the same manner or at the same rate, the diagnosis of AT may not be made until the early school years when the neurologic symptoms (impaired gait, hand incoordination, abnormal eye movements), and the telangiectasia appear or become worse. Ataxia first appears during the toddler stage when children begin to sit and walk, but they then fail to improve much from their initial wobbly gait. In 2006, the average life expectancy was reported to be approximately 25 years. The two most common causes of death are chronic lung disease (1/3 of cases) and cancer (about 1/3 of cases). (Crawford et al., 2006)

Genetic Basis

AT is caused by mutations in the *ATM* (ataxia telangiectasia, mutated) gene located on human chromosome 11q22-q23 and is made up of 66 exons (four non-coding and 62 coding) spanning 150 kb of genomic DNA (Gatti et al., 1988). Mutations include primarily nonsense mutations and frame shifts resulting from insertions and deletions, but also missense and leaky splice-site mutations. The *ATM* gene encodes a large 3056 amino acid protein named ATM kinase whose role is coordinating the cellular response to DNA double-strand breakages. The enzyme also responds to oxidative stress, other forms of genotoxic stress and other stressors that affect cellular homeostasis. (Paz et al., 2011)

Current Therapies

Physical, occupational and speech therapies as well as exercise may help maintain function but will not slow the course of neurodegeneration. Certain anti-Parkinson and anti-epileptic drugs may be useful in the management of symptoms. Commonly prescribed drugs include *trihexyphenidyl* (Artane), *amantadine, baclofen* and BOTOX® injections. Less commonly prescribed drugs that also may be beneficial include *clonazepam*, *gabapentin* and *pregabalin* (Lyrica). Problems with immunity can sometimes be overcome by **immunization**. To slow or prevent the development of chronic lung disease in AT, early intervention for respiratory symptoms is recommended. Liberal use of **antibiotics** should be considered in people with AT who have persistent upper and lower respiratory tract symptoms. Treatments for swallowing problems should be determined following evaluation by an expert in the field of speech-language pathology. Standard cancer treatment regimens need to be modified to avoid the use of **radiation therapy** and radiomimetic drugs, as these are particularly cytotoxic for people with AT. In October 2014, a clinical guidance document on the diagnosis and treatment of AT in children was published by the UK AT Society and covers the key clinical areas of genetics, neurology, pulmonary care, immunology and cancer, plus also physical therapy, dietary management and the implications of surgery in people with AT. (Rothblum-Oviatt et al., 2016; AT Society [UK], 2018)

13.13 PORPHYRIA

Porphyrias are a group of rare metabolic disorders which can either be inherited or acquired affecting the heme pathway. Heme synthesis takes place in several steps requiring a total of 8 different enzymes and the genes encoding them are located on different chromosomes. Heme is essential for the synthesis of haemoglobin, myoglobin, and microsomal cytochromes—all of which play an important role in oxygen transport and/or oxidation–reduction reactions. Each type of porphyria is a result of a specific deficiency in one of the enzymes involved in this biosynthetic pathway. Heme is composed of porphyrin, a large circular molecule made from four rings linked together with an Fe atom at its centre (Figure 1). The disruption of heme production causes overproduction of porphyrins which shows up in the urine as a reddish-purple colour. Individuals with porphyria are unable to complete the synthesis of heme, and thus intermediate products of the biosynthetic pathway, porphyrin or its precursors (e.g. delta-aminolevulinic acid and porpho-bilinogen) accumulate. (Pischik & Kauppinen, 2015; Stein et al., 2017; Edel & Mamet, 2018)

The accumulated precursors are excreted in the urine, in the faeces, or in both depending on their solubility—and measuring their level is the basis of porphyria biochemical diagnosis and typing. The purple colour of porphyrins, causing the dark coloured urine in porphyria patients due to oxidation of porphobilinogen to uroporphyrin and porphobilin, gives the disease its name "porphyria." (Sassa, 2006).

Symptoms

Since the deficiency is inherited from one affected parent, the residual enzyme activity is about 50%, sufficient for regular heme homeostasis, thus keeping the disease latent. Studies indicate that most patients will remain asymptomatic during their whole life without experiencing any porphyria symptoms. An acute attack usually occurs following an exposure to known precipitating factors particularly medications metabolized by the cytochrome P450 system (regarded as unsafe for porphyria patients). Other factors include alcohol use, infections, low caloric intake, and changes in sex hormone balance during the menstrual cycle. Most acute attacks begin as a combination of abdominal pain, mild mental symptoms, such as severe fatigue and inability to concentrate, with or without autonomic dysfunction. The most common symptoms are severe abdominal pain, nausea, vomiting, and constipation. **Tachycardia**, hypertension, and signs of increased sympathetic activity are often associated with abdominal pain. Severe attacks may also present with muscular weakness and or mental disturbance, such as anxiety, disorientation, or hallucinations. (Mayer et a., 1998; Lu et al., 2005; Besur et al., 2015; Edel & Mamet, 2018)

Genetic Basis

Gene mutations in several genes cause deficient production of 7 key enzymes in the biosynthetic pathway: aminolevulinate synthase; aminolevulinic acid dehydratase; hydroxymethylbilane synthase; uroporphyrinogen-III synthase; uroporphyrinogen decarboxylase; coproporphyrinogen oxidase; protoporphyrinogen oxidase; and, ferrochelatase. Six types of porphyrias result from the production disruption of the 7 key enzymes: amino-levulinic acid dehydratase porphyria; acute intermittent porphyria; congenital erythropoietic porphyria; porphyria cutanea tarda; hereditary copro-porphyria; variegate porphyria; and erythropoietic protoporphyria. Acute intermittent porphyria is the most common and is inherited in an autosomal dominant. (Karim et al., 2015; Woolfe et al., 2017; Edel & Mamet, 2018)

Current Therapies

Specific treatment, aimed at stopping the acceleration in heme synthesis which occurs during an acute attack, is targeted to down-regulate aminolevulinate synthase activity. This might be achieved, in very mild cases, by hydration and administration of carbohydrates, or much more effectively, in severe attacks, by blood-derived heme, such as *normosang* in Europe or *hematin* in the United States. **Palliative treatments** given during an acute attack for pain, nausea, and other symptom relief should be performed using only "safe" drugs for porphyria patients. Heme therapy which could be given as needed, and in women suffering from menstrual cycle-related acute attacks, gonadotropin-releasing hormone (GnRH) agonist treatment could be given to avoid the attacks. When no standard treatment is effective, and quality of life is poor, liver transplantation could be an option. (Pischik & Kauppinen, 2015; Stein et al., 2017; Edel & Mamet, 2018)

Disorders of the Human Circulatory System

13.14 LONG QT SYNDROME (LQTS)

Congenital LQTS was first described clinically as Jervell and Lange-Nielsen syndrome and Romano-Ward syndrome in the late 1950s and early 1960s (Johnson & Ackerman, 2010). The electromechanical function of the heart is reflected by electrocardiographic parameters such as the QT interval and is dependent on the coordinated activation and inactivation of inward depolarizing and outward repolarization currents that underlie the major phases of the cardiac action potential (Nerbonne & Kass, 2005; Giudicessi & Ackerman, 2012). Luckily, only a small minority (affects 1 in 2,500 people) of the >250,000 annual sudden deaths in the United States are attributable to LQTS and other heritable arrhythmia syndromes. (Giudicessi & Ackerman, 2013)

Symptoms

Clinically, LQTS is characterized by a prolonged heart rate-corrected QT interval on electrocardiogram (ECG) and a predilection for LQTS-triggered cardiac events including fainting, seizures, and/or sudden cardiac arrest, often during times of emotional or physical duress. LQTS is a potentially lethal genetic disorder of cardiac repolarization that represents a leading cause of sudden cardiac death in children. (Giudicessi & Ackerman, 2013)

Genetic Basis

Congenital LQTS represents a genetically and phenotypically heterogeneous collection of rare, multi-system disorders. LQTS stems from a group of genetically distinct arrhythmogenic disorders resulting from genetic mutations in cardiac potassium and sodium ion channels, termed cardiac channelopathies. Mutations in cell membrane proteins associated with ion channels can also cause LQTS. A total of 13 LQTS susceptibility genes have been discovered to date. Mutations in potassium channel genes *KCNQ1* (LQT1) and *KCNH2* (LQT2) and the sodium channel gene *SCN5A* (LQT3) account for approximately 75% of patients with clinically definite LQTS and encompass over 95% of genetically identifiable LQTS. (Moss, 2003; Ackerman, 2004; Vatta et al., 2006; Tester et al., 2006)

Current Therapies

Since the 1970's, **β-adrenergic receptor antagonists (β-blockers)** have been first-line therapy for the prevention of life-threatening arrhythmias in LQTS that are often triggered by sudden increases in sympathetic activity. These include *propranolol, nadolol* and *metoprolol*. Other therapies include surgery (particularly those with malignant forms of LQTS), implantable **cardioverter-defibrillators**, and avoidance of strenuous physical exercise/other stressors. (Giudicessi & Ackerman, 2013)

13.15 ALPHA-I-ANTITRYPSIN DEFICIENCY (AAT)

In 1963, Laurell and Eriksson, observed an association between alpha-I-antitrypsin and chronic obstructive lung disease, the disease is inherited in an autosomal recessive version. In 1969 an association of AAT and liver disease was reported. (Bearn, 1978) The disorder affects 70,000 to 100,000 individuals in the United States.

Lungs are made of thin sacs called alveoli which allow exchange of gases across their walls into the bloodstream. AAT is a **serum serine protease inhibitor** and functions to protect the lung from the activity of a neutrophil-released protease called elastase. Protease enzymes (such as elastase) are normally released in the lung following activation of neutrophils or macrophages in response to pathogens or tobacco smoke. Elastase can destroy alveoli walls and surrounding tissue, leaving pockets of trapped air. Serum deficiency associated with AAT is caused by an imbalance between proteases and AAT in the lung, leading to slow destruction of the lung alveoli, which can be accelerated by contaminants like cigarette smoke. Although proteases are powerful antimicrobial molecules in the fluid lining the airways, they can damage lung tissue if left unchecked.

Symptoms

The abnormal accumulation of air in the lungs is called **emphysema** and causes shortness of breath. AAT deficiency is also associated with childhood and adult liver **cirrhosis** and, rarely, with hepatocellular carcinoma, inflammation of subcutaneous adipose tissue, and a variety of vascular and autoimmune disorders (Duvoix et al., 2014).

Genetic Basis

The low circulating levels of AAT are the result of mutations in the SERPINA1 gene found on chromosome 14, with more than 120 naturally occurring alleles identified (Schroeder et al., 1985). The normal M allele is present in more than 98% of the population with the most common deficient variants being the severe Z allele, observed with high frequency in Caucasians of Northern European countries and North America. The milder S form is found in high frequency in the Iberian Peninsula in Europe. Most cases of emphysema associated with AAT deficiency are caused by homozygous inheritance of the Z variant which substitutes the amino acid lysine for glutamic acid. The Z mutation causes the AAT protein to polymerize in liver cells (**hepatocytes**) preventing secretion into the blood. The S allele causes a single amino acid substitution of a glutamic acid by a valine resulting in an unstable protein with reduced serum half-life. Most AAT genotypes originate from a combination of the M allele and, to a lesser extent, the mutant S and Z variants. Individuals homozygous for the Z mutation (ZZ) have plasma AAT levels 10 to 15% of the levels of those of the normal M allele and account for more than 95% of cases of clinically recognized AAT deficiency. The three alleles (Z, M and S) are expressed in a codominant manner—that means a person with MZ has levels of AAT that are between the levels of those people who have alleles MM or ZZ. Individuals with at least one normal allele (MZ or MS) or two copies of S (SS) usually produce enough AAT to protect the lungs but have an increased risk of lung disease, especially if they smoke (Bornhorst et al., 2013; de Serres & Blanco, 2014; Chiuchiolo & Crystal, 2016)

Current Therapies

Therapies for AAT antitrypsin deficiency have focused primarily on normalizing AAT levels to protect the lung, standard treatment of emphysema including **bronchodilators** and early use of antibiotics in infections, and to a lesser extent, on strategies to treat the rarer liver diseases (Wang & Perlmutter, 2014). Following preliminary clinical studies in the academic community, and then pharmaceutical company development of large-scale purification of human AAT, the FDA approved the use of weekly AAT **augmentation therapy** for AATD following a clinical trial which demonstrated that weekly infusions would raise to normal plasma and lung epithelial fluid levels of AAT in AAT-deficient individuals. However, AAT augmentation therapy with the human AAT protein is costly, requires weekly to monthly intravenous infusions of purified AAT from pooled human plasma, and has the risk of allergic reactions and viral contamination. Gene therapy is also an option and if effective, it presents a lower risk, fewer issues with limitations in supply, and reduced overall drug cost. The general strategy of AAT gene therapy to augment lung levels of AAT focuses on delivering the normal human M-type AAT complementary DNA (cDNA) under control of a constitutive promoter using a gene transfer vector, so the transduced cells secrete the protein to the blood after a single administration. (Garver et al., 1987; Rosenfeld et al., 1991; Calcedo & Wilson, 2013; Wewers & Crystal, 2013; Crystal, 2014; Chiuchiolo & Crystal, 2016)

CHAPTER SUMMARY

Although mutations are essential for generating variation in living organisms—often, they are responsible for inheritable gene changes which cause disorders due to faults in specific metabolic reaction pathways in the body. Mutations can be inherited from a parent to a child (hereditary) or they can happen during a person's lifetime (acquired). Acquired mutations can be caused by environmental factors such as ultraviolet radiation from the sun. Acquired mutations that humans develop during their lifetime are in cells called somatic cells—the cells that make up most of the body. If a gene mutation is carried by the egg or sperm cell, future generations can inherit the defective gene from their parents.

Genetic disorders can occur due to a defect in a single gene or in a set of genes and as determined by the degree of mutation, they can be categorized into four main types: chromosomal diseases, single-gene disorders, multifactorial disorders and mitochondrial disorders. The names given to genetic conditions are not in one standard way and are often derived from one or a combination of sources including genetic or biochemical defect that causes the condition; one or more major signs or symptoms of the disorder; parts of the body; the geographic area it was first reported; and sometimes the name of a patient or family with the condition.

The body's circulatory system can be compared to the major road networks that connect many states in the United States (often called "freeways") and is made up of the heart and blood vessels (arteries, arterioles, veins, venules, and capillaries). The system carries both blood and lymphatic fluid in two circuits: pulmonary circulation (blood through the lungs for oxygenation) and systemic circuits (from the heart to all body parts).

Sickle cell disease (a chronic red blood cell disorder) caused by defective haemoglobin the protein that carries oxygen in blood. Haemoglobin consists of four protein subunits (2 subunits of alpha-globin and 2 subunits of beta-globin). A gene referred to as HBB (haemoglobin, beta) provides instructions for making beta-globin, and a mutation in this gene produces an abnormal version of beta-globin known as haemoglobin S (HbS). Other mutations in the HBB gene lead to additional abnormal versions of beta-globin such as haemoglobin C (HbC) and haemoglobin E (HbE). Burkitt Lymphoma, a rare form of white blood cell (lymphocyte) cancer caused primarily by mutations on two oncogenes *MYC* (40%) and *ID3* (34%). Chromosomal translocations involving the MYC gene located on chromosome 8 alter the pattern of MYC's expression –disrupting its usual function of controlling cell growth and proliferation.

Gaucher Disease, a rare, autosomal recessive genetic disorder caused by mutations in the *GBA1* gene, located on chromosome 1. Faults in this gene lead to a markedly decreased activity of the lysosomal enzyme, glucocerebrosidase. Without the enzyme, the enzyme's substrate (glucosylceramide) accumulates in macrophages inducing their transformation into Gaucher cells. Gaucher cells mainly infiltrate bone marrow, the spleen, liver, and other organs and are considered the main causes of the disease's symptoms. Chronic Myeloid Leukaemia is characterized by the unrestrained malignancy of pluripotent bone marrow stem cells that circulate in the blood causing enlargement of the spleen, liver, and other organs. The hallmark of CML is the presence of a chromosomal abnormality called the *Philadelphia (Ph) chromosome* named after the city where it was first recorded. It involves translocation between the long arms of chromosomes 9 and 22 bring together two genes: the BCR and ABL.

Niemann-Pick Disease is due to a continuum of phenotypes arising from mutations in the *SMPD1* gene, located within the chromosomal region 11p15.4, a hotspot region for imprinting within the human genome, and studies have shown that the *SMPD1* gene is preferentially expressed from the maternal chromosome (i.e., paternally imprinted). To date, more than 180 mutations have been found within the *SMPD1* gene including point mutations (missense and nonsense), small deletions, and splicing abnormalities. The disease is characterized by a rapidly progressive neurodegenerative course, with reduced pressure in the intraocular fluid in the eyeball (hypotonia), and failure to attain developmental growth milestones.

Paroxysmal nocturnal haemoglobinuria is a rare blood stem cell disorder that manifests with a haemolytic anaemia from uncontrolled complement activation, bone marrow failure, and a propensity for thrombosis It is among the first diseases in which the role the complement system cascade plays in the pathogenesis is well-elucidated. The complement system is a part of the immune system consisting of plasma proteins that interact via three major pathways to form a membrane attack complex which causes membrane lesions, therefore lysing invading pathogenic cells. PNH is caused by a somatic mutation in *PIG-A* gene that leads to a marked deficiency or absence of two important GPI-anchored complement regulatory membrane proteins (CD55 and CD59).

Haemophilia is a sex-linked inherited disorder resulting in a deficiency in certain clotting factors in the blood which affect the intrinsic clotting cascade required to convert prothrombin to thrombin, then fibrin (clot). Haemophilia A has a higher prevalence than haemophilia B. This X-linked bleeding disorder is caused by mutations in coagulation factor VIII (haemophilia A) or factor IX (haemophilia B). Thalassemias are a group of autosomal recessive disorders caused by reduction or lack of production of one or more of the globin chains that make up the haemoglobin (Hb) four subunits. Faulty synthesis of haemoglobin is caused by gene mutations. Adult hemoglobin is composed of two alpha (α) and two beta (β) polypeptide chains. In α-thalassemia, there is deficient synthesis of α-chains—the resulting excess of β-chains bind oxygen poorly, leading to a low concentration of oxygen in tissues. Similarly, in β-thalassemia, there is a lack of β-chains and the excess α-chains can form insoluble aggregates inside red blood cells causing them (and their precursors) to die resulting in severe anaemia.

Atherosclerosis is a progressive inflammatory disorder underlying coronary artery disease (CAD or arteriosclerosis) and stroke, major causes of mortality and morbidity worldwide.

Disorders of the Human Circulatory System

Coronary arteries are small arteries that supply the heart muscle with blood. If those arteries accumulate plaque, the heart muscle is starved of blood and oxygen and can cause heart attack, stroke, and could be fatal. Genome-wide association studies (GWAS) of large human populations have identified numerous novel genes/loci for atherosclerosis and related traits. Several the loci included well-known lipid genes such as LDLR and PCSK9, supporting the importance of low-density lipoproteins (LDL) in atherosclerosis. A few other genes such as CYPA1, CNNM2, and NT5C2, showed evidence of association with hypertension. William's Syndrome is a rare congenital neurogenetic disorder affecting human development and adult cognition. Affected individuals have mild to moderate intellectual disability, or learning difficulties, unique personality characteristics, distinctive facial features, and cardiovascular problems. It is caused by a micro-deletion of chromosome band 7q11.23 involving about 24–28 genes and RNA transcripts. Several genes are included in this loss including CLIP2, ELN, GTF2I, GTF2IRD1, and LIMK1 which are involved in several body conditions.

Ataxia Telangiectasia is characterized by progressive cerebellar degeneration, telangiectasia, immunodeficiency, recurrent sinopulmonary infections, radiation sensitivity, premature aging, and a predisposition to cancer development, especially of lymphoid origin.

AT is caused by mutations in the *ATM* (ataxia telangiectasia, mutated) gene located on human chromosome 11q22-q23 and is made up of 66 exons. Porphyrias are a group of rare metabolic disorders which can either be inherited or acquired affecting the heme pathway. Heme is essential for the synthesis of haemoglobin, myoglobin, and microsomal cytochromes—all of which play an important role in oxygen transport and/or oxidation–reduction reactions. Each type of porphyria is a result of a specific deficiency in one of the enzymes involved in this biosynthetic pathway. Gene mutations in several genes cause deficient production of 7 key enzymes in the biosynthetic pathway producing six types of porphyrias. Long QT Syndrome is characterized by a prolonged heart rate-corrected QT interval on electrocardiogram (ECG) and a predilection for LQTS-triggered cardiac events including fainting, seizures, and/or sudden cardiac arrest, often during times of emotional or physical duress. LQTS stems from a group of genetically distinct arrhythmogenic disorders resulting from genetic mutations in cardiac potassium and sodium ion channels, termed cardiac channelopathies. Mutations have been found in potassium channel genes (LQT1) and (LQT2), and the sodium channel gene (LQT3) accounting for approximately 75% of patients.

Alpha-I-Antitrypsin Deficiency is a serum serine protease inhibitor and functions to protect the lung from the activity of a neutrophil-released protease called elastase. Protease enzymes are normally released in the lung following activation of neutrophils or macrophages in response to pathogens or tobacco smoke. Elastase can destroy alveoli walls and surrounding tissue, leaving pockets of trapped air. The low circulating levels of AAT are the result of mutations in the SERPINA1 gene found on chromosome 14, with more than 120 naturally occurring alleles identified. The three alleles (Z, M and S) are expressed in a codominant manner—that means a person with MZ has levels of AAT that are between the levels of those people who have alleles MM or ZZ. Individuals with at least one normal allele (MZ or MS) or two copies of S (SS) usually produce enough AAT to protect the lungs but have an increased risk of lung disease, especially if they smoke.

End of Chapter Quiz

1. Blood is composed of
   a. Erythrocytes and leukocytes
   b. Thrombocytes and plasma
   c. Erythrocytes and plasma
   d. Thrombocytes and leukocytes
   e. Erythrocytes, leukocytes, thrombocytes and plasma
2. Deoxygenated blood enters the _____ side of the heart via the _____ and _____.
   a. right; superior and inferior vena cava
   b. right; left and right atrium
   c. left; superior and inferior vena cava
   d. left; left and right atrium
   e. right; right atrium and right ventricle
3. Which artery carries deoxygenated blood?
   a. Aorta
   b. Superior vena cava
   c. Inferior vena cava
   d. Pulmonary artery
   e. Right Coronary artery
4. 4. Which vein carries oxygenated blood?
   a. Jugular vein
   b. Superior vena cava
   c. Inferior vena cava
   d. Pulmonary vein
   e. Coronary sinus
5. The passage of blood from the aorta to the rest of the body is referred to as
   a. pulmonary circulation
   b. systemic circulation
   c. cardiac circulation
   d. circulatory circulation
   e. circulatory response
6. Sickle Cell Disease is an autosomal recessive disease. If both parents are carriers for sickle cell, what are the chances their offspring will have sickle cell disease?
   a. 0%
   b. 25%
   c. 50%
   d. 75%
   e. 100%

Disorders of the Human Circulatory System

7. Haemoglobin consists of four protein subunits which are
    a. two subunits of alpha-globin and two subunits of beta globin
    b. four subunits of alpha globin
    c. four subunits of beta globin
    d. heme subunits and iron subunits
    e. none of the above
8. Which statement is false about Burkitt lymphoma?
    a. The three types of Burkitt lymphoma are sporadic, immunodeficiency related and endemic.
    b. Endemic Burkitt lymphoma is often diagnosed in children infected with malaria and/or Epstein-Barr virus
    c. The most frequently mutated genes are oncogenes MYC and ID3
    d. Both young and old patients using high-dose chemotherapy can be cured from Burkitt lymphoma
    e. The symptoms of Burkitt lymphoma vary greatly and depend on the disease's type
9. Chronic myeloid leukaemia is characterized by the unrestrained malignancy of
    a. multipotent adult stem cells
    b. cord blood stem cells
    c. pluripotent bone marrow stem cells
    d. totipotent stem cells
    e. placental cells
10. What procedure is used to open blocked or narrowed coronary arteries?
    a. atrial fibrillation ablation
    b. electrical cardioversion
    c. echocardiogram
    d. computed tomography angiography
    e. coronary angioplasty
11. The symptoms presented in patients with William's Syndrome include
    a. dysmorphic craniofacial features
    b. over-friendliness
    c. high levels of empathy with anxiety
    d. high levels of calcium in the blood
    e. all of the above
12. The symptoms of alpha-I-antitrypsin deficiency include
    a. Emphysema
    b. Liver cirrhosis
    c. Splenomegaly
    d. A and B
    e. All of the above

Thought Questions

1. If one parent is a carrier for SCA and the other parent is normal, what is the probability of having grandchildren with SCA?
2. In thalassemia, describe how excess α or β polypeptide chains affects severity of the condition.
3. Draw and label the electrocardiogram for a patient with LQTS.

REFERENCES

Ackerman, M. J. (2004). Cardiac channelopathies: It's in the genes. *Nature Medicine, 10*(5), 463–464. doi:10.1038/nm0504-463 PMID:15122246

Aerts, J. M., Hollak, C. E., Boot, R. G., Groener, J. E., & Maas, M. (2006). Substrate reduction therapy of glycosphingolipid storage disorders. *Journal of Inherited Metabolic Disease, 29*(2-3), 449–456. doi:10.100710545-006-0272-5 PMID:16763917

American Academy of Pediatrics (AAP). (2018). Health Care Supervision for Children with Williams Syndrome. *Pediatrics, 107*(5), 1192-1204. Retrieved from http://pediatrics.aappublications.org/content/107/5/1192

AT Society (UK). (2018). *Ataxia-Telangiectasia in Children: Guidance on Diagnosis and Clinical Care.* Retrieved from http://www.atsociety.org.uk/data/files/William /A-T_Clinical_Guidance_Document_Final.pdf

Ballas, S. K., Kesen, M. R., Goldberg, M. F., Lutty, G. A., Dampier, C., & Osunkwo, I. (2012). *Malik, P.* Beyond the Definitions of the Phenotypic Complications of Sickle Cell.

Barbon, C. M., Ziegler, R. J., Li, C., Armentano, D., Cherry, M., Desnick, R. J., ... Cheng, S. H. (2005). AAV8-mediated hepatic expression of acid sphingomyelinase corrects the metabolic defect in the visceral organs of a mouse model of Niemann-Pick disease. *Molecular Therapy, 12*(3), 431–440. doi:10.1016/j.ymthe.2005.03.011 PMID:16099409

Bearn, A. G. (1978). Alpha-1-antitrypsin deficiency: A biological enigma. *Gut, 19*(6), 470–473. doi:10.1136/gut.19.6.470 PMID:355063

Besur, S., Schmeltzer, P., & Bonkovsky, H. L. (2015). Acute Porphyrias. *The Journal of Emergency Medicine, 49*(3), 305–312. doi:10.1016/j.jemermed.2015.04.034 PMID:26159905

Bissell, D. M., Anderson, K. E., & Bonkovsky, H. L. (2017). Porphyria. *The New England Journal of Medicine, 377*(9), 862–872. doi:10.1056/NEJMra1608634 PMID:28854095

Bornhorst, J. A., Greene, D. N., Ashwood, E. R., & Grenache, D. G. (2013). α1-Antitrypsin phenotypes and associated serum protein concentrations in a large clinical population. *Chest, 143*(4), 1000–1008. doi:10.1378/chest.12-0564 PMID:23632999

Brodsky, R. A. (2008). Narrative review: paroxysmal nocturnal hemoglobinuria: the physiology of complement-related hemolytic anemia. *Annals of Internal Medicine, 148*(8), 587–595. doi:10.7326/0003-4819-148-8-200804150-00003 PMID:18413620

Brodsky, R. A., Young, N. S., Antonioli, E., Risitano, A. M., Schrezenmeier, H., Schubert, J., ... Hillmen, P. (2008). Multicenter phase 3 study of the complement inhibitor eculizumab for the treatment of patients with paroxysmal nocturnal hemoglobinuria. *Blood, 111*(4), 1840–1847. doi:10.1182/blood-2007-06-094136 PMID:18055865

Calcedo, R., & Wilson, J. M. (2013). Humoral Immune Response to AAV. *Frontiers in Immunology, 4*, 341. doi:10.3389/fimmu.2013.00341 PMID:24151496

Cancer Treatment Centers of America (CTCA). (2018). *Non-Hodgkin lymphoma types.* Retrieved from https://www.cancercenter.com/non-hodgkin-lymphoma/ types/tab/burkitt-lymphoma/

Cao, A., & Kan, Y. W. (2013). The prevention of thalassemia. *Cold Spring Harbor Perspectives in Medicine, 3*(2), a011775. doi:10.1101/cshperspect.a011775 PMID:23378598

Chiuchiolo, M. J., & Crystal, R. G. (2016). Gene Therapy for Alpha-1 Antitrypsin Deficiency Lung Disease. *Annals of the American Thoracic Society, 13*(Suppl 4), S352–S369. doi:10.1513/AnnalsATS.201506-344KV

Collette, J. C., Chen, X., Mills, D. L., Galaburda, A. M., Reiss, A. L., Bellugi, U., & Korenberg, J. R. (2009). William's syndrome: Gene expression is related to parental origin and regional coordinate control. *Journal of Human Genetics, 54*(4), 193–198. doi:10.1038/jhg.2009.5 PMID:19282872

Coronary Artery Disease (C4D) Genetics Consortium. (2011). A genome-wide association study in Europeans and South Asians identifies five new loci for coronary artery disease. *Nature Genetics, 43*(4), 339–344. doi:10.1038/ng.782

Crawford, T. O., Skolasky, R. L., Fernandez, R., Rosquist, K. J., & Lederman, H. M. (2006). Survival probability in ataxia telangiectasia. *Archives of Disease in Childhood, 91*(7), 610–611. doi:10.1136/adc.2006.094268 PMID:16790721

Crystal, R. G. (2014). Adenovirus: The first effective in vivo gene delivery vector. *Human Gene Therapy, 25*(1), 3–11. doi:10.1089/hum.2013.2527 PMID:24444179

Cuenza, L. R., & Adiong, A. A. (2015). Isolated Supravalvar Aortic Stenosis Without William's Syndrome. *Journal of Cardiovascular Echography, 25*(3), 93–95. doi:10.4103/2211-4122.166089 PMID:28465944

Dahl, M., Doyle, A., Olsson, K., Månsson, J.-E., Marques, A. R. A., Mirzaian, M., ... Karlsson, S. (2015). Lentiviral Gene Therapy Using Cellular Promoters Cures Type 1 Gaucher Disease in Mice. *Molecular Therapy, 23*(5), 835–844. doi:10.1038/mt.2015.16 PMID:25655314

Dampier, C., Ely, E., Brodecki, D., & O'Neal, P. (2002). Home management of pain in sickle cell disease: A daily diary study in children and adolescents. *Journal of Pediatric Hematology/Oncology, 24*(8), 643–647. doi:10.1097/00043426-200211000-00008 PMID:12439036

de Serres, F., & Blanco, I. (2014). Role of alpha-1 antitrypsin in human health and disease. *Journal of Internal Medicine, 276*(4), 311–335. doi:10.1111/joim.12239 PMID:24661570

DeZern, A. E., & Brodsky, R. A. (2015). Paroxysmal Nocturnal Hemoglobinuria: A Complement-Mediated Hemolytic Anemia. *Hematology/Oncology Clinics of North America, 29*(3), 479–494. doi:10.1016/j.hoc.2015.01.005 PMID:26043387

(2012). Disease: An Update on Management. *The Scientific World Journal, 949535.* doi:10.1100/2012/949535 PMID:22924029

Dodge, J. C., Clarke, J., Song, A., Bu, J., Yang, W., Taksir, T. V., ... Stewart, G. R. (2005). Gene transfer of human acid sphingomyelinase corrects neuropathology and motor deficits in a mouse model of Niemann-Pick type A disease. *Proceedings of the National Academy of Sciences of the USA, 102*(49), 17822–17827. 10.1073/pnas.0509062102

Dodge, J. C., Clarke, J., Treleaven, C. M., Taksir, T. V., Griffiths, D. A., Yang, W., ... Shihabuddin, L. S. (2009). Intracerebroventricular infusion of acid sphingomyelinase corrects CNS manifestations in a mouse model of Niemann-Pick A disease. *Experimental Neurology, 215*(2), 349–357. doi:10.1016/j.expneurol.2008.10.021 PMID:19059399

Duvoix, A., Roussel, B. D., & Lomas, D. A. (2014). Molecular pathogenesis of alpha-1-antitrypsin deficiency. [PubMed]. *Revue des Maladies Respiratoires, 31*(10), 992–1002. doi:10.1016/j.rmr.2014.03.015

Edel, Y., & Mamet, R. (2018). Porphyria: What Is It and Who Should Be Evaluated? *Rambam Maimonides Medical Journal, 9*(2), e0013. doi:10.5041/RMMJ.10333 PMID:29553924

Faderl, S., Talpaz, M., Estrov, Z., O'Brien, S., Kurzrock, R., & Kantarjian, H. M. (1999). The biology of chronic myeloid leukemia. *The New England Journal of Medicine, 341*(3), 164–172. doi:10.1056/NEJM199907153410306 PMID:10403855

Fang, Y., Shi, C., Manduchi, E., Civelek, M., & Davies, P. F. (2010). MicroRNA-10a regulation of proinflammatory phenotype in athero-susceptible endothelium in vivo and in vitro. *Proceedings of the National Academy of Sciences of the United States of America, 107*(30), 13450–13455. doi:10.1073/pnas.1002120107 PMID:20624982

Fucharoen, S., & Winichagoon, P. (2011). Haemoglobinopathies in southeast Asia. *The Indian Journal of Medical Research, 134*(4), 498–506. PMID:22089614

Garver, R. I. Jr, Chytil, A., Courtney, M., & Crystal, R. G. (1987). Clonal gene therapy: Transplanted mouse fibroblast clones express human alpha 1-antitrypsin gene in vivo. *Science, 237*(4816), 762–764. doi:10.1126cience.3497452 PMID:3497452

Gatti, R. A., Berkel, I., Boder, E., Braedt, G., Charmley, P., Concannon, P., ... Lange, K. (1988). Localization of an ataxia-telangiectasia gene to chromosome 11q22-23. *Nature, 336*(6199), 577–580. doi:10.1038/336577a0 PMID:3200306

Genetic Disease Foundation (GDF). (2018). *Hope through Knowledge.* Retrieved from http://www.geneticdiseasefoundation.org/

Genetics Home Reference (GHR), US Library of Medicine. (2018). *Sickle cell diseases.* Retrieved from https://ghr.nlm.nih.gov/condition/sickle-cell-disease#genes

Giudicessi, J. R., & Ackerman, M. J. (2012). Potassium-channel mutations and cardiac arrhythmias--diagnosis and therapy. *Nature Reviews. Cardiology, 9*(6), 319–332. doi:10.1038/nrcardio.2012.3 PMID:22290238

Giudicessi, J. R., & Ackerman, M. J. (2013). Genotype- and phenotype-guided management of congenital long QT syndrome. *Current Problems in Cardiology, 38*(10), 417–455. doi:10.1016/j.cpcardiol.2013.08.001 PMID:24093767

Grabowski, G. A. (2008). Phenotype, diagnosis, and treatment of Gaucher's disease. *Lancet, 372*(9645), 1263–1271. doi:10.1016/S0140-6736(08)61522-6 PMID:19094956

Gringeri, A. (2011). Factor VIII safety: plasma-derived versus recombinant products. *Blood transfusion = Trasfusione del sangue, 9*(4), 366–370. doi:10.2450/2011.0092-10

Ha, M., & Kim, V. N. (2014). Regulation of microRNA biogenesis. *Nature Reviews. Molecular Cell Biology, 15*(8), 509–524. doi:10.1038/nrm3838 PMID:25027649

Henle, G., & Henle, W. (1966). Immunofluorescence in Cells Derived from Burkitt's Lymphoma. *Journal of Bacteriology, 91*(3), 1248–1256. PMID:4160230

Hill, A., Kelly, R. J., & Hillmen, P. (2013). Thrombosis in paroxysmal nocturnal hemoglobinuria. *Blood, 121*(25), 4985–4996. doi:10.1182/blood-2012-09-311381 PMID:23610373

Hollak, C. E., de Sonnaville, E. S., Cassiman, D., Linthorst, G. E., Groener, J. E., Morava, E., ... Poorthuis, B. J. (2012). Acid sphingomyelinase (Asm) deficiency patients in The Netherlands and Belgium: disease spectrum and natural course in attenuated patients. *Molecular Genetics and Metabolism, 107*(3), 526–533. doi:.2012.06.015 doi:10.1016/j.ymgme

Johnson, J. N., & Ackerman, M. J. (2010). The prevalence and diagnostic/prognostic utility of sinus arrhythmia in the evaluation of congenital long QT syndrome. *Heart Rhythm, 7*(12), 1785–1789. doi:10.1016/j.hrthm.2010.07.030 PMID:20673812

Karim, Z., Lyoumi, S., Nicolas, G., Deybach, J. C., Gouya, L., & Puy, H. (2015). Porphyrias: A 2015 update. *Clinics and Research in Hepatology and Gastroenterology, 39*(4), 412–425. doi:10.1016/j.clinre.2015.05.009 PMID:26142871

Kumar, P., Katoch, M., Bhatia, S., & Gupta, R. (2014). William's syndrome with mitral valve disease. *Medical Journal, Armed Forces India, 70*(2), 189–191. doi:10.1016/j.mjafi.2012.07.011 PMID:24843210

Lebensburger, J. D., Pestina, T. I., Ware, R. E., Boyd, K. L., & Persons, D. A. (2010). Hydroxyurea therapy requires HbF induction for clinical benefit in a sickle cell mouse model. *Haematologica, 95*(9), 1599–1603. doi:10.3324/haematol.2010.023325 PMID:20378564

Lee, Y. T., Lin, H. Y., Chan, Y. W., Li, K. H., To, O. T., Yan, B. P., ... Tse, G. (2017). Mouse models of atherosclerosis: A historical perspective and recent advances. *Lipids in Health and Disease, 16*(1), 12. doi:10.118612944-016-0402-5 PMID:28095860

Levine, G. N., Chodos, A. P., & Loscalzo, J. (1995). Restenosis following coronary angioplasty: Clinical presentations and therapeutic options. *Clinical Cardiology, 18*(12), 693–703. doi:10.1002/clc.4960181203 PMID:8608668

Liu, Y. P., Lien, W. C., Fang, C. C., Lai, T. I., Chen, W. J., & Wang, H. P. (2005). ED presentation of acute porphyria. *The American Journal of Emergency Medicine, 23*(2), 164–167. doi:10.1016/j.ajem.2004.03.013 PMID:15765337

Love, C., Sun, Z., Jima, D., Li, G., Zhang, J., Miles, R., ... Dave, S. (2012). The genetic landscape of mutations in Burkitt lymphoma. *Nature Genetics, 44*(12), 1321–1325. doi:10.1038/ng.2468 PMID:23143597

Lugo, T. G., Pendergast, A. M., Muller, A. J., & Witte, O. N. (1990). Tyrosine kinase activity and transformation potency of bcr-abl oncogene products. *Science, 247*(4946), 1079–1082. doi:10.1126cience.2408149 PMID:2408149

Lukina, E., Watman, N., Arreguin, E. A., Banikazemi, M., Dragosky, M., Iastrebner, M., ... Peterschmitt, M. J. (2010). A phase 2 study of eliglustat tartrate (Genz-112638), an oral substrate reduction therapy for Gaucher disease type 1. *Blood, 116*(6), 893–899. doi:10.1182/blood-2010-03-273151 PMID:20439622

Lusis, A. J. (2012). Genetics of atherosclerosis. *Trends in Genetics: TIG, 28*(6), 267–275. doi:10.1016/j.tig.2012.03.001 PMID:22480919

Marawaha, R. K., Bansal, D., Kaur, S., & Trehan, A. (2004). Wheat grass juice reduces transfusion requirement in patients with thalassemia major: A pilot study. *Indian Pediatrics, 41*(7), 716–720. PMID:15297687

McGann, P. T., Tshilolo, L., Santos, B., Tomlinson, G. A., Stuber, S., Latham, T., ... Ware, R. (2016). Hydroxyurea Therapy for Children with Sickle Cell Anemia in Sub-Saharan Africa: Rationale and Design of the REACH Trial. *Pediatric Blood & Cancer, 63*(1), 98–104. doi:10.1002/pbc.25705 PMID:26275071

McGann, P. T., & Ware, R. E. (2015). Hydroxyurea therapy for sickle cell anemia. *Expert Opinion on Drug Safety, 14*(11), 1749–1758. doi:10.1517/14740338.2015.1088827 PMID:26366626

McGovern, M. M., Wasserstein, M. P., Kirmse, B., Duvall, W. L., Schiano, T., Thurberg, B. L., ... Cox, G. F. (2016). Novel first-dose adverse drug reactions during a phase I trial of olipudase alfa (recombinant human acid sphingomyelinase) in adults with Niemann-Pick disease type B (acid sphingomyelinase deficiency). *Genetics in Medicine, 18*(1), 34–40. doi:10.1038/gim.2015.24 PMID:25834946

Melo, J. V., Hughes, T. P., & Apperley, J. F. (2003). Chronic myeloid leukemia. *Hematology. American Society of Hematology. Education Program, 2003*(1), 132–152.

Meyer, U. A., Schuurmans, M. M., & Lindberg, R. L. (1998). Acute porphyrias: Pathogenesis of neurological manifestations. *Seminars in Liver Disease, 18*(1), 43–52. doi:10.1055-2007-1007139 PMID:9516677

Mignot, C., Doummar, D., Maire, I., & De Villemeur, T. B. (2006). Type 2 Gaucher disease: 15 new cases and review of the literature. *Brain & Development, 28*(1), 39–48. doi:10.1016/j.braindev.2005.04.005 PMID:16485335

Mishra, S., Ray, S., Dalal, J. J., Sawhney, J. P., Ramakrishnan, S., Nair, T., ... Bahl, V. K. (2016). Management standards for stable coronary artery disease in India. *Indian Heart Journal, 68*(Suppl 3), S31–S49. doi:10.1016/j.ihj.2016.11.320

Modell, B., & Darlison, M. (2008). Global epidemiology of haemoglobin disorders and derived service indicators. *Bulletin of the World Health Organization, 86*(6), 480–487. doi:10.2471/BLT.06.036673 PMID:18568278

Mohan, B., & Mittal, C. M. (2011). Supravalvular aortic stenosis in William's syndrome. *Annals of Pediatric Cardiology, 4*(2), 213–214. doi:10.4103/0974-2069.84662 PMID:21976893

Monahan, P. E. (2015). Gene therapy in an era of emerging treatment options for hemophilia B. *Journal of Thrombosis and Homeostasis, 13*(1), S151–S160. doi:10.1111/jth.12957

Moss, A. J. (2003). Long QT Syndrome. *Journal of the American Medical Association, 289*(16), 2041–2044. doi:10.1001/jama.289.16.2041 PMID:12709446

National Heart and Lung and Blood Institute (NIH-NHLBI). (2018). Retrieved from https://www.nhlbi.nih.gov/health-topics/atherosclerosis

Nerbonne, J. M., & Kass, R. S. (2005). Molecular physiology of cardiac repolarization. *Physiological Reviews, 85*(4), 1205–1253. doi:10.1152/physrev.00002.2005 PMID:16183911

Nienhuis, A. W., Nathwani, A. C., & Davidoff, A. M. (2017). Gene Therapy for Hemophilia. *Molecular Therapy: The Journal of the American Society of Gene Therapy, 25*(5), 1163–1167. doi:10.1016/j.ymthe.2017.03.033 PMID:28411016

NIH-GARD. (2018). *Genetics and Rare Diseases Information Centre*. Retrieved from https://rarediseases.info.nih.gov/diseases/7891/disease

Paz, A., Brownstein, Z., Ber, Y., Bialik, S., David, E., Sagir, D., ... Shamir, R. (2010). SPIKE: A database of highly curated human signaling pathways. *Nucleic Acids Research, 39*(Database issue), D793–D799. doi:10.1093/nar/gkq1167 PMID:21097778

Pischik, E., & Kauppinen, R. (2015). An update of clinical management of acute intermittent porphyria. *The Application of Clinical Genetics, 8*, 201–214. doi:10.2147/TACG.S48605 PMID:26366103

Quinn, C. T., Rogers, Z. R., McCavit, T. L., & Buchanan, G. R. (2010). Improved survival of children and adolescents with sickle cell disease. *Blood, 115*(17), 3447–3452. doi:10.1182/blood-2009-07-233700 PMID:20194891

Rogers, G. L., & Herzog, R. W. (2015). Gene therapy for hemophilia. *Frontiers in Bioscience (Landmark Edition), 20*(3), 556–603. doi:10.2741/4324 PMID:25553466

Rosenfeld, M. A., Siegfried, W., Yoshimura, K., Yoneyama, K., Fukayama, M., Stier, L. E., ... Perricaudet, M. (1991). Adenovirus-mediated transfer of a recombinant alpha 1-antitrypsin gene to the lung epithelium in vivo. *Science, 252*(5004), 431–434. doi:10.1126cience.2017680 PMID:2017680

Rothblum, C., Write, J., Lefton-Greif, M. A., McGrath-Morrow, S. A., Crawford, T. O., & Lederman, H. M. (2016). Ataxia telangiectasia: A review. *Orphanet Journal of Rare Diseases, 11*(1), 159. doi:10.118613023-016-0543-7 PMID:27884168

Rowe, M., Fitzsimmons, L., & Bell, A. I. (2014). Epstein-Barr virus and Burkitt lymphoma. *Chinese Journal of Cancer*, *33*(12), 609–619. doi:10.5732/cjc.014.10190 PMID:25418195

Sabatino, D. E., Nichols, T. C., Merricks, E., Bellinger, D. A., Herzog, R. W., & Monahan, P. E. (2012). Animal models of hemophilia. *Progress in Molecular Biology and Translational Science*, *105*, 151–209. doi:10.1016/B978-0-12-394596-9.00006-8 PMID:22137432

Samani, N. J., Erdmann, J., Hall, A. S., Hengstenberg, C., Mangino, M., Mayer, B., ... Schunkert, H. (2007). Genome-wide association analysis of coronary artery disease. *The New England Journal of Medicine*, *357*(5), 443–453. doi:10.1056/NEJMoa072366 PMID:17634449

Sanchez-Martinez, A., Beavan, M., Gegg, M. E., Chau, K.-Y., Whitworth, A. J., & Schapira, A. H. V. (2016). Parkinson disease-linked *GBA* mutation effects reversed by molecular chaperones in human cell and fly models. *Scientific Reports*, *6*(1), 31380. doi:10.1038rep31380 PMID:27539639

Sassa, S. (2006). Modern diagnosis and management of the porphyrias. *British Journal of Haematology*, *135*(3), 281–292. doi:10.1111/j.1365-2141.2006.06289.x PMID:16956347

Savage, D. G., Szydlo, R. M., & Goldman, J. M. (1997). Clinical features at diagnosis in 430 patients with chronic myeloid leukaemia seen at a referral centre over a 16-year period. *British Journal of Haematology*, *96*(1), 111–116. doi:.d01-1982.x doi:10.1046/j.1365-2141.1997

Schmitz, R., Young, R. M., Ceribelli, M., Jhavar, S., Xiao, W., Zhang, M., ... Staudt, L. M. (2012). Burkitt Lymphoma Pathogenesis and Therapeutic Targets from Structural and Functional Genomics. *Nature*, *490*(7418), 116–120. doi:10.1038/nature11378 PMID:22885699

Schroeder, W. T., Miller, M. F., Woo, S. L., & Saunders, G. F. (1985). Chromosomal localization of the human alpha 1-antitrypsin gene (PI) to 14q31-32. *American Journal of Human Genetics*, *37*(5), 868–872. PMID:3876766

Schuchman, E. H., & Desnick, R. J. (2017). Types A and B Niemann-Pick Disease. *Molecular Genetics and Metabolism*, *120*(1-2), 27–33. doi:10.1016/j.ymgme.2016.12.008 PMID:28164782

Schunkert, H., König, I. R., Kathiresan, S., Reilly, M. P., Assimes, T. L., Holm, H., ... Samani, N. J. (2011). Large-scale association analysis identifies 13 new susceptibility loci for coronary artery disease. *Nature Genetics*, *43*(4), 333–338. doi:10.1038/ng.784 PMID:21378990

Sherman, A., Biswas, M., & Herzog, R. W. (2017). Innovative Approaches for Immune Tolerance to Factor VIII in the Treatment of Hemophilia A. *Frontiers in Immunology*, *8*, 1604. doi:10.3389/fimmu.2017.01604 PMID:29225598

Shtivelman, E., Lifshitz, B., Gale, R. P., & Canaani, E. (1985). Fused transcript of abl and bcr genes in chronic myelogenous leukaemia. *Nature*, *315*(6020), 550–554. doi:10.1038/315550a0 PMID:2989692

Sripichai, O., Makarasara, W., Munkongdee, T., Kumkhaek, C., Nuchprayoon, I., Chuansumrit, A., ... Fucharoen, S. (2008). A scoring system for the classification of beta-thalassemia/Hb E disease severity. *American Journal of Hematology*, *83*(6), 482–484. doi:10.1002/ajh.21130 PMID:18186524

Stein, P. E., Badminton, M. N., & Rees, D. C. (2017). Update review of the acute porphyrias. *British Journal of Haematology*, *176*(4), 527–538. doi:10.1111/bjh.14459 PMID:27982422

Steinberg, M. H. (2008). Sickle cell anemia, the first molecular disease: Overview of molecular etiology, pathophysiology, and therapeutic approaches. *The Scientific World Journal*, *8*, 1295–1324. doi:10.1100/tsw.2008.157 PMID:19112541

Steinberg, M. H. (2009). Genetic etiologies for phenotypic diversity in sickle cell anemia. *The Scientific World Journal*, *9*, 46–67. doi:10.1100/tsw.2009.10 PMID:19151898

Stirnemann, J., Belmatoug, N., Camou, F., Serratrice, C., Froissart, R., Caillaud, C., ... Berger, M. G. (2017). A Review of Gaucher Disease Pathophysiology, Clinical Presentation and Treatments. *International Journal of Molecular Sciences*, *18*(2), 441. doi:10.3390/ijms18020441 PMID:28218669

Swift, M., Morrell, D., Cromartie, E., Chamberlin, A. R., Skolnick, M. H., & Bishop, D. T. (1986). The incidence and gene frequency of ataxia-telangiectasia in the United States. *American Journal of Human Genetics*, *39*(5), 573–583. PMID:3788973

Taher, A. T., Viprakasit, V., Musallam, K. M., & Cappellini, M. D. (2013). Treating iron overload in patients with non-transfusion-dependent thalassemia. *American Journal of Hematology*, *88*(5), 409–415. doi:10.1002/ajh.23405 PMID:23475638

Tester, D. J., Will, M. L., Haglund, C. M., & Ackerman, M. J. (2006). Effect of clinical phenotype on yield of long QT syndrome genetic testing. *Journal of the American College of Cardiology*, *47*(4), 764–768. doi:10.1016/j.jacc.2005.09.056 PMID:16487842

The Wellcome Trust Case Control Consortium. (2007). Genome-wide association study of 14,000 cases of seven common diseases and 3,000 shared controls. *Nature*, *447*(7145), 661–678. doi:10.1038/nature05911 PMID:17554300

The World Health Organization (WHO). (2012). *The Top Ten Causes of Death Fact Sheet*. Retrieved from http://www.who.int/mediacentre/factsheets/fs310/en/index2.html

Thompson, P. A., Kantarjian, H. M., & Cortes, J. E. (2015). Diagnosis and Treatment of Chronic Myeloid Leukemia (CML) in 2015. *Mayo Clinic Proceedings*, *90*(10), 1440–1454. doi:10.1016/j.mayocp.2015.08.010 PMID:26434969

Tse, G., Yan, B. P., Chan, Y. W., Tian, X. Y., & Huang, Y. (2016). Reactive Oxygen Species, Endoplasmic Reticulum Stress and Mitochondrial Dysfunction: The Link with Cardiac Arrhythmogenesis. *Frontiers in Physiology*, *7*, 313. doi:10.3389/fphys.2016.00313 PMID:27536244

Tylki-Szymańska, A., Vellodi, A., El-Beshlawy, A., Cole, J. A., & Kolodny, E. (2010). Neuronopathic Gaucher disease: Demographic and clinical features of 131 patients enrolled in the International Collaborative Gaucher Group Neurological Outcomes Subregistry. *Journal of Inherited Metabolic Disease*, *33*(4), 339–346. doi:10.100710545-009-9009-6 PMID:20084461

US National Library of Medicine (USNLM). (2018). Retrieved from https://clinicaltrials.gov/ct2/results?term=acid+sphingomyelinase&Search=Search

Vatta, M., Ackerman, M. J., Ye, B., Makielski, J. C., Ughanze, E. E., Taylor, E. W., ... Towbin, J. A. (2006). Mutant caveolin-3 induces persistent late sodium current and is associated with long-QT syndrome. *Circulation, 114*(20), 2104–2112. doi:10.1161/CIRCULATIONAHA.106.635268 PMID:17060380

Verma, I. C., Saxena, R., & Kohli, S. (2011). Past, present & future scenario of thalassaemic care & control in India. *The Indian Journal of Medical Research, 134*(4), 507–521. PMID:22089615

Wang, J., Kang, Y. X., Pan, W., Lei, W., Feng, B., & Wang, X. J. (2016). Enhancement of Anti-Inflammatory Activity of Curcumin Using Phosphatidylserine-Containing Nanoparticles in Cultured Macrophages. *International Journal of Molecular Sciences, 17*(5), 969. doi:10.3390/ijms17060969 PMID:27331813

Wang, Y., & Perlmutter, D. H. (2013). Targeting intracellular degradation pathways for treatment of liver disease caused by α1-antitrypsin deficiency. *Pediatric Research, 75*(1-2), 133–139. doi:10.1038/pr.2013.190 PMID:24226634

Weatherall, D. J. (2010). The inherited diseases of hemoglobin are an emerging global health burden. *Blood, 115*(22), 4331–4336. doi:10.1182/blood-2010-01-251348 PMID:20233970

Wewers, M. D. & Crystal, R. G. (2013). Alpha-1 antitrypsin augmentation therapy. *Journal of Chronic Obstructive Pulmonary Disease, 10*(1), 64-7. doi:.2013.764402 doi:10.3109/15412555

WHO. (1994). Guidelines for control of hemoglobin disorders. World Health Organization Hereditary Disease Program. Geneva: WHO/HDP/HB/gl/94.1.

Williams Syndrome Association (WSA). (2018). *Therapeutic Interventions.* Retrieved from https://williams-syndrome.org/parent/therapeutic-interventions

Woolf, J., Marsden, J. T., Degg, T., Whatley, S., Reed, P., Brazil, N., ... Badminton, M. (2017). Best practice guidelines on first-line laboratory testing for porphyria. *Annals of Clinical Biochemistry, 54*(2), 188–198. doi:10.1177/0004563216667965 PMID:27555665

KEY TERMS AND DEFINITIONS

Adeno-Associated Viruses (AAV Vectors): These are small viruses widely spread throughout the human population, but do not cause infection. AAVs are often used by scientists as ideal delivery vehicles for gene therapy.

Angina: Chest pain.

Antibiotics: A range of powerful drugs used to treat bacterial infections by killing or slowing down bacteria growth.

Arteriosclerosis: Abnormal thickening or hardening of arteries.

Augmentation Therapy: A treatment used to increase the level of alpha-1 antitrypsin (AAT) protein circulating in the blood and lungs by intravenously transferring AAT from the blood plasma of a healthy donor.

Disorders of the Human Circulatory System

B-Adrenergic Receptor Antagonists (B-Blockers): Drugs that inhibit the binding of norepinephrine and epinephrine by competitively binding to beta-adrenoreceptors.

Blood Transfusion: The process in which donated blood is transferred to another patient through the veins.

Bronchodilators: Substances used to dilate the bronchial tubes and relaxing the lung muscles.

Cardiogram: Produced by a cardiograph, this is a record of heart muscle activity.

Cardioverter-Defibrillators: A small battery powered device that is surgically placed under the skin below the collar bone and is used to detect heart rate abnormalities. If dangerous heart rate is detected, the device sends an electrical shock to correct the rhythm.

Chemotherapy: Therapy designed to treat cancerous diseases through the administration of cytotoxic drugs that aim to regulate and prevent further growth and division of cancerous cells.

Cirrhosis: A degenerative disease that occurs when there is scarring of liver tissue.

Congenital: A condition or trait present from birth.

Coronary Angioplasty: A procedure in which a small balloon is inflated to open blocked or narrowed coronary arteries.

Coronary Artery Bypass Grafting: A surgery that improves blood flow to the heart by creating more pathways for the blood to flow using arteries or veins from other areas of the body.

Dementia: Disease characterized by mental instability, impaired memory, and possible personality changes.

Emphysema: Type A COPD caused by irreversible damage to alveolar walls or the enlargement of distal air sacs.

Enzyme Replacement Therapy (ERT): The process of intravenously replacing enzymes that are deficient in the body.

Erythropoiesis: Production of red blood cells which is initiated by the decreasing levels of oxygen which cause erythropoietin hormone production in the kidneys.

Gene Therapy: The replacement of defective genes with functional genes or the introduction of new genes to prevent or treat diseases.

Genome-Wide Association Studies (GWAS): An observational study that examines if there is an association between genetic variants and traits in different individuals.

HapMap Project: A project aimed to creating a halotype map of the human genome in order to find genetic variants that affect human health and diseases.

Health Disparities: Higher burden of illness, injury, mortality.

Hemoglobinopathies: A group of disorders involving abnormal haemoglobins and anaemia which are caused by defects in the genes that produce haemoglobin.

Hepatocytes: Liver cells.

Immunization: The process in which a person is made to become resistant to an infectious disease.

Ischaemic Cardiomyopathy: Heart muscle disease caused by a reduction in blood flow.

Necrosis: Irreversible cell injury characterized by cell rupture, inflammation, and leakage of contents into extracellular fluid.

Non-Steroidal Anti-Inflammatory Agents (NSAIDS): One of the most common over-the-counter drug used to treat pain and inflammation by preventing an enzyme called cyclooxygenase from producing prostaglandins.

Palliative Treatments: Treatments focused on relieving symptoms and improving the quality of life for patients with serious illnesses.

Physical Therapy: Therapy designed to restore, preserve, or improve physical function through exercise or the use of specially designed equipment.

Pluripotent Bone Marrow Stem Cells: The first step in the developmental process of red blood cells that results in cells that are able to transform into anything.

Protein Replacement Therapy: The process of replacing proteins in the body that are deficient or absent.

Radiation Therapy: A type of cancer treatment that uses beams of high energy to kill cancer cells and shrink tumours.

Serum Serine Protease Inhibitor: Inactivates related proteases in order to regulate protease mediated activities.

Single Nucleotide Polymorphisms (SNPs): A genetic variant of a single nucleotide in a DNA sequence.

Speech Therapy: A clinical program designed to improve speech and language related issues.

Stem Cell Transplantation: The process of replacing unhealthy blood forming cells (stem cells) with healthy cells donated by a healthy individual; also known as a bone marrow transplant.

Stent Insertion: Permanent placement of a wire mesh tube into a clogged artery in order to increase blood flow to the heart.

Substrate Reduction Therapy (SRT): An oral medication that prevents the production of glucocerebroside.

Tachycardia: Term used to describe heart rate that is abnormally rapid.

Chapter 14
Cancers

ABSTRACT

With more than half of all cancer cases occurring in less developed nations of the world, cancer is a source of significant and growing mortality worldwide, with an increase to 19.3 million new cancer cases per year projected for 2025. Standard current treatments for cancer include surgery, radiotherapy, and a host of other systemic treatments comprising cytotoxic chemotherapy, hormonal therapy, immunotherapy, and targeted therapies. Referred to as the "guardian of the genome," the alteration or inactivation of p53 tumour-suppressor gene by mutation or by its interactions with oncogene products or DNA tumour viruses can lead to cancer. The p53 is mutated in about half of almost all types of cancer arising from a wide spectrum of tissues. This chapter focuses on several types of cancer including breast and ovarian, colorectal, small cell lung carcinoma, malignant melanoma, pancreatic, prostate, neurofibromatosis, multiple endocrine neoplasia, and retinoblastoma.

CHAPTER OUTLINE

14.1 Overview
14.2 p53 Tumour Suppressor Gene
14.3 Breast and Ovarian
14.4 Colorectal
14.5 Lung Carcinoma (small cell)
14.6 Malignant Melanoma
14.7 Pancreatic
14.8 Prostate
14.9 Neurofibromatosis
14.10 Multiple Endocrine Neoplasia
14.11 Retinoblastoma
Chapter Summary

DOI: 10.4018/978-1-5225-8066-9.ch014

LEARNING OUTCOMES

- Identify each type of cancer
- Outline the symptoms of each type of cancer
- Explain the genetic basis of each cancer
- Summarize the therapies available to treat each cancer

14.1 OVERVIEW

With more than half of all cancer cases occurring in low- or middle-income nations of the world, cancer is a source of significant and growing mortality worldwide, with an increase to 19.3 million new cancer cases per year projected for 2025 (Block et al., 2015). Standard current treatments for cancer include surgery, **radiotherapy**, and a host of other systemic treatments comprising cytotoxic chemotherapy, **hormonal therapy**, **immunotherapy**, and targeted therapies. Today cancer continues to frustrate clinical treatment efforts, but the search for effective therapies is a never-ending research goal. (Palumbo et al., 2013; Block et al., 2015) Of the global cancer burden, it is estimated that over 20% is attributable to infectious agents. In contrast to virally induced cancers, bacteria have been largely neglected as factors contributing to cancer, with only a few bacterial infections linked to cancer development to date. Bacteria may contribute to cancer development through inflammation, induction of DNA damage by toxins, metabolites, and/or manipulation of host cell signalling pathways during their infection cycle. (Zur, 2009; Samaras et al., 2010; Coghill & Hildesheim, 2014)

Molecular target therapies represent a significant advance in the treatment of cancer and include a variety of drugs like: 1) *Imatinib*, an inhibitor of the tyrosine kinase enzyme BCR-ABL, which has made chronic myelogenous leukaemia a more manageable disease; 2) *Sunitinib, sorafenib* and *bevacizumab* are inhibitors of vascular endothelial growth factor receptor used in renal and colon cancers; 3) *Gefitinib* and *erlotinib* are based on tumour-specific targets and are epidermal growth factor receptor inhibitors used in lung cancer, and the HER2 inhibitor *trastuzumab* used in breast cancer; 4) Another approach is the synthetic lethal model exemplified by the use of poly-ADP ribose polymerase inhibition, in which mutational loss of one or more redundant components of a cell survival pathway in tumourigenic cells confers selective sensitivity to drugs that target remaining pathway components. The expense of the new targeted therapies is a major limitation on their use—for example, 11 of 12 drugs approved by the USFDA in 2012 were priced above $100,000 US per year per patient, primarily because of the accelerating costs of drug development. (Vogelstein et al., 2013; Zahreddine & Borden, 2013; Ciociola et al., 2014)

Genomic sequencing in cancers indicate that they harbour significant genetic heterogeneity, even within a single patient, and based on this heterogeneity, cancers routinely evolve resistance to treatment through switching from one growth pathway to another. The proposed strategy employs the basic principles of rational drug design and aims to curtail cancer growth by precisely targeting many growth pathways simultaneously, although lack of therapeutic success, significant toxicity, and costs make this a challenge (Block et al., 2015; Wang et al., 2015). Block et al. (2015) recommend such an approach to include follow-up maintenance plan to conventional **adjuvant treatment**; situations of rare cancers and disease stages for which no accepted treatments exist; for patients who do not tolerate conventional chemotherapy, hormonal therapy or targeted therapies; for patients who experience relapse or progression after targeted treatment; in hospice or palliative care patients where low- or non-invasive strategies are a legitimate and humane option; and, in situations in which high-cost agents are not feasible.

14.2 THE P53 TUMOUR SUPPRESSOR GENE

The cell cycle (section 4.2, Chapter 4) is composed of a series of steps which ensures cells prepare for the process of cell division through growth, and replication of the DNA and other cellular machinery. The process is negatively or positively regulated by various factors chief among them being the negative regulator coded p53, a tumour-suppressor gene. The earliest model explaining the mechanism that underlies p53 tumour suppressor activity came from studies referring to p53 as the "guardian of the genome" (Lane, 1992). Alteration or inactivation of p53 by mutation, or by its interactions with oncogene products or DNA tumour viruses can lead to cancer. These mutations seem to be the most common genetic change in human cancers. The p53 tumour-suppressor gene is mutated in about half of almost all types of cancer arising from a wide spectrum of tissues. (Levine et al., 1991; Harris & Hoolstein, 1994). A deficiency in p53 enhances the initiation or progression of cancer depending on the type of tumour. Tumours that lack p53 are commonly characterized by more malignant characteristics which include lack of cellular differentiation, genetic instability, and increased invasiveness and metastatic potential (Bieging et al., 2014). The mechanism that underlie p53-mediated tumour suppression is not completely understood; but it is thought to act as a cellular stress sensor that triggers transient cell cycle arrest, permanent cell cycle arrest (cellular senescence) and **apoptosis** in response to a host of diverse stresses. These stressors can include DNA damage, hyperproliferative signals, hypoxia, **oxidative stress**, ribonucleotide depletion and nutrient starvation. It is in response to such stress signals that p53 is displaced from its negative regulators MDM2 and MDM4, thereby allowing its stabilization and activation. (Hu et al., 2012; Bieging et al., 2014)

The p53 gene codes for the p53 protein made up of two amino-terminal transcriptional activation domains, a proline-rich domain, a DNA-binding domain, a tetramerization domain, and a carboxy-terminal region. Inactivation of p53 in human tumours typically occurs through missense mutations in the DNA-binding domain of the p53 protein. p53 promotes the classical cellular functions of cell cycle arrest, senescence and apoptosis primarily through the activation of specific genes. The evidence for p53-dependent regulation of the genes on this list comes from mouse and/or human cells. p53-induced target genes are involved in processes that are important for tumour suppression, including p53-associated responses (apoptosis, cell cycle arrest, senescence and DNA repair), and processes that have recently been associated with p53-dependent tumour suppression (metabolism control, autophagy, tumour microenvironment crosstalk, invasion and metastasis, and stem cell biology). (Vousden & Prives, 2009; Brady & Attardi, 2010)

Six major mechanisms of p53-dependent transcriptional regulation are currently accepted in the literature (Riley et al., 2008; Rinn & Huarte, 2011; Fischer et al., 2014). These include:

1. Direct activation of target genes following p53 binding to a p53 response element (RE);
2. Direct repression of target genes after p53 binding to p53 REs;

3. Direct repression of target genes through p53 binding via **adaptor proteins**;
4. Indirect repression via direct activation of p21 by p53;
5. Indirect repression through interference with transcriptional activators;
6. Indirect repression of target genes via non-coding RNAs (ncRNAs).

14.3 BREAST AND OVARIAN

Breast Cancer

Predictions from the National Cancer Institute (NCI) in 2015 were that 40,290 women and 440 men would die of breast cancer; 231,840 women and 2350 men would be diagnosed with invasive breast cancer; and another 60,290 women would be diagnosed with breast cancer in situ in the US. In early 2016, the NCI estimated that approximately 3,560,570 women in the US are living with a prior diagnosis of breast cancer (DeSantis et al., 2016). Approximately 1.67 million cases of breast cancer are diagnosed annually worldwide (Mavaddat et al., 2015).

Symptoms

Primary or recurrent/local-regional signs and symptoms include a lump in the breast/chest wall/axilla, dimpling of the skin, nipple retraction, clear or bloody nipple discharge (spontaneous), redness, scaling, thickening of nipple-areolar complex, rash on breast, unresponsive to antibiotics (Figure 1). Distant recurrent signs include new-onset localized bone pain lasting longer than 2 weeks (long bones, ribs, spine), persistent chest pain, with or without cough, persistent abdominal pain, unintended weight loss, persistent headache, personality changes, new-onset seizures, and loss of consciousness. (Bodai, 2015)

Figure 1. The growth and metastasis of a malignant breast cancer tumor
Source: Image used under license from Shutterstock.com

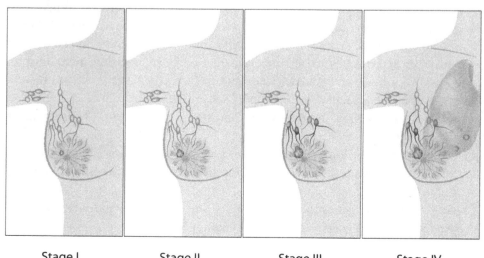

Studies have shown that nearly half of all breast cancer patients have sleep-related problems due to a range of causes including side effects of antineoplastic medications, long hospital stays, or stress (Vincent, 2015). Insomnia, a specific sleep disorder of initiating and maintaining sleep, is most common in cancer patients and often occurs along with anxiety and depression (Savard & Morin, 2001). Cancer treatments, including chemotherapy, may cause damage to the peripheral nerves resulting in neuropathy with symptoms of pain, tingling, numbness, or a pins-and-needles feeling, the inability to feel a hot or cold sensation, or the inability to feel pain. Motor neuropathy can include problems with balance, weak or achy muscles, twitching, cramping or wasting muscles, or swallowing or breathing difficulties. Autonomic nerve damage can cause dizziness or faintness, and digestive, sexual, sweating and urination problems (NCI, 2018). Pain can be caused by cancer therapies, including surgery, radiation therapy, chemotherapy, targeted therapy, supportive care therapies and/or diagnostic procedures (NCI, 2018; IASP, 2018). Lymphedema is a condition following treatment such as surgery or radiation therapy where parts of the lymph system become damaged or blocked, leading to an accumulation of lymph fluid that does not drain properly, builds up in tissues, and causes swelling (Pyszel et al., 2006).

Genetic Basis

Several studies have reported an increased risk for developing breast cancer in women with either *BRCA1* (*on* chromosome 17) or *BRCA2* (on chromosome 13) mutations following exposure to medical radiation, either through mammography or radiation therapy (Bernstein et al., 2013). Both genetic and lifestyle factors are implicated in the aetiology of breast cancer and women with a history of breast cancer in a first-degree relative have about 2X higher risk than women without a family history. Rare high-risk mutations particularly in the *BRCA1* and *BRCA2* genes explain less than 20% of the familial relative risk and account for a small proportion of breast cancer cases in the general population. Low frequency variants conferring intermediate risk, such as those in *CHEK2*, *ATM*, and *PALB2*, explain 2% to 5% of the familial relative risk. **Genome-wide association studies** have led to the discovery of multiple common, low-risk variants (single nucleotide polymorphisms, SNPs) associated with breast cancer risk, many of which are differentially associated by oestrogen receptor status Recently, new risk-associated variants have been identified in large-scale replication studies (Thompson & Easton, 2004; Mavaddat et al., 2010; Garcia-Closas et al., 2013).

Additional research provides evidence associating specific environmental toxicants with an increased risk for developing breast cancer. These toxicants can include: 1) hormones in pharmaceutical agents and personal care products; (2) endocrine disrupting compounds; (3) hormones in food (natural and additives); (4) industrial chemicals; (5) tobacco smoking (active and passive); (6) shift work (light-at-night and melatonin); and (7) radiation (Gray et al., 2017). Breast cancer comprises of several subtype(s) and disease is includes the age of patient (pre- or post-menopausal [assumes a shift in reproductive phases at age 50]); in situ, localized, regional or metastatic presentation; cancer morphological characteristics; histological grade and cellular proliferation rate; and gene expression profile (Alizart et al., 2012).

Current Therapies

The best opportunity to reduce mortality is through early detection. The current standard of care for the treatment of early-stage breast cancer involves giving a patient an informed choice regarding surgical options (Katz & Hawley, 2007). The effectiveness of breast-conserving therapy (BCT), beginning in the 1970s with **quadrantectomy** versus **mastectomy**, has been fully verified in numerous studies with more than two decades of follow-up data (Wood, 2003). These findings have resulted in BCT as the primary surgical option for most patients during the past 2 decades. An integral part of BCT is the mandatory addition of adjuvant radiation therapy; but as with chemotherapy, the risk of death due to radiation therapy starts to rise ten years after treatment and may not be fully manifest until the second decade after therapy (Ewertz & Jensen, 2011).

Several chemotherapeutic agents have been used—although they have the potential for a spectrum of cardiovascular diseases including congestive heart failure, **myocardial ischemia,** hypertension, **arrhythmias**, QT prolongation, **bradycardia, pericarditis**, acute coronary syndrome, and **thromboembolic events** (Bowles et al., 2012). These agents include: 1) a*nthracyclines (doxorubicin and epirubicin);* 2) *alkylating agents (cyclophosphamide); and,* 3) *cytoskeletal disruptors (paclitaxel and docetaxel).* Anthracyclines bind to the DNA of malignant cells, interfering with the replication process and resulting in cellular death. Alkylating agents act by reacting with the proteins of the DNA of cancer cells by adding an alkyl group to them. This disrupts effective DNA replication, resulting in the apoptosis of cancerous cells. Cytoskeletal disruptors inhibit the process of cell division through the interruption of microtubular functions, which are essential for cell division (Youssef & Links, 2005; Bird & Swain, 2008; Yeh & Bickford, 2009; Bodai, 2015).

Hormonal blockade drugs include *tamoxifen* and ***aromatase inhibitors*** *(letrozole, anastrozole, and exemestane)*. *Tamoxifen* is a selective oestrogen receptor modulator and inhibits the growth of breast cancer cells by its antiestrogenic activity through its competitive inhibition of oestrogen binding to oestrogen receptors. Tamoxifen has been heralded as one of the most important advances in the treatment of breast cancer because approximately 70% of these patients have oestrogen receptor-positive cancer. The side effects of tamoxifen therapy are typically those that accompany the onset of menopause. These include hot flashes, mood swings, depression, loss of libido, and vaginal dryness. Aromatase inhibitors work by blocking the enzyme aromatase, which converts adrenal androgens into oestrogens. Aromatase inhibitors have menopausal symptoms including hot flashes, mood swings, vaginal dryness, and loss of libido (Howell et al., 1996. Osborne, 1998; Kennecke et al., 2006; Amir et al., 2011; Bodai, 2015).

Approximately 15% to 30% of all breast cancers are human epidermal growth factor receptor 2-positive (HER2+) and, as such, the HER2 receptor tyrosine kinase pathway has become an important therapeutic target. The main function of the *HER2/neu* oncogene (also called *ERBB2*) is to promote the differentiation, growth, and survival of cells, thereby enhancing the aggressiveness of these breast cancers. Targeted biologic agents include *trastuzumab* and *lapatinib* which are directed at protein kinases and the receptors that activate them (Burstein, 2005; Romond et al., 2005; Perez et al., 2008; Barroso-Sousa et al., 2013; Bodai 2015).

There is a growing body of evidence supporting the use of integrative therapies, especially mind-body therapies, as effective supportive care strategies during breast cancer treatment. There is demonstrated use of integrative therapies for specific clinical indications during and after breast cancer treatment. For example, for anxiety/stress music therapy, meditation, stress management, and yoga are recommended. Meditation, relaxation, yoga, massage, and music therapy are recommended for depression/mood disorders. Meditation and yoga are recommended to improve quality of life. Acupressure and acupuncture are recommended for reducing chemotherapy-induced nausea and vomiting (DeSantis et al., 2016; NCI, 2018).

Diet, physical activity, and weight play a major role in survival among patients with breast cancer. Overall, the long-term survival in breast cancer encompasses a total-health strategy that includes a focus on healthy eating, active living, healthy weight, and emotional resilience (Tuso, 2014; Bodai, 2015).

Ovarian Cancer

With an estimated 21,880 new cases and 13,850 deaths from ovarian cancer in the United States in 2010, ovarian cancer is the 5th leading cause of cancer death among women in the United States (Jemal et al, 2010). NIH-NCI (2018) estimates that in 2018, there will be an estimated 22,240 new cases and 14,070 deaths from ovarian cancer in the United States. Ovarian cancer rates are highest in women aged 55–64 years with the median age of death from ovarian cancer is 70 years (Testa et al., 2018).

Ovarian cancer is a very heterogeneous and occurs as epithelial tumours developing from ovarian epithelial surface cells, which give rise to four main distinct histological subtypes: serous (the most frequent); endometrioid; mucinous, and clear cells. Depending on their invasiveness and aggressiveness ovarian cancers are subdivided into two types: low-grade type-1 and high-grade type-2. Type 1 cancers account for only 10% of the death from ovarian cancer, are slowly growing and include low-grade serous, endometroid and mucinous tumours; while type 2 cancers are rapidly progressing and include high-grade serous carcinomas (Rojas et al., 2016; Testa et al., 2018).

The lifetime ovarian cancer risk for a woman with *BRCA1* mutation is estimated to be 35% to 70%, the ovarian cancer risk lifetime for women with *BRCA2* mutation is between 10 and 30%; the ovarian cancer lifetime risk for the women in the general population is less than 2%. Both types different in three important ways: 1)Type-2 tumours are very frequently associated with *TP53* mutations but not for type-1 tumours; 2) type-1 tumours develop from benign extraovarian lesions that implant on the ovary then subsequently switch to a malignant genotype/phenotype; while type-2 tumours develop from intraepithelial carcinomas originated from Fallopian tube secretory cells; and 3) type-1 tumours have mutations involving these genes: *PIK3CA, PTEN, ARID1A, KRAS* and *BRAF; while* RB1, FOXM1, and NOTCH 3 gene pathways are frequent in type-2 (Coukos et al., 2016; Hamanishi et al., 2016; Kurman & Shih, 2016; Testa et al., 2018).

Surgery remains the cornerstone of treatment for high-grade serous ovarian cancers, preceded or followed by chemotherapy. Several putative ovarian cancer stem cell markers, such as CD24, CD44, CD133, SOX2, SSEA; and the existence of these highly-tumourigenic and chemotherapy-resistant cancer stem cell-specific **biomarkers** might pave the way to the targeting these cells to minimize the drug resistance and the tumour relapse. Immune-based approaches in development for the treatment of ovarian cancer include vaccines, **cytokines**, and adoptive **T-cell therapy**, a treatment involving **lymphocyte transfusion** (Gaillard et al., 2016). Different circulating miRNA signatures have been observed in women with ovarian cancer compared with healthy controls, suggesting that circulating miRNAs may be useful as markers for early detection, as well as for disease prognosis. Therapeutic strategies involving either replacement or suppression of specific miRNAs have been proposed, but no clinical trials investigating miRNAs in ovarian cancer have been initiated to date (Mahdian-Shakib et al., 2016).

Next-generation gene sequencing (NGS) or high-throughput gene sequencing continues to provide insights into the molecular pathways involved in ovarian cancer and several studies are investigating specific pathways like the mutations in the *BRCA1* or *BRCA2* tumour suppressor genes, vascular endothelial growth factor, PI3K/Akt/mTOR signalling network, and tumour suppressor p53 mutations. Hopefully, therapies targeting those pathways will become available as alternative to current ovarian cancer treatment (Siddiqui et al., 2011; Coukos et al., 2016; Hamanishi et al., 2016; Kurman & Shih, 2016; Mittica et al., 2016; Rojas et al., 2016; Testa et al., 2018).

14.4 COLORECTAL

Colorectal cancer is a major cause of cancer morbidity and mortality worldwide (especially for older patients), as its incidence increases markedly after the age of 60 years (Berger et al., 2016). Colorectal cancer is the 3rd leading cause of cancer-related deaths in men and women in the United States. In 2018, ACS estimates that the number of colorectal cancer cases in the US are 97,220 new cases of colon cancer, and 43,030 new cases of rectal cancer; and about 50,630 deaths are expected from this type of cancer in the US (ACS, 2018b).

Symptoms

Due to almost minimal or non-existent symptoms of colorectal cancer symptoms during the early stages of the disease; the signs and symptoms of colorectal cancer may not arise until the disease has progressed into stage II or beyond. There are imaging and laboratory tests available today for colorectal cancer diagnosis including stool tests, other lab tests, a **magnetic resonance imaging (MRI), commuted tomography (CT), or positron emission tomography (PET)** scan, a **colonoscopy**, and other endoscopic procedures. According to the CTCA (2018), colorectal cancer signs and symptoms can be broken down into two general categories: local and systemic. Local symptoms are those that have a direct effect on the colon or rectum and include changes in bowel habits; constipation; diarrhea; alternating diarrhea and constipation; rectal bleeding or blood in stool; abdominal bloating; cramps or discomfort; a feeling that bowels did not empty completely; and stools that are thinner than normal. Systemic colorectal cancer symptoms affect the entire body include unexplained weight loss, unexplained loss of appetite, nausea or vomiting; anemia; jaundice; and weakness or fatigue. (CTCA, 2018)

Genetic Basis

Environmental (chemicals, infectious agents, radiation) and genetic (mutations, immune system and hormone dysfunction) factors can interact in a variety of ways to potentiate **carcinogenesis** (Nakaji et al., 2003). The risk factors for developing colon cancer include age, personal history, family history, racial and ethnic background, lifestyle and diet related factors, and only a minority of cases are associated with underlying genetic disorders. (Triantafillidis et al., 2009; Mishra et al., 2013). Molecular genetic studies have revealed some critical mutations underlying the pathogenesis of the sporadic and inherited forms of colorectal cancer. A number of oncogenes and tumour-suppressor genes are involved in the development of colorectal cancer including the APC, MYC, RAS, p53, MCC and DCC genes (Fearon, 2011). The APC gene alteration (if not inherited), occurs as the earliest molecular alteration in the development of colorectal cancer whereas structural alterations of the rest of the genes are late-occurring events. The *APC* gene is a tumour suppressor gene which normally helps keep cell growth in control. In people with inherited changes in the *APC* gene, this "brake" on cell growth is turned off, causing hundreds of polyps to form in the colon. Over time, cancer will nearly always develop in one or more of these polyps. A mutation in one of the DNA repair enzyme genes like *MYH, MLH1, MSH2, MLH3, MSH6, PMS1*, and *PMS2,* can allow DNA errors to go unfixed. These errors will sometimes affect growth-regulating genes, which may lead to the development of cancer. (Cerda et al., 1993; Zur 2009; Samaras et al., 2010; Coghill & Hildesheim, 2014; ACS, 2018a)

Current Therapies

Despite great improvements in its diagnosis and treatment, invasion and metastasis of colon cancer cells to proximal and distant organs remains the major treatment challenge. The choice of method of colon cancer treatment is crucial since each tumour responds to different methods differentially. Selection of treatment is based on many factors including tumour type, stage of the disease, patient's age, patient's level of health, and patient's attitude towards life. A combination of therapies are available including surgery and polypectomy, radiation therapy, ablation/embolization, cryotherapy, chemotherapy, targeted drug therapy, and immunotherapy. Targeted drug therapy usually uses regimens of 5-fluorouracil (5-FU), *capecitabine* (*Xeloda*), *irinotecan* (*Camptosar*), *oxaliplatin* (*Eloxatin*), *trifluridine* and *tipiracil* (*Lonsurf*). (ACS, 2018a; Mishra et al., 2013). Chemotherapy includes the use of alkylating agents, antimetabolites, plant alkaloids, anti-tumour antibiotics, enzymes, hormones and modifiers of biological response that destroy malignant cells, suppress tumour growth or its cells division. Cancer vaccines are also being tested to boost the immune system to fight against the cancer just like common vaccines do in infection. Gene therapy also provides an opportunity to treat cancer by "active" immunotherapy enabling the transfer of genes encoding antibodies directed against specific oncogenic proteins. Nutritional supplemental therapy includes the use of fermented wheat germ extract, curcumin, quercetin, and garlic (Volate et al., 2005; Altonsy & Andrews, 2011; Mueller & Voigt, 2011; Nautiyal et al., 2011; Mishra et al., 2013).

Research shows that many different miRNAs are aberrantly expressed in various cancer cells, and associated with tumour initiation, promotion, and progression via their interaction with oncogenes and tumour suppressor genes. MiR-126 has been demonstrated to act as a tumour suppressor in colon cancer *in vitro*, and it suppresses colon cancer cell proliferation, invasion and migration. Emerging evidence also shows promise in colorectal treatment using interfering nucleic acids (interfering RNA [RNAi], small-interfering RNA [siRNA], and short-hairpin RNA [shRNA]); **nanoparticles** (like liposomes); and Janus kinase-based therapies (e.g. Jak-1 tyrosine kinase) (Calin & Croce, 2006; Mishra et al., 2013; Yuan et al., 2016).

14.5 LUNG CARCINOMA (SMALL CELL)

Worldwide, lung cancer is currently the leading cause of cancer related deaths in men, and the second leading cause of cancer death in women. According to WCRI (2018) there were an estimated 14.1 million cancer cases around the world in 2012; and of these, 7.4 million cases were in men and 6.7 million in women. This number is expected to increase to 24 million by 2035. Lung cancer has a high incidence coupled with a poor 5-year survival rate of less than 17%. Tobacco smoking still is the main cause for lung cancer development. Lung cancer can be classified histologically as small cell lung carcinoma (SCLC, 20% of cases) and non-small cell lung carcinoma (NSCLC, 80% of cases). NSCLC can further be divided into adenocarcinoma, squamous cell carcinoma, and large cell carcinoma subtypes. (Richter et al., 2017). SCLC is a high-grade tumour of neuroendocrine origin arising from the lower respiratory tract, primarily affecting older adults, nearly always cigarette smokers. It usually occurs in older men (60–70 years old) and is well known to be closely associated with smoking and has the worst prognosis among lung cancers. SCLC is more aggressive than NSCLC and is characterized by a rapid doubling time, high growth fraction (the ratio of proliferating cells to total cells), and greater propensity for early development of widespread metastases (Lee et al., 2016).

Symptoms

Because 90% to 95% of SCLCs arise from lobar or main bronchi, the most common manifestation of SCLC is a large mass centrally located within the lung parenchyma. In early SCLC, this produces symptoms like coughing blood, breathing difficulties, loss of appetite, recurrent bronchitis/emphysema; and in late cancer bone pain, face/arm/neck swellings, jaundice, and lumps in the neck (CTCA, 2018).

Genetic Basis

Based on the observation that the *RB* and *p53* tumour suppressor genes are both mutated in more than 90% of human SCLC, a mouse model for human SCLC was generated by Schaffer et al. (2010). Their studies showed that mutant mice with a deleted *p130, Rb* and *p53* genes in adult lung epithelial cells increased proliferation and significant acceleration of SCLC. Genome-wide expression profiling experiments showed that *Rb/p53/p130* mutant mouse tumours were similar to human SCLC. Richter et al. (2017) found that the *ZAR1* gene in humans (a tumour suppressor gene) is expressed in normal lung but is deactivated in lung cancer cell lines through promoter hypermethylation.

Current Therapies

Despite advances in chemotherapy and radiation therapy regimens, 5-year survival rates for SCLC remain only 5-10%. At the time of diagnosis, SCLC has often already metastasized, and patients almost always relapse following treatment. Lung cancer treatment as well as prognosis differ depending on lung cancer subtypes. Several therapies have muted success including immunotherapy, radiation therapy and use of kinase inhibitors. Since the approval of topotecan in 1996, the USFDA has not approved any new drugs for the treatment of SCLC patients (Bunn et al., 2016). Antibody drug conjugates (ADCs) have been some of the most active agents developed in oncology, combining the specificity of antibodies with the cytotoxicity of traditional chemotherapeutic agents. For instance, the ADC *ovalpituzumab tesirine* binds a ligand on the cell surface and delivers a DNA damaging agent that inhibits the growth of SCLC. Small-cell lung carcinoma (SCLC) has a dismal prognosis in part because of multidrug resistance; and *silibinin* a flavonolignan extracted from milk thistle (*Silybum marianum*) may be useful in the treatment of drug-resistant SCLC (Sadava & Kane, 2013).

14.6 MALIGNANT MELANOMA

Cutaneous malignant melanoma incidence has increased during the past several decades and is the 6th most common cancer accounting for more than 47,000 deaths worldwide annually. The global projected incidence of melanoma for the year 2025 is estimated to be 317,000 new cases compared to the 200,000 cases reported in 2008 (Soura et al., 2016).

Symptoms

According to CTCA (2018), the ABCDE method helps in determining whether an abnormal skin growth is melanoma: Asymmetry (mole is irregular in shape); Border (edge is irregular or notched); Color (mole has uneven shading or dark spots); Diameter (spot is larger than the size of a pencil eraser); Evolving or Elevation (spot is changing in size, shape or texture). Other melanoma symptoms may include: sores that do not heal; pigment, redness or swelling that spreads outside the border of a spot to the surrounding skin; itchiness, tenderness or pain; changes in texture, or scales, oozing or bleeding from an existing mole; and blurry vision or partial loss of sight, or dark spots in the iris (CTCA, 2018).

Genetic Basis

About 7–15% of melanoma cases occur in patients with a family history of melanoma; with most cases due to shared sun exposure experiences among family members with susceptible skin types. About 45% of familial melanomas are associated with germline mutations in the genes *CDKN2A*. The *CDKN2A* gene is located on chromosome 9p21.3 and plays an important role in normal cell cycle progression. *BRAF* gene and *KIT* gene mutations have also been reported in melanoma (Gue at al., 2011; Long et al., 2011). There is a strong relationship between environmental and socioeconomic factors and increased risk of melanoma including UV irradiation, chemicals, air pollution, smoking, chronic inflammation, Western diets, obesity, sedentary lifestyle and increasing age, all associated with increased miRNA-21 signalling (Melnik, 2015). Epigenetic factors may play a role in melanoma genesis with significant changes of microRNA (miRNA) expression in response to environmental exposure of humans (Vrijens et al., 2015). MiRNAs are important posttranscriptional regulators controlling more than 30% of human mRNAs, and certain miRNAs (e.g. miRNA-21) function as potent oncogenes and play an important role in the initiation and progression of cancer. Recent evidence suggests that microRNAs are involved in signalling in cancer initiation and progression. Melnik observed increased miRNA-21 expression during the transition from a benign melanocytic lesion to malignant melanoma, exhibiting highest expression of miR-21. MiRNA-21 inhibits mRNA expression of crucial tumour suppressor proteins (Di Leva et al., 2014; Melnik, 2015).

Current Therapies

Extended resection surgery has been the main treatment option for early melanoma (Guo et al., 2015). Adjuvant therapy using bacille Calmette-Guerin vaccine, interferon, and *talimogene laherparepvec* are recommended for patients with in-transit metastasis. Adjuvant therapy using high-dose interferon is recommended post-operation and for patients at high risk of recurrence. Adjuvant radiotherapy of regional lymph nodes can increase the local control rates but has no impact on the long-term survival. Breakthroughs have been made in the management of stage IV or unresectable melanoma has in recent years. BRAF inhibitor plus MEK inhibitor, anti-CTLA-4 monoclonal antibody (*ipillimumab*), and anti-PD-1 monoclonal antibodies (*nivolumab* and *pembrolizumab*) have been considered standard treatment.

Since 1972, dacarbazine has been the only chemotherapy drug approved by USFDA for treating advanced melanoma. Other chemotherapy drugs include: 1) *Temozolomide* is also used as an oral chemotherapy drug. Platinum-based anti-tumour drugs have also shown certain efficacies in treating melanoma; 2) Taxane compounds include *paclitaxel* (extracted from taxus plants), and *docetaxel* (extracted from the needles of European yew tree); 3) A nanoparticle encapsulated *paclitaxel* (nab-paclitaxel) is a novel anti-tumour compound which uses soluble human albumin to coat paclitaxel and deliver the drug into tumour cells. Personalized targeted therapies including BRAF inhibitors and MEK inhibitors have been used in metastatic melanoma caused by intracellular *BRAF* gene mutation. Two BRAF inhibitors have also been approved by the US FDA (*vemurafenib, dabrafenib* and *trametinib*). *Trametinib* is also used orally as an MEK inhibitor while imatinib has been used as a KIT inhibitor.

Other therapies used include immunotherapy/immune targeted therapy (*ipilimumab, pembrolizumab* and *nivolumab*, and interleukin-2); and anti-angiogenic targeted therapies (recombinant human endostatin and recombinant humanized monoclonal IgG1 antibody injections).

14.7 PANCREATIC

Pancreatic cancer accounts for about 3% of all cancers in the United States and about 7% of all cancer deaths (ACS, 2018). The American Cancer Society's estimates that about 55,440 people (29,200 men/26,240 women) will be diagnosed with pancreatic cancer for pancreatic cancer in the United States for 2018. About 44,330 people (23,020 men/21,310 women) will die of pancreatic cancer (2018). The risk factors for pancreatic cancer include image-detectable pancreatic diseases (pancreatic cysts, pancreatic duct dilation, intraductal papillary mucinous neoplasm, and **chronic pancreatitis**). Lifestyle factors includes smoking, diabetes mellitus, obesity, advancing age, male sex, non-O blood group, occupational exposures, African-American ethnic origin, a high-fat diet, diets high in meat and low in vegetables and folate, and possibly *Helicobacter pylori* infection and periodontal disease. A family history of pancreatic cancer is another known risk, and one that cannot be modified by individual effort or by medicine (Klein at al., 2004; Raimondi et al., 2009; Wolpin et al., 2009; Vincent et al., 2011).

Symptoms

Most pancreatic cancers present non-specifically and are not diagnosed until late in the course of the disease, after the cancer has already spread to other organs. Common symptoms include pain, particularly epigastric pain that radiates to the back, unexplained weight loss, jaundice, clay-coloured stools, nausea, and in ~10% migratory inflammation of a vein (especially in legs).

Genetic Basis

Several genes are implicated in the development of pancreatic cancer. 1) The *TP53* tumour suppressor gene on chromosome 17p is inactivated in 75% of pancreatic cancers. *TP53* codes for the p53 protein, and p53 plays an important role in cellular stress responses, particularly by activating DNA repair, inducing growth arrest and triggering cell death (apoptosis). Loss of p53 function, through mutation of the *TP53* gene, therefore promotes pancreatic neoplasia through the loss of a number of critical cell functions. 2) Inherited mutations in the *BRCA2* gene are associated with a significantly elevated lifetime risk of pancreatic cancer (and also breast, ovarian, and prostate cancers). The prevalence of germline *BRCA2* gene mutations (and possibly *BRCA1* gene mutations) in pancreatic cancer patients varies among different populations and is particularly high in individuals of Ashkenazi Jewish decent. In Ashkenazi Jews with pancreatic cancer, about 4-10% carry a germline *BRCA2* mutation. 3) Germline mutations in the *PALB2* (partner and localizer of *BRCA2*) gene have been reported in 1–3% of familial pancreatic cancer.

Germline mutations in the *p16/CDKN2A* gene are associated with an increased risk of pancreatic cancer. 4) The *p16/CDKN2A* gene, a tumour suppressor gene on chromosome 9p, is inactivated in ~95% of pancreatic cancers. 5) Patients with Lynch Syndrome also have an increased risk of endometrial, gastric, small intestinal, ureteral and pancreatic cancer. Lynch Syndrome is characterized by early onset colon cancer due to germline mutations in one of the DNA mismatch repair genes (*hMSH2*, *hMLH1*, *hPMS1*, *hPMS2* or *hMSH6/GTBP*). Hereditary pancreatitis is a rare inherited form of pancreatitis; and patients with it have a remarkable 58-fold increased risk of developing pancreatic cancer and a lifetime risk (by age 70) of pancreatic cancer of 30–40%. 6) The protein product of the *SMAD4* gene, Smad4 protein functions in the transforming growth factor beta (TGFβ) cell signalling pathway. *SMAD4* gene mutations in pancreatic cancer are associated with poor prognosis and with more widely metastatic disease. 7) Several other genes are somatically mutated in pancreatic cancer at lower frequencies including: *MLL3*, *TGFBR2*, *FBXW7*, *ARID1A*, *AIRID2*, and *ATM*. 8) Several microRNAs are aberrantly expressed in pancreatic cancer, and, because microRNAs tend to be long-lived, these abnormally expressed microRNAs could serve as markers for pancreatic cancer (Hruban et al., 1993; Hahn et al., 1996; Jones et al., 2008; Blackford et al., 2009; Habbe et al., 2009; Roberts et al., 2012).

Current Therapies

Treatment options for people with pancreatic cancer highly depend on the type and stage of the cancer and other factors, can include: surgery, ablation or embolization treatments, radiation therapy, chemotherapy and other drugs. With the greatly improved enhancements in imaging technologies, the radiologic diagnosis of pancreatic cancer is better today. Multi-detector computed tomography, magnetic resonance imaging, and endoscopic ultrasound can all be used to visualize the tumours. Pancreatic cancer is a complex disease, and patients with pancreatic cancer are best treated by a multi-disciplinary team. The optimal treatment greatly depends on careful accurate staging. Patients with Stage I/II disease should undergo surgical resection followed by adjuvant therapy. Neoadjuvant therapy should be considered in this patient population but is controversial, while patients with Stage III borderline resectable cancers should undergo neoadjuvant therapy prior to resection. For patients with stage III locally advanced disease chemotherapy and/or chemoradiotherapy are recommended. A vast majority of these patients eventually develop metastatic disease but select patients can still be considered for surgical resection. In patients with Stage IV and good performance status systemic therapy may be an option, but those with poor overall health should be given supportive therapy (Evans et al., 2008; Pawlik et al., 2008; Tempero et al., 2012. Wolfgang et al., 2013; ACS, 2018).

14.8 PROSTATE

The most prevalent cancer among men globally, prostate cancer, there were 1.4 million new cases reported in 2013; and in 2016, there were approximately 180,890 new cases and 26,120 prostate cancer-related deaths. Prostate cancer cases increased by 217% between 1990 and 2013 as a result of population growth and aging and increased uptake of opportunistic screening, particularly in developing countries. (Raymond et al., 2017) Risk factors to prostate cancer occur in 3 groups: 1) non-modifiable (age, race, geographical location, and known genetic mutations/polymorphisms, and where no specific gene(s) have yet been identified); 2) external factors like exposures to ionizing and UV radiation, plus lifestyle factors when modification might be possible; and 3) blood-based markers, which might be a result of a mixture of the first two.

Symptoms

Although early prostate cancer usually causes no symptoms; more advanced cancers can cause symptoms such as problems urinating (slow or weak urinary stream or the need to urinate more often, especially at night); blood in the urine or semen; trouble getting an erection (erectile dysfunction); pain in the hips, spine, chest, or other areas from cancer that has spread to bones; and weakness or numbness in the legs or feet, or even loss of bladder or bowel control from cancer pressing on the spinal cord (ACS, 2018).

Genetic Basis

The relative risk of developing prostate cancer is higher in men with a family history like that of having a first-degree relative with prostate cancer. BRCA2 mutation confers up to an 8.6-fold increased risk in men below 65 years of age, and such mutations have also been related with aggressive cancer. There are other rare mutations reported in *BRCA1, HOXB13, NBS1* and *CHEK2*. The HOXB13 G84E mutated is the only other identified factor with an appreciative relative risk (3–4-fold) and the abnormal allele frequency is about 1.3 – 1.4%. Other potential familial risk factors for which a genetic basis reported in studies include some types of male pattern baldness, and digit length but their value in risk stratification remains uncertain. Important lifestyle factors include smoking, diet, weight and physical activity. Urinary tract infections might influence the risk for prostate cancer by causing chronic intra-prostatic inflammation, and pathological studies have also suggested that inflammation may be involved in the development of prostate cancer. The possible role of endogenous hormones in the etiology of prostate cancer has been reported; with a positive association between circulating insulin-like growth factors and prostate cancer. Recent expression profiling studies in prostate cancer suggest microRNAs (miRNAs) may serve as potential biomarkers for prostate cancer risk and disease progression (Appleby & Key, 2008; Roddam et al., 2008; Sutcliffe & Platz, 2008; Huncharek et al., 2010; Cuzick et al., 2014; Discacciati & Wolk. 2014; Goh & Eeles, 2014; Karlsson et al., 2014; Luu et al., 2017).

Current Therapies

Treatment options include watchful waiting or active surveillance; surgery; radiation therapy; cryotherapy (cryosurgery); hormone therapy; chemotherapy; vaccine treatment; radical prostatectomy; radiation therapy (with or without androgen deprivation therapy); androgen deprivation therapy, and bone-directed treat-

Cancers

ment. The prostate-specific antigen (PSA) blood test is a widely used test for prostate cancer. Lifestyle modifications like smoking cessation and exercise can decrease the risk of developing prostate cancer. Although associated with an increased number of high-grade prostate cancers, 5α-reductase inhibitors reduce the overall prostate cancer burden. In absence of any detrimental effect on survival, these agents can be cost-effective in prostate cancer prevention. Pharmacological agents like aspirin appear promising but need further evaluation in clinical trials (Cuzick et al., 2014; Discacciati & Wolk; 2014; ACS, 2018).

14.9 NEUROFIBROMATOSIS

Neurofibromatosis, first described in 1882 by Von Recklinghausen, is a genetic defect characterized by a neuroectodermal abnormality and by clinical manifestations of systemic and progressive involvement mainly affecting the skin, nervous system, bones, eyes and possibly other organs. Symptoms manifest differently in each patient, even those within the same family, with a highly variable expression. Carey *et al.* (1986) proposed that NF be classified into only five types, based on distinct clinical features and genetic implications for the patient, as follows: NF1-classical, NF2-acoustic, NF3-segmental, NF4-CALM-familial and NF5-NF-Noonan phenotype. The most common type is NF1 occurring in more than 90% of all cases. Most deaths are related to the malignant transformation of tumours. In patients with NF1, mortality is higher in patients aged from 10 to 40 years and tends to be higher in women than in men (Duong et al., 2011; Antônio et al., 2013).

Symptoms

The clinical manifestations found in NF1 are as follows: neurofibromas, café au lait spots or macules and axillary freckling (in skin); Lisch nodules and optic gliomas (eyes); vascular defects, brain tumours, **macrocephaly** and subsequent learning disabilities, mental retardation, **epilepsy** and headache (in the CNS); **scoliosis**, pectus excavatum, para-spinal tumours, and pseudo-arthrosis (in bones); plus other manifestations such as speech problems, precocious or delayed puberty, hypertension, intestinal neurofibromas and bowel function disorders caused by plexiform neurofibromas (Rauen et al., 2015).

Genetic Basis

Neurofibromatosis type 1, being more frequent in the population, has an estimated incidence of one case per 3,000 inhabitants, and affects all races and both sexes, with half of the cases presenting a family history; while in the other half the disease emerged as the result of a new mutation. The NF1 gene is responsible for neurofibromatosis type 1, located on the long arm of chromosome 17 at 17q11.2 and encodes a protein predominantly expressed in neurons, Schwann cells, oligodendrocytes and astrocytes (DeClue et al., 1991; Daston et al., 1992).

Current Therapies

Patients with neurofibromatosis have a higher mortality rate when compared to the general population. As a result of continued research on the functions of the NF1 and NF2 gene products, today, there have been better diagnoses and monitoring of affected individuals. Despite these advances, no medical treatment is available to prevent or reverse the typical lesions of neurofibromatosis. Medical care is centred on genetic counselling and the early detection of complications that can be treated. Excision of skin tumours and clinical or surgical treatment of injuries to other organ systems affected by the disease are often performed (DeClue et al., 1991; Daston et al., 1992; Duong et al., 2011; Antônio et al., 2013; Rauen et al., 2015).

14.10 MULTIPLE ENDOCRINE NEOPLASIA, MEN

MEN disorders are autosomal dominant and are characterized by the occurrence of tumours involving two or more endocrine glands within the same individual. There are four major forms recognized: MEN1, MEN2 (previously MEN2A), MEN3 (previously MEN2B), and MEN4 (also referred to as MENX).

Symptoms

Each MEN type is associated with the occurrence of specific tumours. MEN1 is characterized by the occurrence of parathyroid, pancreatic islet and anterior pituitary tumours; MEN2 by the occurrence of medullary thyroid carcinoma in association with phaeochromocytoma and parathyroid tumours; MEN3 by the occurrence of medullary thyroid carcinoma and phaeochromocytoma; and MEN4 by the occurrence of parathyroid and anterior pituitary tumours in association with tumours of the adrenals, kidneys, and reproductive organs (Thakker, 1998, 2010; Agarwal, 2014).

Genetic Basis

The gene causing MEN1 is located on the long arm of chromosome 11 (11q13) and codes for a nuclear protein called menin. It is inherited as an autosomal dominant disorder. The *MEN1* gene acts as a tumour-suppressor gene and its abnormalities result in mutations, deletions, and/or truncations of the menin protein. Menin interacts with a large number of proteins many of which have important roles in transcriptional regulation, genomic stability, cell division and cell cycle control. Menin acts as scaffold protein and can increase or decrease gene expression by epigenetic regulation of gene expression via histone methylation. MEN2 is caused by mutations of a tyrosine kinase receptor. MEN3 (previously MEN2B) is due to RET mutations; and MEN4 is due to cyclin-dependent kinase inhibitor (Brandi et al., 1987; Agarwal, 2014; Thakker, 2010, 2014; Norton et al., 2015).

Current Therapies

The FDA has recently approved two agents for the treatment of metastatic pancreatic neuroendocrine tumours, the mTOR inhibitor *everolimus*, and the tyrosine kinase inhibitor *sunitinib*. Two new compounds have recently been FDA-approved for the treatment of metastatic medullary thyroid carcinoma and include *vandetanib* and *cabozantinibz* (Norton et al., 2015).

14.11 RETINOBLASTOMA, RB

Retinoblastoma is a rare cancer of the infant retina, which forms when both retinoblastoma gene (*RB1*) alleles mutate in a susceptible retinal cell (likely a cone photoreceptor precursor) (Dimaras et al., 2015). Retinoblastoma is an aggressive eye cancer that affects 8,000 new babies and children worldwide each year, and of all affected children, 11% reside in high-income, 69% in middle-income, and 20% in low-income countries. There are about 200-300 new cases of retinoblastoma in the United States each year. Retinoblastoma is the most common intraocular malignancy in childhood, comprises 4% of all paediatric cancers and is curable if diagnosed early (Grossniklaus, 2014: Ottaviani & Szijan, 2015; ORW, 2018).

Symptoms

Diagnosis of retinoblastoma is usually clear from presenting signs and clinical examination with the most common sign being white pupil (leukocoria). If parents report a strange reflection in the child's eye, retinoblastoma should be at the top of the differential diagnosis. The second most common sign is misaligned eyes (strabismus) when central vision is lost. Advanced disease may present with iris colour change, enlarged cornea and eye due to increased pressure, or non-infective orbital inflammation. Very late, the eye may bulge from the orbit, a common presentation where awareness and resources are inadequate.

Genetic Basis

Many modifications impair RB1 function, including point mutations, promoter methylation, and small and large deletions. The RB1 protein (pRB) is a cell cycle regulator that binds to E2F transcription factors to repress cell proliferation-related genes. Many patients with retinoblastoma are the first individuals to be diagnosed in a family with only about 6% of patients having a family member previously diagnosed with it. Approximately 45% of people with retinoblastoma carry one *RB1* mutation in their constitutional cells; and of the remaining patients with unilateral retinoblastoma, 98% have somatic biallelic *RB1* loss and 2% have normal *RB1* genes in the tumour, but somatic amplification of the *MYCN* oncogene also. (Rushlow et al., 2013; Dimaras et al., 2015). Recent studies suggest RB can be caused by amplification of the *MYCN* gene. In addition, since *TP53* is rarely mutated in retinoblastoma, other mechanisms of p53 inactivation in these tumours have been discovered, including the genomic gain and overexpression of key inhibitors of p53 activity, *MDM2* and *MDM4* (Thériault et al., 2014). MicroRNAs have been implicated in many crucial cellular pathways in normal and cancer cells, and their role in retinoblastomas has also been explored (Reis et al., 2012).

Current Therapies

The goal of treatment of retinoblastoma is to ensure elimination of the tumour while minimizing collateral injury to other tissues. Globally in frequencies of use, primary treatments for RB include enucleation, intravenous chemotherapy with **focal therapy (laser therapy** and cryotherapy), intra-arterial chemotherapy with focal therapy, and focal therapy alone when tumours are small at diagnosis. External beam radiotherapy (EBRT) imposes a high risk of secondary cancers when the patient carries an *RB1* mutation, but useful in cases that have failed other therapies (Shields, et al. 2013). Secondary treatment options (if primary treatment fails) include focal therapies, repeated systemic chemotherapy, **brachytherap**y (internal radiotherapy), EBRT and whole-eye or proton beam radiation. **Palliation** includes pain management, symptom relief, nutritional support, and psychosocial support for the child and families.

According to Mahinda et al. (2013) and Taich et al. (2014), there are a few translational studies that have resulted in changes in retinoblastoma clinical practice. In this research innovative delivery systems target small-volume subretinal tumours and vitreous seeding include episcleral (outside sclera layer) implants and nanoparticles. Strategies targeting chemotherapy schedules adjustments and dose intensity (metronomic therapy) may particularly play an important role in palliative care for retinoblastoma. (Lybeck & Reid, 2014; Torbidoni et al., 2015) Chemotherapeutic agents used in treating retinoblastoma come from four classes of chemotherapeutic agents: 1) *alkylating agents (*such as *melphalan*); 2) *topoisomerase inhibitors (s*uch as *etoposide* and *topotecan*); *3) platinum based antineoplastic agents* (such as *carboplatin*); 4) *vinca alkyloids (*such as *vincristine*). Adjuvant chemotherapy is utilized for eyes with optic nerve invasion or massive choroidal invasion and include VDC (*vincristine, doxorubicin* or *idorubicin, cyclophosphamide,*), VCE (*vincristine, carboplatin, etoposide*) or hybrid regimens combining chemo-reduction with chemotherapeutic agents (Grossniklaus, 2014; Dimaras et al., 2015; Torbidoni et al., 2015).

CHAPTER SUMMARY

Cancer is a source of significant and growing mortality worldwide, with an increase to 19.3 million new cancer cases per year projected for 2025. Standard current treatments for cancer include surgery, radiotherapy, and a host of other systemic treatments comprising cytotoxic chemotherapy, hormonal therapy, immunotherapy, and targeted therapies. Today cancer continues to frustrate clinical treatment efforts, but the search for effective therapies is a never-ending research goal. Cancers harbour significant genetic heterogeneity, even within a single patient, and based on this heterogeneity, cancers routinely evolve resistance to treatment through switching from one growth pathway to another. Thus, cancer treatment must target many growth pathways simultaneously. Of the global cancer burden, it is estimated that over 20% is attributable to infectious agents including bacteria.

Referred as the "guardian of the genome" the alteration or inactivation of p53 by mutation, or by its interactions with oncogene products or DNA tumour viruses can lead to cancer. The p53 tumour-suppressor gene is mutated in about half of almost all types of cancer arising from a wide spectrum of tissues. p53-induced target genes are involved in processes that are important for tumour suppression. Six major mechanisms of p53-dependent transcriptional regulation have been reported in the literature.

Approximately 1.67 million cases of breast cancer are diagnosed annually worldwide. In 2016 in the US, approximately 3,560,570 women were living with a prior diagnosis of breast cancer, and as many as 40,290 women, and 440 men may have died of breast cancer. Both genetic and lifestyle factors are implicated in the aetiology of breast cancer and women with a history of breast cancer in a first-degree relative have about 2X higher risk than women without a family history. Rare high-risk mutations particularly in the *BRCA1* and *BRCA2* genes explain less than 20%. Genome-wide association studies have led to the discovery of multiple common, low-risk variants (single nucleotide polymorphisms) associated with breast cancer risk, many of which are differentially associated by oestrogen receptor status. Additional research provides evidence associating specific environmental toxicants with an increased risk for developing breast cancer. The best opportunity to reduce mortality is through early detection.

Cancers

The current standard of care for the treatment of early-stage breast cancer involves giving a patient an informed choice regarding surgical options, adjuvant radiation therapy, and chemotherapy. Several chemotherapeutic agents have been used including a*nthracyclines (doxorubicin and epirubicin), alkylating agents (cyclophosphamide), and cytoskeletal disruptors (paclitaxel and docetaxel).* Hormonal blockade drugs include tamoxifen and *aromatase inhibitors (letrozole, anastrozole, and exemestane).* There is a growing body of evidence supporting the use of integrative therapies, especially mind-body therapies, as effective supportive care strategies during breast cancer treatment. Diet, physical activity, and weight may also play a major role in survival among patients with breast cancer.

With an estimated 21,880 new cases and 13,850 deaths from ovarian cancer in the US in 2010, ovarian cancer is the 5th leading cause of cancer death among women in the US especially women aged 55–64 years. Ovarian cancer is a very heterogeneous and occurs as epithelial tumours developing from ovarian epithelial surface cells, which give rise to four main distinct histological subtypes. The lifetime ovarian cancer risk for a woman with *BRCA1* mutation is estimated to be 35% to 70%, the ovarian cancer risk lifetime for women with *BRCA2* mutation is between 10 and 30%; the ovarian cancer lifetime risk for the women in the general population is less than 2%. Surgery remains the cornerstone of treatment for high-grade serous ovarian cancers, preceded or followed by chemotherapy. Immune-based approaches are in development for the treatment of ovarian cancer include vaccines, cytokines, and adoptive T-cell therapy, a treatment involving lymphocyte transfusion.

Colorectal cancer is a major cause of cancer morbidity and mortality worldwide, especially for older patients, as its incidence increases markedly after the age of 60 years. It is the 3rd leading cause of cancer-related deaths in men and women in the United States. Environmental (chemicals, infectious agents, radiation) and genetic (mutations, immune system and hormone dysfunction) factors can interact in a variety of ways to potentiate carcinogenesis. The risk factors for developing colon cancer include age, personal history, family history, racial and ethnic background, lifestyle and diet related factors, and only a minority of cases are associated with underlying genetic disorders. The *APC* gene is a tumour suppressor gene which normally helps keep cell growth in control. People with colorectal cancer have inherited changes in the *APC* gene, causing hundreds of polyps to form in the colon. Over time, cancer will nearly always develop in one or more of these polyps. Despite great improvements in its diagnosis and treatment, invasion and metastasis of colon cancer cells to proximal and distant organs remains the major treatment challenge. The choice of method of colon cancer treatment is crucial since each tumour responds to different methods differentially. Selection of treatment is based on many factors including tumour type, stage of the disease, patient's age, patient's level of health, and patient's attitude towards life. A combination of therapies are available and include surgery and polypectomy, radiation therapy, ablation/embolization, cryotherapy, chemotherapy, targeted drug therapy, and immunotherapy.

Worldwide, lung cancer is currently the leading cause of cancer related deaths in men, and the second leading cause of cancer death in women with an estimated 14.1 million cancer cases around the world (men = 7.4 million; women = 6.7 million). This number is expected to increase to 24 million by 2035. Lung cancer has a high incidence coupled with a poor 5-year survival rate of less than 17% with tobacco smoking still the main cause for lung cancer development. Although small cell lung carcinoma comprises 20% of cases and the non-small cell lung carcinoma 80% of cases, the small cell type has the worst prognosis among lung cancers and is more aggressive characterized by a rapid doubling time, high growth fraction, and greater propensity for early development of widespread metastases. The *RB* and *p53* tumour suppressor genes are both mutated in more than 90% of human small cell type. Studies confirm that mutant mice with a deleted *p130, Rb* and *p53* genes in adult lung epithelial cells increased proliferation and significant acceleration of the small cell type. Despite advances in chemotherapy and radiation therapy regimens, 5-year survival rates for SCLC remain only 5-10%. Several therapies have muted success including immunotherapy, radiation therapy, use of kinase inhibitors and antibody drug conjugates.

Cutaneous malignant melanoma incidence has increased during the past several decades and is the 6th most common cancer accounting for more than 47,000 deaths worldwide annually. The global projected incidence of melanoma for the year 2025 is estimated to be 317,000 new cases compared to the 200,000 cases reported in 2008. About 7–15% of melanoma cases occur in patients with a family history of melanoma; with most cases due to shared sun exposure experiences among family members with susceptible skin types. About 45% of familial melanomas are associated with germline mutations in the genes *CDKN2A*. *BRAF* gene and *KIT* gene mutations have also been reported in melanoma. There is a strong relationship between environmental and socioeconomic factors and increased risk of melanoma including UV irradiation, chemicals, air pollution, smoking, chronic inflammation, Western diets, obesity, sedentary lifestyle and increasing age, all associated with increased miR-21 signalling. Epigenetic factors may play a role in melanoma genesis with significant changes of microRNA (miR) expression in response to environmental exposure of humans. Extended resection surgery has been the main treatment option for early melanoma, although adjuvant therapy using *bacille Calmette-Guerin* vaccine, interferon, and *talimogene laherparepvec* are recommended for patients with in-transit metastasis. Since 1972, dacarbazine has been the only chemotherapy drug approved by US FDA for treating advanced melanoma. Other chemotherapy drugs include *temozolomide*, taxane compounds *paclitaxel* and *docetaxel,* and a nanoparticle encapsulated paclitaxel (a novel anti-tumour compound). Personalized targeted therapies including BRAF^{V600} inhibitors and MEK inhibitors have been used.

Pancreatic cancer accounts for about 3% of all cancers in the United States and about 7% of all cancer deaths. The risk factors for pancreatic cancer include image-detectable pancreatic diseases. Lifestyle factors includes smoking, diabetes mellitus, obesity, advancing age, male sex, non-O blood group, occupational exposures, African-American ethnic origin, a high-fat diet, diets high in meat and low in vegetables and folate, and possibly *Helicobacter pylori* infection and periodontal disease. Several genes are implicated in the development of pancreatic cancer: the *TP53* tumour suppressor gene on chromosome 17p; inherited mutations in the *BRCA2* gene; and germline mutations in the *PALB2* (partner and localizer of *BRCA2*) gene; germline mutations or inactivation in the *p16*/*CDKN2A* tumour suppressor gene; and patients with Lynch Syndrome; *SMAD4* gene mutations; and several other genes somatically mutated at lower frequencies. Treatment options for people with pancreatic cancer highly depend on the type and stage of the cancer and other factors, can include: surgery, ablation or embolization treatments, radiation therapy, chemotherapy and other drugs.

Prostate cancer is the most prevalent cancer among men globally with approximately 180,890 new cases and 26,120 deaths in 2016. Risk factors to prostate cancer occur in 3 groups: non-modifiable, external and blood-based markers. A BRCA2 mutation confers up to an 8.6-fold increased risk in men below 65 years of age, and such mutations have also been related with aggressive cancer. There are other rare mutations reported in *BRCA1, HOXB13, NBS1* and *CHEK2*. Important lifestyle factors include smoking, diet, weight and physical activity. The possible role of endogenous hormones in the aetiology of prostate cancer has been reported; with a positive association between circulating insulin-like growth factors and prostate cancer. Recent expression profiling studies in prostate cancer suggest that microRNAs may serve as potential biomarkers. Treatment options include watchful waiting or active surveillance; surgery; radiation therapy; cryotherapy (cryosurgery); hormone therapy; chemotherapy; vaccine treat-

Cancers

ment; radical prostatectomy; radiation therapy (with or without androgen deprivation therapy); androgen deprivation therapy, and bone-directed treatment. The prostate-specific antigen (PSA) blood test is a widely used test for prostate cancer. Lifestyle modifications like smoking cessation and exercise can decrease the risk of developing prostate cancer.

Neurofibromatosis is a genetic defect characterized by a neuroectodermal abnormality and by clinical manifestations of systemic and progressive involvement mainly affecting the skin, nervous system, bones, eyes and possibly other organs. NF can be classified into five types based on distinct clinical features and genetic implications for the patient and the most common type is NF1 occurring in more than 90% of cases.

The NF1 gene is responsible for neurofibromatosis type 1, located on the long arm of chromosome 17 at 17q11.2 and encodes a protein predominantly expressed in neurons, Schwann cells, oligodendrocytes and astrocytes. Today, as a result of continued research on the functions of the NF1 and NF2 gene products, there have been better diagnoses and monitoring of affected individuals. Despite these advances, no medical treatment is available to prevent or reverse the typical lesions of neurofibromatosis. Medical care is centred on genetic counselling and the early detection of complications that can be treated.

MEN disorders are characterized by the occurrence of tumours involving two or more endocrine glands within the same individual. There are four major forms recognized: MEN1, MEN2, MEN3, and MEN4. The gene causing MEN1 is located on the long arm of chromosome 11 (11q13) and codes for a nuclear protein called menin. It acts as a tumour-suppressor gene and its abnormalities result in mutations, deletions, and/or truncations of the menin protein. Menin acts as scaffold protein and can increase or decrease gene expression by epigenetic regulation of gene expression via histone methylation. The FDA has recently approved two agents for the treatment of metastatic pancreatic neuroendocrine tumours, the mTOR inhibitor *everolimus*, and the tyrosine kinase inhibitor *sunitinib*.

Retinoblastoma is a rare cancer of the infant retina, which forms when both retinoblastoma gene (*RB1*) alleles mutate in retinal cells. The RB1 protein (pRB) is a cell cycle regulator. Retinoblastoma is the most common intraocular malignancy in childhood, and comprises 4% of all paediatric cancers, but curable if diagnosed early. Many modifications impair RB1 function, including point mutations, promoter methylation, and small and large deletions. Recent studies suggest RB can be caused by amplification of the *MYCN* gene, p53 inactivation and currently studies are exploring the role of microRNAs. The goal of treatment of retinoblastoma is to ensure elimination of the tumour while minimizing collateral injury to other tissues. The primary treatments for RB include enucleation, intravenous chemotherapy with focal therapy (laser therapy and cryotherapy), intra-arterial chemotherapy with focal therapy, and focal therapy alone when tumours are small at diagnosis. Secondary treatment options include focal therapies, repeated systemic chemotherapy, brachytherapy (internal radiotherapy), EBRT and whole-eye or proton beam radiation.

End of Chapter Quiz

1. Bacteria may contribute to cancer development through
    a. inflammation
    b. induction of DNA damage by toxins
    c. metabolites
    d. manipulation of host cell signalling pathway during their infection cycle
    e. all of the above

2. The cell cycle is positively or negatively regulated by various factors. The main factor being the negative regulator coded ____, a tumour suppressor gene.
   a. cyclin A
   b. cyclin dependent kinases
   c. p53
   d. cyclin D
   e. E2F

3. Although the mechanism is not completely understood, p53 is thought to act as cellular stress sensor that triggers
   a. cellular senescence
   b. apoptosis
   c. cell proliferation
   d. A and B only
   e. all of the above

4. When parts of the lymph system get damaged or blocked after surgery or radiation therapy, an accumulation of lymph fluids causes swelling. This condition is called
   a. lymphedema
   b. lymphoscintigraphy
   c. mesenteric lymphadenitis
   d. lymphatic filariasis
   e. lymphangioleiomyomatosis

5. Which of the following statements is/are true?
   a. Aromatase inhibitors work by blocking the enzyme aromatase
   b. Aromatase inhibitors have menopausal symptoms such as hot flashes
   c. Aromatase is an enzyme that converts oestrogen into adrenal androgens
   d. A and B only
   e. All of the above

6. Additional research provides evidence associating specific environmental toxicants with an increased risk for developing cancer. This/these toxicants is/are
   a. endocrine disrupting compounds
   b. hormones in food
   c. industrial chemicals
   d. radiation
   e. all of the above

7. Immune-based treatment for ovarian cancer include
   a. vaccines
   b. cytokines
   c. adoptive T-cell therapy
   d. lymphocyte transfusion
   e. all of the above

Cancers

8. Which of the following statements about colorectal cancer is false?
    a. Colorectal cancer is the third leading cause of cancer-related deaths in men and women in the United States.
    b. Systemic colorectal cancer symptoms are those that have a direct effect on the colon or rectum whereas local colorectal cancer symptoms affect the entire body.
    c. There are minimal symptoms of colorectal cancer during the early stages of the disease.
    d. The risk factors for developing colon cancer are age, lifestyle, family history, and diet.
    e. All of the statements above are false
9. What type of imaging and laboratory tests are useful for the diagnosis of colorectal cancer?
    a. MRI and CT
    b. Colonoscopy
    c. PET
    d. A and B only
    e. All of the above
10. Which of the following symptoms are most likely due to prostate cancer?
    a. blood in urine, sharp pain in the abdomen and back, sweating, nausea
    b. chest pain, fatigue, swollen lymph nodes, weight loss
    c. blood in urine or semen, numbness in the legs, erectile dysfunction, pain in the hips, spine and chest, slow urinary stream or frequent urination
    d. chills, malaise, diarrhea, loss of appetite
    e. bloating, indigestion, flatulence, stomach cramps
11. The most common sign of retinoblastoma is _____ and the second most common being _____.
    a. bulging eyes; bloodshot eye
    b. leukocoria; strabismus
    c. conjunctivitis; eye fatigue
    d. dry eyes; uveitis
    e. colour blindness; eye strain

Thought Questions

1. Give examples of the six major mechanisms of p53-dependent transcriptional regulations listed in the chapter.
2. How can interfering nucleic acids (RNAi) and small-interfering RNA (siRNA), and short-hairpin RNA (shRNA) be used to treat colorectal cancer?
3. Describe the chain reaction of events that would result by targeting the NF1 gene in neurofibromatosis.

REFERENCES

ACS. (2018). *Signs and Symptoms of Prostate Cancer*. Retrieved from https://www.cancer.org/cancer/prostate-cancer/detection-diagnosis-staging/signs-symptoms.html

Agarwal, S. K. (2014). Exploring the tumors of multiple endocrine neoplasia type 1 in mouse models for basic and preclinical studies. *International Journal of Endocrine Oncology*, *1*(2), 153–161. doi:10.2217/ije.14.16 PMID:25685317

Alizart, M., Saunus, J., Cummings, M., & Lakhani, S. R. (2012). Molecular classification of breast carcinoma. *Diagnostic Histopathology*, *18*(3), 97–103. doi:10.1016/j.mpdhp.2011.12.003

Altonsy, M. O., & Andrews, S. C. (2011). Diallyl disulphide, a beneficial component of garlic oil, causes a redistribution of cell-cycle growth phases, induces apoptosis, and enhances butyrate-induced apoptosis in colorectal adenocarcinoma cells (HT-29). *Nutrition and Cancer*, *63*(7), 1104–1113. doi:10.1080/01635581.2011.601846 PMID:21916706

American Cancer Society (ACS). (2018a). *What causes colorectal cancer: causes, risk factors and prevention*. Retrieved from https://www.cancer.org/cancer/colon-rectal-cancer/causes-risks-prevention/what-causes.html

American Cancer Society (ACS). (2018b). *Colorectal Cancer*. https://www.cancer.org/cancer/colon-rectal-cancer.html

Amir, E., Seruga, B., Niraula, S., Carlsson, L., & Ocaña, A. (2011). Toxicity of adjuvant endocrine therapy in postmenopausal breast cancer patients: A systematic review and meta-analysis. *Journal of the National Cancer Institute*, *103*(17), 1299–1309. doi:10.1093/jnci/djr242 PMID:21743022

Antônio, J. R., Goloni-Bertollo, E. M., & Trídico, L. A. (2013). Neurofibromatosis: Chronological history and current issues. *Anais Brasileiros de Dermatologia*, *88*(3), 329–343. doi:10.1590/abd1806-4841.20132125 PMID:23793209

Barroso-Sousa, R., Santana, I. A., Testa, L., de Melo Gagliato, D., & Mano, M. S. (2013). Biological therapies in breast cancer: Common toxicities and management strategies. *The Breast*, *22*(6), 1009–1018. doi:10.1016/j.breast.2013.09.009 PMID:24144949

Berger, H., Marques, M. S., Zietlow, R., Meyer, T. F., Machado, J. C., & Figueiredo, C. (2016). Gastric cancer pathogenesis. *Helicobacter*, *21*(Suppl 1), 34–38. doi:10.1111/hel.12338 PMID:27531537

Bernstein, J. L., Thomas, D. C., Shore, R. E., Robson, M., Boice, J. D., Stovall, M., ... WECARE Study Collaborative Group (2013). Contralateral breast cancer after radiotherapy among BRCA1 and BRCA2 mutation carriers: a WECARE study report. *European Journal of Cancer*, *49*(14), 2979–2985. doi:10.1016/j.ejca.2013.04.028

Bieging, K. T., Mello, S. S., & Attardi, L. D. (2014). Unravelling mechanisms of p53-mediated tumour suppression. *Nature Reviews. Cancer*, *14*(5), 359–370. doi:10.1038/nrc3711 PMID:24739573

Bird, B. R., & Swain, S. M. (2008). Cardiac toxicity in breast cancer survivors: Review of potential cardiac problems. *Clinical Cancer Research*, *14*(1), 14–24. doi:10.1158/1078-0432.CCR-07-1033 PMID:18172247

Blackford, A., Serrano, O. K., Wolfgang, C. L., Parmigiani, G., Jones, S., Zhang, X., ... Hruban, R. H. (2009). SMAD4 gene mutations are associated with poor prognosis in pancreatic cancer. *Clinical Cancer Research: An Official Journal of the American Association for Cancer Research*, *15*(14), 4674–4679. doi:10.1158/1078-0432.CCR-09-0227 PMID:19584151

Block, K. I., Block, P. B., & Gyllenhaal, C. (2015). Integrative therapies in cancer: Modulating a broad spectrum of targets for cancer management. *Integrative Cancer Therapies*, *14*(2), 113–118. doi:10.1177/1534735414567473 PMID:25601968

Block, K. I., Gyllenhaal, C., Lowe, L., Amedei, A., Amin, A.R.M. R., Amin, A., ... Zollo, M. (2015). A Broad-Spectrum Integrative Design for Cancer Prevention and Therapy. *Seminars in Cancer Biology*, *35*(Suppl), S276–S304. doi:.2015.09.007 doi:10.1016/j.semcancer

Bodai, B. I., & Tuso, P. (2015). Breast cancer survivorship: A comprehensive review of long-term medical issues and lifestyle recommendations. *The Permanente Journal*, *19*(2), 48–79. doi:10.7812/TPP/14-241 PMID:25902343

Bowles, E. J., Wellman, R., Feigelson, H. S., Onitilo, A. A., Freedman, A. N., Delate, T., ... Wagner, E. H.Pharmacovigilance Study Team. (2012). Risk of heart failure in breast cancer patients after anthracycline and trastuzumab treatment: A retrospective cohort study. *Journal of the National Cancer Institute*, *104*(17), 1293–1305. doi:10.1093/jnci/djs317 PMID:22949432

Brady, C. A., & Attardi, L. D. (2010). p53 at a glance. *Journal of Cell Science*, *123*(Pt 15), 2527–2532. doi:10.1242/jcs.064501 PMID:20940128

Brandi, M. L., Marx, S. J., Aurbach, G. D., & Fitzpatrick, L. A. (1987). Familial multiple endocrine neoplasia type I: A new look at pathophysiology. *Endocrine Reviews*, *8*(4), 391–405. doi:10.1210/edrv-8-4-391 PMID:2891500

Bunn, P. A., Minna, J. D., Augustyn, A., Gazdar, A. F., Ouadah, Y., Krasnow, M. A., ... Hirsch, F. R. (2016). Small Cell Lung Cancer: Can Recent Advances in Biology and Molecular Biology Be Translated into Improved Outcomes? *Journal of Thoracic Oncology*, *11*(4), 453–474. doi:10.1016/j.jtho.2016.01.012

Burstein, H. J. (2005). The distinctive nature of HER2-positive breast cancers. *The New England Journal of Medicine*, *353*(16), 1652–1654. doi:10.1056/NEJMp058197 PMID:16236735

Calin, G. A., & Croce, C. M. (2006). MicroRNA signatures in human cancers. *Nature Reviews. Cancer*, *6*(11), 857–866. doi:10.1038/nrc1997 PMID:17060945

Cancer Treatment Centers of America (CTCA). (2018). *Colorectal cancer symptoms.* Retrieved from https://www.cancercenter.com/colorectal-cancer/symptoms/

Carey, J. C., Baty, B. J., Johnson, J. P., Morrison, T., Skolnik, M., & Kivlin, J. (1986). The genetic aspects of neurofibromatosis. *Annals of the New York Academy of Sciences*, *486*(1 Neurofibromat), 45–46. doi:10.1111/j.1749-6632.1986.tb48061.x PMID:3105404

Ciociola, A. A., Cohen, L. B., & Kulkarni, P. (2014). How drugs are developed and approved by the FDA: Current process and future directions. *The American Journal of Gastroenterology*, *109*(5), 620–623. doi:10.1038/ajg.2013.407 PMID:24796999

Coghill, A. E., & Hildesheim, A. (2014). Epstein-Barr virus antibodies and the risk of associated malignancies: Review of the literature. *American Journal of Epidemiology*, *180*(7), 687–695. doi:10.1093/aje/kwu176 PMID:25167864

Coukos, G., Tanyi, J., & Kandalaft, L. E. (2016). Opportunities in immunotherapy of ovarian cancer. *Annals of Oncology: Official Journal of the European Society for Medical Oncology*, *27*(1suppl 1), 11–15. doi:10.1093/annonc/mdw084 PMID:27141063

CTCA. (2018a). *Lung cancer symptoms*. Retrieved from https://www.cancercenter.com/lung-canc er/symptoms/?

CTCA. (2018b). *Melanoma symptoms*. Retrieved from https://www.cancercenter.com/melanoma/symptoms/?

Cuzick, J., Thorat, M. A., Andriole, G., Brawley, O. W., Brown, P. H., Culig, Z., ... Wolk, A. (2014). Prevention and early detection of prostate cancer. *The Lancet Oncology*, *15*(11), e484–e492. doi:10.1016/S1470-2045(14)70211-6 PMID:25281467

Daston, M. M., Scrable, H., Nordlund, M., Sturbaum, A. K., Nissen, L. M., & Ratner, N. (1992). The protein product of the neurofibromatosis type 1 gene is expressed at highest abundance in neurons, Schwann cells, and oligodendrocytes. *Neuron*, *8*(3), 415–428. doi:10.1016/0896-6273(92)90270-N PMID:1550670

DeClue, J. E., Cohen, B. D., & Lowy, D. R. (1991). Identification and characterization of the neurofibromatosis type 1 protein product. *Proceedings of the National Academy of Sciences of the United States of America*, *88*(22), 9914–9918. doi:10.1073/pnas.88.22.9914 PMID:1946460

DeSantis, C. E., Fedewa, S. A., Goding, S. A., Kramer, J. L., Smith, R. A., & Jemal, A. (2016). Breast cancer statistics, 2015: Convergence of incidence rates between black and white women. *CA: a Cancer Journal for Clinicians*, *66*(1), 31–42. doi:10.3322/caac.21320 PMID:26513636

Di Leva, G., Garofalo, M., & Croce, C. M. (2013). MicroRNAs in cancer. *Annual Review of Pathology*, *9*(1), 287–314. doi:10.1146/annurev-pathol-012513-104715 PMID:24079833

Dimaras, H., Corson, T. W., Cobrinik, D., White, A., Zhao, J., Munier, F. L., ... Gallie, B. L. (2015). Retinoblastoma. *Nature Reviews. Disease Primers*, *1*, 15021. doi:10.1038/nrdp.2015.21 PMID:27189421

Discacciati, A., & Wolk, A. (2014). Lifestyle and dietary factors in prostate cancer prevention. *Recent Results in Cancer Research. Fortschritte der Krebsforschung. Progres Dans les Recherches Sur le Cancer*, *202*, 27–37. doi:10.1007/978-3-642-45195-9_3 PMID:24531774

Duong, T. A., Sbidian, E., Valeyrie-Allanore, L., Vialette, C., Ferkal, S., Hadj-Rabia, S., ... Wolkenstein, P. (2011). Mortality associated with neurofibromatosis 1: a cohort study of 1895 patients in 1980-2006 in France. *Orphanet Journal of Rare Diseases, 6*, 18. doi:10.1186/1750-1172-6-18

Endogenous Hormones and Prostate Cancer Collaborative Group. (2008). Endogenous sex hormones and prostate cancer: A collaborative analysis of 18 prospective studies. *Journal of the National Cancer Institute, 100*(3), 170–183. doi:10.1093/jnci/djm323 PMID:18230794

Evans, D. B., Varadhachary, G. R., Crane, C. H., Sun, C. C., Lee, J. E., Pisters, P. W., ... Wolff, R. A. (2008). Preoperative gemcitabine-based chemoradiation for patients with resectable adenocarcinoma of the pancreatic head. *Journal of Clinical Oncology, 26*(21), 3496–3502. doi:10.1200/JCO.2007.15.8634 PMID:18640930

Ewertz, M., & Jensen, A. B. (2011). Late effects of breast cancer treatment and potentials for rehabilitation. *Acta Oncologica, 50*(2), 187–193. doi:10.3109/0284186X.2010.533190 PMID:21231780

Fearon, E. R. (2011). Molecular genetics of colorectal cancer. *Annual Review of Pathology, 6*(1), 479–507. doi:10.1146/annurev-pathol-011110-130235 PMID:21090969

Fischer, M., Steiner, L., & Engeland, K. (2014). The transcription factor p53: Not a repressor, solely an activator. *Cell Cycle (Georgetown, Tex.), 13*(19), 3037–3058. doi:10.4161/15384101.2014.949083 PMID:25486564

Gaillard, S. L., Secord, A. A., & Monk, B. (2016). The role of immune checkpoint inhibition in the treatment of ovarian cancer. *Gynecologic Oncology Research and Practice, 3*(1), 11. doi:10.118640661-016-0033-6 PMID:27904752

Garcia-Closas, M., Couch, F. J., Lindstrom, S., Michailidou, K., Schmidt, M. K., Brook, M. N., ... Kraft, P. (2013). Genome-wide association studies identify four ER negative-specific breast cancer risk loci. *Nature Genetics, 45*(4), 392–398. doi:10.1038/ng.2561

Goh, C. L., & Eeles, R. A. (2014). Germline genetic variants associated with prostate cancer and potential relevance to clinical practice. *Recent Results in Cancer Research. Fortschritte der Krebsforschung. Progres Dans les Recherches Sur le Cancer, 202*, 9–26. doi:10.1007/978-3-642-45195-9_2 PMID:24531773

Gray, J. M., Rasanayagam, S., Engel, C., & Rizzo, J. (2017). State of the evidence 2017: An update on the connection between breast cancer and the environment. *Environmental Health, 16*(1), 94. doi:10.118612940-017-0287-4 PMID:28865460

Grossniklaus, H. E. (2014). Retinoblastoma. Fifty years of progress. The LXXI Edward Jackson Memorial Lecture. *American Journal of Ophthalmology, 158*(5), 875–891. doi:10.1016/j.ajo.2014.07.025 PMID:25065496

Guo, J., Qin, S., Liang, J., Lin, T., Si, L., & Chen, X., ... Chinese Society of Clinical Oncology (CSCO) Melanoma Panel. (2015). Chinese Guidelines on the Diagnosis and Treatment of Melanoma (2015 Edition). *Annals of Translational Medicine, 3*(21), 322. doi:10.3978/j.issn.2305-5839.2015.12.23 PMID:26734632

Guo, J., Si, L., Kong, Y., Flaherty, K. T., Xu, X., Zhu, Y., ... Qin, S. (2011). Phase II, open-label, single-arm trial of imatinib mesylate in patients with metastatic melanoma harboring c-Kit mutation or amplification. *Journal of Clinical Oncology, 29*(21), 2904–2909. doi:10.1200/JCO.2010.33.9275 PMID:21690468

Habbe, N., Koorstra, J. B., Mendell, J. T., Offerhaus, G. J., Ryu, J. K., Feldmann, G., ... Maitra, A. (2009). MicroRNA miR-155 is a biomarker of early pancreatic neoplasia. *Cancer Biology & Therapy*, *8*(4), 340–346. doi:10.4161/cbt.8.4.7338 PMID:19106647

Hahn, S. A., Schutte, M., Hoque, A. T., Moskaluk, C. A., da Costa, L. T., Rozenblum, E., ... Kern, S. E. (1996). DPC4, a candidate tumor suppressor gene at human chromosome 18q21.1. *Science*, *271*(5247), 350–353. doi:10.1126cience.271.5247.350 PMID:8553070

Hamanishi, J., Mandai, M., & Konishi, I. (2016). Immune checkpoint inhibition in ovarian cancer. *International Immunology*, *28*(7), 339–348. doi:10.1093/intimm/dxw020 PMID:27055470

Harris, C. C., & Hollstein, M. (1994). Clinical Implications of the p53 Tumor-Suppressor Gene. *The New England Journal of Medicine*, *330*(12), 864–865. doi:10.1056/NEJM199403243301215 PMID:8114848

Howell, A., Downey, S., & Anderson, E. (1996). New endocrine therapies for breast cancer. *European Journal of Cancer*, *32A*(4), 576–588. doi:10.1016/0959-8049(96)00032-9 PMID:8695256

Hruban, R. H., van Mansfeld, A. D., Offerhaus, G. J., van Weering, D. H., Allison, D. C., Goodman, S. N., ... Bos, J. L. (1993). K-ras oncogene activation in adenocarcinoma of the human pancreas. A study of 82 carcinomas using a combination of mutant-enriched polymerase chain reaction analysis and allele-specific oligonucleotide hybridization. *American Journal of Pathology*, *143*(2), 545–554. PMID:8342602

Hu, W., Feng, Z., & Levine, A. J. (2012). The Regulation of Multiple p53 Stress Responses is Mediated through MDM2. *Genes & Cancer*, *3*(3-4), 199–208. doi:10.1177/1947601912454734 PMID:23150753

Huncharek, M., Haddock, K. S., Reid, R., & Kupelnick, B. (2010). Smoking as a risk factor for prostate cancer: A meta-analysis of 24 prospective cohort studies. *American Journal of Public Health*, *100*(4), 693–701. doi:10.2105/AJPH.2008.150508 PMID:19608952

International Association for the Study of Pain (IASP). (2018). *Epidemiology of Cancer Pain*. Retrieved from http://www.iasppain.org/AM/Template.cfm?Section=Home& Template=/CM/ ContentDisplay.cfm&ContentID=7395

Jemal, A., Siegel, R., Xu, J., & Ward, E. (2010). Cancer statistics, 2010. *CA: a Cancer Journal for Clinicians*, *60*(5), 277–300. doi:10.3322/caac.20073 PMID:20610543

Jones, S., Zhang, X., Parsons, D. W., Lin, J. C., Leary, R. J., Angenendt, P., ... Kinzler, K. W. (2008). Core signaling pathways in human pancreatic cancers revealed by global genomic analyses. *Science*, *321*(5897), 1801–1806. doi:10.1126cience.1164368 PMID:18772397

Karlsson, R., Aly, M., Clements, M., Zheng, L., Adolfsson, J., Xu, J., ... Wiklund, F. (2014). A population-based assessment of germline HOXB13 G84E mutation and prostate cancer risk. *European Urology*, *65*(1), 169–176. doi:10.1016/j.eururo.2012.07.027 PMID:22841674

Katz, S. J., & Hawley, S. T. (2007). From policy to patients and back: Surgical treatment decision making for patients with breast cancer. *Health Affairs (Project Hope)*, *26*(3), 761–769. doi:10.1377/hlthaff.26.3.761 PMID:17485755

Kennecke, H. F., Ellard, S., O'Reilly, S., & Gelmon, K. A. (2006). New guidelines for treatment of early hormone-positive breast cancer with tamoxifen and aromatase inhibitors. *British Columbia Medical Journal, 48*(3), 121–126.

Klein, A. P., Brune, K. A., Petersen, G. M., Goggins, M., Tersmette, A. C., Offerhaus, G. J., ... Hruban, R. H. (2004). Prospective risk of pancreatic cancer in familial pancreatic cancer kindreds. *Cancer Research, 64*(7), 2634–2638. doi:10.1158/0008-5472.CAN-03-3823 PMID:15059921

Kurman, R. J., & Shih, I. (2016). The Dualistic Model of Ovarian Carcinogenesis: Revisited, Revised, and Expanded. *American Journal of Pathology, 186*(4), 733–747. doi:10.1016/j.ajpath.2015.11.011 PMID:27012190

Lane, D. P. (1992). Cancer. p53, guardian of the genome. *Nature, 358*(6381), 15–16. doi:10.1038/358015a0 PMID:1614522

Lee, D., Rho, J. Y., Kang, S., Yoo, K. J., & Choi, H. J. (2016). CT findings of small cell lung carcinoma: Can recognizable features be found? *Medicine, 95*(47), e5426. doi:10.1097/MD.0000000000005426 PMID:27893684

Levine, A. J., Momand, J., & Finlay, C. A. (1991). The p53 tumour suppressor gene. *Nature, 351*(6326), 453–456. doi:10.1038/351453a0 PMID:2046748

Long, G. V., Menzies, A. M., Nagrial, A. M., Haydu, L. E., Hamilton, A. L., Mann, G. J., ... Kefford, R. F. (2011). Prognostic and clinicopathologic associations of oncogenic BRAF in metastatic melanoma. *Journal of Clinical Oncology, 29*(10), 1239–1246. doi:10.1200/JCO.2010.32.4327 PMID:21343559

Luu, H. N., Lin, H. Y., Sørensen, K. D., Ogunwobi, O. O., Kumar, N., Chornokur, G., ... Di Pietro, G. (2017). miRNAs associated with prostate cancer risk and progression. *BMC Urology, 17*(1), 18. doi:10.118612894-017-0206-6 PMID:28320379

Mahdian-Shakib, A., Dorostkar, R., Tat, M., Hashemzadeh, M. S., & Saidi, N. (2016). Differential role of microRNAs in prognosis, diagnosis, and therapy of ovarian cancer. *Biomedicine and Pharmacotherapy, 84*, 592–600. doi:10.1016/j.biopha.2016.09.087 PMID:27694003

Mahida, J. P., Antczak, C., Decarlo, D., Champ, K. G., Francis, J. H., Marr, B., ... Djaballah, H. (2013). A synergetic screening approach with companion effector for combination therapy: Application to retinoblastoma. *PLoS One, 8*(3), e59156. doi:10.1371/journal.pone.0059156 PMID:23527118

Mavaddat, N., Antoniou, A. C., Easton, D. F., & Garcia-Closas, M. (2010). Genetic susceptibility to breast cancer. *Molecular Oncology, 4*(3), 174–191. doi:10.1016/j.molonc.2010.04.011 PMID:20542480

Mavaddat, N., Pharoah, P. D., Michailidou, K., Tyrer, J., Brook, M. N., Bolla, M. K., ... Garcia-Closas, M. (2015). Prediction of breast cancer risk based on profiling with common genetic variants. *Journal of the National Cancer Institute, 107*(5), djv036. doi:10.1093/jnci/djv036 PMID:25855707

Melnik, B. C. (2015). MiR-21: An environmental driver of malignant melanoma? *Journal of Translational Medicine, 13*(1), 202. doi:10.118612967-015-0570-5 PMID:26116372

Mishra, J., Drummond, J., Quazi, S. H., Karanki, S. S., Shaw, J. J., Chen, B., & Kumar, N. (2012). Prospective of colon cancer treatments and scope for combinatorial approach to enhanced cancer cell apoptosis. *Critical Reviews in Oncology/Hematology, 86*(3), 232–250. doi:10.1016/j.critrevonc.2012.09.014 PMID:23098684

Mittica, G., Genta, S., Aglietta, M., & Valabrega, G. (2016). Immune Checkpoint Inhibitors: A New Opportunity in the Treatment of Ovarian Cancer? *International Journal of Molecular Sciences, 17*(7), 1169. doi:10.3390/ijms17071169 PMID:27447625

Mueller, T., & Voigt, W. (2011). Fermented wheat germ extract--nutritional supplement or anticancer drug? *Nutrition Journal, 10*(1), 89. doi:10.1186/1475-2891-10-89 PMID:21892933

Nakaji, S., Umeda, T., Shimoyama, T., Sugawara, K., Tamura, K., Fukuda, S., ... Parodi, S. (2003). Environmental factors affect colon carcinoma and rectal carcinoma in men and women differently. *International Journal of Colorectal Disease, 18*(6), 481–486. doi:10.100700384-003-0485-0 PMID:12695918

National Cancer Institute (NCI). (2018). *Nerve Problems*. Retrieved from http://www.cancer. gov/about-cancer/treatment/side-effects/nerve-problems

Nautiyal, J., Kanwar, S. S., Yu, Y., & Majumdar, A. P. (2011). Combination of dasatinib and curcumin eliminates chemo-resistant colon cancer cells. *Journal of Molecular Signaling, 6*, 7. doi:10.1186/1750-2187-6-7 PMID:21774804

NIH-NCI. (2018). *Cancer Stat Facts: Ovarian Cancer (At a Glance)*. Retrieved from https://seer.cancer.gov/statfacts/html/ovary.html

Norton, J. A., Krampitz, G., & Jensen, R. T. (2015). Multiple Endocrine Neoplasia: Genetics and Clinical Management. *Surgical Oncology Clinics of North America, 24*(4), 795–832. doi:10.1016/j.soc.2015.06.008 PMID:26363542

One Retinoblastoma World (ORW). (2018). Retrieved from http://map.1rbw.org/

Oronsky, B., Carter, C. A., Mackie, V., Scicinski, J., Oronsky, A., Oronsky, N., ... Reid, T. (2015). The war on cancer: A military perspective. *Frontiers in Oncology, 4*, 387. doi:10.3389/fonc.2014.00387 PMID:25674537

Osborne, C. K. (1998). Tamoxifen in the treatment of breast cancer. *The New England Journal of Medicine, 339*(22), 1609–1618. doi:10.1056/NEJM199811263392207 PMID:9828250

Ottaviani, D., Alonso, C., & Szijan, I. (2015). Uncommon RB1 somatic mutations in a unilateral retinoblastoma patient. *Medicina, 75*(3), 137–141. PMID:26117602

Palumbo, M. O., Kavan, P., Miller, W. H., Panasci, L., Assouline, S., Johnson, N., ... Batist, G. (2013). Systemic cancer therapy: Achievements and challenges that lie ahead. *Frontiers in Pharmacology, 4*, 57. doi:10.3389/fphar.2013.00057 PMID:23675348

Panduro Cerda, A., Lima González, G., & Villalobos, J. J. (1993). Molecular genetics of colorectal cancer and carcinogenesis. *Revista de Investigacion Clinica, 45*(5), 493–504. PMID:8134731

Pawlik, T. M., Laheru, D., Hruban, R. H., Coleman, J., Wolfgang, C. L., Campbell, K., ... Herman, J. M. (2008). Evaluating the impact of a single-day multidisciplinary clinic on the management of pancreatic cancer. *Annals of Surgical Oncology, 15*(8), 2081–2088. doi:10.124510434-008-9929-7 PMID:18461404

Perez, E. A., Koehler, M., Byrne, J., Preston, A. J., Rappold, E., & Ewer, M. S. (2008). Cardiac safety of lapatinib: Pooled analysis of 3689 patients enrolled in clinical trials. *Mayo Clinic Proceedings, 83*(6), 679–686. doi:10.1016/S0025-6196(11)60896-3 PMID:18533085

Pyszel, A., Malyszczak, K., Pyszel, K., Andrzejak, R., & Szuba, A. (2006). Disability, psychological distress and quality of life in breast cancer survivors with arm lymphedema. *Lymphology, 39*(4), 185–192. PMID:17319631

Raimondi, S., Maisonneuve, P., & Lowenfels, A. B. (2009). Epidemiology of pancreatic cancer: An overview. *Nature Reviews. Gastroenterology & Hepatology, 6*(12), 699–708. doi:10.1038/nrgastro.2009.177 PMID:19806144

Rauen, K. A., Huson, S. M., Burkitt-Wright, E., Evans, D. G., Farschrschi, S., Ferner, R. E., ... Upadhyaya, M. (2015). Recent Developments in Neurofibromatoses and RASopathies: Management, Diagnosis and Current and Future Therapeutic Avenues. *American Journal of Medical Genetics. Part A, 167*(1), 1–10. doi:10.1002/ajmg.a.36793 PMID:25393061

Raymond, E., O'Callaghan, M. E., Campbell, J., Vincent, A. D., Beckmann, K., Roder, D., ... Moretti, K. (2017). An appraisal of analytical tools used in predicting clinical outcomes following radiation therapy treatment of men with prostate cancer: A systematic review. *Radiation Oncology (London, England), 12*(1), 56. doi:10.118613014-017-0786-z PMID:28327203

Reis, A. H., Vargas, F. R., & Lemos, B. (2012). More epigenetic hits than meets the eye: microRNAs and genes associated with the tumorigenesis of retinoblastoma. *Frontiers in Genetics, 3*, 284. doi:10.3389/fgene.2012.00284 PMID:23233862

Richter, A. M., Kiehl, S., Köger, N., Breuer, J., Stiewe, T., & Dammann, R. H. (2017). ZAR1 is a novel epigenetically inactivated tumour suppressor in lung cancer. *Clinical Epigenetics, 9*(1), 60. doi:10.118613148-017-0360-4 PMID:28588743

Riley, T., Sontag, E., Chen, P., & Levine, A. (2008). Transcriptional control of human p53-regulated genes. *Nature Reviews. Molecular Cell Biology, 9*(5), 402–412. doi:10.1038/nrm2395 PMID:18431400

Rinn, J. L., & Huarte, M. (2011). To repress or not to repress: This is the guardian's question. *Trends in Cell Biology, 21*(6), 344–353. doi:10.1016/j.tcb.2011.04.002 PMID:21601459

Roberts, N. J., Jiao, Y., Yu, J., Kopelovich, L., Petersen, G. M., Bondy, M. L., ... Klein, A. P. (2011). ATM mutations in patients with hereditary pancreatic cancer. *Cancer Discovery, 2*(1), 41–46. doi:10.1158/2159-8290.CD-11-0194 PMID:22585167

Rojas, V., Hirshfield, K. M., Ganesan, S., & Rodriguez-Rodriguez, L. (2016). Molecular Characterization of Epithelial Ovarian Cancer: Implications for Diagnosis and Treatment. *International Journal of Molecular Sciences, 17*(12), 2113. doi:10.3390/ijms17122113 PMID:27983698

Romond, E. H., Perez, E. A., Bryant, J., Suman, V. J., Geyer, C. E. Jr, Davidson, N. E., ... Wolmark, N. (2005). Trastuzumab plus adjuvant chemotherapy for operable HER2-positive breast cancer. *The New England Journal of Medicine, 353*(16), 1673–1684. doi:10.1056/NEJMoa052122 PMID:16236738

Rushlow, D. E., Mol, B. M., Kennett, J. Y., Yee, S., Pajovic, S., Thériault, B. L., ... Gallie, B. L. (2013). Characterisation of retinoblastomas without RB1 mutations: Genomic, gene expression, and clinical studies. *The Lancet Oncology, 14*(4), 327–334. doi:10.1016/S1470-2045(13)70045-7 PMID:23498719

Sadava, D., & Kane, S. E. (2013). Silibinin reverses drug resistance in human small-cell lung carcinoma cells. *Cancer Letters, 339*(1), 102–106. doi:10.1016/j.canlet.2013.07.017 PMID:23879966

Samaras, V., Rafailidis, P. I., Mourtzoukou, E. G., Peppas, G., & Falagas, M. E. (2010). Chronic bacterial and parasitic infections and cancer: A review. *Journal of Infection in Developing Countries, 4*(5), 267–281. PMID:20539059

Savard, J., & Morin, C. M. (2001). Insomnia in the context of cancer: A review of a neglected problem. *Journal of Clinical Oncology, 19*(3), 895–908. doi:10.1200/JCO.2001.19.3.895 PMID:11157043

Schaffer, B. E., Park, K., Yiu, G., Conklin, J. F., Lin, C., Burkhart, D. L., ... Sage, J. (2010). Loss of p130 accelerates tumor development in a mouse model for human small cell lung carcinoma. *Cancer and Research, 70*(10), 3877–3883. doi:10.1158 /0008-5472.CAN-09-4228

Shields, C. L., Fulco, E. M., Arias, J. D., Alarcon, C., Pellegrini, M., Rishi, P., ... Shields, J. A. (2012). Retinoblastoma frontiers with intravenous, intra-arterial, periocular, and intravitreal chemotherapy. *Eye (London, England), 27*(2), 253–264. doi:10.1038/eye.2012.175 PMID:22995941

Siddiqui, G. K., Maclean, A. B., Elmasry, K., Wong te Fong, A., Morris, R. W., Rashid, M., ... Boxer, G. M. (2011). Immunohistochemical expression of VEGF predicts response to platinum-based chemotherapy in patients with epithelial ovarian cancer. *Angiogenesis, 14*(2), 155–161. doi:10.100710456-010-9199-4 PMID:21221762

Soura, E., Eliades, P. J., Shannon, K., Stratigos, A. J., & Tsao, H. (2016). Hereditary melanoma: Update on syndromes and management: Genetics of familial atypical multiple mole melanoma syndrome. *Journal of the American Academy of Dermatology, 74*(3), 395–407, quiz 408–410. doi:10.1016/j.jaad.2015.08.038 PMID:26892650

Sutcliffe, S., & Platz, E. A. (2008). Inflammation and prostate cancer: A focus on infections. *Current Urology Reports, 9*(3), 243–249. doi:10.100711934-008-0042-z PMID:18765120

Taich, P., Ceciliano, A., Buitrago, E., Sampor, C., Fandino, A., Villasante, F., ... Schaiquevich, P. (2014). Clinical pharmacokinetics of intra-arterial melphalan and topotecan combination in patients with retinoblastoma. *Ophthalmology, 121*(4), 889–897. doi:10.1016/j.ophtha.2013.10.045 PMID:24359624

Tempero, M. A., Arnoletti, J. P., Behrman, S. W., Ben-Josef, E., Benson, A. B., Casper, E. S., ... National Comprehensive Cancer Networks. (2012). Pancreatic Adenocarcinoma, version 2.2012: featured updates to the NCCN Guidelines. *Journal of the National Comprehensive Cancer Network, 10*(6), 703–713.

Testa, U., Petrucci, E., Pasquini, L., Castelli, G., & Pelosi, E. (2018). Ovarian Cancers: Genetic Abnormalities, Tumor Heterogeneity and Progression, Clonal Evolution and Cancer Stem Cells. *Medicines (Basel, Switzerland), 5*(1), 16. doi:10.3390/medicines5010016 PMID:29389895

Thakker, R. V. (1998). Multiple endocrine neoplasia–syndromes of the twentieth century. *The Journal of Clinical Endocrinology and Metabolism, 83*, 2617–2620. PMID:9709920

Thakker, R. V. (2010). Multiple endocrine neoplasia type 1 (MEN1). *Best Practice & Research. Clinical Endocrinology & Metabolism, 24*(3), 355–370. doi:10.1016/j.beem.2010.07.003 PMID:20833329

Thakker, R. V. (2014). Multiple endocrine neoplasia type 1 (MEN1) and type 4 (MEN4). *Molecular and Cellular Endocrinology, 386*(1-2), 2–15. doi:10.1016/j.mce.2013.08.002 PMID:23933118

Thériault, B. L., Dimaras, H., Gallie, B. L., & Corson, T. W. (2013). The genomic landscape of retinoblastoma: A review. *Clinical & Experimental Ophthalmology, 42*(1), 33–52. doi:10.1111/ceo.12132 PMID:24433356

Thompson, D., & Easton, D. (2004). The genetic epidemiology of breast cancer genes. *Journal of Mammary Gland Biology and Neoplasia, 9*(3), 221–236. doi:10.1023/B:JOMG.0000048770.90334.3b PMID:15557796

Torbidoni, A. V., Laurent, V. E., Sampor, C., Ottaviani, D., Vazquez, V., Gabri, M. R., ... Chantada, G. L. (2015). Association of Cone-Rod Homeobox Transcription Factor Messenger RNA with Pediatric Metastatic Retinoblastoma. *JAMA Ophthalmology, 133*(7), 805–812. doi:10.1001/jamaophthalmol.2015.0900 PMID:25928893

Triantafillidis, J. K., Nasioulas, G., & Kosmidis, P. A. (2009). Colorectal cancer and inflammatory bowel disease: Epidemiology, risk factors, mechanisms of carcinogenesis and prevention strategies. *Anticancer Research, 29*(7), 2727–2737. PMID:19596953

Tuso, P. (2014). Physician update: Total health. *The Permanente Journal, 18*(2), 58–63. doi:10.7812/TPP/13-120 PMID:24694316

Vincent, A., Herman, J., Schulick, R., Hruban, R. H., & Goggins, M. (2011). Pancreatic cancer. *Lancet (London, England), 378*(9791), 607–620. doi:10.1016/S0140-6736(10)62307-0 PMID:21620466

Vincent, A. J. (2015). Management of menopause in women with breast cancer. *Climacteric, 18*(5), 690–701. doi:10.3109/13697137.2014.996749 PMID:25536007

Vogelstein, B., Papadopoulos, N., Velculescu, V. E., Zhou, S., Diaz, L. A., & Kinzler, K. W. (2013). Cancer genome landscapes. *Science, 339*(6127), 1546–1558. doi:10.1126cience.1235122 PMID:23539594

Volate, S. R., Davenport, D. M., Muga, S. J., & Wargovich, M. J. (2005). Modulation of aberrant crypt foci and apoptosis by dietary herbal supplements (quercetin, curcumin, silymarin, ginseng and rutin). *Carcinogenesis, 26*(8), 1450–1456. doi:10.1093/carcin/bgi089 PMID:15831530

Vousden, K. H., & Prives, C. (2009). Blinded by the Light: The Growing Complexity of p53. *Cell, 137*(3), 413–431. doi:10.1016/j.cell.2009.04.037 PMID:19410540

Vrijens, K., Bollati, V., & Nawrot, T. S. (2015). MicroRNAs as potential signatures of environmental exposure or effect: A systematic review. *Environmental Health Perspectives, 123*(5), 399–411. doi:10.1289/ehp.1408459 PMID:25616258

Wang, Z., Dabrosin, C., Yin, X., Fuster, M. M., Arreola, A., Rathmell, W. K., ... Jensen, L. D. (2015). Broad targeting of angiogenesis for cancer prevention and therapy. *Seminars in Cancer Biology, 35*(Suppl), S224-S243. doi:10.1016/j.semcancer.2015.01.001

Wolfgang, C. L., Herman, J. M., Laheru, D. A., Klein, A. P., Erdek, M. A., Fishman, E. K., & Hruban, R. H. (2013). Recent progress in pancreatic cancer. *CA: a Cancer Journal for Clinicians, 63*(5), 318–348. doi:10.3322/caac.21190 PMID:23856911

Wolpin, B. M., Chan, A. T., Hartge, P., Chanock, S. J., Kraft, P., Hunter, D. J., ... Fuchs, C. S. (2009). ABO blood group and the risk of pancreatic cancer. *Journal of the National Cancer Institute, 101*(6), 424–331. doi:10.1093/jnci/djp020 PMID:19276450

Wood, W. C. (2003). The future of surgery in the treatment of breast cancer. *The Breast, 12*(6), 472–474. doi:10.1016/S0960-9776(03)00154-1 PMID:14659123

World Cancer Research International (WCRI). (2018). *Worldwide Data*. Retrieved from https://wcrf.org/int/cancer-facts-figures/worldwide-data

Yeh, E. T., & Bickford, C. L. (2009). Cardiovascular complications of cancer therapy: Incidence, pathogenesis, diagnosis, and management. *Journal of the American College of Cardiology, 53*(24), 2231–2247. doi:10.1016/j.jacc.2009.02.050 PMID:19520246

Youssef, G., & Links, M. (2005). The prevention and management of cardiovascular complications of chemotherapy in patients with cancer. *American Journal of Cardiovascular Drugs, 5*(4), 233–243. doi:10.2165/00129784-200505040-00003 PMID:15984906

Yuan, W., Guo, Y. Q., Li, X. Y., Deng, M. Z., Shen, Z. H., Bo, C. B., ... Wang, X. (2016). MicroRNA-126 inhibits colon cancer cell proliferation and invasion by targeting the chemokine (C-X-C motif) receptor 4 and Ras homolog gene family, member A, signaling pathway. *Oncotarget, 7*(37), 60230–60244. doi:10.18632/oncotarget.11176 PMID:27517626

Zahreddine, H., & Borden, K. L. (2013). Mechanisms and insights into drug resistance in cancer. *Frontiers in Pharmacology, 4*, 28. doi:10.3389/fphar.2013.00028 PMID:23504227

Zur Hausen, H. (2009). The search for infectious causes of human cancers: Where and why. *Virology, 392*(1), 1–10. doi:10.1016/j.virol.2009.06.001 PMID:19720205

KEY TERMS AND DEFINITIONS

Adaptor Proteins: Proteins that facilitate the binding of a mRNA transcript by recruiting appropriate signal components such as receptors. They form complexes with other proteins to regulate signal transduction pathways.

Cancers

Adjuvant Treatment: Extra treatments used in addition to the main cancer treatment to prevent cancer from returning.

Apoptosis: Programmed cell death.

Aromatase Inhibitors: A hormone blocking drug that works by blocking the aromatase enzyme, thereby stopping the production of oestrogen.

Arrhythmia: A disorder in which the heart beats with an irregular rate or rhythms.

Biomarkers: A biological characteristic that is used as an indicator of normal biological processes.

Brachytherapy: A type of radiotherapy that involves the placement of radioactive material inside or near the tumour to treat cancers.

Bradycardia: Lower than normal heart rate.

Carcinogenesis: The process of normal cells transforming into cancer cells.

Chronic Pancreatitis: Prolonged inflammation of the pancreas.

Colonoscopy: An exam of the colon and rectum using a small camera to search for polyps and colon cancer.

Computed Tomography (CT): A diagnostic imaging test that uses several x-rays to scan and produce detailed images of areas inside the human body.

Epilepsy: A neurological disorder that causes seizures.

Focal Therapy: The use of focal modalities to treat small tumours.

Genome-Wide Association Studies (GWAS): An observational study that examines if there is an association between genetic variants and traits in different individuals.

Hormonal Therapy: A treatment used to cure or relieve the symptoms of cancer by reducing the amount of oestrogen in the body or blocking the effect of oestrogen on cancer cells.

Immunotherapy: A cancer treatment that helps the body's immune system fight the cancer.

Laser Therapy: A treatment involving lasers to destroy blood vessels that surround and supply a tumour.

Lymphocyte Transfusion: A therapy that involves the introduction of donor lymphocytes to destroy remaining cancer cells.

Macrocephaly: A term used to describe a larger than normal head circumference at birth.

Magnetic Resonance Imaging (MRI): A diagnostic imaging test that uses powerful magnet and radio waves to produce images of body organs and tissues.

Mastectomy: A surgical operation involving the removal of one or both breasts to treat breast cancer.

Myocardial Ischemia: The reduction of blood flow that prevents the heart from receiving enough oxygen.

Nanoparticles: A microscopic particle between 1 and 100 nanometres in size.

Next Generation Sequencing (NGS): A sequencing method used to measure large numbers of individual DNA sequences in a short period of time.

Oxidative Stress: An imbalance between free radicals and antioxidants in charge of homeostasis.

Palliation: To ease the symptoms without curing the underlying cause.

Pericarditis: Inflammation of the pericardium, a sac-like membrane that surrounds the heart.

Positron Emission Tomography (PET): A diagnostic imaging test that uses a special dye that contains radioactive tracers which are absorbed by certain organs and tissues.

Radiotherapy: A treatment used to cure or relieve the symptoms of cancer by using high-energy rays to destroy cancer cells.

Scoliosis: A condition that causes an abnormal curve to the spine.

Single Nucleotide Polymorphisms (SNPs): A genetic variant of a single nucleotide in a DNA sequence.

T-Cell Therapy: A type of cancer treatment involving the alteration of a patient's T-cell in order to attack cancer cells.

Thromboembolic Events: Occurs when a blood clot in an artery or vein breaks off and lodges somewhere else in body and obstructs blood flow of the circulatory system.

Tumorigenic Cells: Cells that are capable of or tend to form tumours.

Chapter 15
Digestive, Ear/Nose/Throat, and Eye Disorders

ABSTRACT

The digestive system includes the structures and organs involved in processing of foods required for growth, development, maintenance, and body repair. Most diseases affecting this system are due to infections from bacteria, viruses, protozoa, and fungi, while others are hereditary. The ear, nose, and throat (ENT) system is a complex set of structures sharing slightly interrelated mechanisms of operation. While some disorders of the ENT are hereditary, environmental influences play a big role. Diseases that affect eyesight primarily centre on three layers of the eye (sclera, choroid, and retina) which make eyesight possible. Disorders of metabolism occur when a crucial enzyme is disabled, or if a control mechanism for a metabolic pathway is affected. The chapter focuses on 14 diseases with suspected genetic causes including cystic fibrosis, diabetes, glucose-galactose malabsorption, hemochromatosis, obesity, Wilson's Disease, Zellweger syndrome, deafness, Pendred syndrome, Best Disease, glaucoma, gyrate atrophy, male pattern baldness, and Alport syndrome.

CHAPTER OUTLINE

15.1 Overview of Digestive, ENT, and Eye Systems
15.2 Cystic Fibrosis
15.3 Diabetes
15.4 Glucose-Galactose Malabsorption
15.5 Hereditary Hemochromatosis
15.6 Obesity
15.7 Wilson's Disease
15.8 Zellweger Syndrome
15.9 Deafness
15.10 Pendred Syndrome
15.11 Best Disease

DOI: 10.4018/978-1-5225-8066-9.ch015

15.12 Glaucoma
15.13 Gyrate Atrophy
15.14 Pattern Baldness
15.15 Alport Syndrome
Chapter Summary

LEARNING OUTCOMES

- Identify each genetic disorder affecting each system
- Outline the symptoms of each disorder
- Explain the genetic basis of each disorder
- Summarize the therapies available to treat each disorder

15.1 OVERVIEW OF THE DIGESTIVE, EAR/NOSE/THROAT, AND EYE SYSTEMS

The digestive system includes the structures and organs involved in processing of foods required for growth, development, maintenance and repair of body tissues. Diseases affecting the digestive system are highly dependent on lifestyle (the foods eaten, amount of physical activity, plus the pace and stress levels each day. Most diseases of the digestive system are due to infections from bacteria, viruses, protozoa, and fungi, while others are hereditary.

The structures of the ear, nose, and throat (ENT) allow a person to make sounds, hear, maintain balance, smell, breathe, and swallow. It is a complex set of structures sharing slightly interrelated mechanisms of operation. While some disorders of the ENT are hereditary, environmental influences play a big role. Eyes allow individuals to perceive their surroundings by forming images at the backs of their eyes. Eyesight is accomplished by a very complex arrangement of layers and structures found in the eye including two sets of transparent fluid (the aqueous and vitreous humors) which nourish eye tissues and help maintain constant ocular shape. The eye is comprised of three layers: sclera (outer protective white coating); choroid (middle layer containing blood vessels which nourish the eye); and the retina (an inner layer containing the photoreceptors (rods and comes) which absorb and convert the light into electrical signals, plus the optic nerve which channels information to the higher centers in the brain for processing. At the front of the eye is the spongy iris, an extension of the choroid and the colored portion of the eye; the pupil (allows light into the eye) and the cornea, which is transparent and bends light rays. The convex lens focuses light rays onto the retina.

The capacity by which cells acquire energy from foods and nutrients eaten and use it to build, store, break and eliminate substances in controlled ways through enzymatic reactions with the help of cofactors (inorganic metal ions and coenzymes or vitamins) is called metabolism. Although the body can tolerate significant errors in metabolic processes, several diseases are caused by errors in metabolic pathways. For example, disorders will occur if a crucial enzyme is disabled, or if a control mechanism for a metabolic pathway is affected. Some diseases arise from inherited traits caused by gene mutations resulting in non-functional enzymes; others involve mutations in regulatory proteins and in transport mechanisms.

Digestive, Ear/Nose/Throat, and Eye Disorders

15.2 CYSTIC FIBROSIS (CF)

CF is considered the most lethal autosomal recessive disorder affecting primarily those of European descent, although it has been reported in all races and ethnicities. CF results in abnormally viscous secretions in the airways of the lungs and in the ducts of the pancreas which cause obstructions leading to inflammation, tissue damage and destruction of both organ systems. Sweat glands, the biliary duct of the liver, the male reproductive tract and the intestine are also often affected. (Cutting, 2015; Deignan & Grody, 2016; Maiuri et al, 2017)

Symptoms

CFTR-related diseases encompass well-known pathological disorders that appear to be influenced by CFTR genotype, including allergic bronchopulmonary fungal infection (by species of the genus *Aspergillus*), unusual stretching and widening of airways, and inflammation of the sinus membranes (**chronic rhinosinusitis**, CRS). When present, the symptoms of CRS often include excessive nasal discharge, nasal obstruction, mouth breathing, headache, poor olfaction, and restless sleep. (Nishioka et al., 1995; Chaaban et al., 2013; Cutting, 2015; Deignan & Grody, 2016; Maiuri et al, 2017)

Genetic Basis

The underlying genetic basis of the disease is related to dysfunction or deficiency of the CF transmembrane conductance regulator (CFTR), an apical membrane anion channel (like chloride or bicarbonate) present in respiratory and exocrine glandular epithelium (Figure 1). Nearly 2,000 variants have been

Figure 1. The genetic basis of cystic fibrosis
Source: Image used under license from Shutterstock.com

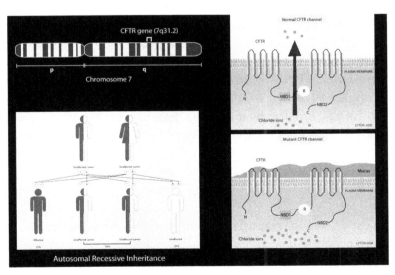

reported, among which 40% are predicted to cause substitution of a single amino acid, 36% are expected to alter RNA processing (including nonsense, frameshift and mis-splicing variants), 3% involve large rearrangements of *CFTR,* and 1% affects promoter regions; 14% seem to be neutral variants, and the effect of the remaining 6% is unclear (Cutting 2015).

There are six categories of mutations described depending on their disease-causing mechanism: 1) Class I mutations result in the absence of CFTR gene synthesis. 2) In class II, the CFTR gene is normally transcribed and translated, but results in a malformed protein folded incorrectly—thus recognized as defective in the endoplasmic reticulum during intracellular trafficking. 3) Class III mutations have the complete CFTR protein present in normal quantities at the cell surface, but it is not transported due to defaults in the chloride transporter (no ion channel activity). 4) Class IV mutations result in abnormalities in chloride conductance. 5) Class V mutations produce decreased quantities of CFTR transcripts and, thus, fewer functional CFTR channels at the cell surface. 6) Lastly, class VI mutations create defects in the stability of the protein leading to accelerated turnover at the cell surface and insufficient quantities of CFTR under steady-state conditions. (Rowe et al., 2005; Chaaban et al., 2013; Cutting, 2015; Deignan & Grody, 2016; Maiuri et al, 2017)

Current Therapies

Medical management consists of nasal saline irrigations as well as medications including antibiotics, decongestants, **antihistamines**, topical and systemic steroids, drugs such as Pulmozyme®, and *N*-acetyl cysteine as well as surfactant lavage. Due to increased knowledge of the CFTR gene, other novel therapies have been developed. For example, small molecules identified by high throughput drug screening can restore activity to the mutant CFTR protein including three drugs available in clinical settings: *ivacaftor*, *lumacaftor*, and *ataluren* (Rowe et al., 2007). Surgical management of the sinuses in CF patients is also thought to improve pulmonary outcomes and is used to justify intervention in asymptomatic individuals. With the growing understanding of the genetic and environmental modifiers of CF, personalized medicine as variant specific therapy is deployed is becoming a possibility. (Chaaban et al., 2013; Cutting, 2015; Deignan & Grody, 2016) Recently, new avenues have been explored to correct the *CFTR* gene defect by genome editing (*DNA and RNA editing* techniques). DNA editing uses engineered nucleases to remove mutated segments of the gene followed by homologous recombination with the wild-type gene. In RNA editing, single-strained antisense RNA-based oligonucleotides replace deleted mRNA segments, leading to the translation of repaired RNA into wild-type CFTR protein. Proteostasis regulators are alternative approaches to circumvent CFTR defect by improving defective proteostasis to avoid unwanted protein-protein interactions and reinstating desirable interactions for misfolded CFTR variants. (Maiuri et al, 2017)

15.3 DIABETES

The American Diabetes Association estimates that about 1.25 million Americans have type-1 diabetes and an estimated 40,000 people will be newly diagnosed each year. Type-1 represents only 5% of people with diabetes have this form of the disease (ADA, 2018a). Type-2 diabetes mellitus affects more than 400 million people around the world and in 2040, there will be more than 640 million people with diabetes worldwide (Marín-Peñalver et al., 2016). Type-1 diabetes (T1D) and type-2 diabetes (T2D) are of have multifactorial causes and genetic predisposition plays a key role. Only about 1–5% of all cases of diabetes result from single-gene mutations (monogenic diabetes) including neonatal diabetes mellitus and maturity-onset diabetes of the young. Neonatal diabetes presents during the first six months of life and can persist throughout life or it can be transient and disappear during infancy, often reappearing later in life. The biological risk factors for type-2 diabetes (which is also referred to as adult-onset diabetes) include body mass index, body fat distribution, brown adipose tissue, metabolic syndrome, adipokines, new biomarkers, imbalance of sex hormones, prediabetes, and gestational diabetes mellitus. The psychosocial risk factors include socioeconomic status, psychosocial stress, sleep deprivation and work stress. Modifiable social factors, like low educational level, occupation, and income, largely contribute to unhealthy lifestyle behaviour and social disparities and thus are related to higher risk of obesity and type-2 particularly in women. The health behavior risks include lifestyle (nutrition and physical activity), sugar-sweetened beverages, alcohol, and cigarette smoking. Research indicates that for reasons less understood, receptors in the cell membrane somehow lose the ability to receive insulin signals in the cell signal-reception process (Figure 2). Initially, the pancreas increases insulin production to try to get glucose into cells, but eventually, it cannot keep up and glucose levels build-up in the blood. (Steck & Winter, 2011; Basile et al., 2014; Yang & Chan, 2016)

Symptoms

According to the American Diabetes Association there are several signs and symptoms of both types of diabetes. These include extreme weakness and/or fatigue, extreme thirst (dehydration), increased urination, abdominal pain, nausea and/or vomiting, blurry vision, slow-healing sores, cuts or wounds, irritability or quick mood changes, changes to (or loss of) menstruation in women, unexplained weight loss, rapid heart rate, reduced blood pressure (falling below 90/60), low body temperature (below 97°F), dry mouth, decreased/blurred vision, headaches, loss of consciousness (rare), numbness and tingling of the hands and feet (ADA, 2018b).

Figure 2. Cell signal-reception process for the hormone insulin
Source: Image used under license from Shutterstock.com

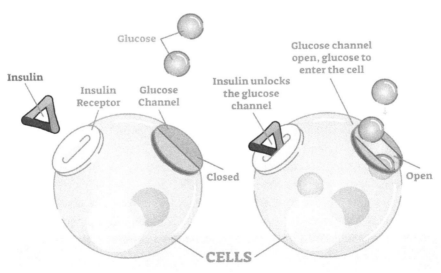

Genetic Basis

Both neonatal and maturity-onset diabetes of the young result from mutations of genes for transcription factors or other proteins that regulate endocrine pancreas development or function. Maturity-onset diabetes of the young presents before the age of 25 years and is caused by mutations in several related genes inherited in an autosomal dominant manner. Mutations in hepatocyte nuclear factor 1A (*HNF1A*) and glucokinase (*GCK) are the* two most common forms. It is estimated that more than 50 and 90 susceptibility gene loci, respectively, are found to be associated with type-1 and type-2 diabetes. The HLA class II region of the major histocompatibility complex confers major genetic risk for type-1. Transcription factor 7-like 2 (*TCF7L2*) is described as among the most susceptible variant for type-2. Despite having similar clinical and pathological similarities, genome-wide association studies (GWAS) indicate that susceptibility loci for both types are different. The shared susceptible loci for both types are: *INS, GLIS3, RASGRP1, COBL, RNLS,* and *BCAR1*. From GWAS studies, approximately one-third of monogenic diabetes genes are associated with type-2, but not type-1 including *HNF1A, HNF4A, PDX1, PAX4, WFS1, ABCC8, SLC2A2, HNF1B, KCNJ11, NEUROD1/BETA2, GCK,* and *PPARG*. *(Marín-Peñalver et al., 2016;* Yang & Chan, 2016)

Current Therapies

Oral agents for the treatment of diabetes include *metformin*, sulfonylureas, meglitinides, alpha-glucosidase inhibitors (*acarbose, miglitol, voglibose*), thiazolidinediones (*sosiglitazone, pioglitazone*), dipeptidyl peptidase-4 inhibitors (*sitagliptin, vildagliptin, saxagliptin, linagliptin, alogliptin*), and sodium glucose co-transporter-2 inhibitors (*dapagliflozin, canagliflozin, empagliflozin*). Injectable agents include RA-GLP1 (*exenatide, liraglutide, lixisenatide, albiglutide, dulaglutide*) and insulin. Other treatments used are *colesevelam, bromocriptine,* and *pramlintide*. The treatment of type-2 in older patients emphasises lifestyle modification (nutrition, physical activity, not smoking). Since all anti-diabetic agents have adverse effects, and expensive, the investigation of novel antidiabetic regimens, with less adverse effects and cheaper, is a major challenge for researchers. Future treatments focus on the use of polyphenols, smart insulin patch, dual-acting peptides, basal insulin analogues, G-protein-coupled receptor 119, oral RA-GLP1, oral insulin, and dual inhibitor *sotagliflozin*, Technosphere insulin, New chitosan formulations of xanthine derivatives, zinc, an insulin mimetic, *Imeglimin,* G protein coupled receptor 40, and hot water extracts of *Salacia chinensis*. (Overton et al., 2008; Pirags et al., 2012; Eldor et al., 2013; Gowda et al., 2013; Morikawa et al., 2015; Marín-Peñalver et al., 2016; Solayman et al., 2016)

15.4 GLUCOSE-GALACTOSE MALABSORPTION (GGM)

The primary transporters that mediate transcellular movements of d-glucose in small intestine have been identified and include the Na^+-d-glucose cotransporter SGLT1 and GLUT2. GGM is a genetic disorder caused by a defect in this sugar transport across the intestinal mucosa. The active process is medicated by a group of genes called Na/glucose co-transporter (SGLT) which are present in the small intestine and renal tubules. (Wright, 1998; Wright et al., 2002; Chedane-Girault et al., 2012)

Digestive, Ear/Nose/Throat, and Eye Disorders

Symptoms

Infants with GGM suffer from chronic, profuse, watery diarrhoea of infectious origin that often leads to hypertonic dehydration. This can be alleviated by oral rehydration therapy.

Genetic Basis

An SGLT1 missense mutation underlies hereditary glucose/galactose malabsorption. Wright et al. (2002) study indicated that the removal of SGLT1 in mice resulted in GGM syndrome that could be cured by a glucose-galactose–reduced diet as has been observed in patients with loss-of-function mutations in SGLT1. (Wright, 1998; Wright et al., 2002)

Current Therapies

Infants can be treated successfully with a carbohydrate-free infant formula with fructose added incrementally to meet energy requirements. Parental education about dietary management of CGGM with specialized formula supplemented with fructose and solid food feedings is also a recommended component of this infant's nutrition therapy. Aggressive nutrition intervention for the infants and judicious dietary counselling for parents can lead to normal growth and neurological development for an infant with GGM. (Abad-Sinden et al., 1997; Wright, 1998; Wright et al., 2002; Al-Lawati & Vargees, 2008; Chedane-Girault et al., 2012)

15.5 HEREDITARY HEMOCHROMATOSIS (HH)

HH define a group of autosomal recessive genetic disorders characterized by iron accumulation in parenchymal organs, primarily the liver, which can potentially result in impaired organ structure and function. It is considered one of the most common hereditary diseases in population of Caucasian origin. The key hormone that controls the saturation of plasma transferrin with iron (transferrin saturation) is hepcidin, and its deficiency causes increased transferrin saturation, a signature unifying feature of all forms of HH. There are other genetic causes of severe iron overload, distinguished from HH by their distinct clinical and pathological presentation which include iron-loading anaemias due to ineffective erythropoiesis or due to **haemolysis**, defects in iron acquisition by the erythroid precursors or repeated blood transfusions (Britton et al., 2002; Porto et al., 2016; Brissot et al., 2017).

Symptoms

Early symptoms of HH are non-specific and may include weakness, lethargy, weight loss and joint pain. Other suggestive signs of more advanced disease include skin pigmentation, liver cirrhosis, calcification of joints, arthritis, diabetes, diminished gonadal activity, and liver cancer.

Genetic Basis

Five types of HH have been identified on the basis of clinical, biochemical, and genetic characteristics. The genes involved are: HFE (encodes HFE protein); HJV (encodes hemojuvelin, a protein involved in iron balance); HAMP (encodes hepcidin); TFR2 (encodes transferrin receptor 2); and SLC40A1 (encodes iron exporter ferroportin). Defects in the hepcidin gene (HAMP) and in genes that stimulate hepcidin production such as HFE, HJV and TFR2 are associated with iron overload (hemochromatosis); with the most common, well-defined and prevalent form being HFE-related. Other less common hereditary forms of iron overload have been recognized and are designated as non-HFE-related HH. Affected HH individuals are usually either homozygous for a single G-to-A mutation resulting in cysteine to tyrosine amino acid substitution at position 282 (C282Y) of the HFE gene identified in up to 80% of study patients. (Brissot & de Bels, 2006; King & Barton, 2006; Zlocha et al., 2006; Brissot et al., 2017)

Current Therapies

Early diagnostic and initiation of iron-depletion therapy prior to the onset of cirrhosis can reduce morbidity and ensure life quality and increase survival times of HH patients. Preventive therapy, focused on family screening, and hepcidin supplementation (for hepcidin deprivation-related HH) remains a key part of the management of HH (Brissot et al., 2017).

15.6 OBESITY

Obesity is defined as the accumulation and storage of excess body fat, while overweight is weight in excess of a weight reference standard (Ogden et al., 2010). Most of the world's population live in countries where overweight and obesity kills more people than underweight. Worldwide obesity has nearly tripled since 1975 and in 2016, more than 1.9 billion adults, 18 years and older, were overweight. Of these over 650 million were obese with 39% of adults aged 18 years and over overweight in 2016, and 13% obese. About 41 million children under the age of 5 were overweight or obese in 2016 and over 340 million children and adolescents aged 5-19 were overweight or obese in 2016. (WHO, 2017) In the United States, approximately 70% of adults and 32% of children are overweight or obese, with a continuing rising trend. To date, existing policies and interventions have not reversed these trends, suggesting that new innovative approaches are needed to transform obesity prevention and control. (Yee et al., 2017) Paediatric overweight and obesity are of concern because of both immediate and later onset health consequences. Obesity in childhood almost always likely leads to adult obesity, and poor health outcomes throughout adulthood; therefore, a major contributor to many preventable chronic diseases and other causes of morbidity. (Daniels, 2009; Freedman et al., 2007; Yanovski, 2015)

Evidence available suggests that multiple factors acting at different scales may be contributing to obesity, including genetics, biology (leptin sensitivity and individual metabolism), individual behaviours (dietary and physical activity choices), social network dynamics (connections to family and friends, which may also influence individual behaviours), the environment (food availability, green spaces for physical activity, and neighbourhood safety), and larger societal forces (economics, policy, education, health awareness, and culture) (Yee et al., 2017). The American Medical Association (AMA) recognizes

Digestive, Ear/Nose/Throat, and Eye Disorders

obesity as a complex, chronic disease that requires medical attention and a disease which is a very public process largely driven by expectation of costs and benefits. The public is slowly getting a broader awareness of factors beyond personal choice influencing obesity and the AMA recognition will help bring more access to care, less blame on people with the condition, and advancing evidence-based approaches for its prevention and treatment (Kyle et al., 2016).

Many medications have been associated with weight gain including insulin secretagogues, glucocorticoids, antipsychotics, mood stabilizers, antidepressants, anticonvulsants, antihypertensives, antihistamines, and chemotherapeutic agents. Sociocultural and ecological environments also play a major role in determining who becomes obese. For example: 1) Arizona Pima Indians who live on a reservation have much higher rates of obesity and diabetes than their genetically-related counterparts in isolated Mexican villages; and, 2) Asian and Hispanic adolescents born in the United States have a higher prevalence of obesity than immigrant members of the same community. The differential response of people to the same environmental conditions can lead to obesity as a result of epigenetic changes which are alterations in gene expression related to disease risk that are modified by the environment during development. (Doche et al., 2012; Maayan & Correll, 2012; Wheeler et al., 2013; van Dijk et al., 2015)

Symptoms

According to the National Heart, Lung and Blood Institute (NHLBI), overweight and obesity have no specific symptoms with the typical signs of overweight and obesity being a high body mass index (BMI) and an unhealthy body fat distribution which raises the waist circumference. Obesity can however cause a variety of complications in many parts of your body which include the following: metabolic syndrome; type-2 diabetes; high blood cholesterol and high triglycerides; high blood pressure; atherosclerosis;, heart attacks and stroke; obstructive sleep apnoea; asthma; obesity hypoventilation syndrome; back pain; non-alcoholic fatty liver disease; osteoarthritis; urinary incontinence; weakened pelvic muscles; gallbladder disease; emotional health issues (such as low self-esteem or depression especially in children); and a variety of cancers (oesophagus, pancreas, colon, rectum, kidney, endometrium, ovaries, gallbladder, breast, or liver). (NHLBI, 2018)

Genetic Basis

Obesity is also just as clearly a genetic disease, because all available data suggest that 60–80% of the observed variance in human body weight can be accounted for by inherited factors. More than 300 genetic loci that are potentially involved in human body weight regulation have been identified through analyses in humans, rodents, and *Caenorhabditis elegans*. Rare inactivating mutations affecting genes in the leptin signalling pathway may account for as much as 3 or 4% of severe, early-onset obesity. Inactivating mutations affecting both alleles of the genes coding for leptin, the leptin receptor, pro-opiomelanocortin (POMC), and enzymes that process POMC such as prohormone convertase-1 have been found in paediatric patients. Abnormalities causing inactivation in genes affecting leptin receptor signal transduction, such as SH2B1 are also associated with obesity and neurocognitive defects. *(*Ashrafi et al., 2003; Rankinen et al., 2005; Farooqi et al., 2006; Wheeler et al., 2013; Yanovski, 2015)

Homozygous and heterozygous inactivating mutations in the melanocortin 4 receptor (MC4R) cause obesity and hyperphagia during childhood, and are the most common known cause of severe, early onset obesity accounting for as many as 3% of such children. Rare mutations in MRAP2, a protein essential for MC4R function, are also associated with paediatric obesity. Some data also support a role for polymorphisms in the MC3R for regulation of body weight, particularly in African American children. Brain derived neurotrophic factor (BDNF) is believed to function downstream from MC4R in the leptin signalling pathway. A heterozygous inactivating mutation in the gene coding for the BDNF receptor has been associated with obesity, seizures, and developmental delay. (Farooqi et al., 2000; Yeo et al., 2004; Savastano et al., 2009; Yanovski, 2015)

Although the mechanisms explaining how identified gene regions might change energy balance are often not fully understood, single nucleotide polymorphisms (SNPs) and copy number in genes and chromosomal regions are reported to be associated with body weight or body composition. An almost universally replicated finding is linkage of the *FTO* gene locus with body weight. *FTO* mRNA is highly expressed in brain areas important for regulation of energy- and reward-driven consumption. Some limited data also suggest that children with less common alleles in *FTO* may have greater food intake, reduced satiety, and greater prevalence of loss of control over their eating. (Ravussin et al., 1994; Savastano et al., 2009; Doche et al., 2012; Wheeler et al., 2013; Yanovski, 2015)

Current Therapies

There are a number of indications that the obesity epidemic is a systems problem, as opposed to a simple problem with linear cause-and-effect relationships, therefore requiring a systems approach (Gortmaker et al., 2011; Bures et al., 2014). Multicomponent interventions that included culturally targeted and tailored components (e.g., health education, diet, physical activity, reduced screen time, behavioural skills, and motivation groups) have shown a greater impact on health behaviours associated with obesity in minority youth than the individual components in isolation (Wilson, 2009). According to WHO (2017), overweight and obesity are largely preventable but supportive environments and communities are fundamental in shaping people's choices, by making the choice of healthier foods and regular physical activity the easiest choice (the choice that is the most accessible, available and affordable), and therefore preventing overweight and obesity. Individuals can limit energy intake from total fats and sugars; increase consumption of fruit and vegetables, as well as legumes, whole grains and nuts; and engage in regular physical activity (60 minutes a day for children and 150 minutes spread through the week for adults).

Sustained implementation of evidence-based and population-based policies which make regular physical activity and healthier dietary choices available, affordable and easily accessible to everyone (particularly to the poorest individuals), makes individual responsibility in creating a healthy lifestyle much easier. The food industry can play a significant role in promoting healthy diets by reducing the fat, sugar and salt content of processed foods; ensuring that healthy and nutritious choices are available and affordable to all consumers; restricting marketing of foods high in sugars, salt and fats, especially those foods aimed at children and teenagers; and ensuring the availability of healthy food choices and supporting regular physical activity practice in the workplace. (WHO, 2017)

Digestive, Ear/Nose/Throat, and Eye Disorders

15.7 WILSON'S DISEASE (WD)

A rare, inherited autosomal recessive disease of copper metabolism, Wilson's disease results in excessive copper deposition in the body first described by Kinnier Wilson in 1912 (Schilsky, 1996). Copper plays an essential role in many metabolic processes including serving as a cofactor for many important enzymes. In the small intestines dietary copper is absorbed through various metal transporters and is exported into the portal circulation by copper-transporting P-type ATPase. It is estimated that approximately 30 individuals per million people are affected worldwide. (Frydman, 1990; Bull et al., 1993; Schilsky, 1996; Kathawala & Hirschfield, 2017)

Symptoms

Wilson's disease may present symptomatically at any age, although most cases are between the ages of 5–35 years. Wilson's disease can present with liver disease varying on a spectrum ranging from acute liver failure, asymptomatic with only biochemical abnormalities, to liver cirrhosis. Individuals with WD can present with several conditions including hepatic (acute hepatitis, haemolysis, and jaundice); neurological (dystonia, pseudosclerosis, **ataxia,** drooling, migraines and insomnia); and psychiatric (academic decline, changes in behaviour with loss of inhibitions, hypersexuality and mild cognitive impairment, depression, anxiety and overt psychosis). Other possible (but infrequent) presentations include **osteoporosis, osteoarthritis,** chondrocalcinosis, hypercalciuria and nephrocalcinosis, **cardiomyopathy** and arrhythmias, gigantism, hypoparathyroidism, infertility and repeated miscarriages. (Chen et al., 2015; Kathawala & Hirschfield, 2017)

Genetic Basis

WD arises from a defective gene identified as *ATP7B* on chromosome 13 (13q14.3), which encodes a copper-transporting P-type ATPase transmembrane protein expressed mainly in hepatocytes. The gene can have over 500 mutations with missense or nonsense mutations very common (60%), followed by insertions/deletions (26%), and splice-site mutations (9%). Modifiers are a group of genes that aggravate or relieve the phenotypes of other virulence genes. Apolipoprotein E (*ApoE*) gene, located in 19q13.2, is a modifier that has the strongest correlation with WD. (Wang et al., 2003; Coffey et al., 2013; Chen et al., 2015)

Current Therapies

Currently, the mainstay of therapy for Wilson's disease remains lifelong drug therapy usually with orally active copper chelators which de-copper the body promoting copper excretion; the use of zinc which reduces intestinal copper absorption in the intestinal mucosa, or both. Chelators used include *dimercaprol, penicillamine, trientine,* and *ammonium tetrathiomolybdate*. For those with acute liver failure, unresponsive to medical therapy, or at end-stage liver disease, liver transplantation must be considered (Kathawala & Hirschfield, 2017).

Dietary management can be combined with other therapies since numerous foods such as chocolate, shellfish, nuts, mushrooms, liver and soy have high concentrations of copper naturally. These foods must be avoided especially in the first year of treatment for Wilson's disease. Antioxidants supplements such as vitamin E or N-acetylcysteine, may have a role as adjunctive therapy. Physiotherapy to maintain muscle function and management is recommended after initiation of anti-copper therapy. Occupational and speech therapies have also been useful in identifying and eliminating environmental barriers to independence and participation in activities of daily living; and optimizing any potential outcomes on recovery following medical therapy. (Litwin et al., 2012; Coffey et al., 2013; Chen et al., 2015)

15.8 ZELLWEGER SYNDROME (ZS)

ZS is an autosomal recessive inherited disorder of the peroxisome, an intracellular organelle composed of a single membrane containing a matrix embedded with over 70 distinct enzymes required for normal lipid metabolism and a host of other biochemical processes critical for normal health and development. The worldwide prevalence of ZS is estimated between 1:50,000 and 1:100,000, with reports of higher incidences some Quebec regions in Canada (Levesque et al., 2012; Lee & Raymond, 2013; Braverman et al., 2016).

Symptoms

Poor development and growth (few developmental milestones achieved), hearing loss, visual impairment (retinal dystrophy, optic nerve abnormalities, cataracts, glaucoma), neurological (hypotonia, neonatal seizures), hepatic dysfunction, skeletal abnormalities, and enamel abnormalities of permanent teeth.

Genetic Basis

The proper assembly of a peroxisome requires a unique set of proteins called peroxins which help incorporate enzymes into the forming peroxisome's matrix. A mutation in a peroxin, (*PEX* gene) produces a reduced or non-functioning peroxin which results in incomplete peroxisomes fail to perform their metabolic duties. There are 16 known human *PEX* genes, and disease-associated mutations have been identified in 13 of these genes. Mutations in *PEX1* account for nearly 70% of all PBD-ZSD cases, with another 26% of cases caused by mutations in *PEX6*, *PEX10*, *PEX12*, or *PEX26*, with the majority of these cases involving *PEX6* mutations (Lee & Raymond, 2013; Braverman et al., 2016).

Current Therapies

Treatment strategies currently being pursued include high-content screening of large chemical libraries for compounds that improve peroxisome assembly and function, as well as gene and cellular therapies. Betaine has potential as a **molecular chaperone** that can improve peroxisome assembly in cultured cells ZS patients with *PEX1* p.G843D mutations. Candidate drug screens identified arginine as another potential molecular chaperone in patient cell lines. Other potential therapeutic opportunities transplant of cell types and cell lineages affected ZS patients.

Digestive, Ear/Nose/Throat, and Eye Disorders

Additional therapies for ZS include dietary restriction of phytanic acid; vitamin K supplements; bile acid therapy (cholic acid) due to altered liver metabolism; use of hearing aids and cochlear implants; periodic ophthalmologic evaluations and use of glasses; regular EEGs screening; and use of common medications to control seizures in children including *levetiracetam, phenobarbital, clonazepam, topiramate,* and *lamotrigine*. Due to skeletal abnormalities, dual-energy X-ray absorptiometry and use of vitamin D is recommended. Future directions in ZS management is coming from research insights gained from genetically engineered mouse models of *PEX* gene defects, and a host of invertebrate models including genetically engineered worms, fruit flies, and zebrafish. (Levesque et al., 2012; Lee & Raymond, 2013; Hiebler et al., 2014; Klouwer et al., 2015; Braverman et al., 2016; Gonzalez et al., 2018)

15.9 DEAFNESS OR HEARING LOSS (HL)

Hearing loss affects about 70 million people worldwide, with about 50%–60% of these cases having a genetic cause; and the remaining 40%–50% of cases attributed to environmental factors such as ototoxic drugs, prematurity, or trauma. Approximately one in every 1,000 children has some form of hearing impairment (Morton, 1990), and one in 2,000 cases is caused by a genetic mutation. A classic example of multi-locus genetic heterogeneity, genetic hearing loss is difficult to study genetically because many different genetic forms have similar clinical phenotypes, plus mutations in the same gene can result in a variety of clinical phenotypes. Cytomegalovirus infection remains the most common environmental cause of congenital HL. However, increasing prevention of environmental and prenatal causes of HL is making the balance shift in favour of genetic origins. (Smith et al., 2005; Mittal et al., 2018)

Syndromic HL is associated with distinctive clinical features and accounts for 30% of hereditary HL, whereas non-syndromic HL accounts for the other 70%. Of the more than 400 syndromes in which HL is a recognized feature, three are the most frequent conditions: Usher syndrome, Pendred syndrome, and Jervell and Lange-Nielson syndrome. (Korver et al., 2017; Mittal et al., 2018)

Symptoms

The American Academy of Pediatrics listed the risk factors for hearing loss to be: hearing, speech, language or developmental delay; family history of hearing loss; neonatal intensive care unit stay >5 days or receiving treatments (extra corporal membrane oxygenation, assisted ventilation, ototoxic drugs [like gentamycin and tobramycin], loop diuretics, or exchange transfusion for **hyperbilirubinaemia**); in utero infections (toxoplasmosis, rubella, cytomegalovirus, herpes simplex or syphilis); craniofacial anomalies; physical findings associated with a syndrome known to cause permanent hearing loss; syndromes associated with congenital hearing loss or progressive or late-onset hearing loss; neurodegenerative disorders or sensorimotor neuropathies; confirmed bacterial or viral meningitis (in particular if caused by mumps, herpes viruses or virus); head trauma, especially of the basal skull, or temporal bone fractures, that require hospitalization; and chemotherapy (AAP, 2007).

The Mayo Clinic reports that the signs and symptoms of hearing loss can include: muffling of speech and other sounds; difficulty understanding words (especially against background noise or in a crowd of people); trouble hearing consonants (frequently asking others to speak more slowly, clearly and loudly); needing to turn up the volume of the television or radio; withdrawal from conversations; and avoidance of some social settings (Mayo Clinic, 2018).

Genetic Basis

Approximately 80% of genetic deafness is non-syndromic (not associated with other clinical features), and autosomal recessive forms account for 60%–75% of the cases. Of the remaining cases, 20%–30% show autosomal dominant inheritance and about 2% are either X-linked or of mitochondrial origin. More than 125 loci have been discovered: 54 autosomal dominant, 71 autosomal recessive, five X-linked, two modifier, and one Y-linked and are all involved in cochlea function. Most deafness genes discovered so far are protein-coding genes which include ion channels, gap junctions, membrane transporters, transcription factors regulating gene expression, adhesion proteins, extracellular matrix proteins, unconventional myosins, and cytoskeletal proteins. Mutations in tRNA- and rRNA-coding genes and intron region of chromosomes have also been found to cause deafness. (Everett et al., 1997; Friedman et al., 2002; Pandya et al., 2003; Wilch et al., 2006; Angeli et al., 2012; Mittal et al., 2018)

For autosomal recessive HL, the most frequent causative genes in order of frequency are *GJB2*, *SLC26A4*, *MYO15A*, *OTOF*, *CDH23*, and *TMC1*. For each of these genes, at least 20 mutations have been reported. Autosomal dominant common mutations include *WFS1*, *MYO7A*, and *COCH*. Several of these genes are also implicated in syndromic HL. (Wilch et al., 2006; Angeli et al., 2012)

Current Therapies

Emerging therapies for hearing loss can be broadly sub-classified as hearing preservation or hearing restoration strategies. Therapeutic non-surgical management of pathogen-associated hearing loss currently is focused on two key areas of intervention: specific antimicrobial therapies, and anti-inflammatory therapies. Both therapies are meant to mitigate the host's immune response to infection, thereby reducing the damage to the cochlea. With increased understanding of infectious disease-related hearing loss, new therapies are on the horizon including the use of free radical scavengers, anti-oxidants, and nanoparticle-based systems. Surgical treatment might be beneficial in select cases in which there is an air-bone gap that is amenable to correction by surgical intervention. Non-medical support including special education and sign language are also options. Restoration of hearing is achieved by implantable or non-implantable hearing devices, including conventional hearing aids, cochlear implants and bone-anchored hearing aids. (Angeli et al., 2012; Korver et al., 2017; Mittal et al., 2018)

15.10 PENDRED SYNDROME

Pendred syndrome is a recessively inherited disorder characterized by diffuse goitre growth associated with sensory-neural deafness first described by Vaughan Pendred in 1896 after observing a significant correlation between goitre and congenital hearing loss. Often, deafness can be extreme following traumatic events, infection, or episode of endolymphatic oedema. The thyroid glands of affected individuals cannot metabolize iodide efficiently, and studies indicate that up to 10–80% of iodine taken up by the gland is discharged after administration of perchlorate (Bigozzi et al., 2005; Sreekar et al., 2013).

Digestive, Ear/Nose/Throat, and Eye Disorders

Symptoms

Pendred syndrome is typically associated with hearing loss and an enlarged thyroid gland—a butterfly-shaped organ at the base of the neck which releases tyrosine-based hormones (**triiodothyronine** or **T3**, and **thyroxine** or **T4**) primarily responsible for regulation of metabolism. People with the Pendred syndrome develop goitre between late childhood and early adulthood. Often, the enlargement does not cause the thyroid to malfunction. Pendred syndrome individuals have severe to profound hearing loss caused by changes in the inner ear at birth. Affected individuals also have problems with balance caused by dysfunction of the vestibular system—the part of the inner ear maintaining balance and orientation. In addition, the vestibular aqueduct, a bony canal that connects the inner ear with the inside of the skull is abnormally enlarged; and the cochlea is abnormally developed and shaped.

Genetic Basis

Pendred syndrome is inherited as an autosomal recessive trait mapped to the gene *SLC26A4* (PDS) coding for a protein "pendrin" on chromosome 7 (7q31). Originally suggested to be a sulphate transporter studies indicate that the protein functions as a membrane-bound chloride-iodine transporter. At least 70 different mutations capable of causing the syndrome have been identified to date. To better understand the role of the pendrin gene in development (SLC26A4), experiments using *"knockout mice"* for the pendrin gene were performed to study mice embryonic and postnatal development. By the 15th day, there was significant expansion of the endo-lymphatic duct and sac and mice showed severe balancing problems running around in circles, in a constant unidirectional sense and displayed severe persistent head tilting. After 30 days of life studies showed a severe cochlear sensory-neural cell degeneration. (Everett et al., 1997; Bigozzi et al., 2005)

Current Therapies

Since Pendred syndrome is incurable, health professionals work together to encourage informed choices about treatment options. Children with Pendred syndrome should start early treatment to gain communication skills, such as learning sign language or cued speech or learning to use a hearing aid. People experiencing significant hearing loss can be evaluated for a cochlear implant, which though cannot restore or create normal hearing, it bypasses injured areas of the ear to provide a sense of hearing in the brain. (NIDCD, 2018)

15.11 BEST DISEASE

The bestrophins were first identified in the human genome as a result of the association of *BEST1* mutations with Best vitelliform macular dystrophy (BVMD). Bestrophins are a family of proteins found throughout the animal kingdom and have been identified in virtually every organism studied. Mutations in the gene *BEST1* are causally associated with as many as five clinically distinct retinal degenerative diseases, which are collectively referred to as the **bestrophinopathies**. The most common of the bestrophinopathies is BVMD, otherwise known as Best disease inherited in an autosomal dominant fashion. (Hartzell et al., 2008; Xiao & Hartzell, 2010; Johnson et al., 2017) The prevalence of BVMD in the United States is currently not known; and in Europe, two studies estimated the prevalence of BVMD to be 2 in 10,000 (Sweden), while a study in Demark estimated this value at 1.5 in 100,000 (Nordström, 1974; Bitner et al., 2012).

Symptoms

Patients with Best disease display impaired vision with decreased visual acuity, abnormal colour vision, macular degeneration, and atrophy of the retinal pigment epithelium.

Genetic Basis

Mutations in *BEST1* have been found in association with clinically distinct retinal degenerative diseases which the HUGO nomenclature committee reassigned gene names as *BEST1* (VMD2), *BEST2* (VMD2L1), *BEST3* (VMD2L2), and *BEST4* (VMD2L3). The four human bestrophins function as calcium-activated anion channels. Over 200 mutations throughout the entire *BEST1* gene have been reported as causing the clinically distinct forms of retinal degeneration including BVMD, adult-onset **vitelliform macular dystrophy** (AVMD) autosomal recessive bestrophinopathy (ARB), autosomal dominant **vitreoretinochoroidopathy** (ADVIRC), and retinitis pigmentosa (RP). (Takahashi & Yamanaka, 2006; Xiao & Hartzell, 2010; Dalvin et al., 2017; Johnson et al., 2017)

Current Therapies

No current therapies or treatments exist for patients suffering from any bestrophinopathy. Treatment is symptomatic and involves the use of low vision aids, and direct laser treatment or **photodynamic therapy**. Newer treatment includes anti-VEGF agents *(bevacizumab)* and trans-corneal electrical retinal stimulation. A number of studies are exploring several treatment possibilities. First, Uggenti et al. (2016) report four autosomal recessive bestrophinopathy (ARB) mutants that failed to conduct Cl⁻ anions. Treatment with proteasome inhibitor 4-phenylbutyrate restored the chloride conductance of these mutant proteins. *Bortezomib* and 4-phenylbutyrate are clinically approved for long-term use as a therapy for ARB patients. Second, Singh et al. (2015) showed that treatment with the drug bafilomycin-A1 decreased photoreceptor outer segment degradation rates and this decrease was completely reversible by valproic acid. Since anomalies in photoreceptor outer segments are involved in the pathogenesis of the bestrophinopathies, these data suggest that valproic acid or similar compounds may be of use in the treatment of BVMD. Third, Guziewicz et al. (2013) showed gene therapy using recombinant adeno-associated virus-mediated gene transfer of *BEST1* is safe and feasible in a canine model. Advanced research in stem-cell transplant technology raises prospects that diseased or damaged tissue in BVMD patients could be replaced with healthy tissue safe from immune rejection. Replacing damaged or dysfunctional retinal pigment epithelium with healthy cells could alleviate or entirely cure BVMD, AVMD, ARB, ADVIRC, and RP. (Kimbrel & Lanza, 2015; Sugita et al., 2016; Dalvin et al., 2017; Johnson et al., 2017)

15.12 GLAUCOMA

Glaucoma is a neurodegenerative disorder of the optic nerve in which retinal ganglion cell death leads to characteristic patterns of visual field loss. Glaucoma is the second leading cause of blindness, accounting for 8% of blindness among the 39 million people who are blind world-wide. The second leading cause of blindness and a major source of morbidity and disability in the United States, glaucoma prevalence in the US was estimated to 2.1%, representing 2.9 million cases in the total population. This includes 1.4 million cases in women, 1.5 million cases in men, 2.3 million cases among people 60 years of age and older, and 0.9 million cases among people of colour. The prevalence of glaucoma was highest in African Americans, followed by non-Caucasian Hispanics, Mexican Americans, and other ethnicities (Gupta et al, 2016). In 2006, the number of individuals estimated to be bilaterally blind from glaucoma was projected to increase from 8.4 million in 2010 to 11.1 million by 2020 (Quigley & Broman, 2006; Rudnicka et al., 2006). While the African populations of the Caribbean, Africa and USA have the highest prevalence of open-angle glaucoma (OAG); it appears there are differences in the prevalence of glaucoma in different African populations in the Caribbean islands and within Africa, attributable to genetic diversity, environmental, and socio-economic factors. (Racette et al., 2003; Leske, 2007; Kyali et al., 2013)

Symptoms

Age, elevated **intraocular pressure** (or ocular hypertension), and family history are major risk factors for glaucoma. The two most common forms of glaucoma are primary **open-angle glaucoma (POAG)** and **angle-closure glaucoma (ACG).** Open-angle glaucoma is often called "the sneak thief of sight" because it has no symptoms until significant vision loss has occurred. There are typically no early warning signs or symptoms of open-angle glaucoma as it develops slowly and sometimes without noticeable sight loss for many years. However, for ACG symptoms include hazy or blurred vision, appearance of rainbow-colored circles around bright lights, severe eye and head pain, nausea or vomiting (accompanying severe eye pain), and sudden loss of sight. (GRF, 2018a)

Genetic Basis

Several genes that cause glaucoma-relevant phenotypes have recently been identified in mice. Mutations in the myocilin gene (*MYOC*) cause **ocular hypertension** and glaucoma. In humans and mice, MYOC is expressed in numerous ocular tissues, most notably in trabecular meshwork cells. (Menaa et al., 2011; Stone et al., 1997). Genetic linkage studies have identified three loci likely to contain genes contributing to congenital glaucoma: GLC3A (2p21), GLC3B (1p36), and GLC3C (14q22). Research in genomics has shown that several genetic loci are linked or associated with glaucoma including genes for congenital glaucoma (*CYP1B1* and *LTBP2*), developmental glaucoma (*PITX2*, *FOXC1*, *PAX6*, and *LMX1B*), juvenile-onset primary open angle glaucoma (*MYOC*), and familial normal-tension glaucoma (*OPTN* and *TBK1*). Although GWAS studies have identified genes/loci associated with glaucoma-relevant phenotypes, in some cases, it is not always clear how a gene or a specific allelic variant impacts the disease. For example: 1) while the gene *CDKN2B-AS1* is strongly associated with glaucoma, its function is unclear; 2) copy number variations in the gene *TBK1* are associated with glaucoma how the copy number variations of the gene impact glaucoma progression is unclear. Using GWAS, chromosome 9p21 has become a major focus of research with multiple genome-wide association studies suggesting associations to OAG. (Burdon et al., 2012; Burdon et al., 2011; Ng et al., 2014; Wang & Wiggs, 2014; Fernandes et al., 2015)

Current Therapies

Visual field loss in glaucoma is irreversible, so detection and treatment are essential to limit the progression of disease and delay additional optic nerve damage. According to the Glaucoma Research Foundation, the only current treatment for glaucoma is the lowering of intraocular pressure, achievable with eyedrops, laser, or incisional surgery. Eye drops are the most common treatment modality and, over time, patients may need to take multiple types of eye drops to stop progression of a disease that typically has no symptoms. (GRF, 2018b)

15.13 GYRATE ATROPHY

Gyrate atrophy of the choroid and retina was first described by Fuchs in 1896. More than 200 individuals with gyrate atrophy have been reported since it was first described in the late 19th century, with cases mainly reported from Finland, the US, Japan, and France (Moloney et al., 2014).

Symptoms

Ophthalmological manifestations such as high **myopia**, marked **astigmatism** and early cataract formation are common. Symptoms such as near-sightedness, night blindness, and loss of peripheral vision develop during childhood, with the field of vision progressively narrowing resulting in tunnel vision. These vision changes may lead to blindness by about the age of 50. While most people with gyrate atrophy of the choroid and retina have no symptoms other than vision loss, neonatal hyperammonemia, neurological abnormalities, intellectual disability, peripheral nerve problems, and muscle weakness may occur. (NCATS, 2018)

Genetic Basis

Gyrate atrophy of the eye layers (choroid and retina) is metabolic disorder secondary to a congenital defect in the gene coding for the mitochondrial matrix enzyme *ornithine-δ-aminotransferase (OAT)*, causing hyperornithinemia. It is found in chromosome 10 and is transmitted in a recessive autosomal fashion. The 10-20-fold increase in levels of the amino acid ornithine lead to a progressive chorio-retinal atrophy. OAT converts the amino acid ornithine from the urea cycle into glutamate. The increase in plasma levels of ornithine greatly affect cellular function. Patients with GA initially complain of night blindness and decreasing of peripheral vision typically in the second decade of life, with total blindness usually occurring between ages 40-60. (Howden et al., 2011; Braham et al., 2018)

Digestive, Ear/Nose/Throat, and Eye Disorders

Current Therapies

Treatment focuses on dietary supplements and/or a specialized diet: low ornithine foods and supplementation of vitamin B6. Patients treated with a low arginine diet showed near normal blood ornithine levels (Kennaway et al., 1980; Wang et al., 2000). Human pluripotent stem cells, which include both human embryonic stem cells and induced pluripotent stem cells (iPS), hold the potential to differentiate into any cell type. Gene-corrected patient-specific iPS cells have been used for gene therapy in patients with gyrate atrophy (Howden et al., 2011; Meyer et al., 2011).

15.14 PATTERN BALDNESS

Pattern baldness (androgenic alopecia) is an androgen-mediated process in genetically susceptible individuals, and environmental and lifestyle risk factors are likely to be trigger. Although often called 'male pattern baldness' this condition can affect up to 70% of men and also 40% of women at some point in their lifetime (Santos et al., 2015). Pattern baldness has a prevalence of about 20% at age 20–30, with the incidence growing at 10% per decade. It affects about 80% of men by the age of 80 years and has substantial psychosocial impacts due to changes in self-consciousness and societal perceptions (Yassa et al., 2011; Hagenaars et al., 2017). Male pattern baldness is often associated with higher levels of dihydrotestosterone and increased expression of androgen receptors; and has also been linked to prostate cancer (Li et al., 2016). Li et al. (2016) found that male-pattern baldness after age 45 was significantly associated with increased risk of both invasive squamous cell carcinoma and basal cell carcinoma, and was associated with melanoma at the head and neck region only. Studies have also positively associated male pattern baldness with androgens as well as insulin-like growth factor-1 (IGF-1) and insulin, all of which are implicated in pathogenesis of colorectal neoplasia (Keum et al., 2016). Lotufo et al. (2000) also found that male pattern baldness appears to be a marker for increased risk of coronary heart disease events, especially among men with hypertension or high cholesterol levels. Santos et al. (2015) note that classic androgenic hair loss in males begins above the temples and vertex, or crown, of the scalp; and as it progresses, only a thin rim of hair at the sides and rear of the head may remain—and it rarely progresses to complete baldness.

Symptoms

According to the Mayo Clinic (2018), the signs and symptoms of pattern baldness are gradual thinning on top of head; circular or patchy bald spots; sudden loosening of hair; full-body hair loss; and patches of scaling that spread over the scalp (ringworm) accompanied by broken hair, redness, swelling and at times, oozing.

Genetic Basis

DNA variants in or near the *AR* gene have been known to increase the risk of patterned hair loss in both men and women, but the exact identity of the causal DNA variant(s) has not been elucidated. Several gene variants have been linked to male-pattern baldness, with the *AR* gene showing the strongest association. A locus on chromosome Xq12 harbouring the androgen receptor gene (*AR*) and its neighbouring ectodysplasin A2 receptor gene (*EDA2R*) is known as the major locus for baldness. GWAS identified 247 independent autosomal loci and 40 independent X chromosome loci linked to baldness. Two genetic loci on chromosome 20p11 (*PAX1/FOXA2*) and 7p21.1 (*HDAC9*) were also identified to be involved in baldness. About 20 loci were found in autosomes and located in or near to genes that have been associated with, for example, hair growth/length in mice (*FGF5*), grey hair (*IRF4*), cancer of the breast (*MEMO1*), bladder cancer (*SLC14A2*), histone acetylation (*HDAC9*), and frontotemporal dementia (*MAPT*). A previous GWAS showed an association of *IRF4* with both hair colour and hair greying, but not with male pattern baldness. *HDAC9* has been identified as a baldness susceptibility gene in a previous study. Two of the top 10 X chromosome SNPs were located in *OPHN1*, a gene previously associated with X-linked poor mental development. (Hébert et al., 1994; Billuart et al., 1998; Rademakers et al., 2004; Sorokin & Chen, 2013; Hagenaars et al., 2017) MicroRNAs have been found to play a role in hair follicle morphogenesis, maintenance and cycling (controlling proliferation, growth arrest and apoptosis) and have been implicated in pattern baldness (Santos et al., 2015).

Current Therapies

Pattern baldness has limited effective therapeutic options—the effectiveness of most treatments relies on how early they are applied. Drug candidates largely fall into the classes of 5-α-R inhibitors, androgen receptor antagonists, ATP-sensitive potassium channel openers, topically growth factors, and a diverse selection of antioxidants and botanical extracts. As of 2016, there are only two FDA-approved current therapies for pattern baldness: topical *minoxidil* (Rogaine) and oral *finasteride* (Propecia). Several others are under testing or investigation including *pinacidil* (Pindac); *dutasteride* (Avodart), *bicalutamide* (Casodex), *flutaide* (Cytomid), *fluridil* (Eucapil) and others. These medications are competitive inhibitors of type II 5-α-R which inhibits the conversion of testosterone to dihydrotestosterone, which is involved in miniaturizing the hair follicles in pattern baldness. (Santos et al., 2015)

15.15 ALPORT SYNDROME (AS)

AS is a rare disease affecting fewer than one in 2000 individuals, although most of these estimates are from hospital-based series of mainly men, for many at-risk women may never have been tested. The syndrome is an inherited disease characterized by progressive renal failure, hearing loss, and ocular abnormalities, including corneal scarring, lenticonus, retinal thinning, and fleck retinopathy. The disease is 85% X-linked and 15% autosomal recessive. The X-linked AS is suspected clinically when men in the family are affected more severely and disease appears to "skip" a generation. Autosomal recessive inheritance is considered when AS affects only one generation, and cases where there is kinship.

Symptoms

The clinical features of AS include a classic triad of renal injury, sensorineural deafness and ocular abnormalities. These include **haematuria, proteinuria**, end-stage renal disease (*ESRD*), lenticonus, retinal thinning, and **retinopathy**. Rare manifestations include soft tissue tumours of the oesophagus and bronchus (leiomyomatosis), aortic **aneurysms**, and giant retinal holes. The "severe" phenotype with renal failure before the age of 30, hearing loss, and often lenticonus and retinopathy is more common with gene rearrangements, indels (insertions/deletions), nonsense, and splicing mutations. (Hertz et al. 2015; Savage et al., 2016)

Genetic Basis

X-linked AS is caused by mutations in the *COL4A5* gene which encodes the collagen IV α5 chain. The autosomal recessive disease is caused by two mutations on different chromosomes in the *COL4A3* or *COL4A4* genes which code for the collagen IV α3 and α4 chains respectively. The collagen IV α3, α4, and α5 chains form a network of the basement membranes of the glomerular filter, the cochlea, cornea, lens capsule, and retina. The collagen IV heterotrimer consists of a long series of glycine-proline-hydroxy-proline repeats, where glycine is present at each third residue. More than 2000 COL4A5 mutations have been described in X-linked Alport syndrome and more than 1000 in COL4A3 and COL4A4 in recessive disease. Mutations in Alport syndrome are different in each family and are missense (40%) or nonsense (40%), most often where a glycine is replaced by another amino acid, or a stop codon, respectively. (Barker et al., 1990; Mochizuki et al., 1994; Hudson et al., 2003; Hertz et al. 2015; Savage et al., 2016)

Current Therapies

Currently, there is no treatment available to cure AS and symptomatic renal protective therapies are currently the mainstay of treatment for this disease. Individuals with **albuminuria** and proteinuria should be treated with **RAAS blockade** (angiotensin-converting-enzyme inhibitors or angiotensin-receptor blockers), **angiotensin converting enzyme (ACE) inhibitors** as first lines of treatment; followed by angiotensin receptor blockers. Other strategies that will delay renal failure progression include treating hypertension, preventing incidental renal damage (from infections, hypotension, nephrotoxic drugs, and preeclampsia), maintaining a healthy weight and lifestyle, optimizing diabetes control, and ceasing smoking. Due to the risk of developing develop hearing loss in middle age, all affected individuals should undergo formal ophthalmologic review at least once with regular check-ups for renal failure. Due to advancing research, newer therapies for AS such as pharmacological interventions, genetic approaches and stem cell therapies are available. (Katayama et al., 2014; Murata et al., 2016; Kashtan et al., 2017)

Pharmacological interventions that have shown efficacy in mice include: vasopeptidase inhibitors, angiotensin-II receptor blockers, ACE inhibitors, 3-hydroxy-3-methylglutaryl-coenzyme-A reductase inhibitors, chemokine receptor-1 inhibitors, tumour necrosis factor-alpha antagonists, renin inhibitors and vitamin-D analogues (Katayama et al., 2014; Savage et al., 2016; Kashtan et al., 2017). In transgenic AS mice, anti-microRNA-21 treatment has been found to reduce glomerular inflammation and impaired renal fibrotic pathways; and a phase II/III study of **bardoxolone therapy** for AS patients with chronic kidney disease has been announced (NIH USNLB, 2018).

CHAPTER SUMMARY

The digestive system includes the structures and organs involved in processing of foods required for growth, development, maintenance and repair of body tissues. Most diseases affecting this system are due to infections from bacteria, viruses, protozoa, and fungi, while others are hereditary. The ear, nose, and throat (ENT) system is a complex set of structures sharing slightly interrelated mechanisms of operation. While some disorders of the ENT are hereditary, environmental influences play a big role. Diseases that affect eyesight primarily center on the three layers of the eye sclera, choroid and retina) which make eyesight possible. Disorders of metabolism occur when a crucial enzyme is disabled, or if a control mechanism for a metabolic pathway is affected. Some diseases arise from inherited traits caused by gene mutations resulting in non-functional enzymes; others involve mutations in regulatory proteins and in transport mechanisms.

Cystic fibrosis, a lethal autosomal recessive disorder affecting primarily people of European descent, results in abnormally viscous secretions in the airways of the lungs and in the ducts of the pancreas which cause obstructions leading to inflammation, tissue damage and destruction of both organ systems. The underlying genetic basis of the disease is related to dysfunction or deficiency of the CF transmembrane conductance regulator (CFTR), an apical membrane anion channel (like chloride or bicarbonate) present in respiratory and exocrine glandular epithelium. Medical management consists of nasal saline irrigations as well as medications including antibiotics, decongestants, antihistamines, topical and systemic steroids, drugs such as Pulmozyme®, and *N*-acetyl cysteine as well as surfactant lavage. Recently, new avenues have been explored to correct the *CFTR* gene defect by genome editing and use of proteostasis regulators.

With an estimated 1.25 million Americans having type-1 and more than 400 million people around the world with type-2, diabetes has multifactorial causes and genetic predisposition plays a key role. The biological risk factors for type-2 diabetes include body mass index, body fat distribution, brown adipose tissue, metabolic syndrome, adipokines, new biomarkers, imbalance of sex hormones, prediabetes, and gestational diabetes mellitus. The psychosocial risk factors include socioeconomic status, psychosocial stress, sleep deprivation and work stress. The health behavior risks include lifestyle (nutrition and physical activity), sugar-sweetened beverages, alcohol, and cigarette smoking. Both neonatal and maturity-onset diabetes of the young result from mutations of genes for transcription factors or other proteins that regulate endocrine pancreas development or function. Maturity-onset diabetes of the young presents before the age of 25 years and is caused by mutations in several related genes inherited in an autosomal dominant manner. Two of the most common mutations are hepatocyte nuclear factor 1A (*HNF1A*) and glucokinase (*GCK*). *There are a variety of oral treatment and i*njectable agents for diabetes. The treatment of type-2 in older patients emphasises lifestyle modification (nutrition, physical activity, not smoking).

Glucose-galactose malabsorption is a genetic disorder caused by a defect in this sugar transport across the intestinal mucosa. The active process is medicated by a group of genes called Na/glucose co-transporter (SGLT) which are present in the small intestine and renal tubules. An SGLT1 missense mutation underlies hereditary glucose/galactose malabsorption. Aggressive nutrition intervention for the infants and judicious dietary counseling for parents can lead to normal growth and neurological development for an infant with GGM.

Hereditary hemochromatosis are autosomal recessive genetic disorders characterized by iron accumulation in parenchymal organs, primarily the liver, which can potentially result in impaired organ structure and function. The key hormone that controls the saturation of plasma transferrin with iron

Digestive, Ear/Nose/Throat, and Eye Disorders

(transferrin saturation) is hepcidin, and its deficiency causes increased transferrin saturation, a signature unifying feature of all forms of HH. The genes involved in HH are HFE (encodes HFE protein); HJV (encodes hemojuvelin, a protein involved in iron balance); HAMP (encodes hepcidin); TFR2 (encodes transferrin receptor 2); and SLC40A1 (encodes iron exporter ferroportin). Early diagnostic and initiation of iron-depletion therapy prior to the onset of cirrhosis can reduce morbidity and ensure life quality and increase survival times of HH patients. Preventive therapy, focused on family screening, and hepcidin supplementation (for hepcidin deprivation-related HH) remains a key part of the management of HH.

Obesity is defined as the accumulation and storage of excess body fat, while overweight is weight in excess of a weight reference standard. Most of the world's population live in countries where overweight and obesity kills more people than underweight. In the United States, approximately 70% of adults and 32% of children are overweight or obese, with a continuing rising trend. To date, existing policies and interventions have not reversed these trends, suggesting that new innovative approaches are needed to transform obesity prevention and control. Evidence available suggests that multiple factors acting at different scales may be contributing to obesity, including genetics, biology (leptin sensitivity and individual metabolism), individual behaviours (dietary and physical activity choices), social network dynamics (connections to family and friends, which may also influence individual behaviours), the environment (food availability, green spaces for physical activity, and neighbourhood safety), and larger societal forces (economics, policy, education, health awareness, and culture).

Data suggests that 60–80% of the observed variance in human body weight can be accounted for by inherited factors; and more than 300 genetic loci that are potentially involved in human body weight regulation have been identified through analyses in humans and other species. Such mutations occur in the melanocortin 4 receptor, and mutations in MRAP2, a protein essential for MC4R function. Single nucleotide polymorphisms (SNPs) and copy number in genes and chromosomal regions are reported to be associated with body weight or body composition. Obesity is an epidemic systems problem with no linear cause-and-effect relationships, thus requiring a systems solution approach. Multicomponent interventions that included culturally targeted and tailored components have shown a greater impact on health behaviours associated with obesity in minority youth than the individual components in isolation.

Wilson's Disease is a rare, inherited autosomal recessive disease of copper metabolism resulting in excessive copper deposition in the body with an estimated 30 individuals per million people affected worldwide. WD arises from a defective gene identified as *ATP7B* on chromosome 13 (13q14.3), which encodes a copper-transporting P-type ATPase transmembrane protein expressed mainly in hepatocytes. Currently, the mainstay of therapy for Wilson's disease remains lifelong drug therapy usually with orally active copper chelators which de-copper the body promoting copper excretion; the use of zinc which reduces intestinal copper absorption in the intestinal mucosa, or both. Dietary management can be combined with other therapies since numerous foods such as chocolate, shellfish, nuts, mushrooms, liver and soy have high concentrations of copper naturally.

Zellweger Syndrome is an autosomal recessive inherited disorder of the peroxisome, an intracellular organelle with over 70 distinct enzymes required for normal lipid metabolism and other biochemical processes critical for normal health and development. The proper assembly of a peroxisome requires a unique set of proteins called peroxins which help incorporate enzymes into the forming peroxisome's matrix. A mutation in a peroxin, (*PEX* gene) produces a reduced or non-functioning peroxin which results in incomplete peroxisomes fail to perform their metabolic duties. Treatment strategies currently being pursued include high-content screening of large chemical libraries for compounds that improve peroxisome assembly and function, as well as gene and cellular therapies. Additional therapies for ZS include dietary restriction of phytanic acid; vitamin K supplements; bile acid therapy (cholic acid) due to altered liver metabolism; use of hearing aids and cochlear implants; periodic ophthalmologic evaluations and use of glasses; regular EEGs screening; and use of common medications to control seizures in children.

Hearing loss affects about 70 million people worldwide, with about 50%–60% of these cases having a genetic cause; and the remaining 40%–50% of cases attributed to environmental factors such as ototoxic drugs, prematurity, or trauma. More than 125 loci have been discovered: 54 autosomal dominant, 71 autosomal recessive, five X-linked, two modifier, and one Y-linked and are all involved in cochlea function. Most deafness genes discovered so far are protein-coding genes which include ion channels, gap junctions, membrane transporters, transcription factors regulating gene expression, adhesion proteins, extracellular matrix proteins, unconventional myosins, and cytoskeletal proteins. Emerging therapies for hearing loss can be broadly sub-classified as hearing preservation or hearing restoration strategies. Therapeutic non-surgical management of pathogen-associated hearing loss currently is focused on two key areas of intervention: specific antimicrobial therapies, and anti-inflammatory therapies. Surgical treatment might be beneficial in select cases in which there is an air-bone gap that is amenable to correction by surgical intervention. Non-medical support including special education and sign language are also options.

Pendred syndrome is a recessively inherited disorder characterized by diffuse goitre growth associated with sensory-neural deafness. The thyroid glands of affected individuals cannot metabolize iodide efficiently, and studies indicate that up to 10–80% of iodine taken up by the gland is discharged after administration of perchlorate. Pendred syndrome has been mapped to the gene *SLC26A4* (PDS) coding for a protein "pendrin" on chromosome 7 (7q31) which functions as a membrane-bound chloride-iodine transporter. At least 70 different mutations capable of causing the syndrome have been identified to date. Since Pendred syndrome is incurable, health professionals work together to encourage informed choices about treatment options. People experiencing significant hearing loss can be evaluated for a cochlear implant, which though cannot restore or create normal hearing, it bypasses injured areas of the ear to provide a sense of hearing in the brain.

The most common of the bestrophinopathies, Best Disease is inherited in an autosomal dominant fashion. Patients with Best Disease display impaired vision with decreased visual acuity, abnormal colour vision, macular degeneration, and atrophy of the retinal pigment epithelium. Mutations in *BEST1* have been found in association with clinically distinct retinal degenerative diseases. Bestrophins function as calcium-activated anion channels and there are over 200 mutations throughout the entire *BEST1* gene causing the clinically distinct forms of retinal degeneration. No current therapies or treatments exist for patients suffering from any bestrophinopathy. Treatment is symptomatic and involves the use of low vision aids, and direct laser treatment or photodynamic therapy. Newer treatment includes anti-VEGF agents (bevacizumab) and transcorneal electrical retinal stimulation.

Glaucoma is a neurodegenerative disorder of the optic nerve in which retinal ganglion cell death leads to characteristic patterns of visual field loss. Glaucoma is the second leading cause of blindness, accounting for 8% of blindness among the 39 million people who are blind world-wide. Age, elevated intraocular pressure (or ocular hypertension), and family history are major risk factors for glaucoma. The two most common forms of glaucoma are primary open-angle glaucoma (POAG) and angle-closure glaucoma (ACG). Several genes that cause glaucoma-relevant phenotypes have recently been identified in mice particularly the genes *MYOC*, GLC3A (2p21), GLC3B (1p36), and GLC3C (14q22). Visual

Digestive, Ear/Nose/Throat, and Eye Disorders

field loss in glaucoma is irreversible, so detection and treatment are essential to limit the progression of disease and delay additional optic nerve damage. The only current treatment for glaucoma is the lowering of intraocular pressure, achievable with eyedrops, laser, or incisional surgery.

Gyrate atrophy affects the choroid and retina layers of the eye causing high myopia, marked astigmatism and early cataract formation. It is metabolic congenital defect in the gene coding for the mitochondrial matrix enzyme *ornithine-δ-aminotransferase (OAT)* found in chromosome 10 and is transmitted in a recessive autosomal fashion. The 10-20-fold increase in levels of the amino acid ornithine lead to a progressive chorio-retinal atrophy. Treatment focuses on dietary supplements and/or a specialized diet: low ornithine foods and supplementation of vitamin B6. Patients treated with a low arginine diet showed near normal blood ornithine levels Human pluripotent stem cells, which include both human embryonic stem cells and induced pluripotent stem cells (iPS) have been used for gene therapy in patients with gyrate atrophy.

Androgenic alopecia (pattern baldness) is an androgen-mediated process in genetically susceptible individuals, and environmental and lifestyle risk factors are likely to be trigger and affects up to 70% of men and 40% of women at some point in their lifetime. Male pattern baldness is often linked to other diseases including prostate cancer, increased risk of both invasive squamous cell carcinoma, basal cell carcinoma, and head and neck region melanoma. Other associated diseases are colorectal neoplasia and coronary heart disease especially among men with hypertension or high cholesterol levels. Several gene variants have been linked to male-pattern baldness, with the *AR* gene showing the strongest association. GWAS identified 247 independent autosomal loci and 40 independent X chromosome loci linked to baldness. Pattern baldness has limited effective therapeutic options—the effectiveness of most treatments relies on how early they are applied. Drug candidates largely fall into the classes of 5-α-R inhibitors, androgen receptor antagonists, ATP-sensitive potassium channel openers, topically growth factors, and a diverse selection of antioxidants and botanical extracts.

Alport Syndrome is characterized by progressive renal failure, hearing loss, and ocular abnormalities, including corneal scarring, lenticonus, retinal thinning, and fleck retinopathy. The disease is 85% X-linked and 15% autosomal recessive. X-linked AS is caused by mutations in the *COL4A5* gene which encodes the collagen IV α5 chain. The autosomal recessive disease is caused by two mutations on different chromosomes in the *COL4A3* or *COL4A4* genes which code for the collagen IV α3 and α4 chains respectively. Currently, there is no treatment available to cure AS and symptomatic renal protective therapies are currently the mainstay of treatment for this disease. Such therapies include angiotensin-converting-enzyme inhibitors or angiotensin-receptor blockers, angiotensin converting enzyme (ACE) inhibitors, and angiotensin receptor blockers. With advancing research, newer therapies for AS such as pharmacological interventions, genetic approaches and stem cell therapies are becoming available.

End of Chapter Quiz

1. Which of the following statement/s are true? Diseases of the digestive system
    a. are dependent on the diet, stress and the amount of physical activity
    b. are due to infections from bacteria, fungi, and protozoa
    c. can sometimes be hereditary
    d. are always hereditary
    e. A, B and C

2. The three layers of the eye starting with the outermost layer to inner layer are called
    a. sclera, choroid, and retina
    b. retina, choroid, and sclera
    c. choroid, retinal, and sclera
    d. iris, pupil, cornea
    e. pupil, iris, cornea
3. The genetic basis of cystic fibrosis is due to the dysfunction of which channel?
    a. sodium potassium ATPase
    b. cystic fibrosis transmembrane conductance regulator (CFTR)
    c. ligand gated calcium channel
    d. g-protein coupled receptor (GPCR)
    e. sodium channel
4. What are the primary transporters that mediate transcellular movements of d-glucose in the small intestine?
    a. SGLT1 and GLUT1
    b. SGLT1 and GLUT2
    c. SGLT1 and GLUT3
    d. SGLT1 and GLUT4
    e. SGLT2 and GLUT1
5. Infants with glucose-galactose malabsorption (GGM) suffer from chronic, profuse, watery diarrhoea that often leads to _____ dehydration.
    a. hypotonic
    b. hypertonic
    c. isotonic
    d. osmotic
    e. none of the above
6. Which hormone controls the saturation of plasma transferrin with iron?
    a. Triiodothyronine (T3)
    b. Erythropoietin
    c. Hepcidin
    d. Thyroxine (T4)
    e. Oestrogen
7. Wilson's disease can present with several conditions. Which of these conditions are incorrectly described?
    a. Ataxia: a neurological condition involving the loss of involuntary muscle movement
    b. Osteoporosis: a disease in which bones become weak
    c. Cardiomyopathy: a disease of the heart muscle in which the heart muscle can become thick, enlarged of stiff
    d. Osteoarthritis: a joint disease caused by the breakdown of cartilage between bones
    e. Arrhythmias: a disorder in which the heart beats with an irregular rate or rhythm

8. The proper assembly of a peroxisome requires a unique set of proteins called
   a. prions
   b. protein C
   c. pepsin
   d. peroxins
   e. p53
9. Around 40-50% of hearing loss cases can be attributed to environmental factors such as
   a. ototoxic drugs
   b. prematurity
   c. tauma
   d. cytomegalovirus infection
   e. all of the above
10. Patients with Best Disease display impaired
    a. motor skills
    b. speech
    c. hearing
    d. vision
    e. C and D
11. The two most common forms of glaucoma are
    a. primary open-angle glaucoma and angle closure glaucoma
    b. normal tension glaucoma and primary open-angle glaucoma
    c. secondary glaucoma and normal tension glaucoma
    d. normal tension glaucoma and angle closure glaucoma
    e. secondary glaucoma and primary open-angle glaucoma

Thought Questions

1. If the signs and symptoms for both types of diabetes are the same, describe how you would distinguish between the two.
2. What role does leptin play in obesity?
3. Draw the layers of the eye and describe the effects of glaucoma.

REFERENCES

Abad-Sinden, A., Borowitz, S., Meyers, R., & Sutphen, J. (1997). Nutrition management of congenital glucose-galactose malabsorption: A case study. *Journal of the American Dietetic Association, 97*(12), 1417–1421. doi:10.1016/S0002-8223(97)00342-8 PMID:9404340

Al-Lawati, T., & Vargees, T. (2008). Glucose Galactose Malabsorption complicated with Rickets and Nephrogenic Diabetes Insipidus. *Oman Medical Journal, 23*(3), 197–198. PMID:22359715

Angeli, S., Lin, X., & Liu, X. Z. (2012). Genetics of hearing and deafness. *The Anatomical Record*, *295*(11), 1812–1829. doi:10.1002/ar.22579 PMID:23044516

Ashrafi, K., Chang, F. Y., Watts, J. L., Fraser, A. G., Kamath, R. S., Ahringer, J., & Ruvkun, G. (2003). Genome-wide RNAi analysis of Caenorhabditis elegans fat regulatory genes. *Nature*, *421*(6920), 268–272. doi:10.1038/nature01279 PMID:12529643

Barker, D. F., Hostikka, S. L., Zhou, J., Chow, L. T., Oliphant, A. R., Gerken, S. C., ... Tryggvason, K. (1990). Identification of mutations in the COL4A5 collagen gene in Alport syndrome. *Science*, *248*(4960), 1224–1227. doi:10.1126cience.2349482 PMID:2349482

Basile, K. J., Guy, V. C., Schwartz, S., & Grant, S. F. (2014). Overlap of genetic susceptibility to type 1 diabetes, type 2 diabetes, and latent autoimmune diabetes in adults. *Current Diabetes Reports*, *14*(11), 550. doi:10.100711892-014-0550-9 PMID:25189437

Bigozzi, M., Melchionda, S., Casano, R., Palladino, T., & Gitti, G. (2005). Pendred syndrome: study of three families. *Acta otorhinolaryngologica Italica: organo ufficiale della Societa italiana di otorinolaringologia e chirurgia cervico-facciale*, *25*(4), 233–239.

Billuart, P., Bienvenu, T., Ronce, N., des Portes, V., Vinet, M. C., Zemni, R., ... Chelly, J. (1998). Oligophrenin-1 encodes a rhoGAP protein involved in X-linked mental retardation. *Nature*, *392*(6679), 923–926. doi:10.1038/31940 PMID:9582072

Bitner, H., Schatz, P., Mizrahi-Meissonnier, L., Sharon, D., & Rosenberg, T. (2012). Frequency, genotype, and clinical spectrum of best vitelliform macular dystrophy: Data from a national center in Denmark. *American Journal of Ophthalmology*, *154*(2), 403–412.e4. doi:10.1016/j.ajo.2012.02.036 PMID:22633354

Braham, I. Z., Ammous, I., Maalej, R., Boukari, M., Boussen, I. M., Errais, K., ... Zhioua, R. (2018). Multimodal imaging of foveoschisis and macular pseudohole associated with gyrate atrophy: A family report. *BMC Ophthalmology*, *18*(1), 89. doi:10.118612886-018-0755-9 PMID:29649987

Braverman, N. E., Raymond, G. V., Rizzo, W. B., Moser, A. B., Wilkinson, M. E., Stone, E. M., ... Bose, M. (2015). Peroxisome biogenesis disorders in the Zellweger spectrum: An overview of current diagnosis, clinical manifestations, and treatment guidelines. *Molecular Genetics and Metabolism*, *117*(3), 313–321. doi:10.1016/j.ymgme.2015.12.009 PMID:26750748

Brissot, P., Cavey, T., Ropert, M., Guggenbuhl, P., & Loréal, O. (2017). Genetic hemochromatosis: Pathophysiology, diagnostic and therapeutic management. *La Presse Medicale*, *46*(12 part 2), e288–e295. doi:10.1016/j.lpm.2017.05.037 PMID:29158016

Brissot, P., & de Bels, F. (2006). Current approaches to the management of hemochromatosis. Hematology: American Society of Hematology Education Program, 36–41.

Britton, R. S., Fleming, R. E., Parkkila, S., Waheed, A., Sly, W. S., & Bacon, B. R. (2002). Pathogenesis of hereditary hemochromatosis: Genetics and beyond. *Seminars in Gastrointestinal Disease*, *13*(2), 68–79. PMID:12064862

Digestive, Ear/Nose/Throat, and Eye Disorders

Bull, P. C., Thomas, G. R., Rommens, J. M., Forbes, J. R., & Cox, D. W. (1993). The Wilson disease gene is a putative copper transporting P-type ATPase similar to the Menkes gene. *Nature Genetics*, 5(4), 327–337. doi:10.1038/ng1293-327 PMID:8298639

Burdon, K. P., Crawford, A., Casson, R. J., Hewitt, A. W., Landers, J., Danoy, P., ... Craig, J. E. (2012). Glaucoma risk alleles at CDKN2B-AS1 are associated with lower intraocular pressure, normal-tension glaucoma, and advanced glaucoma. *Ophthalmology*, 119(8), 1539–1545. doi:10.1016/j.ophtha.2012.02.004 PMID:22521085

Burdon, K. P., Macgregor, S., Hewitt, A. W., Sharma, S., Chidlow, G., Mills, R. A., ... Craig, J. E. (2011). Genome-wide association study identifies susceptibility loci for open angle glaucoma at TMCO1 and CDKN2B-AS1. *Nature Genetics*, 43(6), 574–548. doi:10.1038/ng.824 PMID:21532571

Bures, R. M., Mabry, P. L., Orleans, C. T., & Esposito, L. (2014). Systems science: A tool for understanding obesity. *American Journal of Public Health*, 104(7), 1156. doi:10.2105/AJPH.2014.302082 PMID:24832433

Chaaban, M. R., Kejner, A., Rowe, S. M., & Woodworth, B. A. (2013). Cystic fibrosis chronic rhinosinusitis: A comprehensive review. *American Journal of Rhinology & Allergy*, 27(5), 387–395. doi:10.2500/ajra.2013.27.3919 PMID:24119602

Chedane-Girault, C., Dabadie, A., Maurage, C., Piloquet, H., Chailloux, E., Colin, E., ... Giniès, J. L. (2012). Neonatal diarrhea due to congenital glucose-galactose malabsorption: report of seven cases. *Archives de Pediatrie, 19*(12), 1289–1292. doi:. arcped.2012.09.005 doi:10.1016/j

Chen, C., Shen, B., Xiao, J. J., Wu, R., Duff Canning, S. J., & Wang, X. P. (2015). Currently Clinical Views on Genetics of Wilson's Disease. *Chinese Medical Journal*, 128(13), 1826–1830. doi:10.4103/0366-6999.159361 PMID:26112727

Coffey, A. J., Durkie, M., Hague, S., McLay, K., Emmerson, J., Lo, C., ... Bandmann, O. (2013). A genetic study of Wilson's disease in the United Kingdom. *Brain*, 136(Pt 5), 1476–1487. doi:10.1093/brain/awt035 PMID:23518715

Cutting, G. R. (2014). Cystic fibrosis genetics: From molecular understanding to clinical application. *Nature Reviews. Genetics*, 16(1), 45–56. doi:10.1038/nrg3849 PMID:25404111

Dalvin, L. A., Pulido, J. S., & Marmorstein, A. D. (2016). Vitelliform dystrophies: Prevalence in Olmsted County, Minnesota, United States. *Ophthalmic Genetics*, 38(2), 143–147. doi:10.1080/13816810.2016.1175645 PMID:27120116

Daniels, S. R. (2009). Complications of obesity in children and adolescents. *International Journal of Obesity*, 33(S1Suppl 1), S60–S65. doi:10.1038/ijo.2009.20 PMID:19363511

Deignan, J. L. & Grody, W. W. (2016). Molecular Diagnosis of Cystic Fibrosis. *Current Protocols in Human Genetics, 1,* 88, Unit 9.28. doi:10.1002/0471142905.hg0928s88

Doche, M. E., Bochukova, E. G., Su, H. W., Pearce, L. R., Keogh, J. M., Henning, E., ... Farooqi, I. S. (2012). Human SH2B1 mutations are associated with maladaptive behaviors and obesity. *The Journal of Clinical Investigation, 122*(12), 4732–4736. doi:10.1172/JCI62696 PMID:23160192

Everett, L. A., Glaser, B., Beck, J. C., Idol, J. R., Buchs, A., Heyman, M., ... Green, E. D. (1997). Pendred syndrome is caused by mutations in a putative sulphate transporter gene (PDS). *Nature Genetics, 17*(4), 411–422. doi:10.1038/ng1297-411 PMID:9398842

Farooqi, I. S., Drop, S., Clements, A., Keogh, J. M., Biernacka, J., Lowenbein, S., ... O'Rahilly, S. (2006). Heterozygosity for a POMC-null mutation and increased obesity risk in humans. *Diabetes, 55*(9), 2549–2553. doi:10.2337/db06-0214 PMID:16936203

Fernandes, K. A., Harder, J. M., Williams, P. A., Rausch, R. L., Kiernan, A. E., Nair, K. S., ... Libby, R. T. (2015). Using genetic mouse models to gain insight into glaucoma: Past results and future possibilities. *Experimental Eye Research, 141*, 42–56. doi:10.1016/j.exer.2015.06.019 PMID:26116903

Freedman, D. S., Mei, Z., Srinivasan, S. R., Berenson, G. S., & Dietz, W. H. (2007). Cardiovascular risk factors and excess adiposity among overweight children and adolescents: The Bogalusa Heart Study. *The Journal of Pediatrics, 150*(1), 12–17.e2. doi:10.1016/j.jpeds.2006.08.042 PMID:17188605

Friedman, T. B., Hinnant, J. T., Ghosh, M., Boger, E. T., Riazuddin, S., Lupski, J. R., ... Wilcox, E. R. (2002). DFNB3, spectrum of MYO15A recessive mutant alleles and an emerging genotype-phenotype correlation. *Advances in Oto-Rhino-Laryngology, 61*, 124–130. doi:10.1159/000066824 PMID:12408074

Frydman, M. (1990). Genetic aspects of Wilson's disease. *Journal of Gastroenterology and Hepatology, 5*(4), 483–490. doi:10.1111/j.1440-1746.1990.tb01427.x PMID:2129820

Glaucoma Research Foundation (GRF). (2018a). *What are the Symptoms of Glaucoma?* Retrieved from https://www.glaucoma.org/gleams/what-are-the- symptoms-of-glaucoma.php

Glaucoma Research Foundation (GRF). (2018b). *New Medical Therapies for Glaucoma.* Retrieved from https://www.glaucoma.org/treatment/new-medical- therapies-for-glaucoma.php

Gonzalez, K. L., Ratzel, S. E., Burks, K. H., Danan, C. H., Wages, J. M., Zolman, B. K., & Bartel, B. (2018). A *pex1* missense mutation improves peroxisome function in a subset of *Arabidopsis pex6* mutants without restoring PEX5 recycling. *Proceedings of the National Academy of Sciences of the United States of America, 115*(14), E3163–E3172. doi:10.1073/pnas.1721279115 PMID:29555730

Gortmaker, S. L., Swinburn, B. A., Levy, D., Carter, R., Mabry, P. L., Finegood, D. T., ... Moodie, M. L. (2011). Changing the future of obesity: Science, policy, and action. *Lancet, 378*(9793), 838–847. doi:10.1016/S0140-6736(11)60815-5 PMID:21872752

Gupta, P., Zhao, D., Guallar, E., Ko, F., Boland, M. V., & Friedman, D. S. (2016). Prevalence of Glaucoma in the United States: The 2005-2008 National Health and Nutrition Examination Survey. *Investigative Ophthalmology & Visual Science, 57*(6), 2905–2913. doi:10.1167/iovs.15-18469 PMID:27168366

Guziewicz, K. E., Zangerl, B., Komáromy, A. M., Iwabe, S., Chiodo, V. A., Boye, S. L., ... Aguirre, G. D. (2013). Recombinant AAV-mediated BEST1 transfer to the retinal pigment epithelium: Analysis of serotype-dependent retinal effects. *PLoS One*, *8*(10), e75666. doi:10.1371/journal.pone.0075666 PMID:24143172

Hagenaars, S. P., Hill, W. D., Harris, S. E., Ritchie, S. J., Davies, G., Liewald, D. C., ... Marioni, R. E. (2017). Genetic prediction of male pattern baldness. *PLOS Genetics*, *13*(2), e1006594. doi:10.1371/journal.pgen.1006594 PMID:28196072

Hartzell, H. C., Qu, Z., Yu, K., Xiao, Q., & Chien, L. T. (2008). Molecular physiology of bestrophins: Multifunctional membrane proteins linked to best disease and other retinopathies. *Physiological Reviews*, *88*(2), 639–672. doi:10.1152/physrev.00022.2007 PMID:18391176

Hébert, J. M., Rosenquist, T., Götz, J., & Martin, G. R. (1994). FGF5 as a regulator of the hair growth cycle: Evidence from targeted and spontaneous mutations. *Cell*, *78*(6), 1017–1025. doi:10.1016/0092-8674(94)90276-3 PMID:7923352

Hertz, J. M., Thomassen, M., Storey, H., & Flinter, F. (2014). Clinical utility gene card for: Alport syndrome - update 2014. *European Journal of Human Genetics: EJHG*, *23*(9). doi:10.1038/ejhg.2014.254 PMID:25388007

Hiebler, S., Masuda, T., Hacia, J. G., Moser, A. B., Faust, P. L., Liu, A., ... Steinberg, S. J. (2014). The Pex1-G844D mouse: A model for mild human Zellweger spectrum disorder. *Molecular Genetics and Metabolism*, *111*(4), 522–532. doi:10.1016/j.ymgme.2014.01.008 PMID:24503136

Howden, S. E., Athurva, G., Li, Z., Fung, H., Nisler, B. S., Nie, J., ... Thomson, J. A. (2011). Genetic correction and analysis of induced pluripotent stem cells from a patient with gyrate atrophy. *Proceedings of the National Academy of Sciences of the United States of America*, *108*(16), 6537–6542. 10.1073/pnas.1103388108

Hudson, B. G., Tryggvason, K., Sundaramoorthy, M., & Neilson, E. G. (2003). Alport's syndrome, Goodpasture's syndrome, and type IV collagen. *The New England Journal of Medicine*, *348*(25), 2543–2556. doi:10.1056/NEJMra022296 PMID:12815141

Johnson, A. A., Guziewicz, K. E., Lee, C. J., Kalathur, R. C., Pulido, J. S., Marmorstein, L. Y., & Marmorstein, A. D. (2017). Bestrophin 1 and retinal disease. *Progress in Retinal and Eye Research*, *58*, 45–69. doi:10.1016/j.preteyeres.2017.01.006 PMID:28153808

Joint Committee on Infant Hearing. (2007). Year 2007 position statement: Principles and guidelines for early hearing detection and intervention programs. *American Academy of Pediatrics. Pediatrics*, *120*(4), 898–921. doi:10.1542/peds.2007-2333 PMID:17908777

Kashtan, C. (2017). Alport syndrome: Facts and opinions. *F1000 Research*, *6*, 50. doi:10.12688/f1000research.9636.1 PMID:28163907

Katayama, K., Nomura, S., Tryggvason, K., & Ito, M. (2014). Searching for a treatment for Alport syndrome using mouse models. *World Journal of Nephrology*, *3*(4), 230–236. doi:10.5527/wjn.v3.i4.230 PMID:25374816

Kathawala, M., & Hirschfield, G. M. (2017). Insights into the management of Wilson's disease. *Therapeutic Advances in Gastroenterology*, *10*(11), 889–905. doi:10.1177/1756283X17731520 PMID:29147139

Kennaway, N. G., Weleber, R. G., & Buist, N. R. (1980). Gyrate atrophy of the choroid and retina with hyperornithinemia: Biochemical and histologic studies and response to vitamin B6. *American Journal of Human Genetics*, *32*(4), 529–541. PMID:7395865

Keum, N., Cao, Y., Lee, D. H., Park, S. M., Rosner, B., Fuchs, C. S., ... Giovannucci, E. L. (2016). Male pattern baldness and risk of colorectal neoplasia. *British Journal of Cancer*, *114*(1), 110–117. doi:10.1038/bjc.2015.438 PMID:26757425

Kimbrel, E. A., & Lanza, R. (2015). Current status of pluripotent stem cells: Moving the first therapies to the clinic. *Nature Reviews. Drug Discovery*, *14*(10), 681–692. doi:10.1038/nrd4738 PMID:26391880

King, C., & Barton, D. E. (2006). Best practice guidelines for the molecular genetic diagnosis of Type 1 (HFE-related) hereditary haemochromatosis. *BMC Medical Genetics*, *7*(1), 81. doi:10.1186/1471-2350-7-81 PMID:17134494

Klouwer, F. C., Berendse, K., Ferdinandusse, S., Wanders, R. J., Engelen, M., & Poll-The, B. T. (2015). Zellweger spectrum disorders: Clinical overview and management approach. *Orphanet Journal of Rare Diseases*, *10*(1), 151. doi:10.118613023-015-0368-9 PMID:26627182

Korver, A. M., Smith, R. J., Van Camp, G., Schleiss, M. R., Bitner-Glindzicz, M. A., Lustig, L. R., ... Boudewyns, A. N. (2017). Congenital hearing loss. *Nature Reviews. Disease Primers*, *3*, 16094. doi:10.1038/nrdp.2016.94 PMID:28079113

Kyari, F., Abdull, M. M., Bastawrous, A., Gilbert, C. E., & Faal, H. (2013). Epidemiology of Glaucoma in Sub-Saharan Africa: Prevalence, Incidence and Risk Factors. *Middle East African Journal of Ophthalmology*, *20*(2), 111–125. doi:10.4103/0974-9233.110605 PMID:23741130

Kyle, T. K., Dhurandhar, E. J., & Allison, D. B. (2016). Regarding Obesity as a Disease: Evolving Policies and Their Implications. *Endocrinology and Metabolism Clinics of North America*, *45*(3), 511–520. doi:10.1016/j.ecl.2016.04.004 PMID:27519127

Lee, B. Y., Bartsch, S. M., Mui, Y., Haidari, L. A., Spiker, M. L., & Gittelsohn, J. (2017). A systems approach to obesity. *Nutrition Reviews*, *75*(suppl 1), 94–106. doi:10.1093/nutrit/nuw049 PMID:28049754

Lee, P. R., & Raymond, G. V. (2013). Child neurology: Zellweger syndrome. *Neurology*, *80*(20), e207–e210. doi:10.1212/WNL.0b013e3182929f8e PMID:23671347

Leske, M. C. (2007). Open-angle glaucoma -- an epidemiologic overview. *Ophthalmic Epidemiology*, *14*(4), 166–172. doi:10.1080/09286580701501931 PMID:17896292

Levesque, S., Morin, C., Guay, S. P., Villeneuve, J., Marquis, P., Yik, W. Y., ... Braverman, N. E. (2012). A founder mutation in the PEX6 gene is responsible for increased incidence of Zellweger syndrome in a French Canadian population. *BMC Medical Genetics, 13*(1), 72. doi:10.1186/1471-2350-13-72 PMID:22894767

Li, W. Q., Cho, E., Han, J., Weinstock, M. A., & Qureshi, A. A. (2016). Male pattern baldness and risk of incident skin cancer in a cohort of men. *International Journal of Cancer, 139*(12), 2671–2678. doi:10.1002/ijc.30395 PMID:27542665

Litwin, T., Gromadzka, G., & Czlonkowska, A. (2012). Gender differences in Wilson's disease. *Journal of the Neurological Sciences, 312*(1-2), 31–35. doi:10.1016/j.jns.2011.08.028 PMID:21917273

Lotufo, P. A., Chae, C. U., Ajani, U. A., Hennekens, C. H., & Manson, J. E. (2000). Male pattern baldness and coronary heart disease: The Physicians' Health Study. *Archives of Internal Medicine, 160*(2), 165–171. doi:10.1001/archinte.160.2.165 PMID:10647754

Maayan, L., & Correll, C. U. (2011). Weight gain and metabolic risks associated with antipsychotic medications in children and adolescents. *Journal of Child and Adolescent Psychopharmacology, 21*(6), 517–535. doi:10.1089/cap.2011.0015 PMID:22166172

Maiuri, L., Raia, V., & Kroemer, G. (2017). Strategies for the etiological therapy of cystic fibrosis. *Cell Death and Differentiation, 24*(11), 1825–1844. doi:10.1038/cdd.2017.126 PMID:28937684

Marín-Peñalver, J. J., Martín-Timón, I., Sevillano-Collantes, C., & Del Cañizo-Gómez, F. J. (2016). Update on the treatment of type 2 diabetes mellitus. *World Journal of Diabetes, 7*(17), 354–395. doi:10.4239/wjd.v7.i17.354 PMID:27660695

Marion, R. W. (1991). The genetic anatomy of hearing. A clinician's view. *Annals of the New York Academy of Sciences, 630*(1 Genetics of H), 32–37. doi:10.1111/j.1749-6632.1991.tb19573.x PMID:1952621

Mayo Clinic. (2018). *Hair loss.* Retrieved from https://www.mayoclinic.org/diseases-conditions/hair-loss/symptoms-causes/syc-20372926

Mayo Clinic. (2018). *Hearing Loss.* Retrieved from https://www.mayoclinic.org/diseases-conditions/hearing-loss/symptoms-causes/syc-20373072

Menaa, F., Braghini, C. A., Vasconcellos, J. P., Menaa, B., Costa, V. P., Figueiredo, E. S., & Melo, M. B. (2011). Keeping an eye on myocilin: A complex molecule associated with primary open-angle glaucoma susceptibility. *Molecules (Basel, Switzerland), 16*(7), 5402–5421. doi:10.3390/molecules16075402 PMID:21709622

Meyer, J. S., Howden, S. E., Wallace, K. A., Verhoeven, A. D., Wright, L. S., Capowski, E. E., ... Gamm, D. M. (2011). Optic vesicle-like structures derived from human pluripotent stem cells facilitate a customized approach to retinal disease treatment. *Stem Cells (Dayton, Ohio), 29*(8), 1206–1218. doi:10.1002tem.674 PMID:21678528

Mittal, R., Patel, A. P., Nguyen, D., Pan, D. R., Jhaveri, V. M., Rudman, J. R., ... Liu, X. Z. (2018). Genetic Basis of Hearing Loss in Spanish, Hispanic and Latino Populations. *Gene, 647*, 297–305. doi:10.1016/j.gene.2018.01.027 PMID:29331482

Mochizuki, T., Lemmink, H. H., Mariyama, M., Antignac, C., Gubler, M. C., Pirson, Y., ... Smeets, H. J. (1994). Identification of mutations in the alpha 3(IV) and alpha 4(IV) collagen genes in autosomal recessive Alport syndrome. *Nature Genetics*, *8*(1), 77–81. doi:10.1038/ng0994-77 PMID:7987396

Moloney, T., O'Hagan, S., & Lee, L. (2014). Ultrawide-field fundus photography of the first reported case of gyrate atrophy from Australia. *Clinical Ophthalmology (Auckland, N.Z.)*, (8): 1561–1563. doi:10.2147/OPTH.S64248 PMID:25187693

Morton, N. E. (1990). Genetic linkage and complex diseases: A response. *Genetic Epidemiology*, *7*(1), 33–34. doi:10.1002/gepi.1370070108

Murata, T., Katayama, K., Oohashi, T., Jahnukainen, T., Yonezawa, T., Sado, Y., ... Ito, M. (2016). COL4A6 is dispensable for autosomal recessive Alport syndrome. *Scientific Reports*, *6*(1), 29450. doi:10.1038rep29450 PMID:27377778

National Heart, Lung and Blood Institute (NHLBI). (2018). *Overweight and Obesity*. Retrieved from https://www.nhlbi.nih.gov/health-topics/overweight-and-obesity

National Institute on Deafness and Other Communicable Disorders (NIDCD). (2018). *Pendred Syndrome*. Retrieved from https://www.nidcd.nih.gov/health/pendred-syndrome#treat

Ng, S. K., Casson, R. J., Burdon, K. P., & Craig, J. E. (2013). Chromosome 9p21 primary open-angle glaucoma susceptibility locus: A review. *Clinical & Experimental Ophthalmology*, *42*(1), 25–32. doi:10.1111/ceo.12234 PMID:24112133

Ng, S. K., Casson, R. J., Burdon, K. P., & Craig, J. E. (2014). Chromosome 9p21 primary open-angle glaucoma susceptibility locus: A review. *Clinical & Experimental Ophthalmology*, *42*(1), 25–32. doi:10.1111/ceo.12234 PMID:24112133

NIH US National Library of Medicine. (2018). *Study of RG-012 in Male Subjects with Alport Syndrome (HERA)*. Retrieved from https://clinicaltrials.gov/ct2/show/NCT02855268

Nishioka, G. J., Barbero, G. J., Konig, P., Parsons, D. S., Cook, P. R., & Davis, W. E. (1995). Symptom outcome after functional endoscopic sinus surgery in patients with cystic fibrosis: A prospective study. *Otolaryngology - Head and Neck Surgery*, *113*(4), 440–445. doi:10.1016/S0194-5998(95)70082-X PMID:7567018

Nordström, S. (1974). Hereditary macular degeneration--a population survey in the county of Vsterbotten, Sweden. *Hereditas*, *78*(1), 41–62. doi:10.1111/j.1601-5223.1974.tb01427.x PMID:4448697

Ogden, C. L., Carroll, M. D., Curtin, L. R., Lamb, M. M., & Flegal, K. M. (2010). Prevalence of high body mass index in US children and adolescents, 2007-2008. *Journal of the American Medical Association*, *303*(3), 242–249. doi:10.1001/jama.2009.2012 PMID:20071470

Pandya, A., Arnos, K. S., Xia, X. J., Welch, K. O., Blanton, S. H., Friedman, T. B., ... Nance, W. E. (2003). Frequency and distribution of GJB2 (connexin 26) and GJB6 (connexin 30) mutations in a large North American repository of deaf probands. *Genetics in Medicine*, *5*(4), 295–303. doi:10.1097/01.GIM.0000078026.01140.68 PMID:12865758

Digestive, Ear/Nose/Throat, and Eye Disorders

Porto, G., Brissot, P., Swinkels, D. W., Zoller, H., Kamarainen, O., Patton, S., ... Keeney, S. (2015). EMQN best practice guidelines for the molecular genetic diagnosis of hereditary hemochromatosis (HH). *European Journal of Human Genetics, 24*(4), 479–495. doi:10.1038/ejhg.2015.128 PMID:26153218

Quigley, H. A., & Broman, A. T. (2006). The number of people with glaucoma worldwide in 2010 and 2020. *The British Journal of Ophthalmology, 90*(3), 262–267. doi:10.1136/bjo.2005.081224 PMID:16488940

Racette, L., Wilson, M. R., Zangwill, L. M., Weinreb, R. N., & Sample, P. A. (2003). Primary open-angle glaucoma in blacks: A review. *Survey of Ophthalmology, 48*(3), 295–313. doi:10.1016/S0039-6257(03)00028-6 PMID:12745004

Rademakers, R., Cruts, M., & van Broeckhoven, C. (2004). The role of tau (MAPT) in frontotemporal dementia and related tauopathies. *Human Mutation, 24*(4), 277–295. doi:10.1002/humu.20086 PMID:15365985

Rankinen, T., Zuberi, A., Chagnon, Y. C., Weisnagel, S. J., Argyropoulos, G., Walts, B., ... Bouchard, C. (2006). The human obesity gene map: The 2005 update. *Obesity (Silver Spring, Md.), 14*(4), 529–644. doi:10.1038/oby.2006.71 PMID:16741264

Rowe, S. M., Miller, S., & Sorscher, E. J. (2005). Cystic fibrosis. *The New England Journal of Medicine, 352*(19), 1992–2001. doi:10.1056/NEJMra043184 PMID:15888700

Rowe, S. M., Varga, K., Rab, A., Bebok, Z., Byram, K., Li, Y., ... Clancy, J. P. (2007). Restoration of W1282X CFTR activity by enhanced expression. *American Journal of Respiratory Cell and Molecular Biology, 37*(3), 347–356. doi:10.1165/rcmb.2006-0176OC PMID:17541014

Rudnicka, A. R., Mt-Isa, S., Owen, C. G., Cook, D. G., & Ashby, D. (2006). Variations in primary open-angle glaucoma prevalence by age, gender, and race: A Bayesian Meta-analysis. *Investigative Ophthalmology & Visual Science, 47*(10), 4254–4261. doi:10.1167/iovs.06-0299 PMID:17003413

Santos, Z., Avci, P., & Hamblin, M. R. (2015). Drug discovery for alopecia: Gone today, hair tomorrow. *Expert Opinion on Drug Discovery, 10*(3), 269–292. doi:10.1517/17460441.2015.1009892 PMID:25662177

Savastano, D. M., Tanofsky-Kraff, M., Han, J. C., Ning, C., Sorg, R. A., Roza, C. A., ... Yanovski, J. A. (2009). Energy intake and energy expenditure among children with polymorphisms of the melanocortin-3 receptor. *The American Journal of Clinical Nutrition, 90*(4), 912–920. doi:10.3945/ajcn.2009.27537 PMID:19656839

Savige, J., Colville, D., Rheault, M., Gear, S., Lennon, R., Lagas, S., ... Flinter, F. (2016). Alport Syndrome in Women and Girls. *Clinical Journal of the American Society of Nephrology; CJASN, 11*(9), 1713–1720. doi:10.2215/CJN.00580116 PMID:27287265

Schilsky, M. L. (1996). Wilson disease: Genetic basis of copper toxicity and natural history. *Seminars in Liver Disease, 16*(1), 83–95. doi:10.1055-2007-1007221 PMID:8723326

Singh, R., Kuai, D., Guziewicz, K. E., Meyer, J., Wilson, M., Lu, J., ... Gamm, D. M. (2015). Pharmacological Modulation of Photoreceptor Outer Segment Degradation in a Human iPS Cell Model of Inherited Macular Degeneration. *Molecular Therapy. The Journal of the American Society of Gene Therapy, 23*(11), 1700–1711. doi:10.1038/mt.2015.141 PMID:26300224

Smith, R. J., Bale, J. F. Jr, & White, K. R. (2005). Sensorineural hearing loss in children. *Lancet, 365*(9462), 879–890. doi:10.1016/S0140-6736(05)71047-3 PMID:15752533

Sorokin, A. V., & Chen, J. (2013). MEMO1, a new IRS1-interacting protein, induces epithelial-mesenchymal transition in mammary epithelial cells. *Oncogene, 32*(26), 3130–3138. doi:10.1038/onc.2012.327 PMID:22824790

Sreekar, H., Uppin, V. M., Patil, S., Mutkekar, A., & Tejus, C. (2012). Pendred Syndrome with Retrosternal Goitre- A Rare Case Report. *Indian Journal of Surgery, 75*(S1), 329–330. doi:10.100712262-012-0702-6 PMID:24426607

Steck, A. K., & Winter, W. E. (2011). Review on monogenic diabetes. *Current Opinion in Endocrinology, Diabetes, and Obesity, 18*(4), 252–258. doi:10.1097/MED.0b013e3283488275 PMID:21844708

Stone, E. M., Fingert, J. H., Alward, W. L., Nguyen, T. D., Polansky, J. R., Sunden, S. L., ... Sheffield, V. C. (1997). Identification of a gene that causes primary open angle glaucoma. *Science, 275*(5300), 668–670. doi:10.1126cience.275.5300.668 PMID:9005853

Sugita, S., Iwasaki, Y., Makabe, K., Kamao, H., Mandai, M., Shiina, T., ... Takahashi, M. (2016). Successful Transplantation of Retinal Pigment Epithelial Cells from MHC Homozygote iPSCs in MHC-Matched Models. *Stem Cell Reports, 7*(4), 635–648. doi:10.1016/j.stemcr.2016.08.010 PMID:27641649

Takahashi, K., & Yamanaka, S. (2006). Induction of pluripotent stem cells from mouse embryonic and adult fibroblast cultures by defined factors. *Cell, 126*(4), 663–676. doi:10.1016/j.cell.2006.07.024 PMID:16904174

Uggenti, C., Briant, K., Streit, A. K., Thomson, S., Koay, Y. H., Baines, R. A., ... Manson, F. D. (2016). Restoration of mutant bestrophin-1 expression, localisation and function in a polarised epithelial cell model. *Disease Models & Mechanisms, 9*(11), 1317–1328. doi:10.1242/dmm.024216 PMID:27519691

van Dijk, S. J., Molloy, P. L., Varinli, H., Morrison, J. L., & Muhlhausler, B. S. (2015). Epigenetics and human obesity. *International Journal of Obesity, 39*(1), 85–97. doi:10.1038/ijo.2014.34 PMID:24566855

Wang, R. & Wiggs, J. L. (2014). Common and Rare Genetic Risk Factors for Glaucoma. *Cold Spring Harbor Perspectives in Medicine, 4*(12), a017244. doi:.a017244 doi:10.1101/cshperspect

Wang, T., Steel, G., Milam, A. H., & Valle, D. (2000). Correction of ornithine accumulation prevents retinal degeneration in a mouse model of gyrate atrophy of the choroid and retina. *Proceedings of the National Academy of Sciences of the United States of America, 97*(3), 1224–1229. doi:10.1073/pnas.97.3.1224 PMID:10655512

Wang, X. P., Wang, X. H., Bao, Y. C., & Zhou, J. N. (2003). Apolipoprotein E genotypes in Chinese patients with Wilson's disease. *QJM, 96*(7), 541–542. doi:10.1093/qjmed/hcg093 PMID:12881597

Wheeler, E., Huang, N., Bochukova, E. G., Keogh, J. M., Lindsay, S., Garg, S., ... Farooqi, I. S. (2013). Genome-wide SNP and CNV analysis identifies common and low-frequency variants associated with severe early-onset obesity. *Nature Genetics, 45*(5), 513–517. doi:10.1038/ng.2607 PMID:23563609

Wilch, E., Zhu, M., Burkhart, K. B., Regier, M., Elfenbein, J. L., Fisher, R. A., & Friderici, K. H. (2006). Expression of GJB2 and GJB6 is reduced in a novel DFNB1 allele. *American Journal of Human Genetics*, *79*(1), 174–179. doi:10.1086/505333 PMID:16773579

Wilson, D. K. (2009). New perspectives on health disparities and obesity interventions in youth. *Journal of Pediatric Psychology*, *34*(3), 231–244. doi:10.1093/jpepsy/jsn137 PMID:19223277

World Health Organization (WHO). (2017). *Media Centre: Obesity and Overweight.* Retrieved from http:// www.who.int/en/news-room/fact-sheets/detail/obesity-and-overweight

Wright, E. M. (1998). Glucose galactose malabsorption. *The American Journal of Physiology*, *275*(5), G879–G882. PMID:9815014

Wright, E. M., Turk, E., & Martin, M. G. (2002). Molecular basis for glucose-galactose malabsorption. *Cell Biochemistry and Biophysics*, *36*(2-3), 115–121. doi:10.1385/CBB:36:2-3:115 PMID:12139397

Xiao, Q., Hartzell, H. C., & Yu, K. (2010). Bestrophins and retinopathies. *Pflügers Archiv*, *460*(2), 559–569. doi:10.100700424-010-0821-5 PMID:20349192

Yang, Y., & Chan, L. (2016). Monogenic Diabetes: What It Teaches Us on the Common Forms of Type 1 and Type 2 Diabetes. *Endocrine Reviews*, *37*(3), 190–222. doi:10.1210/er.2015-1116 PMID:27035557

Yanovski, J. A. (2015). Pediatric obesity. An introduction. *Appetite*, *93*, 3–12. doi:10.1016/j.appet.2015.03.028 PMID:25836737

Yassa, M., Saliou, M., De Rycke, Y., Hemery, C., Henni, M., Bachaud, J. M., ... Giraud, P. (1824–1827). ... Giraud, P. (2011). Male pattern baldness and the risk of prostate cancer. *Annals of Oncology: Official Journal of the European Society for Medical Oncology*, *22*(8), 1824–1827. doi:10.1093/annonc/mdq695

Zlocha, J., Kovács, L., Pozgayová, S., Kupcová, V., & Durínová, S. (2006). Molecular genetic diagnostics and screening of hereditary hemochromatosis. *Vnitrni Lekarstvi*, *52*(6), 602–608. PMID:16871764

KEY TERMS AND DEFINITIONS

ACE Inhibitors: A drug used to lower blood pressure and heart failure by reducing the amount of angiotensin II. This allows blood vessels to relax and widen so blood can easily flow through.
Albuminuria: Abnormal amounts of albumin present in the urine.
Aneurysms: A condition that occurs when the artery walls weaken, allowing it to abnormally widen.
Angle-Closure Glaucoma: A rarer form of glaucoma that happens when the iris is not as wide and open as it normally is and blocks the drainage canals causing a quick increase in internal eye pressure.
Antihistamines: A drug used to relieve symptoms of or prevent allergic reactions.
Astigmatism: A refractive error in which the cornea's curvature is abnormal causing blurry vision.
Ataxia: A neurological condition involving the loss of voluntary muscle movement.
Bardoxolone Therapy: A treatment used for kidney disease that activates signalling pathways that reduce inflammation and improves mitochondrial activity.

Bestrophinopathies: A group of five clinically distinct retinal degenerative diseases caused by a mutation in the *BEST1* gene.

Cardiomyopathy: A disease of the heart muscle in which the heart muscle can become thick, enlarged, or stiff.

Chronic Rhinosinusitis: Inflammation of the sinus membranes.

Hyperbilirubinaemia: Elevated levels of bilirubin in the blood.

Intraocular Pressure: Pressure inside the eye.

Myopia: A vision disorder in which the eye is too long causing light to focus in front of instead of on the retina—also known as near-sightedness.

Ocular Hypertension: Pressure inside the eye.

Open-Angle Glaucoma: A type of glaucoma that develops slowly and often goes unnoticed. It is when the angle where the cornea meets the iris is normal but there is an increase in internal eye pressure and damage to the optic nerve due to clogging of the eye's drainage canals.

Osteoarthritis: A joint disease caused by the breakdown of cartilage between bones.

Osteoporosis: A disease in which bones become thin and weak.

Photodynamic Therapy: A therapy used to treat cancer by using a drug called photosensitizing agents and a special kind of light to kill cancer cells.

Proteinuria: Presence of protein in the urine.

RAAS Blockade: Treatment for renal diseases using angiotensin-converting-enzyme inhibitors.

Thyroxine (T4): A hormone produced only by the thyroid gland that regulates basal metabolic rate carrying four iodine atoms.

Triiodothyronine (T3): A thyroid hormone carrying three iodine atoms that regulates basal metabolic rate. T3 is typically formed in the target organs by the removal of an iodine atom from a T4 hormone.

Vitelliform Macular Dystrophy: A genetic form of macular degenerative disease that can cause gradual vision loss.

Vitreoretinochoroidopathy: An ocular disorder that affects the retina, vitreous, and the choroid and can lead to vision impairment.

Chapter 16
Endocrine and Immune System Disorders

ABSTRACT

This chapter discusses nine genetic disorders affecting the endocrine and immune systems of the body. These include congenital adrenal hyperplasia, adrenoleukodystrophy, Cockayne syndrome, diastrophic dysplasia, autoimmune polyglandular syndrome, asthma and allergy, severe combined immunodeficiency (SCID), immunodeficiency with hyper-IgM syndromes, and DiGeorge syndrome. The endocrine system made is a complex collection of hormone-producing glands involved in the control of basic body functions such as metabolism, behaviour, growth, and sexual development. The immune system is a complex and highly developed system which protects the body from pathogenic invaders and other foreign molecules or particles like allergens. The body uses three lines of defence against invaders, two of them non-specific and one highly specific. Most immune disorders result from either an excessive immune response or an "autoimmune attack."

CHAPTER OUTLINE

16.1 Overview of Immune and Endocrine Systems
16.2 Congenital Adrenal Hyperplasia
16.3 Adrenoleukodystrophy
16.4 Cockayne Syndrome
16.5 Diastrophic Dysplasia
16.6 Autoimmune Polyglandular Syndrome
16.7 Asthma
16.8 Severe Combined Immunodeficiency
16.9 Immunodeficiency with Hyper-IgM
16.10 DiGeorge Syndrome
Chapter Summary

DOI: 10.4018/978-1-5225-8066-9.ch016

LEARNING OUTCOMES

- Identify each genetic disorder affecting each system
- Outline the symptoms of each disorder
- Explain the genetic basis of each disorder
- Summarize the therapies currently available to treat each disorder

16.1 OVERVIEW OF ENDOCRINE AND IMMUNE SYSTEMS

The endocrine system made is a complex collection of hormone-producing glands involved in the control of basic body functions such as metabolism, behaviour, growth, and sexual development (Figure 1). These endocrine glands are the pituitary, pineal, thyroid and parathyroid, thymus, adrenals, pancreas, and the gonads (ovaries and testes). Hormones are protein secretions made in glands and used in the body as chemical signals and secreted directly into the bloodstream for transport. Blood carries them to distant tissues and organs where they can bind to specific cell sites embedded in the cell membrane made of integral proteins called receptors. Once they bind, hormones can trigger various responses in the tissues containing the receptors. Steroid hormone signals like those coming from oestradiol or aldosterone freely diffuse through the plasma membrane of cells (no receptors at the surface) to bind molecules in the cell cytoplasm.

Protein hormones like insulin, cortisol, epinephrine, glucagon, etc. use a variety of mechanisms once they bind a receptor at the cell surface initiating a signal transduction process as discussed in Chapter 3 (section 3.3). Specific protein receptors embedded in the cell membrane bind the signals, enabling conformational changes inside channel or transport proteins which transport the signal to the cytoplasm. There are three types of protein receptors for non-steroidal signals: G-protein coupled receptors (GPCR), receptor tyrosine-kinases & ion channel proteins. Non-steroid signals activate other molecular or ionic mediators in the cytoplasm which include GPCRs, G-proteins, effector proteins, and receptor proteins for 2nd messengers (protein kinases A, G, C and calmodulin kinase). Transduction involves several protein molecules in a phosphorylation cascade and other multiple mediators (broadly called second messengers). Such second messengers are ions (like Ca^{2+}), lipid metabolism products (like diacylglycerol [DAG], inositol triphosphate [IP3] and phosphatidyl-inositol-biphosphate [PIP2]), nucleotides (like cyclic AMP or cyclic GMP), and a variety of protein mediators that relay, adapt, transduce, anchor, integrate, modulate or amplifier signals (Chapter 3, section 3.3).

The immune system is a complex and highly developed system which protects the body from pathogenic invaders and other foreign molecules or particles like allergens. Pathogens are viruses, bacteria, fungi, protozoans, and parasitic worms that cause disease or disease symptoms. Individuals with severely defective immune systems can die from infection when pathogens overwhelm the body's ability to seek and attack invaders. The body has three lines of defence against invaders, two of them non-specific, and one highly specific.

The non-specific mechanisms include: 1) External barriers like the skin with a low pH of ~3.5 due to sweat gland secretions; mucous membranes with sticky secretions that trap invaders; secretions like tears, saliva and mucus; and the flushing effects of urine and diarrhoea. 2) Internal phagocytic white blood cells (WBCs, see Figure 1, Chapter 13) including the neutrophils (60-70% of WBCs), monocytes

Figure 1. The major human endocrine organs
Source: Image used under license from Shutterstock.com

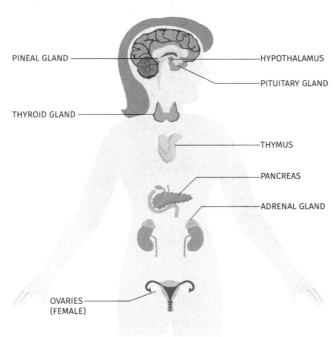

(5% WBCs - develop into macrophages when they reach tissues), eosinophils (1.5% WBCs), and natural killer cells. 3) Inflammatory response initiated by chemical signals released by pathogens or injured cells trigger the release of histamines and prostaglandins which collectively trigger arteriole dilation, venule constriction, and capillary permeability (causing oedema or swelling); while also promoting blood flow to damaged tissue (hence the redness and heat of inflammation) (Figure 2). Often also, toxins from pathogens and the signals from WBCs called pyrogens set the body's thermostat higher—creating a moderate fever which inhibits pathogen growth, facilitates phagocytosis, and speeds up tissue repair. 4) A group of proteins called antimicrobial proteins are also involved which trigger two processes: a) a set of 20 proteins (complement system) that work in a cascade of reactions to form a membrane attack-complex which lyse pathogens is initiated; and b) interferons produced by virus-infected cells diffuse to neighbouring cells alarming them on impeding infection; thereby controlling spread of viral infections like colds or flu.

The specific mechanism of protecting the body is the most important, comprehensive and precise targeting individual foreign invaders. It is differentiated from the non-specific mechanisms by four main features: a) Specificity – the ability to recognize and eliminate specific invaders using antibody-antigen interactions; b) Diversity – ability to respond to a variety of invaders irrespective of geographical location; c) Self/Non-self-recognition – the ability to distinguish the body's own molecules from foreign ones—a process called self-tolerance made possible by two types of markers in body cells called major histocompatibility complexes class I and II (MHCs); d) Memory – the ability of the system to remember previously encountered pathogens, hence responding much faster and more effectively. After an immune response, some B cells remain in body circulation in perpetuity as memory cells.

Figure 2. What happens during inflammation?
Source: Image used under license from Shutterstock.com

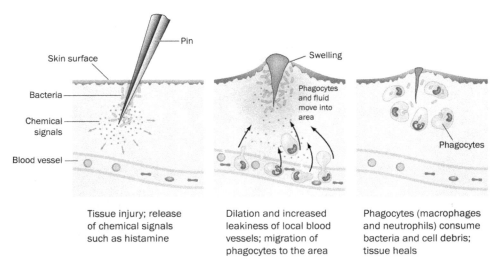

The cells of this final line of defence are the lymphocytes made up in the bone marrow and of two types of cells: B cells (which mature in the bone marrow) and T cells (which migrate to mature in thymus gland). Mature B and T cells concentrate in the lymph nodes, spleen and other lymphatic organs. The effector cells of the B lymphocytes are the plasma cells which produce five types of antibodies—a class of proteins called immunoglobulins or Igs: IgM (develops first in new-borns), IgG (most abundant, in breast milk), IgA (tears, saliva, breast milk) IgD (activates helper-T cells) and IgE (present during allergies). The effector cells of the T lymphocytes include helper T-cells which stimulate B cells to produce antibodies, cytotoxic T-cells which lyse infected cells, and suppressor T-cells which turn off the immune response when pathogen threat diminishes.

Most immune disorders result from either an excessive immune response or an 'autoimmune attack'. Being that a key part of the immune system is to differentiate between invaders and the body's own cells; occasionally this distinction fails triggering a reaction against 'self' cells and molecules (autoimmune attacks).

16.2 CONGENITAL ADRENAL HYPERPLASIA (CAH)

CAH is an umbrella term defining a group of syndromes resulting from inherited defects in one of the five enzymes involved in the biosynthesis of cortisol from cholesterol. These enzymes include 21α-hydroxylase, 11β-hydroxylase, 17α-hydroxylase/17,20-lyase, cholesterol desmolase, and 3β-hydroxysteroid dehydrogenase (Xu et al., 2017). It was initially described in 1865 by Luigi de Crecchio (Delle Piane et al., 2015). The defective cortisol production attenuates the negative feedback to the hypothalamus and pituitary gland, resulting in excessive secretion of corticotropin-releasing hormone (CRH) and adrenocorticotropin (ACTH), respectively. The raised ACTH, in turn, cannot undo the disrupted cortisol synthesis, but its trophic action leads to enlargement of the adrenal gland (**hyperplasia**). The most severe forms

Endocrine and Immune System Disorders

are conventionally called "classic" CAH and are the easiest to recognize. It affects approximately 1 in 16,000 new-borns worldwide. (Finkielstain et al., 2012; Delle Piane et al., 2015; Turcu & Auchus, 2015; Xu et al., 2017)

Symptoms

Diagnosis of all forms of CAH currently relies almost exclusively on non-invasive genetic testing and hormonal testing which has many shortcomings. Symptoms of classic CAH due to 21-hydroxylase deficiency include dehydration, poor feeding, diarrhea, vomiting, heart rhythm problems (arrhythmias), low blood pressure, very low blood sodium levels, low blood glucose, acidosis, weight loss, and shock (due to confusion, irritability, rapid heart rate, and/or coma). In non-classic CAH due to mild 21-hydroxylase deficiency, symptoms observed are related to increased androgens and can include: rapid growth in childhood and early teens but shorter height than both parents, early signs of puberty, acne, irregular menstrual periods (females), fertility problems (in about 10% to 15% of women), excess facial or body hair in women, male-pattern baldness (hair loss near the temples), enlarged penis (males), and small testicles (males). (NICHD, 2018).

Genetic Basis

The most common form of CAH is deficiency of 21-hydroxylase caused by a defective gene CYP21A2, found on chromosome #6 (6p21.3) accounting for over 90% of all cases. A second form of CAH is the deficiency of the enzyme 11β-hydroxylase due to defects in the gene CYP11B1 on chromosome #8 (8q24.3) representing up to 5–8% of all CAH cases. A third form of CAH is deficiency in the enzyme 3β-hydroxysteroid dehydrogenase/isomerase type 2 due to defects in the gene *HSD3B2* on chromosome #1 *(*1p13.1) and is characterised by both mineralocorticoid and glucocorticoid deficiency. Other defective genes described in the literature include CYP17A1 (10q21-q22), STAR (8p11.2), CYP11A1 (15q23–q24) and POR (7q11.2). (Honour, 2009; Delle Piane et al., 2015; Turcu & Auchus, 2015; Xu et al., 2017)

Current Therapies

Treatment goals for all forms of CAH are geared to replace the deficient hormones and to offset the undesirable effects of excessive hormonal production. A commonly used approach has been that of suppressing ACTH by strategic dosing of glucocorticoids. Sustained-released hydrocortisone preparations have been recently developed, with the goal of mimicking the cortisol circadian rhythm and to suppressing ACTH early morning elevation. Continuous subcutaneous hydrocortisone infusion via a pump, replicating a circadian secretory profile, has been used experimentally in young patients with increased cortisol. An alternative method for decreasing ACTH used corticotropin-releasing factor receptor type 1 antagonist. Directly inhibiting androgen synthesis or antagonizing androgen action has also been used as a CAH management strategy. A combination of *flutamide* (an antiandrogen) and *testolactone* (an aromatase inhibitor), has permitted the use of lower doses of hydrocortisone and fludrocortisone acetate and normalized linear growth and bone maturation in children followed for a 2-year period. (White et al., 1987; Laue et al., 1996; Therrell et al., 1998; Merke, 2012; Turcu & Auchus, 2015)

16.3 ADRENOLEUKODYSTROPHY (ALD)

ALD is a progressive "inborn error of metabolism" affecting the brain, spinal cord, peripheral nerves, adrenal cortex and testis. It is an X-linked inherited peroxisomal disorder with a combined incidence of hemizygotes (all phenotypes) and heterozygous female carriers of about 1:16,800 in new-borns (Berger et al., 2014). The peroxisomal matrix degrades saturated, unbranched and very long-chain fatty acids (fatty acyl-chain length of ≥ 22 carbons) by sequential reactions of three enzymes (acyl-CoA oxidase 1, D-bifunctional protein, and either peroxisomal β-ketothiolase 1 or sterol carrier protein x) of the β-oxidation pathway. In ALD patients, saturated long-chain fatty acids accumulate in tissues and body fluids as a diagnostic marker for ALD (Kemp et al., 2012; Wiesinger et al., 2013; Wiesinger et al., 2015). ALD has two predominant phenotypes prevailing: 1) adrenomyeloneuropathy characterised by a slowly progressive axonopathy with the first symptoms usually appearing between 20 and 30 years of age in males and between 40 and 50 years in affected females; 2) The cerebral form usually only affects males and presents with rapidly progressive inflammatory **demyelination** in the brain, leading to rapid cognitive and neurological decline (Wiesinger et al., 2015; Eichler et al., 2017).

Symptoms

A key clinical symptom during aging of ALD patients is a progressive destruction of the axons of peripheral nerves (**axonopathy**) affecting ascending sensory and descending motor spinal cord tracts with 100% of men showing it, and 65% in heterozygous women by the age of 60 years (Berger et al., 2014). The initial symptoms include progressive stiffness and weakness of the legs, impaired vibration sense in the lower limbs, sphincter disturbances and impotence, as well as scarce scalp hair (**alopecia**). About 60% of male ALD patients also present with rapidly progressive, inflammatory cerebral demyelination. (Engelen et al., 2012; Eichler et al., 2017)

Genetic Basis

Mutations affecting the ATP-binding cassette ABC transporter sub-family D member 1 (*ABCD1*) gene, located at chromosome Xq28, have been identified. There are a reported more than 800 different *ABCD1* mutations (ALD Info, 2018). The *ABCD1* gene encodes the peroxisomal half-transporter ABCD1. Studies have shown that he degradation of saturated very long-chain fatty acids by peroxisomal β-oxidation is greatly reduced in cultured ALD fibroblasts (Mosser et al., 1993; Wiesinger et al., 2013; Berger et al., 2014; Wiesinger et al., 2015).

Current Therapies

The available therapies for ALD include replacement therapy with adrenal steroids (mandatory for all patients with impaired adrenal function); dietary therapy with "Lorenzo's Oil" (shows a preventive effect in asymptomatic boys whose brain MRI is normal); and hematopoietic stem cell transplantation in patients in the early stage of the cerebral inflammatory phenotype. Hematopoietic stem cell transplantation or *ex vivo* gene correction of autologous hematopoietic stem cells can arrest the inflammation in early stages providing an efficient treatment for the inflammatory form of ALD. Studies also show that **Lenti-D gene therapy** may be an alternatively safe and effective treatment to **allogeneic stem-cell**

transplantation in boys with early-stage cerebral ALD. Although for the majority of patients with ALD there is currently no curative or preventive treatment, several promising new approaches are being investigated. For instance, research shows that in ALD cells small interfering RNA (siRNA)-mediated inhibition of very long-chain-fatty acids reduces their synthesis and levels. Compounds that can inhibit elongation of very long-chain-fatty acids are treatment candidates. *Bezafibrate* may be effective, safe and well-tolerated compound which lowers the levels of very-long-chain-fatty acids in ALD cells by inhibiting their synthesis. (Mosser, 2006; Engelen et al., 2012; Wiesinger et al., 2015; Eichler et al., 2017)

16.4 COCKAYNE SYNDROME

Cockayne syndrome was first described in 1936 by Edward Cockayne. It has an incidence of 1 in 250,000 live births, and a prevalence of about 2.5 per million (Karikkineth et al., 2017). CS is an autosomal recessive disorder. CS complex are rare autosomal recessive disorders characterised by profound postnatal brain, somatic growth failure and by degeneration of multiple tissues resulting in cachexia, dementia, and premature aging. Average age at death has been estimated at 5, 16, and 31 years in severely, moderately, and mildly affected patients respectively (Natale, 2011). Cockayne syndrome spans a phenotypic spectrum that includes CS type I, CS type II, CS type III, and Xeroderma pigmentosum-Cockayne syndrome (Laugel et al, 2011).

Symptoms

Symptoms of CS include sunken eyes, large ears, thin pointy nose, small chin, dental caries, enamel hypoplasia, photosensitivity, wrinkled and aged appearing skin, thin dry hair, prematurely grey hair, microcephaly usually beginning at age 2, mental retardation with low IQ, delayed milestones, tremors, ataxia, seizures, strokes and subdural haemorrhages, corneal opacification, cataracts [~36–86%], cachectic dwarfism, scoliosis, stooped posture, muscle wasting, renal failure, decreased production of sweat, tears, saliva, micropenis and smaller testicular size (males) and ovarian atrophy (females). (Laugel et al, 2011; Karikkineth et al., 2017)

Genetic Basis

CS is often described as a disorder of transcription and transcription-coupled nucleotide excision repair. It is classified into mainly 2 different complementation groups: 1) CSA, due to a mutation on ERCC8 on chromosome 5q12–q31 and accounts for 20% of CS cases; and 2) CSB, due to a mutation on ERCC6 on chromosome 10q11. CSB accounts for 80% of cases (Natale, 2011; Laugel, 2013; Wilson et al., 2016).

Current Therapies

There is currently no cure for CS and no treatments are available which could slow or halt the progression of the disease. Therefore, the goal of surveillance is to maximize quality of life. The mean age of death is approximately 12 years. However, for symptomatic patients, physical therapy, cochlear implants for hearing, cataract surgeries, sunscreen for photosensitivity, and feeding tubes for malnutrition may be helpful. (Wilson et al., 2016)

16.5 DIASTROPHIC DYSPLASIA (DD)

DD is an autosomal recessive skeletal **dysplasia** caused by SLC26A2 mutation first described by Lamy and Maroteaux in 1960. Individuals with the condition are of short stature with short tubular bones, limited joint motion, **scoliosis**, hypertrophied auricular cartilage, but of normal intelligence. DD is an autosomal recessive trait and due to its phenotypic variability, early diagnosis is challenging but quantitative methods based on radiographic measurements can be used. (Butler et al., 1987). Studies indicate that DTD is overrepresented in Finland and genetic analysis has shown that a specific mutation designated as "DTDST(Fin)," is present in affected members of many Finnish families suggesting that a founder mutation may have occurred in a common ancestor in the past. (Bonafé et al., 2008; Mäkitie et al., 2015). In Finland, 1-2% of the general population are carriers and a total of 183 cases have been diagnosed, with a prevalence ratio of 1 in 30,000.

Symptoms

The symptoms associated with DD are extremely variable, differing in range and severity even among affected family members. However, in all individuals with the disorder, there is abnormal development of bones and joints of the body (skeletal and joint dysplasia). Clinical features of DD include short stature, joint contractures, spinal deformities, and cleft palate. (NORD, 2018)

Genetic Basis

Mutations in the sulphate transporter gene *SLC26A2* (*DTDST*) cause a continuum of skeletal dysplasia phenotypes that includes **achondrogenesis** type 1B (ACG1B), **atelosteogenesis** type 2 (AO2), diastrophic dysplasia (DTD), and milder recessive multiple epiphyseal dysplasia (rMED). The gene family *SLC26/SulP* encodes anion exchangers and channels expressed throughout evolution. SLC26A2 was discovered as a SO_4^{2-} uptake transporter in human chondrocytes and in mice it is expressed in chondrocytes, fibroblasts, and osteoblasts.

Current Therapies

Since DD cannot be cured, therapies available are symptomatic and supportive. Carefully monitoring of affected infants is crucial to ensure prompt detection and appropriate preventive or corrective treatment of respiratory obstruction and distress. As outlined by NORD (2018a) treatments for DD require the coordinated efforts of a team of specialists working together to systematically and comprehensively plan an affected child's treatment. Such specialists may include paediatricians; orthopaedists; surgeons; physical therapists; orthodontists; audiologists; plus other health care professionals. Surgery can be used to correct malformations resulting in breathing and/or feeding difficulties; while orthopaedic techniques, may be performed to help prevent, treat, and/or correct certain skeletal deformities. For children with dental abnormalities, braces, dental surgery, and other corrective procedures are recommended. Genetic counselling is also recommended and may benefit affected individuals and their families. (NORD, 2018a)

16.6 AUTOIMMUNE POLYGLANDULAR SYNDROME (APS)

APS is a rare **polyendocrinopathy** characterised by several endocrine glands and non-endocrine organs failure caused by immune-mediated destruction of tissues (Dittmar & Kahaly, 2003). Autoimmune diseases affecting one organ are frequently followed by the impairment of other glands, resulting in multiple endocrine failure. The hallmark evidence of APS is development of autoantibodies against multiple endocrine and non-endocrine organs causing dysfunction of the affected organ. The most frequently encountered APSs are adrenal insufficiency, autoimmune thyroid disease, insulin-dependent diabetes mellitus, hypoparathyroidism and premature gonadal failure. The classification of APS is based on clinical criteria with four main types: 1) APS-1 with features like chronic candidiasis, chronic **hypoparathyroidism**, and **Addison's disease**. 2) APS-2 presents with Addison's disease, autoimmune thyroid disease, and type 1 diabetes mellitus. 3) APS-3 features autoimmune thyroid diseases associated with other autoimmune diseases (excluding Addison's disease and/or hypoparathyroidism). 4) APS-4 exhibits combinations not included in the previous groups. (Neufeld & Blizzard, 1980; Büyükçelik et al., 2014; Kirmizibekmez et al., 2015; Perniola, 2018)

Symptoms

Symptoms of APS-1 are variable in each patient, exhibiting at least two of the three major conditions that result from the syndrome: chronic mucocutaneous candidiasis, hypoparathyroidism, and adrenocortical insufficiency. Patients with APS-1 can also develop many other autoimmune disorders, including autoimmune liver disease (chronic active hepatitis), ovarian failure (hypogonadism), early onset pernicious anaemia from atrophic gastritis, and a variety of gastro-intestinal problems resulting in chronic malabsorption and diarrhoea. Symptoms of APS-2 may present with shortness of breath, fatigue, weakness, rapid heartbeat, angina, anorexia, abdominal pain, indigestion, and possibly intermittent constipation and diarrhoea. (NORD, 2018a, 2018b)

Genetic Basis

APS is caused by mutations in the autoimmune regulator (AIRE) gene coding for a transcription factor with an autosomal recessive pattern of inheritance. The AIRE gene is located on chromosome 21q22.3, is primarily expressed in thymic medullary epithelial cells, with lesser amounts in lymph nodes and tonsils, and controls induction and maintenance of immune tolerance to self-antigen. Analysis of its multidomain structure reveals that human AIRE belongs to the group of proteins able to bind to chromatin and regulate the process of gene transcription. Over 60 mutations in the AIRE gene have been reported with the two most common mutations being a nonsense mutation and a 13 base-pair deletion in specific exons of the gene. Current research shows that AIRE does not act as a conventional transcription factor by binding to consensus sequences of the target gene promoters; it participates in coordinated events performed by multimolecular complexes. These complexes include CREB-binding protein, DNA-activated protein kinase, DNA-topoisomerase, poly-(ADP-ribose) polymerase 1, bromodomain-containing domain 4, positive transcription elongation factor b, RNA-polymerase II, and a tissue-specific antigen. In the absence of the AIRE protein, many tissue-specific self-antigens are not expressed in the thymus, and multiorgan autoimmunity develops because of autoreactive T cells death. (Cheng et al., 2010; Büyükçelik et al., 2014; Huibregtse et al., 2014; Perniola, 2018)

Current Therapies

Treatment of APS focuses on specific problems include hormones replacement, insulin management for diabetes, and treating the yeast infection. There is no known cure yet for APS. Accurate and timely diagnosis with subsequent appropriate hormone replacement therapy is often life-saving. Therapeutic options are hormone replacement, immunosuppression and avoiding infection. Hormone replacement therapy remains the cornerstone – glucocorticoids and mineralocorticoids for Addison's disease, levothyroxine for hypothyroidism and insulin for type 1 diabetes. (Hargovind et al., 2011) Recently, bone marrow-derived haematopoietic stem cells put into portal circulation with conditioning of *cyclophosphamide*, *bortezomib*, *rituximab* and rabbit-antithymoglobulin have been reported to be helpful therapies (Kirmizibekmez et al., 2015).

16.7 ASTHMA

Asthma and allergy are common conditions that exhibit complex and heterogeneous aetiologies resulting from the complex interplay between genetic factors and environmental exposures that occur at critical times in development. Asthma and allergic disease often co-occur in the same individual or in different individuals within the same families. Both asthma and allergic diseases have significant genetic contributions, with heritability estimates varying between 35% and 95% for asthma, and 33% and 91% for allergic rhinitis (hay fever). Allergy can be defined as a detrimental immune-mediated inflammatory response to often harmless environmental allergens, resulting in one or more allergic diseases such as asthma, allergic rhinitis, atopic dermatitis, and food allergy. The global prevalence of asthma ranges from 1% to 21% of the population in different countries. It is estimated that asthma affects as many as 300 million people worldwide—expected to increase to 400 million by 2025. Asthma is estimated to account for 250,000 deaths each year worldwide. Allergic rhinitis, clinically defined as a symptomatic disorder of the nose induced after allergen exposure by an IgE-mediated inflammation, affects 10% to 20% of the population, with an estimated over 500 million affected people worldwide. Severe asthma accounts for the majority of healthcare costs due to hospitalizations and ED visits, and is associated with the highest asthma related mortality. (Kim et al., 2017)

Epidemiologic studies have highlighted many associations between environmental exposures and subsequent risk for asthma and allergy. Several factors have been reported as modifiers of asthma risk including genetics, race/ethnicity, gender, passive and active tobacco smoke exposure, farm animals and related products, domestic cats and dogs, family size and birth order, day-care attendance in early childhood, respiratory viral infections, microbial exposures, vaccination, antibiotics and antipyretics, mode of delivery, breastfeeding, intake of diet and nutrition, air pollution, obesity, allergens, and occupational exposures. (Renz, 2013; Sadatsafavi et al., 2013; Wenzel et al., 2013; Szefler et al., 2014; Reddel et al., 2015; Loukides, 2016; Kim et al., 2017)

Symptoms

Asthma is not a single disease but rather an umbrella for multiple diseases with similar clinical features, and likely with different genetic and environmental contributor. Its diagnosis is based on clinical features like demonstration of reversible expiratory airflow obstruction and exclusion of alternative diagnoses

Endocrine and Immune System Disorders

that mimic asthma. Asthma is a complex syndrome characterised by the presence of chronic inflammation in the lower airways, resulting in variable airflow obstruction and bronchial hyperresponsiveness and causing recurrent episodes of coughing, wheezing, breathlessness, and chest tightness. respiratory symptoms such as wheeze, shortness of breath, that vary over time and in intensity, together with variable expiratory airflow limitation. (Reddel et al., 2015; GIA, 2018)

Genetic Basis

Recent genome-wide association studies (GWAS), meta-analyses of GWAS, and genome-wide interaction studies (GWISs) have begun to shed light on both common and distinct pathways that contribute to asthma and allergic diseases. GWISs test for interaction effects between each SNP and a specific exposure on asthma risk. At least 15 to 20 asthma risk loci have been identified by GWAS. The great challenges scientists face is to elucidate the effects of associated variation on gene regulation or function and of associated genes on asthma pathogenesis. Several SNPs in or near the following genes have been identified and associated to asthma by GWAS: *HLA-DQ, ORMDL3/GSDML, IL33, IL18R1, IL2RB, SMAD3, SLCA22A5, IL13, RORA, FCER1A, IL13, STAT6, IL4R/IL21R, PYHIN1, TSLP,* and *ADAM33.* (Metzker, 2010; Ober & Yao, 2011; Reddel et al., 2015; Zein et al., 2016; Bønnelykke & Ober, 2017)

Current Therapies

There are now several biomarkers emerging that hold promise for selecting and monitoring therapy including exhaled nitric oxide, serum IgE, periostin, and urinary leukotrienes. Research shows that allergen-specific immunotherapy for paediatric asthma, inhaled corticosteroids and long-acting ß-adrenergic agonist therapy, and leukotriene receptor antagonist therapy are treatment modalities for asthma. There are several new drugs in development or being evaluated in adults: AMG 853, a potent, selective, orally bioavailable, small molecule dual antagonist of D-prostanoid; CRTH2, chemoattractant receptor homologous molecule expressed on TH2 cells; *lebrikizumab*, an anti-IL-13 monoclonal antibody; *dupilumab*, a monoclonal antibody to the alpha subunit of the interleukin-4 receptor; and *benralizumab*, a monoclonal antibody designed to target IL-5Ralpha expressed on eosinophils and basophils. Specific microRNAs have been found to have critical roles in regulating key pathogenic mechanisms in allergic inflammation, including polarization of adaptive immune responses and activation of T cells, regulation of eosinophil development, and modulation of IL-13-driven epithelial responses. MicroRNAs can be useful as biomarkers, as well as help develop microRNA-related novel treatment modalities. Other studies have focused on the use of vitamin D supplementation, and of probiotics in early life. (Moffatt et al., 2010; Tsai et al., 2012; Busse et al., 2013; Elazab et al., 2013; Laviolette et al., 2013; Lu & Rothenberg, 2013; Muehleisen & Gallo, 2013; Noonan et al., 2013; Rebane & Akdis, 2013; Renz, 2013; Sadatsafavi et al., 2013; Wenzel et al., 2013; Szefler et al., 2014; Reddel et al., 2015; Zein et al., 2016; Kim et al., 2017)

16.8 SEVERE COMBINED IMMUNODEFICIENCY (SCID)

SCID is a rare disorder occurring in infancy with life-threatening bacterial, viral or fungal infections; failure to thrive; and diarrhoea. SCID is a collection of individual heritable primary immunodeficiency diseases, resulting from defects in genes controlling the maturation of elements of the adaptive immune system. Popularly known as the "bubble boy disease", it is characterised by severe defects of cellular and humoral immunity that render affected infants susceptible to opportunistic and recurrent infections. Considered the most severe form of primary immunodeficiency, SCID is generally fatal in the first year of life unless recognized early and treated.

Symptoms

Infants born with SCID typically appear normal at birth, are characterised by a severe deficiency of naïve T cells and are at high risk of serious infections after waning of maternal antibody at around 4-6 months of age. If untreated, SCID-affected infants are susceptible to recurrent and opportunistic infections, persistent diarrhoea, faltered growth, and early death.

Genetic Basis

SCID can be caused by mutations in various genes, predominantly affecting T-cell immunity. In SCID, T-cell activation and function are impaired, or T-cell development is hampered causing low or absent peripheral T cells. SCID has traditionally been classified into those patients with residual B cells (T-B+ phenotype) and those whose defects produce an absence of both T cells and B cells (T-B- phenotype). Distinct genetic forms of SCID can be subdivided into T-B+, T-B- or T+B+ SCID, depending on the presence/absence of the respective cell line. The T-B- phenotype accounts for approximately 30% of SCID patients. Mutations in Recombination Activating genes (RAG) 1 and 2, Artemis (DNA cross-link repair enzyme 1C), DNA Ligase IV, DNA-PKcs (DNA-dependent protein kinase catalytic subunit), and Cernunnos-XLF have all been reported to cause SCID in humans. (Schwarz et al., 1996; Li et al., 2002; Noordzij et al., 2003; Kwan & Puck, 2015; Volk et al., 2015)

Many proteins are essential for T cell development, and as many as 20 known genes have been associated with SCID due to deleterious mutations that alter protein expression and prevent the development and maturation of T cells. Almost 50% of SCID cases are caused by *IL2RG* mutations, while all other known SCID defects caused by mutations in autosomal recessive genes. Among the genetic defects that cause T-B- SCID are biallelic mutations in *DCLRE1C* which encodes *Artemis* (a nuclease with intrinsic 5'-3' exonuclease activity on single-stranded DNA). After phosphorylation by and in complex with DNA-dependent protein kinase catalytic subunit, *Artemis* acquires endonuclease activity on 5' and 3' overhangs and hairpins. It is involved in non-homologous end-joining and is essential for opening hairpins, which arise as intermediates during the recombination of the immunoglobulin and T-cell receptor genes in T- and B-cell development. SCID with faulty. (Schwarz et al., 1996; Kwan & Puck, 2015; Volk et al., 2015)

Current Therapies

Established treatments for SCID include restoring the faulty immune system by means of allogeneic hematopoietic cell transplant (HCT) from human leukocyte antigen-matched related or unrelated donors, haploidentical parental donors, or cord blood; enzyme replacement therapy for the adenosine deaminase (ADA)-deficient form of SCID; or experimental gene therapy for X-linked SCID and ADA-SCID. Generally, HCTs have been successful over the past decades, and outcomes following HCT for SCID-affected infants are optimized by earlier HCT and by effective prevention and treatment of infections prior to HCT. Initial gammaretroviral gene therapy trials for other types of SCID have proved effective. (Schwarz et al., 1996; Li et al., 2002; Noordzij et al., 2003; Dvorak & Cowan, 2010; van Til et al., 2012; Kwan & Puck, 2015; Volk et al., 2015)

16.9 IMMUNODEFICIENCY WITH HYPER-IGM

The hyper IgM syndromes are a group of rare genetic disorders leading to loss of T cell-driven immunoglobulin class switch recombination and/or defective somatic hypermutation as well as impaired T cell activation. X-linked hyper IgM is the most common subtype of hyper-IgM syndromes affecting about 70% of the patients (Wang et al., 2014).

Symptoms

As in other primary immune deficiencies, serious infections, autoimmunity, inflammatory complications, and malignancy may develop in hyper IgM individuals. The most common symptoms of X-linked hyper IgM patients were recurrent sinopulmonary infections, neutropenia, oral ulcer, protracted diarrhoea; diffuse problems in respiratory and gastrointestinal systems; skin, bone and joint systems; and central nervous and urinary systems.

Genetic Basis

The most common causes of hyper IgM are mutations in the CD40 Ligand (*CD40LG*) gene located on chromosome X (at Xq26.3-27) in males. The *CD40LG* gene product is the CD40L transmembrane glycoprotein normally expressed on activated CD4+ T lymphocytes. The glycoprotein interacts with CD40 that is constitutively expressed on B cells, macrophages, monocytes, and dendritic cells. This interaction is a critical first step in B cell stimulation for growth and proliferation resulting in the generation of both functional antibodies and memory B cells. Other lesser causes of the hyper IgM syndrome are autosomal recessive mutations in *CD40* which also lead to loss of these B cell functions; mutations in activation-induced cytidine deaminase (*AICDA*) and uracil-DNA glycosylase (*UNG*) affecting males and females equally and leading to different clinical phenotypes. Both AID and UNG act downstream of CD40/CD40L and mutations in these genes lead to subsequently low serum levels of IgG, IgA, and IgE. More complex and involving different mechanisms are the recently described autosomal dominant gain of function mutations in phosphoinositide 3-kinase catalytic delta component (*PIK3CD*) which may lead to **hypogammaglobulinemia** with increased serum IgM and increased susceptibility to infections, autoimmunity, **lymphoid hyperplasia**, and potentially lymphoma. Studies also indicate that males with mutations in NF kappa B essential modulator gene (*IKBKG/NEMO*) may have increased serum IgM levels as well **ectodermal dysplasia** and are sometimes included in the hyper IgM designation. (Aruffo et al., 1993; Zonana et al., 2000; Davies & Thrasher, 2010; Durandy & Kracker, 2012; Hirbod-Mobarakeh et al., 2014; Kracker et al., 2014; Leven et al., 2016)

Current Therapies

Antimicrobials used include *oral trimethoprim/ sulfamethoxazole, acyclovir,* anti-fungal *(itraconazole)* regimen, *dapsone, pentamidine, atovaquone, azithromycin, amoxicillin, thalidomide, cephalexin,* and *amoxicillin/clavulanic acid*, and intravenous (IV) or oral steroids. Immunoglobulin therapy using IV or intramuscular or subcutaneous administration routes has been used. Hematopoietic stem cell and solid organ transplants using matched sibling bone marrow, cord blood, parental bone marrow and also matched unrelated donors or stem cells. Other patients received liver transplants. (Hirbod-Mobarakeh et al., 2014; Kracker et al., 2014; Leven et al., 2016)

16.10 DIGEORGE SYNDROME

DiGeorge syndrome is a complex disorder caused by an embryopathy first described by Dr. Angelo DiGeorge in 1965. The unifying sign of DiGeorge syndrome is a wide spectrum of pathologies of the thymus, ranging from complete athymia, found in complete DiGeorge syndrome patients, to various degrees of thymic dysplasia and disturbed migration of the thymus observed in partial DiGeorge syndrome patients. It occurs in 1 in every 3000–6000 births and is equally distributed between males and females.

Symptoms

The DiGeorge syndrome is a debilitating, multisystemic condition that features (with variable expressivity) cardiac malformations, velopharyngeal insufficiency, hypoparathyroidism with hypocalcemia, and thymic aplasia with immune deficiency. In infants, DiGeorge syndrome typically presents with congenital cardiac anomalies (~75%); immunodeficiency (~75% of patients); and hypocalcaemia due to hypoparathyroidism (~50%). Additional symptoms comprise palatal abnormalities (~75%); gastrointestinal problems such as reflux disease or dysmotility (~30%); genitourinary anomalies such as renal agenesis (~30%); hernia (~1%); neuronal tube defects; polydactyly; anomalies in the spine, face, ear, nose, eyelids or eyes; malignancies or seizures; early-onset Parkinson disease (~30%); and psychiatric disorders such as anxiety disorders and schizophrenia (~60%). (Davies, 2013; Chinn et al., 2013; McDonald-McGinn & Sullivan, 2011; Jonas et al., 2014; McDonald-McGinn et al., 2015; Swillen & McDonald-McGinn, 2015; Lopez-Rivera et al., 2017; Kraus et al., 2018)

Genetic Basis

The DiGeorge syndrome embryopathy is most commonly caused by a deletion on chromosome #22 (22q11.2) but can be caused by other less known mutations. 22q11.2 deletion syndrome is the most common chromosomal microdeletion disorder, estimated to result mainly from non-homologous meiotic recombination events. Today, most (90–95%) newly identified patients with this deletion are found to have *de novo* deletions—that is, neither parent has the deletion. The DiGeorge syndrome involves microdeletions (approximately 0.7–3 million base pairs in size), resulting in a heterogeneous clinical presentation, irrespective of deletion size associated with multi-organ dysfunction. Aberrant inter-chromosomal exchanges involving 8 large, paralogous low copy repeats of DNA span chromosome

Endocrine and Immune System Disorders

22q11.2. These low copy repeats, long nucleotide stretches with extremely high sequence-homology, mispair during meiosis, causing hemizygous chromosomal deletions of variable lengths. The DiGeorge Syndrome Critical Region 8 gene (DGCR8) is also deleted on chromosome 22q11.2. This gene is a pri-microRNA-binding protein required for microRNA biogenesis. Remember that MiRs are a family of small, non-coding RNAs that modulate gene expression by targeting specific messenger RNAs for degradation, translational repression, or both. MiRs affect a wide range of biological responses including proliferation, differentiation, apoptosis, and/or stress responses. In mouse models 22q11.2-deletion, a deficiency of DGCR8 causes a 20–70% reduction in a subset of miRs in the brain causing cognitive impairment. (Landthaler et al., 2004; Bartel, 2009; Fénelon et al., 2011; McDonald-McGinn & Sullivan, 2011; Lopez-Rivera et al., 2017)

Current Therapies

Early diagnosis, preferably prenatally or neonatally, can greatly improve outcomes, emphasizing the importance of universal screening. Management of the syndrome requires an individualized, multidisciplinary and coordinated care plan that takes into account the associated features of the patient. Management requires a multidisciplinary approach involving paediatrics, general medicine, surgery, psychiatry, psychology, interventional therapies (physical, occupational, speech, language and behavioural) and genetic counselling. (Chinn et al., 2013; Davies, 2013; Swillen & McDonald-McGinn, 2015; Lopez-Rivera et al., 2017; Kraus et al., 2018)

CHAPTER SUMMARY

The endocrine system made is a complex collection of hormone-producing glands involved in the control of basic body functions such as metabolism, behaviour, growth, and sexual development. These endocrine glands are the pituitary, pineal, thyroid and parathyroids, thymus, adrenals, pancreas, and the gonads (ovaries and testes). Steroid hormone signals freely diffuse through the plasma membrane of cells (no receptors at the surface) to bind molecules in the cell cytoplasm. Protein hormones use a variety of mechanisms once they bind a receptor at the cell surface initiating a signal transduction process that travels through different cellular components until a response is generated. The immune system is a complex and highly developed system which protects the body from pathogenic invaders and other foreign molecules or particles like allergens. The body uses three lines of defence against invaders, two of them non-specific, and one highly specific. Most immune disorders result from either an excessive immune response or an 'autoimmune attack'.

Congenital Adrenal Hyperplasia is an umbrella term defining a group of syndromes resulting from inherited defects in one of the five enzymes involved in the biosynthesis of cortisol from cholesterol. The most common form of CAH is deficiency of 21-hydroxylase caused by a defective gene CYP21A2, found on chromosome #6 (6p21.3) accounting for over 90% of all cases. There is also a second and third form of CAH affecting different genes. Treatment goals for all forms of CAH are geared to replace the deficient hormones and to offset the undesirable effects of excessive hormonal production.

Adrenoleukodystrophy, an X-linked inherited peroxisomal disorder, is a progressive, "inborn error of metabolism" affecting the brain, spinal cord, peripheral nerves, adrenal cortex and testis. A key clinical symptom during aging of ALD patients is a progressive destruction of the axons of peripheral nerves (axonopathy) affecting ascending sensory and descending motor spinal cord tracts. Mutations affecting the ATP-binding cassette ABC transporter sub-family D member 1 (ABCD1) gene, located at chromosome Xq28, have been identified on this gene which can have more than 800 different mutations. The available therapies for ALD include replacement therapy with adrenal steroids (mandatory for all patients with impaired adrenal function); dietary therapy with "Lorenzo's Oil," (shows a preventive effect in asymptomatic boys whose brain MRI is normal); and hematopoietic stem cell transplantation in patients in the early stage of the cerebral inflammatory phenotype.

Cockayne syndrome was first described in 1936 by Edward Cockayne has an incidence of 1 in 250,000 live births. The CS complex are rare autosomal recessive disorders characterised by profound postnatal brain, somatic growth failure and by degeneration of multiple tissues resulting in cachexia, dementia, and premature aging. CS is a disorder of transcription and transcription-coupled nucleotide excision repair occurring in two forms: CSA, which accounts for 20% of cases CSB, accounting for 80% of cases. There is currently no cure for CS and no treatments are available which could slow or halt the progression of the disease; thus, the goal of surveillance is to maximize quality of life.

Diastrophic Dysplasia is an autosomal recessive skeletal dysplasia first described by Lamy and Maroteaux in 1960. Individuals with the condition are of short stature with short tubular bones, limited joint motion, scoliosis, hypertrophied auricular cartilage, but of normal intelligence. Studies indicate that DTD is overrepresented in Finland and genetic analysis has shown that a specific mutation designated as "DTDST(Fin)," is present in affected members of many Finnish families suggesting that a founder mutation may have occurred in a common ancestor in the past. Mutations in the sulphate transporter gene *SLC26A2* (*DTDST*) cause a continuum of skeletal dysplasia phenotypes. Since DD cannot be cured, therapies available are symptomatic and supportive. Carefully monitoring of affected infants is crucial to ensure prompt detection and appropriate preventive or corrective treatment of respiratory obstruction and distress.

Autoimmune Polyglandular Syndrome is a rare polyendocrinopathy characterised by several endocrine glands and non-endocrine organs failure caused by immune-mediated destruction of tissues. The hallmark evidence of APS is development of autoantibodies against multiple endocrine and non-endocrine organs causing dysfunction of the affected organ. APS is caused by mutations in the autoimmune regulator (AIRE) gene coding for a transcription factor with an autosomal recessive pattern of inheritance. The AIRE gene is located on chromosome 21q22.3, is primarily expressed in thymic medullary epithelial cells, with lesser amounts in lymph nodes and tonsils, and controls induction and maintenance of immune tolerance to self-antigen. There is no known cure yet for APS and treatment focuses on specific problems include hormones replacement, insulin management for diabetes, and treating the yeast infection.

Asthma and allergy are common conditions that exhibit complex and heterogeneous aetiologies resulting from the complex interplay between genetic factors and environmental exposures that occur at critical times in development. Asthma and allergic disease often co-occur in the same individual or in different individuals within the same families. Both asthma and allergic diseases have significant genetic contributions, with heritability estimates varying between 35% and 95% for asthma, and 33% and 91% for allergic rhinitis (hay fever). Epidemiologic studies have highlighted many associations between environmental exposures and subsequent risk for asthma and allergy. Several factors have been reported as modifiers of asthma risk. Recent genome-wide association studies (GWAS), meta-analyses of GWAS, and genome-wide interaction studies (GWISs) have begun to shed light on both common

Endocrine and Immune System Disorders

and distinct pathways that contribute to asthma and allergic diseases. At least 15 to 20 asthma risk loci have been identified by GWAS with several SNPs in or near several genes identified and associated to asthma. There are now several biomarkers emerging that hold promise for selecting and monitoring therapy including exhaled nitric oxide, serum IgE, periostin, and urinary leukotrienes. There are also several new drugs in development or being evaluated in adults.

Severe combined immunodeficiency (SCID) is a rare disorder occurring in infancy with life-threatening bacterial, viral or fungal infections; failure to thrive; and diarrhoea popularly known as the "bubble boy disease". It is characterised by severe defects of cellular and humoral immunity that render affected infants susceptible to opportunistic and recurrent infections. SCID can be caused by mutations in various genes, predominantly affecting T-cell immunity. T-cell activation and function are impaired, or T-cell development is hampered causing low or absent peripheral T cells. Established treatments for SCID include restoring the faulty immune system by means of allogeneic hematopoietic cell transplants. These cells can come from human leukocyte antigen-matched related or unrelated donors, haploidentical parental donors, or cord blood. Enzyme replacement therapy for the adenosine deaminase (ADA)-deficient form of SCID, and experimental gene therapy for X-linked SCID and ADA-SCID are also treatment options.

Immunodeficiency with Hyper-IgM syndromes are a group of rare genetic disorders leading to loss of T cell-driven immunoglobulin class switch recombination and/or defective somatic hypermutation as well as impaired T cell activation. X-linked hyper IgM is the most common subtype of hyper-IgM syndromes affecting about 70% of the patients. The most common causes are mutations in the CD40 Ligand (*CD40LG*) gene located on chromosome X (at Xq26.3-27) in males. The *CD40LG* gene product is the CD40L transmembrane glycoprotein normally expressed on activated CD4+ T lymphocytes. Several treatment options include antimicrobial medications; immunoglobulin therapy using IV or intramuscular or subcutaneous administration; and hematopoietic stem cell and solid organ transplants.

DiGeorge syndrome is a complex disorder caused by an embryopathy first described by Dr. Angelo DiGeorge in 1965. The unifying sign of DiGeorge syndrome is a wide spectrum of pathologies of the thymus, ranging from complete athymia, found in complete DiGeorge syndrome patients, to various degrees of thymic dysplasia and disturbed migration of the thymus observed in partial DiGeorge syndrome patients. The DiGeorge syndrome involves microdeletions of approximately 0.7–3 million base pairs in size associated with multi-organ dysfunction. A pri-microRNA-binding protein gene, required for microRNA biogenesis is also defective in this syndrome. Early diagnosis, preferably prenatally or neonatally, can greatly improve outcomes, emphasizing the importance of universal screening. Management requires a multidisciplinary approach involving paediatrics, general medicine, surgery, psychiatry, psychology, interventional therapies (physical, occupational, speech, language and behavioural) and genetic counselling.

End of Chapter Quiz

1. Which of the following is not part of the endocrine glands?
    a. Pancreas
    b. Spleen
    c. Gonads
    d. Pineal gland
    e. Thyroid gland

2. What are the types of protein receptors for non-steroidal signals?
   a. G-protein coupled receptors
   b. Receptor tyrosine-kinases
   c. Ion channel proteins
   d. A and B only
   e. A, B and C
3. Which of the following are examples of non-specific defence mechanisms?
   a. External barriers such as the skin
   b. Mucous membranes
   c. Internal phagocytic white blood cells
   d. A, B and C
   e. Immunoglobulins
4. What are the four main features of specific defence mechanisms that differ from non-specific defence mechanisms?
   a. Specificity, Diversity, Self/Non-self-recognition, and Memory
   b. Specificity, Sensitivity, Diversity, Memory
   c. Sensitivity, Diversity, Non-self-recognition and Memory
   d. Sensitivity, Diversity, Memory and Self recognition
   e. Specificity, Sensitivity, Self-recognition, Memory
5. Which of the following statements about B and T cells are incorrect?
   a. Mature B and T cells concentrate in the lymph nodes, spleen and other lymphatic organs
   b. B cells mature in the thymus gland and T cells mature in the bone marrow
   c. The effector cells of the B lymphocytes are the plasma cells which produce antibodies
   d. After an immune response, some B cells remain in the body circulation as memory cells
   e. The effector cells of the T lymphocytes include helper T-cells which stimulate B cells to produce antibodies
6. Congenital Adrenal Hyperplasia defines a group of syndromes resulting from inherited defects in one of the five enzymes involved in the biosynthesis of _____ from cholesterol.
   a. vitamin D
   b. bile acid
   c. cortisol
   d. corticotropin-releasing hormone
   e. adrenocorticotropin
7. A key clinical symptom of adrenoleukodystrophy (ALD) in aging patients is
   a. axonopathy
   b. apraxia
   c. agnosia
   d. anomia
   e. aphasia

Endocrine and Immune System Disorders

8. Autoimmune Polyglandular Syndrome-1 has which of the following clinical features
   a. chronic candidiasis
   b. chronic hypoparathyroidism
   c. Addison's disease
   d. A and B only
   e. A, B and C
9. Which of the following are treatment modalities for asthma?
   a. leukotriene receptor antagonist therapy
   b. inhaled corticosteroids and long-acting- ß-adrenergic agonist therapy
   c. allergen-specific immunotherapy
   d. B and C only
   e. A, B and C
10. Which disease is typically known as the "bubble boy disease?"
    a. Severe Combined Immunodeficiency
    b. DiGeorge Syndrome
    c. Immunodeficiency with Hyper-IgM
    d. Autoimmune Polyglandular Syndrome
    e. Diastrophic Dysplasia

Thought Questions

1. Make a flow chart of the immune system's lines of defence.
2. Give an example of an autoimmune attack and how it works.
3. Describe what happens when T-cell activation is impaired.

REFERENCES

Aruffo, A., Farrington, M., Hollenbaugh, D., Li, X., Milatovich, A., Nonoyama, S., ... Neubauer, M. (1993). The CD40 ligand, gp39, is defective in activated T cells from patients with X-linked hyper-IgM syndrome. *Cell, 72*(2), 291–300. doi:10.1016/0092-8674(93)90668-G PMID:7678782

Bain, A., Stewart, M., Mwamure, P., & Nirmalaraj, K. (2015). Addison's disease in a patient with hypothyroidism: Autoimmune polyglandular syndrome type 2. *BMJ Case Reports, 2015*. doi:10.1136/bcr-2015-210506 PMID:26240101

Bartel, D. P. (2009). MicroRNAs: Target recognition and regulatory functions. *Cell, 136*(2), 215–233. doi:10.1016/j.cell.2009.01.002 PMID:19167326

Bonafé, L., Hästbacka, J., de la Chapelle, A., Campos-Xavier, A. B., Chiesa, C., Forlino, A., ... Rossi, A. (2008). A novel mutation in the sulfate transporter gene SLC26A2 (DTDST) specific to the Finnish population causes de la Chapelle dysplasia. *Journal of Medical Genetics, 45*(12), 827–831. doi:10.1136/jmg.2007.057158 PMID:18708426

Bønnelykke, K., & Ober, C. (2016). Leveraging gene-environment interactions and endotypes for asthma gene discovery. *The Journal of Allergy and Clinical Immunology, 137*(3), 667–679. doi:10.1016/j.jaci.2016.01.006 PMID:26947980

Busse, W. W., Wenzel, S. E., Meltzer, E. O., Kerwin, E. M., Liu, M. C., Zhang, N., ... Lin, S. L. (2013). Safety and efficacy of the prostaglandin D2 receptor antagonist AMG 853 in asthmatic patients. *The Journal of Allergy and Clinical Immunology, 131*(2), 339–345. doi:10.1016/j.jaci.2012.10.013 PMID:23174659

Butler, M. G., Gale, D. D., Meaney, F. J., Opitz, J. M., & Reynolds, J. F. (1987). Metacarpophalangeal Pattern Profile Analysis in Diastrophic Dysplasia. *American Journal of Medical Genetics, 28*(3), 685–689. doi:10.1002/ajmg.1320280316 PMID:3425635

Büyükçelik, M., Keskin, M., Keskin, Ö., Bay, A., Kılıç, B. D., Kor, Y., ... Balat, A. (2014). Autoimmune polyglandular syndrome type 3c with ectodermal dysplasia, immune deficiency and hemolytic-uremic syndrome. *Journal of Clinical Research in Pediatric Endocrinology, 6*(1), 47–50. doi:10.4274/Jcrpe.1128 PMID:24637310

Cai, T., Yang, L., Cai, W., Guo, S., Yu, P., Li, J., ... Luo, Z. (2015). Dysplastic spondylolysis is caused by mutations in the diastrophic dysplasia sulfate transporter gene. *Proceedings of the National Academy of Sciences of the United States of America, 112*(26), 8064–8069. 10.1073/pnas.1502454112

Cheng, M. H., Fan, U., Grewal, N., Barnes, M., Mehta, A., Taylor, S., ... Anderson, M. S. (2010). Acquired Autoimmune Polyglandular Syndrome, Thymoma, and an AIRE Defect. *The New England Journal of Medicine, 362*(8), 764–766. doi:10.1056/NEJMc0909510 PMID:20181983

Chinn, I. K., Milner, J. D., Scheinberg, P., Douek, D. C., & Markert, M. L. (2013). Thymus transplantation restores the repertoires of forkhead box protein 3 (FoxP3)+ and FoxP3- T cells in complete DiGeorge anomaly. *Clinical and Experimental Immunology, 173*(1), 140–149. doi:10.1111/cei.12088 PMID:23607606

Davies E. G. (2013). Immunodeficiency in DiGeorge Syndrome and Options for Treating Cases with Complete Athymia. *Frontiers in Immunology, 4*, 322. doi:.2013.00322 doi:10.3389/fimmu

Davies, E. G., & Thrasher, A. J. (2010). Update on the hyper immunoglobulin M syndromes. *British Journal of Haematology, 149*(2), 167–180. doi:10.1111/j.1365-2141.2010.08077.x PMID:20180797

Delle Piane, L., Rinaudo, P. F., & Miller, W. L. (2015). 150 years of congenital adrenal hyperplasia: Translation and commentary of De Crecchio's classic paper from 1865. *Endocrinology, 156*(4), 1210–1217. doi:10.1210/en.2014-1879 PMID:25635623

Dessinioti, C., & Katsambas, A. (2009). Congenital adrenal hyperplasia. *Dermato-Endocrinology, 1*(2), 87–91. doi:10.4161/derm.1.2.7818 PMID:22523607

Dittmar, M., & Kahaly, G. J. (2003). Polyglandular autoimmune syndromes: Immunogenetics and long-term follow-up. *The Journal of Clinical Endocrinology and Metabolism, 88*(7), 2983–2992. doi:10.1210/jc.2002-021845 PMID:12843130

Durandy, A., & Kracker, S. (2012). Immunoglobulin class-switch recombination deficiencies. *Arthritis Research & Therapy, 14*(4), 218. doi:10.1186/ar3904 PMID:22894609

Dvorak, C. C., & Cowan, M. J. (2010). Radiosensitive Severe Combined Immunodeficiency Disease. *Immunology and Allergy Clinics of North America*, *30*(1), 125–142. doi:10.1016/j.iac.2009.10.004 PMID:20113890

Elazab, N., Mendy, A., Gasana, J., Vieira, E. R., Quizon, A., & Forno, E. (2013). Probiotic administration in early life, atopy, and asthma: A meta-analysis of clinical trials. *Pediatrics*, *132*(3), e666–e676. doi:10.1542/peds.2013-0246 PMID:23958764

Fénelon, K., Mukai, J., Xu, B., Hsu, P. K., Drew, L. J., Karayiorgou, M., . . . Macdermott, A. B., ... Gogos, J. A. (2011). Deficiency of Dgcr8, a gene disrupted by the 22q11.2 microdeletion, results in altered short-term plasticity in the prefrontal cortex. *Proceedings of the National Academy of Sciences of the United States of America*, *108*(11), 4447–4452. 10.1073/pnas.1101219108

Finkielstain, G. P., Kim, M. S., Sinaii, N., Nishitani, M., Van Ryzin, C., Hill, S. C., ... Merke, D. P. (2012). Clinical characteristics of a cohort of 244 patients with congenital adrenal hyperplasia. *The Journal of Clinical Endocrinology and Metabolism*, *97*(12), 4429–4438. doi:10.1210/jc.2012-2102 PMID:22990093

Global Initiative for Asthma (GIA). (2018). Retrieved from www.ginasthma.org

Hästbacka, J., de la Chapelle, A., Mahtani, M. M., Clines, G., Reeve-Daly, M. P., Daly, M., ... Weaver, A. (1994). The diastrophic dysplasia gene encodes a novel sulfate transporter: Positional cloning by fine-structure linkage disequilibrium mapping. *Cell*, *78*(6), 1073–1087. doi:10.1016/0092-8674(94)90281-X PMID:7923357

Heneghan, J. F., Akhavein, A., Salas, M. J., Shmukler, B. E., Karniski, L. P., Vandorpe, D. H., & Alper, S. L. (2010). Regulated transport of sulfate and oxalate by SLC26A2/DTDST. *American Journal of Physiology. Cell Physiology*, *298*(6), C1363–C1375. doi:10.1152/ajpcell.00004.2010 PMID:20219950

Hirbod-Mobarakeh, A., Aghamohammadi, A., & Rezaei, N. (2014). Immunoglobulin class switch recombination deficiency type 1 or CD40 ligand deficiency: From bedside to bench and back again. *Expert Review of Clinical Immunology*, *10*(1), 91–105. doi:10.1586/1744666X.2014.864554 PMID:24308834

Honour, J. W. (2009). Diagnosis of diseases of steroid hormone production, metabolism and action. *Journal of Clinical Research in Pediatric Endocrinology*, *1*(5), 209–226. doi:10.4274/jcrpe.v1i5.209 PMID:21274298

Huibregtse, K. E., Wolfgram, P., Winer, K. K., & Connor, E. (2014). Polyglandular autoimmune syndrome type I – a novel AIRE mutation in a North American patient. *Journal of Pediatric Endocrinology & Metabolism*, *27*(0), 1257–1260. doi:10.1515/ jpem-2013-0328 PMID:24945421

Jaring, C. V., Rivera-Arkoncel, M. L., & Lantion-Ang, F. L. (2012). A Filipino woman with autoimmune polyglandular syndrome. *BMJ Case Reports*, *2012*(1), bcr0920114766. doi:10.1136/bcr.09.2011.4766 PMID:22665707

Jonas, R. K., Montojo, C. A., & Bearden, C. E. (2013). The 22q11.2 deletion syndrome as a window into complex neuropsychiatric disorders over the lifespan. *Biological Psychiatry*, *75*(5), 351–360. doi:10.1016/j.biopsych.2013.07.019 PMID:23992925

Karikkineth, A. C., Scheibye-Knudsen, M., Fivenson, E., Croteau, D. L., & Bohr, V. A. (2016). Cockayne syndrome: Clinical features, model systems and pathways. *Ageing Research Reviews*, *33*, 3–17. doi:10.1016/j.arr.2016.08.002 PMID:27507608

Kim, H., Ellis, A. K., Fischer, D., Noseworthy, M., Olivenstien, R., Chapman, K. R., & Lee, J. (2017). Asthma biomarkers in the age of biologics. *Allergy, Asthma, and Clinical Immunology: Official Journal of the Canadian Society of Allergy and Clinical Immunology*, *13*(1), 48. doi:10.118613223-017-0219-4 PMID:29176991

Kim, J. M., Lin, S. Y., Suarez-Cuervo, C., Chelladurai, Y., Ramanathan, M., Segal, J. B., & Erekosima, N. (2013). Allergen-specific immunotherapy for pediatric asthma and rhinoconjunctivitis: A systematic review. *Pediatrics*, *131*(6), 1155–1167. doi:10.1542/peds.2013-0343 PMID:23650298

Kirmizibekmez, H., Mutlu, R. G. Y., Urganzi, N. D., & Öner, A. (2014). Autoimmune Polyglandular Syndrome Type 2: A Rare Condition in Childhood. *Journal of Clinical Research in Pediatric Endocrinology*, *7*(1), 80–82. doi:10.4274/jcrpe.1394 PMID:25800482

Kracker, S., Curtis, J., Ibrahim, M. A., Sediva, A., Salisbury, J., Campr, V., ... Durandy, A. (2014). Occurrence of B-cell lymphomas in patients with activated phosphoinositide 3-kinase δ syndrome. *The Journal of Allergy and Clinical Immunology*, *134*(1), 233–236. doi:10.1016/j.jaci.2014.02.020 PMID:24698326

Kraus, C., Vanicek, T., Weidenauer, A., Khanaqa, T., Stamenkovic, M., Lanzenberger, R., ... Kasper, S. (2018). DiGeorge syndrome: Relevance of psychiatric symptoms in undiagnosed adult patients. *Wiener Klinische Wochenschrift*, *130*(7-8), 283–287. doi:10.100700508-018-1335-y PMID:29671046

Kwan, A., & Puck, J. M. (2015). History and current status of newborn screening for severe combined immunodeficiency. *Seminars in Perinatology*, *39*(3), 194–205. doi:10.1053/j.semperi.2015.03.004 PMID:25937517

Landthaler, M., Yalcin, A., & Tuschl, T. (2004). The human DiGeorge syndrome critical region gene 8 and Its D. melanogaster homolog are required for miRNA biogenesis. *Current Biology*, *14*(23), 2162–2167. doi:10.1016/j.cub.2004.11.001 PMID:15589161

Laue, L., Merke, D. P., Jones, J. V., Barnes, K. M., Hill, S., & Cutler, G. B. Jr. (1996). A preliminary study of flutamide, testolactone, and reduced hydrocortisone dose in the treatment of congenital adrenal hyperplasia. *The Journal of Clinical Endocrinology and Metabolism*, *81*(10), 3535–3539. PMID:8855797

Laugel, V. (2013). Cockayne syndrome: The expanding clinical and mutational spectrum. *Mechanisms of Ageing and Development*, *134*(5-6), 161–170. doi:10.1016/j.mad.2013.02.006 PMID:23428416

Laugel, V., Dalloz, C., Durand, M., Sauvanaud, F., Kristensen, U., Vincent, M. C., ... Dollfus, H. (2010). Mutation update for the CSB/ERCC6 and CSA/ERCC8 genes involved in Cockayne syndrome. *Human Mutation*, *31*(2), 113–126. doi:10.1002/humu.21154 PMID:19894250

Laviolette, M., Gossage, D. L., Gauvreau, G., Leigh, R., Olivenstein, R., Katial, R., ... Molfino, N. A. (2013). Effects of benralizumab on airway eosinophils in asthmatic patients with sputum eosinophilia. *The Journal of Allergy and Clinical Immunology*, *132*(5), 1086–1096.e5. doi:10.1016/j.jaci.2013.05.020 PMID:23866823

Leven, E. A., Maffucci, P., Ochs, H. D., Scholl, P. R., Buckley, R. H., Fuleihan, R. L., ... Cunningham-Rundles, C. (2016). Hyper IgM Syndrome: A Report from the USIDNET Registry. *Journal of Clinical Immunology*, *36*(5), 490–501. doi:10.100710875-016-0291-4 PMID:27189378

Li, L., Moshous, D., Zhou, Y., Wang, J., Xie, G., Salido, E., ... Cowan, M. J. (2002). A founder mutation in Artemis, an SNM1-like protein, causes SCID in Athabascan-speaking Native Americans. *Journal of Immunology (Baltimore, Md.: 1950)*, *168*(12), 6323–6329. doi:10.4049/jimmunol.168.12.6323 PMID:12055248

Lopez-Rivera, E., Liu, Y. P., Verbitsky, M., Anderson, B. R., Capone, V. P., Otto, E. A., ... Sanna-Cherchi, S. (2017). Genetic Drivers of Kidney Defects in the DiGeorge Syndrome. *The New England Journal of Medicine*, *376*(8), 742–754. doi:10.1056/NEJMoa1609009 PMID:28121514

Lu, T. X., & Rothenberg, M. E. (2013). Diagnostic, functional, and therapeutic roles of microRNA in allergic diseases. *The Journal of Allergy and Clinical Immunology*, *132*(1), 3–13, quiz 14. doi:10.1016/j.jaci.2013.04.039 PMID:23735656

Mäkitie, O., Geiberger, S., Horemuzova, E., Hagenäs, L., Moström, E., Nordenskjöld, M., ... Nordgren, A. (2015). SLC26A2 disease spectrum in Sweden - high frequency of recessive multiple epiphyseal dysplasia (rMED). *Clinical Genetics*, *87*(3), 273–278. doi:10.1111/cge.12371 PMID:24598000

McDonald-McGinn, D. M., & Sullivan, K. E. (2011). Chromosome 22q11.2 deletion syndrome (DiGeorge syndrome/velocardiofacial syndrome). *Medicine; Analytical Reviews of General Medicine, Neurology, Psychiatry, Dermatology, and Pediatries*, *90*(1), 1–18. doi:10.1097/MD.0b013e3182060469 PMID:21200182

McDonald-McGinn, D. M., Sullivan, K. E., Marino, B., Philip, N., Swillen, A., Vorstman, J. A., ... Bassett, A. S. (2015). 22q11.2 deletion syndrome. *Nature Reviews. Disease Primers*, *1*, 15071. doi:10.1038/nrdp.2015.71 PMID:27189754

Metzker, M. L. (2010). Sequencing technologies - the next generation. *Nature Reviews. Genetics*, *11*(1), 31–46. doi:10.1038/nrg2626 PMID:19997069

Moffatt, M. F., Gut, I. G., Demenais, F., Strachan, D. P., Bouzigon, E., Heath, S., ... Cookson, W. (2010). A large-scale, consortium-based genomewide association study of asthma. *The New England Journal of Medicine*, *363*(13), 1211–1221. doi:10.1056/NEJMoa0906312 PMID:20860503

Muehleisen, B., & Gallo, R. L. (2013). Vitamin D in allergic disease: Shedding light on a complex problem. *The Journal of Allergy and Clinical Immunology*, *131*(2), 324–329. doi:10.1016/j.jaci.2012.12.1562 PMID:23374263

Natale, V. (2011). A comprehensive description of the severity groups in Cockayne syndrome. *American Journal of Medical Genetics. Part A*, *155A*(5), 1081–1095. doi:10.1002/ajmg.a.33933 PMID:21480477

National Institute of Child Health and Human Development (NICHD). (2018). *What are the symptoms of congenital adrenal hyperplasia (CAH)?* Retrieved from https://www.nichd.nih.gov/health/topics/cah/conditioninfo/symptoms

National Organization for Rare Diseases (NORD). (2018a). *Diastrophic Dysplasia*. Retrieved from https://rarediseases.org/rare-diseases/diastrophic-dysplasia/

National Organization for Rare Diseases (NORD). (2018b). Retrieved from https://rarediseases.org/rare-diseases/autoimmune-polyendocrine-syndrome-type-ii/

National Organization for Rare Diseases (NORD). (2018c). Retrieved from https://rarediseases.org/rare-diseases/autoimmune-polyglandular-syndrome-type-1/

Neufeld, M., & Blizzard, R. M. (1980). Polyglandular autoimmune diseases. In *Symposium on autoimmune aspects of endocrine disorders*. Academic Press.

Noonan, M., Korenblat, P., Mosesova, S., Scheerens, H., Arron, J. R., Zheng, Y., ... Matthews, J. G. (2013). Dose-ranging study of lebrikizumab in asthmatic patients not receiving inhaled steroids. *The Journal of Allergy and Clinical Immunology*, *132*(3), 567–574.e12. doi:10.1016/j.jaci.2013.03.051 PMID:23726041

Noordzij, J. G., Verkaik, N. S., van der Burg, M., van Veelen, L. R., de Bruin-Versteeg, S., Wiegant, W., ... van Dongen, J. J. (2003). Radiosensitive SCID patients with Artemis gene mutations show a complete B-cell differentiation arrest at the pre-B-cell receptor checkpoint in bone marrow. *Blood*, *101*(4), 1446–1452. doi:10.1182/blood-2002-01-0187 PMID:12406895

Ober, C. & Yao, T. (2011). The Genetics of Asthma and Allergic Disease: A 21st Century Perspective. *Immunological Reviews*, *242*(1), 10–30. doi:.1600-065X.2011.01029.x doi:10.1111/j

Perniola, R. (2018). Twenty Years of AIRE. *Frontiers in Immunology*, *9*, 98. doi:10.3389/fimmu.2018.00098 PMID:29483906

Rebane, A., & Akdis, C. A. (2013). MicroRNAs: Essential players in the regulation of inflammation. *The Journal of Allergy and Clinical Immunology*, *132*(1), 15–26. doi:10.1016/j.jaci.2013.04.011 PMID:23726263

Reddel, H. K., Bateman, E. D., Becker, A., Boulet, L., Cruz, A., Drazen, J. M., ... Fitzgerald, J. M. (2015). A summary of the new GINA strategy: A roadmap to asthma control. *The European Respiratory Journal*, *46*(3), 622–639. doi:10.1183/13993003.00853-2015 PMID:26206872

Renz, H. (2013). Advances in in vitro diagnostics in allergy, asthma, and immunology in 2012. *The Journal of Allergy and Clinical Immunology*, *132*(6), 1287–1292. doi:10.1016/j.jaci.2013.08.043 PMID:24139605

Sadatsafavi, M., Lynd, L., Marra, C., Bedouch, P., & Fitzgerald, M. (2013). Comparative outcomes of leukotriene receptor antagonists and long-acting β-agonists as add-on therapy in asthmatic patients: A population-based study. *The Journal of Allergy and Clinical Immunology*, *132*(1), 63–69. doi:10.1016/j.jaci.2013.02.007 PMID:23545274

Schwarz, K., Gauss, G. H., Ludwig, L., Pannicke, U., Li, Z., Lindner, D., ... Bartram, C. R. (1996). RAG mutations in human B cell-negative SCID. *Science*, *274*(5284), 97–99. doi:10.1126cience.274.5284.97 PMID:8810255

Swillen, A., & McDonald-McGinn, D. (2015). Developmental trajectories in 22q11.2 deletion. *American Journal of Medical Genetics. Part C, Seminars in Medical Genetics*, *169*(2), 172–181. doi:10.1002/ajmg.c.31435 PMID:25989227

Szefler, S. J. (2014). Advances in pediatric asthma in 2013: Coordinating asthma care. *The Journal of Allergy and Clinical Immunology*, *133*(3), 654–661. doi:10.1016/j.jaci.2014.01.012 PMID:24581430

Therrell, B. L. Jr, Berenbaum, S. A., Manter-Kapanke, V., Simmank, J., Korman, K., Prentice, L., ... Gunn, S. (1998). Results of screening 1.9 million Texas newborns for 21-hydroxylase-deficient congenital adrenal hyperplasia. *Pediatrics*, *101*(4 Pt. 1), 583–590. doi:10.1542/peds.101.4.583 PMID:9521938

Trivei, H. L., Thakkar, U. G., Vanikar, A. V., & Dave, S. D. (2011). Treatment of polyglandular autoimmune syndrome type 3 using co-transplantation of insulin-secreting mesenchymal stem cells and haematopoietic stem cells. *BMJ Case Reports*, *2011*. doi:10.1136/bcr.07.2011.4436 PMID:22679189

Tsai, C. L., Lee, W. Y., Hanania, N. A., & Camargo, C. A. Jr. (2012). Age-related differences in clinical outcomes for acute asthma in the United States, 2006-2008. *The Journal of Allergy and Clinical Immunology*, *129*(5), 1252–1258.e1. doi:10.1016/j.jaci.2012.01.061 PMID:22385630

Turcu, A. F., & Auchus, R. J. (2015). The Next 150 Years of Congenital Adrenal Hyperplasia. *The Journal of Steroid Biochemistry and Molecular Biology*, *153*, 63–71. doi:10.1016/j.jsbmb.2015.05.013 PMID:26047556

van Til, N. P., de Boer, H., Mashamba, N., Wabik, A., Huston, M., Visser, T. P., ... Wagemaker, G. (2012). Correction of Murine *Rag2* Severe Combined Immunodeficiency by Lentiviral Gene Therapy Using a Codon-optimized *RAG2* Therapeutic Transgene. *Molecular Therapy*, *20*(10), 1968–1980. doi:10.1038/mt.2012.110 PMID:22692499

Volk, T., Pannicke, U., Reisli, I., Bulashevska, A., Ritter, J., Björkman, A., ... Grimbacher, B. (2015). DCLRE1C (ARTEMIS) mutations causing phenotypes ranging from atypical severe combined immunodeficiency to mere antibody deficiency. *Human Molecular Genetics*, *24*(25), 7361–7372. doi:10.1093/hmg/ddv437 PMID:26476407

Wang, L., Zhou, W., Zhao, W., Tian, Z., Wang, W., Wang, X., & Chen, T. (2014). Clinical Features and Genetic Analysis of 20 Chinese Patients with X-Linked Hyper-IgM Syndrome. *Journal of Immunology Research*, *683160*. doi:10.1155/2014/683160 PMID:25215306

Wenzel, S., Ford, L., Pearlman, D., Spector, S., Sher, L., Skobieranda, F., ... Pirozzi, G. (2013). Dupilumab in persistent asthma with elevated eosinophil levels. *The New England Journal of Medicine*, *368*(26), 2455–2466. doi:10.1056/NEJMoa1304048 PMID:23688323

Wilson, B. T., Stark, Z., Sutton, R. E., Danda, S., Ekbote, A. V., Elsayed, S. M., ... Wilson, I. J. (2015). The Cockayne Syndrome Natural History (CoSyNH) study: Clinical findings in 102 individuals and recommendations for care. *Genetics in medicine. Official Journal of the American College of Medical Genetics*, *18*(5), 483–493. doi:10.1038/gim.2015.110 PMID:26204423

Xu, S., Hu, S., Yu, X., Zhang, M., & Yang, Y. (2016). 17α hydroxylase/17,20 lyase deficiency in congenital adrenal hyperplasia: A case report. *Molecular Medicine Reports*, *15*(1), 339–344. doi:10.3892/mmr.2016.6029 PMID:27959413

Zein, J. G., Udeh, B., Teague, W. G., Koroukian, S. M., Schlitz, N. K., Bleecker, E. R., ... Erzurum, S. C. (2016). Impact of Age and Sex on Outcomes and Hospital Cost of Acute Asthma in the United States, 2011-2012. *PLoS One*, *11*(6), e0157301. doi:10.1371/journal.pone.0157301 PMID:27294365

Zonana, J., Elder, M. E., Schneider, L. C., Orlow, S. J., Moss, C., Golabi, M., ... Ferguson, B. M. (2000). A novel X-linked disorder of immune deficiency and hypohidrotic ectodermal dysplasia is allelic to incontinentia pigmenti and due to mutations in IKK-gamma (NEMO). *American Journal of Human Genetics*, *67*(6), 1555–1562. doi:10.1086/316914 PMID:11047757

KEY TERMS AND DEFINITIONS

Achondrogenesis: A group of conditions that affect the development of bone and cartilage.

Addison's Disease: A disorder in which the adrenal gland insufficiently produces cortisol and aldosterone.

Allogeneic Stem-Cell Transplantation: A treatment involving stem cell transfer from a healthy donor to the patient after chemotherapy or radiation.

Alopecia: Hair loss in round patches.

Atelosteogenesis: One of many conditions that affect the development of bone and cartilage of infants that have varying degrees of severity.

Axonopathy: Degradation of the axons of peripheral nerves.

Demyelination: Damage to the protective myelin sheath covering the nerve fibres.

Dysplasia: A phase where cells are not cancerous but look abnormal under a microscope.

Ectodermal Dysplasia: A group of genetic disorders involving the abnormal development of two or more ectodermal structures: skin, sweat glands, nails, teeth, hair, and mucous membranes.

Hyperplasia: Enlargement of an organ caused by an increased number of cells.

Hypogammaglobulinemia: An immune disorder in which there is a reduction in serum gamma globulins.

Hypoparathyroidism: A condition in which there is a deficiency in the production of parathyroid hormone.

Lenti-D Gene Therapy: A form of gene therapy that utilizes the lentivirus as the cloning vehicle.

Lymphoid Hyperplasia: An increase in lymphocytes resulting in the enlargement of lymphoid tissue.

Polyendocrinopathy: A condition when the body develops multiple autoimmune disorders in the endocrine system.

Scoliosis: A condition that causes an abnormal curve to the spine.

Chapter 17
Muscle, Connective Tissue, and Neonatal Disorders

ABSTRACT

The skeleton provides the framework and anchor points against which muscles, attached via tendons, can exert force. Three types of cells are involved in making bone: osteoblasts, osteoclasts, and cartilage. The human muscle system is made up of three types of muscle tissue: skeletal, cardiac, and smooth. The neonate period of life is the first 4 weeks after the birth of an infant. This chapter presents 11 genetic disorders that affect muscles, connective tissue, and newborns. These include achondroplasia, Charcot-Marie tooth syndrome, Duchenne Muscular Dystrophy, Ellis-Van Creveld syndrome, amyotrophic lateral sclerosis, Marfan syndrome, fibrodysplasia ossificans progressive, myotonic dystrophy, Angelman syndrome, Prader-Willi syndrome, fragile-X syndrome, and Waardenburg syndrome.

CHAPTER OUTLINE

17.1 Overview
17.2 Achondroplasia
17.3 Charcot-Marie-Tooth Syndrome
17.4 Duchenne Muscular Dystrophy
17.5 Ellis-Van Creveld Syndrome
17.6 Amyotrophic Lateral Sclerosis
17.7 Marfan Syndrome
17.8 Fibrodysplasia Ossificans Progressiva
17.9 Myotonic Dystrophy
17.10 Angelman Syndrome and Prader-Willi Syndrome
17.11 Fragile-X Syndrome
17.12 Waardenburg Syndrome
Chapter Summary

DOI: 10.4018/978-1-5225-8066-9.ch017

Muscle, Connective Tissue, and Neonatal Disorders

LEARNING OUTCOMES

- Identify each genetic disorder affecting each system
- Outline the symptoms of each disorder
- Explain the genetic basis of each disorder
- Summarize the current therapies available to treat each disorder

17.1 OVERVIEW

Humans have endoskeletons which are rigid internal body parts made up of bone that receive the applied force of muscles. The skeleton provides the framework and anchor points against which muscles, attached via tendons, can exert force (Figure 1). It is made up of 206 bones divided into 2 systems: a) axial skeleton with skull bones, 12 pairs of ribs, the sternum, and the 26 vertebrae; b) appendicular skeleton which is made up of the pectoral girdles (shoulders); pelvic girdle (hips) and pairs of arms, hands, legs and feet. Bones are complex organs that function in movement, providing support by anchoring muscles, protecting internal organs, storage of minerals particularly calcium and phosphorous, and blood cell production (white blood cells, red blood cells and platelets). There are two types of bone tissue: compact bone which is found in the shaft and at both ends of long bones. Compact bone is made up many thin, cylindrical, dense layers around interconnected canals for blood vessels and nerves and it resists mechanical shock. Spongy bone tissue has abundant air spaces but very firm and is found inside both ends and

Figure 1. The structure and bones of the human skeleton
Source: *Image used under license from Shutterstock.com*

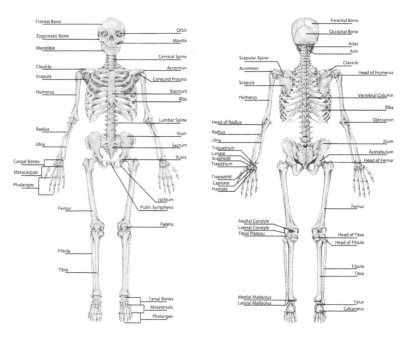

shaft of bone. Spongy tissue contains red bone marrow where blood cells are formed. It also contains yellow bone marrow which is largely fat but can be converted to red marrow in severe blood loss cases.

Three types of cells involved in making bone. 1) Osteoblasts: In the developing embryo cartilage models are first constructed, into which osteoblasts secrete organic substances which mineralize; cartilage then breaks down to form bone marrow cavities. Once osteoblasts are imprisoned by these secretions, they become osteocytes or mature living bone cells. In adults, large amounts of minerals and many osteocytes are removed and replaced routinely in a process called **bone remodeling** which adjusts bone strength and maintains calcium and phosphorous levels in the blood. 2) Osteoclasts secrete enzymes that digest bone in the routine bone remodeling activities in the body. 3) Cartilage are collagen fibres embedded in a rubbery matrix of protein and carbohydrate secreted by cells called chondrocytes. In the adult body cartilage is retained in the flexible parts of our skeleton (ears, nose, intervertebral discs, and windpipe). Physical exercise stresses bone such that the process of mineral deposition exceeds withdrawal—making the bones denser and stronger. With age, bones decline in mass where mineral deposits lag behind withdrawal making the bones hollow and brittle—a condition known as osteoporosis.

The human muscle system is made up of three types of muscle tissue: skeletal, cardiac and smooth. There are over 700 skeletal muscles which help move the body and attach to bone through tendons. Each muscle contains thousands of bundles of long fibre-like muscle cells (with many nuclei from fused embryonic cells) made of individual myofibrils (Figure 2). Each myofibril is made up of two kinds of myofilaments which create muscle striations. Skeletal muscles contract and relax as they move the body. When muscles contract (shortens), many units in their basic units of contraction called sarcomeres shorten. A sarcomere is the region between two (Z) lines of a myofibril and is made up of thin and thick filaments which are made up of actin and myosin proteins respectively. When muscles are resting the actin binding sites are usually blocked by a regulatory protein called tropomyosin—therefore, actin cannot bind myosin. When muscles contract, the actin (short) filaments interact with myosin (thick) to shorten the width of the sarcomeres. When a muscle is stimulated by a neuromuscular junction neurotransmitter

Figure 2. Skeletal muscle structure
Source: Image used under license from Shutterstock.com

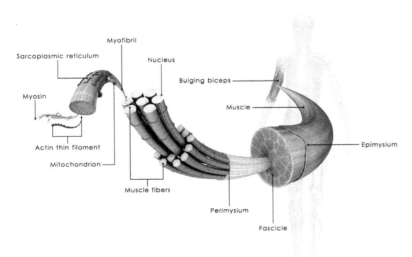

like acetylcholine an action potential is generated and Ca^{2+} ions diffuse from a specialised endoplasmic reticulum covering myofibrils called sarcoplasmic reticulum and reach actin filaments. Here they bind with the regulatory protein troponin (which controls positioning of tropomyosin). This action changes the troponin-tropomyosin complex exposing the myosin binding sites. The myosin head, a centre of the bioenergetic reactions that power muscle contractions, hydrolyses ATP to ADP, binds to Pi and binds the actin forming a cross-bridge. This cross-bridging at about five cross-bridges can form per second contracts the muscle—a process often referred to as the sliding-filament model of muscle contraction. The myosin head is then bound to ATP causing release of actin; Ca^{2+} ions are then pumped out and muscle is at rest again.

The neonate period of life is the first 4 weeks after the birth of an infant. Some human diseases are explained by alterations in a single gene or of a single chromosome; however, most are complex and may involve multiple genes and several metabolic pathways. A myriad of genes, as well as environmental factors control the complex and integrated processes necessary for foetal development. Research using animal models continues to provide important information about the biological and regulatory processes involved in human development.

There are a number of disorders caused by defects in genes important in the formation and function of muscles, connective tissues and in foetal development as discussed below.

17.2 ACHONDROPLASIA (Ach)

Achondroplasia (Ach) is the most common form of dwarfism in humans and is a short stature condition compatible with good general health and normal life expectancy. However, besides the burden of short stature, health issues may arise from complications arising from the particular anatomical features of Ach. It is estimated that it affects approximately 250,000 people worldwide. Ach is related to other chondrodysplasia syndromes including hypochondroplasia (Hch), severe achondroplasia with developmental delay and acanthosis nigricans, and thanatophoric dysplasia. It is an autosomal dominant genetic disease that has 100% expressivity. The shortening of the limbs with proximal segments affected disproportionally produces the short stature, a phenotype referred as rhizomelia. Cartilage growth defects at the skull base produces a large head with frontal bossing and midface hypoplasia. The reduced size of the foramen magnum and **spinal stenosis** are relatively common and often require neurosurgical corrections. The size of the trunk is relatively normal but is often deformed by excessive **lumbar lordosis**. (Baujat et al., 2008; Ornitz & Legeai-Mallet, 2017; Unger et al., 2017)

Symptoms

Symptoms may include prominent forehead (frontal bossing), disproportionately large head-to-body size difference, abnormal hand appearance with persistent space between the long and ring fingers, bowed legs, decreased muscle tone, shortened arms and legs (especially the upper arm and thigh), short stature (significantly below the average height for a person of the same age and sex), narrowing of the spinal column (spinal stenosis), spine curvatures called **kyphosis** and lordosis (MP, 2018).

Muscle, Connective Tissue, and Neonatal Disorders

Genetic Basis

Ach diagnosis is usually made at birth with 80% of cases arising as sporadic mutations in *FGFR3* gene. The condition was mapped to *FGFR3* in 1994 and over 97% of cases result from an autosomal dominant missense single point gain-of-function mutation localized in the transmembrane domain of FGFR3. Genetic linkage studies show that FGFR3 gene is located on the short arm of chromosome 4 and mutation analysis identified an arginine to glycine substitution in fibroblast growth factor receptor 3 (*FGFR3*) in almost all Ach patients. Experiments have shown that mice expressing the FGFR3(p.Gly374Arg) activating mutation, which corresponds to the human FGFR3(p.Gly380Arg) mutation, develop an Ach-like phenotype with reduced chondrocyte proliferation and reduced hypertrophic differentiation and matrix production. The gene FGFR3 signalling activates at least four downstream intracellular signalling pathways including, MAPK, PI3K/AKT, PLCγ, and STATs. FGFR3 signalling also affects surrounding bone, directly and through the regulation of other growth factor signalling pathways in chondrocytes. Similarly to Ach, Hch is caused by a *FGFR3* missense mutation (p.Asn540Lys) localized in tyrosine kinase domain I—the most common Hch mutation, occurring in ~60% of cases. (Shiang et al., 1994; Naski et al., 1998; Wang et al., 1999; Ornitz & Legeai-Mallet, 2017)

Current Therapies

Treatment of the developmental complications of Ach involves symptomatic management, surgical intervention, and lifelong follow-up care. To be effective, therapies for Ach need to be administered throughout from birth to puberty. The primary therapies for patients with Hch include treatment with recombinant human growth hormone (r-hGH) or surgical intervention. Research has focused on FGFR-selective small molecule tyrosine kinase inhibitors (TKI) to directly reduce the high tyrosine kinase activity resulting from mutations in *FGFR3*. Therapeutic efficacy of the TKI CHIR-258 was demonstrated in a mouse model of FGFR3-induced **multiple myeloma**. Two other FGFR TKIs (PD173074 and SU5402) also inhibit the growth and induce apoptosis of multiple myeloma cells. More recently, TKI that is more selective for FGFR3 over others FGFRs (NVP-BGJ398) was used in preclinical rodent models for treating several FGFR-related cancers such as malignant rhabdoid (kidney) tumours. NVP-BGJ398 was shown *in vivo* to reduce FGFR3(p.Tyr367Cys) activation and improve the skeletal phenotype of Ach-like mice. Monoclonal antibodies have been used to inhibit FGFR3 targeting the extracellular part of the receptor to block ligand binding. Soluble FGFR3 decoy receptors (sFGFR3) have also been designed to bind and sequester available FGF to compete with endogenous FGFR3 binding to FGF ligands that functionally regulate chondrogenesis. Drugs currently used for non-skeletal disorders like *meclizine* (an over the-counter H1 receptor inhibitor used to treat motion sickness) promote chondrocyte proliferation and differentiation and attenuates extracellular signal regulated kinases 1/2 phosphorylation. Currently, perhaps the most promising therapy Ach treatment is the use of a stabilized form of C-type natriuretic peptide (CNP) called *vosoritide*. Research has evaluated a 12-amino acid peptide called P3 which bonds with high affinity to the extracellular domain of FGFR3 in mice. P3 treatment restored chondrocyte proliferation and differentiation by decreasing the mitogen-activated protein kinase (MAPK) signalling cascade. Yamashita et al. (2014) have also studied the use of statins to rescue bone growth in achondroplasia and thanatophoric dysplasia type I. (Lorget et al., 2012; Wendt et al., 2015; Gudernova et al., 2016; Miccoli et al., 2016; Ornitz & Legeai-Mallet, 2017; Unger et al., 2017)

17.3 CHARCOT-MARIE-TOOTH SYNDROME

CMT classically refers to inherited motor and sensory neuropathies with a wide range of genotypes and phenotypes characterized primarily by distal muscle weakness, atrophy, and sensory loss. The disease was first described in 1886 by Charcot and Marie in France and independently by Tooth in Great Britain and was named in their honour. Classification of the various types of CMT was originally described by Dyck et al. in 1975 and employed the term "hereditary motor and sensory neuropathy types I–VII", also referred to CMT types 1–7. The most commonly encountered forms of CMT are generally classified as type 1 (demyelinating, 50-80% of CMT cases) and type 2 (axonal, 10-15% of cases). Overall prevalence of CMT is usually reported as 1:2,500, and usually sets in during the first two decades and most frequently in the first 10 years of life. (Reilly, 2007; van Paassen et al., 2014; McCorquodale et al., 2016)

Symptoms

The typical symptoms of CMT1A are difficulty walking or running, due to weakness of the distal leg muscles; calf hypertrophy; foot deformity (usually pes cavus with hammertoes); clawing of the hands; skeletal deformity of the spine (scoliosis); upper extremity symptoms of muscle weakness and atrophy; and decreased manual dexterity. Small fibre sensory loss can be present with complaints of cold feet and decreased temperature discrimination sense; 55-70% of CMT1A patients report pain, severe fatigue, hand tremor, significantly impaired speech perception ability; and predisposal to obstructive sleep apnoea (Pareyson et al., 2006).

Genetic Basis

Since the initial discovery of *PMP22* as the causative gene for CMT1A, more than 45 causative genes have been identified (genetic heterogeneity). CMT can be inherited in an autosomal dominant, autosomal recessive and X-linked manner. The autosomal dominantly inherited demyelinating form of CMT is caused by a duplication on chromosome 17p11.2, containing the gene coding for peripheral myelin protein 22 (PMP22) and thus leading to three copies of the *PMP22* gene. In CMT type 1, *PMP22* duplication (CMT1A) and point mutations account for ~50% or more of CMT1 in North American studies; while international studies are variable and report ranges from 13% to 67%. Other genes known to cause CMT1 are *LITAF*, *EGR2*, and *NEFL and* likely account for ~10% or less of CMT1. In CMT type 2, mutations in *MFN2* (CMT2A) are thought to account for 15%–20% of CMT2 in clinical studies, with other CMT2 genes (*Rab7*, *TRPV4*, *GARS*, *NEFL*, *HSPB1*, *GDAP*, *HSPB8*) accounting for a minority of cases. CMT1A is inherited in an autosomal dominant fashion and offspring of CMT1A patients have a chance of 50% to inherit the *PMP22* duplication from their affected parent.

It is still not understood why an increase or decrease in *PMP22* gene copy number results in CMT1A. PMP22 gene codes for a transmembrane protein and its expression increases upon growth arrest, and in proliferating cells the levels are very low. The highest expression of *PMP22* is detected in the Schwann cells of compact myelin. There seems to be no unifying hypothesis how a *PMP22* copy number variation or mutation leads to disease. In case of point mutations, a toxic effect of misfolded protein is suggested, but this is not the case in duplication and deletion of a normal copy of *PMP22*. Although lacking experimental data, van Paassen et al. (2014) speculate that *PMP22* interacts with other myelin components and

Muscle, Connective Tissue, and Neonatal Disorders

stoichiometry is essential for proper function. (Lupski et al., 1991; Raeymaekers et al., 1991; Pareyson et al., 2006; Fusco et al., 2009; Kohl et al., 2010; Weterman et al., 2010; Videler et al., 2012; van Paassen et al., 2014; McCorquodale et al., 2016)

Current Therapies

Treatment consists of supportive care and may include management by a rehabilitation physician, physiotherapist, occupational therapist, and orthopaedic surgeon. Exercise training, shoe inlays, orthopaedic shoes and orthoses can provide supportive care for the legs. In case of severe skeletal deformities orthopaedic surgery is suggested. Wearing a thumb opposition splint may improve manual dexterity in CMT. To improve thumb opposition of patients with CMT, tendon transfer surgery (moving a tendon from its original attachment to a new one) is an option. Symptomatic drug treatment for positive sensory symptoms and for muscle cramps may be useful. Since CMT patients may also have diabetes mellitus which exacerbates symptoms of the peripheral neuropathy, control of blood sugar is essential. Effective drug treatment is suggested to normalize the myelination process, but the improvement of the axonal function must be the ultimate goal. (Young et al., 2008; McCorquodale et al., 2016)

17.4 DUCHENNE MUSCULAR DYSTROPHY (DMD)

In the US, prevalence of DMD has been reported as 15.9 cases per 100 000 live male births and 19.5 cases per 100 000 live male births in the UK. The absence of dystrophin leads to myofibril membrane fragility and repeated cycles of necrosis and regeneration, leading to eventual muscle atrophy and contractures. Muscle is gradually replaced with fibrous connective tissue and fat, leading to weakness and debilitating contractures. Progressive muscular damage and degeneration occurs in people with DMD, resulting in muscular weakness, associated motor delays, loss of ambulation, respiratory impairment, and cardiomyopathy. Boys with the condition typically die in their 20s or 30s due to either respiratory failure or cardiomyopathy. (Ryder et al., 2017; Birnkrant et al., 2018)

Symptoms

According to the Muscular Dystrophy Association (MDA) many children with DMD begin using a wheelchair sometime between ages 7 and 12. They are often late walkers, and toddlers have enlarged calf muscles known as *pseudohypertrophy,* or "false enlargement," because the muscle tissue is abnormal and may contain scar tissue. Pre-schoolers are clumsy and fall often, have trouble climbing stairs, getting up from the floor or running, difficulty raising their arms. In the teen years, activities involving the arms, legs or trunk may require assistance or mechanical support. Lack of dystrophin can weaken the heart *myocardium*, resulting in **cardiomyopathy**. Over time, sometimes as early as the teen years, the damage done by DMD to the heart can become life-threatening. Beginning at about 10 years of age, children may complain of shortness of breath caused by the weakening of the diaphragm and other muscles that operate the lungs. Problems that indicate poor respiratory function occur and include headaches, mental dullness, difficulty concentrating or staying awake, and nightmares. About 30% of boys with DMD have some degree of *learning disability* (involving attention focusing, verbal learning and memory, and emotional interaction), but rarely do they show serious mental retardation. (MDA, 2018)

Genetic Basis

DMD is a lethal X-linked recessive neuromuscular disorder caused by mutations in the dystrophin gene resulting in absent or insufficient functional dystrophin (a cytoskeletal protein), whose function is to enable the strength, stability, and functionality of muscle myofibrils. The mode of inheritance for DMD was long suspected to be X-linked because of the disease's predominance in males. However, a rare occurrence of a similar clinical syndrome in girls were thought to occur due to disproportionate inactivation of the normal X chromosome in female DMD carriers. Approximately 70% of individuals with DMD have a single-exon or multi-exon deletion or duplication in the dystrophin gene. These mutations include point mutations (nonsense or missense), small deletions, and small duplications or insertions, identified using next-generation sequencing technology. (Ryder et al., 2017; Birnkrant et al., 2018)

Current Therapies

There is currently no cure, and despite exhaustive palliative care, patients are restricted to a wheelchair by the age of 12 years and usually succumb to cardiac or respiratory complications in their late 20s. Physiotherapy, as described in the section on rehabilitation management, and treatment with glucocorticoids, remain the mainstays of DMD treatment and should continue after loss of ambulation. Rehabilitation personnel include physicians, physical therapists, occupational therapists, speech-language pathologists, orthotists, and durable medical equipment providers. Direct physical, occupational, and speech and language therapy should be provided in outpatient and school settings and continue throughout adulthood, augmented by therapies provided during hospital admissions and at home. Endocrine care is necessary to monitor growth and development, identify and diagnose hormone deficiencies, provide endocrine hormone replacement therapy when indicated, and prevent a life-threatening adrenal crisis. The aim of nutritional care is to prevent overweight or obesity and undernutrition or malnutrition through regular assessment of growth and weight; it also aims to promote a healthy, balanced diet, with optimum intake of calories, protein, fluid, and micronutrients, especially calcium and vitamin D. (Guiraud et al., 2015; Gloss et al., 2016; Shimizu-Motohashi et al., 2016; Kornegay, 2017; Ryder et al., 2017)

DMD drugs in use include *ataluren* which was granted conditional marketing authorisation in the European Union in 2014, targeting the approximately 11% of boys with DMD caused by missense mutations in the dystrophin gene. The US Food and Drug Administration (FDA) approved use of *eteplirsen* in 2016 which targets the approximately 13% of boys with a mutation in the dystrophin gene. *Ataluren* and *eteplirsen* are the first of a series of mutation-specific therapies to gain regulatory approval. Clinical trials for DMD include drugs under way target myostatin, anti-inflammatory and antioxidant molecules, compounds to reduce fibrosis, drugs to improve vasodilatation, drugs to improve mitochondrial function, and drugs to regulate utrophin. Current genetically based include dystrophin gene-replacement strategies, genetic modification techniques to restore dystrophin expression, and modulation of the dystrophin homologue, utrophin, as a surrogate to re-establish muscle function. Other innovative therapeutic approaches focusing on exon skipping and read-through of nonsense mutations are in advanced stages, with exon skipping theoretically applicable to a larger number of patients. Exon skipping that targets exons 51, 44, 45, and 53 is being globally investigated including in USA, EU, and Japan. In Japan, a study demonstrated successful dystrophin production in muscles of patients with DMD after treating with exon 53 skipping antisense oligonucleotides. (Guiraud et al., 2015; Shimizu-Motohashi et al., 2016; Ryder et al., 2017)

The FDA has also granted full approval for *deflazacort,* making this the first glucocorticoid specifically for DMD. Studies show after 12 weeks of treatment with the DMD drugs *deflazacort* and *prednisone*, there is improved muscle strength compared with a placebo. *Prednisone* may be offered for improving timed motor function reducing the need for scoliosis surgery, and delaying cardiomyopathy onset by 18 years of age. *Deflazacort* may be offered for improving strength and timed motor function and delaying age at loss of ambulation by 1.4-2.5 years. *Deflazacort* may also be offered for improving pulmonary function, reducing the need for scoliosis surgery, delaying cardiomyopathy onset, and increasing survival at 5-15 years. Additional DMD therapies under investigation include: 1) Inhibition of protein degradation - since DMD is predominantly a muscle-wasting disorder, degradative enzymes of the ubiquitin-proteasome (UPS) and calpain systems have been targeted therapeutically. 2) Myostatin inhibition - inhibition of the myostatin gene (growth and differentiation factor 8), a key negative regulator of muscle growth offers another approach to reverse muscle atrophy. 3) Cell therapies - therapies using muscle-derived or pluripotent stem cells provide a direct means to replace muscle. 4) Genetic therapies have included AAV-mediated insertion of a truncated mini/micro-dystrophin to fit the vector's limited ~4.5 kb carrying capacity, antisense oligonucleotides to induce exon skipping and re-establish the dystrophin reading frame, agents to read-through stop codon mutations, and replacement of dystrophin at the sarcolemma with surrogates such as utrophin. (Moat et al., 2013; Guiraud et al., 2015; Gloss et al., 2016; Shimizu-Motohashi et al., 2016; Kornegay, 2017; Ryder et al., 2017; Birnkrant et al., 2018)

17.5 ELLIS-VAN CREVELD SYNDROME (EVC)

EVC is a is an autosomal recessive skeletal dysplasia (chondral and ectodermal) characterized by short ribs, polydactyly, poor growth and cognitive development, and ectodermal and heart defects. Paediatricians Richard W. B. Ellis of Edinburgh and Simon van Creveld of Amsterdam were the first to describe a case of EVC in 1940 as a rare disease with approximately 150 cases reported worldwide. EVC has a prevalence of approximately 7 per 1,000,000 in the general population, with equal occurrence in males and females. EVC is found at a 1 in 200 incidence rate due to a founder mutation in the Old Order Amish community in Lancaster Country, Pennsylvania, USA. There is parental consanguinity in 30% of the cases with 7/1 000 000 prevalence outside Amish community. Today, the syndrome has been described in other populations and it is known to affect all races. (Stoll et al., 1989; Ibarra-Ramirez et al., 2017; Vona et al., 2017; Naqash et al., 2018)

Symptoms

Prenatal abnormalities (detectable by ultrasound) include narrow thorax, shortening of long bones, hexadactyly and cardiac defects. After birth, cardinal features are short stature, short ribs, polydactyly, and dysplastic fingernails and teeth. Heart defects, especially abnormalities of atrial septation, occur in about 60% of cases. (Digilio et al., 1999; Ibarra-Ramirez et al., 2017)

Genetic Basis

EVC and *EVC2* are oriented in a head-to-head configuration in chromosome 4p16 and encode single-pass type I transmembrane proteins. Mutations of the *EVC1* and *EVC2* genes were identified as the causes of EVC. The *EVC1* gene is found on the distal short arm of chromosome 4 in an area proximal to other chondrodystrophias. Mutations in this gene have been identified in EVC individuals. A second gene, *EVC2*, was identified immediately adjacent *EVC1* which encodes a protein predicted to have one transmembrane segment, three coiled-coil regions, and one RhoGEF domain (SMART). Affected EVC individuals with mutations in *EVC1* or *EVC2* are phenotypically indistinguishable. It must be noted that ECV is genetically heterogeneous; thus, *EVC1* and *EVC2* do not account for the totality of EVC cases. (Polymeropoulos et al., 1996; Galdzicka et al., 2002; Ruiz-Perez et al., 2003; Baujat & Le Merrer, 2007; Tompson et al., 2007; Sasalawad et al., 2013; Naqash et al., 2018)

Current Therapies

Management of EVC is multidisciplinary with symptomatic management mostly required in the neonatal period, including treatment of the respiratory distress due both to narrow chest and heart failure. Bone deformity, especially knee valgus with depression of the lateral tibial plateau and dislocation of the patella will need regular orthopaedic follow-up. Professional dental care should be considered for management of the oral manifestations like neonatal teeth removal to prevent impairment of feeding. Preventive measures include dietary counselling, plaque control, oral hygiene instructions, fluoride varnish application, or daily fluoride mouth rinses. To maintain space and to improve function, aesthetics, and speech, removable or fixed dental prosthesis is recommended depending on age. In infancy and early adulthood, general and specialized paediatric follow-up are also required. The short stature is due to chondrodysplasia of the legs and treatment with growth hormones is suggested. As is typical of chondrodysplasias, rare narrowing of the spinal canal can occur in EVC patients requiring a regular follow-up in adulthood. (Ortiz et al., 1996; Horigome et al., 1997; Digilio et al., 1999; Atasu & Biren, 2000; Baujat & Le Merrer, 2007; Sasalawad et al., 2013; Ibarra-Ramirez et al., 2017; Vona et al., 2017; Naqash et al., 2018)

17.6 AMYOTROPHIC LATERAL SCLEROSIS (ALS)

ALS is a fatal motor neuron disorder characterized by progressive loss of the upper and lower motor neurons at the spinal or bulbar level. It was first described in 1869 by French neurologist Jean-Martin Charcot and it became well known in the United States when baseball player Lou Gehrig was diagnosed with the disease in 1939. ALS is also known as Charcot disease honouring the first person to describe it—Jean-Martin Charcot, and motor neuron disease (MND) as it is one of the five MNDs that affect motor neurons which include primary lateral sclerosis, progressive muscular atrophy, progressive bulbar palsy, and pseudobulbar palsy. The mean age of onset of ALS varies from 50 to 65 years with the median age of onset of 64 years old. Only 5% of the cases have an onset <30 years of age. There is also geographical loci form of ALS where prevalence is 50–100 times higher in certain locations than in any other part of the world. These population include parts of Japan, Guam, Kii Peninsula of Japan, and

South West New Guinea. Scientists believe that the increased incidence of ALS in these regions is due to environmental factors, specifically a neurotoxic non-protein amino acid (β–methylamino-L-alanine, BMAA) in the seeds of the cycad *Cycas micronesica* produced by a symbiotic cyanobacteria in the roots of the cycad that are commonly found in these areas. It is hypothesized that patients in these regions who develop ALS have an inability in preventing BMAA accumulation. Other environmental factors reported to be associated with ALS include: a) occupational exposure to electrical/magnetic fields; b) metals like copper, lead, iron, selenium, copper, zinc, arsenic, cobalt, and cadmium; c) pesticides in air, food and water including organochlorine compounds, pyrethroids, herbicides, and fumigants; d) viruses including polioviruses and other enteroviruses. Several medical conditions like head trauma, cancer, metabolic diseases, and neuroinflammation have also been associated with ALS. (Rowland & Shneider, 2001; Ingre et al., 2015; Zarei et al., 2015; Martin et al., 2017)

Symptoms

The established risk factors for ALS are older age, male gender, and a family history of the disease. Symptoms of ALS are divided into primary and secondary symptoms. Primary symptoms include muscle weakness and atrophy, spasticity, speech disturbances, poor management of oral secretions, difficulty swallowing, and respiratory complications that result in death. Secondary symptoms usually accompany primary symptoms, and they can significantly reduce the quality of life of patients, such as pain or difficulty performing daily tasks. The most common symptoms that appear in both types of ALS are muscle weakness, twitching, and cramping, which eventually often leads to muscle impairment. In the most advanced stages, ALS patients will develop symptoms of **dyspnoea** and **dysphagia**. Hyper-metabolism, together with abnormalities in lipid and carbohydrate metabolism, is also prevalent in patients with ALS (Leigh & Ray-Chaudhuri, 1994; Zarei et al., 2015; Mariosa et al., 2017)

Genetic Basis

The most common form of ALS is sporadic (90–95%) and has no obvious genetically inherited component; but the remaining 5–10% of the cases are familial-type ALS (fALS) with associated genetic dominant inheritance factors. Previous epidemiologic studies suggest exposure to environmental toxins like cigarette smoke, agriculture chemicals, heavy metals, solvents, electrical magnetic fields, type of diet, dust/fibres/fumes, and physical activity could be associated with ALS.

The most common cause of ALS is a mutation of the gene encoding the antioxidant enzyme superoxide dismutase 1 (SOD1). The SOD1 gene, which codes for copper/zinc ion-binding SOD, 18 other genes have been identified in association with familial ALS. Additional genes that are known to cause fALS include: TARDBP, encodes TAR DNA-binding protein 43 (TDP-43); FUS, which codes for fusion in sarcoma; ANG, which codes for angiogenin, ribonuclease, and the RNAase A family 5; OPTN, which codes for optineurin; and C9ORF72 which harbours an expansion mutation of a hexanucleotide repeat sequence, GGGGCC, in a non-coding part of the gene. In ALS patients, the number of GGGGCC repeats can be more than a thousand, in contrast with 2–30 repeats in control populations. Mutant SOD1 has a structural instability that causes a misfold in the mutated enzyme, which can lead to aggregation in the motor neurons within the central nervous system (CNS). Several hypotheses have been proposed in regard to the mechanism underlying the mode of action of mutant SOD and the subsequent neurodegeneration seen in ALS. The most important proposed hypothesis for the pathogenesis of ALS includes glutamate excitotoxicity structural and functional abnormalities of mitochondria, impaired axonal structure or transport defects, and free radical-mediated oxidative stress. SOD1 mutations, which account for 20% of cases of fALS and 5% of sporadic ALS; TARDBP mutations represent 5–10% of fALS mutations. TDP-43 and FUS, which represents 5% of fALS mutations, are part of the process of gene expression and regulation including transcription, RNA splicing, transport, and translation, as well as processing small regulatory RNAs. ANG, responsible for the remaining 1% of fALS, is a gene, coding for an angiogenic factor that responds to hypoxia. (Leigh & Ray-Chaudhuri, 1994; Forsberg et al., 2011; Kabashi et al., 2011; Dangoumau et al., 2014; Deivasigamani et al., 2014; Zarei et al., 2015; Martin et al., 2017)

Current Therapies

ALS is currently incurable, and the emphasis has focused on treatments and interventions that prolong survival. Since there are no drugs that can halt or reverse the progressive loss of neurons, ALS management strategies feature optimizing the quality of life and helping maintain the patient's autonomy for as long as possible. A well concerted effort from multiple specialists: neurologists, physical therapists, speech therapists, occupational therapists, respiratory therapists, social workers, dietitians, and nursing care managers is suggested in the management of patients. Attention must be paid to important issues such as concerns about the course of the disease, the nutritional and respiratory management during late stages of life, and the patients advanced directives and end-of-life decisions. ALS recommended dietary supplements include vitamins E and A, creatine and *pu-erh* tea extract. (Berger et al., 2000; Groeneveld et al., 2003; Van den Berg et al., 2005; Simmons, 2005; Morozova et al., 2008; Yu et al., 2014)

17.7 MARFAN SYNDROME (MFS)

MFS was first described by French paediatrician Antoine Marfan in 1896 in a five and half-year-old girl. Its prevalence is estimated to be 1 in 5,000 in the US, affecting both genders equally. MFS is an inherited, autosomal dominant disorder in ~75% of families (remaining 25% of cases caused by de-novo mutations) with a high degree of clinical variability that affects many parts of the body like skeletal, ocular and cardiovascular systems. MFS affects males and females equally and the mutation shows no ethnic or geographical or gender bias. (Van de Velde et al., 2006; Haneline & Lewkovich, 2007; Kumar & Agarwal, 2014; Gehle et al., 2017)

Symptoms

Manifestations of MFS affect multiple organ systems including skeletal muscle, skin, the cardiovascular system (e.g. aortic dissection or tear) and the eye. MFS displays several abnormalities of the thoracic and abdominal aorta, including abnormal aortic stiffness, aortic aneurysm and dissection, and mitral valve prolapse and regurgitation. Mitral valve prolapse is the most common valvular abnormality in MFS. Histologic tissue analyses display elastic lamellae fragmentation, disorganization of the aortic architecture with excessive collagen and mucopolysaccharide accumulation, and a relative decrease of vascular smooth muscle cells. Systemic features include chest, vertebral, and feet skeletal deformities; excessive elongation of upper and lower limbs; and altered ratios among the body's segments. MFS is

Muscle, Connective Tissue, and Neonatal Disorders

associated with chronic fatigue, chronic pain, and psychological despair that compromise the quality of life and impose restrictions on the autonomy of affected persons. The most serious complication in patients with MFS is aortic dissection and the risk is substantially increased during pregnancy and the post-partum period. (Ramirez & Dietz, 2007; Groth et al., 2015; Gehle et al., 2017)

Genetic Basis

MFS is an autosomal dominant inherited connective tissue disorder primarily caused by mutations in *FBN1 gene* on chromosome 15 (15q21.1), *which* encodes an extracellular matrix protein fibrillin 1, a structural component of the extracellular matrix which is also involved in the regulation of β bioavailability. More than 2,000 mutations in this gene have been described. Mutations in many other genes coding for proteins of the transforming growth factor -β signalling (like *TGFBR*s, and *SMAD*s), vascular smooth muscle cell contractility (*MYH11 and ACTA2*), and other ECM proteins (*COL3 and FBN2*) cause clinical phenotypes overlapping with MFS. These mutations comprise of nonsense, in-frame and out-of-frame deletion/insertion, splice site, and missense. (Ramirez & Dietz, 2007; von Kodolitsch & Robinson, 2007; Hoffjan, 2012; Doyle et al., 2015; Goland & Elkayam, 2017)

Current Therapies

Since MFS is a severe, chronic, life-threatening disease with multiorgan involvement, it requires optimal multidisciplinary care to normalize both prognosis and quality of life including paediatricians, paediatric cardiologists, cardiologists, vascular surgeons, heart surgeons, vascular interventionists, orthopaedic surgeons, ophthalmologists, human geneticists, and nurses; and of auxiliary disciplines including forensic pathologists, orthodontists, dentists, radiologists, rhythmologists, pulmonologists, sleep specialists, neurologists, obstetric surgeons, psychiatrists, psychologists, and rehabilitation specialists. Drugs that block transforming growth factor -β (TGFβ) signalling (also used to treat high blood pressure) can reduce the symptoms of MFS. Examples include *propranolol, atenolol*, TGFβ neutralizing antibody, and the angiotensin-II type 1 receptor blocker *losartan*, and angiotensin converting enzyme (ACE) inhibitors (like *perindopril*). However, due to poor tolerance in some people other medications like calcium channel blockers (also used to lower blood pressure) which block movement of calcium into cells from the extracellular space are good alternatives. There are several sub-classes of these Ca^{2+} blockers based on chemical structure, including **dihydropyridines** (like *amlodipine*) and non-dihydropyridines, which include both **phenylalkylamines** (like *verapamil*) and **benzothiazepines** (like *diltiazem*). (Doyle et al., 2015; Groth et al., 2015; Kodolitsch et al., 2016; Pepe et al., 2016; Goland & Elkayam, 2017)

17.8 FIBRODYSPLASIA OSSIFICANS PROGRESSIVA (FOP)

FOP, the most catastrophic of the progressive heterotopic endochondral ossification (HEO) disorders in humans, is a rare and disabling genetic condition characterized by congenital malformations of the great toes. The worldwide prevalence of FOP is approximately one in two million individuals with no ethnic, racial, gender, or geographic predisposition to the disorder. (Shore & Kaplan, 2008; Morales-Piga & Kaplan, 2010; Pignolo et al., 2013; Morales-Piga & Kaplan, 2015; Qi et al., 2017)

Symptoms

The two main clinical features which define classic FOP are: 1) malformations of the great toes; and, 2) progressive heterotopic endochondral ossification in characteristic anatomic patterns. Individuals with FOP appear normal at birth except for malformations of the great toes. During the first decade of life, most children develop episodic, painful inflammatory soft tissue swellings (or flare-ups). Although some flare-ups regress spontaneously, several of them transform soft connective tissues including aponeuroses, fascia, ligaments, tendons, and skeletal muscles into mature heterotopic bone. Through endochondral ossification, skeletal muscles and connective tissues are replaced by ribbons, sheets, and plates of heterotopic bone leading to an armament-like encasement of bone and permanent immobility. Minor trauma such as intramuscular immunizations, mandibular blocks for dental work, muscle fatigue, blunt muscle trauma from bumps, bruises, falls, or influenza-like viral illnesses can trigger painful new flare-ups of FOP leading to progressive heterotopic ossification. Worse still, surgical removal of heterotopic bone may provoke explosive and painful new bone growth. (Hebela et al., 2005; Deirmengian et al., 2008; Shore & Kaplan, 2008; Kaplan et al., 2012; Kilmartin et al., 2014; Qi et al., 2017)

Genetic Basis

The FOP defective gene has been mapped to chromosome 2 (2q23-24 region) and occurs in all sporadic and familial cases of the disorder. Missense mutations have been found in the glycine-serine (GS) activation domain of the FOP gene. The gene encodes a bone morphogenetic protein (BMP) signalling pathway type I receptor called activin receptor type IA/activin-like kinase 2 (ACVR1/ALK2). Most cases of FOP arise as a result of a spontaneous new mutation, but a paternal age effect has been reported following an autosomal dominant genetic transmission. The ACVR1 mutation is fully penetrant, and all individuals reported as carrying this mutation develop FOP. (Kaplan et a., 1993; Shen et al., 2009; Petrie et al., 2009; Pignolo et al., 2013; Kilmartin et al., 2014; Qi et al., 2017)

Current Therapies

Currently, there is no proven efficacy with any therapy in altering the natural history of the disease. Effective treatment for FOP will likely be based on interventions that modulate overactive ACVR1/ALK2 signalling, or that specifically block postnatal heterotopic endochondral ossification. Current management is focused on early diagnosis, assiduous avoidance of injury or harm, symptomatic amelioration of painful flare-ups, and optimization of residual function. Glucocorticoids have successfully been used to manage new flare-ups affecting the function of major joints in the appendicular skeleton. Non-steroidal anti-inflammatory medications, cyclo-oxygenase-2 inhibitors, leukotriene inhibitors and mast cell stabilizers are useful anecdotally in managing chronic discomfort and ongoing flare-ups. Individuals with FOP experience the thoracic insufficiency syndrome—a life-threatening complication of cardiopulmonary function causing pneumonia and right-sided heart failure. Hearing impairment occurs in about half of FOP patients, submandibular swelling, and approximately a two-fold higher prevalence of kidney stones.

Clinical management of FOP remains symptomatic. Emerging insights into the pathophysiology of ACVR1/ALK2-mediated heterotopic ossification, and better understanding of the FOP gene have provided impetus to at least four long-term approaches to the treatment and/or prevention of FOP: 1) Blocking activity of the mutant FOP receptor; 2) Diverting the responding mesenchymal stem cells to a soft tissue fate; 3) Inhibiting inflammatory and neuro-inflammatory triggers of FOP flare-ups; and, 4) Altering the inductive and/or conducive microenvironments that promote the formation of FOP lesions. Signal transduction inhibitors have shown potential for development into powerful therapeutic agents—for example, the small molecule signal transduction inhibitor—*dorsomorphin*, blocks all bone morphogenetic protein BMP-specific signalling in cells in which the FOP gene has been overexpressed. Retinoic acid receptor gamma (RARγ) agonists have also been found to inhibit BMP-induced chondrogenesis required for the cartilaginous scaffold of heterotopic endochondral ossification.

Research is on-going on genetically engineered and pharmacologically-manipulated heterotopic endochondral ossification forming mice. Studies show that blocking any major control point in the sensory nerve pathway—theTRPV1 ion channel; the dorsal root ganglion cells; the pre-protachykinin gene that encodes Substance P; the receptor for Substance P (neurokinin 1 receptor); or the c-kit gene (required for mast cell development); all slows down or completely abolishes heterotopic endochondral ossification. Substance P triggers the release of inflammation-inducing and oedema-causing modulators that amplify the innate inflammatory response, further stimulating heterotopic endochondral ossification. Kaplan et al. (2012), demonstrated that inhibitory RNA (RNAi) duplexes are capable of specifically suppressing the mutant copy of the ACVR1/ALK2 gene in connective tissue progenitor cells from FOP patients. The RNAi duplexes decreased elevated BMP signalling in FOP cells to levels observed in control cells. (Shore et al., 2005; Kaplan et al., 2008; Zaghloul et al., 2008; Mullins & Shore, 2009; ICC-FOP, 2011; Kan et al., 2011; Shimono et al., 2011; Kaplan et al., 2012; Pignolo et al., 2013; Kilmartin et al., 2014; Qi et al., 2017)

17.9 MYOTONIC DYSTROPHY (DM)

DM is one of the most prevalent and probably also one of the most difficult to understand genetic disorders, due to its heterogeneity and its highly complex and variable clinical manifestation and molecular aetiology. DM is the collective name for a disease with two genetic subtypes, DM1 and DM2. DM shows an autosomal dominant inheritance pattern with progressive multi-systemic disorders that affect skeletal and smooth muscle as well as the eye, heart, endocrine system, and central nervous systems (CNS). DM1 is the most common form of muscular dystrophy in adults. Both types are caused by untranslated nucleotide repeat expansions in two distinct genes which leads to an RNA pathology of aberrant alternative splicing of several genes. Although both DM1 and DM2 are characterized by skeletal muscle dysfunction and share other clinical features, the diseases differ in the muscle groups that are affected. In DM1, distal muscles are mainly affected, whereas in DM2 problems are mostly found in proximal muscles. DM1 shows the highest prevalence, ranging from 0.5 and 18 cases per 100,000 individuals among different ethnic populations. (Day & Ranum, 2005; Echenne & Bassez, 2013; Thornton, 2014; Thornton et al., 2017)

Symptoms

Both disorders have multisystem features including myotonic myopathy, cataract, and cardiac conduction disease. CNS features are highly variable between patients, DM1 is commonly associated with sleep disturbance, behavioural effects, and cognition changes. The most common CNS symptom, effecting around 80% of patients, is daytime hyper-somnolence with a global disorganization of sleep habits and diurnal rhythm. Studies have shown sleep-onset REM in 26-54% of patients. DM1 is also associated with a variable constellation of behavioural and cognitive changes, that may include anxiety, avoidant behaviour, apathy, memory impairment, executive dysfunction, and problems with visuospatial processing. Cognitive impairment is one of the most common manifestations and challenging management aspects of childhood DM1. Approximately 50% of children with DM1 have at least one psychiatric diagnosis with internalising disorders (phobia, depression, anxiety) and attention deficit hyperactivity disorders are common. Cataracts before the age of 55, or family history of premature cataracts, suggests the diagnosis of DM1 or DM2 in patients with muscle symptoms.

Cardiac dysrhythmia, particularly heart block, is the second leading cause of death after respiratory failure. Other health conditions found are respiratory manifestations (due to inspiratory and expiratory muscle weakness), significant cardiac disorders, and diabetes mellitus. For many patients the first symptom is grip myotonia. Gastrointestinal symptoms are highly prevalent in DM1 with the frequency of cholelithiasis increased, likely reflecting gallbladder smooth muscle involvement. Intestinal dysmotility is common, producing symptoms of bowel urgency and diarrhoea, often alternating with constipation. DM1 is associated with higher risk of cancer, most notably involving the thyroid gland, ovary, colon, endometrium, brain, and eye (choroidal melanoma). Primary hypogonadism is common in men with DM1, and to a lesser extent in DM2—resulting in testicular atrophy, reduced fertility, erectile dysfunction, and low testosterone. DM1 is associated with metabolic derangements including insulin resistance, increased cholesterol, and hypertriglyceridemia. Abnormal liver function tests are common in DM1 and DM2. Modest elevations of alanine and aspartate aminotransferase levels, gamma-glutamyltransferase, and alkaline phosphatase may occur. Balding can occur in men and women with DM1. (Day & Ranum, 2005; Minnerop et al., 2011; Campbell et al., 2013; Theadom et al., 2014; Ho et al., 2015; Gourdon & Meola, 2017; Thornton et al., 2017)

Genetic Basis

DM1 is attributed to the expansion of a CTG trinucleotide repeat in the 3' non-coding region of the dystrophia myotonica *protein kinase* gene *(DMPK)* on chromosome 19 (19q13.3). DM1 has autosomal dominant inheritance and expressivity (penetrance) is variable. Individuals with DM1 have at least 50 and in some cases upwards of 3,000 CTG repeats in *DMPK*. Congenital myotonic dystrophy is characterised by severe hypotonia and weakness at birth, often with respiratory insufficiency. The incidence of CDM is up to 1 in 47619 live births and the mortality in the neonatal period may be 30%-40%. In DM2, the number of CCTG repeats in the first intron of *zinc finger 9* (*ZNF9*) gene. *ZNF9* occurs in several different forms (polymorphic) in the general population, ranging from 10 to 33 repeats. On average, the CTG expansion increases by more than 200 repeats when transmitted from one generation to the next leading to anticipation—the genetic phenomenon where symptoms begin at an earlier age in successive generations. Studies in DM1 and DM2 genes were perplexing because *DMPK* and *ZNF9* have no obvious functional connections and the repeat expansions are in genomic segments that do not encode proteins. Current evidence an RNA-mediated pathogenesis theory in which the shared clinical features of DM1 and DM2 involve a novel genetic mechanism in which repetitive RNA exerts a toxic effect. This RNA toxicity likely involves the activation of signalling pathways by mutant *DMPK* RNA. Newer therapeutic tools target different steps of the DM pathological cascade including the DNA mutation itself, toxic RNAs, and mediator proteins. The most current technologies are targeting DM using modified stem cells; nanotechnology; gene therapy using CRISPR-Cas and TALE nucleases to reduce CTG repeats

Muscle, Connective Tissue, and Neonatal Disorders

and viral approaches using adenovirus associated virus vectors; miRNAs and GSK3ß inhibitor drugs like *tideglusib*. (Brook et al., 1992; Todd & Paulson, 2010; Theadom et al., 2014; Ho et al., 2015; Lee et al., 2016; Gourdon & Meola, 2017; Thornton et al., 2017)

Current Therapies

No treatments are currently available that would fundamentally alter the course of DM1 or DM2. A multidisciplinary team approach is critical in providing supportive care to manage manifestations, reduce complications, optimise function and undertake health surveillance. This includes involvement of genetic counsellors, nurses, educators, physiotherapists, speech therapists, occupational therapists, social workers, and dieticians in addition to medical specialists. The management of DM is based on genetic counselling, preserving function and independence, preventing cardiopulmonary complications, and providing symptomatic treatment for myotonia, hyper-somnolence, and pain. Research shows that daytime hyper-somnolence in DM1 can be successfully managed with stimulant medications, such as *methylphenidate* and *modafinil*. The use of various anticonvulsant or antiarrhythmic drugs like *mexiletine* showed reduction of grip myotonia in DM1 by up to 50% in patients after 7 weeks of treatment. Anabolic agents were tested in DM1 to find their effectiveness in overcoming the muscle wasting. Some improvement of muscle mass with testosterone or recombinant insulin-like growth factor was evident but functional ability was not achieved. Research in gene therapy holds promise for the treatment of myotonic dystrophy. Most studies focus on the RNA mediated pathways of disease using antisense therapy. This therapy synthesizes strands of nucleic acid (called **antisense oligonucleotides**), complimentary to target mutations hoping that the target mutant sequence will be silenced.

Other therapies include drug therapy (mexiletine, anti-epileptics, amino acids, antidepressants); dehydroepiandrosterone; insulin-like growth factor; non-steroidal anti-inflammatory drugs (NSAIDs); and nutritional supplements like vitamin D and calcium. The use of nocturnal non-invasive ventilation (bi-level positive airway pressure or BiPAP and continuous positive airway pressure or CPAP has proved useful. Antibiotics for management of acute infections, prophylactic vaccinations, use of pacemaker or defibrillator insertions if indicated, total tonsillectomy or adenoidectomy have been useful. (MacDonald et al., 2002; Trip et al., 2006; Logigian et al., 2010; Bennett & Thornton, 2012; Wheeler et al., 2012; Gao & Cooper, 2013; Thornton, 2014; Ho et al., 2015; Lee et al., 2016; Gourdon & Meola, 2017; Thornton et al., 2017)

17.10 ANGELMAN SYNDROME (AS) AND PRADER-WILLI SYNDROME (PWS)

AS was first described by Harry Angelman, an English paediatrician in 1965. The defect is on a gene found on chromosome 15. AS has an estimated incidence of approximately 1 in 12,000–20,000 live births. PWS was first described by Prader, Labhart and Willi in 1956 and it represents the most common genetic cause of obesity with an estimated incidence of 1 in 15,000-25,000 live births. A maternal-only contribution is diagnostic of PWS, while an exclusively paternal contribution indicates AS. The two conditions are caused by a gene imprinting process whereby genes are expressed in a tissue-specific manner depending on parent of origin. (Abramowitz & Bartolomei, 2012; Peters et al., 2012; Bird, 2014; Larson et al., 2015; Wheeler et al., 2017)

Symptoms

The typical symptoms of AS are usually evident within the first year of life, with delayed attainment of gross motor, fine motor, receptive language, expressive language, and social skills. Seizures occur in 80%–95% of children with AS and usually start in childhood. The onset of seizures is before age 3 years in 75% of affected individuals. Behaviourally, AS features include a happy demeanour, easily provoked laughter, short attention span, hyper-motoric behaviour, mouthing of objects, and an affinity for water. Most children have an apparently reduced need for sleep (as little as 5–6 hours per night) and abnormalities of the sleep–wake cycle, with long or frequent periods of wakefulness during the night. **Microcephaly** is common in 80% of AS individuals and subtle craniofacial phenotype develops with time, consisting of midface recession, **prognathism**, and broad mouth (perhaps as a result of the latter two are possibly consequences of tongue thrusting, mouthing behaviours and increased smiling).

The main features of PWS are neonatal hypotonia and failure to thrive, childhood onset hyperphagia and obesity, small hands and feet, short stature, hypogonadism, and cognitive impairment. Infants with PWS present with neonatal hypotonia, hypoplasia of the clitoris/labia minora in girls and small penis and undescended testis in boys. Hypotonia is associated with poor sucking and feeding, often resulting in failure to thrive. Mothers may report decreased foetal activity and infants are often found in the breech position at the time of delivery. Clinical features include increased neonatal head/chest circumference ratio, narrow bifrontal diameter, **dolichocephaly**, almond shaped eyes, downturned angles of the mouth with abundant and thick saliva, small hands and feet with straight borders of the ulnar side of the hands and inner side of the legs. The presence of some of these features associated with neonatal hypotonia should alert physicians for early diagnosis of PWS during infancy. These features may become more prominent by age 2-3 years. Excessive eating and obsession with food generally begins in the preschool age group and will lead to morbid obesity if not controlled. (Thibert et al., 2013; Angulo et a., 2015; Larson et al., 2015; Wheeler et al., 2017)

Genetic Basis

AS is caused by a lack of expression of the maternally inherited *UBE3A* gene in the brain. The *UBE3A* gene is one of a small subset of human genes that go through imprinting. **Genetic imprinting** is the process of conferring functional differences onto specific genes such that their expression occurs from only one parent's allele. In most body tissues, the *UBE3A* gene appears to be expressed from both alleles (but unequally favouring the maternal allele in the brain), while the paternally derived *UBE3A* gene is silenced, leaving only the maternally inherited copy active. In mice, *UBE3A* is expressed from both paternal and maternal chromosomes in most tissues except in neurons—where only the maternally-inherited copy of the gene is expressed. AS results from deficient expression of the maternal copy of the *UBE3A* gene due to one of four molecular aetiologies: 1) deletion of the AS critical region on maternal chromosome 15q11-q13 (70% of cases); 2) paternal uniparental disomy for chromosome 15 (2% of cases); 3) an imprinting defect causing lack of expression of the maternal copy of *UBE3A* (3% of cases); and 4) point mutations in the maternally inherited copy of *UBE3A* (10% of cases). Several epigenetic mechanisms overlay the information contained within the DNA sequence and can include gene expression controllers like DNA elements preventing nearby chromatin domains from interacting (DNA insulators); histone modifications (acetylation, phosphorylation, and methylation); DNA methylation; and promoters of linked imprinted genes competing for access to enhancers (transcriptional enhancer competition). Deficits in the imprinted gene cluster on chromosome 15q11-13 cause Prader–Willi syndrome (PWS) and AS. Loss of expression of maternally derived *UBE3A* causes AS, while loss of expression of paternally derived gene(s) causes PWS. *UBE3A* encodes E6-associated protein (E6-AP), an E3 ubiquitin ligase. E6-AP's main function is believed to be participation in protein degradation in proteasomes via the ubiquitin pathway. The ubiquitin-proteasome system targets cellular proteins for destruction by covalently attaching ubiquitin to one or more lysine residues of proteins destined for degradation. The defective UBE3A activity results in failure to degrade its substrates because of impaired ubiquitination. (Kishino et al., 1997; Tan et al., 2011; Abramowitz & Bartolomei, 2012; Peters et al., 2012; Allen et al., 2013; Chamberlain, 2013; Ishida & Moore, 2013; Bird, 2014; Tan et al., 2014; Angulo et a., 2015; Larson et al., 2015; Wheeler et al., 2017)

Current Therapies

The multiple domains of healthcare in AS and PWS are best served by a comprehensive approach and an interdisciplinary team, working towards the goals of health and wellness, safety, social inclusion, and autonomy. The primary areas of clinical management include the following: seizures, sleep, aspiration risk, constipation, dental care, vision, obesity, scoliosis, bone density, mobility, communication, behaviour, and anxiety. Studies on treatments for movement disorders in AS and PWS are few, most focusing on case reports. For example, a case study found *levodopa* decreased tremor and other Parkinsonism symptoms in 2 adults with AS, and intensive physiotherapy showed a nearly 60% increase in total scores on gross motor function over the course of 36 months. *Minocycline* was found to bring about some minor improvements in fine motor skills and improvement in auditory comprehension in children with AS aged 4 to 12; but no systematic studies have examined the extent to which these therapies may help the majority of individuals with AS. Standard of care for movement disorders typically involves assessment and intervention by physical therapists and occupational therapists. Physiotherapy or bracing is often recommended for scoliosis and to improve movement, while increased activity, full range of joint movement, and monitoring of diets are encouraged to prevent obesity, and an effective intervention for increasing mobility in adults with AS—but long-term benefits are unknown. Studies exploring the use of functional analysis of behaviour in children with AS and PWS suggest that escape, tangible obtainment, and/or social engagement seeking are the primary functions of some behaviours. One reported that individually provided functional communication training decreased challenging behaviours. Stimulants for hyperactivity, antipsychotics for aggression have been used. Risperidone and methylphenidate have been prescribed for the management of hyperactivity in patients—but with no evidence-based studies indicating efficacy of these treatments in AS and PWS. Treatment of central sleep apnoea via sustained-release melatonin improved sleep rounds and reduced insomnia in a case study - melatonin and *dipiperon*. The gold standard for the treatment of sleep apnoea in adults is CPAP or BiPAP. In children adenoidectomy, tonsillectomy or adenotonsillectomy are often first lines of treatment. Supplemental oxygen therapy may be added in the presence of obesity hypoventilation syndrome. The management of other sleep disturbances may include implementation of adequate sleep hygiene, sleep wake schedule regulation and even circadian rhythm modification. (Cataletto et al., 2011; Tan et al., 2011; Allen et al., 2013; Bird, 2014; Angulo et al., 2015; Larson et al., 2015; Wheeler et al., 2017)

17.11 FRAGILE-X SYNDROME (FSX)

FXS is an X-inherited genetic disease first described in 1943 by Martin and Bell. FXS is considered the most common inherited cause of intellectual disability and the second most prevalent cause after Down syndrome. While most cases of Down syndrome are *de novo*, FXS is always inherited from an affected or carrier parent and affects as many as 1 in 4,000 males and 1 in 8,000 females. It has been estimated that as many as 50% of individuals with FSX have co-occurring autism. (Bagni et al., 2012; Saldarriaga et al., 2014; Thurman et al., 2017)

Symptoms

The FXS cognitive and emotional phenotype and physical characteristics depend on the amount of FMRP produced, which depends on the number of trinucleotide repeats and the methylation degree of the *FMR1*. Depending on the FMRP protein concentration, a clinical spectrum therefore develops. When FMRP levels are not very low the symptoms are less severe, there's a moderate emotional and learning difficulties and a normal intelligence quotient. If the production of the FMRP protein decreases or stops, a severe cognitive deficit develops causing mental impairment. About 80% of the patients with FXS will have one or more of their common facial characteristics of long face, large and protruded ears, and abnormally large testes (**macroorchidism**). Studies report that 84% of males and 67% of females with the full mutation in a large national survey of parents had attention problems including attention, impulsivity and inhibition control. Studies also show that 70% of boys and 56% of girls have been treated or diagnosed with an anxiety disorder. Individuals with FXS can exhibit several behaviours commonly associated with autism, including difficulties with social communication, self-injurious behaviour, perseverative or restricted behaviour, motor stereotypies, poor eye contact, and odd or delayed speech. Autism spectrum disorder (ASD) is identifiable as a cluster of symptoms, rather one specific condition, with many currently unknown causes. Scientists think there may be considerably more overlap between FSX and ASD (Howes et al., 2018). Social withdrawal (like avoidance and indifference) and adaptive socialization behaviours (like recognizing emotions or appropriate social interactions) are often independent predictors of autism spectrum disorder in individuals with FXS. Behaviour problems, including tactile defensiveness, hand flapping, poor eye contact, hyperactivity, tantrums, perseveration, hyperarousal to sensory stimuli, and impulsivity are hallmark features of FXS. Self-injury (often the biting of hands and fingers) is also common in individuals with FXS, with prevalence rates as high as 79% of boys. (Symons et al., 2003; Kaufmann et al., 2004; Hatton et al., 2006; Brock & Hatton, 2010; Smith et al., 2012; Hagerman & Hagerman, 2013; Leigh et al., 2013; Saldarriaga et al., 2014; Ciaccio et al., 2017; Thurman et al., 2017)

Genetic Basis

Affected men have a classic phenotype characterized by long face, large and protruding ears and macroorchidism. FXS is caused by an abnormal expansion (triplet expansion) in the number of the CGG repeats located in the 5' UTR in the fragile X mental retardation 1 gene (*FMR1*) located at Xq27.3. There is a dynamic expansion of the CGG repeat in each generation from the premutation range and expanding to a full mutation when passed on by a woman to her children. The transcriptional silencing of *FMR1* means the encoded protein FMRP, which regulates mRNA metabolism in the brain and con-

Muscle, Connective Tissue, and Neonatal Disorders

trolling the expression of key molecules involved in receptor signalling and spine morphology, is lost. Premutation carriers though not showing the classic FXS phenotype, develop other medical, psychiatric and neurological problems. Depending on the number of repetitions, 4 types of alleles are defined with different clinical manifestations; a) normal alleles (0-44 CGG repeats); b) premutation alleles (55-200 repeats); c) intermediate allele (45 and 54 repeats serving as a precursor for full mutation alleles); and d) full mutation alleles (>200 repeats). The silencing of the *FMR1* gene is the result of a series of complex epigenetic modifications following the expansion of the trinucleotide repeat. Although FXS is usually caused by the methylation and gene silencing associated with the full mutation, other mechanisms can also be at play. Though accounting for less than 1% of all FXS cases described, deletions of the coding region of the gene can lead to loss of FMRP; or point mutations and reading frame shifts leading to a functional deficit of the protein FMRP and the consequent phenotype can occur. Worldwide prevalence as determined by molecular assays is estimated at 1 per 5,000 men, and 1 per 4,000-6,000 women. (Sutcliffe et al., 1992; Hirst et al., 1993; Stöger et al., 1997; Bassell & Warren, 2008; Delahunty et al., 2009; Bagni et al., 2012; Hagerman & Hagerman, 2013; Leigh et al., 2013; Saldarriaga et al., 2014; Raspa et al., 2017; Ciaccio et al., 2017; Thurman et al., 2017)

Current Therapies

Multiple studies have been carried out in the attempt to develop target treatments for FXS focus on the activation of the *FMR1* gene or the treatment of the symptoms associated to this disease. Restoration of the *FMR1* gene activity by decreasing the DNA methylation and altering the H3 and H4 histones acetylation code has produced some good results. The use of 5-azad C (a methyltransferase inhibitor) and histone deacetylase inhibitors (like TSA, butyrate and 4-phenylbutyrate) have been proven to act synergistically in FMR1 restoration. Both drugs have been tested solely *in vitro*, due to the risk of inducing cellular apoptosis *in vivo*. The mechanism of action of some of the medications used focus on epigenetic modulation, glutamatergic system and regulation of the translation of FMRP target mRNAs. Valproic acid, known as a silent gene reactivator has been used as a weak reactivator of the FM alleles in FXS. It has been reported that the administration of this drug to patients with FXS improves their attention deficit hyperactivity disorder. L-acetyl-carnitine also seems to inhibit the formation of fragile sites in the X chromosome. Research has also showed that compound GSK6A given to mice lacking the FMR1 gene in an established animal model of fragile X syndrome, relieves symptomatic behaviours, such as impaired social interactions and inflexible decision making, which can be displayed by humans with FSX. GSK6A inhibits one particular form of a cellular signalling enzyme: the p110β form of PI3 (phosphoinositide-3) kinase. (Gross et al., 2018)

Minocycline, mGluR5 antagonists, GABA A and B agonists have been demonstrated to be effective in clinical treatment of children and adults with FXS by preventing overproduction of metalloproteinase-9 produced in response to the normal synaptic stimulations in this disease. To treat ADHD symptoms, the best medications available in FXS are the stimulants including preparations of methylphenidate or mixed amphetamine salts. Clonidine can also be used to treat the sleep disturbance that is a common problem for young children with FXS. Due to the high frequency of mood, anxiety and behaviour disorders in patients with FXS, serotonin selective reuptake inhibitors (SSRI) have been used to treat aggressive behaviour, anxiety, depression effectively. Mood stabilizer medication such as aripiprazole or risperidone can be used in SSRI-induced aggressive behaviour. Lithium and sertraline are also excellent mood stabilizers and targeted treatment for FXS. Children with FXS are also typically very good with the computer so applications that focus on language, reading, math, or social deficits can be used with the help of the teacher, therapist or parent. Developing cognitive therapies, as well as educational and behavioural intervention in FXS patients to allow strengthening of their social abilities, reading abilities, and adaptive behaviour is recommended. (Wirojanan et al., 2009; Moskowitz et al., 2011; Rueda et al., 2011; Bagni et al., 2012; Ciaccio et al., 2017; Thurman et al., 2017)

17.12 WAARDENBURG SYNDROME (WS)

Petrus Johannes Waardenburg, a German ophthalmologist was the first person to describe WS in 1951. Fundamentally a disorder of the abnormal neural crest cell differentiation and migration (**neurocristopathy**), it accounts for 2% of congenital deafness most commonly described in western populations and affects ~1 in 42,000 people. There are >400 types of syndromic hearing loss, and WS is the most common, accounting for 2–5% of congenital deafness. It equally affects both male and females and all races with an incidence of one in 40,000. (Zaman et al., 2015; Chen et al., 2016; Rawlani et al., 2018; Wang et al., 2018)

Symptoms

WS is a type of auditory-pigmentary syndrome, the clinical manifestations of which include congenital neurosensory deafness; change in iris pigmentation; telecanthus; abnormal distribution of hair and skin pigmentation (prematurely white hair, eyebrows and eyelashes below age 30; facial freckles; skin depigmentation leukoplakia); wide root of the nose; and synophrys. Waardenburg syndrome (WS) is the most common type of autosomal dominant syndromic hearing loss. The clinical spectrum of the oral manifestations of patients affected by WS include a number of oral alterations such as dental agenesis of the lower lateral incisors, taurodontism and conical teeth; cleft lip and/or palate and undefined jaw malformations and mandibular prognathism; irregular teeth, abnormalities of the tooth enamel, a high palate and a fissured tongue. WS is characterized by physical absence of melanocytes affecting skin, hair, eyes, or stria vascularis of cochlea producing congenital deafness, partial or total heterochromia iridis, hypertrichosis of the medial part of the eyebrows, broad and elevated nasal root and dystopia canthorum. (Yang et al., 2013; Zaman et al., 2015; Chen et al., 2016; Sólia-Nasser et al., 2016; Rawlani et al., 2018; Wang et al., 2018)

Genetic Basis

WS is classified into four major types and 10 subtypes depending on the additional symptoms present: WS1, WS2, WS3 and WS4. WS1 is the classic form of WS with dystopia canthorum (lateral displacement of the inner canthi), whereas WS2 is characterized by the presence of unpigmented tissue and deafness without dystopia canthorum. WS3 is similar to WS1 with additional musculoskeletal abnormalities, and WS4 is characterized by the presence of aganglionic megacolon. The diagnosis of WS1 requires the presence of at least two major criteria, or one major and two minor criteria. WS1 and WS2 are more common than WS3 and WS4. WS is largely attributed to the mutations in seven genes: paired box 3 *(PAX3)*, sex-determining region Y-box 10 *(SOX10)*, melanogenesis associated transcription factor *(MITF)*, endothelin-2, endothelin receptor type-B *(EDNRB)*, snail family zinc finger-2 transcriptional repressor *(SNAI2)*, endothelin-3 *(EDN3)* and a gene encoding *KIT* ligand *(KITLG)*. Malfunctioning of these genes caused by missense and nonsense mutations, frame shifts, small in-frame insertions or deletions and splice alterations causes abnormal development of neural crest cells, change in iris pigmentation, deafness, hair greying, and abnormal skin pigmentation in pigment-producing cells called melanocytes. Melanocytes make a pigment called melanin, which contributes to skin, hair, and eye colour and plays an essential role in the normal function of the inner ear. The seven genes are involved in a network of interactions, and mutations in these genes determine faults in embryogenesis derived from neural crest cells and characterise the phenotypic diversity of WS. All or most WS1 cases are caused by mutations at the PAX3 locus on chromosome 2 (2q37). Approximately 70 different PAX3 point mutations have been identified. WS1 caused by mutations in the *PAX3* gene, is characterized by evidence of dystopia canthorum and the full symptomatology of the disease with a narrow nose, marked hypoplasia of the nasal bone, short philtrum, and short and retropositioned maxilla. WS2 is due to mutations in the *MITF* and *SNAI2* genes and is heterogeneous group with normally located canthi (without dystopia canthorum). Sensorineural hearing loss (77%) and heterochromia iridium (47%) are the two most important diagnostic indicators for this type. WS3 is similar to WS1 but is also associated with musculoskeletal abnormalities. WS4 is the association of WS with congenital aganglionic megacolon (**Hirschsprung disease**) and is caused by mutations in the *SOX10, EDN3,* or *EDNRB* genes. (Khong & Rosenberg, 2002; Pingault et al., 2010; Yang et al., 2013; Zaman et al., 2015; Chen et al., 2016; Chen et al., 2017; Rawlani et al., 2018; Wang et al., 2018)

Current Therapies

There is no specific treatment for WS and early diagnosis will enable the treatment of deafness. A very individualized and meticulous anaesthetic approach in managing these patients is suggested requiring preoperative evaluation, co-existence of other system abnormalities, airway management, and perioperative nutrition strategies. A consortium of professionals must work together including anaesthesiologists, paediatricians and paediatric-surgeons to manage the multiple surgery needs for these patients. Genetic counselling for families experiencing the disorder has also been described as helpful. More recently advancing research has given hope to new remedies. The role for SOX10 in maintaining cell cycle control in melanocytes suggests a rational new direction for targeted treatment or prevention of melanoma (melanoma immunotherapy). Studies have identified the presence of *de novo* cellular immune reactivity against the transcription factor *SOX10* using tumour-infiltrating lymphocytes obtained from a patient. The patient experienced a dramatic clinical response to immunotherapy. The feasibility of cochlear implantation for sensorineural hearing loss in patients is increasingly being recommended. (Peker et al., 2015; Sharma & Arora, 2015; Koyama et al., 2016; Wang et al., 2018)

CHAPTER SUMMARY

The skeleton provides the framework and anchor points against which muscles, attached via tendons, can exert force. It is made up of 206 bones divided into 2 systems: a) axial skeleton with skull bones, 12 pairs of ribs, the sternum, and the 26 vertebrae; b) appendicular which is made up of the pectoral girdles (shoulders); pelvic girdle (hips) and pairs of arms, hands, legs and feet. Three types of cells involved in making bone: osteoblasts, osteoclasts and cartilage. Physical exercise stresses bone such that the process of mineral deposition exceeds withdrawal—making the bones denser and stronger. With age, bones decline in mass where mineral deposits lag behind withdrawal making the bones hollow and brittle—a condition known as osteoporosis. The human muscle system is made up of three types of muscle tissue: skeletal, cardiac and smooth. There are over 700 skeletal muscles which help move the body and attach to bone through tendons. Each muscle contains thousands of bundles of long fibre-like muscle cells (with many nuclei from fused embryonic cells) made of individual myofibrils. Each myofibril is made up of two kinds of myofilaments which create muscle striations.

The neonate period of life is the first 4 weeks after the birth of an infant. Some human diseases are explained by alterations in a single gene or of a single chromosome; however, most are complex and may involve multiple genes and several metabolic pathways. A myriad of genes, as well as environmental factors control the complex and integrated processes necessary for foetal development. Research using animal models continues to provide important information about the biological and regulatory processes involved in human development. There are a number of disorders caused by defects in genes important in the formation and function of muscles, connective tissues and in foetal development.

Achondroplasia is the most common form of dwarfism in humans and is a short stature condition compatible with good general health and normal life expectancy. However, besides the burden of short stature, health issues may arise from complications arising from the particular anatomical features of Ach. Ach diagnosis is usually made at birth with 80% of cases arising as sporadic mutations in *FGFR3* gene. The condition was mapped to *FGFR3* in 1994 and over 97% of cases result from an autosomal dominant missense single point gain-of-function mutation localized in the transmembrane domain of FGFR3. Genetic linkage studies show that FGFR3 gene is located on the short arm of chromosome 4 and mutation analysis identified an arginine to glycine substitution in fibroblast growth factor receptor 3 (*FGFR3*) in almost all Ach patients. Treatment of the developmental complications of Ach involves symptomatic management, surgical intervention, and lifelong follow-up care. To be effective, therapies for Ach need to be administered throughout from birth to puberty. Currently, perhaps the most promising therapy Ach treatment is the use of a stabilized form of C-type natriuretic peptide (CNP) called *vosoritide*.

CMT classically refers to inherited motor and sensory neuropathies with a wide range of genotypes and phenotypes characterized primarily by distal muscle weakness, atrophy, and sensory loss. The disease was first described in 1886 by Charcot and Marie in France and independently by Tooth in Great Britain and was named in their honour. The most commonly encountered forms of CMT are generally classified as type 1 (demyelinating, 50-80% of CMT cases) and type 2 (axonal, 10-15% of cases). Since the initial discovery of *PMP22* as the causative gene for CMT1A, more than 45 causative genes have been identified (genetic heterogeneity). CMT can be inherited in an autosomal dominant, autosomal recessive and X-linked manner. The autosomal dominantly inherited demyelinating form of CMT is caused by a duplication on chromosome 17p11.2, containing the gene coding for peripheral myelin protein 22 (PMP22) and thus leading to three copies of the *PMP22* gene. CMT treatment consists of supportive care and may include management by a rehabilitation physician, physiotherapist, occupational therapist, and orthopaedic surgeon.

The prevalence of DMD in the US has been reported as 15.9 cases per 100 000 live male births and 19.5 cases per 100 000 live male births in the UK. The absence of dystrophin leads to myofibril membrane fragility and repeated cycles of necrosis and regeneration, leading to eventual muscle atrophy and contractures. Muscle is gradually replaced with fibrous connective tissue and fat, leading to weakness and debilitating contractures. Progressive muscular damage and degeneration occurs in people with DMD, resulting in muscular weakness, associated motor delays, loss of ambulation, respiratory impairment,

Muscle, Connective Tissue, and Neonatal Disorders

and cardiomyopathy. DMD is a lethal X-linked recessive neuromuscular disorder caused by mutations in the dystrophin gene resulting in absent or insufficient functional dystrophin (a cytoskeletal protein), whose function is to enable the strength, stability, and functionality of muscle myofibrils. There is currently no cure, and despite exhaustive palliative care, patients are restricted to a wheelchair by the age of 12 years and usually succumb to cardiac or respiratory complications in their late 20s. Rehabilitation personnel include physicians, physical therapists, occupational therapists, speech-language pathologists, orthotists, and durable medical equipment providers. DMD drugs in use include *ataluren*, *eteplirsen*, *myostatin*, *deflazacort* and *prednisone*.

EVC is a is an autosomal recessive skeletal dysplasia (chondral and ectodermal) characterized by short ribs, polydactyly, poor growth and cognitive development, and ectodermal and heart defects. It was first described by pediatricians Richard W. B. Ellis and Simon van Creveld in 1940. Mutations of the *EVC1* and *EVC2* genes were identified as the causes of EVC. *EVC1* and *EVC2* are oriented in a head-to-head configuration in chromosome 4p16 and encode single-pass type I transmembrane proteins. Management of EVC is multidisciplinary with symptomatic management mostly required in the neonatal period, including treatment of the respiratory distress due both to narrow chest and heart failure.

ALS is a fatal motor neuron disorder characterized by progressive loss of the upper and lower motor neurons at the spinal or bulbar level. It was first described in 1869 by French neurologist Jean-Martin Charcot and it became well known in the United States when baseball player Lou Gehrig was diagnosed with the disease in 1939. The most common form of ALS is sporadic (90–95%) and has no obvious genetically inherited component; but the remaining 5–10% of the cases are familial-type ALS (fALS) with associated genetic dominant inheritance factors. Previous epidemiologic studies suggest exposure to environmental toxins like cigarette smoke, agriculture chemicals, heavy metals, solvents, electrical magnetic fields, type of diet, dust/fibres/fumes, and physical activity could be associated with ALS. The most common cause of ALS is a mutation of the gene encoding the antioxidant enzyme superoxide dismutase 1 (SOD1). The SOD1 gene, which codes for copper/zinc ion-binding SOD, 18 other genes have been identified in association with familial ALS. ALS is currently incurable, and the emphasis has focused on treatments and interventions that prolong survival. Since there are no drugs that can halt or reverse the progressive loss of neurons, ALS management strategies feature optimizing the quality of life and helping maintain the patient's autonomy for as long as possible. A well concerted effort from multiple specialists: neurologists, physical therapists, speech therapists, occupational therapists, respiratory therapists, social workers, dietitians, and nursing care managers is suggested in the management of patients.

MFS was first described by French paediatrician Antoine Marfan in 1896 in a five and half-year-old girl as an inherited, autosomal dominant disorder in ~75% of families (remaining 25% of cases caused by de-novo mutations) with a high degree of clinical variability that affects many parts of the body like skeletal, ocular and cardiovascular systems. MFS is a connective tissue disorder primarily caused by mutations in *FBN1 gene* on chromosome 15 (15q21.1), *which* encodes an extracellular matrix protein fibrillin 1, a structural component of the extracellular matrix which is also involved in the regulation of β bioavailability. More than 2,000 mutations in this gene have been described. Since MFS is a severe, chronic, life-threatening disease with multiorgan involvement, it requires optimal multidisciplinary care to normalize both prognosis and quality of life including paediatricians, paediatric cardiologists, cardiologists, vascular surgeons, heart surgeons, vascular interventionists, orthopaedic surgeons, ophthalmologists, human geneticists, and nurses; and of auxiliary disciplines including forensic pathologists, orthodontists, dentists, radiologists, rhythmologists, pulmonologists, sleep specialists, neurologists, obstetric surgeons, psychiatrists, psychologists, and rehabilitation specialists. Drugs like *propranolol, atenolol, losartan, perindopril, amlodipine, verapamil,* and *diltiazem* can reduce the symptoms of MFS.

FOP, the most catastrophic of the progressive heterotopic endochondral ossification (HEO) disorders in humans, is a rare and disabling genetic condition characterized by congenital malformations of the great toes. The worldwide prevalence of FOP is approximately one in two million individuals with no ethnic, racial, gender, or geographic predisposition to the disorder. The FOP defective gene has been mapped to chromosome 2 (2q23-24 region) and occurs in all sporadic and familial cases of the disorder. Missense mutations have been found in the glycine-serine (GS) activation domain of the FOP gene. The gene encodes a bone morphogenetic protein (BMP) signalling pathway type I receptor called activin receptor type IA/activin-like kinase 2 (ACVR1/ALK2). Most cases of FOP arise as a result of a spontaneous new mutation, but a paternal age effect has been reported following an autosomal dominant genetic transmission. Currently, there is no proven efficacy with any therapy in altering the natural history of the disease. Effective treatment for FOP will likely be based on interventions that modulate overactive ACVR1/ALK2 signalling, or that specifically block postnatal heterotopic endochondral ossification. Current management is focused on early diagnosis, assiduous avoidance of injury or harm, symptomatic amelioration of painful flare-ups, and optimization of residual function. Glucocorticoids, non-steroidal anti-inflammatory medications, cyclo-oxygenase-2 inhibitors, leukotriene inhibitors and mast cell stabilizers are useful anecdotally in managing chronic discomfort and ongoing flare-ups. Signal transduction inhibitors like *dorsomorphin* have shown potential for development into powerful therapeutic agents. Research is on-going on genetically engineered and pharmacologically-manipulated heterotopic endochondral ossification forming mice.

DM is one of the most prevalent and probably also one of the most difficult to understand genetic disorders, due to its heterogeneity and its highly complex and variable clinical manifestation and molecular aetiology. DM is the collective name for a disease with two genetic subtypes, DM1 and DM2. DM shows an autosomal dominant inheritance pattern with progressive multi-systemic disorders that affect skeletal and smooth muscle as well as the eye, heart, endocrine system, and central nervous systems (CNS). DM1 is attributed to the expansion of a CTG trinucleotide repeat in the 3' non-coding region of the dystrophia myotonica p*rotein kinase* gene *(DMPK)* on chromosome 19 (19q13.3). DM1 has autosomal dominant inheritance and expressivity (penetrance) is variable. Individuals with DM1 have at least 50 and in some cases upwards of 3,000 CTG repeats in *DMPK*.

No treatments are currently available that would fundamentally alter the course of DM1 or DM2. A multidisciplinary team approach is critical in providing supportive care to manage manifestations, reduce complications, optimise function and undertake health surveillance. This includes involvement of genetic counsellors, nurses, educators, physiotherapists, speech therapists, occupational therapists, social workers, and dieticians in addition to medical specialists. Other therapies include drug therapy (mexiletine, anti-epileptics, amino acids, antidepressants); dehydroepiandrosterone; insulin-like growth factor; non-steroidal anti-inflammatory drugs (NSAIDs); and nutritional supplements like vitamin D and calcium.

AS was first described by Harry Angelman, an English paediatrician, in 1965. The defect is on a gene found on chromosome 15. PWS was first described by Prader, Labhart and Willi in 1956 and it represents the most common genetic cause of obesity. A maternal-only contribution is diagnostic of PWS, while an exclusively paternal contribution indicates AS. The two conditions are caused by a gene imprinting process whereby genes are expressed in a tissue-specific manner depending on parent of origin. AS is caused by a lack of expression of the maternally inherited *UBE3A* gene in the brain. The *UBE3A* gene

Muscle, Connective Tissue, and Neonatal Disorders

is one of a small subset of human genes that go through imprinting. tissues except in neurons—where only the maternally-inherited copy of the gene is expressed. Deficits in the imprinted gene cluster on chromosome 15q11-13 cause Prader–Willi syndrome (PWS) and AS. Loss of expression of maternally derived *UBE3A* causes AS, while loss of expression of paternally derived gene(s) causes PWS. *UBE3A* encodes E6-associated protein (E6-AP), an E3 ubiquitin ligase.

The multiple domains of healthcare in AS and PWS are best served by a comprehensive approach and an interdisciplinary team, working towards the goals of health and wellness, safety, social inclusion, and autonomy. The primary areas of clinical management include the following: seizures, sleep, aspiration risk, constipation, dental care, vision, obesity, scoliosis, bone density, mobility, communication, behaviour, and anxiety. Studies on treatments for movement disorders in AS and PWS are few, most focusing on case reports where use of certain drugs like *levodopa* decreased tremor and other Parkinsonism symptoms in 2 adults with AS.

FXS is an X-inherited genetic disease first described in 1943 by Martin and Bell. FXS is considered the most common inherited cause of intellectual disability and the second most prevalent cause after Down syndrome. It has been estimated that as many as 50% of individuals with fragile X syndrome have co-occurring autism. FXS is caused by an abnormal expansion (triplet expansion) in the number of the CGG repeats located in the 5' UTR in the fragile X mental retardation 1 gene (*FMR1*) located at Xq27.3. There is a dynamic expansion of the CGG repeat in each generation from the premutation range and expanding to a full mutation when passed on by a woman to her children. The transcriptional silencing of *FMR1* means the encoded protein FMRP, which regulates mRNA metabolism in the brain and controlling the expression of key molecules involved in receptor signalling and spine morphology, is lost. Multiple studies have been carried out in the attempt to develop target treatments for FXS focus on the activation of the *FMR1* gene or the treatment of the symptoms associated to this disease. Restoration of the *FMR1* gene activity by decreasing the DNA methylation and altering the H3 and H4 histones acetylation code has produced some good results. The use of 5-azad and histone deacetylase inhibitors have been proven to act synergistically in FMR1 restoration. Minocycline, mGluR5 antagonists, GABA A and B agonists have been demonstrated to be effective in clinical treatment of children and adults with FXS by preventing overproduction of metalloproteinase-9 produced in response to the normal synaptic stimulations in this disease.

WS was first described in 1951 by a German ophthalmologist—Petrus Johannes Waardenburg as a disorder of the abnormal neural crest cell differentiation and. There are >400 types of syndromic hearing loss, and WS is the most common, accounting for 2–5% of congenital deafness. WS) is classified into four major types and 10 subtypes depending on the additional symptoms present. WS is largely attributed to the mutations in seven genes whose malfunctioning is caused by missense and nonsense mutations, frame shifts, small in-frame insertions or deletions and splice alterations causing abnormal development of neural crest cells, change in iris pigmentation, deafness, hair greying, and abnormal skin pigmentation in pigment-producing cells called melanocytes. There is no specific treatment for WS and early diagnosis will enable the treatment of deafness. A very individualized and meticulous anaesthetic approach in managing these patients is suggested requiring preoperative evaluation, co-existence of other system abnormalities, airway management, and perioperative nutrition strategies.

End of Chapter Quiz

1. The axial skeleton includes all of the following except:
   a. skull bones
   b. pectoral girdles
   c. sternum
   d. 12 pairs of ribs
   e. 26 vertebrae
2. What are the two types of bone tissue?
   a. compact bone and spongy bone tissue
   b. long bone and short bone tissue
   c. hard bone and soft bone tissue
   d. smooth bone and striated bone tissue
   e. none of the above
3. The three types of cells involved in making bone are
   a. squamous, cuboidal, columnar
   b. keratinocytes, melanocytes, Langerhans
   c. osteoblasts, osteocytes, and osteoclasts
   d. merkel, chondrocytes, granulocytes
   e. agranulocytes, osteoblasts, osteocytes
4. When muscles are at rest, the actin binding sites are blocked by a regulatory protein called
   a. acetylcholine
   b. troponin
   c. tropomyosin
   d. myofibril
   e. sarcomeres
5. When a muscle is stimulated by acetylcholine, an action potential is generated and ____ ions diffuse from the sarcoplasmic reticulum and reach the actin filaments.
   a. Potassium
   b. Chloride
   c. Sodium
   d. Calcium
   e. Phosphorus
6. What is the most common form of dwarfism in humans?
   a. Achondroplasia
   b. Spondyloepiphyseal dysplasias
   c. Diastrophic dysplasia
   d. Campomelic dysplasia
   e. Cartilage Hair Hypoplasia

Muscle, Connective Tissue, and Neonatal Disorders

7. The absence of this protein leads to myofibril membrane fragility leading to muscle atrophy and contractures.
   a. Tau
   b. Dystrophin
   c. Keratin
   d. Coronin
   e. Spectrin
8. Which of the following statements about Duchenne Muscular Dystrophy (DMD) is incorrect?
   a. DMD is a lethal Y-linked recessive neuromuscular disorder caused by mutation in the dystrophin gene
   b. DMD is predominant in males
   c. Due to the lack of dystrophin, patients with DMD have weak heart myocardium resulting in cardiomyopathy
   d. Children with DMD may complain of shortness of breath due the weakening of the diaphragm and other muscles
   e. There is currently no cure for DMD
9. Which of the following statements about Angelman Syndrome (AS) and Prader-Willi Syndrome (PWS) is false?
   a. Microcephaly is common in 80% of AS individuals whereas, dolichocephaly is common in individuals with PWS
   b. The main features of PWS are neonatal hypertonia, large hands and feet, and hypergonadism.
   c. AS and PWS are caused by a gene imprinting process whereby genes are expressed in a tissue-specific manner depending on parent origin
   d. Behaviourally, patients with AS have a happy demeanour, easily provoked laughter, and affinity of water.
   e. A maternal only contribution is diagnostic of PWS, while an exclusively paternal contribution indicates AS
10. What are the clinical manifestations associated with Waardenburg Syndrome?
    a. Congenital neurosensory deafness
    b. Change in iris pigmentation
    c. Telecanthus
    d. Synophrys
    e. All of the above

Thought Questions

1. Describe the stages of muscle contraction and relaxation in detail.
2. Which parts of the nervous system does Amyotrophic Lateral Sclerosis (ALS) affect, and how can that eventually lead to paralysis?
3. How do you think environmental factors play a part in the neurodegeneration of motor neurons?

REFERENCES

Abramowitz, L. K., & Bartolomei, M. S. (2011). Genomic imprinting: Recognition and marking of imprinted loci. *Current Opinion in Genetics & Development, 22*(2), 72–78. doi:10.1016/j.gde.2011.12.001 PMID:22195775

Allen, K. D., Kuhn, B. R., DeHaai, K. A., & Wallace, D. P. (2013). Evaluation of a behavioral treatment package to reduce sleep problems in children with Angelman Syndrome. *Research in Developmental Disabilities, 34*(1), 676–686. doi:10.1016/j.ridd.2012.10.001 PMID:23123881

Angulo, M. A., Butler, M. G., & Cataletto, M. E. (2015). Prader-Willi syndrome: A review of clinical, genetic, and endocrine findings. *Journal of Endocrinological Investigation, 38*(12), 1249–1263. doi:10.100740618-015-0312-9 PMID:26062517

Atasu, M., & Biren, S. (2000). Ellis-van Creveld syndrome: Dental, clinical, genetic and dermatoglyphic findings of a case. *The Journal of Clinical Pediatric Dentistry, 24*(2), 141–145. PMID:11314324

Bagni, C., Tassone, F., Neri, G., & Hagerman, R. (2012). Fragile X syndrome: Causes, diagnosis, mechanisms, and therapeutics. *The Journal of Clinical Investigation, 122*(12), 4314–4322. doi:10.1172/JCI63141 PMID:23202739

Bassell, G. J., & Warren, S. T. (2008). Fragile X syndrome: Loss of local mRNA regulation alters synaptic development and function. *Neuron, 60*(2), 201–214. doi:10.1016/j.neuron.2008.10.004 PMID:18957214

Baujat, G., Legeai-Mallet, L., Finidori, G., Cormier-Daire, V., & Le Merrer, M. (2008). Achondroplasia. *Best Practice & Research. Clinical Rheumatology, 22*(1), 3–18. doi:10.1016/j.berh.2007.12.008 PMID:18328977

Baujat, G., & Merrer, M. L. (2007). Ellis-Van Creveld syndrome. *Orphanet Journal of Rare Diseases, 2*(1), 27. doi:10.1186/1750-1172-2-27 PMID:17547743

Berger, M. M., Kopp, N., Vital, C., Redl, B., Aymard, M., & Lina, B. (2000). Detection and cellular localization of enterovirus RNA sequences in spinal cord of patients with ALS. *Neurology, 54*(1), 20–25. doi:10.1212/WNL.54.1.20 PMID:10636120

Bird, L. M. (2014). Angelman syndrome: Review of clinical and molecular aspects. *The Application of Clinical Genetics, 7*, 93–104. doi:10.2147/TACG.S57386 PMID:24876791

Birnkrant, D. J., Bushby, K., Bann, C. M., Apkon, S. D., Blackwell, A., Brumbaugh, D., ... Weber, D. R. (2018). Diagnosis and management of Duchenne muscular dystrophy, part 1: Diagnosis, and neuromuscular, rehabilitation, endocrine, and gastrointestinal and nutritional management. *Lancet Neurology, 17*(3), 251–267. doi:10.1016/S1474-4422(18)30024-3 PMID:29395989

Brock, M., & Hatton, D. (2010). Distinguishing features of autism in boys with fragile X syndrome. *Journal of Intellectual Disability Research, 54*(10), 894–905. doi:10.1111/j.1365-2788.2010.01315.x PMID:20704635

Brook, J. D., McCurrach, M. E., Harley, H. G., Buckler, A. J., Church, D., Aburatani, H., ... Hudson, T. (1992). Molecular basis of myotonic dystrophy: Expansion of a trinucleotide (CTG) repeat at the 3' end of a transcript encoding a protein kinase family member. *Cell, 68*(4), 799–808. doi:10.1016/0092-8674(92)90154-5 PMID:1310900

Campbell, C., Levin, S., Siu, V. M., Venance, S., & Jacob, P. (2013). Congenital myotonic dystrophy: Canadian population-based surveillance study. *The Journal of Pediatrics, 163*(1), 120–125. doi:10.1016/j.jpeds.2012.12.070

Cataletto, M., Angulo, M., Hertz, G., & Whitman, B. (2011). Prader-Willi syndrome: A primer for clinicians. *International Journal of Pediatric Endocrinology, 12*(1), 12. doi:10.1186/1687-9856-2011-12 PMID:22008714

Chamberlain, S. J. (2012). RNAs of the human chromosome 15q11-q13 imprinted region. *Wiley Interdisciplinary Reviews. RNA, 4*(2), 155–166. doi:10.1002/wrna.1150 PMID:23208756

Chen, D., Zhao, N., Wang, J., Li, Z., Wu, C., Fu, J., & Xiao, H. (2017). Whole-exome sequencing analysis of Waardenburg syndrome in a Chinese family. *Human Genome Variation, 4*, 17027. doi:10.1038/hgv.2017.27 PMID:28690861

Chen, Y., Yang, F., Zheng, H., Zhou, J., Zhu, G., Hu, P., & Wu, W. (2016). Clinical and genetic investigation of families with type II Waardenburg syndrome. *Molecular Medicine Reports, 13*(3), 1983–1988. doi:10.3892/mmr.2016.4774 PMID:26781036

Ciaccio, C., Fontana, L., Milani, D., Tabano, S., Miozzo, M., & Esposito, S. (2017). Fragile X syndrome: A review of clinical and molecular diagnoses. *Italian Journal of Pediatrics, 43*(1), 39. doi:10.118613052-017-0355-y PMID:28420439

Dangoumau, A., Verschueren, A., Hammouche, E., Papon, M. A., Blasco, H., Cherpi-Antar, C., ... Vourc'h, P. (2014). Novel SOD1 mutation p.V31A identified with a slowly progressive form of amyotrophic lateral sclerosis. *Neurobiology of Aging, 35*(1), 266.e1–266.e4. doi:10.1016/j.neurobiolaging.2013.07.012 PMID:23954173

Day, J. W., & Ranum, L. P. (2005). RNA pathogenesis of the myotonic dystrophies. *Neuromuscular Disorders, 15*(1), 5–16. doi:10.1016/j.nmd.2004.09.012 PMID:15639115

Deirmengian, G. K., Hebela, N. M., O'Connell, M., Glaser, D. L., Shore, E. M., & Kaplan, F. S. (2008). Proximal tibial osteochondromas in patients with fibrodysplasia ossificans progressiva. *The Journal of Bone and Joint Surgery, 90*(2), 366–374. doi:10.2106/JBJS.G.00774 PMID:18245597

Deivasigamani, S., Verma, H. K., Ueda, R., Ratnaparkhi, A., & Ratnaparkhi, G. S. (2014). A genetic screen identifies Tor as an interactor of VAPB in a Drosophila model of amyotrophic lateral sclerosis. *Biology Open, 3*(11), 1127–1138. doi:10.1242/bio.201410066 PMID:25361581

Digilio, M. C., Marino, B., Ammirati, A., Borzaga, U., Giannotti, A., & Dallapiccola, B. (1999). Cardiac malformations in patients with oral-facial-skeletal syndromes: Clinical similarities with heterotaxia. *American Journal of Medical Genetics, 84*(4), 350–356. doi:10.1002/(SICI)1096-8628(19990604)84:4<350::AID-AJMG8>3.0.CO;2-E PMID:10340650

Doyle, J. J., Doyle, A. J., Wilson, N. K., Habashi, J. P., Bedja, D., Whitworth, R. E., ... Dietz, H. C. (2015). A deleterious gene-by-environment interaction imposed by calcium channel blockers in Marfan syndrome. *eLife*, *4*, e08648. doi:10.7554/eLife.08648 PMID:26506064

Echenne, B., & Bassez, G. (2013). Congenital and infantile myotonic dystrophy. *Handbook of Clinical Neurology*, *113*, 1387–1393. doi:10.1016/B978-0-444-59565-2.00009-5 PMID:23622362

Forsberg, K., Andersen, P. M., Marklund, S. L., & Brännström, T. (2011). Glial nuclear aggregates of superoxide dismutase-1 are regularly present in patients with amyotrophic lateral sclerosis. *Acta Neuropathologica*, *121*(5), 623–634. doi:10.100700401-011-0805-3 PMID:21287393

Fusco, C., Frattini, D., Scarano, A., & Giustina, E. D. (2009). Congenital pes cavus in a Charcot-Marie-tooth disease type 1A newborn. *Pediatric Neurology*, *40*(6), 461–464. doi:10.1016/j.pediatrneurol.2008.12.010 PMID:19433282

Galdzicka, M., Patnala, S., Hirshman, M. G., Cai, J. F., Nitowsky, H., Egeland, J. A., & Ginns, E. I. (2002). A new gene, EVC2, is mutated in Ellis-van Creveld syndrome. *Molecular Genetics and Metabolism*, *77*(4), 291–295. doi:10.1016/S1096-7192(02)00178-6 PMID:12468274

Gao, Z., & Cooper, T. A. (2012). Antisense oligonucleotides: Rising stars in eliminating RNA toxicity in myotonic dystrophy. *Human Gene Therapy*, *24*(5), 499–507. doi:10.1089/hum.2012.212 PMID:23252746

Gehle, P., Goergen, B., Pilger, D., Ruokonen, P., Robinson, P. N., & Salchow, D. J. (2017). Biometric and structural ocular manifestations of Marfan syndrome. *PLoS One*, *12*(9), e0183370. doi:10.1371/journal.pone.0183370 PMID:28931008

Gloss, D., Moxley, R. T. III, Ashwal, S., & Oskoui, M. (2016). Practice guideline update summary: Corticosteroid treatment of Duchenne muscular dystrophy: Report of the Guideline Development Subcommittee of the American Academy of Neurology. *Neurology*, *86*(5), 465–472. doi:10.1212/WNL.0000000000002337 PMID:26833937

Goland, S., & Elkayam, U. (2017). Pregnancy and Marfan syndrome. *Annals of Cardiothoracic Surgery*, *6*(6), 642–653. doi:10.21037/acs.2017.10.07 PMID:29270376

Gourdon, G., & Meola, G. (2017). Myotonic Dystrophies: State of the Art of New Therapeutic Developments for the CNS. *Frontiers in Cellular Neuroscience*, *11*, 101. doi:10.3389/fncel.2017.00101 PMID:28473756

Groeneveld, G. J., Veldink, J. H., van der Tweel, I., Kalmijn, S., Beijer, C., de Visser, M., ... van den Berg, L. H. (2003). A randomized sequential trial of creatine in amyotrophic lateral sclerosis. *Annals of Neurology*, *53*(4), 437–445. doi:10.1002/ana.10554 PMID:12666111

Gross, C., Banerjee, A., Tiwari, D., Longo, F., White, A. R., & Allen, A. G. ... Bassell, G. J. (2018). Isoform-selective phosphoinositide 3-kinase inhibition ameliorates a broad range of fragile X syndrome-associated deficits in a mouse model. *Neuropsychopharmacology*. doi:10.103841386-018-0150-5

Growth, K. A., Hove, H., Kyhl, K., Folkestad, L., Gaustadnes, N. V., Stochhol, K., ... Gravholt, C. H. (2015). Prevalence, incidence, and age at diagnosis in Marfan Syndrome. *Orphanet Journal of Rare Diseases*, *10*(1), 153. doi:10.118613023-015-0369-8 PMID:26631233

Gudernova, I., Vesela, I., Balek, L., Buchtova, M., Dosedelova, H., Kunova, M., ... Krejci, P. (2016). Multikinase activity of fibroblast growth factor receptor (FGFR) inhibitors SU5402, PD173074, AZD1480, AZD4547 and BGJ398 compromises the use of small chemicals targeting FGFR catalytic activity for therapy of short-stature syndromes. *Human Molecular Genetics, 25*(1), 9–23. doi:10.1093/hmg/ddv441 PMID:26494904

Guiraud, S., Chen, H., Burns, D. T., & Davies, K. E. (2015). Advances in genetic therapeutic strategies for Duchenne muscular dystrophy. *Experimental Physiology, 100*(12), 1458–1467. doi:10.1113/EP085308 PMID:26140505

Hagerman, R., & Hagerman, P. (2013). Advances in clinical and molecular understanding of the FMR1 premutation and fragile X-associated tremor/ataxia syndrome. *Lancet Neurology, 12*(8), 786–798. doi:10.1016/S1474-4422(13)70125-X PMID:23867198

Hagerman, R. J., Berry-Kravis, E., Kaufmann, W. E., Ono, M. Y., Tartaglia, N., Lachiewicz, A., ... Tranfaglia, M. (2009). Advances in the treatment of fragile X syndrome. *Pediatrics, 123*(1), 378–390. doi:10.1542/peds.2008-0317 PMID:19117905

Haneline, M., & Lewkovich, G. N. (2007). A narrative review of pathophysiological mechanisms associated with cervical artery dissection. *Journal of the Canadian Chiropractic Association, 51*(3), 146–157. PMID:17885677

Hatton, D. D., Sideris, J., Skinner, M., Mankowski, J., Bailey, D. B. Jr, Roberts, J., & Mirrett, P. (2006). Autistic behavior in children with fragile X syndrome: Prevalence, stability, and the impact of FMRP. *American Journal of Medical Genetics. Part A, 140A*(17), 1804–1813. doi:10.1002/ajmg.a.31286 PMID:16700053

Hebela, N., Shore, E. M., & Kaplan, F. S. (2005). Three pairs of monozygotic twins with fibrodysplasia ossificans progressiva: The role of environment in the progression of heterotopic ossification. *Clinical Reviews in Bone and Mineral Metabolism, 3*(3-4), 205–208. doi:10.1385/BMM:3:3-4:205

Hirst, M. C., Knight, S. J., Christodoulou, Z., Grewal, P. K., Fryns, J. P., & Davies, K. E. (1993). Origins of the fragile X syndrome mutation. *Journal of Medical Genetics, 30*(8), 647–650. doi:10.1136/jmg.30.8.647 PMID:8411050

Ho, G., Cardamone, M., & Farrar, M. (2015). Congenital and childhood myotonic dystrophy: Current aspects of disease and future directions. *World Journal of Clinical Pediatrics, 4*(4), 66–80. doi:10.5409/wjcp.v4.i4.66 PMID:26566479

Hoffjan, S. (2012). Genetic dissection of marfan syndrome and related connective tissue disorders: An update 2012. *Molecular Syndromology, 3*(2), 47–58. doi:10.1159/000339441 PMID:23326250

Horigome, H., Hamada, H., Sohda, S., Oyake, Y., & Kurosaki, Y. (1997). Prenatal ultrasonic diagnosis of a case of Ellis-van Creveld syndrome with a single atrium. *Pediatric Radiology, 27*(12), 942–944. doi:10.1007002470050277 PMID:9388288

Howes, O. D., Findon, J. L., Wichers, R. H., Charman, T., King, B. H., Loth, E., ... Murphy, D. G. (2018). Autism Spectrum Disorder: Consensus guidelines on assessment, treatment and research from the British Association for Psychopharmacology. *Journal of Psychopharmacology (Oxford, England)*, *32*(1), 3–29. doi:10.1177/0269881117741766 PMID:29237331

Ibarra-Ramirez, M., Campos-Acevedo, L. D., Lugo-Trampe, J., Martínez-Garza, L. E., Martinez-Glez, V., Valencia-Benitez, M., Lapunzina, P., ... Ruiz-Peréz, V. (2017). Phenotypic Variation in Patients with Homozygous c.1678G>T Mutation in EVC Gene: Report of Two Mexican Families with Ellis-van Creveld Syndrome. *The American Journal of Case Reports, 18*, 1325–1329. doi:10.12659/AJCR.905976

Ingre, C., Roos, P. M., Piehl, F., Kamel, F., & Fang, F. (2015). Risk factors for amyotrophic lateral sclerosis. *Clinical Epidemiology*, *7*, 181–193. doi:10.2147/CLEP.S37505 PMID:25709501

Ishida, M., & Moore, G. E. (2013). The role of imprinted genes in humans. *Molecular Aspects of Medicine*, *34*(4), 826–840. doi:10.1016/j.mam.2012.06.009 PMID:22771538

Kabashi, E., Bercier, V., Lissouba, A., Liao, M., Brustein, E., Rouleau, G. A., & Drapeau, P. (2011). FUS and TARDBP but not SOD1 interact in genetic models of amyotrophic lateral sclerosis. *PLOS Genetics*, *7*(8), e1002214. doi:10.1371/journal.pgen.1002214 PMID:21829392

Kan, L., Lounev, V. Y., Pignolo, R. J., Duan, L., Liu, Y., Stock, S. R., ... Kessler, J. A. (2011). Substance P signaling mediates BMP-dependent heterotopic ossification. *Journal of Cellular Biochemistry*, *112*(10), 2759–2772. doi:10.1002/jcb.23259 PMID:21748788

Kaplan, F. S., Chakkalakal, S. A., & Shore, E. M. (2012). Fibrodysplasia ossificans progressiva: Mechanisms and models of skeletal metamorphosis. *Disease Models & Mechanisms*, *5*(6), 756–762. doi:10.1242/dmm.010280 PMID:23115204

Kaplan, F. S., Le Merrer, M., Glaser, D. L., Pignolo, R. J., Goldsby, R. E., Kitterman, J. A., ... Shore, E. M. (2008). Fibrodysplasia ossificans progressiva. *Best Practice & Research. Clinical Rheumatology*, *22*(1), 191–205. doi:10.1016/j.berh.2007.11.007 PMID:18328989

Kaplan, F. S., McCluskey, W., Hahn, G., Tabas, J. A., Muenke, M., & Zasloff, M. A. (1993). Genetic transmission of fibrodysplasia ossificans progressiva. Report of a family. *The Journal of Bone and Joint Surgery*, *75*(8), 1214–1220. doi:10.2106/00004623-199308000-00011 PMID:8354680

Kaplan, F. S., Shore, E. M., & Pignolo, R. J.The International Clinical Consortium on FOP (ICC-FOP). (2011). The Medical Management of Fibrodysplasia Ossificans Progressiva: Current Treatment Considerations. *Clin Proc Intl Clin Consort FOP*, *4*, 1–100.

Kaplan, J., Kaplan, F. S., & Shore, E. M. (2011). Restoration of normal BMP signaling levels and osteogenic differentiation in FOP mesenchymal progenitor cells by mutant allele-specific targeting. *Gene Therapy*, *19*(7), 786–790. doi:10.1038/gt.2011.152 PMID:22011642

Kaufmann, W. E., Cortell, R., Kau, A. S., Bukelis, I., Tierney, E., Gray, R. M., ... Standard, P. (2004). Autism spectrum disorder in fragile X syndrome: Communication, social interaction, and specific behaviors. *American Journal of Medical Genetics. Part A*, *129A*(3), 225–234. doi:10.1002/ajmg.a.30229 PMID:15326621

Keane, G. M., & Pyeritz, E. R. (2008). Medical management of Marfan syndrome. *Circulation*, *117*(21), 2802–2813. doi:10.1161/CIRCULATIONAHA.107.693523 PMID:18506019

Khong, H. T., & Rosenberg, S. A. (2002). The Waardenburg syndrome type 4 gene, SOX10, is a novel tumor-associated antigen identified in a patient with a dramatic response to immunotherapy. *Cancer Research*, *62*(11), 3020–3023. PMID:12036907

Kilmartin, E., Grunwald, Z., Kalan, F. S., & Nussbaum, B. L. (2014). General Anesthesia for Dental Procedures in Patients with Fibrodysplasia Ossificans Progressiva: A Review of 42 Cases in 30 Patients. *Anesthesia and Analgesia*, *118*(2), 298–301. doi:10.1213/ANE.0000000000000021 PMID:24361843

Kishino, T., Lalande, M., & Wagstaff, J. (1997). UBE3A/E6-AP mutations cause Angelman syndrome. *Nature Genetics*, *15*(1), 70–73. doi:10.1038/ng0197-70 PMID:8988171

Kohl, B., Groh, J., Wessig, C., Wiendl, H., Kroner, A., & Martini, R. (2010). Lack of evidence for a pathogenic role of T-lymphocytes in an animal model for Charcot-Marie-Tooth disease 1A. *Neurobiology of Disease*, *38*(1), 78–84. doi:10.1016/j.nbd.2010.01.001 PMID:20064611

Kornegay, J. N. (2017). The golden retriever model of Duchenne muscular dystrophy. *Skeletal Muscle*, *7*(1), 9. doi:10.118613395-017-0124-z PMID:28526070

Koyama, H., Kashio, A., Sakata, A., Tsutsumiuchi, K., Matsumoto, Y., & Karino, S. (2016). ... Yamasoba, T. (2016). The Hearing Outcomes of Cochlear Implantation in Waardenburg Syndrome. *BioMed Research International*, *2854736*. doi:10.1155/2016/2854736

Kumar, A., & Agarwal, S. (2014). Marfan syndrome: An eyesight of syndrome. *Meta Gene*, *2*, 96–105. doi:10.1016/j.mgene.2013.10.008 PMID:25606393

Larson, A. M., Shinnick, J. E., Shaaya, E. A., Thiele, E. A., & Thibert, R. L. (2014). Angelman syndrome in adulthood. *American Journal of Medical Genetics. Part A*, *167A*(2), 331–344. doi:10.1002/ajmg.a.36864 PMID:25428759

Lee, H. B., Sundberg, B. N., Sigafoos, A. N., & Clark, K. J. (2016). Genome Engineering with TALE and CRISPR Systems in Neuroscience. *Frontiers in Genetics*, *7*, 47. doi:10.3389/fgene.2016.00047 PMID:27092173

Legeai-Mallet, L., Benoist-Lasselin, C., Munnich, A., & Bonaventure, J. (2004). Overexpression of FGFR3, Stat1, Stat5 and p21Cip1 correlates with phenotypic severity and defective chondrocyte differentiation in FGFR3-related chondrodysplasias. *Bone*, *34*(1), 26–36. doi:10.1016/j.bone.2003.09.002 PMID:14751560

Leigh, M. J., Nguyen, D. V., Mu, Y., Winarni, T. I., Schneider, A., Chechi, T., ... Hagerman, R. J. (2013). A randomized double-blind, placebo-controlled trial of minocycline in children and adolescents with fragile x syndrome. *Journal of Developmental and Behavioral pediatrics. Journal of Developmental and Behavioral Pediatrics*, *34*(3), 147–155. doi:10.1097/DBP.0b013e318287cd17 PMID:23572165

Leigh, P. N., & Ray-Chaudhuri, K. (1994). Motor neuron disease. *Journal of Neurology, Neurosurgery, and Psychiatry*, *57*(8), 886–896. doi:10.1136/jnnp.57.8.886 PMID:8057109

Logigian, E. L., Martens, W. B., Moxley, R. T., McDermott, M. P., Dilek, N., Wiegner, A. W., ... Moxley, R. T. (2010). Mexiletine is an effective antimyotonia treatment in myotonic dystrophy type 1. *Neurology, 74*(18), 1441–1448. doi:10.1212/WNL.0b013e3181dc1a3a PMID:20439846

Lorget, F., Kaci, N., Peng, J., Benoist-Lasselin, C., Mugniery, E., Oppeneer, T., ... Legeai-Mallet, L. (2012). Evaluation of the therapeutic potential of a CNP analog in a Fgfr3 mouse model recapitulating achondroplasia. *American Journal of Human Genetics, 91*(6), 1108–1014. doi:10.1016/j.ajhg.2012.10.014 PMID:23200862

Lupski, J. R., De Oca-Luna, R. M., Slaugenhaupt, S., Pentao, L., Guzzetta, V., Trask, B. J., ... Patel, P. I. (1991). DNA duplication associated with Charcot-Marie-Tooth disease type 1A. *Cell, 66*(2), 219–232. doi:10.1016/0092-8674(91)90613-4 PMID:1677316

MacDonald, J. R., Hill, J. D., & Tarnopolsky, M. A. (2002). Modafinil reduces excessive somnolence and enhances mood in patients with myotonic dystrophy. *Neurology, 59*(12), 1876–1880. doi:10.1212/01.WNL.0000037481.08283.51 PMID:12499477

Mariosa, D., Beard, J. D., Umbach, D. M., Bellocco, R., Keller, J., Peters, T. L., ... Kamel, F. (2017). Body Mass Index and Amyotrophic Lateral Sclerosis: A Study of US Military Veterans. *American Journal of Epidemiology, 185*(5), 362–371. doi:10.1093/aje/kww140 PMID:28158443

Martin, S., Khleifat, A. A., & Al-Chalabi, A. (2017). What causes amyotrophic lateral sclerosis? *Version 1. F1000 Research, 6*, 371. doi:10.12688/f1000research.10476.1 PMID:28408982

McCorquodale, D., Pucillo, E. M., & Johnson, N. E. (2016). Management of Charcot-Marie-Tooth disease: Improving long-term care with a multidisciplinary approach. *Journal of Multidisciplinary Healthcare, 9*, 7–19. doi:10.2147/JMDH.S69979 PMID:26855581

Medline Plus (MP), US National Library of Medicine. (2018). *Achondroplasia*. Retrieved from https://medlineplus.gov/ency/article/001577.htm

Miccoli, M., Bertelloni, S., & Massart, F. (2016). Height Outcome of Recombinant Human Growth Hormone Treatment in Achondroplasia Children: A Meta-Analysis. *Hormone Research in Paediatrics, 86*(1), 27–34. doi:10.1159/000446958 PMID:27355624

Minnerop, M., Weber, B., Schoene-Bake, J. C., Roeske, S., Mirbach, S., Anspach, C., ... Kornblum, C. (2011). The brain in myotonic dystrophy 1 and 2: Evidence for a predominant white matter disease. *Brain, 134*(Pt 12), 3530–3546. doi:10.1093/brain/awr299 PMID:22131273

Moat, S. J., Bradley, D. M., Salmon, R., Clarke, A., & Hartley, L. (2013). Newborn bloodspot screening for Duchenne muscular dystrophy: 21 years experience in Wales (UK). *European journal of human genetics. European Journal of Human Genetics, 21*(10), 1049–1053. doi:10.1038/ejhg.2012.301 PMID:23340516

Morales-Piga, A., & Kaplan, F. S. (2010). Osteochondral diseases and fibrodysplasia ossificans progressiva. *Advances in Experimental Medicine and Biology, 686*, 335–348. doi:10.1007/978-90-481-9485-8_19 PMID:20824454

Morozova, N., Weisskopf, M. G., McCullough, M. L., Munger, K. L., Calle, E. E., Thun, M. J., & Ascherio, A. (2008). Diet and amyotrophic lateral sclerosis. *Epidemiology (Cambridge, Mass.)*, *19*(2), 324–337. doi:10.1097/EDE.0b013e3181632c5d PMID:18300717

Moskowitz, L. J., Carr, E. G., & Durand, V. M. (2011). Behavioral intervention for problem behavior in children with fragile X syndrome. *American Journal on Intellectual and Developmental Disabilities*, *116*(6), 457–478. doi:10.1352/1944-7558-116.6.457 PMID:22126659

Muscular Dystrophy Association (MDA). (2018). *Duchenne Muscular Dystrophy*. Retrieved from https://www.mda.org/disease/duchenne-muscular-dystrophy/signs-and-symptoms

Naqash, T. A., Alshahrani, I., & Simasetha, S. (2018). Ellis-van Creveld Syndrome: A Rare Clinical Report of Oral Rehabilitation by Interdisciplinary Approach. *Case Reports in Dentistry*, *8631602*. doi:10.1155/2018/8631602 PMID:29607224

Naski, M. C., Wang, Q., Xu, J., & Ornitz, D. M. (1996). Graded activation of fibroblast growth factor receptor 3 by mutations causing achondroplasia and thanatophoric dysplasia. *Nature Genetics*, *13*(2), 233–237. doi:10.1038/ng0696-233 PMID:8640234

Ornitz, D. M., & Legeai-Mallet, L. (2016). Achondroplasia: Development, Pathogenesis, and Therapy. *Developmental Dynamics*, *246*(4), 291–309. doi:10.1002/dvdy.24479 PMID:27987249

Pareyson, D., Scaioli, V., & Laurà, M. (2006). Clinical and electrophysiological aspects of Charcot-Marie-Tooth disease. *Neuromolecular Medicine*, *8*(1-2), 3–22. doi:10.1385/NMM:8:1-2:3 PMID:16775364

Peker, K., Ergil, J., & Öztürk, I. (2015). Anaesthesia Management in a Patient with Waardenburg Syndrome and Review of the Literature. *Turkish Journal of Anaesthesiology and Reanimation*, *43*(5), 360–362. doi:10.5152/TJAR.2015.52714 PMID:27366529

Pepe, G., Giusti, B., Sticchi, E., Abbate, R., Gensini, G., & Nistri, S. (2016). Marfan syndrome: Current perspectives. *The Application of Clinical Genetics*, *9*, 55–65. doi:10.2147/TACG.S96233 PMID:27274304

Peters, S. U., Horowitz, L., Barbieri-Welge, R., Taylor, J. L., & Hundley, R. J. (2012). Longitudinal follow-up of autism spectrum features and sensory behaviors in Angelman syndrome by deletion class. *Journal of Child Psychology and Psychiatry, and Allied Disciplines*, *53*(2), 152–159. doi:10.1111/j.1469-7610.2011.02455.x PMID:21831244

Petrie, K. A., Lee, W. H., Bullock, A. N., Pointon, J. J., Smith, R., Russell, R. G., ... Triffitt, J. T. (2009). Novel mutations in ACVR1 result in atypical features in two fibrodysplasia ossificans progressiva patients. *PLoS One*, *4*(3), e5005. doi:10.1371/journal.pone.0005005 PMID:19330033

Pignolo, R. J., Shore, E. M., & Kaplan, F. S. (2013). Fibrodysplasia ossificans progressiva: diagnosis, management, and therapeutic horizons. *Pediatric Endocrinology Reviews*, *10*(2), 437–448.

Pingault, V., Ente, D., Dastot-Le Moal, F., Goossens, M., Marlin, S., & Bondurand, N. (2010). Review and update of mutations causing Waardenburg syndrome. *Human Mutation*, *31*(4), 391–406. doi:10.1002/humu.21211 PMID:20127975

Polymeropoulos, M. H., Ide, S. E., Wright, M., Goodship, J., Weissenbach, J., Pyeritz, R. E., ... Francomano, C. A. (1996). The gene for the Ellis-van Creveld syndrome is located on chromosome 4p16. *Genomics*, *35*(1), 1–5. doi:10.1006/geno.1996.0315 PMID:8661097

Qi, Z., Luan, J., Zhou, X., Cui, Y., & Han, J. (2017). Fibrodysplasia ossificans progressiva: Basic understanding and experimental models. *Intractable & Rare Diseases Research*, *6*(4), 242–248. doi:10.5582/irdr.2017.01055 PMID:29259851

Raeymaekers, P., Timmerman, V., Nelis, E., De Jonghe, P., Hoogendijk, J. E., Baas, F., & Bolhuis, P. A. (1991). Duplication in chromosome 17p11.2 in Charcot-Marie-Tooth neuropathy type 1a (CMT 1a). The HMSN Collaborative Research Group. *Neuromuscular Disorders: NMD*, *1*(2), 93–97. doi:10.1016/0960-8966(91)90055-W PMID:1822787

Ramirez, F., & Dietz, H. C. (2007). Marfan syndrome: From molecular pathogenesis to clinical treatment. *Current Opinion in Genetics & Development*, *17*(3), 252–258. doi:10.1016/j.gde.2007.04.006 PMID:17467262

Raspa, M., Wheeler, A. C., & Riley, C. (2017). Public Health Literature Review of Fragile X Syndrome. *Pediatrics*, *139*(Suppl 3), S153–S171. doi:10.1542/peds.2016-1159C PMID:28814537

Rawlani, S. M., Ramtake, R., Dhabarde, A., & Rawlani, S. S. (2018). Waardenburg syndrome: A rare case. *Oman Journal of Ophthalmology*, *11*(2), 158–160. doi:10.4103/ojo.OJO_51_2014 PMID:29930451

Reik, W., & Walter, J. (2001). Genomic imprinting: Parental influence on the genome. *Nature Reviews. Genetics*, *2*(1), 21–32. doi:10.1038/35047554 PMID:11253064

Reilly, M. M. (2007). Sorting out the inherited neuropathies. *Practical Neurology*, *7*(2), 93–105. PMID:17430873

Rowland, L. P., & Shneider, N. A. (2001). Amyotrophic lateral sclerosis. *The New England Journal of Medicine*, *344*(22), 1688–1700. doi:10.1056/NEJM200105313442207 PMID:11386269

Rueda, J. R., Ballesteros, J., Guillen, V., Tejada, M. I., & Solà, I. (2011). Folic acid for fragile X syndrome. *Cochrane Database of Systematic Reviews*, *5*, CD008476. doi:10.1002/14651858.CD008476.pub2 PMID:21563169

Ruiz-Perez, V. L., Tompson, S. W., Blair, H. J., Espinoza-Valdez, C., Lapunzina, P., Silva, E. O., ... Goodship, J. A. (2003). Mutations in two nonhomologous genes in a head-to-head configuration cause Ellis-van Creveld syndrome. *American Journal of Human Genetics*, *72*(3), 728–732. doi:10.1086/368063 PMID:12571802

Ryder, S., Leadley, R. M., Armstrong, N., Westwood, M., de Kock, S., Butt, T., ... Kleijnen, J. (2017). The burden, epidemiology, costs and treatment for Duchenne muscular dystrophy: An evidence review. *Orphanet Journal of Rare Diseases*, *12*(1), 79. doi:10.118613023-017-0631-3 PMID:28446219

Saldarriaga, W., Tassone, F., Gonzalez-Teshima, L. Y., Forero-Forero, J. V., Ayala-Zapata, S., & Hagerman, R. (2014). Fragile X Syndrome. *Colombia Médica (Cali, Colombia)*, *45*(4), 190–198. PMID:25767309

Sasalawad, S. S., Hugar, S. M., Poonacha, K. S., & Mallikarjuna, R. (2013). Ellis-van Creveld syndrome. *BMJ Case Reports, 2013*(2), bcr2013009463. doi:10.1136/bcr-2013-009463 PMID:23843404

Sharma, K., & Arora, A. (2015). Waardenburg Syndrome: A Case Study of Two Patients. *Indian Journal of Otolaryngology and Head and Neck Surgery: Official Publication of the Association of Otolaryngologists of India, 67*(3), 324–328. doi:10.100712070-015-0870-3 PMID:26405672

Shen, Q., Little, S. C., Xu, M., Haupt, J., Ast, C., Katagiri, T., ... Shore, E. M. (2009). The fibrodysplasia ossificans progressiva R206H ACVR1 mutation activates BMP-independent chondrogenesis and zebrafish embryo ventralization. *The Journal of Clinical Investigation, 119*(11), 3462–3472. doi:10.1172/JCI37412 PMID:19855136

Shiang, R., Thompson, L. M., Zhu, Y. Z., Church, D. M., Fielder, T. J., Bocian, M., ... Wasmuth, J. J. (1994). Mutations in the transmembrane domain of FGFR3 cause the most common genetic form of dwarfism, achondroplasia. *Cell, 78*(2), 335–342. doi:10.1016/0092-8674(94)90302-6 PMID:7913883

Shimizu-Motohashi, Y., Miyatake, S., Komaki, H., Takeda, S., & Aoki, Y. (2016). Recent advances in innovative therapeutic approaches for Duchenne muscular dystrophy: From discovery to clinical trials. *American Journal of Translational Research, 8*(6), 2471–2489. PMID:27398133

Shimono, K., Tung, W. E., Macolino, C., Chi, A. H., Didizian, J. H., Mundy, C., ... Iwamoto, M. (2011). Potent inhibition of heterotopic ossification by nuclear retinoic acid receptor-γ agonists. *Nature Medicine, 17*(4), 454–460. doi:10.1038/nm.2334 PMID:21460849

Shore, E. M., Feldman, G. J., Xu, M., & Kaplan, F. S. (2005). The genetics of fibrodysplasia ossificans progressiva. *Clinical Reviews in Bone and Mineral Metabolism, 3*(3-4), 201–204. doi:10.1385/BMM:3:3-4:201

Shore, E. M., & Kaplan, F. S. (2008). Insights from a rare genetic disorder of extra-skeletal bone formation, fibrodysplasia ossificans progressiva (FOP). *Bone, 43*(3), 427–433. doi:10.1016/j.bone.2008.05.013 PMID:18590993

Simmons, Z. (2005). Management strategies for patients with amyotrophic lateral sclerosis from diagnosis through death. *The Neurologist, 11*(5), 257–270. doi:10.1097/01.nrl.0000178758.30374.34 PMID:16148733

Smith, L. E., Barker, E. T., Seltzer, M. M., Abbeduto, L., & Greenberg, J. S. (2012). Behavioral phenotype of fragile X syndrome in adolescence and adulthood. *American Journal on Intellectual and Developmental Disabilities, 117*(1), 1–17. doi:10.1352/1944-7558-117.1.1 PMID:22264109

Sólia-Nasser, L., de Aquino, S. N., Paranaíba, L. M., Gomes, A., Dos-Santos-Neto, P., Coletta, R. D., ... Martelli-Junior, H. (2016). Waardenburg syndrome type I: Dental phenotypes and genetic analysis of an extended family. *Medicina Oral, Patologia Oral y Cirugia Bucal, 21*(3), e321–e327. doi:10.4317/medoral.20789 PMID:27031059

Stöger, R., Kajimura, T. M., Brown, W. T., & Laird, C. D. (1997). Epigenetic variation illustrated by DNA methylation patterns of the fragile-X gene FMR1. *Human Molecular Genetics, 6*(11), 1791–1801. doi:10.1093/hmg/6.11.1791 PMID:9302255

Sutcliffe, J. S., Nelson, D. L., Zhang, F., Pieretti, M., Caskey, C. T., Saxe, D., & Warren, S. T. (1992). DNA methylation represses FMR-1 transcription in fragile X syndrome. *Human Molecular Genetics, 1*(6), 397–400. doi:10.1093/hmg/1.6.397 PMID:1301913

Symons, F. J., Clark, R. D., Hatton, D. D., Skinner, M., & Bailey, D. B. Jr. (2003). Self-injurious behavior in young boys with fragile X syndrome. *American Journal of Medical Genetics. Part A, 118A*(2), 115–121. doi:10.1002/ajmg.a.10078 PMID:12655491

Tan, W. H., Bacino, C. A., Skinner, S. A., Anselm, I., Barbieri-Welge, R., Bauer-Carlin, A., ... Bird, L. M. (2011). Angelman syndrome: Mutations influence features in early childhood. *American Journal of Medical Genetics. Part A, 155A*(1), 81–90. doi:10.1002/ajmg.a.33775 PMID:21204213

Tan, W. H., Bird, L. M., Thibert, R. L., & Williams, C. A. (2014). If not Angelman, what is it? A review of Angelman-like syndromes. *American Journal of Medical Genetics. Part A, 164A*(4), 975–992. doi:10.1002/ajmg.a.36416 PMID:24779060

Theadom, A., Rodrigues, M., Roxburgh, R., Balalla, S., Higgins, C., Bhattacharjee, R., ... Feigin, V. (2014). Prevalence of muscular dystrophies: A systematic literature review. *Neuroepidemiology, 43*(3-4), 259–268. doi:10.1159/000369343 PMID:25532075

Thibert, R. L., Larson, A. M., Hsieh, D. T., Raby, A. R., & Thiele, E. A. (2013). Neurologic manifestations of Angelman syndrome. *Pediatric Neurology, 48*(4), 271–279. doi:10.1016/j.pediatrneurol.2012.09.015 PMID:23498559

Thornton C. A. (2014). Myotonic dystrophy. *Neurologic Clinics, 32*(3), 705–719. doi:10.1016/j.ncl.2014.04.011

Thornton, C. A., Wang, E., & Carrell, E. (2017). Myotonic dystrophy: Approach to therapy. *Current Opinion in Genetics & Development, 44*, 135–140. doi:10.1016/j.gde.2017.03.007 PMID:28376341

Thurman, A., McDuffie, A., Hagerman, R. J., Josol, C. K., & Abbeduto, L. J. (2017). Language Skills of Males with Fragile X Syndrome or Nonsyndromic Autism Spectrum Disorder. *Journal of Autism and Developmental Disorders, 47*(3), 728–743. doi:10.100710803-016-3003-2 PMID:28074353

Todd, P. K., & Paulson, H. L. (2010). RNA-mediated neurodegeneration in repeat expansion disorders. *Annals of Neurology, 67*(3), 291–300. doi:10.1002/ana.21948 PMID:20373340

Tompson, S. W., Ruiz-Perez, V. L., Blair, H. J., Barton, S., Navarro, V., Robson, J. L., ... Goodship, J. A. (2007). Sequencing EVC and EVC2 identifies mutations in two-thirds of Ellis-van Creveld syndrome patients. *Human Genetics, 120*(5), 663–670. doi:10.100700439-006-0237-7 PMID:17024374

Trip, J., Drost, G., van Engelen, B. G., & Faber, C. G. (2006). Drug treatment for myotonia. *Cochrane Database of Systematic Reviews, 1*, CD004762. PMID:16437496

Unger, S., Bonafé, L., & Gouze, E. (2017). Current Care and Investigational Therapies in Achondroplasia. *Current Osteoporosis Reports, 15*(2), 53–60. doi:10.100711914-017-0347-2 PMID:28224446

Van de Velde, S., Fillman, R., & Yandow, S. (2006). Protrusio acetabuli in Marfan syndrome. History, diagnosis, and treatment. *The Journal of Bone Joint Surgery, 88*(3), 639–646. PMID:16510833

Van den Berg, J. P., Kalmijn, S., Lindeman, E., Veldink, J. H., de Visser, M., Van der Graaff, M. M., ... Van den Berg, L. H. (2005). Multidisciplinary ALS care improves quality of life in patients with ALS. *Neurology, 65*(8), 1264–1267. doi:10.1212/01.wnl.0000180717.29273.12 PMID:16247055

van Paassen, B. W., van der Kooi, A. J., van Spaendonck-Zwarts, K. Y., Verhamme, C., Baas, F., & de Visser, M. (2014). PMP22 related neuropathies: Charcot-Marie-Tooth disease type 1A and Hereditary Neuropathy with liability to Pressure Palsies. *Orphanet Journal of Rare Diseases, 9*(1), 38. doi:10.1186/1750-1172-9-38 PMID:24646194

Videler, A., Eijffinger, E., Nollet, F., & Beelen, A. (2012). A thumb opposition splint to improve manual dexterity and upper-limb functioning in Charcot-Marie-Tooth disease. *Journal of Rehabilitation Medicine, 44*(3), 249–253. doi:10.2340/16501977-0932 PMID:22366728

von Kodolitsch, Y., & Robinson, P. N. (2007). Marfan syndrome: An update of genetics, medical and surgical management. *Heart (British Cardiac Society), 93*(6), 755–760. doi:10.1136/hrt.2006.098798 PMID:17502658

von Kodolitsch, Y., Rybczynski, M., Vogler, M., Mir, T. S., Schüler, H., Kutsche, K., ... Pyeritz, R. E. (2016). The role of the multidisciplinary health care team in the management of patients with Marfan syndrome. *Journal of Multidisciplinary Healthcare, 9*, 587–614. doi:10.2147/JMDH.S93680 PMID:27843325

Vona, B., Maroofian, R., Mendiratta, G., Croken, M., Peng, S., Ye, X., ... Shi, L. (2017). Dual Diagnosis of Ellis-van Creveld Syndrome and Hearing Loss in a Consanguineous Family. *Molecular Syndromology, 9*(1), 5–14. doi:10.1159/000480458 PMID:29456477

Wang, L., Qin, L., Li, T., Liu, H., Ma, L., Li, W., ... Liao, S. (2017). Prenatal diagnosis and genetic counseling for Waardenburg syndrome type I and II in Chinese families. *Molecular Medicine Reports, 17*(1), 172–178. doi:10.3892/mmr.2017.7874 PMID:29115496

Wang, Y., Spatz, M. K., Kannan, K., Hayk, H., Avivi, A., Gorivodsky, M., ... Givol, D. (1999). A mouse model for achondroplasia produced by targeting fibroblast growth factor receptor 3. *Proceedings of the National Academy of Sciences of the United States of America, 96*(8), 4455–4460. 10.1073/pnas.96.8.4455

Wendt, D. J., Dvorak-Ewell, M., Bullens, S., Lorget, F., Bell, S. M., Peng, J., ... Bunting, S. (2015). Neutral endopeptidase-resistant C-type natriuretic peptide variant represents a new therapeutic approach for treatment of fibroblast growth factor receptor 3-related dwarfism. *The Journal of Pharmacology and Experimental Therapeutics, 353*(1), 132–149. doi:10.1124/jpet.114.218560 PMID:25650377

Weterman, M. A., van Ruissen, F., de Wissel, M., Bordewijk, L., Samijn, J. P., van der Pol, W. L., ... Baas, F. (2009). Copy number variation upstream of PMP22 in Charcot-Marie-Tooth disease. *European Journal of Human Genetics: EJHG, 18*(4), 421–428. doi:10.1038/ejhg.2009.186 PMID:19888301

Wheeler, A. C., Sacco, P., & Cabo, R. (2017). Unmet clinical needs and burden in Angelman syndrome: A review of the literature. *Orphanet Journal of Rare Diseases, 12*(1), 164. doi:10.118613023-017-0716-z PMID:29037196

Wheeler, T. M., Leger, A. J., Pandey, S. K., MacLeod, A. R., Nakamori, M., Cheng, S. H., ... Thornton, C. A. (2012). Targeting nuclear RNA for in vivo correction of myotonic dystrophy. *Nature, 488*(7409), 111–115. doi:10.1038/nature11362 PMID:22859208

Williams, C. A. (2010). The behavioral phenotype of the Angelman syndrome. *American Journal of Medical Genetics. Part C, Seminars in Medical Genetics, 154C*(4), 432–437. doi:10.1002/ajmg.c.30278 PMID:20981772

Wirojanan, J., Jacquemont, S., Diaz, R., Bacalman, S., Anders, T. F., Hagerman, R. J., & Goodlin-Jones, B. L. (2009). The efficacy of melatonin for sleep problems in children with autism, fragile X syndrome, or autism and fragile X syndrome. *Journal of Clinical Sleep Medicine: JCSM: Official Publication of the American Academy of Sleep Medicine, 5*(2), 145–150. PMID:19968048

Yamashita, A., Morioka, M., Kishi, H., Kimura, T., Yahara, Y., Okada, M., ... Tsumaki, N. (2014). Statin treatment rescues FGFR3 skeletal dysplasia phenotypes. *Nature, 513*(7519), 507–511. doi:10.1038/nature13775 PMID:25231866

Yang, S., Dai, P., Liu, X., Kang, D., Zhang, X., Yang, W., ... Yuan, H. (2013). Genetic and Phenotypic Heterogeneity in Chinese Patients with Waardenburg Syndrome Type II. *PLoS One, 8*(10), e77149. doi:10.1371/journal.pone.0077149 PMID:24194866

Young, P., De Jonghe, P., Stögbauer, F., & Butterfass-Bahloul, T. (2008). Treatment for Charcot-Marie-Tooth disease. *Cochrane Database of Systematic Reviews*, (1): CD006052. doi:10.1002/14651858.CD006052.pub2 PMID:18254090

Yu, Y., Hayashi, S., Cai, X., Fang, C., Shi, W., Tsutsui, H., & Sheng, J. (2014). Pu-erh tea extract induces the degradation of FET family proteins involved in the pathogenesis of amyotrophic lateral sclerosis. *BioMed Research International, 254680*. doi:10.1155/2014/254680 PMID:24804206

Zaghloul, K. A., Heuer, G. G., Guttenberg, M. D., Shore, E. M., Kaplan, F. S., & Storm, P. B. (2008). Lumbar puncture and surgical intervention in a child with undiagnosed fibrodysplasia ossificans progressiva. *Journal of Neurosurgery. Pediatrics, 1*(1), 91–94. doi:10.3171/PED-08/01/091 PMID:18352811

Zaman, A., Capper, R., & Baddoo, W. (2018). Waardenburg syndrome: More common than you think! *Clinical Otolaryngology, 40*(1), 44–48. doi:10.1111/coa.12312 PMID:25200653

Zarei, S., Carr, K., Reiley, L., Diaz, K., Guerra, O., Altamirano, P. F., ... Chinea, A. (2015). A comprehensive review of amyotrophic lateral sclerosis. *Surgical Neurology International, 6*(1), 171. doi:10.4103/2152-7806.169561 PMID:26629397

Muscle, Connective Tissue, and Neonatal Disorders

KEY TERMS AND DEFINITIONS

Antisense Oligonucleotides: Short synthetic polymers that can bind to specific RNA molecules to alter protein expression.

Benzothiazepines: A subclass of calcium channel blocking drugs that can be used to reduce the force of cardiac contractions as well as a vasodilator.

Bone Remodeling: A highly regulated cycle where osteoclasts destroy old bone so that osteoblasts make new bone.

Cardiomyopathy: A disease of the heart muscle in which the heart muscle can become thick, enlarged, or stiff.

Dihydropyridines: A subclass of calcium channel blocking drugs that is often used to reduce peripheral resistance and arterial pressure.

Dolichocephaly: A condition characterized by an abnormally elongated head relative to head width.

Dysphagia: Difficulty swallowing.

Dyspnoea: Shortness of breath or difficulty breathing.

Hirschsprung Disease: A birth defect in which nerve cells that line the intestine do not properly form during development causing the inability to pass stool.

Kyphosis: An abnormally excessive outward curvature or forward rounding of the spine.

Lumbar Lordosis: Inward curvature of the lower back.

Macroorchidism: A condition affecting males that is characterized by abnormally large testes relative to age.

Microcephaly: A birth defect in which an infant's head is smaller than other infants of the same age and sex.

Multiple Myeloma: Cancer of the plasma cells.

Neurocristopathy: A group disorders involving abnormal neural crest cell differentiation and migration.

Phenylalkylamines: A subclass of calcium channel blocking drugs that is used to target cells of the heart.

Prognathism: A facial deformity characterized by the protrusion of either the upper or lower jaw.

Pseudohypertrophy: Enlargement of calf muscles.

Spinal Stenosis: The narrowing of the spinal canals.

Stoichiometry: Expression of the quantitative relationship between products and reactants in a chemical reaction.

Chapter 18
Human Nervous System Disorders

ABSTRACT

The nervous system (NS) is comprised of nerve cells (neurons), which transfer and process information, and neuroglia (or glial cells), which provide the supportive framework neurons need to function effectively. There are two divisions of the nervous system: central (CNS) and peripheral (PNS). The CNS consists of the brain and spinal cord and forms an intricate network of specialised cells that are responsible for coordinating all bodily functions. The PNS delivers sensory information from peripheral sensory tissues and systems to the CNS and carries motor commands from the CNS to peripheral tissues. This chapter discusses 15 diseases that directly affect the nervous system mostly caused by mutations in a single gene, with others having more complex modes of inheritance. They include Alzheimer's Disease, epilepsy, essential tremor, familial Mediterranean fever, Friedreich's ataxia, Huntington's disease, maple syrup syndrome, Menkes disease, narcolepsy, Parkinson's Disease, phenylketonuria, Refsum disease, spinal muscular atrophy, tangier disease, and spinocerebellar ataxia.

CHAPTER OUTLINE

18.1 Overview of the Nervous System
18.2 Alzheimer's Disease
18.3 Epilepsy
18.4 Essential Tumor
18.5 Familial Mediterranean Fever
18.6 Friedreich's Ataxia
18.7 Huntington's Disease
18.8 Maple Syrup Urine Disease
18.9 Menkes Disease
18.10 Narcolepsy
18.11 Parkinson's Disease

DOI: 10.4018/978-1-5225-8066-9.ch018

Human Nervous System Disorders

18.12 Phenylketonuria
18.13 Refsum Disease
18.14 Spinal Muscular Atrophy
18.15 Tangier Disease
18.16 Spinocerebellar Ataxia
Chapter Summary

LEARNING OUTCOMES

- Identify each genetic disorder affecting each system
- Outline the symptoms of each disorder
- Explain the genetic basis of each disorder
- Summarize the current therapies available to treat each disorder

18.1 OVERVIEW OF THE NERVOUS SYSTEM

The nervous system (NS) is comprised of nerve cells (neurons) which transfer and process information; and neuroglia (or glial cells) which, twenty times more numerous, provide the supportive framework neurons need to function effectively. There are two divisions of the nervous system: central (CNS) and peripheral (PNS) (Figure 1). The CNS consists of the brain and spinal cord and forms an intricate network of specialised cells that are responsible for coordinating all bodily functions. It is the seat of higher functions like intelligence, memory, learning and emotions and also integrates, processes, and coordinates sensory data and motor commands. The PNS is all neural tissue outside the CNS and delivers sensory information from peripheral sensory tissues and systems like eyes to the CNS. It also carries our motor commands from the CNS to peripheral tissues. Sensory information in the PNS comes from sensory receptors in the eyes, nose, ears, mouth and also from receptors in the skin, muscles, and joints. Motor command information is delivered via the somatic NS which controls voluntary and involuntary skeletal muscle contractions; and the visceral or autonomic NS which controls smooth muscle, cardiac muscles and glandular activity (without conscious control). There several diseases that directly affect the NS with a genetic component—some are due to a mutation in a single gene, while others have more complex modes of inheritance.

18.2 ALZHEIMER'S DISEASE (AD)

According to Alzheimer's Disease International, it is estimated that there were 46.8 million people worldwide living with dementia in 2015 expected to reach 131.5 million in 2050 (ADI, 2018). The World Health Organization (WHO) states that dementia is a global public health priority as the 7th leading cause of death with approximately 50 million people worldwide with an estimated cost of $818 billion (WHO, 2018). AD accounts for 60–70% of dementia cases worldwide estimated to quadruple by 2050. Currently, approved drugs only temporarily alleviate some of the disease's symptoms to a limited extent. Global leaders have set a deadline of 2025 for finding an effective way to treat or prevent AD. However, despite

Figure 1. The central and peripheral nervous systems in humans
Source: Image used under license from Shutterstock.com

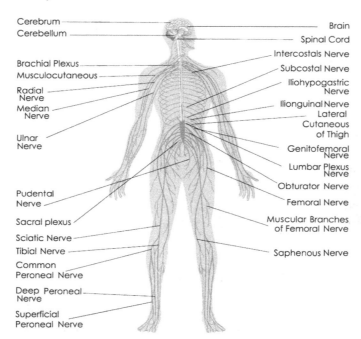

the evaluation in clinical trials of numerous potential treatments, only four **cholinesterase inhibitors** and memantine have shown sufficient safety and efficacy to allow marketing approval at an international level. These five therapeutic agents are nevertheless only symptomatic treatments, temporarily ameliorating memory and thinking problems, with a modest clinical effect, without treating the underlying cause of AD or slowing the rate of decline. It has to be emphasized that for a disease-modifying therapy of an AD agent to be approved after preclinical development and initial characterization, it takes a total development time of more than 9 years. Phase 1 takes approximately 13 months, Phase 2 approximately 28 months, and Phase 3 approximately 51 months, followed by regulatory review of approximately 18 months. The cost of developing a therapy for AD, including the cost of failures, is estimated at $5.7 billion in the current environment. (Vradenburg, 2015; Šimić et al., 2017; Hampel et al., 2018)

AD is associated with several comorbidities including: 1) **Lewy body disease**, a subset of diseases which includes Parkinson's disease and dementia with Lewy bodies which have abnormal accumulation of α-synuclein in neurons; 2) cerebrovascular diseases that cause vascular brain injury, including atherosclerosis, arteriolosclerosis, and **cerebral amyloid angiopathy**; 3) hippocampal sclerosis; 4) **argyrophilic grain disease**; 5) TDP-43 proteinopathy; and 6) cerebral autosomal dominant arteriopathy and **leukoencephalopathy** (CADASIL) as well as many other neuropathologic changes.

Symptoms

The clinical diagnosis of AD has been generally based on the original 1984 protocol of the National Institute of Neurological and Communicative Disorders and Stroke – Alzheimer's Disease and Related Disorders Association. The presence of cognitive impairment and a suspected dementia syndrome must be confirmed by neuropsychological testing for a diagnosis of possible or probable AD. The American Psychiatric Association Diagnostic and Statistical Manual for Mental Disorders 5th edition (DSM-5) criteria, requires that along with a memory disorder and impairment in at least one additional cognitive domain, the criteria also required both impairments to interfere with social functioning or activities of daily living (McKhann et al., 1984; APA, 2018). According to the Alzheimer's Association (AA), the signs and symptoms of AD include memory loss, difficulty in completing familiar tasks, challenges in planning or solving problems, confusion with time and place, trouble comprehending visual images and spatial relationships, problems with writing or speaking words, misplacing things and being unable to retrace steps, decreased and poor judgement, changes in mood and personality, and withdrawal from work and social situations (AA, 2018).

Genetic Basis

Studies showed that cerebrovascular amyloid (caused by cerebral amyloid angiopathy) and senile plaques are composed of **amyloid β protein (Aβ)**, the antigenic determinants shared in both AD and Down's syndrome. A *Val717Ile* missense mutation in the amyloid precursor protein (*APP*) gene on chromosome 21 was found to be causally related to the early-onset autosomal-dominant familial AD. According to the amyloid theory, excessive production of Aβ, existing in monomeric, oligomeric, and aggregated forms as senile plaques cleaved from a larger amyloid precursor protein (APP) molecule by β-secretase enzyme (encoded by the *BACE1* gene) and γ-secretase is the key pathological event which drives all other pathological changes. These pathological changes include altered calcium homeostasis, microglial activation/inflammation, astrocytosis, an upregulated production of nitric oxide and DNA damage, dysregulation of energy metabolism and cell cycle control, a significant increase in the full-length mitochondrial DNA (mtDNA) accompanied by extensive fragmentation of the unamplified mtDNA, the development of neurofibrillary tangles (NFT), synaptic loss, excitotoxicity, neuronal death, and dementia, not only in early-onset AD cases but also in late-onset cases. (Williamson et al., 2009; Cox et al., 2016; Šimić et al., 2017: Hampel et al., 2018)

Collectively, the genetic aetiology of AD is very complex with early onset AD being less than 1% of cases, and often familial (fAD). It has an autosomal dominant and fully penetrant inheritance caused by any of more than 200 pathogenic mutations in *APP* (33 duplication mutations), *PSEN1* (185 mutations) and *PSEN2* (13 mutations). Most AD cases (over 99%) however are sporadic, late-onset and have few evident genetic components. At least 14 genes with the highest association with late onset AD identified by positional mapping, targeted gene analysis and genome-wide association studies are: 1) *BIN1* which encodes several isoforms of a nucleoplasmic adaptor protein, one of which was identified as MYC-interacting protein; 2) *CLU* which encodes apolipoprotein J; 3) *ABCA7* for ATP-binding cassette transporter; 4) *CR1* for complement component receptor 1; 5) *PICALM* for phosphatidylinositol binding clathrin assembly protein; 6) *MS4A6A*; 7) *CD33* for a transmembrane receptor expressed on cells of myeloid lineage; 8) *MS4A4E* coding for protein membrane-spanning 4-domains; 9) *CD2AP* which codes for a scaffolding molecule that regulates the actin cytoskeleton; 10) *MAPT* which encodes the microtubule-associated protein tau (*MAPT*) found in AD neurofibrillary tangles; 11) *GAB2* and *APOE* is biologically relevant to AD neuropathology and was found to be overexpressed in pathologically vulnerable neurons. The GAB2 protein was detected in neurons, tangle-bearing neurons, and dystrophic neurites; and interference with *GAB2* gene expression increased tau phosphorylation. Studies suggest *GAB2* modifies LOAD risk in *APOE* ε4 carriers and may influence Alzheimer neuropathology; 12) *IDE* gene, which produces the insulin-degrading enzyme, is a possible candidate enzyme responsible for the degradation and clearance of Aβ; 13) There is significant genetic association between single-nucleotide polymorphisms in the *UBQLN1* gene and the age at onset of AD; 14) The AlzGene online database (www.alzgene.org) has recently associated sortilin-related receptor (*SORL1*) to AD pathology. (Williamson et al., 2009; Dunlop et al., 2013; Iacono et al., 2014; Calderón-Garcidueñas et al., 2015; Brockmeyer & D'Angiulli, 2016; Cox et al., 2016; Šimić et al., 2017; Hampel et al., 2018)

Research in twins suggest that other factors, likely epigenetic and environmental, are related to AD pathogenesis. Perhaps the most researched epigenetic mark is DNA methylation to cytosine located in CpG dinucleotides forming 5-methylcytosine. Environmental toxins, pollutants and metals are reported to negatively affect global DNA methylation patterns. For example, prenatal methylmercury exposure resulted in long-lasting depression-like behaviour and hypermethylation of brain-derived neurotropic factor gene in mouse hippocampus. Air pollution exposure especially damages the blood-brain barrier in the brainstem and can trigger an autoimmune response contributing to the neuroinflammatory and AD pathology present in children in large metropolitan centres. Some environmental toxins, such as β-methylamino-l-alanine (BMAA) produced by cyanobacteria cause misfolding and aggregation of various proteins. Chronic dietary exposure to BMAA has been shown to trigger the formation of both neurofibrillary tangles and Aβ deposits in the brain of vervet monkeys. (Goate et al., 1991; Onishchenko et al., 2008; Williamson et al., 2009; Dunlop et al., 2013; Iacono et al., 2014; Calderón-Garcidueñas et al., 2015; Brockmeyer & D'Angiulli, 2016; Cox et al., 2016; Šimić et al., 2017 Hampel et al., 2018)

Current Therapies

There are numerous potential AD therapeutics targeting the serotonergic system that have been tested in preclinical and clinical trials. The 5-HT$_{1A}$, 5-HT$_4$, and 5-HT$_6$ receptors are considered novel therapeutic targets in AD since 5-HT was found to have indirect influences on neuronal degeneration and memory deficits. Partial agonists of the 5-HT$_{1A}$ receptor, *tandospirone* and *buspirone* are useful in the treatment of behavioural and psychological symptoms of dementia. The 5-HT$_4$ receptor agonists have been shown to increase the release of acetylcholine and decreased production and deposition of Aβ in a mouse model of AD. Currently, several different 5-HT$_6$ receptor antagonists are in clinical trials as drug candidates for AD including *escitalopram, buspirone, tandospirone, reboxetine, atomoxetine, levodopa, risperidone, dextroamphetamine, haloperidol, circadin, melatonin* and nanotherapeutics. Two new agents, PRX-03140 (a 5-HT$_4$ antagonist) and SB-742457 (a 5-HT$_6$ antagonist), have completed phase II trials, while LuAE58054, an antagonist of the 5-HT$_6$ receptor, is in phase III trial with 930 mild to moderate AD patients, in combination with the AChEI donepezil. (Salzman, 2001; Sato et al., 2007; Jia et al., 2014; Pimenova et al., 2014; Ramirez et al., 2014; Šimić et al., 2017; Hampel et al., 2018)

Drugs like the α_1-adrenoreceptor agonist *prazosin* target the noradrenergic system and have mainly been used in ameliorating behavioural symptoms in AD and reducing amyloid burden and neuroinflammation; the β-adrenoreceptor antagonist *propranolol* for aggression and agitation; and the antidepressant *imipramine* and *venlafaxine* for depression, as well as β-blockers for slowing cognitive decline. Other drugs like the α_2-adrenoreceptor antagonists like *piperoxane* improved memory deficits in aged mice; and *fluparoxan* in amyloid precursor protein mice; while the β3-adrenoreceptor agonist *CL316243* improved learning in chicks impaired by the injection of Aβ_{42}. The noradrenaline precursor L-threo-dihydroxyphenylserine improved memory of amyloid precursor protein transgenic mice and amyloid precursor protein double-mutant mice and in reducing Aβ burden in 5xFAD mice. The efficacy of noradrenaline reuptake inhibitors *atomoxetine* (used in the treatment of attention-deficit disorder) and *reboxetine* (used in depression) in the treatment of demented patients is under investigation in clinical trials. (Khachaturian et al., 2006; Rosenberg et al., 2008; Gibbs et al., 2010; Heneka et al., 2010; Scullion et al., 2011; Kalinin et al., 2012; Chalermpalanupap et al., 2013; Hammerschmidt et al., 2013; Šimić et al., 2017; Hampel et al., 2018)

Cognitive impairment in AD can be partially restored by *levodopa* and dopamine agonists like *rotigotine* are shown to be effective in restoring LTP-like cortical plasticity in AD patients. In mice, the non-selective dopamine agonist *apomorphine* is effective in reverting perturbed behavioural tasks (such as water-maze) due to oxidopamine-induced parkinsonism in rodents, promoting Aβ degradation and protecting hippocampal neurons from oxidative stress. Many dopaminergic drugs sought in the treatment of AD are used to reduce apathy, a behavioural symptom in about 70% of patients. Dopamine uptake inhibitors *dextroamphetamine* and *methylphenidate*, used to treat attention deficit hyperactivity disorder and the antidepressant *bupropion* have been tested for treatment of apathy in AD. *Haloperidol*, an antipsychotic D_2 receptor antagonist has been tested in clinical trials for treatment of psychosis and agitation in AD. The antipsychotic *risperidone*, a dopamine receptor antagonist, was tested for the treatment of hallucinations and delusions, and agitation and aggression in AD patients. Phase IV clinical trial testing the effect of repetitive transcranial magnetic stimulation (rTMS) for treatment of apathy in AD has not concluded. (Tariot et al., 1987; Burke et al., 1993; Freedman et al., 1998; Brusa et al., 2003; Herrmann et al., 2008; Himeno et al., 2011; Mitchell et al., 2011; Martorana et al., 2013; Koch et al., 2014; Šimić et al., 2017; Hampel et al., 2018)

Potential therapeutics targeting histaminergic system in AD have focused on histamine H_3R antagonists and seven showed beneficial effects on cognition in preclinical models: *thioperamide*, BF2.649, ABT-239, ABT-288, GSK189254, JNJ-10181457 and PF-03654746. In clinical studies, the H_3R antagonist ABT-288 was shown to be safe in healthy elderly subjects and demonstrated efficacy across several cognitive domains. Therapeutic effects of **melatonin** in AD have been tested in 14 trials including AD patients and 8 trials including mild cognitive impairment patients. Melatonin supplementation slows the progression of cognitive impairment in AD and mild cognitive impairment patients, ameliorates sundowning and improves sleep. Current focus is on the development of prolonged-release melatonin analogues that could prolong the effects of melatonin. Cholinomimetics (*tacrine, rivastigmine, donepezil,* and *galantamine*) have been found to temporarily alleviate some of the disease's symptoms by enhancing cholinergic neurotransmission in AD patients; while *memantine* (a non-competitive antagonist of *N*-methyl-d-aspartate receptors) may have protective activity against glutamate-induced excitotoxic neuronal death. (Cardinali et al., 2010; Brioni et al., 2011; Yiannopoulou & Papageorgiou, 2013; Haig et al., 2014; Šimić et al., 2017; Hampel et al., 2018)

According to Vradenburg (2015), the key areas by which AD drug development could produce viable drug candidates by 2025 are: a) improvement in trial design; b) better trial infrastructure; c) disease registries of well-characterized patient cohorts to help with fast/timely enrollment of appropriate study population; d) validated biomarkers to better detect disease, determine risk and monitor disease as well as predict disease response; e) more sensitive clinical assessment tools; and, f) faster regulatory review.

18.3 EPILEPSY

Epilepsy includes a group of heterogeneous syndromes defined by clinical, electroencephalographic (EEG), and brain imaging criteria. Epilepsy is a network disorder in which the normal physiologic connections between cortical and subcortical pathways/regions are interrupted or disturbed. At the very basic level, seizures are a result of imbalance between excitatory and inhibitory inputs to cells (that is, increased excitation or decreased inhibition). This brings about an abnormal synchronization of electrical activity in a group of active neurons. Depending on the site of origin and the subsequent brain structures and networks affected, seizures may produce a variety of clinical features and symptoms, and can remain localized or generalize across the entire brain. Epilepsy genetics encompasses two broad categories: a) genes and loci discovered in association with primary epilepsy syndromes, where epilepsy is a primary presenting feature; and b) genes found to be associated with disorders of brain development related to epilepsy. It is estimated that about 65 million people worldwide have epilepsy, with >80% living in developing nations of the world. In the US, >3 million people are reported to have this disorder. Research indicates that the peak incidence is higher in the older population, rising from the age of 65; and >25% of new-onset epilepsies are diagnosed after 65 years of age. Considering that the global population aged >65 will increase by ~400 million reaching almost 1 billion by 2030, epilepsy cases in older adults is expected to rise substantially. The population of older adults with epilepsy >65 years consists of two main groups: those who have had epilepsy for many years and, owing to improvements in healthcare, are now living to older age; and those who develop epilepsy *de novo* in later life. While several underlying causes may contribute to new-onset epilepsy in the elderly, cerebrovascular disease accounts for 50–70% of cases and is the single most common cause. (Brodie et al., 2009; Birbeck, 2010; Ngugi *et al.*, 2010; Brodie et al., 2012; Choi et al., 2017; Sen et al., 2018)

Symptoms

Symptoms vary depending on the type of seizure and in most cases, a person with epilepsy will tend to have the same type of seizure each time—so, future episodes will present symptoms similarly. Since epilepsy results from abnormal activity in the brain, seizures can affect any process the brain coordinates. The signs and symptoms of a seizure can include: temporary confusion, staring straight ahead, repetitive swallowing, uncontrollable jerking movements of the arms and legs (often spreading to whole body), repetitive lip smacking, aimless fiddling movements, loss of consciousness or awareness, and psychic symptoms such as fear, anxiety or de javu. (Mayo Clinic, 2018; WebMD, 2018)

Genetic Basis

Causative mutations in many genes, including some genes coding for ion channel subunits and others involved in synaptic function or brain development, have been reported. The region of the sodium channel subunit genes (*SCN1A, SCN1B, and SCN1A*) was found in GWAS studies to be associated with mesial temporal lobe epilepsy and hippocampal sclerosis with febrile seizures. A second locus on chromosome 4 (4p15.1) with the gene *PCDH7* has newly been associated with epilepsy. It encodes a calcium-dependent adhesion protein and a member of the cadherin gene family expressed in the CNS (specifically in thalamocortical circuits and the hippocampus). PCDH7 is implicated as possible cause of common forms of epilepsy, because mutations in another protocadherin gene, *PCDH19*, have been confirmed to cause epilepsy and mental retardation in female patients. For genetic generalized epilepsy, an association at 2p16.1 with genes *VRK2* and *FANCL* within close proximity was reported. Decreased brain glutamine synthetase and manganese concentrations have been found in epileptic cases. A region on chromosome 3 (3q26.2) with the gene *GOLIM4 which* encodes Golgi internal membrane protein 4, (involved in manganese homoeostasis) is closely related to epileptic episodes. The internal membrane protein 4 is degraded when manganese increases above normal concentrations. All brain manganese is bound to glutamine synthetase, an enzyme playing a key part in production or degradation of the neurotransmitters glutamate, glutamine, and GABA. Decreased brain glutamine synthetase and manganese concentrations have been reported in epilepsy. The 4p12 region contains two GABA receptor subunit genes (*GABRG2* and *GABRD*); and mutations in other GABA receptors have been reported to cause epilepsy. The 2p16.1 interval contained genes encoding vaccinia-related kinase 2 (*VRK2*) and Fanconi anaemia, complementation group L (*FANCL*) *found to be associated with* genetic generalised epilepsy. VRK2 is a serine-threonine protein kinase involved in signal transduction and apoptosis. *FANCL*, codes for a ring-type E3 ubiquitin ligase of the Fanconi anaemia pathway involved in DNA repair and homologous recombination. Other genes associated with syndromes and cerebral cortical malformations highly associated with epilepsy include *LIS1, DCX, ARX, FLNA, GPR56*, and *MECP2*. Mutations in *TUBA1A and TBC1D24* have recently been identified in association with focal onset epilepsy. Early infantile epileptic encephalopathy with suppression-burst, a severe form of early onset epilepsy, has also recently been associated with mutations in the gene *STXBP1*. Epigenetic factors including chromatin modifications and DNA methylation have been implicated in epilepsy. They can alter neural network dynamics by interfering with signalling pathways and enzyme and receptor expression. Such epigenetic changes may explain inter-individual differences in the emergence of epilepsy, such as after brain trauma. (Brodie et al., 2009; Birbeck, 2010; Ngugi et al., 2010; Poduri & Lowenstein, 2011; Brodie et al., 2012; Li et al., 2012; Lerche et al., 2013; Machnes et al., 2013; ILAECC, 2014; Choi et al., 2017; Sen et al., 2018)

Researchers have identified structural genomic variations associated with epilepsy conducting large-scale studies of copy number variation in patients studied with idiopathic epilepsies without autism, intellectual disabilities, or dysmorphic features. In one group studied in Northern European, subjects with idiopathic generalized epilepsy by high-density SNP arrays, microdeletions in almost 2% at 15q11.2 and 16p13.11 were reported. In another study, subjects with idiopathic focal or generalized onset epilepsy were analysed by whole-genome oligonucleotide array CGH, and nearly 3% had deletions at loci previously associated with autism and intellectual disability (located at 15q11.2, 15q13.3, or 16p13.11). Studies also found deletions in 16p13.11. (Choi et al., 2017; Sen et al., 2018)

Current Therapies

The pathophysiology of epilepsy is multifaceted, involving several neurotransmitter systems and many receptors, ion channels, intracellular signalling cascades, genes, and epigenetic modifications. Antiepileptic drugs (AEDs) have many benefits but also many side effects, including aggression, agitation, and irritability, in some patients with epilepsy. There have been more than 15 antiepileptic drugs introduced in the last 20 years, many of which work in very unique mechanisms; yet, more than 30% of adolescent and adult patients with the common epilepsies continue to have seizures, despite receiving treatment with many of these drugs used either singly or in combination. The majority of current AEDs target primarily either the classic voltage-gated ion channels (Na^+, K^+, and Ca^{2+}) or the GABA system, but many AEDs have mixed mechanisms and others selectively inhibit other targets (e.g., neurotransmitter release by *levetiracetam* and AMPA receptors by *perampanel*). Many AEDs are also used to treat aggression in psychiatric populations, such as patients with schizophrenia, schizoaffective disorders, bipolar disorders, and autism spectrum disorders. AEDs such as valproic acid, *topiramate, gabapentin,* and *lamotrigine* are frequently used, including those targeting ion channels like *carbamazepine, oxcarbazepine, lamotrigine, brivaracetam,* and *phenytoin*.

Children and adolescents who are treated with newer AEDs like *gabapentin, lamotrigine, tiagabine, vigabatrin, levetiracetam, perampanel* (especially at higher doses), *phenobarbital, sodium valproate, topiramate,* and *zonisamide* should be monitored closely for possible behavioural adverse effects. The atypical **antipsychotics drugs** that are high-affinity antagonists at 5-HT$_{2A}$ and/or 5-HT$_{1A}$ receptors and D$_2$ receptor antagonists like *clozapine, olanzapine, quetiapine,* and *risperidone* and the conventional antipsychotic drugs like *haloperidol*, are widely used "off-label" to attenuate aggressive behaviour. The rationale derives from the abnormalities of 5-HT and DA systems in the pathophysiology of aggressive behaviour. *Clozapine* has been demonstrated to be superior to *olanzapine* and *haloperidol* in the control of aggression. In more resistant cases, the combination of *clozapine* and an AED such as valproic acid remains the most effective treatment. Based on current research on these drugs, some AEDs seem to be associated with higher risk than others, including *clobazam, clonazepam, levetiracetam, perampanel, phenobarbital, tiagabine, topiramate, vigabatrin,* and *zonisamide*. The potential for aggressive behaviour should be explained to every patient starting treatment with any of these drugs, particularly patients with known anger management issues. The AEDs with strongest evidence for a risk of aggressive behaviours are *levetiracetam, perampanel,* and possibly *topiramate*, but most patients taking these, and any other AEDs, will have no problems with aggressive behaviours. Involvement of partners and families is important, since many people are not aware that they have a short temper or that their demeanour could be perceived as aggressive. The choice of AED therapy for patients with newly diagnosed and chronic epilepsy must be taken very carefully due to the potential of exacerbating adverse effects like aggressive behaviour. To prevent problems with aggression in patients with epilepsy, on-going and future research will clarify the neurobiology of aggression and epilepsy and may help clinical decision making and treatment selection. (Rogawski & Löscher, 2004; Scharfman, 2007; Brodie, 2010; Brodie et al., 2012; Comai et al., 2012a; Comai et al., 2012b; Löscher et al., 2013; Moshé et al., 2015; Brodie et al., 2016)

18.4 ESSENTIAL TREMOR (ET)

ET is a chronic, progressive neurologic disease with the hallmark motor feature of a tremor that occurs during voluntary movements such as writing or eating involving the hands and arms, but which may also eventually spread to involve the head (i.e., neck), voice, jaw, and other body regions. Due to pathologic heterogeneity, there is increasing support for the notion that ET may be a family of diseases whose central defining feature is kinetic tremor of the arms, and which might more appropriately be referred to as "the essential tremors". Among the most prevalent adult-onset movement disorders, ET can occur at any age, and paediatric cases have been reported, although most cases arise later in life. Studies indicate that ET worldwide was 0.9%; but prevalence among persons aged 65 years and older was 4.6%, and in some studies, the prevalence among persons aged 95 and older reached values more than 20%. Research indicates that ET misdiagnosis rate in two studies has been reported to range from 37-50% (Jain et al., 2006; Bermejo-Pareja, 2011; Elble, 2013; Louis, 2014; Espay et al., 2017)

Symptoms

Abnormal oscillations in the central nervous system are crucial to the pathogenesis of ET and the clinical presentation of tremor involves a rhythmic motor activity. The main feature of ET is **kinetic tremor** which is typically mildly asymmetric; and in approximately 5% of patients, the tremor is markedly asymmetric or even unilateral. In about 50% of patients, the tremor has an intentional component. Postural tremor also occurs in ET, and is generally worse when arms are folded when held straight in front of the patient. Studies note that the amplitude of kinetic tremor exceeds that of postural tremor. In addition to tremor, another motor feature of ET is gait ataxia, which is more than that seen in similarly aged controls. This is often mild in most patients, but may be of moderate severity in others with the associated functional problems. Mild saccadic eye movement abnormalities have also been detected in ET patients in some studies. The presence of non-motor features in ET has including diminished hearing, a mild olfactory deficit in some, and mild cognitive changes like mild cognitive impairment and dementia. Psychiatric manifestations include both secondary anxiety and depression. (Helmchen et al., 2003; Earhart et al., 2009; Kronenbuerger et al., 2009; Bermejo-Pareja, 2011; Thenganatt & Louis, 2012; Gitchel et al., 2013; Louis, 2013)

Genetic Basis

Both genetic and environmental (toxic) factors are likely contributors to disease aetiology. The genetic architecture of familial and sporadic ET is likely to be explained by several modes of inheritance and transmission, including both Mendelian and complex disease pattern. Aggregation studies indicate that 30–70% of ET patients have a family history with the vast majority (>80%) of young-onset (≤40 years old) cases reporting ≥1 affected first-degree relative. Studies suggest an autosomal-dominant mode of inheritance with reduced penetrance. Two published GWAS variably identified single-nucleotide polymorphisms (SNPs) in the leucine-rich repeat and Ig domain containing 1 (*LINGO1*) gene or an intronic variant in the *SLC1A2* gene, which reached genome wide significance and are associated with increased risk for ET. A more recent rare variant (p.R47H) in *TREM2* was identified as a risk factor for ET and the gene may turn out to be a promising candidate. A rare amino acid substitution (p.R47H) in the TREM2 protein (triggering receptor expressed on myeloid cells) has been identified as a risk factor for several neurodegenerative diseases, including Alzheimer's disease. The genetic loci harbouring ET genes have been found on chromosomes 2, 3, and 6 [3q13 (ETM1), 2p22-p25 (ETM2), and 6p23 (ETM3)]. Using linkage and whole-exome sequencing, a p.Q290X mutation in the gene *FUS/TLS*, was linked to ET. In a more recent study examining both ET and Parkinson disease, the mitochondrial serine protease *HTRA2* variant was shown to be present in patients with both diseases. All affected individuals in the family were either heterozygous or homozygous and homozygosity was associated with earlier age at onset of tremor, more severe postural tremor, and more severe kinetic tremor. Many environmental toxins associated with ET are under investigation, including β-carboline alkaloids (like the dietary toxin harmane) and lead. (Higgins et al., 1997; Gulcher et al., 1997; Shatunov et al., 2006; Louis & Dogu, 2007; Louis, 2008; Merner et al., 2012; Unal Gulsuner et al., 2014)

Current Therapies

Despite new insights into ET pathogenesis, its therapy remains purely symptomatic, and virtually all medications used for the reduction of tremor have initially been developed and approved for other pathologies. ET therapies include the following broad classes of medications: anti-convulsants, beta-adrenergic blockers, GABAergic agents, calcium channel blockers, and atypical neuroleptics. Only *propranolol* has been approved by the US FDA for the treatment of ET. It is a nonselective β-adrenergic receptor antagonist. Other drugs in use for ET and well-reviewed by Hedera et al., (2013) include the following: 1) *Primidone* is an anticonvulsant that is metabolized to phenobarbital and phenylethylmalonamide. 2) *Gabapentin* is a structural analogue of gamma-aminobutyric acid [GABA] and does not directly bind to $GABA_A$ or $GABA_B$ receptors. It likely interacts with auxiliary subunits of voltage-gated calcium channels. 3) *Pregabalin* is another structural derivative of GABA, and like gabapentin, it does not display any affinity to GABA-ergic receptors. It acts like gabapentin with a high affinity binding to the alpha2-delta site (an auxiliary subunit) of voltage-sensitive channels. 4) *Topiramate* has a complex mechanism of action including blocking of voltage-gated sodium channels, augmenting of GABA activity at the $GABA_A$ receptors, antagonizing the AMPA/kainate glutamate receptors, and inhibiting the carbonic anhydrase enzyme). 5) *Atenolol* is a competitive, β-1 selective (cardio-selective) adrenergic antagonist. 6) *Nimodipine* belongs to a calcium channel blockers class of medications, binding to the L-type voltage-gated calcium channels. 7) *Clozapine* belongs to the group of atypical antipsychotics and mainly blocks dopamine D_1 and D_4 receptors. 8) The benzodiazepines (*clonazepam* and *alprazolam*) directly bind to the $GABA_A$ receptor complex, and their presence triggers more influx of Cl⁻ ions through the increased binding of GABA. This results in hyperpolarization of the cell membrane thereby inhibition of action potential firing—hence their anxiolytic, anticonvulsant, sedative, muscle relaxant, and likely also antitremorogenic effects. Additional pharmacological agents tried in ET include *levetiracetam, lacosamide, mirtazapine, amantadine, memantine isoniazid, methazolamide, acetalozamide*, and *clonidine*, but at present there is no sufficient evidence to support their use on a trial and error basis even in medically refractory cases. (Zesiewicz et al., 2011, 2013; Elble, 2013; Hedera et al., 2013)

18.5 FAMILIAL MEDITERRANEAN FEVER (FMF)

FMF was first described by Reimann in 1948, and seven years later, the disease was defined as FMF in 1955 (Sohar et al., 1967). Throughout the world, FMF is most frequently seen in Turkey with a prevalence varying between 1:150 and 1:10,000. The second most frequently affected ethnic group is Armenians with studies showing a prevalence of 1:500. Disease onset is prior to 20 years of age in 90% of cases and in 60% of cases, the age at onset is under 10 years. However, the disease may develop after the first few years of life and the sex distribution shows a slight male dominance. (Sohar et al., 1967; Onen, 2006; Özen et al., 2017)

Symptoms

FMF is characterised by recurrent fever and **serositis** (for instance, peritonitis, pleuritis, and synovitis) symptoms. There are reported individual and ethnic differences in both the frequency and course of the clinical symptoms. Often, a single sign may sometimes accompany high fever, at other times, more than one symptom can occur simultaneously. Clinical symptoms may differ over time even in the same patient. Attacks develop quite spontaneously and continue for at least 12 hours and most symptoms resolve within 3–4 days, with the interval between attacks clinically relatively symptom-free. However, arthritis and myalgia may have a prolonged course. (Padeh et al., 2003; Sönmez et al., 2016; Özen et al., 2017)

Genetic Basis

FMF is associated with mutations in the *Mediterranean Fever* (*MEFV*) gene, located on the short arm of chromosome 16 (16p13.3). *MEFV* encodes a 781-amino acid protein termed pyrin responsible for the regulation of apoptosis, inflammation and cytokines, and is mainly expressed in neutrophils, eosinophils, dendritic cells and fibroblasts. The epigenetic mechanisms such as histone modification, methylation, and microRNAs may play role in the pathogenesis of FMF. Studies show a slightly higher methylation level of exon 2 of *MEFV* in FMF patients when compared to healthy controls. Though considered as an autosomal recessive disease, research indicates that a significant portion of the patients had only one mutation in the *MEFV* gene (heterozygotes). The expectation would then be that heterozygote individuals will be carriers lacking the FMF phenotype. Studies show that the disease in heterozygotes could not be distinguished from that of homozygous patients, hence FMF could be considered a dominant condition with low penetrance. Microorganisms may affect FMF since pyrin is a component of NLRP3. Pyrin has also been shown to detect virulent pathogenic activity. Pyrin is a part of the inflammasome complex NLRP3 (a pathogen recognition receptor) serving as an intracellular organelle required to produce interleukin-1β—a potent pro-inflammatory cytokine crucial for host-defence responses to infection and injury. Pyrin was found to interact with an inflammasome adaptor protein which activates caspase-1 leading to the cleavage and activation of IL-1β. Research demonstrated a significant increase in pyrin expression in FMF patients compared to healthy controls (Touitou, 2001; Booty et al., 2009; Marek-Yagel et al., 2009; Kirectepe et al., 2011; Lopez-Castejon & Brough, 2011Cakır et al., 2012; Sari et al., 2014; Özen & Batu, 2015; Onen, 2016; Özen et al., 2016; Sönmez et al., 2016; Özen et al., 2017)

Current Therapies

The European League Against Rheumatism (EULAR) recommendations emphasize that the treatment aim of FMF is to control acute attacks, minimize the chronic and subclinical inflammation, prevent complications, and provide an acceptable quality of life. The drug *colchicine* has been the main treatment for FMF since 1972. It inhibits the release of the cytokine IL-1β release from bone-marrow-derived macrophages preventing microtubule elongation by binding to tubulin monomers and inhibiting polymer formation. *Colchicine* treatment is lifelong in FMF and the most common side effect seen in up to 10% of patients during the first month of the treatment is gastrointestinal disturbance. Other rare side effects include vitamin B12 deficiency, reversible peripheral neuritis and myopathy, bone marrow suppression, and alopecia. Anti-IL-1 therapy appears to be a promising second-line therapy in refractory or intolerant patients. Three types of anti-IL-1 agents in clinical use all administered subcutaneously: a) *Anakinra*, a recombinant homolog of the human IL-1 receptor; b) *Canakinumab*, a fully human immunoglobulin G1 monoclonal antibody; and c) *Rilonacept*, a dimeric fusion protein which captures IL-1. Anti-tumour necrosis factor (TNF) drugs were effective in reducing episode frequency in patients unresponsive to *colchicine*. The drugs *thalidomide, adalimumab, etanercept,* and *infliximab* whose mechanism of action is inhibiting TNF, were found to have positive effects. Research suggests that *colchicine* treatment must be maintained while administering alternative treatments to *colchicine*-resistant patients. Anti-TNF treatment can have beneficial effects for controlling FMF attacks in FMF patients with chronic arthritis and/or sacroiliitis. (Yilmaz et al., 2011; Özen et al., 2011; Ozgocmen & Akgul, 2011; Dinarello et al., 2012; Hashkes et al., 2012; Padeh et al., 2012; Sönmez et al., 2016; Özen et al., 2017)

18.6 FRIEDREICH'S ATAXIA (FRDA)

FRDA is the most common autosomal recessive form of ataxia (loss of voluntary control of bodily movements) accounting for half of the inherited progressive ataxias and three-quarters of those with onset before age 25. The incidence of FRDA is close to 1 in 50 000 individuals, with the carrier frequency estimated to be 1 in 100 individuals. It affects people in most parts of the world especially peoples of European, Middle Eastern, North African, and Indian descent. In East Asia (China, Japan, Korea and Southeast Asia), sub-Saharan Africa, and Amerindians FRDA has never been reported in the indigenous populations. Approximately 1:85 people carry a mutant copy of the frataxin gene. (Tsou et al., 2011; Ashley et al., 2012; Bürk, 2017; Napierala et al., 2017; Cotticelli et al., 2018)

Symptoms

A very devastating disease, the typical age of FRDA onset is 10 years with many patients wheelchair-bound by 19 years of age, and the average life expectancy is only 40 years of age. There appears to be an inverse relationship between the age of FRDA onset and the size of the GAA repeat expansion—thus, the size of the GAA repeat expansion directly correlates with the severity of disease. The disorder is characterized by progressive gait and limb **ataxia**, ataxic speech, abnormal proprioception and vibratory sense, and loss of reflexes, with a slowly progressive course that culminates in reliance on hands-on assistance for self-care and wheelchair dependency. Children become increasingly clumsy with physical activity, an absence of lower limb deep tendon reflexes, scoliosis, foot deformities, diabetes, visual disturbance, **dysphasia, dysarthria**, eye movement abnormalities, hearing and cognitive deficits, sexual problems, sleep apnoea, urological disturbances, psychological issues, and cardiomyopathy. (Fogel & Perlman, 2007; Tsou et al., 2011; Bürk, 2017; Napierala et al., 2017; Cotticelli et al., 2018)

Genetic Basis

FRDA results from a mutation of the frataxin gene identified in 1996 and located on chromosome 9 (9q13). It encodes a protein product which regulates mitochondrial iron metabolism. A defective frataxin causes oxidative stress and cell damage. About 98% of the people with FRDA are homozygous for a GAA triplet repeat expansion in the first intron of the frataxin gene, and the other 2% have point mutations. Friedreich ataxia is a chromatin disease where GAA repeats induce epigenetic heterochromatin-mediated gene silencing through histone modifications. Three physiological consequences of decreased levels of frataxin are: mitochondrial dysfunction, production of free radicals, and changes in iron metabolism. People without the disorder have normally 3-33 GAA repeats, while those affected have between 60-1500 GAA repeats on both alleles (homozygous)—true for 95% of patients. The other 5% of individuals are compound heterozygous. Molecular genetic diagnostic testing is done by examining GAA repeat expansions and then sequencing the gene for point mutations. (Pandolfo, 2008; Ashley et al., 2012; Bürk, 2017; Napierala et al., 2017; Cotticelli et al., 2018)

Current Therapies

Nursing and rehabilitative interventions are the mainstays of treatment, as there are no curative therapies. There is currently no US FDA-approved treatment for FRDA, but advances in research of its pathogenesis have led to clinical trials of potential treatments. Therapeutic developments centre around the three biochemical defects of decreased frataxin gene, mitochondrial iron accumulation, and oxidative stress and include: 1) Erythropoietin, a hormone secreted by the kidneys controls erythropoiesis (red blood cell production) and can potentially increase frataxin expression. Due to the increased levels of iron required during **erythropoiesis**, upregulation of iron metabolism-related genes, which include frataxin, is required. Recombinant human erythropoietin has been found to increase frataxin in cellular models. Preclinical trials on erythropoietin and analogues to pinpoint the property that increases frataxin expression without causing erythrocytosis in going on. 2) Resveratrol, a compound studied for antioxidant effects, increased frataxin expression in cells in culture and in mouse models with GAA repeat expansions. 3) A protein-based therapy has also been developed by attaching the cationic cell-penetrating TAT peptide from HIV to frataxin. TAT-frataxin complex crosses into the mitochondrial matrix and increases the level of frataxin in cells that are in culture. Studies show that knockout mouse models injected with the TAT-frataxin, had a life expectancy increase of 50% or more and had corrections in diseases of the heart muscle (cardiomyopathy). 4) Deferiprone, currently being studied is a small lipid-soluble molecule that chelates iron. 5) Antioxidants are being used as treatment for oxidative stress—for example, in the US, *idebenone* has reached phase-3 trial as a double-blind, multi-dose, placebo-controlled trial. 6) Pioglitazone, a drug for diabetes, activates genes that are involved in mitochondrial biogenesis and antioxidant defences and is in clinical trials in France. 7) EGb761is a free radical scavenger from *gingko biloba* and is also in clinical trials. 8) Ox1, an indole compound, acts as an antioxidant and may be able to weakly chelate iron. 9) A polyunsaturated fatty acid with hydrogen molecules that are susceptible to radical reactions with deuterium. 10) Deuterated polyunsaturated fatty acids stop the chain reaction of reactive oxygen species and has been developed into a dietary supplement. 11) p38 inhibitors and a synthetic shRNA (gFA11) which reverses the growth defect of FRDA cells in culture are both being investigated as possible therapeutic agents. 12) Other potential agents such as coenzyme Q10, vitamin B3, L-carnitine, *deferiprone, physostigmine, riluzole, amantadine*, and frataxin-inducing agents (like histone deacetylase inhibitors and interferon gamma), plus gene therapy are currently under investigation. (Hart et al., 2005; Fogel & Perlman, 2007; Rai et al., 2008; Lynch et al., 2010; Myers et al., 2010; Ashley et al., 2012; Brigatti et al., 2012; Aranca et al., 2016; Bürk, 2017; Napierala et al., 2017; Cotticelli et al., 2018)

18.7 HUNTINGTON'S DISEASE (HD)

HD is an autosomal dominant neurodegenerative disease characterized by motor impairment, cognitive decline, and psychiatric manifestations and behavioural abnormalities. The first published accounts of the disorder that would come to bear his name occurred in 1872 by a young physician from Connecticut—George Huntington. HD has a prevalence of 5-10 per 100,000 in South America, North America, Australia, and most European countries and countries of European descent, but significantly lower in Africa and Asia, with an estimated prevalence of 0.5:100,000 in Japan and China, and even lower in South Africa. HD affects males and females at the same frequency, and the mean age of onset is around 40 years although it can be as early as 4 and as late as 80 years of age. Epidemiologic studies show that there are about 30,000 HD patients and that there are about 150,000 people at risk of developing the disease in the US. There has been a reported increase in HD prevalence in Australia, the Americas and the UK over the last few decades, and a number of reasons for the increased prevalence are suggested including more accurate diagnosis (definitive genetic tests), improved supportive care, and general increase in overall life expectancy. Aging has been associated with a decline in a variety of pathways relevant to neurodegenerative disease especially those that are critical for handling misfolded proteins. Thus, aging may further reduce the overall ability of cells to withstand random stresses. (Walker, 2007; Finkbeiner, 2011; Rawlins et al., 2016; Chaganti et al., 2017)

Symptoms

The best-known symptoms are excessive uncontrolled jerky motor movements, called chorea, and gait disturbances. Some patients experience **dystonia**—an increase in muscle tone. Neuropsychiatric symptoms especially depression and anxiety are common. Some patients report obsessive-compulsive symptoms and effects on cognition. Deficits in executive function appear early; a global dementia often follows. Unusually long CAG stretches (>50) produce earlier symptoms; the youngest person with HD developed symptoms at age 2. Children often display symptoms so different that the syndrome is called juvenile HD (JHD), reflecting the earlier onset. Unlike adult-onset HD, JHD is associated with a paucity of movements (bradykinesia) and an increased incidence of seizures. The most prominent neuropathology in HD occurs within the striatal part of the basal ganglia, in which gross atrophy is accompanied by extensive neuronal loss and **astrogliosis**, both of which become more severe as the disease progresses, with the atrophy leading to great enlargement of the lateral ventricles). Some of these dying neurons show nuclear fragmentation and marker expression characteristic of apoptotic cell death. (Margolis & Ross, 2003; Walker, 2007; Finkbeiner, 2011; Chaganti et al., 2017)

Genetic Basis

HD is caused by a pathologic cytosine-adenine-guanine (CAG) repeat expansion, on the 5' end of the *huntingtin (HTT)* gene (Figure 2). The ubiquitous expression of mutant huntingtin protein (mHTT), is thought to be the predominant toxic agent in HD. The normal (wild-type) huntingtin gene is ubiquitously expressed and has many interaction partners performing many functions including vesicular trafficking; the mediation of endocytosis, vesicular recycling and endosomal trafficking; coordination of cell division; transcriptional regulation; and metabolism. The age of HD onset strongly correlates with expanded allele

Figure 2. Huntington's disease genetic mechanisms
Source: Image used under license from Shutterstock.com

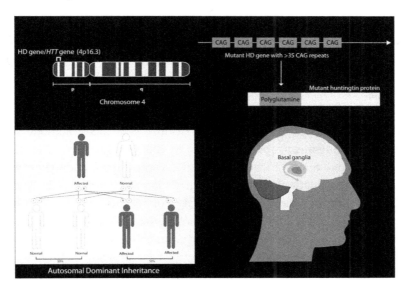

CAG repeat length, which correlates negatively with the age of onset, and accounts for approximately 70% of the variation in age of onset in HD.

Clinical HD always develops in individuals with CAG repeat length of 40 or greater, and there is reduced penetrance among individuals with repeat lengths from 36–39. CAG repeat lengths of 27–35 are considered intermediate alleles, but the CAG repeat size can expand into the pathogenic HD range in the next generation. Studies show that more than 90% of individuals with repeat sizes of ≥44 would present with HD before the age of 60. Surprisingly, individuals who have two HD-causing alleles (homozygous for mutant *HTT*) appear to develop symptoms about the same age as people with a single allele and the same CAG expansion. However, homozygosity for the CAG mutation has been reported to lead to a more severe clinical course of the disorder. The CAG expansions might mediate neurodegeneration through abnormal polyQ expansions. The polyQ expansions confer to proteins the propensity to aggregate and the rate of aggregation dramatically increases as the polyQ expansion grows. The genetic mutation that causes all HD leads to an abnormal expansion of a polyQ tract in the HTT protein. In turn, the polyQ expansion confers to HTT one or more toxic functions and the propensity to aggregate and accumulate intracellularly. Research shows that aggregates of HTT with disease-associated polyQ expansions develop relatively insoluble amyloid-like structures in vitro. Several lines of evidence indicate that HD is fundamentally a **proteopathy** and that measures to mitigate protein misfolding or to promote the clearance of misfolded proteins may reduce behavioural deficits and neuropathological abnormalities. Sub-microscopic molecular assemblies called oligomers or misfolded monomers are toxic and trigger neurodegeneration. Oligomers might cause toxicity by creating pores in membranes or by binding to specific cellular proteins, such as receptors, and altering their function. Cells contain complex interconnected molecular networks to prevent proteins from misfolding and to correct or degrade those that become misfolded. Those networks might sense mutant HTT as a threat and initiate beneficial coping responses which might delay HD. It is not clear how a polyQ expansion in mutant HTT induces neurodegeneration, but aggregated protein species might form structures that confer toxicity, and with sufficient non-native folds intracellularly, the molecular chaperone system that maintains cellular functional normality may be overwhelmed. In the vast majority of cases (>80%) of the hereditary transmissions from HD parents, the expanded CAG repeat is only mildly altered by one or a few CAG repeats, usually decreasing if transmitted maternally, and increasing if transmitted paternally. (Bates et al., 1997; Scherzinger et al., 1997; Langbehn et al., 2004; Finkbeiner, 2011; Reiner et al., 2011; Keum et al., 2016; Chaganti et al., 2017)

Current Therapies

Since all cases of HD are caused by the same basic mutation, it is an ideal candidate for the development of therapeutics targeting pathogenic processes close to its root genetic cause. The disorder is caused by toxic gains of function, so reducing mHTT production should alleviate its pathogenesis. Novel therapeutic strategies target the mHTT production pathway. Recent advances to improve the design and delivery of targeted therapeutics will improve their efficacy, safety, tolerability and duration of effect. Therapies designed to interact with *HTT* mRNA include antisense oligonucleotides (ASOs) and RNA interference (RNAi) compounds which accelerate degradation of the mRNA transcript. Both are nucleotide-based therapeutic molecules that bind to mRNA selectively. ASOs are synthetic single-stranded DNA molecules that principally bind pre-mRNA in the nucleus, targeting it for degradation by RNase H. RNAi uses RNA-based therapeutic molecules—including siRNA, shRNA or miRNA depending on structure and sequence. These bind to mature, spliced cytosolic mRNA, targeting it for removal. As of December 2017, the drug coded IONIS-HTT$_{RX}$, was the first ASO designed to lower huntingtin protein (HTT) to be tested in people with HD targeting human huntingtin in clinical trials. IONIS-HTT$_{Rx}$ is expected to reduce expression of both mutant and wild-type *HTT* alleles equally but is not clear whether lowering the wild-type allele is safe in humans. The FDA has approved *nusinersen*, a lumbar intrathecally-administered ASO that was found to extend survival in spinal muscular atrophy via targeted modulation of gene expression. Other coded drugs tested are PRECISION-HD1 and PRECISION-HD2. Clinical trials for *deutetrabenazine*, a vesicular monoamine transporter type 2 selective inhibitor just recently completed.

Virally-delivered RNA therapeutics which permanently transduce CNS cells, effectively turning them into factories making a drug that suppresses manufacture of mHTT are also being investigated. The potency of huntingtin reduction and phenotypic benefits from AAV-delivered siRNA is being tested in several HD rodent models. Non-viral vectors are also receiving increased attention in HD therapy. Small molecule approaches using a brain-penetrant, orally bioavailable small molecules that reduce *HTT* through altering mRNA splicing and processing are also under investigation. The two DNA-targeting gene therapies currently under investigation are zinc finger proteins (ZFPs) and CRISPR/Cas9. Zinc-finger transcriptional repressors may reduce huntingtin by targeting DNA without altering it, while CRISPR-Cas9 therapeutics bring the promise of permanently correcting the CAG expansion mutation that causes HD. Zinc fingers are naturally-occurring structural motifs that can bind specific DNA sequences. ZFPs tooled for therapeutic use typically contain a zinc finger array specific to the DNA sequence of interest—one 'finger' per three bases fused to a functional domain intended to act upon the DNA. Some examples are zinc finger nucleases that cleave DNA, and zinc finger transcription factors that modulate gene expression. The use of *HTT*-directed CRISPR/Cas9 therapeutic strategies has become a possibility recently. The system can restore alleles to harmless length by excising CAGs and can inactivate the mutant allele by inserting stop codons or missense mutations. (Walker, 2007; Klug, 2010; Finkbeiner, 2011; Chaganti et al., 2017; Wild & Tabrizi, 2017; Rodrigues & Wild, 2018)

18.8 MAPLE SYRUP URINE DISEASE (MUSD)

MSUD is an autosomal recessive disorder characterized by disruption of the normal activity of the branched-chain α-ketoacid dehydrogenase (BCKAD) complex. This metabolic pathway's first step involves the conversion of branched-chain amino acids (BCAAs)—leucine, isoleucine, and valine into their relevant α-ketoacids by branch-chain aminotransferase within the mitochondria. During the second step in the BCAA catabolism, the BCKAD complex initiates oxidative decarboxylation of the α-ketoacids converting them into acetoacetate, acetyl-CoA, and succinyl-CoA. The metabolic disorder occurs when there is decreased function of the BCKAD enzyme complex since BCAA catabolism is essential for normal physiological functions. MSUD is amenable to treatment through dietary restriction of BCAAs, and with early treatment, patients typically have good clinical outcomes. MSUD is therefore part of the Recommended Uniform Screening Panel (RUSP), which compiles a list of actionable, early onset disorders for which screening is recommended for all new-borns in the United States. Increased use of tandem mass spectrometry in new-born screening has greatly improved clinical outcomes in affected individuals by helping facilitate early detection and timely medical intervention for patients with MSUD. MSUD has had a worldwide incidence of 1:185,000 live births, but is much more frequent in certain founder populations including the Old Order Mennonites of Pennsylvania, where occurrence rates may be as high as 1:200 live births. MSUD is amenable to treatment through dietary restriction of BCAAs and outcomes are positive if it is detected and treated within the first few days of life—hence the need for new-born screening. Individuals with the classic neonatal form have <2% of BCKAD enzymatic activity and present with maple syrup odour in cerumen (earwax) shortly after birth and in urine during the first week of life. If untreated, the neonate may develop irritability, lethargy, poor feeding, apnoea, muscle rigidity (opisthotonus), "bicycling" movements, followed by coma and early death due to brain oedema. The intermediate form of MSUD is characterized by up to 30% of BCKAD residual activity; thus, these individuals may appear healthy during the neonatal period, although maple syrup odour in cerumen may be present. During year 1 of their life, they may have feeding problems, poor growth, and intellectual disability, and are susceptible to similar neurologic features as individuals with the classic form. Individuals with intermittent MSUD are completely asymptomatic, having normal growth and neurological development, even on an unrestricted diet. During catabolic states, clinical and biochemical features of the classic form may arise in patients and should be controlled similarly to individuals with the severe form. (Blackburn et al., 2017; Grünert et al., 2018; Kathait et al., 2018)

Symptoms

MUSD is characterized by neurological and developmental delay, encephalopathy, feeding problems, and a maple syrup odour to the urine. Patients with this disorder have elevations of branch chain ketoacids in the urine in addition to elevated BCAAs in the plasma (Blackburn et al., 2017; Grünert et al., 2018; Kathait et al., 2018)

Genetic Basis

Genetic studies have determined that MSUD is an autosomal recessive disease caused by pathogenic variants in genes encoding any of the 4-branched chain α-keto acid dehydrogenase (BCKAD) subunits. The BCKDC catalyses the rate-limiting step in the catabolism of the BCAA. The enzyme complex consists of three catalytic components: a decarboxylase (E1) composed of two E1α and two E1β subunits, a transacylase (E2) core of 24 identical lipoate bearing subunits and a dehydrogenase (E3) existing as a homodimer. Mutations in the genes encoding the E1α, E1β and E2 subunits result in an MSUD phenotype, while mutations in the E3 subunit cause a more complex phenotype with lactic acidosis. BCAAs can be found in protein-rich diets and are among the nine amino acids essential in humans, playing important roles in protein synthesis and function, cellular signalling, and glucose metabolism. Mutation in and varying residual enzymatic activity defines five MSUD variants. (Brosnan & Brosnan, 2006; Lynch & Adams, 2014; Blackburn et al., 2017; Grünert et al., 2018)

Current Therapies

One of the primary goals in treating MSUD is to manage diet by reducing BCAAs and provide adequate macronutrients to prevent catabolism and help maintain plasma BCAAs within targeted treatment ranges. Nutrition management consists of BCAA-free medical food specially formulated to provide 80%–90% of protein needs, and most of energy and micronutrient necessities throughout life. Individuals with MSUD are at risk for metabolic decompensation. The goal in acute management is to suppress catabolism and promote protein anabolism. Patients must have a "sick day" prescription to manage illness at home, although this practice is not universal. Sick day protocol usually entails guidelines on increasing BCAA-free amino acid formula intake to 120% of the usual intake, decreasing leucine intake by 50%–100%, and providing small but frequent feedings throughout a 24-hour period. The monitoring of BCAA plasma concentrations is necessary to guide appropriate diet adjustments during the illness. The sick day protocol is usually initiated upon the first signs of an illness and for minor illnesses can be managed at home. In acute cases, more aggressive approaches like dialysis, hemofiltration, parenteral nutrition, and/or tube feedings must be taken. Acute dietary treatment needs to be aggressive and include sufficient energy (up to 150% of the normal energy consumption), based on BCAA-free formula and fluid administration (up to 150 mL/kg). Liver transplantation from unrelated deceased individuals or living related donor is an option (when available) but associated with high costs, severe complications including death, and a need for lifelong immunosuppression. Thus, hepatocyte transplantation is becoming a better and viable cell therapy for metabolic liver disease. Sodium phenylbutyrate (a nitrogen scavenging medication), is commonly used for the treatment of patients with urea cycle disorders. Placental-derived stem cell transplantation lengthened survival and corrected many amino acid imbalances in a mouse model of iMSUD. This highlights the potential for their use as a viable alternative clinical therapy for MSUD and other liver-based metabolic diseases. (Brunetti-Pierri et al., 2011; Skvorak et al., 2013; Blackburn et al., 2017; Grünert et al., 2018; Kathait et al., 2018)

18.9 MENKES DISEASE (MD)

MD was discovered when an association was made between copper deficiency and a demyelinating disease of the brain in sheep's offspring grazing in copper-deficient pastures around 1937 in Australia. Then in 1962, Menkes et al. (1962), described a distinctive clinical syndrome comprising neurological degeneration in patients of English–Irish heritage who showed unusual hair quality and failure to thrive. The infants appeared normal at birth through early infancy and then developed epileptic seizures and regressed developmental milestones dying between the age of 7 months and 3.5 years. MD normally presents in males at 2–3 months of age, and progresses to loss of previously obtained developmental milestones and the onset of hypotonia, seizures and failure to thrive. MD has an incidence of 1/140,000 to 1/300,000. MD markedly decreases a cells' ability to absorb copper ions with an estimated reduction of copper transport of 0–17% of that exhibited under healthy conditions. This causes severe cerebral degeneration and arterial changes, resulting in death in infancy. Copper is a trace metal with a ready capacity to gain and donate electrons—a property which makes it highly desirable as a cofactor for numerous enzymes—including those that are critical for proper neurological function. It is also a potential generator of toxic free radicals attributed to neurodegeneration via oxidative stress. The regulation of copper metabolism is critical to normal cell function and an elaborate system of chaperones and transporters has evolved that enables simultaneous utilization of and protection from it. (Menkes et al., 1967; Kaler et al., 2008; Tümer & Møller, 2010; Zlatic et al., 2015; Ojha & Prasad, 2016)

Symptoms

MD is characterized by infantile-onset cerebral and cerebellar neurodegeneration (widespread atrophy of the grey and white matter), focal and generalized seizures, connective tissue abnormalities, hypothermia, developmental delay, impaired cognitive development, growth failure, and very low serum copper, and unusual hair quality (sparse, tangled, and lustreless). (Kaler et al., 2008; Prasad et al., 2011; Zlatic et al., 2015; Peng et al., 2018)

Genetic Basis

A diverse range of mutations in *ATP7A* gene found on the X chromosome causes three distinct X-linked recessive disorders: occipital horn syndrome, spinal muscular atrophy, distal X-linked 3, and MD. The gene encodes a trans-Golgi copper-transporter (ATPase 1)—a major component of the intra-cellular copper homeostasis. ATP7A is a member of a large family of P-type ATPases that are energy-utilizing membrane proteins functioning as cation pumps. It is involved in the delivery of copper to the secreted copper enzymes and in the export of surplus copper from cells. They are called 'P-type' ATPases, as they form a phosphorylated intermediate during the transport of cations across a membrane. More than 350 different mutations affecting the *ATP7A* gene have been described which cause a profound reduction in the quantity and/or functional capacity of ATP7A molecules. Genetic defects are caused by small deletions or insertions (22% of cases), nonsense mutations (18% of cases), splice junction mutations (18% of cases), large gene deletions (17% of cases), and missense mutations. MD neurological manifestations is ascribed to five enzymes expressed in brain that require copper for their function which include mitochondrial cytochrome oxidase C, and four enzymes that acquire copper in the Golgi apparatus: PAM, DBH, LOX, and tyrosinase. (Kim et al., 2008; Banci et al., 2010; Tümer & Møller, 2010; Kaler, 2011; Moizard et al., 2011; Zlatic et al., 2015; Peng et al., 2018)

Current Therapies

MD is a multisystem disorder with no cure requiring a multidisciplinary approach to improve the quality of life and holistic care for patients. Neonatal nurses or neonatologists can play an important role in the early recognition of subtle signs and symptoms such as hypothermia, unusual hair, and dysmorphic facial features. This can facilitate early referral and diagnosis, supporting the initiation of copper supplementation as required. Interventions involve the institution of copper replacement therapy and the potential use of chaperones in the future. Failure to thrive should be assessed by a feeding team in association with dietitians, occupational therapists, physiotherapists, nurse educators, genetic counsellors, and social workers. Various small-molecule copper complexes—including copper chloride, copper gluconate, copper histidine and copper sulphate have been used for treatment of MD, with variable clinical outcomes. Copper replacement treatment seems especially relevant for paediatric patients with a family history of ATP7A-related distal motor neuropathy if such individuals are known to possess the mutant *ATP7A* allele but have not developed neurological symptoms. In the future, gene therapy that restores at least low levels of functional ATP7A is perhaps the best hope for patients with *ATP7A* mutations who are unresponsive to conventional treatment. Preliminary results of brain-directed *ATP7A* gene therapy using AAV serotype 5 (AAV5) in a mouse model of MD appear highly promising. (Kaler et al., 2008; Donsante & Kaler, 2010; Kim et al., 2010; Tümer & Møller, 2010; Ojha & Prasad, 2016; Peng et al., 2018)

18.10 NARCOLEPSY

Narcolepsy is a neurological disorder that afflicts 1 in 2000 individuals and is characterized by excessive daytime sleepiness and **cataplexy**—a sudden loss of muscle tone triggered by positive emotions. In the US, narcolepsy affects an estimated 250,000 individuals, with only 20% of them correctly diagnosed. The difficulties in narcolepsy diagnosis occur because symptoms remain undiagnosed or misdiagnosed with other conditions such as sleep apnoea, idiopathic hypersomnia, schizophrenia, depression or sleep deprivation. (Merkle et al., 2015; Black et al., 2017; Calik, 2017)

Symptoms

Narcolepsy is the disruption of the brain wake-promoting system caused by the loss of wake-promoting orexinergic neurons in the lateral hypothalamus. There is overwhelming daytime sleepiness or persistent need to sleep during the day, emotionally-triggered muscle paralysis, onset of sleep attacks possibly associated with head trauma, sleep attacks during physical activity, night-time sleeplessness, hallucinations, mental sluggishness and obesity. Cataplexy is a symptom unique to narcolepsy and refers to the sudden loss of skeletal muscle tone without the loss of consciousness, often triggered by strong emotions such as laughter, surprise, or anger (Liblau et al., 2015; Merkle et al., 2015; Black et al., 2017).

Genetic Basis

Risk factors for narcolepsy can be found in genes that encode the major histocompatibility complex proteins, also known as human leukocyte antigen (HLA) genes. Components of the HLA class II encoding HLA-DRB1-DQA1-DQB1 haplotype have been associated with several autoimmune diseases, including rheumatoid arthritis and type 1 diabetes. A strong genetic association of narcolepsy with HLA-DR2 and HLA-DQ1 in the major histocompatibility (MHC) region has been established. Studies have identified HLA DQA1*01:02 and DQB1*06:02 subtypes as the primary candidate susceptibility genes for narcolepsy in the HLA class II region and reported that complex HLA-DR and -DQ interactions contribute to the genetic predisposition to human narcolepsy. Research in canine narcolepsy attributed it to disruption of the hypocretin (orexin) receptor 2 gene. Hypocretin (orexin) neurons play a critical role in the regulation of sleep and wakefulness, and disturbances of the hypocretin system have been directly linked to narcolepsy in animals and humans. An excitatory neuropeptide hormone produced in the hypothalamus region of the brain, hypocretin promotes wakefulness, food intake, and energy expenditure. It has now been established the cause of narcolepsy in humans is the selective loss of hypocretin neurons that are normally found as a mere few thousand cluster of cells in the lateral hypothalamus. This selective loss of neurons may be due to an autoimmune disorder, with genetic and environmental factors as additional contributors. Based on the tight association of narcolepsy with a specific HLA subtype (DQB1*06:02), many studies postulate the disorder may be autoimmune in nature. supported a connection between the loss of hypocretin neuropeptides and narcolepsy with cataplexy. Studies further showed that hypocretin deficiency in humans diagnosed with narcolepsy with cataplexy was not due to mutations in hypocretin system genes but rather a secondary loss of hypocretin neurons in the dorso-lateral hypothalamus. Although the majority of patients with the sporadic form of narcolepsy are HLA DQB1*06:02 positive, there have been consistent reports of patients with defined narcolepsy-cataplexy, without the HLA DQB1*06:02 allele. GWAS in three ethnic groups, indicate a strong association between narcolepsy (mostly sporadic cases with DQB1*06:02) and polymorphisms in the TCRα (T-cell receptor alpha) locus. Additional SNP in the 3′ untranslated region of P2RY11, the purinergic receptor subtype $P2Y_{11}$ gene. Recently, variants in two additional loci, cathepsin H and tumour necrosis factor super family member 4 (TNFSF4), attained genome-wide significance. (Liu et al., 1999; Thannickal et al., 2000; Brown et al., 2001; Mignot et al., 2001; Hallmayer et al., 2009; Kornum et al., 2011; Faracoet al., 2013; Singh et al., 2013)

Current Therapies

There is currently no known cure for narcolepsy and treatment is primarily symptomatic. Drugs treat individual symptoms alone like excessive daytime sleepiness (for example, *modafinil/armodafinil, methylphenidate,* and *amphetamine*); cataplexy (use of antidepressants like such as *venlafaxine* and *clomipramine*); or both excessive daytime sleepiness and cataplexy (sodium oxybate). Unfortunately, these drugs have abuse, tolerability, and adherence issues, necessitating other drug options. Only sodium oxybate is approved by the FDA to treat both excessive daytime sleepiness and cataplexy. It is often well tolerated and improves the quality of life of narcoleptic patients. ADX-N05 is a phenylalanine derivative with dopaminergic and noradrenergic activity that has been assessed for treatment of excessive daytime sleepiness in narcolepsy. *Pitolisant* was found to increase wakefulness more than placebo and to a level indistinguishable from *modafinil*. Although *pitolisant* is a useful wake-promoting therapeutic in practice, it does not show anti-cataplectic effects—histaminergic cells remain active during cataplexy. Histaminergic neurons in the posterior hypothalamus are one of the targets of hypocretin replacement therapy (HCRT) innervation that mediates HCRT-induced wakefulness. $GABA_B$ agonists with R-baclofen used in the treatment of muscle spasticity, have recently been evaluated for narcolepsy therapy in a preclinical study using two mouse models of HCRT neuron ablation. Replacement of the HCRT that is lost from HCRT neurodegeneration is an obvious therapeutic approach, but there are number of obstacles to be overcome before being realized including exacerbation of cataplexy and sleep fragmentation found in narcoleptic *orexin/ataxin-3* mice treated with the dual HCRT antagonist *almorexant*. Another potential approach is viral vector-based delivery of the *prepro-HCRT* gene. Expression of *HCRT* using recombinant adeno-associated virus in neurons of the lateral hypothalamus effectively prevented cataplexy in narcoleptic *orexin/ataxin-3* mice. Novel approaches using pluripotent stem cells for treatment of narcolepsy are also under development. Immuno- or neuro-protective therapies are aimed at treating autoimmune attack on HCRT neurons and is being investigated. (Huang et al., 2001; Liu et al., 2011; Black et al., 2013; Dauvilliers et al., 2013; Black et al., 2014; Flygare & Parthasarathy, 2015; Liblau et al., 2015; Merkle et al., 2015; Black et al., 2017; Calik, 2017)

18.11 PARKINSON'S DISEASE (PD)

PD is increasingly recognized as a heterogeneous multisystem disorder involving other neurotransmitter systems, such as the serotonergic, noradrenergic and cholinergic circuits. A wide variety of nonmotor symptoms (NMS) linked with these neurotransmitters are commonly observed in patients with PD and as a result of this variability, subtyping of PD has been proposed, including a system based on time of onset and ongoing rate of cognitive decline. People with PD exhibit more rapid decline in a number of cognitive domains: executive, attentional, and visuospatial domains, and memory. Cognitive decline in PD is a continuous process affecting nearly all patients over time, and the demarcations between the four cognitive groups—cognitively normal, severe cognitive decline (SCD), PD with mild cognitive impairment (PD-MCI) and PD with dementia (PDD) are not strict. The following mechanisms are proposed to contribute to cognitive decline in Parkinson disease: protein misfolding (α-synuclein, amyloid and tau); neurotransmitter activity; synaptic dysfunction and loss; neuroinflammation and diabetes; mitochondrial dysfunction and retrograde signaling; microglial and astroglial changes; genetics; epigenetics; adenosine receptor activation; and cerebral network disruption (Svenningsson et al., 2012; Sauerbier et al., 2016; Ffytche et al., 2017; van Stiphout et al., 2018).

Symptoms

PD is one of the most common age-related brain disorders—defined primarily as a movement disorder, with the typical symptoms being resting tremor, rigidity, bradykinesia and postural instability, and is pathologically characterized by degeneration of nigrostriatal dopaminergic neurons and the presence of Lewy bodies (misfolded α-synuclein) in the surviving neurons. The combined National Institute of Neurological Disorders and Stroke (NINDS) and National Institute of Mental Health (NIMH) work group concluded that instead of representing distinct symptom classes, the various symptoms of PD formed a continuum progressing over the course of time (Sauerbier et al., 2016; Aarsland et al., 2017; Ffytche et al., 2017; van Stiphout et al., 2018).

Genetic Basis

Potential genetic risk factors for cognitive impairment in PD have been reported in various studies: *1) GBA* (glucosylceramidase) mutations are associated with greater cognitive impairment and risk of dementia in patients with Parkinson disease (PD). 2) *MAPT* (microtubule-associated protein tau) - there are mixed findings for cognitive impairment: the H1 haplotype is associated with an increased risk of dementia and modulation of temporal-parietal activation, as measured by functional MRI (fMRI). 3) *APOE* (apolipoprotein E) - The *APOE*$*\varepsilon$4 allele was associated with cognition in most (but not all) studies. 4) *LRRK2* (leucine-rich repeat serine/threonine-protein kinase 2) - some larger studies report a reduced prevalence of cognitive impairment and dementia in carriers of *LRRK2* mutations. 5) *SNCA* (α-synuclein) - multiplications and disease-causing mutations in this gene are associated with cogni-

Human Nervous System Disorders

tive decline and dementia in monogenic PD. 6) *COMT* (catechol *O* methyltransferase) – there is some evidence that the allele is associated with impairment on frontal-dependent tasks but no evidence for dementia risk. 7) *BDNF* (brain-derived neurotrophic factor) – some studies report an association between cognitive impairment and this factor. 8) *UBQLN1* (ubiquilin-1) – suggested but no association has been found with cognitive impairment. 9) *FMR1* (fragile X mental retardation protein 1) also suggested but no association with cognitive impairment found. 10) Immune/inflammatory genes: *IL10, IL17A, IL-18, IFNG* – reported mixed findings where *IL10* was associated with lower risk of cognitive impairment and IL17A was associated with greater risk in one study, but not in subsequent studies (Svenningsson et al., 2012; Halliday et al., 2014; Collins & Williams-Gray, 2016; Sauerbier et al., 2016; Aarsland et al., 2017; van Stiphout et al., 2018).

Current Therapies

No disease-modifying treatments with effects on cognition in PD are available, and there is no robust evidence that progression to dementia can be impeded. Epidemiological studies indicate that green tea and coffee can influence the risk of PD and, thus, a potential preventive role has been explored, although the effects on cognitive decline are unclear. *Atomoxetine*, a selective noradrenaline reuptake inhibitor, was associated with some cognitive benefit in patients with PD. Empirical evidence supports the hypothesis that *rasagiline*, a selective monoamine oxidase B inhibitor with effects on motor symptoms in PD, also has a positive cognitive effect. A rivastigmine patch offered some benefit, although the primary outcome measure (the clinical impression of change) was not significantly affected. *Vortioxetine,* a novel agent that acts on a number of serotonergic receptors in addition to inhibiting serotonin reuptake, has produced cognitive improvement in elderly patients with depression. $5-HT_6$ receptor antagonists and *donepezil* for PD-MCI might exert effects on cognition and mood owing to enhancement of cholinergic, glutamatergic, noradrenergic and dopaminergic neurotransmission. Among the non-pharmacological approaches available for patients with PD are cognitive training, physical exercise, and neuro-stimulation. Systematic cognitive training to improve cognition in PD research showed small but statistically significant improvements in working memory, processing speed and executive functioning after cognitive training. Aerobic physical exercise has a range of beneficial effects on the brain, with potential relevance to cognition. Deep brain stimulation of the subthalamic nucleus is an established practice in PD and there is some preliminary evidence of positive cognitive effects after stimulation of the cholinergic nucleus basalis of Meynert in patients with AD (Postuma et al., 2012; David et al., 2015; Kuhn et al., 2015; Leung et al., 2015; Jurado-Coronel et al., 2016; Aarsland et al., 2017; Ffytche et al., 2017; van Stiphout et al., 2018).

18.12 PHENYLKETONURIA (PKU)

PKU is a recessively inherited disease caused by mutations in the gene encoding the enzyme phenylalanine hydroxylase (PAH). It is one of the most common inherited inborn errors of metabolism that impairs postnatal cognitive development, and the first metabolic disorder in which a toxic agent, phenylalanine (Phe), causes mental retardation, and in which treatment was found to prevent clinical symptoms. The incidence of various PAH variations differs by race and ethnicity. The hepatic enzyme PAH catalyses the conversion of the essential amino acid phenylalanine tyrosine, a precursor of the neurotransmitters dopamine, noradrenaline and adrenaline. The resulting elevated levels of Phe (a condition known as

hyperphenylalaninaemia, HPA) is the primary biochemical marker of PKU. PKU was described by A. Følling in 1934 and it has a current estimated average prevalence of approximately 1 in 10,000 Caucasian births (Bélanger-Quintana et al., 2011; Hafid & Christodoulou, 2015; Liu et al., 2017; Romani et al., 2017).

Symptoms

Although children with untreated PKU appear normal at birth, by the time they are 3 to 6 months, they begin to lose interest in their surroundings. Developmentally delayed can be seen by age 1 and infant skin has less pigmentation than is normal. Without diet therapy, children with PKU develop severe intellectual and developmental disabilities. The other symptoms of the disorder include: behavioural or social problems; seizures, shaking, or jerking movements in the arms and legs; stunted or slow growth; skin rashes, like eczema; microcephaly; a musty odour in urine, breath, or skin due to excessive Phe in the body; fair skin and blue eyes (caused by the inability to metabolize phenylalanine into melanin) (NIH, 2018).

Genetic Basis

Mutations in the *PAH* gene accounts for 98% of cases of PKU mapped to chromosome 12 (region 12q23.2). There are currently over 560 known *PAH* mutations. The remaining cases are caused by mutations affecting either the synthesis or the regeneration of tetrahydrobiopterin (BH_4), a co-factor involved in the enzyme PAH metabolic roles. The genes involved in the BH_4-deficiency type HPA include guanosine triphosphate cyclohydrolase I (*GTPCH*), 6-pyruvoyl-tetrahydropeterin synthase (*PTPS*), pterin-4a-carbinolamine dehydratase (*PCD*), and quinoid dihydropteridine reductase (*DHPR*). Sepiapterin reductase (*SR*) is also a genetic factor in BH_4-deficiency disorders, but mutations in *SR* do not lead to HPA (Scriver et al., 1994; Liu et al., 2017; Romani et al., 2017).

Current Therapies

The dietary management of PKU was established many decades ago and consists of a restriction of dietary natural protein to minimize Phe intake. It requires supplementation with special medical formulas that supply sufficient essential amino acids, energy, vitamins and minerals. To avoid mental retardation in infants, the diet should be started in the first weeks of life emphasizing the critical role of neonatal screening programs in the early identification of these patients. Treatment is lifelong as HPA in adulthood has been associated with attention problems, mood instability and white matter degeneration leading to seizures and gait disturbances. In expectant mothers, moderately high Phe levels in the mother can cause microcephaly, mental retardation and congenital heart defects in the foetus. Though in use for a long time, the PKU dietary treatment has been attributed to growth retardation and specific deficits such as calcium, iron, selenium, zinc or vitamin D and B12 deficiencies, including signs of osteoporosis at an early age. In addition, the PKU diet imposes a heavy burden to the patient and their families economically and socially—hence reports of non-compliance, particularly among adolescents and young adults. There are new dietary therapies including more palatable medical formulas, large neutral amino acid supplementation, and the development of medical foods based upon glycol-macropeptide, a naturally

Human Nervous System Disorders

low-Phe protein (Bélanger-Quintana et al., 2011; Ho & Christodoulou, 2014; Hafid & Christodoulou, 2015; Liu et al., 2017; Romani et al., 2017).

Novel pharmacologic therapies to directly ameliorate the effects of a mutant enzyme like pharmacologic chaperones including the naturally occurring cofactor (5,6,7,8-tetrahydrobiopterin, BH4), help stabilize misfolded mutant enzymes and prevent **proteolisis**. BH4 has been used to treat a subset of PKU patients for over ten years under experimental conditions and in 2014, BH4 was commercialized in the form of sapropterin dihydrochloride, a synthetic form of BH4. BH4 treatment is likely to act via this method, although the exact mechanism is uncertain. During *in vitro* expression of mutant PAH protein, the presence of BH_4 leads to increased protein levels and thus catalytic activity, supporting a chaperone-like role of the BH_4, increasing the half-life of the mutant PAH protein and protecting it from targeted degradation in the ubiquitin-dependent proteolytic pathway. Although a single case, orthotopic liver transplantation successfully cured PKU in a single patient who was transplanted because of cryptogenetic cirrhosis. Therapeutic liver repopulation following hepatocyte transplant continues to be investigated in preclinical models. Clinical gene therapy trials are underway in several inborn errors of metabolism, and preclinical gene transfer experiments continue in PKU animal models. A newer therapy involving a novel enzyme substitution that utilizes subcutaneous injection of phenylalanine ammonia lyase to metabolize circulating blood Phe, is currently in clinical trial. Recombinant adeno-associated virus vectors have also been used to deliver the *PAH* gene to the liver in a murine PKU model, allowing correction of HPA of up to one year. Another novel type of therapy of genetic disorders is the use of the nonsense read-through agents, such as the aminoglycoside antibiotic gentamicin, for treating individuals with nonsense mutations. *In vitro* testing of two aminoglycosides against four *PAH* nonsense mutations have demonstrated their ability to restore PAH enzyme activity. Another promising therapy is enzyme replacement and substitution therapies. Enzyme replacement via introduction of wild-type, functional PAH protein has been hampered by the instability of the protein produced *in vitro*, rendering large-scale production and purification of the protein costly and inefficient. However, phenylalanine ammonia lyase (PAL), an enzyme normally found in plants and fungi and catalyses the deamination of phenylalanine to ammonia and *trans*-cinnamic acid, the latter of which is then quickly converted into hippurate and excreted in urine. The use of recombinant PAL (with polyethylene glycol polymers covalently linked to lysine residues, so-called PEGylation) avoids the immune-mediated degradation of the enzyme associated with the wild-type, functional PAH protein therapy (Dobbelaere et al., 2003; Ney et al., 2009; Enns et al., 2010; MacDonald et al., 2010; Bélanger-Quintana et al., 2011; Ho & Christodoulou, 2014; Hafid & Christodoulou, 2015; Liu et al., 2017; Romani et al., 2017).

18.13 REFSUM DISEASE (RD)

RD was first described by Sigvald Refsum in 1946. It is a rare, autosomal, recessively-inherited disorder of peroxisome branched-chain lipid metabolism due to a defect in the initial step in the alfa-oxidation of phytanic acid—a C-16 saturated fatty acid. RD belongs to peroxisomal biogenesis disorders resulting from a generalized peroxisomal function impairment. Phytanic acid metabolism was recognized to be vital for human health with the identification of peroxisomal disorders, such as RD. Phytanic acid accumulates in retinal pigment epithelium and other tissues and causes cellular death through calcium deregulation, free radical formation and apoptosis. RD prevalence is estimated to be about 1 in 3000 to 4000 in the population (van den Brink et al., 2003; Jayaram & Downes, 2008; Bompaire et al., 2015).

Symptoms

The adolescent onset of clinical symptoms is due to a gradual accumulation of phytanic acid, the only **pathognomic** biochemical marker of RD. Clinical symptoms usually begin in late childhood before the age of 20. Patients often experience severe clinical complications that range neurological impairment to cardiovascular anomalies. Defining characteristics include progressive adult retinitis pigmentosa, peripheral neuropathy, anosmia, and cerebellar ataxia. Other symptoms include nerve deafness, skeletal dysplasia, ichthyosis, cataracts, and cardiac arrhythmias (van den Brink et al., 2003; Jayaram & Downes, 2008).

Genetic Basis

Two causative genes have been identified: the gene *PHYH* located on chromosome 6 encodes phytanoyl-CoA hydroxylase; and located on chromosome 10 is the gene *PEX7* encoding peroxin 7, a receptor required to import several proteins into the peroxisomal matrix (Mihalik et al., 1997; van den Brink et al., 2003).

Current Therapies

The treatment of RD may consist of a special diet avoiding phytanic acid-rich food (dairy products, fish, meat, and fat of ruminant animals). **Plasmapheresis** or **lipopheresis** can be used in the event of acute arrhythmias or extreme weakness. Where dietary control has been inadequate, these treatments have been shown to help improve clinical outcomes. LDL **apheresis** (removal of low-density lipoprotein cholesterol from a patient's blood) can be used in addition to the diet when phytanic acid plasma levels are critically high (>300 μmoles/litre) (van den Brink et al., 2003; Jayaram & Downes, 2008).

18.14 SPINAL MUSCULAR ATROPHY (SMA)

SMA and amyotrophic lateral sclerosis (ALS, covered in Chapter 17) are two devastating neurological conditions with the common pathological hallmark of motor neuron degeneration, ultimately leading to muscle wasting and death. SMA is a monogenic neuromuscular disorder with a carrier frequency of 1 in 40 to 67 and affects about 1 in 8,500–12,500 new-borns—and the most common genetic cause of infant mortality. Patients present with severe muscle weakness and atrophy, predominantly in proximal (e.g., trunk) muscles, due to degeneration of lower motor neurons of the spinal cord ventral horn. SMA has a broad range of age of onset, severity, rate of progression, and variability between and within the five types (0-IV) subtypes. Type I SMA patients display the most severe symptoms, with death in infancy (median age: 13.5 months) if invasive ventilation (for up to 16 hours a day) is not implemented. Types II and III have a later childhood onset and are associated with survival into adulthood (with 93% surviving to 25 years) and the potential for a normal lifespan, albeit with considerable physical disability (Scoto et al., 2017; Tosolini & Sleigh, 2017; Verhaart et al., 2017).

Symptoms

Respiratory complications are the major cause of morbidity and mortality in SMA—the major complications being impaired cough (resulting in reduced clearance of lower airway secretions); chest wall and lung underdevelopment, and recurrent infections that exacerbate muscle weakness. Scoliosis is a common complication of SMA (especially in types I and II), and is present in 60% to 95% of patients, secondary to progressive muscle weakness (Arnold et al., 2015; Farrar et al., 2017).

Genetic Basis

Humans have two nearly identical inverted *SMN* genes on chromosome 5q13, and the most common form of SMA is caused by mutations in the survival motor neuron 1 (*SMN1*) gene, resulting in SMN protein deficiency. An almost identical survival motor neuron 2 (*SMN2*) gene produces a small amount of functional SMN protein. A single functioning *SMN1* allele produces sufficient protein for LMNs to remain healthy. When protein production from *SMN1* is impaired, as it is in SMA patients, *SMN2* can only partially compensate because the gene is aberrantly spliced 90% of the time (creating truncated, non-functional SMN2 protein) capable of producing only about 10% of the full-length SMN made by *SMN1*. It is recognized that the SMN2 copy number is a modulator of the SMA phenotype. The SMN protein found in the nucleus and cytoplasm of almost all cells, and amongst its other functions, it plays a vital housekeeping role in spliceosome assembly. In the multi-protein SMN complex, SMN directs the efficient cytoplasmic assemblage of small nuclear RNAs (snRNAs) with a group of seven proteins (known as SM ribonucleoproteins) that collectively make up the extremely stable SM core of small nuclear ribonucleoproteins (snRNPs). After nuclear-import, the snRNPs role is the catalytic removal of introns from pre-mRNA transcripts in the process of splicing (Lefebvre et al., 1995; Gruss et al., 2017; Singh et al., 2017).

Current Therapies

The pipeline of therapies for SMA encompasses four different strategies, including *SMN1* gene replacement, modulation of *SMN2* encoded full-length protein levels, neuroprotection, and targeted improvements of muscle strength and function. Another muscle target is the *myostatin-follistatin* pathway, in which *myostatin* acts as a negative regulator of muscle growth. Administration of recombinant *follistatin* to *SMN* mice resulted in significant improvement in muscle mass, gross motor function and lifespan. Appropriate nutritional management of SMA patients is also critical for improving quality of life and optimizing survival. Studies indicate that gene therapy and ASO approaches to increase SMN levels are getting into clinical trials. ASOs are short (15–25 nucleotides), synthetic, single-stranded DNA or RNA sequences that specifically bind to target pre-mRNA or mRNA sequences, impacting gene expression. An ASO gene therapy and perhaps the most advanced compound known to increase production of fully functional SMN protein is *nusinersen*, approved in the US by the FDA in 2016, and subsequently marketed in the EU in 2017. Delivered via single intrathecal injections *nusinersen* enters the cerebral spinal fluid by lumbar puncture. Small molecule, *SMN2* splice-modifying drugs, such as RG7916 and LMI070 which augment SMN and SMN-independent neuroprotection strategies are being investigated. A second gene therapy called AVXS-101 has shown significant pre-clinical potential. It delivers the

SMN1 gene using non-replicating self-complementary adeno-associated virus serotype 9 (scAAV9). One major advantage of AVXS-101 over *nusinersen* is that AAV9 can cross the blood-brain-barrier in mice, cats, and non-human primates, permitting intravenous delivery. Following demonstration of its neuroprotective properties of motor neurons in cell culture in SMA mice, *olesoxime* has entered clinical trials in SMA patients. Pharmacological compounds aimed at specifically targeting skeletal muscle are presently the only non-SMN, non-CNS drugs in clinical trials for SMA patients. Among those is the fast-skeletal troponin activator (CK-2127107), which slows calcium release from fast skeletal muscle troponin and sensitizes the sarcomere to calcium increasing contractile response to nerve signalling. Research is also focused on novel small-molecule compounds identified from high-throughput screening and medicinal chemistry optimization such as RG7800, LMI070, and RG3039. SMN independent therapeutic targets have been identified in preclinical studies including the compounds, *fasudil* and Y-27632, which regulate actin cytoskeleton integrity; the antioxidant flavonoid called *quercetin*, that suppresses beta-catenin signalling; and BAY 55-9837, that indirectly stabilizes *SMN* mRNA. Studies also find that PTEN a tumour suppressor gene is an important disease modifier in SMN7 mice, and therapies aimed at lowering PTEN expression may offer a potential therapeutic strategy for SMA (Millasseau & Zeviani, 1995; Garcia-Cao et al., 2012; Mulcahy et al., 2014; Arnold et al., 2015; Calder et al., 2016; Bowerman et al., 2017; Farrar et al., 2017; Gruss et al., 2017; Singh et al., 2017; Scoto et al., 2017; Tosolini & Sleigh, 2017).

18.15 TANGIER DISEASE (TD)

First identified in two siblings from Tangier Island off the Chesapeake Bay (Virginia, USA), TD was first reported by Fedrickson et al. in 1967 as a new disease of HDL-deficiency. Individuals with TD have severe HDL deficiency with less than 5% of normal plasma HDL levels and higher incidence of premature cardiovascular disease, extremely enlarged yellow tonsil, clouding of the cornea, and enlarged spleen and liver (Fredrickson et al., 1961; Fredrickson, 1964; Brunham et al., 2006; Negi et al., 2013; Nagappa et al., 2016).

Symptoms

TD is characterized by very low plasma levels of HDL and Apolipoprotein A-I (ApoA-I), low total cholesterol and normal or high levels of triglycerides. Classical clinical symptoms include hyperplastic orange tonsils, peripheral neuropathy (more than 50% of TD patients), **hepatosplenomegaly**, enlarged lymph nodes, ischemic heart disease or stroke, corneal clouding and premature atherosclerosis. In addition, there is an accumulation of cholesterol in macrophages and related cells (Sechi et al., 2014; Nagappa et al., 2016).

Genetic Basis

TD is a rare autosomal recessive disorder, resulting from mutations in the ATP binding cassette transporter gene, *ABCA1*, mapped to chromosome 9 (9q22-q31). The gene encodes a multiple trans-membrane domain protein (ABCA1) involved in the efflux of free cholesterol from peripheral cells to ApoA-I generating nascent high-density lipoprotein (HDL). With the ABCA1 defect, patients present with a characteristic severe deficiency or absence of HDL in the plasma, rapid catabolism of ApoAI, and an accumulation of cholesterol esters in macrophages and other reticuloendothelial cells in multiple tissues. ABCA1 is an integral membrane protein consisting of 12 transmembrane domains and two ATP binding cassette domains, playing diverse roles in transmembrane lipid and ion transport. Approximately 180 mutations have been reported in the *ABCA1* gene, and these are associated with a range of clinical, biochemical, and cellular phenotypes, with most mutations residing in the two large extracellular domains. At the cellular level, several studies have suggested a homeostatic link between ABCA1 function and NPC1/NPC2, the two proteins involved in the transport of cholesterol and other lipids from late endosomes/lysosomes to other cellular compartments, including the endoplasmic reticulum. The deficit of either NPC1 or NPC2 causes Niemann-Pick disease (covered in Chapter 13) (Bodzioch et al., 1999; Negi et al., 2013; Choi et al., 2003; Brunham et al., 2006; Cziraky et al., 2008; Sechi et al., 2014; Wang & Smith, 2014; Brunham et al., 2015; Nagappa et al., 2016).

Current Therapies

There are no drug therapies that raise HDL cholesterol levels in TD patients—thus treatment of TD patients is supportive and based on specific disease manifestations. Surgery to remove enlarged tissues like the spleen and tonsils may be necessary in some patients. TD patients need regular consultation with their primary care providers to assess cardiovascular risk and perform neurological and ophthalmological examinations. It is also recommended that families of patients with TD seek genetic counselling. Macrophage-directed gene therapy with ABC1 is a suggested therapeutic approach. Bone marrow stem cells would also most likely to ameliorate the symptoms of this disorder, but no proof-of-principle studies have yet been reported in animals. Niacin is the most effective agent for raising HDL-C levels, and pharmacoeconomic modelling suggests that niacin/statin combination therapy may promote the cost-effective achievement of optimal lipid values in several at-risk patient populations (NORD, 2018).

18.16 SPINOCEREBELLAR ATAXIA (SCA)

SCA2 has a wide geographical distribution across the world, being the second most common subtype of SCA worldwide, after only to SCA3; although in some parts of the world, it represents the most common subtype. SCA population-based surveys performed in isolated geographical regions hinder the unbiased evaluation of global prevalence for SCA2. Large populations of SCA2 families have been described in Mexico, South Africa, India, Italy, and Venezuela, and in Cuba high prevalence is attributed to a prominent founder effect on the island. SCA2 is characterized by a broad group of progressive features, including gait ataxia (97%), postural instability, cerebellar **dysarthria, dysmetria,** and **dysdiachokinesia,** and rarely hypotonia. The age of onset of the cerebellar syndrome in SCA2 is variable, although in most subjects, it appears in the second or third decade of life. Like other SCAs caused by polyQ expansions, the age of onset is closely correlated to the expanded CAG repeats in the *ATXN2* gene, which explains between 60 and 80% of its variability (Shakkottai & Fogel, 2013; Matilla-Dueñas et al., 2014; Monte et al., 2018).

Symptoms

Cerebellar cell degeneration and loss is a major neuropathological feature in spinocerebellar ataxias. In fact, by the time the patients demonstrate ataxia, the most prominent motor symptoms of SCA, brain atrophy is already detected in most cases. As disease progresses, substantial loss of the Purkinje cell layer and all four deep cerebellar nuclei is evident, and the remaining neurons are atrophied and/or misplaced (heterotopy) within the cerebellum, as shown by neuroimaging and histological analyses. Painful disabling muscle cramps are reported by 88% of the SCA2 patients and usually affect the lower limbs, followed by abdominal and trunk muscles. They also have sleep disorders including restless legs syndrome, periodic legs movements syndrome, REM sleep behaviour disorders, insomnia, and nocturnal leg cramps. Cognitive performance of SCA2 patients is characterized by early frontal-executive dysfunctions, verbal memory impairments and attentional deficits. The most frequent symptoms include depression and anxiety states (up to 22% of cases), with rare **psychosis**. Notable olfactory dysfunctions include impaired olfactory threshold, quality, identification and discrimination. The most salient feature of oculomotor disturbance in SCA2 is the slowing of horizontal saccadic eye movements, with 90% of cases showing mixed sensori-motor peripheral neuronopathy. Movement disorders beyond cerebellar ataxia are prominently observed in some forms of SCAs including resting tremor, dystonia, myoclonus, rigidity, and chorea/dyskinesia (Auburger, 2012; Shakkottai & Fogel, 2013; Matilla-Dueñas et al., 2014; Monte et al., 2018).

Genetic Basis

Spinocerebellar ataxias (SCAs) comprise a large heterogeneous group of autosomal dominant cerebellar ataxias caused by a large variety of genetic defects including repeat expansions, conventional mutations, and large rearrangements in genes. SCAs have been classified into at least 43 subtypes depending on their genetic locus. Groups of disorders may be recognized with shared molecular mechanisms of disease which include the polyglutamine ataxias, ataxias associated with ion-channel dysfunction, mutations in signal transduction molecules, and disease associated with non-coding repeats. Polyglutamine ataxias disorders include SCA1, SCA2, SCA3, SCA6, SCA7, SCA12, SCA17, and DRPLA, where expansion within a glutamine encoding CAG repeat accounts for disease. The CAG/CTG repeat in genes is often translated to a polyglutamine domain in proteins. Elongated repeat regions result in elongated polyglutamine tracts. The first protein proven to be involved in neurodegenerative disorders and contain an elongated polyglutamine domain was huntingtin, the factor responsible for Huntington's disease (covered earlier in section 18.7, this chapter). Direct ion-channel mutations or secondary ion-channel dysfunction has been implicated in the pathogenesis of SCA5, SCA6, SCA13, SCA15/16, SCA19/22, and SCA27. Mutations in signal transduction molecules are the direct cause of disease in SCA11, SCA12, SCA14 and SCA23. Non-coding repeats/ RNA toxicity is the likely mechanism of pathogenesis in SCA8, SCA10, SCA31 and SCA36.

SCA2 is caused by the abnormal expansion of *cytosine–adenine–guanine* (CAG) repeats in a coding region of the *ATXN2* gene (12q23-q24.1). This leads to the expression of abnormally long polyglutamine (polyQ) sequences in the protein. Normal alleles vary from 13-31 triplet repeats, and alleles with 22 trinucleotide repeats are the most common. Alleles carrying 28–33 repeats are considered intermediate expansions predisposing an elevated risk for ALS or Parkinson Disease. Expanded alleles present ≥32 triplet repeats with a large range of full penetrance above 35 repeats, which usually exhibit a pure CAG tract. The presence of CAA interruptions in expanded alleles appears to predispose to a phenotype with Parkinson or with motor neuron disease. Ataxin-2 protein is a ubiquitously expressed polypeptide involved in the regulation of several RNA processing pathways, endocytosis, modulation of calcium signalling pathways, as well as control of metabolism and energy balance. The polyQ-expanded ataxin-2 protein exhibits toxic properties as it loses its biological functions causing dysfunction and death of a large population of neurons in the cerebellum, brainstem, spinal cord, and brain cortex. This results in progressive cerebellar syndrome including the extra-cerebellar features which clinically characterizes the disease. Currently, multiple evidence from clinical, electrophysiological, and imaging approaches indicate that the toxic damage could start up to 15 years before the ataxic onset, with a SCA2 precursory stage characterized by subtle motor and unspecific non-motor features (Orr et al., 1993; Honti & Vécsei, 2005; Dueñas et al., 2006; Shakkottai & Paulson, 2009; Durr, 2010; Auburger, 2012; Seidel et al., 2012; Matilla-Dueñas et al., 2014; Smeets & Verbeek, 2016; Monte et al., 2018).

Current Therapies

There are currently no cures or effective treatments for genetic or idiopathic (spontaneous) ataxias and treatment is therefore wholly symptomatic. The identification of earlier molecular dysfunction in SCAs helps to direct the study of mechanisms of neurotoxicity to earlier stages of the disease. Applying effective treatments to SCAs in the window of reversibility of early neuronal damage is now the major challenge, and the key barrier to effective therapy is that SCAs present clinically when neuronal loss is well advanced and thus irreversible. Current treatments are almost all directed at modifying symptoms; few address underlying pathogenic mechanisms and are inevitably delivered too late to rescue dying neurons. The therapeutic approaches for SCA2 are limited to supportive care that partially improves some cerebellar and non-cerebellar manifestations but fail to halt the progression of the disease, and specific factors limit the successful of effective clinical trials. 1) SCA2 is still considered rare disease representing a not enough attractive disease for most pharmaceutical companies. 2) The majority of clinical trials have enrolled small and heterogeneous samples of subjects, most of them in advanced clinical stages of the disease, when the therapy is difficult by the extended neuronal degeneration. 3) There is a marked absence of predictive and progression biomarkers to assess the efficacy of therapeutics. There are currently two kinds of therapeutic interventions evaluated in SCA2 patients consisting of pharmacological and physiotherapeutic interventions. A number of clinical trials have been assessed in SCA2 patients, although future clinical trials in prodromal disease stages need to confirm the findings. These include the use of lithium carbonate, *riluzole* (a potassium channel activator), zinc sulphate, *lisuride,* and B-vitamins. Recent findings based on the reduction of ataxin-2 expression by means anti-sense oligonucleotide (ASO) therapy revealed the efficacy of this therapeutical approach in the motor performance and Purkinje cells firing rate in SCA2 mouse models representing promising options for future clinical trials in humans. Physiotherapy is a very important strategy in the rehabilitation of SCA2 patients and Cuban patients treated for six hours daily for two months showed a significant improvement in coordination, postural stability, saccade latency, and antioxidant defences (Orr et al., 1993; Dueñas et al., 2006; Durr, 2010; Auburger, 2012; Seidel et al., 2012; Shakkottai & Fogel, 2013; Matilla-Dueñas et al., 2014; Smeets & Verbeek, 2016; Monte et al., 2018).

Since many SCAs (including the polyglutamine ataxias), are caused by autosomal-dominant gain-of-function mutations; gene suppression strategies such as RNA interference and antisense oligonucleotides, are an attractive option for the treatment of gain-of-function mutations. Strategies which target disease-causing alleles have been effective in multiple SCA models. Adeno-associated viral delivery of short hairpin-RNAs (shRNAs) against the mutant human ataxin-1 gene improved Purkinje neuron morphology and motor function in a mouse model of SCA1, while both AAV-mediated overexpression of wild-type ataxin-1 and miRNA against ataxin-1 have recently been shown to be effective. In several of the SCAs, aberrant protein folding is a hallmark feature in the disease pathogenesis. In polyglutamine ataxias, protein folding is relevant since expanded glutamine repeat sequences are particularly prone to misfolding events and seem to have toxic effects in certain neuronal populations. Molecular chaperone pathways play a large role in protein homeostasis and are altered in several neurodegenerative protein misfolding diseases. These pathways include the heat shock proteins (Hsp), which are likely involved in the pathogenesis of neurodegenerative disorders. SCA14 is associated with an increase in Hsp70, which may be a protective response to reduce elevated cell stress due to the presence of mutant protein oligomers. A knockdown of Hsp70 increased protein aggregation and decreased survival in studies, suggesting that the Hsp70 pathway may be important for protein quality control in SCAs. Since neurodegeneration improves upon Hsp70 activation or overexpression in some models of protein folding disease, Hsp70 activators may be therapeutic options for the treatment of human SCAs. Gene targeting strategies may be best for the specific targeting of dominant mutations like those seen in SCAs—but gene delivery options are currently associated with significant health risks and are complicated (Fujikake et al., 2008; Williams & Paulson, 2008; Ogawa et al., 2013; Pratt et al., 2015; Bushart et al., 2016).

CHAPTER SUMMARY

The nervous system (NS) is comprised of nerve cells (neurons) which transfer and process information; and neuroglia (or glial cells) which, twenty times more numerous, provide the supportive framework neurons need to function effectively. There are two divisions of the nervous system: central (CNS) and peripheral (PNS). The CNS consists of the brain and spinal cord and forms an intricate network of specialised cells that are responsible for coordinating all bodily functions. The PNS is all neural tissue outside the CNS and delivers sensory information from peripheral sensory tissues and systems like eyes to the CNS. It also carries our motor commands from the CNS to peripheral tissues. There several diseases that directly affect the NS with a genetic component—some are due to a mutation in a single gene, while others have more complex modes of inheritance.

An estimated 46.8 million people worldwide living with dementia in 2015 expected to reach 131.5 million in 2050 with AD accounting for 60–70% of dementia cases worldwide. AD is associated with several comorbidities including: 1) Lewy body disease, a subset of diseases which includes Parkinson's disease and dementia with Lewy bodies which have abnormal accumulation of α-synuclein in neurons; 2) cerebrovascular diseases that cause vascular brain injury, including atherosclerosis, arteriolosclerosis, and cerebral amyloid angiopathy; 3) hippocampal sclerosis; 4) argyrophilic grain disease; 5) TDP-43 proteinopathy; and 6) cerebral autosomal dominant arteriopathy and leukoencephalopathy (CADASIL) as well as many other neuropathologic changes. Studies showed that cerebrovascular amyloid (caused by cerebral amyloid angiopathy) and senile plaques are composed of amyloid β protein (Aβ), the antigenic determinants shared in both AD and Down's syndrome. A *Val717Ile* missense mutation in the amyloid precursor protein (*APP*) gene on chromosome 21 was found to be causally related to the early-onset autosomal-dominant familial AD.

Human Nervous System Disorders

At least 14 genes with the highest association with late onset AD identified by positional mapping, targeted gene analysis and genome-wide association studies, although in over 99% of cases it is sporadic, late-onset with little evidence of genetic involvement. Research in twins suggest that other factors, likely epigenetic and environmental, are related to AD pathogenesis. There are numerous potential AD therapeutics targeting the serotonergic system that have been tested in preclinical and clinical trials. Partial agonists of the $5-HT_{1A}$ receptor, tandospirone and buspirone are useful in the treatment of behavioural and psychological symptoms of dementia. The $5-HT_4$ receptor agonists have been shown to increase the release of acetylcholine and decreased production and deposition of Aβ in a mouse model of AD. Currently, several different $5-HT_6$ receptor antagonists are in clinical trials as drug candidates for AD including *escitalopram, buspirone, tandospirone, reboxetine, atomoxetine, levodopa, risperidone, dextroamphetamine, haloperidol, circadin, melatonin* and nanotherapeutics. Several other drug therapies have been tested.

Epilepsy is a network disorder in which the normal physiologic connections between cortical and subcortical pathways/regions are interrupted or disturbed. At the very basic level, seizures are a result of imbalance between excitatory and inhibitory inputs to cells (that is, increased excitation or decreased inhibition). The brings about an abnormal synchronization of electrical activity in a group of active neurons. It is estimated that about 65 million people worldwide have epilepsy, with >80% living in developing nations of the world. In the US, >3 million people are reported to have this disorder. Causative mutations in many genes, including some genes coding for ion channel subunits (sodium, calcium, potassium, and manganese), and others involved in synaptic function, signal transduction or brain development have been reported. Antiepileptic drugs (AEDs) have many benefits but also many side effects, including aggression, agitation, and irritability, in some patients with epilepsy. There have been more than 15 antiepileptic drugs introduced in the last 20 years, many of which work in very unique mechanisms; yet, more than 30% of adolescent and adult patients with the common epilepsies continue to have seizures, despite receiving treatment with many of these drugs used either singly or in combination.

ET is a chronic, progressive neurologic disease with the hallmark motor feature of a tremor that occurs during voluntary movements such as writing or eating involving the hands and arms, but which may also eventually spread to involve the head (i.e., neck), voice, jaw, and other body regions. Due to pathologic heterogeneity, there is increasing support for the notion that ET may be a family of diseases whose central defining feature is kinetic tremor of the arms, and which might more appropriately be referred to as "the essential tremors". Both genetic and environmental (toxic) factors are likely contributors to disease aetiology. Studies suggest an autosomal-dominant mode of inheritance with reduced penetrance. The genetic loci harbouring ET genes have been found on chromosomes 2, 3, and 6. Many environmental toxins associated with ET are under investigation, including β-carboline alkaloids (like the dietary toxin harmane) and lead. Despite new insights into ET pathogenesis, its therapy remains purely symptomatic, and virtually all medications used for the reduction of tremor have initially been developed and approved for other pathologies. ET therapies include the following broad classes of medications: anti-convulsants, beta-adrenergic blockers, GABAergic agents, calcium channel blockers, and atypical neuroleptics.

FMF was first described by Reimann in 1948, but Sohar et al. (1967) defined the disease as FMF in 1955 and most frequently seen in Turkey, and Armenia. FMF is associated with mutations in the *Mediterranean Fever (MEFV)* gene, located on the short arm of chromosome 16 (16p13.3). *MEFV* encodes a 781-amino acid protein termed pyrin responsible for the regulation of apoptosis, inflammation and cytokines, and is mainly expressed in neutrophils, eosinophils, dendritic cells and fibroblasts. The epigenetic mechanisms such as histone modification, methylation, and microRNAs may play role in the pathogenesis of FMF. The European League Against Rheumatism (EULAR) recommendations emphasize that the treatment aim of FMF is to control acute attacks, minimize the chronic and subclinical inflammation, prevent complications, and provide an acceptable quality of life. The drug *colchicine* has been the main treatment for FMF since 1972. Other drugs used include *canakinumab, rilonacept, thalidomide, adalimumab, etanercept,* and *infliximab*.

FRDA is the most common autosomal recessive form of ataxia (loss of voluntary control of bodily movements) accounting for half of the inherited progressive ataxias and three-quarters of those with onset before age 25. The incidence of FRDA is close to 1 in 50 000 individuals, with the carrier frequency estimated to be 1 in 100 individuals. FRDA results from a mutation of the frataxin gene located on chromosome 9 and encoding a protein product which regulates mitochondrial iron metabolism. About 98% of the people with FRDA are homozygous for a GAA triplet repeat expansion in the first intron of the frataxin gene, and the other 2% have point mutations. Nursing and rehabilitative interventions are the mainstays of treatment, as there are no curative therapies. There is currently no US FDA-approved treatment for FRDA, but advances in research of its pathogenesis have led to clinical trials of potential treatments. Therapeutic developments centre around the three biochemical defects of decreased frataxin gene, mitochondrial iron accumulation, and oxidative stress and include: erythropoietin, resveratrol, a protein-based TAT-frataxin complex therapy, deferiprone, pioglitazone, EGb761, Ox1, p38 inhibitors/synthetic shRNA plus other potential agents.

HD is an autosomal dominant neurodegenerative disease characterized by motor impairment, cognitive decline, and psychiatric manifestations and behavioural abnormalities. HD affects males and females at the same frequency, and the mean age of onset is around 40 years although it can be as early as 4 and as late as 80 years of age. Epidemiologic studies show that there are about 30,000 HD patients and that there are about 150,000 people at risk of developing the disease in the US. Aging has been associated with a decline in a variety of pathways relevant to neurodegenerative disease especially those that are critical for handling misfolded proteins. HD is caused by a pathologic cytosine-adenine-guanine (CAG) repeat expansion, on the 5' end of the *Huntingtin (HTT)* gene. The ubiquitous expression of mutant huntingtin protein (mHTT), is thought to be the predominant toxic agent in HD. The normal (wild-type) huntingtin gene is ubiquitously expressed and has many interaction partners performing many functions including vesicular trafficking; the mediation of endocytosis, vesicular recycling and endosomal trafficking; coordination of cell division; transcriptional regulation; and metabolism. Since all cases of HD are caused by the same basic mutation, it is an ideal candidate for the development of therapeutics targeting pathogenic processes close to its root genetic cause. Novel therapeutic strategies target the mHTT production pathway. Therapies designed to interact with *HTT* mRNA include antisense oligonucleotides (ASOs) and RNA interference (RNAi) compounds which accelerate degradation of the mRNA transcript. The FDA has approved *nusinersen*, a lumbar intrathecally-administered ASO found to extend survival in spinal muscular atrophy via targeted modulation of gene expression. Other coded drugs tested are PRECISION-HD1, PRECISION-HD2 and *deutetrabenazine*. Non-viral vectors are also receiving increased attention in HD therapy. Small molecule approaches using a brain-penetrant, orally bioavailable small molecules that reduce *HTT* through altering mRNA splicing and processing are also under investigation. The two DNA-targeting gene therapies currently under investigation are zinc finger proteins (ZFPs) and CRISPR/Cas9.

MSUD is an autosomal recessive disorder characterized by disruption of the normal activity of the branched-chain α-ketoacid dehydrogenase (BCKAD) complex. This metabolic pathway's first step involves the conversion of branched-chain amino acids (BCAAs)—leucine, isoleucine, and valine into their relevant α-ketoacids by branch-chain aminotransferase within the mitochondria. MSUD has had a worldwide incidence of 1:185,000 live births but is much more frequent in certain founder populations including the Old Order Mennonites of Pennsylvania, where occurrence rates may be as high as 1:200 live births. Genetic studies have determined that MSUD is an autosomal recessive disease caused by pathogenic variants in genes encoding any of the 4-branched chain α-keto acid dehydrogenase (BCKAD) subunits. The BCKDC catalyses the rate-limiting step in the catabolism of the BCAA. The enzyme complex consists of three catalytic components: a decarboxylase (E1) composed of two E1α and two E1β subunits, a transacylase (E2) core of 24 identical lipoate bearing subunits and a dehydrogenase (E3) existing as a homodimer. Mutations in the genes encoding the E1α, E1β and E2 subunits result in an MSUD phenotype. One of the primary goals in treating MSUD is to manage diet by reducing BCAAs and provide adequate macronutrients to prevent catabolism and help maintain plasma BCAAs within targeted treatment ranges. Nutrition management consists of BCAA-free medical food specially formulated to provide 80%–90% of protein needs, and most of energy and micronutrient necessities throughout life. Liver transplantation and hepatocyte transplantation are becoming viable cell therapies, and so is placental-derived stem cell transplantation.

MD was discovered when an association was made between copper deficiency and a demyelinating disease of the brain in sheep's offspring grazing in copper-deficient pastures around 1937 in Australia. Then in 1962, Menkes et al. (1962), described a distinctive clinical syndrome comprising neurological degeneration in patients of English–Irish heritage who showed unusual hair quality and failure to thrive. MD has an incidence of 1/140,000 to 1/300,000. MD markedly decreases a cells' ability to absorb copper ions with an estimated reduction of copper transport of 0–17% of that exhibited under healthy conditions. This causes severe cerebral degeneration and arterial changes, resulting in death in infancy. A diverse range of mutations in *ATP7A* gene found on the X chromosome causes three distinct X-linked recessive disorders: occipital horn syndrome, spinal muscular atrophy, distal X-linked 3, and MD. The gene encodes a trans-Golgi copper-transporter (ATPase 1)—a major component of the intra-cellular copper homeostasis. ATP7A is a member of a large family of P-type ATPases that are energy-utilizing membrane proteins functioning as cation pumps. It is involved in the delivery of copper to the secreted copper enzymes and in the export of surplus copper from cells. Genetic defects are caused by small deletions or insertions (22% of cases), nonsense mutations (18% of cases), splice junction mutations (18% of cases), large gene deletions (17% of cases), and missense mutations. MD is a multisystem disorder with no cure requiring a multidisciplinary approach to improve the quality of life and holistic care for patients. Neonatal nurses or neonatologists can play an important role in the early recognition of subtle signs and symptoms such as hypothermia, unusual hair, and dysmorphic facial features. Interventions involve the institution of copper replacement therapy and the potential use of chaperones in the future. Various small-molecule copper complexes—including copper chloride, copper gluconate, copper histidine and copper sulphate have been used for treatment of MD, with variable clinical outcomes. In the future, gene therapy that restores at least low levels of functional ATP7A is perhaps the best hope for patients with *ATP7A* mutations who are unresponsive to conventional treatment.

Narcolepsy is a neurological disorder that afflicts 1 in 2000 individuals and is characterized by excessive daytime sleepiness and cataplexy—a sudden loss of muscle tone triggered by positive emotions. In the US, narcolepsy affects an estimated 250,000 individuals, with only 20% of them correctly diagnosed. The difficulties in narcolepsy diagnosis occur because symptoms remain undiagnosed or misdiagnosed with other conditions such as sleep apnoea, idiopathic hypersomnia, schizophrenia, depression or sleep deprivation. Risk factors for narcolepsy can be found in genes that encode the major histocompatibility complex proteins, also known as human leukocyte antigen (HLA) genes. Components of the HLA class II encoding HLA-DRB1-DQA1-DQB1 haplotype have been associated with several autoimmune diseases, including rheumatoid arthritis and type 1 diabetes. A strong genetic association of narcolepsy with HLA-DR2 and HLA-DQ1 in the major histocompatibility (MHC) region has been established. Studies further showed that hypocretin deficiency in humans diagnosed with narcolepsy with cataplexy was not due to mutations in hypocretin system genes but rather a secondary loss of hypocretin neurons in the dorso-lateral hypothalamus. GWAS in three ethnic groups, indicate a strong association between narcolepsy and polymorphisms in the TCRα locus. Recently, variants in two additional loci, cathepsin H and tumour necrosis factor super family member 4 (TNFSF4), attained genome-wide significance. There is currently no known cure for narcolepsy and treatment is primarily symptomatic. Drugs treat individual symptoms alone like excessive daytime sleepiness (for example, *modafinil/armodafinil, methylphenidate*, and *amphetamine*); cataplexy (use of antidepressants like such as *venlafaxine* and *clomipramine*); or both excessive daytime sleepiness and cataplexy (sodium oxybate). $GABA_B$ agonists with R-baclofen used in the treatment of muscle spasticity, have recently been evaluated for narcolepsy therapy in a preclinical study using two mouse models of HCRT neuron ablation. Replacement of the HCRT that is lost from HCRT neurodegeneration is being evaluated as a potential therapy. Another potential approach is viral vector-based delivery of the *prepro-HCRT* gene. Novel approaches using pluripotent stem cells for treatment of narcolepsy are also under development. Immuno- or neuro-protective therapies are aimed at treating autoimmune attack on HCRT neurons and is being investigated.

PD is increasingly recognized as a heterogeneous multisystem disorder involving other neurotransmitter systems, such as the serotonergic, noradrenergic and cholinergic circuits. A wide variety of nonmotor symptoms (NMS) linked with these neurotransmitters are commonly observed in patients with PD and as a result of this variability, subtyping of PD has been proposed, including a system based on time of onset and ongoing rate of cognitive decline. Several mechanisms have been proposed as contributors of cognitive decline in Parkinson disease: protein misfolding (α-synuclein, amyloid and tau); neurotransmitter activity; synaptic dysfunction and loss; neuroinflammation and diabetes; mitochondrial dysfunction and retrograde signaling; microglial and astroglial changes; genetics; epigenetics; adenosine receptor activation; and cerebral network disruption. Potential genetic risk factors for cognitive impairment in PD have been reported in various studies: *1) GBA* (glucosylceramidase) mutations; *MAPT* (microtubule-associated protein tau); *APOE* (apolipoprotein E); *LRRK2* (leucine-rich repeat serine/threonine-protein kinase 2); *SNCA* (α-synuclein); *COMT* (catechol *O* methyltransferase); *BDNF* (brain-derived neurotrophic factor); *UBQLN1* (ubiquilin-1; *FMR1* (fragile X mental retardation protein 1); and immune/inflammatory genes. No disease-modifying treatments with effects on cognition in PD are available, and there is no robust evidence that progression to dementia can be impeded. Epidemiological studies indicate that green tea and coffee can influence the risk of PD; the drugs *atomoxetine, rasagiline, vortioxetine,* and *donepezil* are also helpful in reducing symptoms. Systematic cognitive training, aerobic physical exercise and deep brain stimulation have all shown preliminary evidence of positive cognitive benefits.

PKU is a recessively inherited disease caused by mutations in the gene encoding the enzyme phenylalanine hydroxylase (PAH). It is one of the most common inherited inborn errors of metabolism that impairs postnatal cognitive development, and the first metabolic disorder in which a toxic agent, phenylalanine (Phe), causes mental retardation, and in which treatment was found to prevent clinical symptoms. Mutations in the *PAH* gene accounts for 98% of cases of PKU mapped to chromosome 12 (region 12q23.2). There are currently over 560 known *PAH* mutations. The remaining cases are caused by mutations affecting either the synthesis or the regeneration of tetrahydrobiopterin (BH_4), a co-factor involved in the enzyme PAH metabolic roles. The dietary management of PKU was established many decades ago and consists of a restriction of dietary natural protein to minimize Phe intake. It requires supplementation with special medical formulas that supply sufficient essential amino acids, energy, vitamins and minerals. There are new dietary therapies including more palatable medical formulas, large neutral amino acid supplementation, and the development of medical foods based upon glycolmacropeptide, a naturally low-Phe protein. Novel pharmacologic therapies to directly ameliorate the effects of a mutant enzyme like pharmacologic chaperones including the naturally occurring cofactor (5,6,7,8-tetrahydrobiopterin, BH4), help stabilize misfolded mutant enzymes and prevent proteolisis. Orthotopic liver transplantation, therapeutic liver repopulation following hepatocyte transplant, and clinical gene therapy trials have been used or are underway in several inborn errors of metabolism. A newer therapy involving a novel enzyme substitution that utilizes subcutaneous injection of phenylalanine ammonia lyase to metabolize circulating blood Phe, is currently in clinical trial. Recombinant adeno-associated virus, use of the nonsense read-through agents, enzyme replacement and substitution therapies have been tested and evaluated as therapy options.

RD was first described by Sigvald Refsum in 1946 as a rare, autosomal, recessively-inherited disorder of peroxisome branched-chain lipid metabolism caused by a defect in the initial step in the alfa-oxidation of phytanic acid—a C-16 saturated fatty acid. RD belongs to peroxisomal biogenesis disorders resulting from a generalized peroxisomal function impairment. It has a prevalence is estimated to be about 1 in 3000 to 4000 in the population. Two causative genes have been identified: the gene *PHYH* located on chromosome 6 encodes phytanoyl-CoA hydroxylase; and located on chromosome 10 is the gene *PEX7* encoding peroxin 7, a receptor required to import several proteins into the peroxisomal matrix. The treatment of RD may consist of a special diet avoiding phytanic acid-rich food (dairy products, fish, meat, and fat of ruminant animals). Plasmapheresis or lipopheresis can be used in the event of acute arrhythmias or extreme weakness. Where dietary control has been inadequate, these treatments have been shown to help improve clinical outcomes. LDL apheresis (removal of low-density lipoprotein cholesterol from a patient's blood) can be used in addition to the diet when phytanic acid plasma levels are critically high.

SMA is a monogenic neuromuscular disorder with a carrier frequency of 1 in 40 to 67 and affects about 1 in 8,500–12,500 new-borns—and the most common genetic cause of infant mortality. Patients present with severe muscle weakness and atrophy, predominantly in proximal (e.g., trunk) muscles, due to degeneration of lower motor neurons of the spinal cord ventral horn. SMA has a broad range of age of onset, severity, rate of progression, and variability between and within the five recognized subtypes (0-IV). Humans have two nearly identical inverted *SMN* genes on chromosome 5q13, and the most common form of SMA is caused by mutations in the survival motor neuron 1 (*SMN1*) gene, resulting in SMN protein deficiency. An almost identical survival motor neuron 2 (*SMN2*) gene produces a small amount of functional SMN protein. The pipeline of therapies for SMA encompasses four different strategies, including *SMN1* gene replacement, modulation of *SMN2* encoded full-length protein levels, neuroprotection, and targeted improvements of muscle strength and function. Another muscle target is the *myostatin-follistatin* pathway, in which *myostatin* acts as a negative regulator of muscle growth. Studies indicate that gene therapy and ASO approaches to increase SMN levels are getting into clinical trials. Research is also focused on novel small-molecule compounds identified from high-throughput screening and medicinal chemistry optimization such as RG7800, LMI070, and RG3039. SMN independent therapeutic targets have been identified in preclinical studies including the compounds, *fasudil* and Y-27632, which regulate actin cytoskeleton integrity; the antioxidant flavonoid called *quercetin*, that suppresses beta-catenin signalling; and BAY 55-9837, that indirectly stabilizes *SMN* mRNA. Studies also find that

PTEN a tumour suppressor gene is an important disease modifier in SMN7 mice, and therapies aimed at lowering PTEN expression may offer a potential therapeutic strategy for SMA.

First identified in two siblings from Tangier Island off the Chesapeake Bay (Virginia, USA), TD was first reported by Fedrickson et al. in 1967 as a new disease of HDL-deficiency. Individuals with TD have severe HDL deficiency with less than 5% of normal plasma HDL levels and higher incidence of premature cardiovascular disease, extremely enlarged yellow tonsil, clouding of the cornea, and enlarged spleen and liver. TD is a rare autosomal recessive disorder, resulting from mutations in the ATP binding cassette transporter gene, *ABCA1,* mapped to chromosome 9 (9q22-q31). The gene encodes a multiple trans-membrane domain protein (ABCA1) involved in the efflux of free cholesterol from peripheral cells to ApoA-I generating nascent high-density lipoprotein (HDL). With the ABCA1 defect, patients present with a characteristic severe deficiency or absence of HDL in the plasma, rapid catabolism of ApoAI, and an accumulation of cholesterol esters in macrophages and other reticuloendothelial cells in multiple tissues. There are no drug therapies that raise HDL cholesterol levels in TD patients—thus treatment of TD patients is supportive and based on specific disease manifestations. Surgery to remove enlarged tissues like the spleen and tonsils may be necessary in some patients. Macrophage-directed gene therapy, bone marrow stem cells and niacin/statin combination therapy may promote the cost-effective achievement of optimal lipid values in several at-risk patient populations.

SCAs comprise a large heterogeneous group of autosomal dominant cerebellar ataxias caused by a large variety of genetic defects including repeat expansions, conventional mutations, and large rearrangements in genes. SCAs have been classified into at least 43 subtypes depending on their genetic locus. Disorders have a shared molecular mechanism of disease which include the polyglutamine ataxias, ataxias associated with ion-channel dysfunction, mutations in signal transduction molecules, and disease associated with non-coding repeats.

SCA2 is the second most common disorder and one of the most severe subtypes caused by the abnormal expansion of *cytosine–adenine–guanine* (CAG) repeats in a coding region of the *ATXN2* gene. This leads to the expression of abnormally long polyglutamine (polyQ) sequences in the protein. Normal alleles vary from 13-31 triplet repeats, with expanded alleles of ≥32 triplet repeats a signature mark for SCA. There are currently no cures or effective treatments for genetic or idiopathic (spontaneous) ataxias and treatment is therefore wholly symptomatic. The identification of earlier molecular dysfunction in SCAs helps to direct the study of mechanisms of neurotoxicity to earlier stages of the disease. Applying effective treatments to SCAs in the window of reversibility of early neuronal damage is now the major challenge, and the key barrier to effective therapy is that SCAs present clinically when neuronal loss is well advanced and thus irreversible. A number of clinical trials have been assessed in SCA2 patients, although future clinical trials in prodromal disease stages need to confirm the findings. These include the use of lithium carbonate, *riluzole* (a potassium channel activator), zinc sulphate, *lisuride,* and B-vitamins. Recent findings based on the reduction of ataxin-2 expression by means anti-sense oligonucleotide (ASO) therapy revealed the efficacy of this therapeutical approach in the motor performance and Purkinje cells firing rate in SCA2 mouse models. Physiotherapy has also shown promise with a significant improvement in coordination, postural stability, saccade latency, and antioxidant defences. Gene suppression strategies such as RNA interference and antisense oligonucleotides, are also being investigated as treatment options. Use of heat shock proteins (Hsp), is receiving increased attention due to its involvement in the pathogenesis of neurodegenerative disorders. A knockdown of Hsp70 increased protein aggregation and decreased survival in studies, suggesting that the Hsp70 pathway may be important for protein quality control in SCAs.

Human Nervous System Disorders

End of Chapter Quiz

1. What is the total development time of a disease-modifying therapy for Alzheimer's Disease (AD) to be approved after preclinical development and initial characterization?
   a. Six months
   b. One year
   c. Three years
   d. Six years
   e. More than nine years
2. Which of the following statements about Alzheimer's Disease treatment is false?
   a. The FDA approved drugs effectively cure patients with AD over time
   b. Memantine may have protective activity against glutamate-induced excitotoxic neuronal death
   c. The therapeutic agents are only symptomatic treatments that temporarily alleviate memory problems
   d. Melatonin supplementation slows the progression of cognitive impairment in AD
   e. Tacrine, rivastigmine, donepezil and galantamine are cholinomimetic drugs that enhance cholinergic neurotransmission in AD patients.
3. There is an inverse relationship between the age of Friedreich's Ataxia onset and the size of the ___ repeat expansion.
   a. GAU
   b. GAG
   c. GAC
   d. GAA
   e. GGA
4. One of the symptoms of Huntington's Disease is excessive uncontrolled jerky motor movements called _____.
   a. proteopathy
   b. astrogliosis
   c. chorea
   d. zinc finger
   e. cas9
5. What is the primary biochemical marker of Phenylketonuria?
   a. Hyperphenylalaninemia
   b. Hyperphenylalaninemia
   c. Tetrahydrobiopterin deficiency
   d. Sepiapterin reductase
   e. Quinoid dihydropteridine reductase
6. Which of the following are treatments for Refsum Disease?
   a. Avoiding phytanic acid rich food
   b. Plasmapheresis
   c. Lipopheresis
   d. LDL apheresis
   e. All of the above

7. What are the respiratory complications that causes morbidity and mortality in patients with Spinal Muscular Atrophy?
   a. Impaired cough
   b. Chest wall and lung underdevelopment
   c. Reduced clearance of lower airway secretions
   d. A and C only
   e. A, B and C
8. Individuals with Tangier Disease will have which symptoms?
   a. Severe HDL deficiency
   b. Enlarged yellow tonsil
   c. Clouding of the cornea
   d. Hepatosplenomegaly
   e. All of the above
9. Spinocerebellar Ataxia type 2 is characterized by a group of progressive features that include:
   a. dysarthria
   b. dysmetria
   c. dysdiadochokinesia
   d. gait ataxia
   e. all of the above
10. Menkes Disease decreases a cells' ability to absorb _____ ions
    a. Manganese
    b. Potassium
    c. Copper
    d. Sodium
    e. Calcium

Thought Questions

1. Describe the comorbidities of Alzheimer's Disease.
2. There have been improvements in CRISPR/Cas9 in treating diseases. Describe how you could use the latest CRISPR/Cas9 method to treat Huntington's Disease.
3. Make and describe a flow chart of the human nervous system.

REFERENCES

Aarsland, D., Creese, B., Politis, M., Chaudhuri, K. R., Ffytche, D. H., Weintraub, D., & Ballard, C. (2017). Cognitive decline in Parkinson disease. *Nature Reviews. Neurology, 13*(4), 217–231. doi:10.1038/nrneurol.2017.27 PMID:28257128

Al Hafid, N., & Christodoulou, J. (2015). Phenylketonuria: a review of current and future treatments. *Translational Pediatrics, 4*(4), 304–317. doi:.issn.2224-4336.2015.10.07 doi:10.3978/j

Alzheimer's Association (AA). (2018). *Ten early signs and symptoms of Alzheimer's Disease*. Retrieved from https://www.alz.org/10-signs-symptoms-alzheimers-dementia.asp

Alzheimer's Disease International (ADI). (2018). *World Alzheimer Report 2016*. Retrieved from https://www.alz.co.uk/research/world-report-2016

American Psychiatric Association (APA). (2018). *Diagnostic and Statistical Manual for Mental Disorders 5th edition (DSM-5)*. Retrieved from https://www.psychiatry.org/psychiatrists/ practice/dsm

Aranca, T. V., Jones, T. M., Shaw, J. D., Staffetti, J. S., Ashizawa, T., Kuo, S., ... Ying, S. H. (2016). Emerging therapies in Friedreich's ataxia. *Neurodegenerative Disease Management*, *6*(1), 49–65. doi:10.2217/nmt.15.73 PMID:26782317

Arnold, W. D., Kassar, D., & Kissel, J. T. (2014). Spinal Muscular Atrophy: Diagnosis and Management in a New Therapeutic Era. *Muscle & Nerve*, *51*(2), 157–167. doi:10.1002/mus.24497 PMID:25346245

Ashley, C. N., Hoang, K. D., Lynch, D. R., Perlman, S. L., & Maria, B. L. (2012). Childhood Ataxia: Clinical Features, Pathogenesis, Key Unanswered Questions, and Future Directions. *Journal of Child Neurology*, *27*(9), 1095–1120. doi:10.1177/0883073812448840 PMID:22859693

Auburger, G. W. (2012). Spinocerebellar ataxia type 2. *Handbook of Clinical Neurology*, *103*, 423–436. doi:10.1016/B978-0-444-51892-7.00026-7 PMID:21827904

Banci, L., Bertini, I., Cantini, F., & Ciofi-Baffoni, S. (2010). Cellular copper distribution: A mechanistic systems biology approach. *Cellular and Molecular Life Sciences*, *67*(15), 2563–2589. doi:10.100700018-010-0330-x PMID:20333435

Bélanger-Quintana, A., Burlina, A., Harding, C. O., & Muntau, A. C. (2011). Up to date knowledge on different treatment strategies for phenylketonuria. *Molecular Genetics and Metabolism*, *104*(0), S19–S25. doi:10.1016/j.ymgme.2011.08.009 PMID:21967857

Bermejo-Pareja, F. (2011). Essential tremor--a neurodegenerative disorder associated with cognitive defects? *Nature Reviews. Neurology*, *7*(5), 273–282. doi:10.1038/nrneurol.2011.44 PMID:21487422

Bezard, E., Gross, C. E., & Brotchie, J. M. (2003). Presymptomatic compensation in Parkinson's disease is not dopamine-mediated. *Trends in Neurosciences*, *26*(4), 215–221. doi:10.1016/S0166-2236(03)00038-9 PMID:12689773

Birbeck, G. L. (2010). Epilepsy care in developing countries: Part I of II. *Epilepsy Currents*, *10*(4), 75–79. doi:10.1111/j.1535-7511.2010.01362.x PMID:20697498

Black, S. W., Morairty, S. R., Chen, T. M., Leung, A. K., Wisor, J. P., Yamanaka, A., & Kilduff, T. S. (2014). $GABA_B$ agonism promotes sleep and reduces cataplexy in murine narcolepsy. *The Journal of Neuroscience*, *34*(19), 6485–6494. doi:10.1523/JNEUROSCI.0080-14.2014 PMID:24806675

Black, S. W., Morairty, S. R., Fisher, S. P., Chen, T. M., Warrier, D. R., & Kilduff, T. S. (2013). Almorexant promotes sleep and exacerbates cataplexy in a murine model of narcolepsy. *Sleep*, *36*(3), 325–336. doi:10.5665leep.2442 PMID:23449602

Black, S. W., Yamanaka, A., & Kilduff, T. S. (2017). Challenges in the development of therapeutics for narcolepsy. *Progress in Neurobiology*, *152*, 89–113. doi:10.1016/j.pneurobio.2015.12.002 PMID:26721620

Blackburn, P. R., Gass, J. M., Vairo, F. P., Farnham, K. M., Atwal, H. K., Macklin, S., ... Atwal, P. S. (2017). Maple syrup urine disease: Mechanisms and management. *The Application of Clinical Genetics*, *10*, 57–66. doi:10.2147/TACG.S125962 PMID:28919799

Bodzioch, M., Orsó, E., Klucken, J., Langmann, T., Böttcher, A., Diederich, W., ... Schmitz, G. (1999). The gene encoding ATP-binding cassette transporter 1 is mutated in Tangier disease. *Nature Genetics*, *22*(4), 347–351. doi:10.1038/11914 PMID:10431237

Bompaire, F., Marcaud, V., Trionnaire, E. L., Sedel, F., & Levade, T. (2015). Refsum Disease Presenting with a Late-Onset Leukodystrophy. *JIMD Reports*, *19*, 7–10. doi:10.1007/8904_2014_355 PMID:25604618

Booty, M. G., Chae, J. J., Masters, S. L., Remmers, E. F., Barham, B., Le, J. M., ... Aksentijevich, I. (2009). Familial Mediterranean fever with a single MEFV mutation: Where is the second hit? *Arthritis and Rheumatism*, *60*(6), 1851–1861. doi:10.1002/art.24569 PMID:19479870

Bowerman, M., Becker, C. G., Yáñez-Muñoz, R. J., Ning, K., Wood, M. J. A., & Gillingwater, T. H. (2017). Therapeutic strategies for spinal muscular atrophy: SMN and beyond. Disease Models & Mechanisms, 10(8), 943–954. doi:10.1242/dmm.030148

Brigatti, K. W., Deutsch, E. C., Lynch, D. R., & Farmer, J. M. (2012). Novel Diagnostic Paradigms for Friedreich Ataxia. *Journal of Child Neurology*, *27*(9), 1146–1151. doi:10.1177/0883073812448440 PMID:22752491

Brioni, J. D., Esbenshade, T. A., Garrison, T. R., Bitner, S. R., & Cowart, M. D. (2011). Discovery of histamine H_3 antagonists for the treatment of cognitive disorders and Alzheimer's disease. *The Journal of Pharmacology and Experimental Therapeutics*, *336*(1), 38–46. doi:10.1124/jpet.110.166876 PMID:20864505

Brockmeyer, S., & D'Angiulli, A. (2016). How air pollution alters brain development: The role of neuroinflammation. *Translational Neuroscience*, *7*(1), 24–30. doi:10.1515/tnsci-2016-0005 PMID:28123818

Brodie, M. J. (2010). Antiepileptic drug therapy the story so far. *Seizure*, *19*(10), 650–655. doi:10.1016/j.seizure.2010.10.027 PMID:21075011

Brodie, M. J., Barry, S. J., Bamagous, G. A., Norrie, J. D., & Kwan, P. (2012). Patterns of treatment response in newly diagnosed epilepsy. *Neurology*, *78*(20), 1548–1554. doi:10.1212/WNL.0b013e3182563b19 PMID:22573629

Brodie, M. J., Besag, F., Ettinger, A. B., Mula, M., Gobbi, G., Comai, S., ... Steinhoff, B. J. (2016). Epilepsy, Antiepileptic Drugs, and Aggression: An Evidence-Based Review. *Pharmacological Reviews*, *68*(3), 563–602. doi:10.1124/pr.115.012021 PMID:27255267

Brodie, M. J., Elder, A. T., & Kwan, P. (2009). Epilepsy in later life. *Lancet Neurology*, *8*(11), 1019–1030. doi:10.1016/S1474-4422(09)70240-6 PMID:19800848

Brosnan, J. T., & Brosnan, M. E. (2006). Branched-chain amino acids: Enzyme and substrate regulation. *The Journal of Nutrition*, *136*(1Suppl), 207S–211S. doi:10.1093/jn/136.1.207S PMID:16365084

Brown, R. E., Sergeeva, O., Eriksson, K. S., & Haas, H. L. (2001). Orexin A excites serotonergic neurons in the dorsal raphe nucleus of the rat. *Neuropharmacology*, *40*(3), 457–459. doi:10.1016/S0028-3908(00)00178-7 PMID:11166339

Brunetti-Pierri, N., Lanpher, B., Erez, A., Ananieva, E. A., Islam, M., Marini, J. C., ... Lee, B. (2010). Phenylbutyrate therapy for maple syrup urine disease. *Human Molecular Genetics*, *20*(4), 631–640. doi:10.1093/hmg/ddq507 PMID:21098507

Brunham, L. R., Kang, M. H., Karnebeek, C. V., Sadananda, S. N., Collins, J. A., Zhang, L., ... Hayden, M. R. (2015). Clinical, Biochemical, and Molecular Characterization of Novel Mutations in *ABCA1* in Families with Tangier Disease. *JIMD Reports*, *18*, 51–62. doi:10.1007/8904_2014_348 PMID:25308558

Brunham, L. R., Singaraja, R. R., & Hayden, M. R. (2006). Variations on a gene: Rare and common variants in ABCA1 and their impact on HDL cholesterol levels and atherosclerosis. *Annual Review of Nutrition*, *26*(1), 105–129. doi:10.1146/annurev.nutr.26.061505.111214 PMID:16704350

Brusa, L., Bassi, A., Stefani, A., Pierantozzi, M., Peppe, A., Caramia, M. D., ... Stanzione, P. (2003). Pramipexole in comparison to l-dopa: A neuropsychological study. *Journal of Neural Transmission (Vienna, Austria)*, *110*(4), 373–380. doi:10.100700702-002-0811-7 PMID:12658365

Burke, W. J., Roccaforte, W. H., Wengel, S. P., Bayer, B. L., Ranno, A. E., & Willcockson, N. K. (1993). L-deprenyl in the treatment of mild dementia of the Alzheimer type: Results of a 15-month trial. *Journal of the American Geriatrics Society*, *41*(11), 1219–1125. doi:10.1111/j.1532-5415.1993.tb07306.x PMID:8227897

Bushart, D. D., Murphy, G. G., & Shakkottai, V. G. (2008). Precision medicine in spinocerebellar ataxias: Treatment based on common mechanisms of disease. *Annals of Translational Medicine*, *4*(2), 25. doi:10.3978/j.issn.2305-5839.2016.01.06 PMID:26889478

Cakir, N., Pamuk, Ö. N., Derviş, E., Imeryüz, N., Uslu, H., Benian, Ö., ... Senocak, M. (2012). The prevalences of some rheumatic diseases in western Turkey: Havsa study. *Rheumatology International*, *32*(4), 895–908. doi:10.100700296-010-1699-4 PMID:21229358

Calder, A. N., Androphy, E. J., & Hodgetts, K. J. (2016). Small Molecules in Development for the Treatment of Spinal Muscular Atrophy. *Journal of Medicinal Chemistry*, *59*(22), 10067–10083. doi:10.1021/acs.jmedchem.6b00670 PMID:27490705

Calderón-Garcidueñas, L., Vojdani, A., Blaurock-Busch, E., Busch, Y., Friedle, A., Franco-Lira, M., ... D'Angiulli, A. (2015). Air pollution and children: Neural and tight junction antibodies and combustion metals, the role of barrier breakdown and brain immunity in neurodegeneration. *Journal of Alzheimer's Disease*, *43*(3), 1039–1058. doi:10.3233/JAD-141365 PMID:25147109

Calik, M. W. (2017). Update on the treatment of narcolepsy: Clinical efficacy of pitolisant. *Nature and Science of Sleep*, *9*, 127–133. doi:10.2147/NSS.S103462 PMID:28490912

Cardinali, D. P., Vigo, D. E., Olivar, N., Vidal, M. F., & Brusco, L. I. (2014). Melatonin Therapy in Patients with Alzheimer's Disease. *Antioxidants*, *3*(2), 245–277. doi:10.3390/antiox3020245 PMID:26784870

Chaganti, S. S., McCusker, E. A., & Loy, C. T. (2017). What do we know about Late Onset Huntington's Disease? *Journal of Huntington's Disease*, *6*(2), 95–103. doi:10.3233/JHD-170247 PMID:28671137

Chalermpalanupap, T., Kinkead, B., Hu, W. T., Kummer, M. P., Hammerschmidt, T., Heneka, M. T., ... Levey, A. I. (2013). Targeting norepinephrine in mild cognitive impairment and Alzheimer's disease. *Alzheimer's Research & Therapy*, *5*(2), 21. doi:10.1186/alzrt175 PMID:23634965

Choi, H., Pack, A., Elkind, M. S., Longstreth, W. T. Jr, Ton, T. G., & Onchiri, F. (2017). Predictors of incident epilepsy in older adults: The Cardiovascular Health Study. *Neurology*, *88*(9), 870–877. doi:10.1212/WNL.0000000000003662 PMID:28130470

Choi, H. Y., Karten, B., Chan, T., Vance, J. E., Greer, W. L., Heidenreich, R. A., ... Francis, G. A. (2003). Impaired ABCA1-dependent lipid efflux and hypoalphalipoproteinemia in human Niemann-Pick type C disease. *The Journal of Biological Chemistry*, *278*(35), 32569–32577. doi:10.1074/jbc.M304553200 PMID:12813037

Clark, L. N., & Louis, E. D. (2018). Essential tremor. *Handbook of Clinical Neurology*, *147*, 229–239. doi:10.1016/B978-0-444-63233-3.00015-4 PMID:29325613

Collins, L. M., & Williams-Gray, C. H. (2016). The Genetic Basis of Cognitive Impairment and Dementia in Parkinson's Disease. *Frontiers in Psychiatry*, *7*, 89. doi:10.3389/fpsyt.2016.00089 PMID:27242557

Comai, S., Tau, M., & Gobbi, G. (2012). The psychopharmacology of aggressive behavior: a translational approach: part 1: neurobiology. *Journal of Clinical Psychopharmacology*, *32*(1), 83–94. doi:10.1097/JCP.0b013e31823f8770 PMID:22198449

Cotticelli, M. G., Xia, S., Kaur, A., Lin, D., Wang, Y., Ruff, E., ... Wilson, R. B. (2018). Identification of p38 MAPK as a novel therapeutic target for Friedreich's ataxia. *Scientific Reports*, *8*(1), 5007. doi:10.103841598-018-23168-x PMID:29568068

Cox, P. A. (2012). The psychopharmacology of aggressive behavior: a translational approach: part 2: clinical studies using atypical antipsychotics, anticonvulsants, and lithium. *Journal of Clinical Psychopharmacology*, *32*(2), 237–260. doi:10.1097/JCP.0b013e31824929d6 PMID:22367663

Cox, P. A., Davis, D. A., Mash, D. C., Metcalf, J. S., & Banack, S. A. (2016). Dietary exposure to an environmental toxin triggers neurofibrillary tangles and amyloid deposits in the brain. *Proceedings of the Royal Society B: Biological Sciences*, *283*(1823). 10.1098/rspb.2015.2397

Cummings, J., Aisen, P. S., DuBois, B., Frolich, L., Jack, C. R. Jr, Jones, R. W., ... Scheltens, P. (2016). Drug development in Alzheimer's disease: The path to 2025. *Alzheimer's Research & Therapy*, *8*(1), 39. doi:10.118613195-016-0207-9 PMID:27646601

Cziraky, M. J., Watson, K. E., & Talbert, R. L. (2008). Targeting low HDL-cholesterol to decrease residual cardiovascular risk in the managed care setting. *Journal of Managed Care Pharmacy*, *14*(8Suppl), S3–S28. PMID:19891279

Dauvilliers, Y., Bassetti, C., Lammers, G. J., Arnulf, I., Mayer, G., Rodenbeck, A., ... Schwartz, J.-C. (2013). Pitolisant versus placebo or modafinil in patients with narcolepsy: A double-blind, randomised trial. *Lancet Neurology, 12*(11), 1068–1075. doi:10.1016/S1474-4422(13)70225-4 PMID:24107292

David, F. J., Robichaud, J. A., Leurgans, S. E., Poon, C., Kohrt, W. M., Goldman, J. G., Comella, C. L., Vaillancourt, D. E., ... Corcos, D. M. (2015). Exercise improves cognition in Parkinson's disease: The PRET-PD randomized, clinical trial. *Movement Disorders, 30*(12), 1657–1663. doi:10.1002/mds.26291

Dinarello, C. A., Simon, A., & van der Meer, J. W. (2012). Treating inflammation by blocking interleukin-1 in a broad spectrum of diseases. *Nature Reviews. Drug Discovery, 11*(8), 633–652. doi:10.1038/nrd3800 PMID:22850787

Dobbelaere, D., Michaud, L., Debrabander, A., Vanderbecken, S., Gottrand, F., Turck, D., & Farriaux, J. P. (2003). Evaluation of nutritional status and pathophysiology of growth retardation in patients with phenylketonuria. *Journal of Inherited Metabolic Disease, 26*(1), 1–11. doi:10.1023/A:1024063726046 PMID:12872834

Donsante, A., & Kaler, S. G. (2010). Patterns of motor function recovery in a murine model of severe Menkes disease rescued by brain-directed AAV5 gene therapy plus copper. *Molecular Therapy, 18*(Suppl. 1), S10–S11. doi:10.1016/S1525-0016(16)37465-2

Dueñas, A. M., Goold, R., & Giunti, P. (2006). Molecular pathogenesis of spinocerebellar ataxias. *Brain, 129*(Pt 6), 1357–1370. doi:10.1093/brain/awl081 PMID:16613893

Dunlop, R. A., Cox, P. A., Banack, S. A., & Rodgers, K. J. (2013). The non-protein amino acid BMAA is misincorporated into human proteins in place of L-serine causing protein misfolding and aggregation. *PLoS One, 8*(9), e75376. doi:10.1371/journal.pone.0075376 PMID:24086518

Durr, A. (2010). Autosomal dominant cerebellar ataxias: Polyglutamine expansions and beyond. *Lancet Neurology, 9*(9), 885–894. doi:10.1016/S1474-4422(10)70183-6 PMID:20723845

Elble, R. J. (2013). What is essential tremor? *Current Neurology and Neuroscience Reports, 13*(6), 353. doi:10.100711910-013-0353-4 PMID:23591755

Enns, G. M., Koch, R., Brumm, V., Blakely, E., Suter, R., & Jurecki, E. (2010). Suboptimal outcomes in patients with PKU treated early with diet alone: revisiting the evidence. *Molecular Genetics and Metabolism, 101*(2-3), 99–109. doi:.ymgme.2010.05.017 doi:10.1016/j

Espay, A. J., Lang, A. E., Erro, R., Merola, A., Fasano, A., Berardelli, A., & Bhatia, K. P. (2017). Essential pitfalls in "essential" tremor. *Movement Disorders, 32*(3), 325–331. doi:10.1002/mds.26919 PMID:28116753

Faraco, J., Lin, L., Kornum, B. R., Kenny, E. E., Trynka, G., Einen, M., ... Mignot, E. (2013). ImmunoChip study implicates antigen presentation to T cells in narcolepsy. *PLOS Genetics, 9*(2), e1003270. doi:10.1371/journal.pgen.1003270 PMID:23459209

Farrar, M. A., Park, S. B., Vucic, S., Carey, K., Turner, B. J., Gillingwater, T. H., ... Kiernan, M. C. (2016). Emerging therapies and challenges in spinal muscular atrophy. *Annals of Neurology, 81*(3), 355–368. doi:10.1002/ana.24864 PMID:28026041

Ffytche, D. H., Creese, B., Politis, M., Chadhuri, K. R., Weintraub, D., Ballard, C., & Aarsland, D. (2017). The psychosis spectrum in Parkinson disease. *Nature Reviews. Neurology, 13*(2), 81–95. doi:10.1038/nrneurol.2016.200 PMID:28106066

Finkbeiner, S. (2011). Huntington's Disease. *Cold Spring Harbor Perspectives in Biology, 3*(6), a007476. doi:10.1101/cshperspect.a007476 PMID:21441583

Flygare, J., & Parthasarathy, S. (2015). Narcolepsy: Let the Patient's Voice Awaken Us! *The American Journal of Medicine, 128*(1), 10–13. doi:10.1016/j.amjmed.2014.05.037 PMID:24931392

Fogel, B. L., & Perlman, S. (2007). Clinical features and molecular genetics of autosomal recessive cerebellar ataxias. *Lancet Neurology, 6*(3), 245–257. doi:10.1016/S1474-4422(07)70054-6 PMID:17303531

Fredrickson, D. S. (1964). The Inheritance of High Density Lipoprotein Deficiency (Tangier Disease). *The Journal of Clinical Investigation, 43*(2), 228–236. doi:10.1172/JCI104907 PMID:14162531

Fredrickson, D. S., Altrocchi, P. H., Avioli, L. V., Goodman, D. S., & Goodman, H. C. (1961). Tangier disease: Combined clinical staff conference at the National Institutes of Health. *Annals of Internal Medicine, 55*(6), 1016–1031. doi:10.7326/0003-4819-55-6-1016

Freedman, M., Rewilak, D., Xerri, T., Cohen, S., Gordon, A. S., Shandling, M., & Logan, A. G. (1998). L-deprenyl in Alzheimer's disease: Cognitive and behavioral effects. *Neurology, 50*(3), 660–608. doi:10.1212/WNL.50.3.660 PMID:9521253

Fujikake, N., Nagai, Y., Popiel, H. A., Okamoto, Y., Yamaguchi, M., & Toda, T. (2008). Heat shock transcription factor 1-activating compounds suppress polyglutamine-induced neurodegeneration through induction of multiple molecular chaperones. *The Journal of Biological Chemistry, 283*(38), 26188–26197. doi:10.1074/jbc.M710521200 PMID:18632670

Garcia-Cao, I., Song, M. S., Hobbs, R. M., Laurent, G., Giorgi, C., de Boer, V. C. J., ... Pandolfi, P. P. (2012). Systemic elevation of PTEN induces a tumor suppressive metabolic state. *Cell, 149*(1), 49–62. doi:10.1016/j.cell.2012.02.030 PMID:22401813

Gibbs, M. E., Maksel, D., Gibbs, Z., Hou, X., Summers, R. J., & Small, D. H. (2010). Memory loss caused by Aβ protein is rescued by a $β_3$-adrenoceptor agonist. *Neurobiology of Aging, 31*(4), 614–624. doi:10.1016/j.neurobiolaging.2008.05.018 PMID:18632189

Gitchel, G. T., Wetzel, P. A., & Baron, M. S. (2013). Slowed saccades and increased square wave jerks in essential tremor. *Tremor and Other Hyperkinetic Movements (New York, N.Y.), 3*. doi:10.7916/D8251GXN PMID:24116343

Goate, A., Chartier-Harlin, M. C., Mullan, M., Brown, J., Crawford, F., Fidani, L., ... James, L. (1991). Segregation of a missense mutation in the amyloid precursor protein gene with familial Alzheimer's disease. *Nature, 349*(6311), 704–706. doi:10.1038/349704a0 PMID:1671712

Grünert, S. C., Rosenbaum-Fabian, S., Schumann, A., Schwab, K. O., Mingirulli, N., & Spiekerkoetter, U. (2018). Successful pregnancy in maple syrup urine disease: A case report and review of the literature. *Nutrition Journal, 17*(1), 51. doi:10.118612937-018-0357-7 PMID:29753318

Gruss, O. J., Meduri, R., Schilling, M., & Fischer, U. (2017). UsnRNP biogenesis: Mechanisms and regulation. *Chromosoma*, *126*(5), 577–593. doi:10.100700412-017-0637-6 PMID:28766049

Gulcher, J. R., Jónsson, P., Kong, A., Kristjánsson, K., Frigge, M. L., Kárason, A., ... Stefánsson, K. (1997). Mapping of a familial essential tremor gene, FET1, to chromosome 3q13. *Nature Genetics*, *17*(1), 84–87. doi:10.1038/ng0997-84 PMID:9288103

Haig, G. M., Bain, E., Robieson, W., Othman, A. A., Baker, J., & Lenz, R. A. (2014). A randomized trial of the efficacy and safety of the H3 antagonist ABT-288 in cognitive impairment associated with schizophrenia. *Schizophrenia Bulletin*, *40*(6), 1433–1442. doi:10.1093chbulbt240 PMID:24516190

Halliday, G. M., Leverenz, J. B., Schneider, J. S., & Adler, C. H. (2014). The neurobiological basis of cognitive impairment in Parkinson's disease. *Movement Disorders: Official Journal of the Movement Disorder Society*, *29*(5), 634–650. doi:10.1002/mds.25857 PMID:24757112

Hallmayer, J., Faraco, J., Lin, L., Hesselson, S., Winkelmann, J., Kawashima, M., ... Mignot, E. (2009). Narcolepsy is strongly associated with the T-cell receptor alpha locus. *Nature Genetics*, *41*(6), 708–711. doi:10.1038/ng.372 PMID:19412176

Hammerschmidt, T., Kummer, M. P., Terwel, D., Martinez, A., Gorji, A., Pape, H. C., ... Heneka, M. T. (2012). Selective loss of noradrenaline exacerbates early cognitive dysfunction and synaptic deficits in APP/PS1 mice. *Biological Psychiatry*, *73*(5), 454–463. doi:10.1016/j.biopsych.2012.06.013 PMID:22883210

Hampel, H., Toschi, N., Babiloni, C., Filippo, B., Black, K. L., & Bokde, A. L. W. (2018, June 12). ... for the Alzheimer Precision Medicine Initiative (APMI). Revolution of Alzheimer Precision Neurology: Passageway of Systems Biology and Neurophysiology. *Journal of Alzheimer's Disease*, *64*(s1Suppl 1), S47–S105. doi:10.3233/JAD-179932 PMID:29562524

Hart, P. E., Lodi, R., Rajagopalan, B., Bradley, J. L., Crilley, J. G., Turner, C., ... Cooper, J. M. (2005). Antioxidant treatment of patients with Friedreich ataxia: Four-year follow-up. *Archives of Neurology*, *62*(4), 621–626. doi:10.1001/archneur.62.4.621 PMID:15824263

Hashkes, P. J., Spalding, S. J., Giannini, E. H., Huang, B., Johnson, A., Park, G., ... Lovell, D. J. (2012). Rilonacept for colchicine-resistant or -intolerant familial Mediterranean fever: A randomized trial. *Annals of Internal Medicine*, *157*(8), 533–541. doi:10.7326/0003-4819-157-8-201210160-00003 PMID:23070486

Hedera, P., Cibulčík, F., & Davis, T. L. (2013). Pharmacotherapy of essential tremor. *Journal of Central Nervous System Disease*, *5*, 43–55. doi:10.4137/JCNSD.S6561 PMID:24385718

Helmchen, C., Hagenow, A., Miesner, J., Sprenger, A., Rambold, H., Wenzelburger, R., ... Deuschl, G. (2003). Eye movement abnormalities in essential tremor may indicate cerebellar dysfunction. *Brain*, *126*(Pt 6), 1319–1332. doi:10.1093/brain/awg132 PMID:12764054

Heneka, M. T., Nadrigny, F., Regen, T., Martinez-Hernandez, A., Dumitrescu-Ozimek, L., & Terwel, D., ... Kummer, M. P. (2010). Locus ceruleus controls Alzheimer's disease pathology by modulating microglial functions through norepinephrine. *Proceedings of the National Academy of Sciences of the United States of America*, *107*(13), 6058–6063. 10.1073/pnas.0909586107

Herrmann, N., Rothenburg, L. S., Black, S. E., Ryan, M., Liu, B. A., Busto, U. E., & Lanctôt, K. L. (2008). Methylphenidate for the treatment of apathy in Alzheimer disease: Prediction of response using dextroamphetamine challenge. *Journal of Clinical Psychopharmacology, 28*(3), 296–301. doi:10.1097/JCP.0b013e318172b479 PMID:18480686

Higgins, J. J., Pho, L. T., & Nee, L. E. (2004). A gene (ETM) for essential tremor maps to chromosome 2p22-p25. *Movement Disorders, 12*(6), 859–864. doi:10.1002/mds.870120605 PMID:9399207

Himeno, E., Ohyagi, Y., Ma, L., Nakamura, N., Miyoshi, K., Sakae, N., ... Kira, J. (2011). Apomorphine treatment in Alzheimer mice promoting amyloid-b degradation. *Annals of Neurology, 69*, 248–256. doi:10.1002/ana.22319 PMID:21387370

Ho, G., & Christodoulou, J. (2014). Phenylketonuria: Translating research into novel therapies. *Translational Pediatrics, 3*(2), 49–62. doi:10.3978/j.issn.2224-4336.2014.01.01 PMID:26835324

Honti, V., & Vécsei, L. (2005). Genetic and molecular aspects of spinocerebellar ataxias. *Neuropsychiatric Disease and Treatment, 1*(2), 125–133. doi:10.2147/nedt.1.2.125.61044 PMID:18568057

Huang, Z. L., Qu, W. M., Li, W. D., Mochizuki, T., Eguchi, N., & Watanabe, T., ... Hayaishi, O. (2001). Arousal effect of orexin A depends on activation of the histaminergic system. *Proceedings of the National Academy of Sciences of the United States of America, 98*(17), 9965–9970. 10.1073/pnas.181330998

Iacono, D., Volkman, I., Nennesmo, I., Pedersen, N. L., Fratiglioni, L., Johansson, B., ... Gatz, M. (2014). Neuropathologic assessment of dementia markers in identical and fraternal twins. *Brain Pathology (Zurich, Switzerland), 24*(4), 317–333. doi:10.1111/bpa.12127 PMID:24450926

International League Against Epilepsy Consortium on Complex Epilepsies. (2014). Genetic determinants of common epilepsies: a meta-analysis of genome-wide association studies. *The Lancet. Neurology, 13*(9), 893–903. doi:10.1016/S1474-4422(14)70171-1

Jain, S., Lo, S. E., & Louis, E. D. (2006). Common misdiagnosis of a common neurological disorder: How are we misdiagnosing essential tremor? *Archives of Neurology, 63*(8), 1100–1104. doi:10.1001/archneur.63.8.1100 PMID:16908735

Jayaram, H., & Downes, S. M. (2008). Midlife diagnosis of Refsum Disease in siblings with Retinitis Pigmentosa – the footprint is the clue: A case report. *Journal of Medical Case Reports, 2*(1), 80. doi:10.1186/1752-1947-2-80 PMID:18336720

Jia, Q., Deng, Y., & Qing, H. (2014). Potential therapeutic strategies for Alzheimer's disease targeting or beyond β-amyloid: Insights from clinical trials. *BioMed Research International, 837157*. doi:10.1155/2014/837157 PMID:25136630

Jun, G., Ibrahim-Verbaas, C. A., Vronskaya, M., Lambert, J.-C., Chung, J., Naj, A. C., ... Farrer, L. A. (2016). A Novel Alzheimer disease locus located near the gene encoding tau protein. *Molecular Psychiatry, 21*(1), 108–117. doi:10.1038/mp.2015.23 PMID:25778476

Kaler, S. G. (2011). ATP7A-related copper transport diseases-emerging concepts and future trends. *Nature Reviews. Neurology, 7*(1), 15–29. doi:10.1038/nrneurol.2010.180 PMID:21221114

Kaler, S. G., Holmes, C. S., Goldstein, D. S., Tang, J., Godwin, S. C., Donsante, A., ... Patronas, N. (2008). Neonatal diagnosis and treatment of Menkes disease. *The New England Journal of Medicine*, *358*(6), 605–614. doi:10.1056/NEJMoa070613 PMID:18256395

Kalinin, S., Polak, P. E., Lin, S. X., Sakharkar, A. J., Pandey, S. C., & Feinstein, D. L. (2011). The noradrenaline precursor L-DOPS reduces pathology in a mouse model of Alzheimer's disease. *Neurobiology of Aging*, *33*(8), 1651–1663. doi:10.1016/j.neurobiolaging.2011.04.012 PMID:21705113

Kathait, A. S., Puac, P., & Castillo, M. (2018). Imaging Findings in Maple Syrup Urine Disease: A Case Report. *Journal of Pediatric Neurosciences*, *13*(1), 103–105. doi:10.4103/JPN.JPN_38_17 PMID:29899783

Katrin, B. (2017). Friedreich Ataxia: Current status and future prospects. *Cerebellum & Ataxias*, *4*(1), 4. doi:10.118640673-017-0062-x PMID:28405347

Keum, J. W., Shin, A., Gillis, T., Mysore, J. S., Elneel, K. A., Lucente, D., ... Lee, J. (2016). The *HTT* CAG-Expansion Mutation Determines Age at Death but Not Disease Duration in Huntington Disease. *American Journal of Human Genetics*, *98*(2), 287–298. doi:10.1016/j.ajhg.2015.12.018 PMID:26849111

Khachaturian, A. S., Zandi, P. P., Lyketsos, C. G., Hayden, K. M., Skoog, I., Norton, M. C., ... Breitner, J. C. (2006). Antihypertensive medication use and incident Alzheimer disease: The Cache County Study. *Archives of Neurology*, *63*(5), 686–692. doi:10.1001/archneur.63.5.noc60013 PMID:16533956

Kim, B. E., Nevitt, T., & Thiele, D. J. (2008). Mechanisms for copper acquisition, distribution and regulation. *Nature Chemical Biology*, *4*(3), 176–185. doi:10.1038/nchembio.72 PMID:18277979

Kim, B. E., Turski, M. L., Nose, Y., Casad, M., Rockman, H. A., & Thiele, D. J. (2010). Cardiac copper deficiency activates a systemic signaling mechanism that communicates with the copper acquisition and storage organs. *Cell Metabolism*, *11*(5), 353–363. doi:10.1016/j.cmet.2010.04.003 PMID:20444417

Kirectepe, A. K., Kasapcopur, O., Arisoy, N., Celikyapi, E. G., Hatemi, G., Ozdogan, H., & Tahir Turanli, E. (2011). Analysis of MEFV exon methylation and expression patterns in familial Mediterranean fever. *BMC Medical Genetics*, *12*(1), 105. doi:10.1186/1471-2350-12-105 PMID:21819621

Klug, A. (2010). The discovery of zinc fingers and their applications in gene regulation and genome manipulation. *Annual Review of Biochemistry*, *79*(1), 213–231. doi:10.1146/annurev-biochem-010909-095056 PMID:20192761

Koch, G., Di Lorenzo, F., Bonnì, S., Giacobbe, V., Bozzali, M., Caltagirone, C., & Martorana, A. (2014). Dopaminergic modulation of cortical plasticity in Alzheimer's disease patients. *Neuropsychopharmacology: Official Publication of the American College of Neuropsychopharmacology*, *39*(11), 2654–2661. doi:10.1038/npp.2014.119 PMID:24859851

Kornum, B. R., Kawashima, M., Faraco, J., Lin, L., Rico, T. J., Hesselson, S., ... Mignot, E. (2010). Common variants in P2RY11 are associated with narcolepsy. *Nature Genetics*, *43*(1), 66–71. doi:10.1038/ng.734 PMID:21170044

Kuhn, J., Hardenacke, K., Lenartz, D., Gruendler, T., Ullsperger, M., Bartsch, C., ... Sturm, V. (2015). Deep brain stimulation of the nucleus basalis of Meynert in Alzheimer's dementia. *Molecular Psychiatry*, *20*(3), 353–360. doi:10.1038/mp.2014.32 PMID:24798585

Langbehn, D. R., Brinkman, R. R., Falush, D., Paulsen, J. S., & Hayden, M. R. (2004). A new model for prediction of the age of onset and penetrance for Huntington's disease based on CAG length. *Clinical Genetics, 65*(4), 267–277. doi:10.1111/j.1399-0004.2004.00241.x PMID:15025718

Lefebvre, S., Bürglen, L., Reboullet, S., Clermont, O., Burlet, P., Viollet, L., ... Zeviani, M. (1995). Identification and characterization of a spinal muscular atrophy-determining gene. *Cell, 80*(1), 155–165. doi:10.1016/0092-8674(95)90460-3 PMID:7813012

Lerche, H., Shah, M., Beck, H., Noebels, J., Johnston, D., & Vincent, A. (2013). Ion channels in genetic and acquired forms of epilepsy. *The Journal of Physiology, 591*(4), 753–764. doi:10.1113/jphysiol.2012.240606 PMID:23090947

Leung, I. H., Walton, C. C., Hallock, H., Lewis, S. J., Valenzuela, M., & Lampit, A. (2015). Cognitive training in Parkinson disease: A systematic review and meta-analysis. *Neurology, 85*(21), 1843–1851. doi:10.1212/WNL.0000000000002145 PMID:26519540

Li, M., Wang, Y., Zheng, X. B., Ikeda, M., Iwata, N., Luo, X. J., ... Su, B. (2012). Meta-analysis and brain imaging data support the involvement of VRK2 (rs2312147) in schizophrenia susceptibility. *Schizophrenia Research, 142*(1-3), 200–205. doi:10.1016/j.schres.2012.10.008 PMID:23102693

Liblau, R. S., Vassalli, A., Seifinejad, A., & Tafti, M. (2015). Hypocretin (orexin) biology and the pathophysiology of narcolepsy with cataplexy. *Lancet Neurology, 14*(3), 318–328. doi:10.1016/S1474-4422(14)70218-2 PMID:25728441

Lin, L., Faraco, J., Li, R., Kadotani, H., Rogers, W., Lin, X., ... Mignot, E. (1999). The sleep disorder canine narcolepsy is caused by a mutation in the hypocretin (orexin) receptor 2 gene. *Cell, 98*(3), 365–376. doi:10.1016/S0092-8674(00)81965-0 PMID:10458611

Liu, M., Blanco-Centurion, C., Konadhode, R., Begum, S., Pelluru, D., Gerashchenko, D., ... Shiromani, P. J. (2011). Orexin gene transfer into zona incerta neurons suppresses muscle paralysis in narcoleptic mice. *The Journal of Neuroscience, 31*(16), 6028–6040. doi:10.1523/JNEUROSCI.6069-10.2011 PMID:21508228

Liu, N., Huang, Q., Li, Q., Zhao, D., Li, X., Cui, L., ... Kong, X. (2017). Spectrum of *PAH* gene variants among a population of Han Chinese patients with phenylketonuria from northern China. *BMC Medical Genetics, 18*(1), 108. doi:10.118612881-017-0467-7 PMID:28982351

Lopez-Castejon, G., & Brough, D. (2011). Understanding the mechanism of IL-1β secretion. *Cytokine & Growth Factor Reviews, 22*(4), 189–195. doi:10.1016/j.cytogfr.2011.10.001 PMID:22019906

Löscher, W., Klitgaard, H., Twyman, R. E., & Schmidt, D. (2013). New avenues for anti-epileptic drug discovery and development. *Nature Reviews. Drug Discovery, 12*(10), 757–776. doi:10.1038/nrd4126 PMID:24052047

Louis, E. D. (2008). Environmental epidemiology of essential tremor. *Neuroepidemiology, 31*(3), 139–149. doi:10.1159/000151523 PMID:18716411

Louis, E. D. (2012). The primary type of tremor in essential tremor is kinetic rather than postural: Cross-sectional observation of tremor phenomenology in 369 cases. *European Journal of Neurology, 20*(4), 725–727. doi:10.1111/j.1468-1331.2012.03855.x PMID:22925197

Louis, E. D. (2014). Essential tremor' or 'the essential tremors': Is this one disease or a family of diseases? *Neuroepidemiology, 42*(2), 81–89. doi:10.1159/000356351 PMID:24335621

Louis, E. D., & Dogu, O. (2007). Does age of onset in essential tremor have a bimodal distribution? Data from a tertiary referral setting and a population-based study. *Neuroepidemiology, 29*(3-4), 208–212. doi:10.1159/000111584 PMID:18043006

Louis, E. D., Gerbin, M., & Galecki, M. (2013). Essential tremor 10, 20, 30, 40: Clinical snapshots of the disease by decade of duration. *European Journal of Neurology, 20*(6), 949–954. doi:10.1111/ene.12123 PMID:23521518

Lynch, C. J., & Adams, S. H. (2014). Branched-chain amino acids in metabolic signalling and insulin resistance. *Nature Reviews. Endocrinology, 10*(12), 723–736. doi:10.1038/nrendo.2014.171 PMID:25287287

Lynch, D. R., Perlman, S. L., & Meier, T. (2010). A phase-3, double-blind, placebo-controlled trial of idebenone in Friedreich ataxia. *Archives of Neurology, 67*(8), 941–947. doi:10.1001/archneurol.2010.168 PMID:20697044

MacDonald, A., Gokmen-Ozel, H., van Rijn, M., & Burgard, P. (2010). The reality of dietary compliance in the management of phenylketonuria. *Journal of Inherited Metabolic Disease, 33*(6), 665–670. doi:10.100710545-010-9073-y PMID:20373144

Machnes, Z. M., Huang, T. C., Chang, P. K., Gill, R., Reist, N., Dezsi, G., ... Szyf, M. (2013). DNA methylation mediates persistent epileptiform activity in vitro and in vivo. *PLoS One, 8*(10), e76299. doi:10.1371/journal.pone.0076299 PMID:24098468

Marek-Yagel, D., Berkun, Y., Padeh, S., Abu, A., Reznik-Wolf, H., Livneh, A., ... Pras, E. (2009). Clinical disease among patients heterozygous for familial Mediterranean fever. *Arthritis and Rheumatism, 60*(6), 1862–1866. doi:10.1002/art.24570 PMID:19479871

Margolis, R. L., & Ross, C. A. (2003). Diagnosis of Huntington disease. *Clinical Chemistry, 49*(10), 1726–1732. doi:10.1373/49.10.1726 PMID:14500613

Martorana, A., Di Lorenzo, F., Esposito, Z., Lo Giudice, T., Bernardi, G., Caltagirone, C., & Koch, G. (2013). Dopamine D_2-agonist rotigotine effects on cortical excitability and central cholinergic transmission in Alzheimer's disease patients. *Neuropharmacology, 64*, 108–113. doi:10.1016/j.neuropharm.2012.07.015 PMID:22863599

Matilla-Dueñas, A., Ashizawa, T., Brice, A., Magri, S., McFarland, K. N., Pandolfo, M., ... Sanchez, I. (2014). Consensus Paper: Pathological Mechanisms Underlying Neurodegeneration in Spinocerebellar Ataxias. *Cerebellum (London, England), 13*(2), 269–302. doi:10.100712311-013-0539-y PMID:24307138

Mayo Clinic. (2018). *Epilepsy*. Retrieved from https://www.mayoclinic.org/diseases-conditions/epilepsy/symptoms-causes/syc-20350093

Menkes, J. H., Alter, M., Steigleder, G. K., Weakley, D. R., & Sung, J. H. (1962). A sex-linked recessive disorder with retardation of growth, peculiar hair, and focal cerebral and cerebellar degeneration. *Pediatrics, 29*, 764–779. PMID:14472668

Merkle, F. T., Maroof, A., Wataya, T., Sasai, Y., Studer, L., Eggan, K., & Schier, A. F. (2015). Generation of neuropeptidergic hypothalamic neurons from human pluripotent stem cells. *Development (Cambridge, England), 142*(4), 633–643. doi:10.1242/dev.117978 PMID:25670790

Merner, N. D., Girard, S. L., Catoire, H., Bourassa, C. V., Belzil, V. V., Rivière, J. B., ... Rouleau, G. A. (2012). Exome sequencing identifies FUS mutations as a cause of essential tremor. *American Journal of Human Genetics, 91*(2), 313–319. doi:10.1016/j.ajhg.2012.07.002 PMID:22863194

Mignot, E., Lin, L., Rogers, W., Honda, Y., Qiu, X., Lin, X., ... Risch, N. (2001). Complex HLA-DR and -DQ interactions confer risk of narcolepsy-cataplexy in three ethnic groups. *American Journal of Human Genetics, 68*(3), 686–699. doi:10.1086/318799 PMID:11179016

Mihalik, S. J., Morrell, J. C., Kim, D., Sacksteder, K. A., Watkins, P. A., & Gould, S. J. (1997). Identification of PAHX, a Refsum disease gene. *Nature Genetics, 17*(2), 185–189. doi:10.1038/ng1097-185 PMID:9326939

Mitchell, R. A., Herrmann, N., & Lanctôt, K. L. (2011). The role of dopamine in symptoms and treatment of apathy in Alzheimer's disease. *CNS Neuroscience & Therapeutics, 17*(5), 411–427. doi:10.1111/j.1755-5949.2010.00161.x PMID:20560994

Moizard, M. P., Ronce, N., Blesson, S., Bieth, E., Burglen, L., Mignot, C., ... Raynaud, M. (2011). Twenty-five novel mutations including duplications in the ATP7A gene. *Clinical Genetics, 79*(3), 243–253. doi:10.1111/j.1399-0004.2010.01461.x PMID:21208200

Monte, T. L., Reckziegel, E. D. R., Augustin, M. C., Locks-Coelho, L. D., Santos, A. S. P., Furtado, G. V., ... Jardim, L. B. (2018). The progression rate of spinocerebellar ataxia type 2 changes with stage of disease. *Orphanet Journal of Rare Diseases, 13*(1), 20. doi:10.118613023-017-0725-y PMID:29370806

Moshé, S. L., Perucca, E., Ryvlin, P., & Tomson, T. (2015). Epilepsy: New advances. *Lancet, 385*(9971), 884–898. doi:10.1016/S0140-6736(14)60456-6 PMID:25260236

Mulcahy, P. J., Iremonger, K., Karyka, E., Herranz-Martín, S., Shum, K. T., Tam, J. K., & Azzouz, M. (2014). Gene therapy: A promising approach to treating spinal muscular atrophy. *Human Gene Therapy, 25*(7), 575–586. doi:10.1089/hum.2013.186 PMID:24845847

Myers, L., Farmer, J. M., Wilson, R. B., Friedman, L., Tsou, A., Perlman, S. L., ... Lynch, D. R. (2008). Antioxidant use in Friedreich ataxia. *Journal of the Neurological Sciences, 267*(1-2), 174–176. doi:10.1016/j.jns.2007.10.008 PMID:17988688

Nagappa, M., Taly, A. B., Mahadevan, A., Pooja, M., Bindu, P. S., Chickabasaviah, Y. T., ... Sinha, S. (2016). Tangier's disease: An uncommon cause of facial weakness and non-length dependent demyelinating neuropathy. *Annals of Indian Academy of Neurology, 19*(1), 137–139. doi:10.4103/0972-2327.175436 PMID:27011649

Napierala, J. S., Li, Y., Lu, Y., Lin, K., Hauser, L. A., Lynch, D. R., & Napierala, M. (2017). Comprehensive analysis of gene expression patterns in Friedreich's ataxia fibroblasts by RNA sequencing reveals altered levels of protein synthesis factors and solute carriers. *Disease Models & Mechanisms*, *10*(11), 1353–1369. doi:10.1242/dmm.030536 PMID:29125828

National Institutes of Health (NIH). (2018). *Phenylketonuria.* Retrieved from https://www.nichd.nih.gov/health/topics/pku/conditioninfo/symptoms

National Organization for Rare Disorders (NORD). (2018). *Tangier Disease.* Retrieved from https://rarediseases.org/rare-diseases/tangier-disease/

Negi, S. I., Brautbar, A., Virani, S. S., Anand, A., Polisecki, E., Asztalos, B. F., ... Jones, P. H. (2013). A novel mutation in the ABCA1 gene causing an atypical phenotype of Tangier disease. *Journal of Clinical Lipidology*, *7*(1), 82–87. doi:10.1016/j.jacl.2012.09.004 PMID:23351586

Ney, D. M., Gleason, S. T., van Calcar, S. C., MacLeod, E. L., Nelson, K. L., Etzel, M. R., ... Wolff, J. A. (2009). Nutritional management of PKU with glycomacropeptide from cheese whey. *Journal of Inherited Metabolic Disease*, *32*(1), 32–39. doi:10.100710545-008-0952-4 PMID:18956251

Ngugi, A. K., Bottomley, C., Kleinschmidt, I., Sander, J. W., & Newton, C. R. (2010). Estimation of the burden of active and life-time epilepsy: A meta-analytic approach. *Epilepsia*, *51*(5), 883–890. doi:10.1111/j.1528-1167.2009.02481.x PMID:20067507

Ogawa, K., Seki, T., Onji, T., Adachi, N., Tanaka, S., Hide, I., ... Sakai, N. (2013). Mutant γPKC that causes spinocerebellar ataxia type 14 upregulates Hsp70, which protects cells from the mutant's cytotoxicity. *Biochemical and Biophysical Research Communications*, *440*(1), 25–30. doi:10.1016/j.bbrc.2013.09.013 PMID:24021284

Ojha, R., & Prasad, A. N. (2016). Menkes disease: What a multidisciplinary approach can do. *Journal of Multidisciplinary Healthcare*, (9): 371–385. doi:10.2147/JMDH.S93454 PMID:27574440

Onen, F. (2006). Familial Mediterranean fever. *Rheumatology International*, *26*(6), 489–496. doi:10.100700296-005-0074-3 PMID:16283319

Onishchenko, N., Karpova, N., Sabri, F., Castrén, E., & Ceccatelli, S. (2008). Long-lasting depression-like behavior and epigenetic changes of BDNF gene expression induced by perinatal exposure to methylmercury. *Journal of Neurochemistry*, *106*(3), 1378–1387. doi:10.1111/j.1471-4159.2008.05484.x PMID:18485098

Orr, H. T., Chung, M. Y., Banfi, S., Kwiatkowski, T. J. Jr, Servadio, A., Beaudet, A. L., ... Zoghbi, H. Y. (1993). Expansion of an unstable trinucleotide CAG repeat in spinocerebellar ataxia type 1. *Nature Genetics*, *4*(3), 221–226. doi:10.1038/ng0793-221 PMID:8358429

Ozen, S., & Batu, E. D. (2015). The myths we believed in familial Mediterranean fever: What have we learned in the past years? *Seminars in Immunopathology*, *37*(4), 363–369. doi:10.100700281-015-0484-6 PMID:25832989

Özen, S., Batu, E. D., & Demir, S. (2017). Familial Mediterranean Fever: Recent Developments in Pathogenesis and New Recommendations for Management. *Frontiers in Immunology*, *8*(3), 253. doi:10.3389/fimmu.2017.00253 PMID:28386255

Ozen, S., Bilginer, Y., Aktay Ayaz, N., & Calguneri, M. (2011). Anti-interleukin 1 treatment for patients with familial Mediterranean fever resistant to colchicine. *The Journal of Rheumatology*, *38*(3), 516–518. doi:10.3899/jrheum.100718 PMID:21159830

Özen, S., Demirkaya, E., Erer, B., Livneh, A., Ben-Chetrit, E., Giancane, G., ... Carmona, L. (2016). EULAR recommendations for the management of familial Mediterranean fever. *Annals of the Rheumatic Diseases*, *75*(4), 644–651. doi:10.1136/annrheumdis-2015-208690 PMID:26802180

Ozgocmen, S., & Akgul, O. (2011). Anti-TNF agents in familial Mediterranean fever: Report of three cases and review of the literature. *Modern Rheumatology*, *21*(6), 684–690. doi:10.310910165-011-0463-2 PMID:21567247

Padeh, S., Gerstein, M., & Berkun, Y. (2012). Colchicine is a safe drug in children with familial Mediterranean fever. *The Journal of Pediatrics*, *161*(6), 1142–1146. doi:10.1016/j.jpeds.2012.05.047 PMID:22738946

Padeh, S., Shinar, Y., Pras, E., Zemer, D., Langevitz, P., Pras, M., & Livneh, A. (2003). Clinical and diagnostic value of genetic testing in 216 Israeli children with Familial Mediterranean fever. *The Journal of Rheumatology*, *30*(1), 185–190. PMID:12508410

Pandolfo, M. (2008). Friedreich Ataxia. *Archives of Neurology*, *65*(10), 1296–1303. doi:10.1001/archneur.65.10.1296 PMID:18852343

Peng, C., Hsu, C., Wang, N., Lee, H., Lin, S., Chan, W., ... Jiang, C. (2018). Spontaneous retroperitoneal hemorrhage in Menkes disease: A rare case report. *Medicine; Analytical Reviews of General Medicine, Neurology, Psychiatry, Dermatology, and Pediatries*, *97*(6), e9869. doi:10.1097/MD.0000000000009869 PMID:29419699

Pimenova, A. A., Thathiah, A., De Strooper, B., & Tesseur, I. (2014). Regulation of amyloid precursor protein processing by serotonin signaling. *PLoS One*, *9*(1), e87014. doi:10.1371/journal.pone.0087014 PMID:24466315

Poduri, A., & Lowenstein, D. (2011). Epilepsy genetics--past, present, and future. *Current Opinion in Genetics & Development*, *21*(3), 325–332. doi:10.1016/j.gde.2011.01.005 PMID:21277190

Postuma, R. B., Lang, A. E., Munhoz, R. P., Charland, K., Pelletier, A., Moscovich, M., ... Barreto, G. E. (2015). Implication of Green Tea as a Possible Therapeutic Approach for Parkinson Disease. *CNS & Neurological Disorders - Drug Targets*, *15*(3), 292–300. doi:10.2174/18715273156661602021255 19 PMID:26831259

Postuma, R. B., Lang, A. E., Munhoz, R. P., Charland, K., Pelletier, A., Moscovich, M., ... Shah, B. (2012). Caffeine for treatment of Parkinson disease: A randomized controlled trial. *Neurology*, *79*(7), 651–658. doi:10.1212/WNL.0b013e318263570d PMID:22855866

Prasad, A. N., Levin, S., Rupar, C. A., & Prasad, C. (2011). Menkes disease and infantile epilepsy. *Brain & Development*, *33*(10), 866–876. doi:10.1016/j.braindev.2011.08.002 PMID:21924848

Pratt, W. B., Gestwicki, J. E., Osawa, Y., & Lieberman, A. P. (2015). Targeting Hsp90/Hsp70-based protein quality control for treatment of adult onset neurodegenerative diseases. *Annual Review of Pharmacology and Toxicology*, *55*(1), 353–371. doi:10.1146/annurev-pharmtox-010814-124332 PMID:25292434

Rai, M., Soragni, E., Jenssen, K., Burnett, R., Herman, D., Coppola, G., ... Pandolfo, M. (2008). HDAC inhibitors correct frataxin deficiency in a Friedreich ataxia mouse model. *PLoS One*, *3*(4), e1958. doi:10.1371/journal.pone.0001958 PMID:18463734

Ramirez, M. J., Lai, M. K., Tordera, R. M., & Francis, P. T. (2014). Serotonergic therapies for cognitive symptoms in Alzheimer's disease: Rationale and current status. *Drugs*, *74*(7), 729–736. doi:10.100740265-014-0217-5 PMID:24802806

Rawlins, M. D., Wexler, N. S., Wexler, A. R., Tabrizi, S. J., Douglas, I., Evans, S. J., & Smeeth, L. (2016). The Prevalence of Huntington's Disease. *Neuroepidemiology*, *46*(2), 144–153. doi:10.1159/000443738 PMID:26824438

Reiner, A., Dragatsis, I., & Dietrich, P. (2011). Genetics and neuropathology of Huntington's disease. *International Review of Neurobiology*, *98*, 325–372. doi:10.1016/B978-0-12-381328-2.00014-6 PMID:21907094

Rodrigues, F. B., & Wild, E. J. (2018). Huntington's Disease Clinical Trials Corner: February 2018. *Journal of Huntington's Disease*, *7*(1), 89–98. doi:10.3233/JHD-189001 PMID:29480210

Rogawski, M. A., & Löscher, W. (2004). The neurobiology of antiepileptic drugs. *Nature Reviews. Neuroscience*, *5*(7), 553–564. doi:10.1038/nrn1430 PMID:15208697

Romani, C., Palermo, L., MacDonald, A., Limback, E., Hall, S. K., & Geberhiwot, T. (2017). The impact of phenylalanine levels on cognitive outcomes in adults with phenylketonuria: Effects across tasks and developmental stages. *Neuropsychology*, *31*(3), 242–254. doi:10.1037/neu0000336 PMID:28240926

Rosenberg, P. B., Mielke, M. M., Tschanz, J., Cook, L., Corcoran, C., Hayden, K. M., ... Lyketsos, C. G. (2008). Effects of cardiovascular medications on rate of functional decline in Alzheimer disease. *The American Journal of Geriatric Psychiatry: Official Journal of the American Association for Geriatric Psychiatry*, *16*(11), 883–892. doi:10.1097/JGP.0b013e318181276a PMID:18978249

Salzman, C. (2001). Treatment of the agitation of late-life psychosis and Alzheimer's disease. *European Psychiatry*, *1*, 25s–28s. doi:10.1016/S0924-9338(00)00525-3 PMID:11520475

Sarı, İ., Birlik, M., & Kasifoğlu, T. (2014). Familial Mediterranean fever: An updated review. *European Journal of Rheumatology*, *1*(1), 21–33. doi:10.5152/eurjrheum.2014.006 PMID:27708867

Sathasivam, K., Hobbs, C., Mangiarini, L., Mahal, A., Turmaine, M., Doherty, P., ... Bates, G. P. (1999). Transgenic models of Huntington's disease. *Philosophical Transactions of the Royal Society of London. Series B, Biological Sciences*, *354*(1386), 963–969. doi:10.1098/rstb.1999.0447 PMID:10434294

Sato, S., Mizukami, K., & Asada, T. (2007). A preliminary open-label study of 5-HT1$_A$ partial agonist tandospirone for behavioural and psychological symptoms associated with dementia. *The International Journal of Neuropsychopharmacology*, *10*(02), 281–283. doi:10.1017/S1461145706007000 PMID:16817981

Sauerbier, A., Jenner, P., Todorova, A., & Chaudhuri, K. R. (2016). Non-motor subtypes and Parkinson's disease. *Parkinsonism & Related Disorders*, *22*(Suppl 1), S41–S46. doi:10.1016/j.parkreldis.2015.09.027 PMID:26459660

Scharfman, H. E. (2007). The neurobiology of epilepsy. *Current Neurology and Neuroscience Reports*, *7*(4), 348–354. doi:10.100711910-007-0053-z PMID:17618543

Scherzinger, E., Lurz, R., Turmaine, M., Mangiarini, L., Hollenbach, B., Hasenbank, R., ... Wanker, E. E. (1997). Huntingtin-encoded polyglutamine expansions form amyloid-like protein aggregates in vitro and in vivo. *Cell*, *90*(3), 549–558. doi:10.1016/S0092-8674(00)80514-0 PMID:9267034

Scoto, M., Finkel, R. S., Mercuri, E., & Muntoni, F. (2017). Therapeutic approaches for spinal muscular atrophy (SMA). *Gene Therapy*, *24*(9), 514–519. doi:10.1038/gt.2017.45 PMID:28561813

Scriver, C. R., Eisensmith, R. C., Woo, S. L., & Kaufman, S. (1994). The hyperphenylalaninemias of man and mouse. *Annual Review of Genetics*, *28*(1), 141–165. doi:10.1146/annurev.ge.28.120194.001041 PMID:7893121

Scullion, G. A., Kendall, D. A., Marsden, C. A., Sunter, D., & Pardon, M. C. (2011). Chronic treatment with the alpha2-adrenoceptor antagonist fl uparoxan prevents age-related deficits in spatial working memory in APPxPS1 transgenic mice without altering Aβ plaque load or astrocytosis. *Neuropharmacology*, *60*(2-3), 223–234. doi:10.1016/j.neuropharm.2010.09.002 PMID:20850464

Sechi, A., Dardis, A., Zampieri, S., Rabacchi, C., Zanoni, P., Calandra, S., ... Bembi, B. (2014). Effects of miglustat treatment in a patient affected by an atypical form of Tangier disease. *Orphanet Journal of Rare Diseases*, *9*(1), 143. doi:10.118613023-014-0143-3 PMID:25227739

Seidel, K., Siswanto, S., Brunt, E. R., den Dunnen, W., Korf, H. W., & Rüb, U. (2012). Brain pathology of spinocerebellar ataxias. *Acta Neuropathologica*, *124*(1), 1–21. doi:10.100700401-012-1000-x PMID:22684686

Sen, A., Valentina, C., & Husain, M. (2018). Cognition and dementia in older patients with epilepsy. *Brain*, *141*(6), 1592–1608. doi:10.1093/brain/awy022 PMID:29506031

Shakkottai, V. G., & Fogel, B. L. (2013). Clinical Neurogenetics: Autosomal Dominant Spinocerebellar Ataxia. *Neurologic Clinics*, *31*(4), 987–1007. doi:10.1016/j.ncl.2013.04.006 PMID:24176420

Shakkottai, V. G., & Paulson, H. L. (2009). Physiologic alterations in ataxia: Channeling changes into novel therapies. *Archives of Neurology*, *66*(10), 1196–1201. doi:10.1001/archneurol.2009.212 PMID:19822774

Shatunov, A., Sambuughin, N., Jankovic, J., Elble, R., Lee, H. S., Singleton, A. B., ... Goldfarb, L. G. (2006). Genomewide scans in North American families reveal genetic linkage of essential tremor to a region on chromosome 6p23. *Brain*, *129*(Pt 9), 2318–2331. doi:10.1093/brain/awl120 PMID:16702189

Simic, G., Leko, M. B., Wray, S., Harrington, C. R., Delalle, I., Jovanov-Milosevi, N., ... Hof, P. R. (2017). Monoaminergic Neuropathology in Alzheimer's disease. *Progress in Neurobiology*, *151*, 101–138. doi:10.1016/j.pneurobio.2016.04.001 PMID:27084356

Singh, A. K., Mahlios, J., & Mignot, E. (2013). Genetic association, seasonal infections and autoimmune basis of narcolepsy. *Journal of Autoimmunity*, *43*, 26–31. doi:10.1016/j.jaut.2013.02.003 PMID:23497937

Singh, R. N., Howell, M. D., Ottesen, E. W., & Singh, N. N. (1860). Diverse role of survival motor neuron protein. *Biochimica et Biophysica Acta*, (3): 299–315. doi:10.1016/j.bbagrm.2016.12.008 PMID:28095296

Skvorak, K. J., Dorko, K., Marongju, F., Tahan, V., Hansel, M. C., Gramignoli, R., ... Strom, S. C. (2013). Placental stem cell correction of murine intermediate maple syrup urine disease. *Hepatology (Baltimore, Md.)*, *57*(3), 1017–1023. doi:10.1002/hep.26150 PMID:23175463

Smeets, C. J., & Verbeek, D. S. (2016). Climbing fibers in spinocerebellar ataxia: A mechanism for the loss of motor control. *Neurobiology of Disease*, *88*, 96–106. doi:10.1016/j.nbd.2016.01.009 PMID:26792399

Sohar, E., Gafni, J., Pras, M., & Heller, H. (1967). Familial Mediterranean fever. A survey of 470 cases and review of the literature. *The American Journal of Medicine*, *43*(2), 227–253. doi:10.1016/0002-9343(67)90167-2 PMID:5340644

Sönmez, H. E., Batu, E. D., & Özen, S. (2016). Familial Mediterranean fever: Current perspectives. *Journal of Inflammation Research*, *9*, 13–20. doi:10.2147/JIR.S91352 PMID:27051312

Svenningsson, P., Westman, E., Ballard, C., & Aarsland, D. (2012). Cognitive impairment in patients with Parkinson's disease: Diagnosis, biomarkers, and treatment. *Lancet Neurology*, *11*(8), 697–707. doi:10.1016/S1474-4422(12)70152-7 PMID:22814541

Tariot, P. N., Sunderland, T., Weingartner, H., Murphy, D. L., Welkowitz, J. A., Thompson, K., & Cohen, R. M. (1987). Cognitive effects of L-deprenyl in Alzheimer's Disease. *Psychopharmacology*, *91*(4), 489–495. doi:10.1007/BF00216016 PMID:3108930

Thannickal, T. C., Moore, R. Y., Nienhuis, R., Ramanathan, L., Gulyani, S., Aldrich, M., ... Siegel, J. M. (2000). Reduced number of hypocretin neurons in human narcolepsy. *Neuron*, *27*(3), 469–474. doi:10.1016/S0896-6273(00)00058-1 PMID:11055430

Thenganatt, M. A., & Louis, E. D. (2012). Personality profile in essential tremor: A case-control study. *Parkinsonism & Related Disorders*, *18*(9), 1042–1044. doi:10.1016/j.parkreldis.2012.05.015 PMID:22703869

Tosolini, A. P., & Sleigh, J. N. (2017). Motor Neuron Gene Therapy: Lessons from Spinal Muscular Atrophy for Amyotrophic Lateral Sclerosis. *Frontiers in Molecular Neuroscience*, *10*, 405. doi:10.3389/fnmol.2017.00405 PMID:29270111

Touitou, I. (2001). The spectrum of Familial Mediterranean Fever (FMF) mutations. *European Journal of Human Genetics*, *9*(7), 473–483. doi:10.1038j.ejhg.5200658 PMID:11464238

Tsou, A. Y., Paulsen, E. K., Lagedrost, S. J., Perlman, S. L., Mathews, K. D., Wilmot, G. R., ... Lynch, D. R. (2011). Mortality in Friedreich ataxia. *Journal of the Neurological Sciences*, *307*(1–2), 46–49. doi:10.1016/j.jns.2011.05.023 PMID:21652007

Tümer, Z., & Møller, L. B. (2010). Menkes disease. *European Journal of Human Genetics*, *18*(5), 511–518. doi:10.1038/ejhg.2009.187 PMID:19888294

Unal Gulsuner, H., Gulsuner, S., Mercan, F. N., Onat, O. E., Walsh, T., Shahin, H., ... Tekinay, A. B. (2014). Mitochondrial serine protease HTRA2 in a kindred with essential tremor and Parkinson disease. *Proceedings of the National Academy of Sciences of the United States of America*, *111*(51), 18285–18290. doi:10.1073/pnas.1419581111

van den Brink, D. M., Brites, P., Haasjes, J., Wierzbicki, A. S., Mitchell, J., Lambert-Hamill, M., ... Wanders, R. J. (2003). Identification of PEX7 as the second gene involved in Refsum disease. *American Journal of Human Genetics*, *72*(2), 471–477. doi:10.1086/346093 PMID:12522768

van Stiphout, M. A. E., Marinus, J., van Hilten, J. J., Lobbezoo, F., & de Baat, C. (2018). Oral Health of Parkinson's Disease Patients: A Case-Control Study. *Parkinson's Disease*, *9315285*. doi:10.1155/2018/9315285 PMID:29854385

Verhaart, I. E. C., Robertson, A., Leary, R., McMacken, G., König, K., Kirschner, J., ... Lochmüller, H. (2017). A multi-source approach to determine Spinal Muscular Atrophy incidence and research ready population. *Journal of Neurology*, *264*(7), 1465–1473. doi:10.100700415-017-8549-1 PMID:28634652

Vradenburg, G. (2015). A pivotal moment in Alzheimer's disease and dementia: How global unity of purpose and action can beat the disease by 2025. *Expert Review of Neurotherapeutics*, *15*(1), 73–82. doi:10.1586/14737175.2015.995638 PMID:25576089

Walker, F. O. (2007). Huntington's disease. *Lancet*, *369*(9557), 218–228. doi:10.1016/S0140-6736(07)60111-1 PMID:17240289

Wang, S., & Smith, J. D. (2014). ABCA1 and nascent HDL biogenesis. *BioFactors (Oxford, England)*, *40*(6), 547–554. doi:10.1002/biof.1187 PMID:25359426

WebMD. (2018). *Symptoms of epilepsy*. Retrieved from https://www.webmd.com/epilepsy/understanding-epilepsy- symptoms

Wild, E. J., & Tabrizi, S. J. (2017). Therapies targeting DNA and RNA in Huntington's disease. *Lancet Neurology*, *16*(10), 837–847. doi:10.1016/S1474-4422(17)30280-6 PMID:28920889

Williams, A. J., & Paulson, H. L. (2008). Polyglutamine neurodegeneration: Protein misfolding revisited. *Trends in Neurosciences*, *31*(10), 521–528. doi:10.1016/j.tins.2008.07.004 PMID:18778858

Williamson, J., Goldman, J., & Marder, K. S. (2009). Genetic Aspects of Alzheimer Disease. *The Neurologist*, *15*(2), 80–86. doi:10.1097/NRL.0b013e318187e76b PMID:19276785

World Health Organization (WHO). (2018). *Mental Health*. Retrieved from http://www.who.int/mental_health/neurology/dementia/en/

Yiannopoulou, K. G., & Papageorgiou, S. G. (2013). Current and future treatments for Alzheimer's disease. *Therapeutic Advances in Neurological Disorders*, *6*(1), 19–33. doi:10.1177/1756285612461679 PMID:23277790

Yilmaz, R. O. S., Ozyurt, H., & Erkorkmaz, U. (2011). Serum vitamin B12 status in children with familial Mediterranean fever receiving colchicine treatment. *Hong Kong Journal of Paediatrics, 16*, 3–8.

Zanatta, D., Rios Romenets, S., Altman, R., Chuang, R., ... Shah, B. (2012). Caffeine for treatment of Parkinson disease: A randomized controlled trial. *Neurology, 79*(7), 651–658. doi:10.1212/WNL.0b013e318263570d PMID:22855866

Zesiewicz, T. A., Elble, R. J., Louis, E. D., Gronseth, G. S., Ondo, W. G., Dewey, R. B., ... Weiner, W. J. (1752–1755). ... Weiner, W. J. (2011). Evidence-based guideline update: treatment of essential tremor: report of the Quality Standards subcommittee of the American Academy of Neurology. *Neurology, 77*(19), 1752–1755. doi:10.1212/WNL.0b013e318236f0fd

Zesiewicz, T. A., Shaw, J. D., Allison, K. G., Staffetti, J. S., Okun, M. S., & Sullivan, K. L. (2013). Update on treatment of essential tremor. *Current Treatment Options in Neurology, 15*(4), 410–423. doi:10.100711940-013-0239-4 PMID:23881742

Zlatic, S., Comstra, H. S., Gokhale, A., Petris, M. J., & Faudez, V. (2015). Molecular Basis of Neurodegeneration and Neurodevelopmental Defects in Menkes Disease. *Neurobiology of Disease, 81*, 154–161. doi:10.1016/j.nbd.2014.12.024 PMID:25583185

KEY TERMS AND DEFINITIONS

Antipsychotics Drugs: A group of drugs used to treat mental health conditions.

Apheresis: A medical procedure in which an individual component is removed from a patient's blood using an apparatus that separates the components and returns the remaining blood into the patient.

Argyrophilic Grain Disease: A neurodegenerative disease marked by the presence of argyrophilic grains.

Astrogliosis: A chemical reaction involving an increase of astrocytes due to damage after CNS injuries.

Ataxia: A neurological condition involving the loss of voluntary muscle movement.

Cataplexy: A condition characterized by sudden and uncontrollable voluntary muscle weakness caused by strong emotion or laughter and often results in physical collapse while conscious.

Cerebral Amyloid Angiopathy: A condition caused by a build-up of amyloid on the walls of the arteries in the brain which can lead to a brain haemorrhage and dementia.

Cholinesterase Inhibitors: A class of drugs that prevent the breakdown of acetylcholine.

Dysarthria: A speech disorder characterized by poor articulation of speech.

Dysdiachokinesia: The inability to perform quick, alternating movements.

Dysmetria: A condition that occurs as a result the development of lesions in the cerebellum which then affects coordinated movement, thought processing, and behaviour.

Dysphasia: A condition in which the ability to produce and understand language is affected.

Dystonia: A neurological disorder marked by involuntary muscle contractions.

Erythropoiesis: Production of red blood cells which is initiated by the decreasing levels of oxygen which cause erythropoietin hormone production in the kidneys.

Hepatosplenomegaly: Abnormal enlargement of both the spleen and liver.

Hyperphenylalaninaemia: Abnormally elevated levels of plasma phenylalanine.

Kinetic Tremor: A rhythmic, involuntary movement of a body part that is associated with voluntary movements.

Leukoencephalopathy: Neurological disorders of white matter in the brain.

Lewy Body Disease: One of the most common causes of dementia which results from the abnormal accumulation of alpha-synuclein in the brain.

Lipopheresis: Removal of low-density lipoprotein cholesterol from a patient's blood.

Pathognomic: A characteristic that is indicative of a particular disease.

Plasmapheresis: Removal of plasma from a patient's blood.

Proteolisis: The process of breaking down proteins into smaller polypeptides or into amino acids.

Proteopathy: Diseases involving the disruption of cell function as a result of the production of abnormal proteins.

Psychosis: Disorders of mental illnesses that result in abnormal thinking, perceptions, behaviour, and emotions.

Chapter 19
Genetics and Public Health

ABSTRACT

This chapter wraps up by discussing the crucial role played by public health specialists who must reconcile traditional public health concerns of health inequality and equity with safe and effective health interventions and diagnostics that meet individual health needs. Since most genetic diseases in the realm of public health are an interplay of different genetic, lifestyle, and environmental factors, genomic science has given greater emphasis to the importance of molecular and cellular mechanisms in health and disease. New biological knowledge must be integrated with the social and environmental models to improve health at individual and population levels. Public health specialists must now be able to integrate genome-based knowledge into public health in a responsible, ethical, and effective way and anticipate the increase in the health service requirements likely to occur in the future. The foundational pillars of bioethics (beneficence, non-maleficence, autonomy, and justice) must be protected by all public health stakeholders.

CHAPTER OUTLINE

19.1 The Role of Public Health and Public Policy
19.2 Big Data, Personalized Medicine, and Biomarkers
19.3 Ethical, Legal and Social Considerations
Chapter Summary

LEARNING OUTCOMES

- Characterize the role of public health and public policy in the use of genomic information
- Explain the meaning of personalized medicine and its potential applications
- Discuss the social, legal and ethical issues arising in the use of genetic information

DOI: 10.4018/978-1-5225-8066-9.ch019

19.1 THE ROLE OF PUBLIC HEALTH AND PUBLIC POLICY

Connecting genes and their phenotypic manifestation is complicated necessitating the translation of information about genes and DNA sequences into facts about genetic susceptibility to disease, the interaction between these susceptibilities, modifiable risk factors, and the impact of this knowledge on population health. From a public health perspective, systematic, evidence-based evaluations and technology assessments are critical to the incorporation and use of genomics in **clinical** and **public health practice**. *Genetics* is the science of inheritance focusing on how Mendelian inherited traits with single-mutations cause a disease. The more recent term *genomics* focuses on studying complex set of genes, their expression, and how they interact with other genes and the environment to affect disease development. Genomic science has given greater emphasis to the importance of molecular and cellular mechanisms in health and disease. Since most genetic diseases in the realm of public health are an interplay of different genetic, lifestyle and environmental factors, **public health specialists** must reconcile traditional public health concerns (like **health disparities** and **health equity**) with safe and effective health interventions and diagnostics that meet individual health needs. Once the genetic and environmental factors involved in the causation of disease and how they interact (referred to as epigenomics) is understood, public health genomics' goal is to devise effective preventive interventions targeted at individuals with susceptible genotypes. New biological knowledge must be integrated with the social and environmental models to improve health both at the individual and population levels. Public health genomics is clearly a very interdisciplinary research field bringing together different disciplines that will include medical sciences, statistics, biotechnology, engineering, pharmaceutical research/industry, health policy, ethics, law, sociology, public health practitioners, genetic centres, governments, non-governmental organizations, and representatives of patient groups (Marzuillo et al., 2014; Zimmern & Khoury, 2017).

Following the completion of the Human Genome Project in 2003, the utility of genomics in improving public health education, outreach, and interventions has come into the limelight as public health practitioners in the public and private arenas reposition to provide more proactive guidance and leadership to improve population and community health outcomes. Public health specialists must now be able to integrate genome-based knowledge into public health in a responsible and effective way and anticipate the increase in the health service requirements likely to occur in the future. Public health agencies must work out feasible ways to access and analyse the results of genome-based research and technologies and identify information gaps at both individual and population levels to allow the formulation of appropriate policies to guide **evidence-based interventions**. Public health aims at improving the health of the entire population by implementing preventive strategies, but behind the entire population is the individual.

In a most inspiring characterization of the complexity of today's biomedical research problems, Zerhouni (2003) noted over 15 years ago that scientists must reorient in the new millennium to envelope the interdisciplinarity of knowledge to new organizational models emphasizing team science. This will enable a holistic understanding of the interplay between genetics, diet, infectious agents, environment, behaviour, and social structures. Ultimately, public health successes and/or failures will be measured by the level of involvement in four main areas: 1) **Translational research** to conduct systemic reviews of genomic studies, develop evidence-based policy and practice guidelines, and monitor impacts on people's health; 2) **Epidemiological studies**, to extend **genetic biobanks** and disease registries with genomics information; 3) Harmonization of genomics data with the persistent health disparities and equity concerns; 4) Evaluating the social, ethical and cultural issues arising from expanding realms of genomic data.

Genetics and Public Health

As human genomics rapidly advances, appropriate and timely policies governing genomic data and clinical technologies are necessary to guide decision making and best practice in research, public health, and clinical care. Due to the rapidity of genomics technological change, new policy decisions to address new knowledge are urgently needed and must keep pace with the rate of change. Genomics policies at federal, state, organization, and institutional levels address a wide spectrum of policy issues, from how genomics research samples are stored and shared to whether and how to utilize new genomics technologies in clinical practice (Lemke & Harris-Wai, 2015). Key stakeholders in genomics include diverse groups of patients, research participants, the general public, health providers, researchers, advocacy groups, taxpayers, policy makers, among others. Stakeholders may support or oppose decisions and may be influential in the organization or within the community in which they operate. Stakeholder engagement is a process by which an organization involves people who may be affected by the decisions it makes or who can influence the implementation of decisions. The International Association of Public Participation's spectrum of participation defines five broad levels of increasing involvement in the public engagement process: (i) inform through fact sheets, websites, or open houses; (ii) consult by public comment, focus groups, surveys, or public meetings; (iii) involve through workshops or deliberative polling; (iv) collaborate through citizen advisory committees, consensus building, or participatory decision making; and (v) empower people through citizen juries or delegated decisions (Lemke & Harris-Wai, 2015; IAPP, 2018).

Khoury et al. (2011) listed three priorities for how public health specialists should translate the role of genomics in meeting the health needs of the population. 1) Serve as informed advocates for emerging genomic applications in practice. 2) Implement current evidence-based genomic applications to improve health and prevent disease, and being cognizant of premature use, misuse, or overuse of genomic applications. 3) Use genomics tools to evaluate the health impact of public health interventions. Data obtained from genome-disease associations and genome-environment interactions which elucidates genomic variants to common diseases, is becoming critical to the formulation and assessment of genome-based public health strategies to be incorporated into existing recommendations for disease prevention. The challenges facing public health specialists is the development of a skilled public health workforce competent in the differentiating consumable genomic technology ready for use in population health; and technology that is not quite ready for direct transfer into the public health arena for health promotion and disease prevention. System-wide organizational changes are called for within the health care system to provide these services effectively and efficiently. Such changes will centre on three main pillars: a) workforce capacity building and development of structure; b) promotion of genomic education in public health professionals including clinicians; c) outreach activities in the community to promote genomic health literacy (Brand & Probst-Hensch, 2007; Brand et al., 2008; Frenk et al., 2010; CDC, 2011; Khoury et al., 2011; Marzuillo et al., 2014).

A four-phased framework of moving promising genomic applications to clinical and public health practice for population health benefit was presented by Khoury et al. (2011): a) gene discovery and how it affects health; b) development of guidelines to support evidence-based practice; c) translate guidelines into actual practice; and d) assessment of the impact of the practice to population health. Translation guidelines to health practice is purported to be the most challenging problem in healthcare and disease prevention dealing with increasing the spread of knowledge about evidence-based interventions (dissemination research), integrating these interventions into existing programs and structures, and widespread adoption of these interventions by the whole range of stakeholders (diffusion research). Other challenges include workforce training, public health literacy, information systems and public participation. The last phase evaluated impacts evidence-based recommendations and guidelines on real-world health outcomes.

According to Fried et al. (2014) the whole-scale incorporation of genomics in the public health arena will undoubtedly be greatly influenced by large-scale macrosocial forces of globalization, urbanization, population aging, and health disparities. The authors define globalization as the increasing interconnectedness of ideas, behaviours, resources, and human capital globally creating opportunities for innovative, effective public health interventions including individual-based solutions and concerted efforts towards shared health challenges/threats. Urbanization and migration pose a unique challenge to public health research and practice since more than half of the world's population is now living in urban areas and increasing urbanization in most developing nations. These forces will provide challenges and opportunities in public health practice as illustrated by the following three examples: 1) Migration in many countries poses a big challenge as it causes massive movement of people into cities causing severe strain to urban planning and aging urban infrastructures. 2) Implementation of large-scale environmental modifications that influence people's health (especially vulnerable populations) in cities can become more manageable. 3) Interconnectedness of urban populations provides opportunities for behavioural interventions at a truly large scale (Fried et al., 2014; Marzuillo et al., 2014; Zimmern & Khoury, 2017).

Population aging has characterized global populations in the 20th and 21st centuries as people live longer greatly altering the population pyramids of countries worldwide. Studies indicate that approximately 11% of the world's population is currently older than 60 years, with about 27% younger than 14 years; and by 2030, 16.5% of the world's population will be 60 years and older, and about 23% will be 14 years or younger. By the year 2050, people older than 60 years population will surpass those 14 years or younger. Since increased longevity raises the chronic disease burden and shifts the population distribution of disease, mitigative strategies of disease prevention must be developed. On the positive side an aging population creates opportunities that can optimize societal well-being and lifestyle habits intergenerational transfers within families. Intergroup health disparities in health are increasing throughout the United States and globally. Health disparities in many parts of the world are continuous, graded, and cumulative across the life course across racial/ethnicity, socioeconomic status, and gender. The long-lasting effects of these intergroup differences in health is an ever-increasing challenge to public health requiring creative and innovative approaches to address (Fried et al., 2014).

19.2 BIG DATA, PERSONALIZED MEDICINE AND BIOMARKERS

By applying emerging technologies to measure and understand disease, pathogens, health behaviours, exposures and population disease susceptibility, precision public health is driven by big data to prevent disease, promote health, reduce health disparities and develop policies and interventions to improve public health. **Big data** are very elaborate and extensive stored data sets that can be manipulated and analysed to drive insights, innovation and develop new interventions. Big data has been used in public health studies of disease surveillance and signal detection, predicting risk, targeting interventions, and understanding disease. Examples of public health disciplines that benefit from big data use are community health, environmental health science, epidemiology, infectious disease/vector control, maternal and child health, occupational health and safety, and nutrition. The use of big data has been shown to improve precision in select disciplines of public health (Khoury et al., 2016; Dolley, 2018).

Genetics and Public Health

Personalized medicine is a term with increased recent use. It means incorporating the science of pharmacogenomics and other individualized interventions into clinical practice. It involves preventive, diagnostic, and therapeutic interventions, as they relate to associated risk defined through genetics as well as clinical and family histories. It requires genetic, lifestyle, and environmental data to meet goals of more customized and potentially individualized clinical treatments. Ginsburg and Phillips (2018) make a clear distinction between personalized medicine and **precision medicine**. Personalized medicine is a clinical approach to patient health that factors their genetic make-up while also paying attention to their preferences, beliefs, attitudes, knowledge and social context. On the other hand, precision medicine depicts a model for health care delivery that relies heavily on data, analytics, and information (Ginsburg & Phillips, 2018).

There are high expectations about the future impact of the discoveries through GWAS on preventive and clinical health care practice. Nevertheless, the predictive value of genetic profiling is limited for most complex disorders particularly because for some known markers, it is not definite how much they contribute to disease incidence. For example, people who are homozygotes for the APOE ε4 gene variant, have an increased risk of Alzheimer's disease; but not everyone who tests positive for this variant will develop Alzheimer's disease. Seshadri et al. (1995) noted that fewer than 30% of people with the APOE ε4 polymorphism develop Alzheimer's disease. Despite the growing availability of commercial, **direct-to-consumer (DTC) genetic testing** via the Internet, there is still a lot of territory remaining to be covered in evidence-based applications of genetic profiling in clinical and public health care practice. Unravelling the genomic understanding of common diseases in all its complexity, and all its interactions will keep scientists very engaged in the years ahead. The increased utilization of Big data in precision public health has big risks—summed here as the duty of protecting the dignity, privacy, security of citizens and patients, while pursuing interventional strategies that will deliver more meaningful public health outcomes in a reasonable timeframe (Roberts et al., 2014; NIST, 2015; Khoury & Galea, 2016; Khoury et al., 2016; Dolley, 2018).

The existence of large interindividual differences on drug reactions has prompted the application of genomics in public health (especially based on genome-wide association studies [GWAS]), in the fields of pharmacogenetics and pharmacogenomics. Pharmacogenomics is the study of the role of inherited and acquired genetic variation in drug response. It is shifting from focusing on individual candidate genes to GWAS which are based on a rapid scan of biomarkers across the genome of persons affected by a particular disorder or drug-response phenotype, and people who are not affected, with tests for association that compare genetic variation in case-controlled settings. In addition, pharmacogenomics can provide new insights into mechanisms of drug action and as a result can contribute to the development of new therapeutic agents. The focus on biomarkers and their applicability in drug therapy arose from the observation that some drugs are ineffective in some people, others cause adverse health effects in people, while others only achieve sub-effective results in others. Using clinical and genomics criteria, pharmacogenomics underlines the principle of optimizing drug dose while preventing adverse side-effects—the right drug should be used in the right dose in the right patient at the right time (Fackler & McGuire, 2009).

In its labelling of drugs, the FDA is now utilizing **genomic biomarkers**—including germline or somatic gene variants (polymorphisms or mutations), functional deficiencies with a genetic aetiology, gene expression differences, and chromosomal abnormalities—all from GWAS (FDA, 2018). For instance, the use of the cardiovascular prodrug *clopidogrel* has been demonstrated to be less effective in patients with the gene variant *CYP2C19* genetically polymorphic to *CYP2C9*. The *CYP2C19 gene* encodes a truncated version of the cytochrome P-450 enzyme required in the activation of *clopidogrel*. This enzyme inhibits ADP–stimulated platelet activation by binding irreversibly to a specific platelet receptor of ADP, $P2Y_{12}$, thus inhibiting platelet aggregation decreasing the risk of subsequent ischemic vascular events. This results in decreased *clopidogrel* metabolic activation, a decreased antiplatelet effect, and an increased likelihood of a cardiovascular event—observations that are confirmed by GWAS. It is important to note that although pharmacogenomics acknowledges the crucial role played by genetic biomarkers affecting drug metabolism, there are numerous factors that can alter drug responses. These include other medications, plus environmental and physiological factors such as nutrition, ageing, liver and kidney function and patient compliance—all of which provide a better, more holistic understanding of individual pathways in systems biomedicine. Pharmacogenomics facilitates the identification of biomarkers that can help physicians optimize drug selection, dose, and treatment duration and avert adverse drug reactions.

19.3 ETHICAL, LEGAL AND SOCIAL CONSIDERATIONS

Ethical, legal and social issues are often better viewed as arbitrary categories, mostly because these issues are almost always complex and intertwined. Advanced research in genomics and advent of new like next-generation sequencing, GWAS and GWIS has crucial implications for the understanding, prevention, and treatment of human diseases (Roberts et al., 2014). Public health genomics then will require a more elaborate set of structures and processes to access, evaluate, and integrate knowledge from genomics-based research and ultimately translate these findings for improved population health outcomes. A conceptual framework suggested by Burke et al. (2006) highlights the key components that must be developed to support core activities including informing public policy; development and evaluation of preventive and clinical health services; communication and stakeholder engagement; and education and training.

Next generation sequencing (NGS) is expected to improve the quality of new-born screening and increase the predictive value of the generated results. However, this technology raises a number of programmatic and ethical issues in these screening programs. For example, additional unanticipated findings from NGS like predictive genetic information that may not be directly related to the onset of adult conditions; and which may not be presently actionable for the child in question, may raise social and ethical issues in parents—therefore requiring closer attention before NGS protocols become mainstream. Some health conditions arising from screening programs are shifting the health focus towards a life course perspective raising several public health, ethical and policy challenges. One example is given by Roberts et al (2014) of phenylketonuria (PKU) to illustrate this perspective shift. PKU treatment calls for consumption of a diet that is low in phenylalanine during childhood. Information from international longitudinal studies like that of Pratt et al. (2000) then showed that elevated phenylalanine levels in mothers with PKU were associated with increased risk of having a child with birth defects and cognitive impairment. This meant that a phenylalanine-restricted diet had to be expanded to include girls and women of childbearing age (CGMP, 2008). Evidence shows improved outcomes in adolescents and adults treated with this diet necessitating individuals to remain on a phenylalanine-restricted diet for life (life course) (Macleod & Ney, 2010) which raises some ethical and policy challenges. For example, who is responsible for the expense of the special diet especially for parents who cannot afford it out-ofpocket? Could coverage expansion be made part of government screening programs policy? Are there resources and funds to conduct extensive long-term follow-up surveillance of these children lifelong?

Another example highlighting the policy shifts and the screening information obtained about family members who may be carriers of a disease condition that will impact the health of their offspring. Such conditions are like sickle-cell anaemia and cystic fibrosis. It is beneficial for individuals to get this information for future reproductive planning. Should such information gained from childhood screening programs be made available to other family members who may benefit from knowing about it? Who is responsible for telling family members—health professionals, genetic counsellors, social workers, etc.? What about if health professionals decide not to share the information, are there legal recourses if family members later know the information was withheld? Another complicated situation is a case where a health condition is found in childhood but does not manifest disease symptoms until later in life. As noted by Kwon and Steiner (2011), this situation creates a group of individuals, called "patients-in-waiting", who are diagnosed with a disorder but are "waiting" for symptoms of their disorder to develop. Who is responsible for the damage caused by the psychological and emotional effects of living with a looming diagnosis of a disorder? Who is responsible for the follow-up procedures and costs involved as infants move into adulthood?

The interactions between genes and their environment, modify how gene expression occurs. Environmental factors are influential predictors of subsequent phenotypes and disease risk in later life, and include dietary (total caloric intake, specific micro- and macronutrient levels, phytochemicals); physical (temperature, species density); social (stress, behaviour, socioeconomic); chemical (toxins, endocrine disruptors, drugs), and unknown/unpredictable (stochastic like geological events) effects. In public health, epigenetic changes hold promise as targets for preventive and therapeutic interventions because they are potentially reversible. For instance, studies show that in mice models, epigenetic alterations associated with perinatal chemical exposure can be counteracted by maternal nutritional supplementation with methyl donors like folic acid or components of soy (Dolinoy et al., 2007). Bekdash et al., (2013) also showed that in rats, *in utero* choline supplementation negated alcohol-induced alterations in epigenetic modification.

Unlike genetic mutations, which are static and non-modifiable, epigenetic marks are dynamic and potentially modifiable. According to Roberts et al. (2014), epigenetic evaluation has become an essential tool in public health interventions for two reasons: 1) Environmental exposures are good candidates for modifiable risk factors as they can be effectively regulated at the personal behavioural level (with nutritional and/or pharmacological approaches) as well as regulatory policy level (by reducing the prevalence of toxic exposures in a given state or municipality). 2) Through **epigenomic profiling,** biomarkers of exposure can be identified enabling clinicians to identify at-risk individuals prior to disease onset. Once the specific epigenetic biomarkers that are associated with health outcomes are established, epigenomics become a promising tool for individualized reproductive health care, and population-wide disease diagnostic, screening, and prevention strategies (Roberts et al., 2014).

Despite the progress made in epigenetics research, Roberts et al. (2014) note that a better integration of the field into public health planning is hampered by lack of data in four main areas: 1) studies on exposure prior to conception to ensure coverage of the initial epigenetic reprogramming events; (2) studies that characterize exposures and health outcomes over time, with corresponding measures of potential epigenetic drift; (3) investigations on tissue and cell-specific differences in epigenetic responses to environmental exposures; and (4) examination of exposure mixtures that more closely approximate real-world exposures that people face.

The potential future application of findings from health screening programs, epigenetics and epigenomics studies raise very delicate issues including health policies that determine access to genomic information; challenges in providing health education and understanding behavioural responses to genetic test results; and procedural and distributive issues in public health genomics. Health policy challenges in public health genomics involve the regulation of access to genomic information generated by emerging technologies. Clinicians have traditionally been the gatekeepers of people's health data. Although many individuals, particularly those with family histories of certain diseases, are not only curious about their genetic profiles, but also have a "right-to-know" their personal health information; it must be emphasized that data generated by recent genetic technology is limited in its predictive (therefore utilitarian value). There will be difficulties in conveying information in accurate and easily understood terms; plus, the possible psychological and social harms in individuals receiving such information remains unknown. Decisions regarding the involvement of the potential disease-risk of genetic information "owners" has received a lot of mixed reviews in the literature. Leading professional organizations in clinical genetics hold the position that genetic testing should be deferred until adulthood unless there is information supporting medical benefit of this information in childhood. Some authors note the value of preserving autonomy, suggesting that genetic risk information could infringe on the child's right to an "open future" and the corresponding opportunity to make testing decisions as a consenting adult (Davis, 1997). Others support a more flexible stance against testing minors for adult-onset conditions, noting that in the absence of proven (versus speculative) harms from genetic risk disclosure, parental authority in such decision-making should be respected (Wilfond & Ross, 2009; Roberts et al., 2014).

The increased availability and use of DTC genetic testing without a lot of government oversight has raised ethical, legal and social implications by claiming exaggerated benefits of testing without fully disclosing the associated risks of testing including privacy and familial implications. Most companies do not know the psychological vulnerability of consumers requesting genetic information, and there is no genetic counselling associated with tests. Others note that such genetic information may lead to more (perhaps unnecessary), health care utilization and screening. Proponents of consumer genomics view direct access to one's genome as an individual right, noting potential benefits of learning more about one's predisposition to disease and likelihood of response to specific medications. Some countries have severely restricted such DTC marketers (like Germany) while others (US) have issued warnings. Roberts and Ostergren (2013), note that neither the health benefits promoted by such companies like improvements in positive health behaviours nor the "worst-case scenarios" envisioned by its critics like catastrophic psychological distress, misunderstanding of test results, or undue burden on the healthcare system have materialized to date. Potential discrimination from insurers and employers is a big concern from genetic susceptibility test results. In the U.S., the federal Genetic Information Non-discrimination Act (GINA) was passed in 2008, prohibiting health insurers and employers from using genetic information (including family history) to inform decisions about coverage, premiums, or hiring. However, GINA does not cover life, disability, or long-term care insurance. (Korobkin & Rajkumar, 2008). Since scientists are still determining the relevance of genomic information to human health, educating the public about it poses numerous challenges. The complexity of most medical disorders with many potential multiple contributors such as health behaviours, environmental exposures, comorbid conditions, and social determinants of health makes it difficult to determine how these factors influence disease expression. Besides, there are also other genetic factors like gene expressivity (penetrance), variable expressivity, and genotype-phenotype correlations can impact the expression of gene mutations that may not be distinctly clear. There are also issues of differences in racial and ethnic groups, individual genetic variability, geographical location, and other lifestyle factors (Bustamante et al., 2011; Roberts & Ostergren, 2013; Roberts et al., 2014; FDA, 2018).

Disclosure of genetic information is also riddled with complications of translating complex findings to individuals who often lack the advanced skills in health literacy, numeracy, and genetic literacy that would be required to fully appreciate the meaning and implications of results. Ideally, genomic information would be delivered by a trained genetic counsellor with expertise in human genetics, health education, and interpersonal practice. To meet anticipated increased future demands for genetic counselling, studies call for the development of alternative models of genetic service delivery with recognized need for increased involvement of non-genetics health care professionals, use of educational media, and briefer protocols (Guttmacher, 2001; Lipkus, 2007). Genomic information potential lies in its ability to promote healthy behaviours among high-risk subpopulations to reduce chances of disease onset. Although some behaviours are complex and difficult to change (like smoking cessation and improved diet/exercise habits), genetic susceptibility testing may enhance preferences for biological interventions (like using medications) over health behaviour changes (like lifestyle change) when both are viable options (Senior & Marteau, 2007).

Concerned that health benefits and burdens be distributed fairly across the population, public health ethics has long placed an emphasis on issues of justice. There is increased reflection by scholars that how applications of genomics technologies could exacerbate, rather than ameliorate, health disparities between racial and ethnic groups by focusing on biological causes of disease instead of more compelling social and environmental risk factors. On the other hand, public health ethics may be viewed through the lens of procedural justice, where notions of responsible stewardship would require transparency to the general public regarding policy changes and research initiatives. Since fundamental rights of the individual such as privacy issues, the person's autonomy, confidentiality, privacy and the 'right (not) to know' are hallmarks of ethical medical research practice, they need deliberate consideration and protection in public health genomics (Cleeren et al., 2011; Zimmern & Khoury, 2017).

With the ultimate goal of improving public health, genomic biobanks are now increasingly being used by government, academic, and research organizations to fuel biomedical research. They store large amounts of personalized genomic DNA and other health data and policies that govern the use of participant data lack oversight and uniformity. Biobanks make large-scale efforts to recruit from the public (locally, nationally and internationally) and they are not specific to a disease category. For example, the Mayo Clinic biorepository houses data from Minnesota residents. Kaiser Permanente's Research Program on Genes, Environment, and Health has data from large numbers of Kaiser Permanente's Californian health plan members. There are also smaller-scale biobanks collecting and storing disease-specific data from the general public (Lemke & Harris-Wai, 2015).

Genomic biobanking is a privacy issue of concern since genetic data from one person extends to other immediate or extended family members—raising the problem of informed consent. People affected by the contribution of genetic information by a family member may not have explicitly consented to (future) studies on the relationship between the human genome and susceptibility to the disorders being studied in that database. In addition, the sharing of genetic data—which is a central feature of "open access" science practice becomes problematic as personal information is not designed for wide-scale sharing or for public health purposes. Policy makers must ready to protect consumers and to monitor the implications of genome-based knowledge and technologies for health, social and environmental policy goals. Since genetic testing is considered somewhat predictive in nature, issues of stigmatization and discrimination arise with the most commonly expressed fear being that genetic information will be used in ways that conflict with people's expectations. Further, not everyone will have access to genetic testing due to socioeconomic considerations—creating social inequality binaries of "haves" and "have nots".

Another issue of concern is what genetics tests show in some individuals. For instance, if an individual has a genetic predisposition to a certain disease like heart disease or stroke, do health professionals cease treatments (or health advice) to such individuals, preferring to focus on individuals where public health efforts will produce better outcomes? Yet another issue bringing about social inequality is degree of genomic literacy in test takers. Not everyone can navigate through the various "must-do-actions" that genetic susceptibility to a disease entails. These can include behaviour modifications like increased physical activity, better diets, relocating to "safer" neighbourhoods away from polluting factories, or adopting a better healthy lifestyle. Affordability of the modifying behaviours called for by tests is another matter. Cleeren et al., (2011) call this the "double-edged sword" of genetics, since it appears like implementing genomics in public health can result in either the widening or the narrowing of health disparities among the population (Sankar et al., 2004; Gostin & Powers, 2006; Evans et al., 2011; Roberts et al., 2014; Zimmern & Khoury, 2017).

While scientific innovations like genomics must be praised and rewarded due to their important contribution to the understanding human health and disease; ethical questions addressing the right of individuals and the good of the overall community, honest and open democratic processes of decision making, the prevention of conditions that lead to poor health outcomes, and the protection of disenfranchised or marginalized populations must continually be raised, evaluated, and solutions found. Public health practitioners, policy makers, health professionals, and researchers will all benefit by duly being cognizant of the foundational pillars of bioethics—beneficence, non-maleficence, autonomy and justice. Unless the competing interests of different stakeholders in the age of genomics are carefully managed and reconciled—in a collaborative framework that encourages scientific innovations and appropriate clinical and public health applications—it will be difficult to affirm that genomics, cell and molecular biology are a blessing to future public health practice. As Zimmern and Khoury note, public health practitioners have an integral role to play in educating consumers, empowering individual autonomy, abstaining from **genetic exceptionalism**, and providing an honest broker role for evaluating evidence of genomic approaches to health services delivery (Beskow et al., 2001; Clayton, 2003; Gebbie et al., 2003; Khoury et al., 2009; Kaye et al., 2010; Knoppers et al., 2010; Cleeren et al., 2011; Burton et al., 2014; Zimmern & Khoury, 2017).

The increased potential of integrating current new technologies like GWAS, GWIS, genomics, epigenomics and NGS creates great optimism amongst stakeholders in public health practice. Continued research and engagement is needed to harness the potential of genomic discoveries to improve human health and to secure public trust in ongoing public health genomic initiatives. As more genetic information becomes available, public health practice interdisciplinarity involving social sciences, health policy, psychology, biostatistics, biology, chemistry, environmental science, epidemiology, and ethics becomes even more critical in meeting community health needs and reducing the public health burdens in the 21st century.

CHAPTER SUMMARY

Genomic science has given greater emphasis to the importance of molecular and cellular mechanisms in health and disease. Since most genetic diseases in the realm of public health are an interplay of different genetic, lifestyle and environmental factors, public health specialists must reconcile traditional public health concerns (like health disparities and equity) with safe and effective health interventions and diagnostics that meet individual health needs. New biological knowledge must be integrated with the social and environmental models to improve health both at the individual and population levels; and public health specialists must now be able to integrate genome-based knowledge into public health in a responsible and effective way and anticipate the increase in the health service requirements likely to occur in the future. Public health genomics is a very interdisciplinary research field bringing together different disciplines that will include medical sciences, statistics, biotechnology, engineering, pharmaceutical research/industry, health policy, ethics, law, sociology, public health practitioners, genetic centres, governments, non-governmental organizations, and representatives of patient groups. Public health successes and/or failures will be measured by the level of involvement in translational research; epidemiological studies; harmonization of genomics data with the persistent health disparities and equity concerns; and the evaluation of the social, ethical and cultural issues arising from expanding realms of genomic data.

In translating the role of genomics in meeting the health needs of the population public health specialists must serve as informed advocates for emerging genomic applications in practice; be implementers of current evidence-based genomic applications to improve health and prevent disease; and use genomics tools to evaluate the health impact of public health interventions. There will be a greater need for workforce capacity building and development of structure; promotion of genomic education in public health professionals including clinicians; and outreach in the community to promote genomic health literacy. The whole-scale incorporation of genomics in the public health is not immune to large-scale macrosocial forces of globalization, urbanization, population aging, and health disparities. Thus, public health specialists must be cognizant of how these forces modify, challenge, and benefit public health practice. Personalized medicine means incorporating the science of pharmacogenomics and other individualized interventions into clinical practice. It involves preventive, diagnostic, and therapeutic interventions, as they relate to associated risk defined through genetics as well as clinical and family histories.

Ethical, social and legal issues (ESLI) are almost always complex and intertwined and increased availability of genomic information will have broad societal implications in these areas. Public health genomics will require a more elaborate set of structures and processes to access, evaluate, and integrate knowledge from genomics-based research and ultimately translate these findings for improved population health outcomes. Conceptually, the key components that must be developed to support core activities including informing public policy; development and evaluation of preventive and clinical health services; communication and stakeholder engagement; and education and training. With increased use of genomic data, some contentious ESLI issues that must be addressed are: genomic biobanking, sharing of genomic data, possible stigmatization and discrimination, access to genetic tests due to socioeconomic conditions, differential treatment in individuals arising from the discovery of specific genetic conditions, and differences in genomic literacy amongst the public. The foundational pillars of bioethics (beneficence, non-maleficence, autonomy and justice) must be protected by public health practitioners, policy makers, health professionals, and researchers to enhance the collaborative framework that encourages scientific innovations and appropriate clinical and public health applications. Continued research and engagement is needed to harness the potential of genomic discoveries to improve human health and to secure public trust in ongoing public health genomic initiatives.

End of Chapter Quiz

1. Public health successes and/or failures will be measured by the level of involvement in which main areas?
    a. Translational research
    b. Epidemiological studies
    c. Harmonization of genomics data with health disparities and equity concern
    d. Evaluating the social, ethical and culture issues
    e. All of the Above
2. The whole-scale incorporation of genomics in the public health arena will be influenced by which macrosocial forces?
    a. Globalization
    b. Urbanization
    c. Population Aging
    d. Health disparities
    e. All of the Above
3. _____ is a clinical approach to patient health that factors their genetic make-up while also paying attention to their preferences, beliefs, attitudes, knowledge and social context.
    a. Precision medicine
    b. Pharmacogenomics
    c. Personalized medicine
    d. Genetic testing
    e. Health disparities
4. The study of the role of inherited and acquired genetic variation in drug response is called
    a. Pharmacokinetics
    b. Pharmacogenomics
    c. Pharmacodynamics
    d. Pharmacovigilance
    e. Pharmacopeia
5. This term refers to the statement that the implementation of genomics in public health can result in either the widening or the narrowing of health disparities in the population.
    a. Double-edged sword of genetics
    b. Yin and Yang of genetics
    c. Genetic exceptionalism
    d. Strengths and weakness of genetics
    e. None of the Above

Genetics and Public Health

6. Which of the following statements about public health is incorrect?
   a. Public health practitioners reposition to provide more proactive guidance and leadership to improve population and community health outcomes.
   b. Public health agencies must work out feasible ways to access and analyse the results of genome-based research and technologies and identify information gaps at both individual and population levels.
   c. Public health genomics is a very interdisciplinary research field bringing together different disciplines such as medical science, statistics and sociology.
   d. Public health aims to improve only the health of one individual by implementing preventative strategies.
   e. All of the statements are correct
7. _____ are very elaborate and extensive stored data sets that can be manipulated and analysed to drive insights, innovation and develop new interventions.
   a. Biomarkers
   b. Big Data
   c. Personalized Medicine
   d. Precision Medicine
   e. Pharmacogenomics
8. In the United States, the federal _____ was passed, prohibiting health insurers and employers from using genetic information to inform decisions about coverage, premiums or hiring.
   a. Genetic Information Non-discrimination Act
   b. Genome Wide Privacy Act
   c. Whole Genome Privacy Act
   d. Private Sequencing Information Act
   e. Global Genetic Non-discrimination Act
9. Next generation sequencing is expected to improve the quality of new-born screening and increase the predictive value of the generated results; however, this technology raises a lot of concerns such as
   a. ethical issues
   b. social issues
   c. legal issues
   d. psychological and emotional effects
   e. all of the above
10. Public health practitioners must
    a. abstain from genetic exceptionalism
    b. accept genetic exceptionalism
    c. educate consumers
    d. A and C only
    e. B and C only

Thought Questions

1. Can you think of any downfalls to precision or personalized medicine?
2. Describe the role of pharmacogenetics in personalized medicine.
3. How would you improve public health and public policy, given what you already know?

REFERENCES

Antoniou, A., Pharoah, P. D., Narod, S., Risch, H. A., Eyfjord, J. E., Hopper, J. L., ... Easton, D. F. (2003). Average risks of breast and ovarian cancer associated with BRCA1 or BRCA2 mutations detected in case Series unselected for family history: A combined analysis of 22 studies. *American Journal of Human Genetics, 72*(5), 1117–1130. doi:10.1086/375033 PMID:12677558

Bekdash, R. A., Zhang, C., & Sarkar, D. K. (2013). Gestational choline supplementation normalized fetal alcohol-induced alterations in histone modifications, DNA methylation, and proopiomelanocortin (POMC) gene expression in β-endorphin-producing POMC neurons of the hypothalamus. *Alcoholism, Clinical and Experimental Research, 37*(7), 1133–1142. doi:10.1111/acer.12082 PMID:23413810

Beskow, L. M., Khoury, M. J., Baker, T. G., & Thrasher, J. F. (2001). The Integration of Genomics into Public Health Research, Policy and Practice in the United States. *Community Genetics, 4*(1), 2–11. PMID:11493747

Brand, A., Brand, H., & Schulte in den Bäumen, T. (2008). The impact of genetics and genomics on public health. *European Journal of Human Genetics, 16*(1), 5–13. doi:10.1038j.ejhg.5201942 PMID:17957226

Brand, A. M., & Probst-Hensch, N. M. (2007). Biobanking for epidemiological research and public health. *Pathobiology, 74*(4), 227–238. doi:10.1159/000104450 PMID:17709965

Burke, W., Khoury, M. J., Stewart, A., & Zimmern, R. L. (2006). The path from genome-based research to population health: Development of an international public health genomics network. *Genetics in Medicine, 8*(7), 451–458. doi:10.1097/01.gim.0000228213.72256.8c PMID:16845279

Burton, H., Jackson, C., & Abubakar, I. (2014). The impact of genomics on public health practice. *British Medical Bulletin, 112*(1), 37–46. doi:10.1093/bmb/ldu032 PMID:25368375

Chen, L. S., Kwok, O. M., & Goodson, P. (2008). US health educators' likelihood of adopting genomic competencies into health promotion. *American Journal of Public Health, 98*(9), 1651–1657. doi:10.2105/AJPH.2007.122663 PMID:18633090

Clayton, E. W. (2003). Ethical, legal, and social implications of genomic medicine. *The New England Journal of Medicine, 349*(6), 562–569. doi:10.1056/NEJMra012577 PMID:12904522

Cleeren, E., Van der Heyden, J., Brand, A., & Van Oyen, H. (2011). Public health in the genomic era: Will Public Health Genomics contribute to major changes in the prevention of common diseases? *Archives of Public Health, 69*(1), 8. doi:10.1186/0778-7367-69-8 PMID:22958637

Committee on Genetics. (2008). Maternal phenylketonuria. *Pediatrics, 122*(2), 445–449. doi:10.1542/peds.2008-1485 PMID:18245437

Davis, D. S. (1997). Genetic dilemmas and the child's right to an open future. *The Hastings Center Report, 27*(2), 7–15. doi:10.2307/3527620 PMID:9131346

Disparities and Inequalities Report. (2011). Morbidity and Mortality Weekly Report. *Centers for Disease Control and Prevention, 60*(suppl), 1–116.

Dolinoy, D. C., Huang, D., & Jirtle, R. L. (2007). Maternal nutrient supplementation counteracts bisphenol A-induced DNA hypomethylation in early development. *Proceedings of the National Academy of Sciences of the United States of America, 104*(32), 13056–13061. doi:10.1073/pnas.0703739104 PMID:17670942

Dolley, S. (2018). Big Data's Role in Precision Public Health. *Frontiers in Public Health, 6*, 68. doi:10.3389/fpubh.2018.00068 PMID:29594091

Dowell, S. F., Blazes, D., & Desmond-Hellmann, S. (2016). Four steps to precision public health. *NAT-News, 540*(7632), 189. doi:10.1038/540189a

Fackler, J. L., & McGuire, A. L. (2009). Paving the Way to Personalized Genomic Medicine: Steps to Successful Implementation. *Current Pharmacogenomics and Personalized Medicine, 7*(2), 125–132. doi:10.2174/187569209788653998 PMID:20098629

Frenk, J., Chen, L., Bhutta, Z. A., Cohen, J., Crisp, N., Evans, T., ... Zurayk, H. (2010). Health professionals for a new century: Transforming education to strengthen health systems in an interdependent world. *Lancet, 376*(9756), 1923–1958. doi:10.1016/S0140-6736(10)61854-5 PMID:21112623

Fried, L. P., Begg, M. D., Bayer, R., & Galea, S. (2013). MPH Education for the 21st Century: Motivation, Rationale, and Key Principles for the New Columbia Public Health Curriculum. *American Public Health Association, 104*(1), 23–30. doi:10.2105/AJPH.2013.301399

Ginsburg, G. S., & Phillips, K. A. (2018). Precision Medicine: From Science to Value. *Health Affairs (Project Hope), 37*(5), 694–701. doi:10.1377/hlthaff.2017.1624 PMID:29733705

Institute of Medicine (US). (2003). Committee on Educating Public Health Professionals for the 21st Century. In Who Will Keep the Public Healthy? Educating Public Health Professionals for the 21st Century. Washington, DC: National Academies Press (US).

International Association of Public Participation (IAPP). (2018*). Advancing Public Practice Participation*. Retrieved from https://www.iap2.org/page/ethics?

Kaye, J., Boddington, P., de Vries, J., Hawkins, N., & Melham, K. (2009). Ethical implications of the use of whole genome methods in medical research. *European Journal of Human Genetics: EJHG, 18*(4), 398–403. doi:10.1038/ejhg.2009.191 PMID:19888293

Khoury, M. J. (2009). Interview: Dr. Muin J. Khoury Discusses the future of public health genomics and why it matters for personalized medicine and global health. *Current Pharmacogenomics and Personalized Medicine, 7*(3), 158–163. doi:10.2174/1875692110907030158

Khoury, M. J., Bowen, M. S., Burke, W., Coates, R. J., Dowling, N. F., Evans, J. P., ... St Pierre, J. (2011). Current priorities for public health practice in addressing the role of human genomics in improving population health. *American Journal of Preventive Medicine*, *40*(4), 486–493. doi:10.1016/j.amepre.2010.12.009 PMID:21406285

Khoury, M. J., & Galea, S. (2016). Will Precision Medicine Improve Population Health? *Journal of the American Medical Association*, *316*(13), 1357–1358. doi:10.1001/jama.2016.12260 PMID:27541310

Khoury, M. J., Iademarco, M. F., & Riley, W. T. (2015). Precision Public Health for the Era of Precision Medicine. *American Journal of Preventive Medicine*, *50*(3), 398–401. doi:10.1016/j.amepre.2015.08.031 PMID:26547538

Knoppers, B. M., Leroux, T., Doucet, H., Godard, B., Laberge, C., Stanton-Jean, M., ... Avard, D. (2010). Framing genomics, public health research and policy: Points to consider. *Public Health Genomics*, *13*(4), 224–234. doi:10.1159/000279624 PMID:20395691

Kwon, J. M., & Steiner, R. D. (2011). "I'm fine; I'm just waiting for my disease": The new and growing class of presymptomatic patients. *Neurology*, *77*(6), 522–523. doi:10.1212/WNL.0b013e318228c15f PMID:21753177

Lemke, A. A., & Harris-Wai, J. N. (2015). Stakeholder engagement in policy development: Challenges and opportunities for human genomics. *Genetics in Medicine*, *17*(12), 949–957. doi:10.1038/gim.2015.8 PMID:25764215

Macleod, E. L., & Ney, D. M. (2010). Nutritional Management of Phenylketonuria. *Annales Nestle*, *68*(2), 58–69. doi:10.1159/000312813

Marzuillo, C., De Vito, C., D'Addario, M., Santini, P., D'Andrea, E., Boccia, A., & Villari, P. (2014). Are public health professionals prepared for public health genomics? A cross-sectional survey in Italy. *BMC Health Services Research*, *14*(1), 239. doi:10.1186/1472-6963-14-239 PMID:24885316

National Institute of Standards and Technology (NIST). (2015). *NIST Big Data Interoperability Framework: Volume 1, Definitions* (NIST Special Publication 1500-1). Available from: http://nvlpubs.nist.gov/nistpubs/SpecialPublications/NIST.SP.1500-1.pdf

Platt, L. D., Koch, R., Hanley, W. B., Levy, H. L., Matalon, R., Rouse, B., ... Friedman, E. G. (2000). The international study of pregnancy outcome in women with maternal phenylketonuria: Report of a 12-year study. *American Journal of Obstetrics and Gynecology*, *182*(2), 326–333. doi:10.1016/S0002-9378(00)70219-5 PMID:10694332

Roberts, J. S., Dolinoy, D. C., & Tarini, B. A. (2014). Emerging issues in public health genomics. *Annual Review of Genomics and Human Genetics*, *15*(1), 461–480. doi:10.1146/annurev-genom-090413-025514 PMID:25184533

Seshadri, S., Drachman, D. A., & Lippa, C. F. (1995). Apolipoprotein E epsilon 4 allele and the lifetime risk of Alzheimer's disease. What physicians know, and what they should know. *Archives of Neurology*, *52*(11), 1074–1079. doi:10.1001/archneur.1995.00540350068018 PMID:7487559

Simone, B., Mazzucco, W., Gualano, M. R., Agodi, A., Coviello, D., Dagna Bricarelli, F., ... Boccia, S. (2013). The policy of public health genomics in Italy. *Health Policy (Amsterdam)*, *110*(2-3), 214–219. doi:10.1016/j.healthpol.2013.01.015 PMID:23466031

US Food and Drug Administration (FDA). (2018). *23andMe, Inc. 11/22/13*. Retrieved from https://www.fda.gov/ICECI/EnforcementActions/WarningLetters/2013/ucm376296.htm

US Food and Drug Administration (FDA). (2018). *Table of Pharmacogenomic Biomarkers in Drug Labeling*. Retrieved from https://www.fda.gov/Drugs/ScienceResearch/ ucm572698.htm

Wang, L., McLeod, H. L., & Weinshilboum, R. M. (2011). Genomics and Drug Response. *The New England Journal of Medicine*, *364*(12), 1144–1153. doi:10.1056/NEJMra1010600 PMID:21428770

Wilfond, B., & Ross, L. F. (2009). From genetics to genomics: Ethics, policy, and parental decision-making. *Journal of Pediatric Psychology*, *34*(6), 639–647. doi:10.1093/jpepsy/jsn075 PMID:18647793

Zerhouni, E. (2003). Medicine. The NIH Roadmap. *Science*, *302*(5642), 63–72. doi:10.1126cience.1091867 PMID:14526066

Zimmern, R. L., & Khoury, M. J. (2012). The impact of genomics on public health practice: The case for change. *Public Health Genomics*, *15*(3-4), 118–124. doi:10.1159/000334840 PMID:22488453

KEY TERMS AND DEFINITIONS

Big Data: Stored data sets that can be manipulated and analysed to develop new interventions.

Clinical Health Practice: The application of health sciences in interventions designed to enhance the well-being on an individual.

Direct-to-Consumer (DTC) Genetic Testing: Genetic testing marketed to consumers without the direct involvement of a healthcare professional.

Epidemiological Studies: The study of distribution of disease in populations and the application to improve the health outcomes of those populations.

Epigenomic Profiling: A process used for disease stratification and for personalized medicine made possible by the epigenetic changes that occur in the genome. These include DNA methylation and post-translational histone modifications or chromatin restructuring.

Evidence-Based Intervention: Practices or programs that have documented empirical evidence of effectiveness.

Genetic Exceptionalism: The notion that genetic information is unique and should be treated differently from other medical information.

Genomic Biobanks: A collection of biological samples of genetic data for use in research.

Genomic Biomarkers: DNA or RNA characteristic that can be used as an indicator of normal biological processes, pathogenic processes, as well as response to interventions.

Health Disparities: Higher burden of illness, injury, mortality, or violence experienced of socially disadvantaged populations.

Health Equity: Occurs when every person has the opportunity to reach their fullest health potential.

Precision Medicine: Customized medical care that is tailored to each patient's individual characteristic.

Public Health Practice: The application of public health sciences to improve the overall health of the public.

Public Health Specialist: Health professionals that focus on prevention of disease and work towards creating conditions where populations can be healthy.

Translational Research: Improvement of health outcomes through conducting systemic reviews of genomic studies to address medical needs.

Appendix

END OF CHAPTER QUIZZES KEYS

Chapter 1

1. D
2. C
3. E
4. B
5. A
6. D
7. B
8. A
9. E
10. A

Chapter 2

1. E
2. A
3. B
4. C
5. E
6. B
7. B
8. C
9. E
10. A

APPENDIX

Chapter 3

1. E
2. C
3. A
4. B
5. C
6. E
7. D
8. C
9. B
10. E
11. C
12. D

Chapter 4

1. E
2. B
3. A
4. C
5. C
6. D
7. C
8. A
9. B
10. D

Chapter 5

1. E
2. A
3. E
4. B
5. C
6. D
7. C
8. A
9. B
10. B

APPENDIX

Chapter 6

1. D
2. E
3. E
4. A
5. A
6. E
7. D
8. C
9. D
10. B

Chapter 7

1. D
2. A
3. B
4. E
5. D
6. E
7. C
8. B
9. C
10. A

Chapter 8

1. A
2. C
3. B
4. A
5. A
6. C
7. E
8. B
9. D
10. E

Chapter 9

1. D
2. C
3. C
4. B
5. B
6. A
7. D
8. D
9. E
10. D

Chapter 10

1. B
2. C
3. B
4. D
5. D
6. D
7. C
8. A
9. A
10. B

Chapter 11

1. C
2. A
3. E
4. B
5. E
6. C
7. C
8. A
9. B
10. B

APPENDIX

Chapter 12

1. B
2. E
3. A
4. D
5. A
6. C
7. E
8. B
9. E
10. D

Chapter 13

1. E
2. A
3. D
4. D
5. B
6. B
7. A
8. D
9. C
10. E
11. E
12. D

Chapter 14

1. E
2. C
3. D
4. A
5. D
6. E
7. E
8. B
9. E
10. C
11. B

APPENDIX

Chapter 15

1. E
2. A
3. B
4. B
5. B
6. C
7. A
8. D
9. E
10. D
11. A

Chapter 16

1. B
2. E
3. D
4. A
5. B
6. C
7. A
8. E
9. E
10. A

Chapter 17

1. B
2. A
3. C
4. C
5. D
6. A
7. B
8. A
9. B
10. E

APPENDIX

Chapter 18

1. E
2. A
3. D
4. C
5. A
6. E
7. E
8. E
9. E
10. C

Chapter 19

1. E
2. E
3. C
4. B
5. A
6. D
7. B
8. A
9. E
10. D

Glossary

1000 Genomes Project: An international research effort launched between 2008 and 2015 that was responsible for generating the largest public catalogue of genetic variation and genotypic data.

2:8:8:8 Octet Rule: A rule in electron configuration that focuses on the tendency of atoms to bond in a way that mimics the number of valence electrons in inert gases.

23 Chromosome Pairs: The normal number of chromosome pairs that human cells contain.

3' OH End: Refers to the 3' carbon of DNA and RNA sugar backbone that has a hydroxyl group attached to it.

40S Ribosomal Subunit: In eukaryotes, the smaller subunit of the ribosome complex that is responsible for decoding mRNA.

5' Carbon End: Refers to the 5' carbon of DNA and RNA backbone that has a phosphate group attached to it.

60S Ribosomal Subunit: In eukaryotes, the larger subunit of the ribosome complex that is responsible for peptide formation.

7-Methylguanosine Cap: A cap that is added to the 5' end of the hRNA molecule that protects the mRNA from degradation.

A (Aminoacyl) Site: The binding site in a ribosome that holds the incoming aminoacyl-tRNA complex, or the next amino acid that is to be adding to the growing polypeptide chain.

ACE Inhibitors: A drug used to lower blood pressure and heart failure by reducing the amount of angiotensin II. This allows blood vessels to relax and widen so blood can easily flow through.

Achondrogenesis: A group of conditions that affect the development of bone and cartilage.

Activation Energy: The minimum amount of energy required for a chemical reaction to proceed.

Glossary

Activators: Molecules that recognize and bind to enhancer sites located upstream of the promoter region.

Active Site: The specific site of an enzyme, in which binding of a certain substrate catalyzes a particular reaction.

Adaptor Proteins: Proteins that facilitate the binding of a mRNA transcript by recruiting appropriate signal components such as receptors. They form complexes with other proteins to regulate signal transduction pathways.

Addison's Disease: A disorder in which the adrenal gland insufficiently produces cortisol and aldosterone.

Adeno-Associated Viruses (AAV Vectors): These are small viruses widely spread throughout the human population, but do not cause infection. AAVs are often used by scientists as ideal delivery vehicles for gene therapy.

Adenoviruses (AD): A class of double-stranded DNA viruses that have been known to cause eye and respiratory diseases.

Adhering Junctions: A type of junction that helps cells stay in place by joining with neighboring cells through cytoskeletal proteins.

Adhesion: The attraction of molecules of one kind for molecules of a different kind.

Adipose Tissue: Fat stored as loose connective tissue composed of adipocytes that is found under the skin, around internal organs between muscles, in bone marrow and in breast tissue.

Adjuvant Treatment: Extra treatments used in addition to the main cancer treatment to prevent cancer from returning.

Albuminuria: Abnormal amounts of albumin present in the urine.

Allele: Alternative forms of the same gene responsible for a given trait.

Allogeneic Stem-Cell Transplantation: A treatment involving stem cell transfer from a healthy donor to the patient after chemotherapy or radiation.

Allosteric Activation: A type of activation in which binding to an allosteric site, increases the formation of more product.

Allosteric Site: A site, other than the active site, located on the enzyme.

Alopecia: Hair loss in round patches.

Alpha Rays: A form of radiation with two protons and two neutrons (like helium nucleus) emitted when an atom undergoes radioactive decay; considered as the most destructive form of radiation.

Alternative Splicing: A phenomenon of splicing that results in different options or patterns for a given gene; a process mediated by introns, exons, and proteins that increases the complexity of gene expression.

Amino Acids: The building blocks or monomers of proteins.

Aminoacyl-tRNA Synthase: Enzyme that catalyzes the addition of a corresponding amino acid to its tRNA.

Amphipathic: Having both hydrophilic and hydrophobic elements.

Amplicon: A piece of DNA that is the product of PCR amplification.

Amyloid-B Protein: Aggregates of misfolded proteins, a major hallmark of Alzheimer's Disease.

Anabolism: Metabolic reactions that require energy to synthesize biomolecules.

Analysis of Variance (ANOVA): A statistical testing method that can be used to determine differences between the means of three or more variables.

Anaphase I: Phase in meiosis where homologous chromosomes move away from the equator of the cell toward the spindle poles.

Anaphase II: During this phase, the chromosomes pull away from the metaphase plate and move toward the spindle poles.

Aneurysms: A condition that occurs when the artery walls weaken, allowing it to abnormally widen.

Angina: Chest pain.

Angle-Closure Glaucoma: A rarer form of glaucoma that happens when the iris is not as wide and open as it normally is and blocks the drainage canals causing a quick increase in internal eye pressure.

Animalia: Multicellular, heterotrophic eukaryotes that lack a cell wall; includes both invertebrates and vertebrates.

Antibiotics: A range of powerful drugs used to treat bacterial infections by killing or slowing down bacteria growth.

Glossary

Antibodies: A protein that responds to the presence of specific antigens and evokes an immune response.

Anticodon: A set of 3 nucleotides on a tRNA that correspond to a complementary codon in mRNA.

Antigen: A molecule that binds to an antibody.

Antihistamines: A drug used to relieve symptoms of or prevent allergic reactions.

Antipsychotics Drugs: A group of drugs used to treat mental health conditions.

Antisense Oligonucleotides: Short synthetic polymers that can bind to specific RNA molecules to alter protein expression.

Apheresis: A medical procedure in which an individual component is removed from a patient's' blood using an apparatus that separates the components and returns the remaining blood into the patient.

Apoptosis: Programmed cell death.

Aqueous Solution: A type of homogeneous mixture in which water is the dissolving medium.

Archaebacteria: The oldest Kingdom of unicellular prokaryotes that thrive in harsh conditions.

Argyrophilic Grain Disease: A neurodegenerative disease marked by the presence of argyrophilic grains.

Aromatase Inhibitors: A hormone blocking drug that works by blocking the aromatase enzyme, thereby stopping the production of oestrogen.

Arrhythmia: A disorder in which the heart beats with an irregular rate or rhythms.

Arteriosclerosis: Abnormal thickening or hardening of arteries.

Artificial Selection: Breeding of plants or animals to obtain preferable traits.

Asexual Reproduction: A type of reproduction in which a single parent organism reproduces itself, resulting in the formation of a genetically identical offspring; associated with mitosis.

Astigmatism: A refractive error in which the cornea's curvature is abnormal causing blurry vision.

Astrogliosis: A chemical reaction involving an increase of astrocytes due to damage after CNS injuries.

Ataxia: A neurological condition involving the loss of voluntary muscle movement.

Atelosteogenesis: One of many conditions that affect the development of bone and cartilage of infants that have varying degrees of severity.

Atherosclerosis: A disease caused by a buildup of plaque in the arteries.

Atoms: Smallest unit of matter.

Attenuation: A modulation system in which transcription is halted before the operon genes are transcribed.

Attenuator: A regulator region within the leader region of the trp operon that will halt transcription.

Augmentation Therapy: A treatment used to increase the level of alpha-1 antitrypsin (AAT) protein circulating in the blood and lungs by intravenously transferring AAT from the blood plasma of a healthy donor.

Autoradiography: A technique that utilizes X-rays to visualize radioactively labeled molecules.

Autosomes: 22 pairs of chromosomes that exist in the human genome and are not sex chromosomes.

Axonopathy: Degradation of the axons of peripheral nerves.

B-Adrenergic Receptor Antagonists (B-Blockers): Drugs that inhibit the binding of norepinephrine and epinephrine by competitively binding to beta-adrenoreceptors.

Bacteria: Single-celled organisms that lack a nucleus; also referred to as prokaryotes.

Bardoxolone Therapy: A treatment used for kidney disease that activates signalling pathways that reduce inflammation and improves mitochondrial activity.

Benzothiazepines: A subclass of calcium channel blocking drugs that can be used to reduce the force of cardiac contractions as well as a vasodilator.

Bestrophinopathies: A group of five clinically distinct retinal degenerative diseases caused by a mutation in the *BEST1* gene.

Beta Rays: A form of radiation with a -1 charge and a mass of about 1 electron.

Big Data: Stored data sets that can be manipulated and analysed to develop new interventions.

Bilayers: A membrane structure composed of two layers, often formed through the aggregation of phospholipids.

Glossary

Binary Fission: A type of asexual reproduction in which a single-celled organism reproduces itself via DNA replication and division.

Binding Proteins: Proteins that act as a link to bring two molecules together.

Bioeconomy: An economy in which biological processes and biological based products are used to reduce environmental impacts and improve overall global sustainability.

Biomarkers: A biological characteristic that is used as an indicator of normal biological processes.

Bioprocessing: The process and application of biotechnology-based tools as a means for commercial use.

Bioremediators: Biological organisms such as bacteria and fungi—microorganisms that are used to reduce the amount of environmental pollutants.

Biotechnology: The utilization and application of knowledge, techniques, and procedures from a variety of biology-related disciplines.

Biotin (Vitamin B7): A vitamin B-complex that aids in carbohydrate and fat metabolism and in fatty acid and glucose synthesis.

Blood Transfusion: The process in which donated blood is transferred to another patient through the veins

Bone Remodeling: A highly regulated cycle where osteoclasts destroy old bone so that osteoblasts make new bone.

Brachytherapy: A type of radiotherapy that involves the placement of radioactive material inside or near the tumour to treat cancers.

Bradycardia: Lower than normal heart rate.

Bronchodilators: Substances used to dilate the bronchial tubes and relaxing the lung muscles.

Buffers: Substances used to resist fluctuations in pH upon the addition of an acid or base.

cAmp Receptor Protein (CRP): A regulatory protein that binds cAMP and causes a conformational change in CAP.

Carcinogenesis: The process of normal cells transforming into cancer cells.

Cardiogram: Produced by a cardiograph, this is a record of heart muscle activity.

Cardiomyopathy: A disease of the heart muscle in which the heart muscle can become thick, enlarged or stiff.

Cardiomyopathy: A disease of the heart muscle in which the heart muscle can become thick, enlarged or stiff.

Cardioverter-Defibrillators: A small battery powered device that is surgically placed under the skin below the collar bone and is used to detect heart rate abnormalities. If dangerous heart rate is detected, the device sends an electrical shock to correct the rhythm.

Carnivore: An animal that consumes other animals.

Carotenoids: Pigment molecules found in chloroplasts that aid in photosynthesis and protection.

Catabolism: Metabolic reactions that breakdown biomolecules for energy.

Catabolite Activator Protein (CAP): The cAMP-CRP complex that allows CRP to bind tightly to the promoter to begin gene transcription.

Cataplexy: A condition characterized by sudden and uncontrollable voluntary muscle weakness caused by strong emotion or laughter and often results in physical collapse while conscious.

CDK-Cyclin Complex: The complex that forms when cyclin binds to CDK.

Cell Division: Process of a cell splitting into two daughter cells with identical genetic material.

Cell Theory: All living things are made up of one or more cells.

Cell Wall: Rigid outer layer surrounding the cell membrane that gives a cell protection and shape.

Cell: The basic unit of structure and function in all living organisms.

Cellular Machinery: A network of organelles and structures that aid in cellular functions.

Cellular Metabolism: The sum of chemical processes in an organism.

Cellular Respiration: The process in which carbohydrates are broken down into carbon dioxide and water through the use of oxygen.

Central Vacuole: Large structure inside plant cells that occupies majority of the volume.

Centromere: A structure that links two chromatids together to form a chromosome.

Glossary

Cerebral Amyloid Angiopathy: A condition caused by a build-up of amyloid on the walls of the arteries in the brain which can lead to a brain haemorrhage and dementia.

Chemical Bonds: A union between atoms in a molecule.

Chemotherapy: Therapy designed to treat cancerous diseases through the administration of cytotoxic drugs that aim to regulate and prevent further growth and division of cancerous cells.

Chiasmata: A cross-shaped structure that connects homologues together, serving as a link.

Chloroplasts: A type of plastid that is used in photosynthesis.

Cholesterol: A compound composed of a four-ring carbon structure with a carbon side chain.

Cholinesterase Inhibitors: A class of drugs that prevent the breakdown of acetylcholine.

Chromatids: Half of an identical copy of a replicated chromosome.

Chromatin Remodelling: Regulated modification of chromatin to allow DNA to be compressed into the nucleus of a cell.

Chromatin: A DNA and protein complex that forms a chromosome.

Chromoplasts: A type of plastid that synthesizes and stores mainly yellow and red photosynthetic pigments.

Chromosomal Mutations: Larger-scale structural mutations that involve changes in whole segments of DNA.

Chromosomes: Units that contain chromatin specific for each species (e.g. humans have 46 chromosomes).

Chronic Pancreatitis: Prolonged inflammation of the pancreas.

Chronic Rhinosinusitis: Inflammation of the sinus membranes.

Chylomicrons: A class of lipoproteins that transport lipids from the intestines to other tissues in the body.

Cirrhosis: A degenerative disease that occurs when there is scarring of liver tissue.

Clinical Health Practice: The application of health sciences in interventions designed to enhance the well-being on an individual.

Cloning: The process of making an identical copy.

Codominance: A genetic effect in which both alleles of are phenotypically expressed in an organism.

Codons: A set of 3 nucleotides or triplets on the mRNA that codes for a specific anticodon.

Coenzymes: Organic compounds that work with enzymes to facilitate their activity.

Coevolution: The process by which two or more species influence each other's evolution.

Cofactors: Small organic, or inorganic compounds that work with enzymes to facilitate their activity.

Cohesion: The attraction of molecules for other molecules of the same kind.

Colonoscopy: An exam of the colon and rectum using a small camera to search for polyps and colon cancer.

Competitive DNA Binding: A competition to bind DNA when there is overlap between enhancer and silencer sequences.

Competitive Inhibition: A type of reversible inhibition in which an inhibitor binds to the active site, increasing the Km of the reaction.

Complementary DNA (cDNA): DNA synthesized from mature mRNA transcripts that can be used to analyze the structure, organization, and expression of eukaryotic genes.

Complex Organization: Levels of organization classified from the simplest living organisms to the most complex organisms.

Compound Light Microscope: A microscope with more than one lens and its own light source to send the image to your eye.

Computed Tomography (CT): A diagnostic imaging test that uses several x-rays to scan and produce detailed images of areas inside the human body.

Condensation (Dehydration): A reaction involving the use of water to form a chemical bond.

Congenital: A condition or trait present from birth.

Conjugation: DNA transfer via cell-to-cell contact.

Connective Tissue: One of four basic tissue types in the body; supportive tissue such as bone, cartilage, and fat tissue that serve protection and support functions.

Glossary

Consumers: Organisms that consume other organisms for energy. They are also known as heterotrophs.

Control Checkpoints: Points in the cell cycle that prevent the cell from progressing to the next phase until conditions are favorable.

Control Group: A group in a study or experiment that receives no treatment, or a placebo, and is used as a comparison for the experimental group.

Cooperative Lattices: Sets of similarly functioning structures organized in bundles as seen in human taste buds.

Coronary Angioplasty: A procedure in which a small balloon is inflated to open blocked or narrowed coronary arteries.

Coronary Artery Bypass Grafting: A surgery that improves blood flow to the heart by creating more pathways for the blood to flow using arteries or veins from other areas of the body.

Correlation Analysis: A statistical tool used to understand the relationship of two variables.

Covalent Bond: A type of chemical bond involving the sharing of electrons between atoms.

CRISPR Arrays: Short segments of DNA that allow bacteria to remember and recognize when an infectious agent invades again.

cRMI: Adaptor proteins that assemble nuclear export receptors for different classes of mRNA.

Crossing Over: The exchanging of genes between two non-sister chromatids that occurs during prophase I.

Cyclic AMP (cAmp): An intracellular protein activator or second messenger in signal transduction of various cellular processes.

Cyclin: Regulatory proteins that bind and activate CDK.

Cyclin-Dependent Protein Kinase (CDK): A group of enzymes that phosphorylate specific proteins.

Cytokinesis: The cellular process of cytoplasmic division that results in the formation of the cleavage furrow and two daughter cells.

Cytoplasm: Intracellular fluid that holds all contents of the cell (except the nucleus in eukaryotes).

Cytosol: The aqueous component of a cell that make up the cytoplasm, proteins and other cellular structures.

Decomposers: Organisms that break down organic material.

Decomposition: The breakdown of organic material into simpler substances.

Dementia: Disease characterized by mental instability, impaired memory, and possible personality changes.

Demyelination: Damage to the protective myelin sheath covering the nerve fibres.

Deoxyribonucleic Acid (DNA): A type of nucleic acid that contains an organism's hereditary information; composed of adenine, guanine, thymine, and cytosine bases.

Designer Babies: A baby whose genetic makeup is selected for to include desirable traits.

Detritus Feeders: Organisms that consume dead plants and animals.

Development: The process of growing and become more mature or advanced.

Dihybrid Inheritance: Inheritance of two different characteristic traits.

Dihydropyridines: A subclass of calcium channel blocking drugs that is often used to reduce peripheral resistance and arterial pressure.

Diploid: Term used to describe two sets of chromosomes in a cell.

Direct-To-Consumer (DTC) Genetic Testing: Genetic testing marketed to consumers without the direct involvement of a healthcare professional.

Disulphide Bridge: A bond formed between the thiol groups of 2 cysteine amino acids.

DNA Fingerprint: A unique pattern of repeated sequences, or RFLPs, obtained via gel electrophoresis, that can be used to identify an individual's DNA.

DNA Helicase: An enzyme that initiates the unwinding (or unfurling) of DNA during DNA replication.

DNA Library: An assortment of DNA fragments that have carefully been cloned into vectors.

DNA Ligase: Enzyme that joins the fragmented nucleotides together via a phosphodiester bond.

DNA Microarrays (Nucleic Acid Arrays): Specific DNA sequences that are deposited and bound to a solid surface such as a glass microscope slide.

Glossary

DNA Polymerase I: Enzyme that has proofreading ability, where it removes nucleotides from the 5' end and replace with the correct DNA.

DNA Polymerase III: A holoenzyme that catalyzes the synthesis of the complementary strand of DNA in the 5' to 3' end direction.

DNA Primase: Enzymes that create RNA primers complementary to DNA strands.

Dolichocephaly: A condition characterized by an abnormally elongated head relative to head width.

Domains: A taxonomy that classifies three different cellular life forms; Archaea, Bacteria and Eukarya.

Dominant: Term used to describe the allele whose effects completely mask the expression of the other; traits which are abbreviated by upper-case letters.

Double-blind Experiment: A type of experiment in which the subject nor the administrator knows the treatment or test being received; it is designed to eliminate or reduce potential bias in the experiment.

Dry Weight: The normal weight of an organism that excludes excess fluid.

Dysarthria: A speech disorder characterized by poor articulation of speech.

Dysdiachokinesia: The inability to perform quick, alternating movements.

Dysmetria: A condition that occurs as a result the development of lesions in the cerebellum which then affects coordinated movement, thought processing, and behaviour.

Dysphagia: Difficulty swallowing.

Dysphasia: A condition in which the ability to produce and understand language is affected.

Dysplasia: A phase where cells are not cancerous but look abnormal under a microscope.

Dyspnoea: Shortness of breath or difficulty breathing.

Dystonia: A neurological disorder marked by involuntary muscle contractions.

Dystrophin: A protein that plays a role in skeletal and cardiac muscle movement.

E (Exit) Site: The binding site in a ribosome where the deacylated tRNA will exit the ribosome.

Ectoderm: One of three primary germ layers, responsible for forming external structures such as the skin, hair, nails, and nervous system.

Ectodermal Dysplasia: A group of genetic disorders involving the abnormal development of two or more ectodermal structures: skin, sweat glands, nails, teeth, hair and mucous membranes.

Egg: Female reproductive cell.

Electron Microscope: A microscope that uses electrons to create an image and has higher magnification and resolving power.

Electron Transfer: A type of reaction involving the loss or gain of electrons in the formation of ions from atoms.

Electrons: Negatively charged subatomic particles that orbit an atomic nucleus.

Embryo: Forms in the first trimester following fertilization of an egg by a sperm cell during the developmental stage of an organism.

Emphysema: Type A COPD caused by irreversible damage to alveolar walls or the enlargement of distal air sacs.

Encyclopaedia of DNA Elements (ENCODE) Project: A public research association that started in 2003 with a main goal of identifying all functional elements in the genomes of humans and mice.

Endocrine Disruptor: Exogenous chemicals that alter or interfere with the body's normal endocrine function, thus further impairing homeostasis.

Endocytosis: Movement or uptake of extracellular material into a cell.

Endoderm: One of three primary germ layers, responsible for forming internal structures such as glandular organs and the gut lining.

Endoplasmic Reticula: A large folded membrane system consisting of the rough ER and smooth ER.

Endosymbiosis: A form of relationship in which one cell lives inside another to form and acts as a single organism.

Energy: The capacity to do work.

Enhancesome: A type of enhancer containing multiple sites for activator binding.

Enzyme Replacement Therapy (ERT): The process of intravenously replacing enzymes that are deficient in the body.

Glossary

Enzyme: A biological catalyst that acts to lower the activation energy of a reaction, often composed of proteins.

Epidemiological Studies: The study of distribution of disease in populations and the application to improve the health outcomes of those populations.

Epigenetics: The study of changes in an organism that influences gene expression but are not mediated by the original DNA sequence.

Epigenomic Profiling: A process used for disease stratification and for personalized medicine made possible by the epigenetic changes that occur in the genome. These include DNA methylation and post-translational histone modifications or chromatin restructuring.

Epigenomics: The identification of the complete set of epigenetic modifications implicated in gene expression.

Epilepsy: A neurological disorder that causes seizures.

Epinephrine: Hormone secreted by the medulla of the adrenal gland, primarily during stress.

Epithelial Cells: Tightly packed cells that line the surface of the body and act as a protective barrier.

Epithelial Tissue: One of four basic tissue types in the body; covering or lining tissue that serves to protect, support, secrete, and absorb (i.e., skin).

Erythropoiesis: Production of red blood cells which is initiated by the decreasing levels of oxygen which cause erythropoietin hormone production in the kidneys.

Essential Amino Acids: Amino acids that cannot be produced by the organism and must be supplied by its diet.

Ester Bond: A bond formed from a dehydration reaction between a carboxylic acid and an alcohol.

Ethics: The study of ethical and moral theories and their relation to different social systems and cultures.

Eubacteria: The most common type of asexual, unicellular prokaryotes; eubacteria can be autotrophic or heterotrophic and move by cilia, flagella and pseudopods.

Eukaryotes: Multicellular or unicellular organisms with membrane bound organelles.

Evidence-Based Intervention: Practices or programs that have documented empirical evidence of effectiveness.

Evolution: The idea of change in the genetic composition over time.

Exons: Coding regions of mRNA that are spliced together by spliceosomes during post-translational modification and will exit the nucleus.

Exoskeleton: The structural component of insects that provide support and protection for the body.

Experimental Group: A group in a study that receives treatment.

Facilitated Diffusion: A form of passive transport in which a solute moves down its concentration gradient via channels or pores in the membrane.

Family Tree: A genealogical diagram showing familial relationships of many generations.

Fat: Lipids that are solid at room temperature; a hydrophobic molecule usually composed of long hydrocarbon chains.

Fatty Acids: Long hydrocarbon chains with either single or double bonds between carbon atoms and a carboxyl functional group at one end.

Feedback Inhibition: A type of inhibition in which product formation decreases enzyme activity.

First Messenger: An extracellular signal that initiates intracellular activity through receptors on the cell surface (e.g. hormone).

Fitness: A measure of one's ability to pass on genes and reproduce.

Foamy Viral Vectors: Vectors that contain nonpathogenic foamy viruses that can be used in the delivery of genes to a number of different cell types.

Focal Therapy: The use of focal modalities to treat small tumours.

Foetus: An unborn, developing offspring that occurs between the embryonic stage and birth.

Forensic Evidence: Evidence that is obtained by scientific methods and is often used in criminal and legal proceedings.

Fossils: Preserved remains or remnants of a living organism.

Frameshift Mutations: Mutations that occur from the insertion or deletion of nucleotides and results in a shift in the reading frame.

Glossary

Functional Group: Group of atoms that are recognized for specific characteristics within a molecule.

Functional Group Transfer: A type of reaction involving the transfer of a functional group.

Fungi: Eukaryotic organism that include: yeasts, molds, and mushrooms. Most fungi are unicellular and are considered decomposers. They lack chlorophyll and have chitin in their cell walls.

G_1 Phase: The first stage of the cell division cycle in which most of the cell growth is occurring; also referred to as Gap 1.

G_2 Phase: The third stage of the cell division cycle that results in the formation of condensed and shortened chromosomes, organelles, and a spindle.

Gametes: Sex cells (eggs in female or sperms in male).

Gametogenesis: The process of the gamete production.

Gamma Rays: A form of radiation having no charge or mass.

Gap Junctions: Pore-like connections between adjacent cells that allow cytoplasmic mixing and the exchange of small molecules such as ions and amino acids.

Gel Electrophoresis: A method used in the lab to separate DNA fragments based on their size.

Gene Therapy: The replacement of defective genes with functional genes or the introduction of new genes to prevent or treat diseases.

Genes: Units of information about specific traits on DNA.

Genetic Adaptation: Traits that result from natural selection and random variation, which give rise to favorable genetic changes in an organism.

Genetic Code: Standard set of rules and meanings of nucleotide triplets in which all life forms share.

Genetic Engineering: The process of altering the characteristics of an organism by insertion of modified genes.

Genetic Exceptionalism: The notion that genetic information is unique and should be treated differently from other medical information.

Genetic Material: Components that must contain information, replicate accurately, and are capable of change.

Genetic Mutations: Heritable alterations in one or more base pairs that damage DNA and can affect a single nucleotide pair or larger segments of a chromosome.

Genetically Modified (GM) Crops: Crops that have one or more genes coding for desirable traits have been inserted through the process of genetic engineering

Genetically Modified Organisms (GMOs): Organisms whose genomes have been genetically engineered to incorporate genes that are not normally found in their genomes.

Genome Mutations: Mutations that affect the normal chromosome number.

Genome-Wide Analyses: The process of measuring and identifying gene features of entire genomes.

Genome-Wide Association Studies (GWAS): An observational study that examines if there is an association between genetic variants and traits in different individuals.

Genomic Biobanks: A collection of biological samples of genetic data for use in research.

Genomic Biomarkers: DNA or RNA characteristic that can be used as an indicator of normal biological processes, pathogenic processes as well as response to interventions.

Genomics: A study of how bases on DNA determine whole sets of genes and how they function together.

Genotype: The actual alleles present in an individual.

Genotypic Ratio: The ratio used to describe differences in allele frequency from a cross or breeding event.

Germ Cells: Reproductive cells (Ovaries in females and testes in males).

Glucagon: A peptide hormone produced and secreted pancreatic α-cells; the release stimulates the breakdown of glycogen in the liver, thus increasing blood glucose levels.

Glucose: A simple sugar that humans use as a major source of energy.

Glycerol: An alcohol composed of a three-carbon chain.

Glycogen: A polysaccharide composed of glucose residues; acts as energy storage for animals.

Glycolipid: A lipid with a carbohydrate covalently attached to it.

Glossary

Glycoprotein: A protein with a carbohydrate covalently attached to it.

Glycosidic Bond: A covalent bond between the anomeric carbon of a sugar and an alcohol or amine of another molecule formed by a dehydration reaction.

Golgi Complex: A eukaryotic organelle that sorts and packages proteins before they are transported to their final destination.

G-Proteins: A signaling molecule that binds GTP embedded in the plasma membrane.

Growth Factors: Molecular signals that prioritize and promote rapid cell division.

Growth: The developmental process by which an organism matures or increases in size.

Guide RNA: An RNA sequence used in CRISPR Cas9 technology to direct enzymes to specific DNA sequences.

Haematuria: Presence of blood in the urine.

Haemolysis: Rupture of red blood cells.

Halogen: Non-metal and gas elements that compose Group 17 in the periodic table.

Haploid: Term used to describe one set of chromosomes in a cell.

HapMap Project: A project aimed to creating a halotype map of the human genome in order to find genetic variants that affect human health and diseases.

Health Disparities: Higher burden of illness, injury, mortality, or violence experienced of socially disadvantaged populations.

Health Equity: Occurs when every person has the opportunity to reach their fullest health potential.

Hemoglobinopathies: A group of disorders involving abnormal haemoglobins and anaemia which are caused by defects in the genes that produce haemoglobin.

Hepatocytes: Liver cells.

Hepatosplenomegaly: Abnormal enlargement of both the spleen and liver

Herbicides: A pesticide used to kill unwanted vegetation.

Herbivore: An animal that consumes plants.

Hereditary Material: DNA (Deoxyribonucleic Acid) and RNA (Ribonucleic Acid).

Herpes Simplex Virus (HSV): Virus responsible for causing herpes and herpes-related infections.

Heterozygous: Terminology used to describe different alleles.

High Heat Capacity: A property used to describe a substance's ability to absorb a great amount of heat or energy for a given increase in temperature.

High-Density Lipoproteins (HDL): Responsible for the transportation of cholesterol back to the liver.

Hirschsprung Disease: A birth defect in which nerve cells that line the intestine do not properly form during development causing the inability to pass stool.

Histidine Operon: A bacterial operon that regulates transcription for histidine biosynthesis via attenuation mechanisms.

Histone: Water-soluble proteins found in chromatin that coil DNA and form nucleosomes.

Homeostasis: The ability of the body to maintain stable conditions internally as the external environment changes.

Homeotherms: Warm-blooded animal that maintains a constant body temperature.

Homeotic Genes: Highly conserved DNA sequences that code for specific transcription factors that regulate gene expression of specific anatomical structures.

Homologous: Term used to describe the same set of genes in a chromosome pair.

Homozygous: Terminology used to describe the same alleles.

Hormonal Therapy: A treatment used to cure or relieve the symptoms of cancer by reducing the amount of oestrogen in the body or blocking the effect of oestrogen on cancer cells.

hRNA: The primary transcript of high-molecular weight formed from DNA that will undergo post-transcriptional modification to become mRNA.

Hugo Gene Nomenclature Committee: A committee that sets the official standard for human genome nomenclature and associated genomic information. It aims to assign a unique yet meaningful name or symbol for each known human gene.

Human Genome Project: An international project created in 1990 with the goal of sequencing and identifying the entire human genome.

Glossary

Human Microbiome Project: A research initiative established in 2008 under the United States National Institutes of Health that enabled the study of human microbiota and its role in the development of human health and disease.

Hybrid Vectors: Genetically modified vectors that contain elements from more than one virus.

Hydrocarbon: A compound consisting of hydrogen and carbon atoms.

Hydrogen Bond: A weak chemical bond involving the interaction of a hydrogen atom.

Hydrogen: An element in the periodic table that contains only one shell, with a single electron and a single proton in the nucleus.

Hydrogenation: A chemical reaction involving the addition of a hydrogen to monounsaturated or polyunsaturated fatty acids.

Hydrolysis: A chemical breakdown of a bond due to a reaction with water.

Hydrolytic Enzyme: Enzyme that uses water to break down molecules into their simplest units.

Hydrophilic: Water loving.

Hydrophobic Interaction: An interaction between water insoluble substances.

Hydrophobic: Non-water loving.

Hyperbilirubinaemia: Elevated levels of bilirubin in the blood.

Hyperphenylalaninaemia: Abnormally elevated levels of plasma phenylalanine.

Hyperplasia: Enlargement of an organ caused by an increased number of cells.

Hypogammaglobulinemia: An immune disorder in which there is a reduction in serum gamma globulins.

Hypoparathyroidism: A condition in which there is a deficiency in the production of parathyroid hormone.

Hypothesis: An assumption or a possible explanation that can later be scientifically tested.

Immunization: The process in which a person is made to become resistant to an infectious disease.

Immunotherapy: A cancer treatment that helps the body's immune system fight the cancer.

Imprinting: A phenomenon where of the two inherited copies of each gene (alleles), one of them is silenced by epigenetic processes such that only one of them (either from mother or father) is expressed.

Inborn: Natural or innate.

Induced-Fit Model: A theory that states that when a proper substrate comes in contact with an enzyme's active site, the enzyme will undergo some conformational change on substrate binding.

Inducer: Protein that binds to the repressor, thus preventing the repressor from binding to the operator.

Industrial Biotechnology: The implementation of biotechnology-based tools into traditional industrial processes as a means to improve and sustain global sustainability.

Inert Gases: Unreactive elements that make up Group 18 of the periodic table; also referred to as noble gases.

Infant: Refers to the offspring after birth that will continue to grow and develop.

Initiator Proteins: Proteins that bind to the replicator and signal to begin replication.

Innermost Shell: The lowest energy shell with a maximum carrying capacity of 2 electrons.

Insecticides: A pesticide used to harm, kill or repel insects.

Insulator/Insulator Protein: A group of elements found between the enhancer and the promoter.

Insulin: A peptide hormone produced and secreted by pancreatic β-cells; promotes the uptake of glucose from liver and muscle cells, thus lowering blood glucose levels.

Integral Proteins: Transmembrane proteins that are embedded in the membrane.

Interdependency: Organisms that rely on resources in their environment to survive.

International Cancer Genome Consortium: An international association established in 2007 with the aim of identifying, defining, and understanding the genomic changes of 25,000 untreated cancer cases.

International HapMap Project: An international project created in 2002 that contributed to the development of a haplotype map of the human genome.

Interphase: The phase when the cell has grown in size, replicated its DNA, and prepared for cell division.

Intraocular Pressure: Pressure inside the eye.

Glossary

Introns: Non-coding regions of mRNA that are removed by spliceosomes during post-translational modification.

Invertebrates: Soft bodied animals lacking a spine.

Ionic Bond: A type of chemical bond formed between oppositely charged particles and electrostatic attraction.

Ionic: Of or relating to ions.

Ions: An electrically charged atom or molecule.

Ischaemic Cardiomyopathy: Heart muscle disease caused by a reduction in blood flow.

Isomer: The product resulting from a molecular rearrangement.

Isotopes: Atoms that share the same number of protons but differ in the number of neutrons.

Junk DNA: Non-coding DNA sequences in the genome.

Karyotype Analysis: A process used to observe the number and appearance of chromosomes in order to detect birth disorders.

Karyotypes: Images of human chromosomes as they line up in pairs according to size.

Kinetic Tremor: A rhythmic, involuntary movement of a body part that is associated with voluntary movements.

Kinetochore: A protein bundle that serves as an attachment point for spindle fibres on a centromere.

Kingdoms: The second highest taxonomy under Domain, that classifies and separates the following into smaller groups called phyla: Eubacteria, Archaebacteria, Protista, Fungi, Plantae and Animalia.

Kyphosis: An abnormally excessive outward curvature or forward rounding of the spine.

Lac Repressor: Protein that acts as a negative regulator and prevents transcription by RNA polymerase when lactose levels are low.

Lactose Operon: An inducible system that requires an inducer (lactose) to remove a repressor protein from the operator site in order to begin transcription of a gene.

Lagging Strand: A new DNA strand that is synthesized discontinuously in the 5' to 3' direction.

Laser Therapy: A treatment involving lasers to destroy blood vessels that surround and supply a tumour.

Leader: A regulatory region of the trp operon that precedes the start of the coding region for the trpE gene.

Leading Strand: A new DNA strand that is synthesized continuously in the 5' to 3' direction.

Lenti-D Gene Therapy: A form of gene therapy that utilizes the lentivirus as the cloning vehicle.

Lentiviruses: A type of retrovirus that contains two regulatory genes and is known for their slow and long incubation period.

Leucoplasts: A type of plastid that lack photosynthetic pigments and are commonly found in storage organs.

Leukoencephalopathy: Neurological disorders of white matter in the brain.

Lewy Body Disease: One of the most common causes of dementia which results from the abnormal accumulation of alpha-synuclein in the brain.

Life: A combination of characteristics of an organism that consist of homeostasis, organization, metabolism, growth, response to stimuli, adaptation, and reproduction.

Ligand: A molecule that serves as the key for a given receptor, usually a hormone or neurotransmitter.

Linkage: Genes that are closer together on the same chromosome are more likely inherited together.

Lipopheresis: Removal of low-density lipoprotein cholesterol from a patient's blood.

Lipoproteins: A lipid and protein complex that aids in the transport of lipids throughout the lymph and blood.

Locus: The location of a gene on a chromosome.

Logarithmic Scale: A non-linear scale used to simplify graphs or tables of large numeric values.

Low-Density Lipoproteins (LDLs): Transport triglycerides and cholesterol from liver cells to tissue cells.

Lumbar Lordosis: Inward curvature of the lower back.

Glossary

Lymphocyte Transfusion: A therapy that involves the introduction of donor lymphocytes to destroy remaining cancer cells.

Lymphoid Hyperplasia: An increase in lymphocytes resulting in the enlargement of lymphoid tissue.

Macrocephaly: A term used to describe a larger than normal head circumference at birth.

Macromolecular Antigens: A class of antigens derived from complex macromolecules that can be used to revolutionize pharmaceutical development.

Macromolecules: Large complex molecules made up of repeating units of the same or different molecules.

Macroorchidism: A condition affecting males that is characterized by abnormally large testes relative to age.

Magnetic Resonance Imaging (MRI): A diagnostic imaging test that uses powerful magnet and radio waves to produce images of body organs and tissues.

Major Elements: A group of central elements required by living organisms; includes carbon, nitrogen, oxygen, and hydrogen.

Male-Specific Region: A region on the Y chromosome that contains many genes involved in turning an embryo into a male.

Mastectomy: A surgical operation involving the removal of one or both breasts to treat breast cancer.

Mature mRNA: mRNA that has been spliced, processed, and is ready for exportation to the cytoplasm to begin the process of translation.

Mean: The average of a set of numbers calculated by getting the sum of all the numbers and dividing it by the total numbers in the set.

Median: The middle value in a set of numbers in organized in a numerical order.

Meiosis: A type of cell division that results in germ cells for sexual reproduction.

Meiosis I: The first half of meiosis in which the cell divides into two non-identical haploid daughter cells..

Meiosis II: The second half of meiosis in which the cell divides into two identical haploid daughter cells.

Membrane Proteins: Proteins that have a hydrophobic surface and are located at the cell membrane.

Membrane: Enclosing boundary consisting of a phospholipid bilayer.

Mesoderm: One of three primary germ layers, responsible for forming "middle structures" such as bones, muscles, kidneys, blood vessels, and the heart.

Metaphase I: Phase in which tetrads randomly align themselves in pairs (or homologues) at the equator of the cell.

Metaphase: Longest phase within the mitotic phase that involves the attachment of spindle microtubules to the kinetochores of chromatids and the formation of the metaphase plate.

Metaphase II: During this phase, chromosomes align themselves along the metaphase plate.

Metaphase Plate: The alignment of individual chromatids at the equator of the cell during metaphase.

Microcephaly: A birth defect in which an infant's head is smaller than other infants of the same age and sex.

Microfilaments: Rod-like structures formed in the cytoplasm, responsible for gross movements of the cell.

Microtubules: A tube-like structure composed of alpha-tubulin and beta-tubulin, that function as support and transportation inside cells.

Minor Elements: A group of elements required in minor proportions by living organisms (i.e., iron, calcium, copper, zinc).

Mismatch Repair: A repair system that corrects any errors in base pairing.

Missense Mutation: A type of point mutation that results in the substitution of one amino acid for another in the final polypeptide.

Mitochondrial DNA: Genetic material (DNA) in the mitochondria that carries the code to convert chemical energy into ATP.

Mitosis: A type of cell division that results in two identical copies of the original cell.

Mitotic (Meiotic) Phase: The phase of a cell when it replicates to form two daughter cells in mitosis and four non-identical daughter cells in meiosis.

Glossary

Mitotic Promoting Factor (MPF): A protein (cyclin-CDK) factor that regulates the transition of cells from G2 to M phase.

Mode: The value that appears most frequently in a set of numbers.

Molecular Rearrangement: A chemical reaction involving the movement of substituents in the same molecule and preservation of the original chemical formula.

Monohybrid Inheritance: Inheritance of one characteristic trait.

Monosomy X: Having only one X chromosome in the female cell due to nondisjunction.

Monounsaturated Fatty Acids: Fatty acids with single double bonds.

Morula: A "ball" of cells containing blastomeres, following the division of a zygote.

mRNA Ribonucleoprotein (mRNP) Complex: Protein complex containing the appropriate export receptors that is translocated through the nuclear pore to the cytoplasm.

Multiple Myeloma: Cancer of the plasma cells.

Muscle Tissue: One of four basic tissue types in the body; the three type include skeletal, cardiac, and smooth.

Mutation: An alteration in the DNA sequence due to an insertion, deletion or rearrangement of a single base pair or fragments.

Myocardial Ischemia: The reduction of blood flow that prevents the heart from receiving enough oxygen.

Myopia: A vision disorder in which the eye is too long causing light to focus in front of instead of on the retina—also known as near-sightedness.

Nanoparticles: A microscopic particle between 1 and 100 nanometres in size.

Nanotechnology: Field of research designed for the implementation of nanomaterials for new biological or biochemical applications.

National Institute of Health (NIH) Roadmap Epigenomics Program: An international consortium with the goal of creating accessible human epigenomic data that can be used to further integrate and expand on current research in the scientific community.

Natural Elements: Naturally occurring elements on Earth.

Natural Selection: The process that favors the survival and reproductive success of organisms that are best adjusted to the environment.

Necrosis: Irreversible cell injury characterized by cell rupture, inflammation, and leakage of contents into extracellular fluid.

Negative Feedback: A regulatory mechanism in the body that reduces a stimulus in order to maintains stable conditions.

Neonate: A fully developed foetus ready for birth.

Nervous Tissue: One of four basic tissue types in the body; composed of specialized cells that function to receive stimuli and conduct impulses throughout the body (i.e., nerve cells, neurons).

Neurocristopathy: A group disorders involving abnormal neural crest cell differentiation and migration.

Neutrons: Uncharged subatomic particles that reside within an atomic nucleus.

Next Generation Sequencing (NGS): A sequencing method used to measure large numbers of individual DNA sequences in a short period of time.

Nicotinamide Adenine Dinucleotide (NAD): An important electron carrier involved in various redox reactions of living systems—for example, aerobic respiration.

Non-Competitive Inhibition: A type of reversible inhibition in which the inhibitor binds to an allosteric site, decreasing the Vmax of the reaction.

Nondisjunction: When homologous chromosomes or sister chromatids fail to separate during meiosis.

Non-Histone Proteins: Protein that are found in chromatins after histones are removed.

Non-Kinetochore: Microtubules that do not interact with the kinetochore of the chromatid, but instead with each other from opposite sides of the cell.

Non-Polar: Chemical property used to describe the equal distribution of electrons in a chemical bond.

Nonsense Mutation: A type of point mutation that results in the substitution of a stop codon for an amino acid in the final polypeptide.

Non-Steroidal Anti-Inflammatory Agents (NSAIDS): One of the most common over-the-counter drug used to treat pain and inflammation by preventing an enzyme called cyclooxygenase from producing prostaglandins.

Non-Viral Vectors: Synthetically produced biological particles that have been modified to carry specific factors of gene expression; often used for therapeutic purposes.

Nucleic Acids: Polymers that are made of nucleotides and form either deoxyribonucleic acid (DNA) or ribonucleic acid (RNA).

Nuclein: A term which was previously used to describe a nucleic acid.

Nucleolus: A structure within the nucleus where ribosomal RNA transcription, processing and ribosome assembly happens.

Nucleoplasm: A liquid substance that supports and holds the nucleus and structures within the nuclear membrane.

Nucleosome: DNA wrapped around histones, forming a repeating unit of chromatin.

Nucleus: An organelle found in eukaryotes that contains the cell's genetic material.

Ocular Hypertension: Pressure inside the eye.

Oestrous: A recurring period of fertility in more advanced mammals that make females more receptive to males.

Oils: Lipids that are liquid at room temperature—usually unsaturated.

Okazaki Fragments: Term used to label the synthesized DNA fragments on the lagging strand.

Oligosaccharides: A saccharide polymer composed of typically 2 to 9 monosaccharides or simple sugars.

Omnivore: An animal that consumes plants and other animals.

Oogenesis: The process of the egg or ovum production.

Open-Angle Glaucoma: A type of glaucoma that develops slowly and often goes unnoticed. It is when the angle where the cornea meets the iris is normal but there is an increase in internal eye pressure and damage to the optic nerve due to clogging of the eye's drainage canals.

Operator: The key regulatory element of transcription initiation in operons that is composed of a specific DNA sequence.

Operon: Structure consisting of a cluster of co-regulated genes transcribed as a single mRNA and plays a role in regulation when binding to repressors or inducers.

Orbital: The region of space around a molecule where an electron is likely to be found; also commonly referred as a shell.

Organelles: Specialized structures found in the cytoplasm of a living cell that carry out specific functions.

Organs: A group of tissues that have a specific function in the body.

Organ-Systems: A group of organs that work together to perform multiple functions in the body.

Ossification: The process of creating bone from cartilage.

Osteoarthritis: A joint disease caused by the breakdown of cartilage between bones.

Osteoporosis: A disease in which bones become thin and weak.

Outermost Shell: The shell that determines the chemical properties of an atom.

Ovaries: Female reproductive organ in which eggs are formed.

Oxidative Stress: An imbalance between free radicals and antioxidants in charge of homeostasis.

"p" arm: The shorter section of the chromosome when it is divided by the centromere.

P (Peptidyl) Site: The binding site in a ribosome where methionine binds and holds the tRNA containing the growing polypeptide chain.

Palliation: To ease the symptoms without curing the underlying cause.

Palliative Treatments: Treatments focused on relieving symptoms and improving the quality of life for patients with serious illnesses.

Particulate Matter: The mixture of solid and liquid particles suspended in the air.

Pathognomic: A characteristic that is indicative of a particular disease.

Pedigree: A diagram used to determine the lineage and probability of inheriting genes in humans and direct ancestors.

Peer-Reviewed: An evaluation of work by experts in the appropriate field to assess the quality before publication.

Peptide Bonds: A type of covalent bond formed through a condensation reaction of two amino acids.

Glossary

Peptidyl Transferase: Enzyme that catalyzes the formation of peptide bonds in a growing polypeptide chain.

Pericarditis: Inflammation of the pericardium, a sac-like membrane that surrounds the heart.

Periodic Table: A table of chemical elements arranged by respective characteristics/properties, including atomic number, mass, electronegativity, etc.

Peripheral Proteins: Proteins that are located on one side of the membrane.

Persistent Organic Pollutants: Toxic organic chemicals that are resistant to environmental degradation and that persist and accumulate in adipocytes of living organisms.

Personal Wellness Profile: A questionnaire that is used to assess an individual's health lifestyle.

Personalized Medicine: The application of genomic advances and gene sequencing knowledge towards the delivery of personalized medical care to improve health outcome and reduce patient mortality and morbidity.

pH: A logarithmic scale that quantifies the acidity or basicity of an aqueous solution.

Phagocytosis: Engulfment of large particles into a cell.

Pharmacogenomics: A sector within biomedicine and biotechnology that aims to revolutionize modern medicine by advancing pharmaceutical development in vaccines, therapies, and biological molecules.

"Pharm" Animals: Animals in which certain genes have been incorporated or altered into their genome through genetic engineering.

Phenotype: The physically observable traits in an individual.

Phenotypic Ratio: The ratio used to describe differences in observable traits from a cross or breeding event.

Phenylalkylamines: A subclass of calcium channel blocking drugs that is used to target cells of the heart.

Phosphodiester Bond: The bond that forms between two phosphate groups of nucleotides.

Phosphorylate: The process of adding phosphate to proteins.

Photodynamic Therapy: A therapy used to treat cancer by using a drug called photosensitizing agents and a special kind of light to kill cancer cells.

Photosynthesis: The process by which all plants, algae and some microorganisms synthesize food using sunlight energy plus the raw materials carbon dioxide and water.

Physical Therapy: Therapy designed to restore, preserve, or improve physical function through exercise or the use of specially designed equipment.

Phytochemicals: Biologically active chemicals present in plants that play a role in human health and disease.

Pinocytosis: Uptake of extracellular fluid into a cell.

Placebo: Inactive pill that has the same appearance as the active drug.

Plantae: Eukaryotic plants that contain chlorophyll and obtain energy through photosynthesis.

Plasma Membrane: A lipid bilayer that separates the internal components of the cell from the extracellular space; also referred to as a cell membrane.

Plasmapheresis: Removal of plasma from a patient's blood.

Plasmids: Small circular pieces of double stranded DNA that can replicate independently of the host's chromosomal DNA, usually found in bacteria.

Plasmodesmata: Narrow channels connected between plant cells that allow the passage of certain molecules.

Plastids: Organelles responsible for photosynthesis and food storage in plant cells and algae.

Pluripotent Bone Marrow Stem Cells: The first step in the developmental process of red blood cells that results in cells that are able to transform into anything.

Poikilotherms: Cold-blooded animal whose body temperature changes with the environment

Point Mutations: Mutations that occur from one nucleotide change for another in DNA.

Polarity: Measure of the unequal or equal distribution of electrons in a covalent bond.

Poly-A Polymerase: Enzyme that catalyzes the addition of adenine nucleotides during post-transcriptional modification.

Poly-A Tail: A set of approximately 200 adenine nucleotides at the 3' OH end of a pre-mRNA.

Polyadenylation: The process by which poly-A polymerase adds on adenine nucleotides are adding onto the 3'OH end of pre-mRNA.

Polyendocrinopathy: A condition when the body develops multiple autoimmune disorders in the endocrine system.

Polymerase Chain Reaction (PCR): A technique used to make copies of a target piece of DNA.

Polyribosomes: A complex of several ribosomes that attach to mRNA and translate in sequence.

Polyunsaturated Fatty Acids: Fatty acids with more than two double bonds.

Positive Gene Regulation: Gene regulation that results in increased enzymatic activity in the presence of product; in contrast to negative gene regulation.

Positron Emission Tomography (PET): A diagnostic imaging test that uses a special dye that contains radioactive tracers which are absorbed by certain organs and tissues.

Poxviruses (PV): Double-stranded DNA viruses responsible for causing smallpox, cowpox, and other related infectious diseases.

Precision Medicine: Customized medical care that is tailored to each patient's individual characteristic.

Precursor Messenger RNA (Pre-mRNA): A primary transcript that contains both exons and introns, and thus precedes the formation of mRNA.

Pre-Initiation Complex: Complex of transcription factors and RNA Polymerase II that initiates transcription.

Pre-mRNA Splicing: Post-transcriptional modification that involve splicing out introns and splicing together exons.

Primer: Short RNA sequences created from DNA primase.

Primosome: A protein complex involved with DNA primases to synthesize RNA primers.

Probability: The likelihood of an event to occur.

Producers: Organisms that produce their own food and obtain energy from chemicals or the sun. They are also known as autotrophs.

Prognathism: A facial deformity characterized by the protrusion of either the upper or lower jaw.

Prokaryotes: Single-celled organisms that lack membrane-bound organelles.

Promoter: A specific region upstream of the DNA sequence that regulates transcription by binding to RNA polymerase II.

Proofreading: Editing and correcting the errors that enzymes may have made while they were creating daughter strands.

Prophase: First phase within the mitotic phase that involves the formation of mitotic spindles, the migration of centrioles to opposite sides of the cell, and disappearance of the nucleolus and nuclear membrane.

Prophase I: At this phase, chromosomes condense and perform a crossover between other chromosomes.

Prophase II: During this phase, the nuclear envelope breaks and the chromosomes condense.

Protein Replacement Therapy: The process of replacing proteins in the body that are deficient or absent.

Protein-Based Therapeutics: A type of biotechnology-based product that aims to prevent and treat a variety of medical issues that have yet to be addressed.

Protein-DNA Complex: A structure of protein and DNA that have bound together to regulate cell replication.

Proteinuria: Presence of protein in the urine.

Proteolisis: The process of breaking down proteins into smaller polypeptides or into amino acids.

Proteomics: A study and characterization of proteins expressed by a genome at one given time in a cell.

Proteopathy: Diseases involving the disruption of cell function as a result of the production of abnormal proteins.

Protista: Unicellular eukaryotic organisms (though there are known relatively simple multicellular forms) that include both autotrophic and heterotrophic forms of the following organisms: algae, slime, molds, and protozoa.

Protists: Unicellular eukaryotes that fall under the kingdom Protista.

Protons: Positively charged subatomic particles that reside within an atomic nucleus, the number of which determines its periodic element.

Glossary

Protozoans: Paraphyletic grouping of "animal-like" single-celled eukaryotic organisms.

Pseudohypertrophy: Enlargement of calf muscles.

Psychosis: Disorders of mental illnesses that result in abnormal thinking, perceptions, behaviour and emotions.

Public Health Practice: The application of public health sciences to improve the overall health of the public.

Public Health Specialist: Health professionals that focus on prevention of disease and work towards creating conditions where populations can be healthy.

Punnett Square: Visual diagram used to predict the offspring, genotype and phenotype ratio, of a particular cross or breeding event.

Purines: The larger, double carbon-nitrogen rings that are found in DNA, such as adenine and guanine.

Pyrimidines: The smaller, single carbon-nitrogen rings that are found in DNA, such as cytosine and thymine.

"q" arm: The longer section of the chromosome when it is divided by the centromere.

RAAS Blockade: Treatment for renal diseases using angiotensin-converting-enzyme inhibitors.

Radiation Therapy: A type of cancer treatment that uses beams of high energy to kill cancer cells and shrink tumours.

Radiotherapy: A treatment used to cure or relieve the symptoms of cancer by using high-energy rays to destroy cancer cells.

Real-Time Polymerase Chain Reaction (RT-PCR): A newer, faster technique of PCR amplification that does not require post-PCR analysis.

Receptors: A protein on the surface of a cell that binds to signaling molecules to facilitate specific functions.

Recessive: Term used to describe the allele whose effects show up only when the dominant allele is absent; traits which are abbreviated by lower-case letters.

Recombinant DNA Technology: Biomedical technology that focuses on altering genetic material outside an organism as a means to enhancing and improving characteristics inside living organisms or their products.

Recombinants: Novel genotypes that result from the reshuffling of genes and the crossover of chromosomes during meiosis.

Reduction Division: The first cell division in which the number of chromosomes is reduced by half.

Regression: A statistical procedure that estimates the relationship between dependent and independent variables.

Regulatory Sequences: DNA segments that regulate the expression of genes in an organism.

Repetitive DNA: Short repeating sequences of DNA.

Replication Origins: The starting point for DNA replication to begin its process.

Repressors: Molecules that bind to the operator to halt transcription of a particular gene.

Reproduction: The ability of organisms to produce offspring.

Response to Stimuli: The ability of a living organism to detect a stimulus and respond accordingly.

Restriction Endonucleases (REases): DNA-cleaving enzymes that biologically function to protect the host genome against foreign DNA.

Restriction Length Fragment Polymorphisms (RFLPs): A set of DNA sequence fragments of varying length.

Retroviruses: A group of RNA viruses that replicate by producing a DNA copy of their genome and inserts it into a host cell.

Reverse Transcriptase: An enzyme that produces a copy of DNA from mRNA.

Reversible Reactions: Chemical reactions that are reversible.

R-Group: Shorthand designation for alkyl groups.

Ribosomes: Organelles that synthesize proteins.

RNA Polymerase Ii: Enzyme complex that catalyzes the transcription of DNA by binding to the promoter.

Rough ER: Site of protein synthesis/modification for proteins targeted to enter the secretory pathway.

Glossary

rRNA: Ribosomal RNA synthesized from nucleolar DNA; the ribosome component responsible for catalyzing protein synthesis.

Rutting: The period of sexual desire and mating in mammals.

S Phase: The synthesis phase where the cell duplicates DNA via semiconservative replication.

Salts: Ionic compounds that result from the neutralization reaction of an acid and base in water.

Sanger Sequencing: The original sequencing technology used to sequence short stretches of DNA.

Saturated Fatty Acids: Fatty acids with all single carbon bonds.

Scaffold Complex: Complex that actively transcribes genes at the start site by by-passing the slow process of recruiting factors required for transcription initiation.

Scientific Process: The hypothetical-deductive stepwise procedure used in science to make inquiries, formulate hypothesis, design experiments, collect and analyze data, and make conclusions.

Scoliosis: A condition that causes an abnormal curve to the spine.

Second Messenger: A chemical signal that relays instructions from the cell surface to enzymes in the cytoplasm.

Segregate: The separation of alleles during the meiosis.

Semi-Conservative Replication: A type of DNA replication in which the two new strands of DNA produced each contain an original strand and a new strand.

Sensory Cells: Cells that detect smell, light, touch, sound, taste, and temperature through sensory receptors.

Sertoli Cells: Somatic cells that promote the formation of testis and spermatogenesis from germ cells.

Serum Serine Protease Inhibitor: Inactivates related proteases in order to regulate protease mediated activities.

Sex Chromosomes: The 23rd pair of chromosomes, X and Y, that determines sex and sexual characteristics of an organism.

Sex Determining Region: Also known as testis-determining factor, it is a gene on the Y chromosome that gives rise to male phenotypes, such as the testes.

Sexual Reproduction: A type of reproduction in which two parent organisms (female and male) mate and produce offspring; associated with meiosis and genetic variation.

SHOX: A gene that is essential in bone development and growth.

Sickle-Cell Anaemia: An incomplete dominant genetic disorder caused by a defect in haemoglobin synthesis resulting in haemoglobin insolubility and instability.

Signal Transduction Pathway: A cascade of chemical reactions that eventually lead to a target reaction or molecule.

Significance Level: A statistical predetermined value that allows one to reject or not reject a null hypothesis—usually 0.05 in biological data.

Silencers: Specific sequences upstream of the promoter region that prevent RNA polymerase from binding to the promoter, thus curtailing gene transcription.

Single Nucleotide Polymorphisms (SNPs): A genetic variant of a single nucleotide in a DNA sequence.

Small Nuclear Ribonucleoprotein Particles (snRNPs): Uridine-rich RNA-proteins that interact with unmodified pre-mRNA and other proteins to form a larger complex known as the spliceosome.

Small Ubiquitin-Related Modifier (SUMO): A polypeptide of less than 100 amino acids that attach or detach to other proteins to modify their functions

Smooth ER: Site of steroid hormone synthesis and the degradation of environmental toxins.

Solute: The substance being dissolved like sugar or salt.

Solvent: The solution in which the solute is dissolved.

Species: A group of organisms that share similar characteristics and have the ability to interbreed and produce fertile offspring.

Speech Therapy: A clinical program designed to improve speech and language related issues.

Sperm: Male reproductive cell.

Spermatogenesis: The process of sperm or spermatozoa production.

Spinal Stenosis: The narrowing of the spinal canals.

Spliceosome: An enzyme that catalyzes esterification reactions in the removal of introns and the joining of exons.

Splicing Regulatory Elements: Elements of gene expression between transcription and translation that play a role in protein processing and degradation.

Standard Deviation: A quantification of the variations or dispersions around the mean within a dataset.

Standard Error: A quantification of the variations of different means from multiple datasets – it measures how precise the means are by comparing and testing their differences.

Stem Cell Transplantation: The process of replacing unhealthy blood forming cells (stem cells) with healthy cells donated by a healthy individual; also known as a bone marrow transplant.

Stent Insertion: Permanent placement of a wire mesh tube into a clogged artery in order to increase blood flow to the heart.

Stoichiometry: Expression of the quantitative relationship between products and reactants in a chemical reaction.

Student t-Tests: A hypothesis testing method that can determine the statistical difference between two samples by dividing the difference between group means by the group variability.

Substrate Reduction Therapy (SRT): An oral medication that prevents the production of glucocerebroside.

Sugars: Monosaccharides or disaccharides primarily used for energy storage.

Synapsis: The process of homologous chromosome pairing during the initial phases of meiosis.

Synthetic Elements: Man-made elements.

Tachycardia: Term used to describe heart rate that is abnormally rapid.

Tap: Adaptor proteins that assemble nuclear export receptors for different classes of mRNA.

TATA Binding Protein: Protein that recognizes and binds to a specific region of repeating thymine and adenine bases.

TBP: Gene that codes for a TATA binding protein.

T-Cell Therapy: A type of cancer treatment involving the alteration of a patient's T-cell in order to attack cancer cells.

Telomerase: An enzyme that extends the telomeres in chromosomes to reverse shortened telomeres.

Telomeres: Protective chemical structures located at the end of a chromosome to prevent the loss of genes.

Telophase: The last phase within the mitotic phase that involves the disappearance of kinetochores and the reformation of a nuclear envelope around each set of chromosomes.

Telophase I: Phase that represents the complete separation of homologues and prepares for the second half of meiosis.

Telophase II: The completion phase of meiosis, in which four non-identical daughter cells are produced.

Territories: An area that is controlled by a community of animals of the same species.

Testcross: A genetic cross between a known homozygous recessive parent and a suspected heterozygous parent to determine specific genotypes (Pp or PP).

Testes: Oval shaped male reproductive gland in which sperm is produced.

Testosterone: A male sex hormone that helps develop the secondary sexual characteristics.

Tetrad: A group of four chromatids that are formed in meiosis.

Tetraploid: When cells have two additional sets of homologous chromosomes, totaling to four.

Theory of Natural Selection: A theory proposed by Charles Darwin that explained that natural selection is based on four main principles; variation, adaptive, selection, and increase in frequency.

Thermoreceptors: Receptor of a sensory neuron that responds to a rise in temperature.

Thromboembolic Events: Occurs when a blood clot in an artery or vein breaks off and lodges somewhere else in body and obstructs blood flow of the circulatory system.

Thyroxine (T4): A hormone produced only by the thyroid gland that regulates basal metabolic rate carrying four iodine atoms.

Tight Junctions: Tight band-like structures that form a seal between the membranes of adjacent cells, thus blocking flow of molecules across the cell layer.

Tissues: A group of cells that have a specific function.

Tracers: A substance that can be readily followed by its distinctive property.

Glossary

Transamination: A reaction involving the transfer of an amino group from one amino acid to a keto acid.

Transcription and Export (TREX) Complex: Multi-protein complex that is assembled upon the pre-mRNA during transcription and plays an important role in the efficiency of mature mRNA export.

Transcription Factors: Family of proteins that regulate (activate or repress) gene expression and thus transcription.

Transcription: The process in which RNA polymerase uses a single-stranded DNA template to make mRNA.

Transduction: Introduction of foreign DNA to a cell by a virus or a viral vector.

Transfats: A form of unsaturated fat that has been modified by the chemical process of adding hydrogen ions that changes the H-atoms orientation from a *cis* to a *trans* position.

Transformation: Uptake and expression of naked DNA by bacteriophage.

Transgenics: The process of isolating genetic material from one organism and introducing it into another organism's genome.

Transition Metals: A group of elements in the periodic table containing partially filled d orbitals.

Transitions: A type of mutation in which pyrimidines are changed to another (i.e., C to T) or purines are changed to another (i.e., A to G). The transition rate is higher than the transversion rate in animal genomes.

Translation: The process in which a mRNA is translated into a sequence of amino acids during specific protein synthesis.

Translational Research: Improvement of health outcomes through conducting systemic reviews of genomic studies to address medical needs.

Transposable Elements: Mobile genetic DNA sequences consisting of repeats that are found in the genomes of nearly all eukaryotes.

Transposons: Repetitive DNA sequences that can insert or remove themselves from one location to another in the genome.

Transversions: A type of mutation in which pyrimidines are changed to purines, and vice versa.

Triiodothyronine (T3): A thyroid hormone carrying three iodine atoms that regulates basal metabolic rate. T3 is typically formed in the target organs by the removal of an iodine atom from a T4 hormone.

Triphosphate Molecules: Chain of three phosphates associated with the base adenosine (ATP) or guanosine (GTP) and stores the cell's energy.

Triplets: Sets of three bases on a mature messenger RNA strand.

Triploid: When cells have one additional set of homologous chromosomes totaling to three.

Trisomy 21: An additional autosome on chromosome #21 as a result of nondisjunction.

tRNA: Transfer RNA synthesized from regular DNA; RNA molecule that carries the anticodon sequence for polypeptide chain formation.

Tryptophan Operon: A regulatory system that, in the presence of tryptophan, will turn off transcription by binding to the trp repressor, thus preventing RNA polymerase from accessing the promoter.

Tumorigenic Cells: Cells that are capable of or tend to form tumours.

Universal Donors: A person whose blood can be accepted by all blood types without causing an immune response (i.e., Type O).

Universal Recipients: A person who can receive all blood types during a blood transfusion, without causing an immune response (i.e., AB+).

Universal Solvent: A solvent capable of dissolving many solutes (i.e., water).

Utilization of Energy: The ability of organisms to consume and transform energy into work.

Variable Number of Tandem Repeats (VNTRs or Microsatellite DNA): Specific regions of DNA that demonstration variations in the number of tandem repeats.

Vasoconstriction: Narrowing of blood vessels.

Vasodilation: Widening of blood vessels.

Vertebrates: Animals with a spine.

Vesicles: Membrane-bound organelles that transport substances between cells.

Viruses: Obligate intracellular parasites that cannot replicate without a host; lacks nucleus and organelles.

Glossary

Vitamins: Organic, essential nutrients that regulate bodily process that support growth and maintain life.

Vitelliform Macular Dystrophy: A genetic form of macular degenerative disease that can cause gradual vision loss.

Vitreoretinochoroidopathy: An ocular disorder that affects the retina, vitreous, and the choroid and can lead to vision impairment.

Wet Weight: The normal weight of an organism that includes watery fluids.

Whole Exome Sequencing: A sequencing method used to determine all the exons in a genome of an organism.

Whole Genome Sequencing: A sequencing method used to determine whole genomes of an organism.

X-Ray Diffraction: A method that determines the molecular structure of a substance using a beam of X-rays which diffract or spread out in specific patterns.

Zygote: The first stage in the formation of a new organism after fertilization of an egg by a sperm.

About the Author

Oscar Wambuguh is a professor at California State University, East Bay (CSUEB)—part of the California State University system with its 23 campuses all over the State of California, and the largest 4-year public university education system in the United States. A biologist by training, he has since 2000 been teaching courses in biological and health sciences including general biology, human biology, genes and human health, and environmental health. Before joining the CSU system, Dr. Wambuguh served as a graduate student instructor in biological sciences for five years when completing his doctoral studies at the University of California, Berkeley (UCB). This textbook is a result of the many years he instructed general biology at UCB and CSUEB—and for the last 15 years, genes and human health at CSUEB. Dr. Wambuguh is active in research and writing in the field of environmental health management and in the scholarship of teaching and learning. Today, it is firmly established that past and on-going anthropogenic activities have greatly contributed to increased global warming and climate change over the last three decades. These changes will affect all aspects of human existence and our life-support systems including our biodiversity conservation efforts, food production, disease patterns and associated public health measures, water and energy resources, sea-level rise, and with increased human migration—the national security of many nations. With such a myriad of changes, finding ways in which humans can live more sustainably on this planet has become crucial not only for our very own survival and that of future human generations, but also for all other organisms that share this planet with us. Dr. Wambuguh has published many peer-reviewed articles in scientific journals and general public readership magazines and bulletins. His two previous books include Conservation of Biological Diversity in Developing Countries and Green Energy Development in Rural Landscapes.

Index

"p" Arm 130, 141, 144
"q" Arm 130, 141, 144
1000 Genomes Project 270, 286
2:8:8:8 Octet Rule 55
3' OH End 152, 156, 160, 168, 179, 185, 217, 227
5' Carbon End 147, 160
60S Ribosomal Subunits 184
7-Methylguanosine Cap 168, 179, 184

A

ACE Inhibitors 381, 397
Achondrogenesis 406, 424
Activators 194, 197, 203, 245-246, 259, 500
Active Site 80-81, 85-86, 88, 91, 167
Adaptor Proteins 169, 179, 184, 253, 358
Addison's Disease 407-408, 424
Adeno-Associated Viruses (AAV Vectors) 322
Adjuvant Treatment 326, 359
Albuminuria 381, 397
Allele 118, 122, 124, 127, 136, 141, 303, 308, 311, 338, 442, 445, 482-484, 488-491, 495
Allogeneic Stem-Cell Transplantation 404, 424
Alopecia 379, 385, 404, 424, 479
Alternative Splicing 168, 195, 197, 247, 268-270, 278, 439
Aminoacyl-tRNA Synthase 172, 184
Aneurysms 381, 397
Angina 301, 322, 407
Angle-Closure Glaucoma 377, 384, 397
Antibiotics 11, 97, 210, 293, 305, 309, 322, 328, 333, 364, 382, 408, 441
Anticodon 172-173, 184, 249
Antihistamines 364, 369, 382, 397
Antipsychotics Drugs 476, 527
Antisense Oligonucleotides 432-433, 441, 467, 484, 499, 502, 506
Apheresis 494, 505, 527
Apoptosis 65, 153, 160, 196-197, 243-244, 246, 327, 330, 336, 359, 380, 413, 429, 445, 475, 479, 494, 501
Argyrophilic Grain Disease 470, 500, 527
Aromatase Inhibitors 330, 343, 359
Arrhythmia 307, 359
Arteriosclerosis 42-43, 301, 310, 322
Astigmatism 378, 385, 397
Astrogliosis 482, 527
Ataxia 288, 295, 304-305, 311, 371, 397, 405, 468, 477, 480, 494, 497-498, 502, 527
Atelosteogenesis 406, 424
Atoms 19-20, 23-24, 26-31, 34, 39, 41-42, 47, 50, 52, 55, 88, 244, 398
Attenuation 196, 203
Augmentation Therapy 309, 322
Autoradiography 160
Autosomes 128, 130, 132, 137, 141-142, 144, 178, 180, 380
Axonopathy 404, 414, 424

B

Bardoxolone Therapy 381, 397
Benzothiazepines 437, 467
Bestrophinopathies 375-376, 384, 398
Big Data 532-533, 545
Binding Proteins 160, 241
Bioeconomy 223, 234
Biomarkers 272, 331, 338, 344, 359, 365, 382, 409, 415, 473, 499, 532-535, 545
Bioprocessing 223-224, 227, 234
Bioremediators 223, 234
Biotechnology 205, 208, 214-215, 220-221, 223-227, 234, 240, 287, 530, 539
Biotin (Vitamin B7) 197, 268
Blood Transfusion 293, 301, 323
Bone Remodeling 427, 467

Brachytherapy 342, 345, 359
Bradycardia 330, 359
Bronchodilators 309, 323

C

cAmp Receptor Protein (CRP) 203
Carcinogenesis 197, 246, 332, 343, 359
Cardiogram 302, 323
Cardiomyopathy 301, 323, 371, 398, 431, 433, 449, 467, 480-481
Cardioverter-Defibrillators 307, 323
Catabolite Activator Protein (CAP) 195, 203
Cataplexy 488-489, 503-504, 527
Cell 2-3, 5, 7, 36, 39, 43, 50, 52, 56-60, 62-63, 65-67, 70-76, 80, 86-87, 91, 93-94, 97-102, 104-106, 109-111, 114, 120, 125, 137-138, 145, 149, 151-156, 160-161, 166, 168, 174, 180, 186-187, 191-193, 195-197, 208, 210, 212-213, 218-221, 223, 227, 234-235, 240-241, 244, 246-247, 249-251, 254, 256, 259, 268, 271, 274-275, 288, 291-296, 298, 300-301, 307, 309-310, 323-327, 330-337, 340-341, 343-345, 359, 364-365, 372, 375, 377, 379, 381, 384-385, 400, 404, 410-415, 424, 426, 433, 437-439, 446-447, 450-451, 467, 471, 478, 480-482, 486-487, 496, 498, 500, 502-503, 528, 538
Cell Theory 57, 86, 91, 97, 114
Centromere 102-104, 114, 130, 141, 144, 253
Cerebral Amyloid Angiopathy 470-471, 500, 527
Chaperonins 47, 55
Chemical Bonds 19, 21, 24-25, 34, 52, 55
Chemotherapy 213, 294, 296, 323, 325, 327, 329-331, 333-334, 336-338, 341-345, 373, 424
Cholinesterase Inhibitors 470, 527
Chromatids 102-103, 106, 111, 114, 131
Chromatin 62, 98, 114, 130, 144, 155-156, 160-161, 166-167, 179, 193, 196-197, 239-246, 251, 253-256, 259, 268, 407, 442, 475, 480, 545
Chromatin Remodelling 193, 241-242, 251, 268
Chromosomes 65, 93, 97-98, 101-103, 105-106, 110-111, 114, 119-120, 124, 127-128, 130-131, 134-135, 137-138, 141-142, 144, 152-154, 156, 161, 178, 180, 270, 274, 278, 296, 305, 310, 374, 381, 385, 442, 477, 501
Chronic Pancreatitis 336, 359
Chronic Rhinosinusitis 363, 398
Cirrhosis 308, 323, 367-368, 371, 383, 493
Clinical Health Practice 545
Codominance 115, 122, 125, 127
Codons 170-172, 176, 180, 184, 214, 249, 484

Colonoscopy 332, 359
Competitive DNA Binding 203, 247, 268
Complementary DNA (cDNA) 215, 234, 309
Complex Organization 15, 18
Computed Tomography (CT) 359
Congenital 274, 292, 303, 306-307, 311, 323, 373-374, 377-378, 385, 399, 402, 412-413, 437, 440, 446-447, 450-451, 492
Connective Tissue 48, 77, 79, 87, 91, 303, 425, 431, 437, 439, 448-449, 487
Coronary Angioplasty 302, 323
Coronary Artery Bypass Grafting 302, 323
Covalent Bond 27-28, 36, 39, 55
CRISPR Arrays 211, 226, 234
Crossing Over 106, 114, 120, 133
Cyclic AMP (Camp) 203
Cytosol 57-60, 87, 91

D

Dementia 100, 276, 280, 295, 323, 380, 405, 414, 469-472, 477, 482, 490-491, 500-501, 504, 527-528
Demyelination 404, 424
Deoxyribonucleic Acid (DNA) 3, 50, 114, 146, 160
Development 3-4, 15-16, 18, 80, 87, 93-96, 110-111, 138, 166, 178, 187, 195, 206, 216, 221-223, 234, 240-241, 244, 249, 251, 254, 259-260, 272, 279, 286-287, 295, 297, 303-305, 309, 311, 326, 331-333, 336, 338, 343-344, 361-362, 366-367, 369, 372, 375, 380, 382-383, 399-400, 406-410, 413-415, 424, 428, 432-433, 439, 447-451, 467, 470-471, 473-474, 484-485, 487, 489, 491-492, 501-502, 504-505, 527, 530-531, 533-534, 537, 539
Dihybrid Inheritance 119, 124, 127
Dihydropyridines 437, 467
Direct-to-Consumer (DTC) Genetic Testing 277, 533, 545
DNA Fingerprint 219, 227, 234
DNA Helicase 160, 167
DNA Library 213, 226, 234
DNA Ligase 151, 156, 160, 214, 410
DNA Microarrays (Nucleic Acid Arrays) 234
DNA Polymerase I 151, 156, 160
DNA Polymerase III 150, 152, 156, 160
DNA Primase 160-161
Dolichocephaly 442, 467
Domains 7-8, 16, 18, 196, 327, 442-443, 451, 473, 490, 497
Dysarthria 480, 497, 527

Index

Dysdiachokinesia 497, 527
Dysmetria 497, 527
Dysphagia 435, 467
Dysphasia 480, 527
Dysplasia 399, 406, 411-412, 414-415, 424, 428-429, 433, 449, 494
Dyspnoea 435, 467
Dystonia 371, 482, 498, 527

E

Ectodermal Dysplasia 411, 424
Emphysema 308-309, 323, 334
Encyclopaedia of DNA Elements (ENCODE) Project 270, 286
Endocrine Disruptor 268
Endosymbiosis 153, 160
Enhancesome 203, 246, 268
Enzyme Replacement Therapy (ERT) 295, 323
Enzymes 45, 47, 52, 56, 64-65, 80-85, 88, 91, 98, 100, 153, 156, 160-161, 163-164, 166, 170, 179, 188, 190, 195-196, 206, 208-211, 213, 217, 223, 226, 234-235, 242-243, 245-246, 248, 250-251, 254, 259, 290, 296, 298, 305-306, 308, 311, 323, 333, 362, 369, 371-372, 382-383, 402, 404, 413, 427, 433, 487, 493, 503, 505
Epidemiological Studies 491, 504, 530, 539, 545
Epigenetics 192, 239-240, 253-254, 258-260, 268, 490, 504, 535-536
Epigenomic Profiling 254, 535, 545
Epigenomics 240-241, 259, 268, 270, 287, 530, 535-536, 538
Epilepsy 255, 260, 295, 339, 359, 468, 474-476, 501
Epithelial Tissue 77, 91
Erythropoiesis 300-301, 323, 367, 481, 527
Ester Bond 39, 43, 55
Eukaryotes 7-8, 15, 18, 57, 149, 152, 177, 180, 184, 187, 190, 192, 196, 249
Evidence-Based Intervention 545
Evolution 1, 10-12, 16, 104, 114, 151, 154, 163, 170, 178, 272, 279, 406
Exons 168-169, 179, 184-185, 217, 226, 235, 268, 305, 311, 407, 432

F

Fatty Acids 31-32, 39-41, 43, 55, 80, 245, 404, 481
Focal Therapy 341, 345, 359
Forensic Evidence 224, 234
Frameshift Mutations 162, 176-177, 180, 184

G

Gametogenesis 93, 109, 112, 114
Gel Electrophoresis 227, 234
Gene Therapy 205-206, 210-211, 220-221, 225, 227, 234, 291, 296-300, 309, 322-323, 333, 376, 379, 385, 404, 410-411, 415, 424, 440-441, 481, 488, 493, 495, 497, 503, 505-506
Genes 73, 87, 98, 104-106, 111, 114-116, 120-121, 124, 127-128, 132-135, 138, 141, 144-146, 152-154, 156, 161-167, 171, 175, 178-179, 184, 186-190, 192-193, 195-197, 203-208, 210-220, 223, 226-227, 234, 240, 244-246, 249, 251, 254-256, 260, 269-275, 277-280, 288, 294, 296, 300, 302-307, 309-311, 323, 327, 329, 331-337, 341-344, 366, 368-372, 374, 377, 380-385, 403, 409-411, 413, 415, 428, 430, 434-435, 437, 439-443, 446-451, 471, 474-475, 477, 481, 485-486, 488-489, 491-492, 494-495, 498, 501, 503-506, 530, 533, 535, 537
Genetic Adaptation 163, 179, 184
Genetic Code 170-171, 174, 176, 180, 184, 211
Genetic Engineering 205, 208, 221-223, 226, 234
Genetic Exceptionalism 538, 545
Genetic Mutations 162, 175, 180, 184, 274, 279, 307, 311, 338, 535
Genome-Wide Analyses 248, 268
Genome-Wide Association Studies (GWAS) 302, 311, 323, 359, 366, 409, 414
Genomic Biobanks 537, 545
Genomic Biomarkers 533, 545
Genomics 218-219, 227, 234, 269-270, 272, 276, 278-280, 377, 530-534, 536-539
Genotype 118-119, 122, 124, 127, 331, 363
Genotypic Ratio 118, 127
Germ Cells 93, 98, 104, 111, 114, 132, 153, 241, 252, 258
Glycosidic Bond 36-37, 55
Growth 3, 15, 18, 93-94, 99, 101, 110-111, 114, 163, 188, 206, 219, 294-297, 303-304, 310, 322-323, 326-328, 330-338, 342-344, 361-362, 367, 372, 374, 379-380, 382, 384-385, 399-401, 403, 405, 410-411, 413-414, 428-430, 432-434, 437-438, 441, 448-450, 481, 485, 487, 492, 495, 505
Growth Factors 99-101, 111, 114, 338, 344, 380, 385
Guide RNA 212, 234

H

HapMap Project 270, 286, 302, 323
Health Disparities 323, 530, 532, 537-539, 545
Health Equity 530, 546
Hemoglobinopathies 213, 291, 323
Hepatocytes 308, 323, 371, 383
Hepatosplenomegaly 297, 496, 527
Herbicides 239, 258, 260, 268, 435
Heterozygous 115, 118, 121, 124, 127, 292, 295, 369-370, 404, 477, 481
Hirschsprung Disease 447, 467
Homeostasis 5, 15, 18, 174, 268, 297, 305-306, 359, 471, 487, 500, 503
Homeotic Genes 167, 184
Homozygous 115, 118, 121, 124, 127, 295, 308, 368-369, 477, 479-481, 483, 502
Hormonal Therapy 325-327, 342, 359
Hugo Gene Nomenclature Committee 273, 286
Human Genome Project 164-165, 179, 184, 269-270, 272, 278, 286, 302, 530
Human Microbiome Project 270, 286
Hydrocarbon 39, 55
Hydrogen Bond 28, 55
Hyperbilirubinaemia 373, 398
Hyperphenylalaninaemia 492, 527
Hyperplasia 399, 402, 411, 413, 424
Hypogammaglobulinemia 411, 424
Hypoparathyroidism 371, 407, 412, 424
Hypothesis 18, 164, 179, 304, 430, 436, 491

I

Immunization 305, 323
Immunotherapy 325-326, 333-334, 336, 342-344, 359, 409, 447
Imprinting 253-254, 268, 297, 303, 310, 441-442, 450-451
Inducer 190, 196, 203
Industrial Biotechnology 223, 234
Initiator Proteins 160
Insecticides 239, 258, 260, 268
Insulator 203, 268
Interdependency 5, 15, 18
International Cancer Genome Consortium 270, 286
International HapMap Project 270, 286
Intraocular Pressure 377-378, 384-385, 398
Introns 164, 168-169, 179, 184-185, 268-271, 278, 495
Ionic Bond 26, 55
Ischaemic Cardiomyopathy 301, 323
Isotopes 23-24, 52, 55

J

Junk DNA 164, 179, 185, 269-270, 278

K

Karyotype Analysis 130, 141, 144
Kinetic Tremor 476-477, 501, 528
Kingdoms 7-8, 16, 18
Kyphosis 428, 467

L

Lac Repressor 203
Lactose Operon 186-188, 195, 203
Lagging Strand 151-152, 156, 160-161
Laser Therapy 341, 345, 359
Leading Strand 151-152, 156, 160
Lenti-D Gene Therapy 404, 424
Leukoencephalopathy 470, 500, 528
Lewy Body Disease 470, 500, 528
Life 1-2, 4, 7-8, 11, 15, 18-19, 24, 29, 33-34, 52, 56-57, 82, 86, 94, 104, 109-110, 137, 184, 187, 213, 258, 272-273, 277, 293, 295, 297-298, 300, 304, 306, 324, 330, 333, 343, 365, 368, 375, 378, 383, 405, 409-410, 414, 425, 428, 430, 435-438, 442, 448-449, 474, 476, 478-482, 485-487, 489, 492, 495, 497, 502-503, 532, 534-536
Linkage 36, 128, 132, 134, 141, 144, 192, 370, 377, 429, 448, 477
Lipopheresis 494, 505, 528
Lipoproteins 38, 41-43, 52, 55, 302, 311
Locus 124, 127, 273, 279, 302, 370, 380, 447, 474, 489, 498, 504, 506
Lumbar Lordosis 428, 467
Lymphocyte Transfusion 331, 343, 359
Lymphoid Hyperplasia 411, 424

M

Macrocephaly 339, 359
Macromolecules 19, 32, 34, 36, 38, 43, 45, 50, 52-53, 55, 65, 80, 165
Macroorchidism 444, 467
Magnetic Resonance Imaging (MRI) 332, 359
Mastectomy 330, 359
Mature mRNA 168-169, 171-172, 179-180, 185, 214, 234
Meiosis 93-94, 98, 104-111, 114, 118-120, 124, 127, 243, 413
Microcephaly 405, 442, 467, 492

Index

Mismatch Repair 145, 152, 156, 160, 337
Missense Mutation 185, 367, 382, 429, 471, 500
Mitochondrial DNA 145, 153, 156, 160, 471
Mitosis 93-94, 98, 100-102, 106, 108-112, 114, 130, 243
Mitotic Promoting Factor (MPF) 101, 114
Monohybrid Inheritance 117, 124, 127
mRNA Ribonucleoprotein (mRNP) Complex 169, 179, 185
Multiple Myeloma 429, 467
Muscle Tissue 77, 79, 87, 91, 425, 427, 431, 448
Mutation 11, 160, 175-176, 180, 185, 213, 292, 295, 298, 308-310, 325, 327, 331-332, 336-339, 341-344, 367-368, 370, 372-373, 382-383, 398, 405-407, 414, 429-430, 432-433, 435-436, 438, 440, 444-445, 448-451, 469, 471, 477, 479-480, 483-484, 486, 500, 502
Myocardial Ischemia 330, 359
Myopia 378, 385, 398

N

Nanoparticles 303, 333, 342, 359
Nanotechnology 15, 268, 440
National Institute of Health (NIH) Roadmap Epigenomics Program 270, 287
Natural Selection 1, 8, 10-11, 16, 18, 175-176, 184, 292
Necrosis 301, 323, 381, 431, 448, 480, 489, 504
Nervous Tissue 77, 79, 87, 91, 100
Neurocristopathy 446, 467
Next Generation Sequencing (NGS) 214, 234, 359, 534
Nicotinamide Adenine Dinucleotide 246, 268
Nonsense Mutation 185, 407
Non-Steroidal Anti-Inflammatory Agents (NSAIDS) 293, 323
Nucleic Acids 50-51, 53, 145-147, 155, 160, 215, 333
Nuclein 146, 161
Nucleosome 62, 154, 156, 161, 254

O

Ocular Hypertension 377, 384, 398
Okazaki Fragments 150-151, 156, 161
Open-Angle Glaucoma 377, 384, 398
Operator 187-190, 195-196, 203-204
Operon 186-191, 195-196, 203-204
Organelles 18, 57-58, 60, 62-63, 65-66, 87, 91, 97-98, 106, 145, 149, 153, 155, 204
Osteoarthritis 369, 371, 398
Osteoporosis 300, 371, 398, 427, 448, 492
Oxidative Stress 245, 302, 305, 327, 359, 436, 473, 480-481, 487, 502

P

Palliation 342, 359
Palliative Treatments 306, 324
Particulate Matter 55, 257, 260, 268
Pathognomic 494, 528
Pedigree 128, 141, 144
Peptide Bonds 45-47, 52, 55, 185
Peptidyl Transferase 180, 185
Pericarditis 330, 359
Periodic Table 20-21, 23-24, 31, 50, 55
Persistent Organic Pollutants 239, 258, 260, 268
Personal Wellness Profile 141, 144
Personalized Medicine 272, 277, 280, 287, 364, 532-533, 539, 545
pH 5, 19, 29, 31-32, 52-53, 55, 65, 82-85, 88, 296-297, 310, 400
Pharmacogenomics 221, 234, 272, 279, 287, 533-534, 539
Phenotype 115, 122, 124-125, 127, 216, 254, 259-260, 303, 331, 339, 381, 404, 410, 414, 428-429, 442, 444-445, 479, 486, 495, 499, 503, 533
Phenotypic Ratio 118, 127
Phenylalkylamines 437, 467
Phosphodiester Bond 51, 55, 147, 160, 167, 209
Photodynamic Therapy 376, 384, 398
Physical Therapy 304-305, 324, 405
Phytochemicals 260, 268, 535
Plasmapheresis 494, 505, 528
Plasmids 206, 209-210, 212, 214, 225-226, 234
Pluripotent Bone Marrow Stem Cells 296, 310, 324
Point Mutations 162, 176, 180, 185, 297, 310, 341, 345, 430, 432, 442, 445, 447, 480-481, 502
Poly-A Polymerase 168, 185
Poly-A Tail 168, 179, 185, 249
Polyadenylation 168, 185, 195, 197, 247-248
Polyendocrinopathy 407, 414, 424
Polymerase Chain Reaction (PCR) 217-218, 234
Polyribosomes 185
Positive Gene Regulation 195, 204, 246, 268
Positron Emission Tomography (PET) 332, 359
Precision Medicine 272, 533, 546
Precursor Messenger RNA (Pre-mRNA) 167, 179, 185
Primer 150, 152, 155-156, 161
Primosome 161
Prognathism 442, 446, 467
Prokaryotes 7, 15, 18, 57, 187, 190
Promoter 166-167, 179, 185, 188-190, 192, 194-197, 203-204, 210, 246-247, 251, 253, 256, 268, 271, 309, 334, 341, 345, 364
Proofreading 145, 160-161

Protein Replacement Therapy 299, 324
Protein-Based Therapeutics 222, 234
Protein-DNA Complex 161
Proteinuria 380-381, 398
Proteolisis 493, 505, 528
Proteomics 218-219, 227, 234
Proteopathy 483, 528
Pseudohypertrophy 431, 467
Psychosis 240, 371, 473, 498, 528
Public Health Practice 530-532, 538-539, 546
Public Health Specialist 546
Punnett Square 127

R

RAAS Blockade 381, 398
Radiation Therapy 305, 324, 329-330, 333-334, 337-338, 343-345
Radiotherapy 52, 325-326, 335, 341-342, 345, 359-360
Real-Time Polymerase Chain Reaction (RT-PCR) 217, 235
Recombinant DNA Technology 205-206, 209-210, 225, 235
Recombinants 119-120, 127
Reduction Division 93, 106, 111, 114
Repetitive DNA 164, 177, 185, 270, 278
Replication Origins 161
Repressors 194, 197, 204, 247, 259, 484
Reproduction 2, 5, 15-16, 18, 93-94, 97-98, 104, 106, 111, 114, 175
Response to Stimuli 18
Restriction Endonucleases (REases) 235
Restriction Length Fragment Polymorphisms (RFLPs) 235
Retroviruses 206, 210, 214, 225-226, 235, 299
Reverse Transcriptase 214, 235
Ribosomes 62-63, 87, 98, 162, 171, 179-180, 185, 204, 247, 249
RNA Polymerase Ii 166-167, 179-180, 185, 204

S

Sanger Sequencing 215, 226, 235
Scientific Process 18
Scoliosis 339, 360, 405-406, 414, 424, 430, 433, 443, 451, 480, 495
Semi-Conservative Replication 151, 161
Serum Serine Protease Inhibitor 308, 311, 324
Sex Chromosomes 128, 130, 138, 141-142, 144, 178, 180

Sex Determining Region 132, 141, 144
Silencers 194, 204, 247, 268
Single Nucleotide Polymorphisms (SNPs) 271, 302, 324, 360, 370, 383
Small Nuclear Ribonucleoprotein Particles (snRNPs) 179, 185
Small Ubiquitin-Related Modifier (SUMO) 244, 268
Speech Therapy 304, 324
Spinal Stenosis 428, 467
Spliceosome 168-169, 179, 185, 495
Splicing Regulatory Elements 204, 247, 268
Stem Cell Transplantation 293, 298, 324, 404, 414, 486, 503
Stent Insertion 302, 324
Stoichiometry 431, 467
Substrate Reduction Therapy (SRT) 295, 324
Synapsis 106, 114

T

Tachycardia 306, 324
TATA Binding Protein 166, 179, 185
T-Cell Therapy 331, 343, 360
Telomerase 145, 153, 156, 161, 195, 197, 251, 253
Telomeres 145, 153, 156, 161
Testcross 118, 127
Thromboembolic Events 330, 360
Thyroxine (T4) 398
Transcription Factors 132, 166-167, 179, 184-185, 193-194, 197, 204, 216, 242, 245-246, 251, 259, 341, 366, 374, 382, 384, 484
Transgenics 206, 222, 235
Transitions 162, 176, 178, 180, 185
Translational Research 530, 539, 546
Transposons 162, 164, 176-178, 180, 185
Transversions 162, 176, 178, 180, 185
Triiodothyronine (T3) 398
Tryptophan Operon 186-187, 190, 195, 204
Tumorigenic Cells 360

U

Utilization of Energy 3, 15, 18

V

Variable Number of Tandem Repeats (VNTRs or Microsatellite DNA) 235
Vitelliform Macular Dystrophy 375-376, 398
Vitreoretinochoroidopathy 376, 398

Index

W

Whole Exome Sequencing 215, 217, 226, 235
Whole Genome Sequencing 215-217, 226, 235

X

X-Ray Diffraction 146-147, 161
B-Adrenergic Receptor Antagonists (B-Blockers) 307, 323

Purchase Print, E-Book, or Print + E-Book

IGI Global books are available in three unique pricing formats:
Print Only, E-Book Only, or Print + E-Book. Shipping fees apply.

www.igi-global.com

Recommended Reference Books

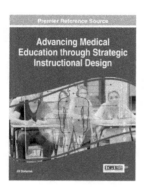

Advancing Medical Education through Strategic Instructional Design

ISBN: 978-1-5225-2098-6
© 2017; 349 pp.
List Price: $205

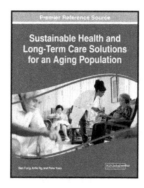

Sustainable Health and Long-Term Care Solutions for an Aging Population

ISBN: 978-1-5225-2633-9
© 2018; 441 pp.
List Price: $225

Workforce Development Theory and Practice in the Mental Health Sector

ISBN: 978-1-5225-1874-7
© 2017; 378 pp.
List Price: $190

Exploring the Nutrition and Health Benefits of Functional Foods

ISBN: 978-1-5225-0591-4
© 2017; 523 pp.
List Price: $200

Internet of Things and Advanced Application in Healthcare

ISBN: 978-1-5225-1820-4
© 2017; 349 pp.
List Price: $210

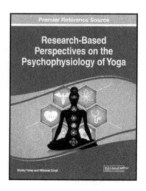

Research-Based Perspectives on the Psychophysiology of Yoga

ISBN: 978-1-5225-2788-6
© 2018; 456 pp.
List Price: $225

Do you want to stay current on the latest research trends, product announcements, news and special offers?
Join IGI Global's mailing list today and start enjoying exclusive perks sent only to IGI Global members.
Add your name to the list at **www.igi-global.com/newsletters.**

Publisher of Peer-Reviewed, Timely, and Innovative Academic Research

www.igi-global.com | Sign up at www.igi-global.com/newsletters | facebook.com/igiglobal | twitter.com/igiglobal | linkedin.com/igiglobal

Ensure Quality Research is Introduced to the Academic Community

Become an IGI Global Reviewer for Authored Book Projects

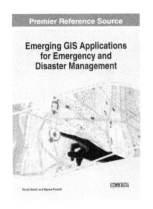
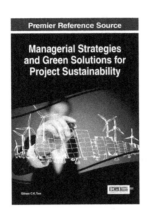

The overall success of an authored book project is dependent on quality and timely reviews.

In this competitive age of scholarly publishing, constructive and timely feedback significantly expedites the turnaround time of manuscripts from submission to acceptance, allowing the publication and discovery of forward-thinking research at a much more expeditious rate. Several IGI Global authored book projects are currently seeking highly qualified experts in the field to fill vacancies on their respective editorial review boards:

Applications may be sent to:
development@igi-global.com

Applicants must have a doctorate (or an equivalent degree) as well as publishing and reviewing experience. Reviewers are asked to write reviews in a timely, collegial, and constructive manner. All reviewers will begin their role on an ad-hoc basis for a period of one year, and upon successful completion of this term can be considered for full editorial review board status, with the potential for a subsequent promotion to Associate Editor.

If you have a colleague that may be interested in this opportunity, we encourage you to share this information with them.

www.igi-global.com

Celebrating 30 Years of Scholarly Knowledge Creation & Dissemination

InfoSci®-Books

A Collection of 4,000+ Reference Books Containing Over 87,000 Full-Text Chapters Focusing on Emerging Research

This database is a collection of over 4,000+ IGI Global single and multi-volume reference books, handbooks of research, and encyclopedias, encompassing groundbreaking research from prominent experts worldwide. These books are highly cited and currently recognized in prestigious indices such as: Web of Science™ and Scopus®.

Librarian Features:
- No Set-Up or Maintenance Fees
- Guarantee of No More Than A 5% Annual Price Increase
- COUNTER 4 Usage Reports
- Complimentary Archival Access
- Free MARC Records

Researcher Features:
- Unlimited Simultaneous Users
- No Embargo of Content
- Full Book Download
- Full-Text Search Engine
- No DRM

To Find Out More or To Purchase This Database:
www.igi-global.com/infosci-books

eresources@igi-global.com • Toll Free: 1-866-342-6657 ext. 100 • Phone: 717-533-8845 x100

www.igi-global.com

IGI Global Proudly Partners with

Enhance Your Manuscript with eContent Pro International's Professional
Copy Editing Service

Expert Copy Editing

eContent Pro International copy editors, with over 70 years of combined experience, will provide complete and comprehensive care for your document by resolving all issues with spelling, punctuation, grammar, terminology, jargon, semantics, syntax, consistency, flow, and more. In addition, they will format your document to the style you specify (APA, Chicago, etc.). All edits will be performed using Microsoft Word's Track Changes feature, which allows for fast and simple review and management of edits.

Additional Services

eContent Pro International also offers fast and affordable proofreading to enhance the readability of your document, professional translation in over 100 languages, and market localization services to help businesses and organizations localize their content and grow into new markets around the globe.

IGI Global Authors Save 25% on eContent Pro International's Services!

Scan the QR Code to Receive Your 25% Discount

The 25% discount is applied directly to your eContent Pro International shopping cart when placing an order through IGI Global's referral link. Use the QR code to access this referral link. eContent Pro International has the right to end or modify any promotion at any time.

Email: customerservice@econtentpro.com

econtentpro.com

CPSIA information can be obtained
at www.ICGtesting.com
Printed in the USA
BVHW010218170722
642286BV00003B/31